Theoretical Astrophysics
Volume II: Stars and Stellar Systems

The study of stellar physics plays a central role in the broader study of astrophysics. This authoritative textbook, which is the second volume of a comprehensive three-volume course on theoretical astrophysics, tackles stars and stellar systems. Designed to help graduate students and researchers develop an understanding of the key physical processes governing stars and stellar systems, it teaches the fundamentals, and then builds on them to give the reader an in-depth understanding of advanced topics. The book's modular design allows the chapters to be approached individually, yet the transitions are seamless, creating a coherent and connected whole. It can be used alone or in conjunction with Volume I, which covers a wide range of astrophysical processes, and the forthcoming Volume III, on galaxies and cosmology.

After reviewing the key observational results and nomenclature used in stellar astronomy, the book develops a solid understanding of central concepts including stellar structure and evolution, the physics of stellar remnants (such as white dwarfs, neutron stars and black holes), pulsars, binary stars, the Sun and the planetary system, the interstellar medium, and globular clusters. Throughout, the reader's grasp of all of the topics is developed and tested with more than seventy-five exercises.

This indispensible volume provides graduate students with a self-contained introduction to stellar physics and will allow them to master the material sufficiently to read and engage in research with heightened understanding.

THANU PADMANABHAN is a Professor at Inter-University Center for Astronomy and Astrophysics in Pune, India. His research interests are Gravitation, Cosmology, and Quantum Theory. He has published over one hundred technical papers in these areas and has written four books: *Structure Formation in the Universe, Cosmology and Astrophysics Through Problems, After the First Three Minutes*, and, together with J.V. Narlikar, *Gravity, Gauge Theories and Quantum Cosmology*.

Professor Padmanabhan is a member of the Indian Academy of Sciences, National Academy of Sciences, and International Astronomical Union. He has received numerous awards, including the Shanti Swarup Bhatnagar Prize in Physics (1996) and the Millenium Medal (2000) awarded by the Council of Scientific and Industrial Research, India.

He has also written more than one hundred popular science articles, a comic strip serial, and several regular columns on astronomy, recreational mathematics, and history of science that have appeared in international journals and papers. He is married, has one daughter, and lives in Pune, India. His hobbies include chess, origami, and recreational mathematics.

THEORETICAL ASTROPHYSICS
Volume II: Stars and Stellar Systems

T. PADMANABHAN

Inter-University Centre for Astronomy and Astrophysics
Pune, India

PUBLISHED BY THE PRESS SYNDICATE OF THE UNIVERSITY OF CAMBRIDGE
The Pitt Building, Trumpington Street, Cambridge, United Kingdom

CAMBRIDGE UNIVERSITY PRESS
The Edinburgh Building, Cambridge CB2 2RU, UK
40 West 20th Street, New York, NY 10011-4211, USA
10 Stamford Road, Oakleigh, VIC 3166, Australia
Ruiz de Alarcón 13, 28014 Madrid, Spain
Dock House, The Waterfront, Cape Town 8001, South Africa

http://www.cambridge.org

© Cambridge University Press 2001

This book is in copyright. Subject to statutory exception
and to the provisions of relevant collective licensing agreements,
no reproduction of any part may take place without
the written permission of Cambridge University Press.

First published 2001

Printed in the United States of America

Typeface Times Roman 11/13 pt.   *System* LaTeX 2_ε [TB]

A catalog record for this book is available from the British Library.

Library of Congress Cataloging in Publication Data
Padmanabhan, T. (Thanu), 1957–
Theoretical astrophysics : stars and stellar systems / T. Padmanabhan.
p.   cm.
Includes bibliographical references and index.
ISBN 0-521-56241-4 – ISBN 0-521-56631-2 (pb)
1. Stars.   2. Galaxies.   I. Title: Stars and stellar systems.   II. Title.
QB801.P23   2001
523.8 – dc21

00-058586

ISBN 0 521 56241 4 hardback
ISBN 0 521 56631 2 paperback

Dedicated to the memory of my father, Shri. N. Thanu Iyer,
who insisted on Excellence

THEORETICAL ASTROPHYSICS

– in three volumes –

VOLUME I: ASTROPHYSICAL PROCESSES

1: Order–of–Magnitude Astrophysics; 2: Dynamics; 3: Special Relativity, Electrodynamics, and Optics; 4: Basics of Electromagnetic Radiation; 5: Statistical Mechanics; 6: Radiative Processes; 7: Spectra; 8: Neutral Fluids; 9: Plasma Physics; 10: Gravitational Dynamics; 11: General Theory of Relativity; 12: Basics of Nuclear Physics.

VOLUME II: STARS AND STELLAR SYSTEMS

1: Overview: Stars and Stellar Systems; 2: Stellar Structure; 3: Stellar Evolution; 4: Supernova (Type II); 5: White Dwarfs, Neutron Stars, and Black Holes; 6: Pulsars; 7: Binary Stars and Accretion; 8: The Sun and the Solar System; 9: The Interstellar Medium; 10: Globular Clusters.

VOLUME III: GALAXIES AND COSMOLOGY

1: Overview: Galaxies and Cosmology; 2: Galactic Structure and Dynamics; 3: Friedmann Model; 4: Thermal History of the Universe; 5: Structure Formation; 6: Cosmic Microwave Background Radiation; 7: Formation of Baryonic Structures; 8: Active Galactic Nuclei; 9: Intergalactic Medium and Absorption Systems; 10: Cosmological Observations.

Contents

Preface		xiii
1	**Overview: Stars and Stellar Systems**	**1**
1.1	Introduction	1
1.2	Stars	2
1.3	Stellar Magnitudes and Colours	7
1.4	Overview of Stellar Evolution	14
1.5	Pulsars	25
1.6	Stellar Binaries	28
1.7	Interstellar Medium	33
1.8	Theoretical Constraints on Astronomical Observations	37
	1.8.1 Limitations Due to Earth's Atmosphere	38
	1.8.2 Limitation on Resolution	40
	1.8.3 Sensitivity and Signal-to-Noise Ratio	44
	1.8.4 Dynamic Range of Observations	48
	1.8.5 Luminosity Bias	52
	1.8.6 Confusion Limit	55
2	**Stellar Structure**	**57**
2.1	Introduction	57
2.2	Equations of Stellar Structure	57
2.3	Solutions to Equations of Stellar Structure	66
2.4	Toy Stellar Models	82
	2.4.1 Homologous Stellar Models	83
	2.4.2 The Radiative Stellar Envelope	90
	2.4.3 Fully Convective Stars with H^- Opacity	96
2.5	Observational Aspects of Stellar Atmospheres	101
	2.5.1 Continuum Radiation	102
	2.5.2 Lines	105

3 Stellar Evolution — 113

- 3.1 Introduction — 113
- 3.2 Overview — 114
- 3.3 Pre-Main-Sequence Collapse — 116
  - 3.3.1 Gravitational Instability and Mass Scales — 116
  - 3.3.2 Collapse of a Spherical Cloud — 118
  - 3.3.3 Contraction onto the Main Sequence — 124
  - 3.3.4 Brown Dwarfs — 129
  - 3.3.5 General Discussion of Homologous Collapse — 136
- 3.4 Evolution of High-Mass Stars — 139
  - 3.4.1 Initial Stages in Main Sequence — 140
  - 3.4.2 Shell Burning of Hydrogen — 140
  - 3.4.3 Core Contraction and Envelope Expansion — 142
  - 3.4.4 Red Giant Phase and Core Helium Ignition — 149
  - 3.4.5 Horizontal and Asymptotic Giant Branches — 150
- 3.5 Evolution of Low-Mass Stars — 151
  - 3.5.1 Core and Shell Hydrogen Burning — 151
  - 3.5.2 Red Giant Phase — 154
  - 3.5.3 Helium Flash — 157
  - 3.5.4 Evolution after the Helium Flash — 160
- 3.6 Late-Stage Evolution of Stars — 161
- 3.7 Stellar Oscillations and Stability — 171
  - 3.7.1 Adiabatic Spherical Pulsations — 172
  - 3.7.2 Nonadiabatic Effects in Radial Pulsations — 178
  - 3.7.3 Adiabatic Nonradial Oscillations — 181

4 Supernova (Type II) — 188

- 4.1 Introduction — 188
- 4.2 Overview — 188
- 4.3 Formation of Iron Cores and Neutrino Cooling — 190
- 4.4 Photodisintegration and Neutronisation — 196
- 4.5 Neutrino Opacity and Trapping — 198
- 4.6 Core Collapse — 202
- 4.7 Bounce and Possible Shock Formation — 206
- 4.8 Supernova Luminosity and Light Curves — 210
- 4.9 Evolution of Supernova Remnants — 216
- 4.10 Shock Acceleration in Supernova Remnants — 224
- 4.11 Stellar Nucleosynthesis — 230

5 White Dwarfs, Neutron Stars, and Black Holes — 236

- 5.1 Introduction — 236
- 5.2 Structure of White Dwarfs — 236

	5.2.1 White Dwarfs Supported by Degenerate Electron Pressure	238
	5.2.2 Coulomb Corrections at Low Densities	243
	5.2.3 Corrections at High Densities	246
	5.2.4 General Relativistic Corrections	249
	5.2.5 Effect of Rotation and Magnetic Field	257
5.3	Surface Structure and Thermal Evolution of White Dwarfs	259
5.4	Equation of State at Higher Densities	267
5.5	Neutron Star Models	271
5.6	Mass Bounds for Neutron Stars	274
5.7	Internal Structure of Neutron Stars	277
	5.7.1 Layers Inside a Neutron Star	277
	5.7.2 Superfluidity in Neutron Stars	278
	5.7.3 Microscopic Origin of a Bandgap	283
5.8	Gravitational Collapse and Black Holes	287
5.9	Rotating Black Holes	293

6 Pulsars — 296

6.1	Introduction	296
6.2	Overview	296
6.3	Electromagnetic Field Around the Pulsar	299
	6.3.1 The Aligned Rotator	304
	6.3.2 Pair Production in a Pulsar Atmosphere	314
6.4	Atomic Structures in Strong Magnetic Fields	318
6.5	Glitches in Pulsars	323
6.6	Pulsar Timing	326
6.7	Pulsar Scintillation	334
	6.7.1 Weak Scattering ($r_{\text{diff}} \gg r_F$)	338
	6.7.2 Strong Scattering ($r_{\text{diff}} \ll r_F$)	339

7 Binary Stars and Accretion — 343

7.1	Introduction	343
7.2	Overview	343
7.3	Example of an Evolving Binary System	349
7.4	Low-Mass and High-Mass X-Ray Binaries	354
7.5	Accretion Disks	358
	7.5.1 Model for Thin Disks	358
	7.5.2 Nature of Disk Viscosity	362
	7.5.3 Emergent Spectrum from a Thin Disk	365
7.6	Magnetic Effects in Accretion	371
	7.6.1 Magnetic Torques and Pulsar Spin-Up	375
	7.6.2 Millisecond Pulsars	379
7.7	General Relativistic Effects in Binary Systems	382

	7.7.1	Gravitational Radiation from Binary Pulsars	382
	7.7.2	Black Holes in Binary Systems	386
	7.7.3	Gravitational Radiation from Coalescing Binaries	389
7.8	Varieties of Accreting Binary Systems	392
	7.8.1	Novas	394
	7.8.2	Supernova Type I	396
	7.8.3	Cataclysmic Variables	397
	7.8.4	X-Ray Transients and Be Stars	399
	7.8.5	X-Ray Bursters	401

8 The Sun and the Solar System — 404

- 8.1 Introduction — 404
- 8.2 The Standard Solar Model — 404
- 8.3 Solar Neutrinos — 411
- 8.4 Solar Oscillations — 415
- 8.5 The Atmosphere and the Corona of the Sun — 420
- 8.6 Solar Flares — 425
  - 8.6.1 Magnetic Buoyancy — 426
  - 8.6.2 Magnetic Reconnection — 427
- 8.7 Generation of the Solar Magnetic Field — 431
- 8.8 Solar Wind — 436
- 8.9 Brief Description of the Solar System — 445
- 8.10 Aspects of Solar System Dynamics — 449
  - 8.10.1 Perturbation Theory — 451
  - 8.10.2 Example: Precession of Mercury — 452
  - 8.10.3 Precession of the Equinoxes — 456
  - 8.10.4 Tidal Friction — 462
  - 8.10.5 Long-Term Evolution of the Solar System — 466

9 The Interstellar Medium — 469

- 9.1 Introduction — 469
- 9.2 Overview — 469
- 9.3 Ionisation of the ISM Around a Star — 472
- 9.4 Propagation of Ionisation Fronts — 478
- 9.5 Heating and Cooling of the ISM — 482
- 9.6 Global Structure of the ISM — 490
- 9.7 Interstellar Electron Density and Magnetic Field — 498
  - 9.7.1 Electron Density from the Dispersion Measure — 498
  - 9.7.2 Faraday Rotation and the Interstellar Magnetic Field — 500
- 9.8 Radiation from Ionised Interstellar Gas — 502
- 9.9 21-cm Observations of Neutral Hydrogen — 509
- 9.10 Molecular Lines from the ISM — 514
- 9.11 Maser Action in the ISM — 518

9.12	Interstellar Dust Grains	522
9.13	Giant Molecular Clouds and Star Formation	527

10 Globular Clusters — 532

10.1	Introduction	532
10.2	Stellar Distribution and Ages of Globular Clusters	532
10.3	Time Scales in the Evolution of Globular Clusters	536
10.4	Fokker–Planck Description of Globular Cluster Dynamics	542
10.5	Aspects of Globular Cluster Evolution	552
10.6	Binary Stars in Globular Clusters	556
10.7	Interaction Between Clusters and Disk	559
10.8	Open Clusters	561
	Notes and References	565
	Index	571

Preface

"Yadhyadh vibuthimatsthwam srimadhurjithameva va
thaththadevava gachchatwam mama tejoamsa sambhavam"
("...Whatever that is glorious, prosperous or powerful anywhere,
know that to be a manifestation of a part of My splendour...")

— Bhagawad-Gita, Chapter 10, verse 41.

During the past decade or so, theoretical astrophysics has emerged as one of the most active research areas in physics. This advance has also reflected the greater interdisciplinary nature of the research that has been carried out in this area in recent years. As a result, those who are learning theoretical astrophysics with the aim of making a research career in this subject need to assimilate a considerable amount of concepts and techniques, in different areas of astrophysics, in a short period of time. Every area of theoretical astrophysics, of course, has excellent textbooks that allow the reader to master that *particular* area in a well-defined way. Most of these textbooks, however, are written in a traditional style that focusses on one area of astrophysics (say stellar evolution, galactic dynamics, radiative processes, cosmology, etc.) Because different authors have different perspectives regarding their subject matter, it is not very easy for a student to understand the key unifying principles behind several different astrophysical phenomena by studying a plethora of separate textbooks, as they do not link up together as a series of core books in theoretical astrophysics covering everything that a student would need. A few books, which *do* cover the whole of astrophysics, deal with the subject at a rather elementary (first-course) level.

What we require is clearly something analogous to the famous Landau-Lifshitz course in theoretical physics, but focussed to the subject of theoretical astrophysics at a fairly advanced level. In such a course, one could present all the key physical concepts (e.g., radiative processes, fluid mechanics, plasma physics, etc.)

from a unified perspective and then apply them to different astrophysical situations.

This book is the second of a set of three volumes intended to do exactly that. The three volumes form one single coherent unit of study, using which a student can acquire mastery over all the traditional astrophysical topics. What is more, these volumes will emphasise the unity of concepts and techniques in different branches of astrophysics. The interrelationship among different areas and common features in the analysis of different theoretical problems will be stressed throughout. Because many of the basic techniques need to be developed only once, it is possible to achieve significant economy of presentation and crispness of style in these volumes.

Needless to say, there are some basic "boundary conditions" one has to respect in such an attempt to cover the whole of theoretical astrophysics in approximately 3×560 pages. Not much space is available to describe the nuances in greater length or to fill in details of algebra. For example, I have made conscious choices as to which parts of the algebra can be left to the reader and which parts need to be worked out explicitly in the text, and I have omitted detailed discussions of elementary concepts and derivations. However, I do *not* expect the reader to know anything about astrophysics. All astrophysical concepts are developed *ab initio* in these volumes. The approach used in these three volumes is similar to that used by Gengis Khan, namely (*i*) cover as much area as possible, (*ii*) capture the important points, and (*iii*) be utterly ruthless!

To cut out as much repetition as possible, the bulk of the physical principles are presented at one go in Vol. I and are applied in the other two volumes to different situations. These three volumes also concentrate on *theoretical* aspects. Observation and phenomenology are, of course, discussed in Vols. II and III to the extent necessary to make the motivation clear. However, I do not have the space to discuss how these observations are made, the errors, reliability, etc., of the observations or the astronomical techniques. (Maybe there should be a fourth volume describing observational astrophysics!)

The target audience for this three-volume work will be fairly large and comprises (1) students in the first year of their Ph.D. program in theoretical physics, astronomy, astrophysics, and cosmology; (2) research workers in various fields of theoretical astrophysics, cosmology etc.; and (3) teachers of graduate courses in theoretical astrophysics, cosmology, and related subjects. In fact, anyone working or interested in some area of astronomy or astrophysics will find something useful in these volumes. They are also designed in such a way that parts of the material can be used in modular form to suit the requirements of different people and different courses.

Let me briefly highlight the features that are specific to Vol. II. The reader is assumed to be familiar with the material covered in Vol. I, having either studied that volume (which is the recomended procedure!) or through independent courses in basic physics. The spirit of the three coherent volumes is to avoid repetition as much as possible, and hence I have merely referred to the relevant parts of Vol. I whenever some input is required. Given the familiarity with basic physical processes, it was fairly easy to order the topics of Vol. II in a logical sequence. The fundamentals of stellar structure, stellar evolution, and stellar remnants – treated as isolated systems – are covered in Chaps. 2–6. The behaviour of binary stellar systems is different enough to warrant a separate chapter, Chap. 7. Chapters 8–10 are in some sense special topics: Chap. 8 deals with the Sun and the solar system, which deserve a detailed discussion in any course of astrophysics; Chap. 9 covers the interstellar medium and the cross talk between stars and the rest of the contents of the galaxy; finally, the short Chap. 10 describes some aspects of globular clusters.

This volume provided a tough challenge as regards the discussion of phenomenological input, and a few words regarding my policy are in order. Stellar astronomy is probably one of those areas in which observations lead the theory and the availability of accurate data allows one to recognise the complexity of several phenomena. These volumes, however, are intended to be a course on *theoretical* astrophysics, and hence the emphasis naturally is on the theoretical aspects rather than on observational and phenomenological issues. Given this dichotomy, it is easy to fall into one of the two traps: (1) Drown the reader in an accurate but unclassified sea of astronomical data just because accurate data are available or (2) ignore the phenomenological input and treat the subject as a branch of applied mathematics. I have tried hard to avoid both these pitfalls by adopting the following approach. I describe the necessary observational issues (but not observational techniques) and provide a minimum of observational data whenever they are relevant. I have also tried to motivate theoretical developments based on specific observational inputs, especially when a more fundamental approach would be unwarranted or facetious. At the same time I have tried to bring some amount of method and order in the presentation of the topics so that the reader will be able to grasp how a theoretical astrophysicist goes about the task of developing the models. One major problem in this approach was the interdependency of concepts (and even jargon) that prevents a fully streamlined development of topics. I have attempted to solve this difficulty by providing an overview in Chap. 1 that develops the necessary astronomical jargon, introduces the *dramatis personae*, and summarises the observational data that are general enough to be presented right at the outset. Chapter 1 also discusses several key issues of observational astronomy that are generic and reasonably independent of the technology available at a given time. (I plead guilty of not having yet learnt how to write my first chapters; this is the worst chapter in the book.)

All this required the exercise of my judgement in deciding the choice of topics, their emphasis, and the proper blend of phenomenology, observations, and theoretical rigour. It is impossible to satisfy everyone as regards the "correctness" of such decisions and I have tried to do some optimisation so as to provide the maximum benefit to the reader.

Any one of these topics is fairly vast and often requires a full textbook to do justice to it, whereas I have devoted approximately 60 pages to discussing each of them! I would like to emphasise that such a crisp, condensed discussion is not only possible but also constitutes a basic matter of policy in these volumes. After all, the idea *is* to provide the student with the essence of several textbooks in one place. It should be clear to lecturers that these materials can be easily regrouped to serve different graduate courses at different levels, especially when complemented by other textbooks.

Because of the highly pedagogical nature of the material covered in this volume, I have not given detailed references to original literature except on rare occasions when a particular derivation is not available in standard textbooks. The annotated list of references given at the end of the book cites several other textbooks that I found very useful. Some of these books, of course, contain extensive bibliographies and references to original literature. The selection of core books cited here clearly reflects the personal bias of the author, and I apologise to anyone who feels their work or contribution has been overlooked.

Several people have contributed to the making of these volumes and especially to Vol. II. The idea for these volumes originated over a dinner with J.P. Ostriker in late 1994, while I was visiting Princeton. I was lamenting to Jerry about the lack of a comprehensive set of books covering all of theoretical astrophysics, and Jerry said, "Why don't *you* write them?" He was very enthusiastic and supportive of the idea and gave extensive comments on and suggestions for the original outline I produced the next week. I am grateful to him for the comments and for the moral support that I needed to launch into such a project. I sincerely hope the volumes do not disappoint him.

Adam Black of Cambridge University Press took up the proposal with his characteristic enthusiasm and initiative. I should also thank him for choosing six excellent (anonymous) referees for this proposal whose support and comments helped to mould it into the proper framework. The processing of this volume was handled by Ellen Carlin of CUP and I thank her for the effort she has put in.

Many of my friends and colleagues carried out the job of reading the earlier drafts and providing comments. Of these, M. Vivekanand has gone through most of the book with meticulous care and has made extensive comments. H.M. Antia, D. Bhattacharya, S. Bhavsar, J. Chengalur, Nissim Kanekar, S. Konar, D. Narasimha, J.V. Narlikar, A.R. Ramprakash, S. Srianand, K. Subramanian, and F. Sutaria made detailed comments on selected chapters. Some of the figures and data were provided by H.M. Antia, Ranjan Gupta, Yashwant Gupta, Balchand Joshi, and A.R. Ramprakash. I thank all of them for their help.

I have been a regular visitor to the Astronomy department of Caltech during the past several years, and the work on the volumes has benefitted tremendously through my discussions and interactions with the students and staff of the Caltech Astronomy department. I would like to specially thank Roger Blandford, Peter Goldreich, Shri Kulkarni, Sterl Phinney, and Tony Readhead for several useful discussions and for sharing with me their insights and experience in the teaching of astrophysics.

This project would not have been possible but for the dedicated support from Vasanthi Padmanabhan, who not only did the entire TEXing and formatting but also produced most of the figures – often writing the necessary programs for the same. I thank her for the help and look forward to receiving the same for the last volume as well! I also thank Sunu Engineer, who was resourceful in solving several computer-related problems that cropped up periodically. It is a pleasure to acknowledge the library and other research facilities available at IUCAA, which were useful in this task.

T. Padmanabhan

1
Overview: Stars and Stellar Systems

1.1 Introduction

Stellar physics provides a natural starting point for the study of astrophysics for several reasons. To begin with, this is probably the best understood area of astrophysics. Second, there is a vast amount of reliable data dealing with stellar physics. This observational input motivates accurate and sophisticated theoretical modelling as well as provides the opportunity for a detailed comparison between models and observations. Finally, stellar physics also forms the basis for the study of several other related areas, even in the domain of extragalactic astronomy and cosmology. For example, measurements of cosmic distances and the ages of different structures – which are very important in cosmology – cannot be done without accurate modelling of the stellar phenomena that are used as tools; the study of formation and evolution of galactic systems requires an understanding of star formation and stellar evolution, etc. This volume deals with different aspects of the astrophysics of stellar systems.

The evolution of stars differs significantly, depending on whether the star is isolated or is a member of a binary system. The bulk of the chapters in the book (from Chap. 3 to Chap. 6) deal with stellar evolution and stellar remnants in isolated contexts. Chapter 7 is devoted to the study of evolution of binary stars, and Chap. 10 covers the dynamics of systems like globular clusters that have a very large number of stars. Because stars reside in – and are strongly coupled to – the interstellar medium (ISM), one full chapter (Chap. 9) is devoted to discussing the physics of the ISM.

Because of the interdependence of many of the concepts involved in this study, it is difficult to provide a completely logical and structured approach to stellar physics. It is necessary to take observational and phenomenological inputs within different contexts in order to guide the theoretical concepts appropriately. The purpose of this first chapter is to provide the necessary background so that the latter chapters can be developed in a streamlined fashion. In this chapter, most of the astronomical jargon, units, and a certain amount of observational and phenomenological inputs are introduced that will be required throughout the book. The observational input is kept at a general level, and aspects that are very specific to a particular topic will be discussed *in situ* in the relevant chapters.

It is also important to appreciate certain issues of observational astronomy that are very special to this subject and do not exist in other branches of applied physics, which draw inputs from laboratory experiments. Section 1.8 of this chapter highlights some of these issues and their relevance.[1–6] (The superscript numbers refer to the notes and the references given at the end of the book.)

1.2 Stars

There is sufficient evidence to believe that stars form when the local condensations of gas in a galaxy, contracting under self-gravity, reach sufficiently high central temperatures to ignite self-sustained nuclear reactions in the core. Although a fundamental theory that takes it into account all the relevant factors (rotation of the gas, magnetic and other nongravitational forces) and predicts the properties of the initial distribution of stars does not exist, the basic idea – that stars originate when sustained nuclear reactions take place in the core region of a self-gravitating cloud – seems to be well borne out, both theoretically and observationally. This idea will form the basis for the description of stellar structure and evolution in Chaps. 2 and 3.

In fact, this idea has sufficient predictive power to allow us to estimate several key properties of stars in an approximate manner. We shall briefly recall these results from Vol. I, Subsection 1.5.3. Consider a spherically symmetric cloud of, say, hydrogen gas, with mass M, radius R, and containing N protons, which is contracting under self-gravity. As the cloud contracts, the gravitational pressure $P_g \propto (GM^2/R^4)$ increases and needs to be balanced either by the thermal pressure of the gas or by the electron degeneracy pressure. This requires that the thermal energy or the Fermi energy of electrons (per particle) be comparable with the gravitational potential energy of the system (per particle); the latter is given by

$$\epsilon_g \equiv \frac{E_{\text{grav}}}{N} = \left(\frac{Gm_p^2}{R}\right) N = \left(\frac{4\pi}{3}\right)^{1/3} Gm_p^2 N^{2/3} n^{1/3}, \qquad (1.1)$$

where m_p is the mass of the proton and $n = (3N/4\pi R^3)$ is the number density of particles. To take into account both thermal and quantum degeneracy contributions to the energy, we take the total energy per particle to be $(k_B T + \epsilon_F)$, where $\epsilon_F = (\hbar^2/2m_e)(3\pi^2 n)^{2/3}$ is the Fermi energy of the electrons in the nonrelativistic limit. This energy will be comparable with gravitational energy if $(k_B T + \epsilon_F) \simeq Gm_p^2 N^{2/3} n^{1/3}$. Using the expression for ϵ_F for the nonrelativistic electrons, we get

$$k_B T \simeq G m_p^2 N^{2/3} n^{1/3} - \frac{(3\pi^2)^{2/3}}{2} \frac{\hbar^2}{m_e} n^{2/3}. \qquad (1.2)$$

As the radius R of the system is reduced, the second term on the right-hand side ($\propto n^{2/3}$) grows faster than the first ($\propto n^{1/3}$); hence the temperature of the system will first increase, then reach a maximum, and finally decrease again. The maximum temperature $T_{\max}$ is reached when $n = n_c$, with

$$n_c^{1/3} \cong \frac{\alpha_G}{(3\pi^2)^{2/3}} \left(\frac{N^{2/3}}{\lambda_e} \right), \quad k_B T_{\max} \simeq \frac{\alpha_G^2}{2(3\pi^2)^{2/3}} (N^{4/3} m_e c^2), \qquad (1.3)$$

where $\lambda_e \equiv (\hbar/m_e c) \approx 3.8 \times 10^{-11}$ cm is the Compton wavelength of the electron and $\alpha_G \equiv (G m_p^2/\hbar c) \approx 6 \times 10^{-39}$ is the gravitational equivalent of the fine-structure constant.

At temperatures higher than $\sim 10^3$ K, hydrogen will be ionised and we will have a plasma of electrons and protons. If the maximum temperature $T_{\max}$ of the plasma is sufficiently high to trigger nuclear fusion in the system, then we obtain a gravitationally bound, self-sustained nuclear reactor. For two protons to fuse together and undergo nuclear reaction, it is necessary that their de Broglie wavelengths $\lambda_{\text{deB}} \equiv (h/m_p v)$ overlap. Because this requires overcoming the Coulomb repulsion, such direct interaction can take place only if the kinetic energy of colliding particles is at least of the order of the electrostatic potential energy at the separation λ_{deB}. This requires kinetic energies of the order of $\epsilon \approx (e^2/\lambda_{\text{deB}}) \approx (\alpha^2/2\pi^2) m_p c^2 \approx 1$ keV, where $\alpha \equiv (e^2/\hbar c)$ is the fine-structure constant. It is conventional to write this expression as $\epsilon_{\text{nucl}} \approx \eta \alpha^2 m_p c^2$, with $\eta \simeq 0.1$. The quantity ϵ_{nucl} sets the scale for triggering nuclear reactions in astrophysical contexts. The energy corresponding to the maximum temperature $k_B T_{\max}$ [obtained in expression (1.3) above] will be larger than ϵ_{nucl} when

$$N > N_* \equiv (2\eta)^{3/4} (3\pi^2)^{1/2} \left(\frac{m_p}{m_e} \right)^{3/4} \left(\frac{\alpha}{\alpha_G} \right)^{3/2} \approx 4 \times 10^{56} \qquad (1.4)$$

for $\eta \simeq 0.1$. The corresponding condition on mass is $M > M_*$, with

$$M_* \approx (2\eta)^{3/4} (3\pi^2)^{1/2} \left(\frac{m_p}{m_e} \right)^{3/4} \left(\frac{\alpha}{\alpha_G} \right)^{3/2} m_p \approx 4 \times 10^{32} \text{ gm}, \qquad (1.5)$$

which is comparable with the mass of the smallest stars observed in our universe. The radius is $R_* \cong (G M_* m_p / k_B T_{\max}) \simeq 3 \times 10^{10}$ cm. The mass and the radius of the Sun, for example, are $M_\odot = 2 \times 10^{33}$ gm and $R_\odot \simeq 7 \times 10^{10}$ cm, respectively. Most of the stars in the universe powered by nuclear reactions have masses in the range $(0.1$–$60) M_\odot$.

According to the above description, stars form in overdense regions of primordial gas in the galaxy. To understand the relationship between the stars and the galaxy, it is necessary to model the origin of the galaxy itself. This is somewhat more uncertain and will be discussed in detail in Vol. III. It is, however, possible to understand the characteristic mass and size of a galaxy by analysing the cooling processes operating in a protogalactic cloud. Such an analysis in Vol. I, Subsection 1.5.1, showed that the size and the mass of a typical galaxy are

$$R_g \simeq \alpha^3 \alpha_G^{-1} \lambda_e \left(\frac{m_p}{m_e}\right)^{1/2} \simeq 74 \, \text{kpc},$$

$$M_g \simeq \alpha^5 \alpha_G^{-2} \left(\frac{m_p}{m_e}\right)^{1/2} m_p \simeq 3 \times 10^{44} \, \text{gm},$$

(1.6)

where 1 kpc $\simeq 3 \times 10^{21}$ cm. A comparison of M_g with expression (1.5) for M_* shows that the number of stars $N_{\text{star}} \simeq (M_g/M_*)$ in a typical galaxy will be given by a combination of fundamental constants $N_{\text{star}} = \alpha^{7/2} \alpha_G^{-1/2} (m_e/m_p)^{1/4} \simeq 10^{12}$. Typical galaxies indeed have approximately 10^{11}–10^{12} stars, although there is a fair amount of spread in this number. Most of the visible mass in the galaxy is contained in a region somewhat smaller than the size estimated above, at $R_{\text{gal}} \approx 20$ kpc.

Given the radius of the galaxy and the number of stars in it, we can estimate the mean distance between the stars to be $d_{\text{star}} \approx (R_{\text{gal}}/N_*^{1/3}) \approx 3 \times 10^{18}$ cm $\equiv 1$ pc. Thus we expect to see stars in our galaxy to be distributed at distances varying from a few parsecs to a few tens of kiloparsecs. A star like the Sun, with a radius of $\sim 10^{11}$ cm and located at a distance of 10 pc from us, will subtend an angle of ~ 1 milliarcsecond; it is clear that most stars will look like point objects.

One of the important questions in observational astronomy is the determination of spatial distribution of stars in the galaxy, which – to a large extent – is independent of the structure and the dynamics of the stars. Thus the simplest observation we can make regarding a star is to measure its position in the sky, which requires the specification of two suitable angular coordinates. However, because most of the astronomical data are either gathered from Earth or from satellites, which have systematic motion at short time scales with respect to a fixed cosmic reference frame, it is necessary to define accurately the coordinate system used in any given astronomical observation. Although several coordinate systems are possible, each having its own domain of applicability and utility, there is one coordinate system called *the equatorial system* that appears to be natural to the observations based on Earth. We shall now briefly describe how such a coordinate system is defined, as the measurement of the position of an object in the sky is of central importance in any branch of astronomy.

The rotation of the Earth about its axis defines two unique directions in the sky called the north celestial pole (NCP) and the south celestial pole (SCP), which are obtained by the extension of the Earth's axis of rotation to the celestial sphere (see Fig. 1.1). The great circle in the celestial sphere, formed by the plane perpendicular to this axis, defines the celestial equator. Treating the Earth

1.2 Stars

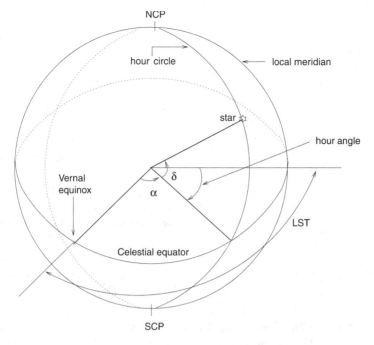

Fig. 1.1. The equatorial coordinate system.

as a sphere of radius R, we can define a local normal at any given location; the intersection of this normal with the celestial sphere defines the direction of the zenith at any given location on Earth. The great circle on the celestial sphere passing through the celestial poles and the zenith is called the local meridian.

As the Earth revolves around the Sun annually, the Sun appears to move from west to east on the celestial sphere in a path called the ecliptic, inclined at approximately $23°27'$ to the celestial equator. The two great circles – the ecliptic and the celestial equator – intersect at two different points called the vernal and the autumn equinoxes. The Sun passes through the vernal equinox (VE) on approximately 21 March, moving from south to north of the celestial equator.

Given the celestial equator and the VE, it is possible to define the position of any celestial object – say, a star – by the following procedure. We first draw the great circle (called the hour circle) passing through the celestial poles and the star. We can now specify the coordinates of the star by giving two angles: (1) the *declination* δ, which is the angular distance measured from the celestial equator to the star along the star's hour circle and (2) the *right ascension* (RA) α, which is the angular distance along the celestial equator from the VE to the star's hour circle. (In terms of the standard spherical polar coordinate system, $\alpha \equiv \phi$ and $\delta \equiv 90° - \theta$.) By convention, the RA is usually expressed in terms of hours, minutes, and seconds rather than in degrees, with the convention of 24 h corresponding to 360°. RA increases from west to east so that the stars with larger RA rise later than those with smaller ones. The declination is taken to be positive northward of the celestial equator and negative southward.

The motivation for using time units rather than angular units to measure RA comes from the fact that a star's RA is very nearly equal to the time between the meridian transit of the star and the meridian transit of the VE. The *local sidereal time* (LST) is the RA of the meridian expressed in units of time. It is also usual to define a quantity called the *hour angle*, which is the angle along the celestial equator between the meridian and the hour circle that, by convention, is measured in the sense opposite to α. It follows that a star's hour angle (1) is the difference between the LST and the star's RA and (2) will be equal to the time since the star crossed the meridian. Obviously, a star will be at the meridian when the LST is equal to the star's RA.

The coordinate system based on the NCP and the VE suffers from the difficulty that neither of these directions remains static in the celestial sphere because of the complicated dynamical process acting in the solar system (see Chap. 8.) Careful corrections have to be applied in order to define the coordinate system with respect to a hypothetical mean NCP and mean VE. Further, it is also necessary to define a particular instant in time with respect to which α and δ are measured. Several such epochs are used, with the most popular ones being those based on the Julian year 1950 or 2000.

Exercise 1.1
Practice with coordinate systems I: An astronomer wants to observe a star with RA and dec$(X, Y) = (23^h 20^m 39^s, +18°08'33'')$, say. It would be best if the source is as high as possible in the sky at the time of observation. Which part of the year is best suited for this? [Answer: If the declination is positive, it is best to do the observation from the northern hemisphere whereas for $Y < 0$ it is better to use a telescope in the southern hemisphere. What we really need to determine is the time of the year at which the observation can best be carried out. To have the star as high as possible in the sky, an observer would like to choose the time of the year such that the star crosses the observer's meridian near the middle of the night. We begin by noting the two relations: (1) The LST when a star crosses the meridian is equal to the RA of the star and (2) the LST at midnight on any particular day is the RA of the Sun on that particular day (RA$_\odot$) plus the hour angle of the Sun at midnight. Because the latter is 12^h, it follows that LST (midnight) = RA$_\odot + 12^h$. Because we want the LST at the time of the star's crossing the meridian to be the same as the LST at midnight, we can equate the two expressions for the LST. This gives the RA$_\odot$ = RA$_* - 12^h = (X - 12)^h$. This is the expression we need. Because the RA of the Sun is zero on the spring equinox (roughly March 21) and varies by 2 h/month, the RA of the Sun will have the value $(X - 12)^h$ after approximately $(X - 12)/2$ months since the spring equinox. In this specific case mentioned in the question, $(X - 12)/2 \approx 6$ months, and hence the second or third week of September will be the best time for the indicated observation.]

Exercise 1.2

Practice with coordinate systems II: A star has RA ($+6^h$) and dec $+30°$. What is its altitude above the horizon for an observer at $30°$ north when the LST is 9^h? At what LST will it be overhead? At what time of the year will the star be directly overhead at local midnight?

Exercise 1.3

Galactic coordinate system: Another coordinate system frequently used in extragalactic astronomy is based on our galaxy. In this scheme the galactic equator is chosen to be that great circle on the celestial sphere that closely approximates the plane of the Milky Way – which, in turn, is inclined at an angle of $62.87°$ to the celestial equator. The north galactic pole is located at $(\alpha_{GP}, \delta_{GP}) = (192.85948°, 27.12825°) \simeq (12^h51^m, +27°7.7')$. The galactic latitude b of an object is the angle from the galactic equator to the star along the great circle through the star and galactic poles. The galactic longitude l is measured along the galactic equator from the direction of the galactic centre. This direction corresponding to $l = 0, b = 0$ has the equatorial coordinates $(\alpha, \delta) = (266.405°, -28.936°) \simeq (17^h45.6^m, -28°56.2')$. Show that the galactic coordinates (l, b) are related to the equatorial coordinates (α, δ) by

$$\sin b = \sin \delta_{GP} \sin \delta + \cos \delta_{GP} \cos \delta \cos(\alpha - \alpha_{GP}),$$
$$\cos b \sin(l_{CP} - l) = \cos \delta \sin(\alpha - \alpha_{GP}), \quad (1.7)$$
$$\cos b \cos(l_{CP} - l) = \cos \delta_{GP} \sin \delta - \sin \delta_{GP} \cos \delta \cos(\alpha - \alpha_{GP}),$$

where $l_{CP} = 123.932°$ is the longitude of the NCP.

1.3 Stellar Magnitudes and Colours

Once the nuclear reactions occur in the hot central region of the gas cloud, its structure changes significantly. If the transport of this energy to the outer regions is through photon diffusion, then the opacity of matter will play a vital role in determining the stellar structure. In particular, the opacities determine the relation between the luminosity and the mass of the star.

A photon with mean free path $l = (n\sigma)^{-1}$, randomly walking through the hot plasma in the interior of the star, will have $N_{coll} \simeq (R/l)^2$ collisions in traversing the radius R. This will take the time $t_{esc} \simeq (lN_{coll}/c) \simeq (R/c)(R/l)$ for the photon to escape. The luminosity of a star L will be proportional to the ratio between the radiant energy content of the star, E_γ, and t_{esc}. Because $E_\gamma \simeq (aT^4)R^3 \propto T^4R^3$, we find that

$$L \propto \frac{R^3 T^4 l}{R^2} \propto RT^4 l \propto \frac{RT^4}{n\sigma}. \quad (1.8)$$

For a wide class of stars, we may assume that the central temperature $k_B T \simeq (GMm_p/R)$ is reasonably constant because nuclear reactions – which depend strongly on T – act as a thermostat. If Thomson scattering dominates, then

$\sigma = \sigma_T \equiv [(8\pi/3)(e^2/m_e c^2)^2]$ and we get

$$L \propto \frac{RT^4}{\sigma_T n} \propto \frac{T^4 R^4}{\sigma_T N} \propto \frac{M^4}{M} \propto M^3. \qquad (1.9)$$

The situation is different if interaction of photons with partially ionised atoms provide the opacity. The cross section for bound–free and bound–bound opacity in thermal equilibrium has been obtained in Vol. I, Chap. 1, Subsection 1.4.4, where it was shown that $l \propto T^{7/2} n^{-2} \propto T^{7/2} R^6 M^{-2}$. [The bound–free opacity can be understood as follows: In equilibrium, the photoionisation rate, which removes energy from the radiation field, should match the recombination rate. The amount of energy removed by photoionisation is proportional to $dE_{\rm ion} \propto n_{\rm atom} \sigma_{\rm bf} T^4$, where $\sigma_{\rm bf}$ is the photoionisation cross section. The energy supplied by recombination scales as $dE_{\rm rec} \propto n_e n_i v \propto n_e n_i T^{1/2}$. Equating $dE_{\rm ion}$ to $dE_{\rm rec}$, we get $n_{\rm atom} \sigma_{\rm bf} T^4 \propto n_e n_i T^{1/2}$. Introducing the bound–free opacity $\kappa_{\rm bf}$ by the definition $\kappa_{\rm bf} = (n_{\rm atom} \sigma_{\rm bf}/\rho)$ and taking $n_e \propto \rho$, $n_i \propto \rho$, we find that $\kappa_{\rm bf} \propto \rho T^{-3.5}$.] In this case, we have

$$L \propto RT^4 l \propto R^7 T^{15/2} M^{-2} \propto M^{11/2} R^{-1/2}. \qquad (1.10)$$

Taking $(GM/R) \approx$ constant so that $R \propto M$ now gives $L \propto M^5$. Taken along with expression (1.9), we expect the luminosity of a star to be related to its mass by $L \propto M^\alpha$, with $\alpha \simeq 3\text{--}5$.

If we imagine the surface of the star to be at some effective temperature T_s, then the total blackbody luminosity from the star will be $L = (4\pi R^2)(\sigma T_s^4)$, where $\sigma = (\pi^2 k^4/60 \hbar^3 c^2)$ is the Stefan–Boltzmann constant. It is convenient to use this relation to define the surface temperature T_s of the star with a given luminosity L and radius R. If the radiation from the star is approximated as that of a blackbody, then the intensity f_ν (which is the energy per unit area per unit time per solid angle per frequency) of thermal radiation emitted by the star will be

$$f_\nu = \frac{dE}{dA\,dt\,d\Omega\,d\nu} = B_\nu \equiv \frac{2h\nu^3}{c^2} \frac{1}{e^{h\nu/k_B T_s} - 1}. \qquad (1.11)$$

The quantity νB_ν (which gives the intensity per logarithmic band in frequency) reaches a maximum value near $h\nu \approx 4 k_B T$, which translates to the fact that a blackbody at 6000 K will have the maximum for νB_ν at 6000 Å. The maximum intensity is $(\nu B_\nu)_{\rm max} \approx (T/100\ {\rm K})^4$ W m^{-2} sr^{-1}.

This description provides a few more useful observed characteristics of a star. In principle, we can fit the spectrum of a star to a blackbody spectrum (approximately) and obtain T_s. The total energy flux l received from the star (called the *apparent luminosity*) can also be measured directly. Because $l = L/(4\pi d^2)$, where d is the distance to the star, we can determine the *absolute luminosity* L of the star if the distance to the star is known. Assuming that the distance can be independently measured, we will be able to determine L and T_s and plot the location of the stars in a two-dimensional (L–T_s) plane.

1.3 Stellar Magnitudes and Colours

From our definition, it follows that $T_s \propto L^{1/4}R^{-1/2} \propto L^{1/4}M^{-1/2}$. Combining with $M \propto L^{1/5}$, valid when the interior is only partially ionised, we get $T_s \propto L^{1/4}L^{-1/10} \propto L^{3/20}$. On the other hand, if Thomson scattering dominates with $L \propto M^3$, we get $T_s \propto L^{1/12}$. Thus, if the stars are plotted on a log T_s–log L plane, (called the Hertzsprung–Russel diagram or the H-R diagram) we expect them to lie within the lines with slopes $3/20 = 0.15$ and $1/12 \simeq 0.08$. The observed slope is ~ 0.13, giving reasonable support to the basic ideas developed above. Observationally, it is found that stellar surface temperatures vary from approximately 3×10^3 K to 3×10^4 K as the mass varies from approximately $0.1\,M_\odot$ to $60\,M_\odot$. The corresponding variation in the radius is in the range of $(0.8–70) \times 10^{10}$ cm, and the luminosity ranges from $10^{-3}\,L_\odot$ to $10^{5.7}L_\odot$, where $L_\odot = 3 \times 10^{33}$ ergs s^{-1} is the luminosity of the Sun.

The intensity f_ν, defined above as the amount of energy received per second per unit area per unit frequency interval, is a fundamental quantity characterising the radiation from *any* celestial object. It is therefore important to stress some practical issues related to its measurement. Actual measurements in optical, IR, and UV bands do not measure the f_ν of a source directly. The observed intensity f can be usually expressed in the form

$$f \equiv \int_0^\infty f_\nu \mathcal{T}_\nu \mathcal{F}_\nu \mathcal{R}_\nu \, d\nu \equiv \int_0^\infty f_\lambda S_\lambda \, d\lambda, \qquad (1.12)$$

where each of the factors signify the following features: (1) $\mathcal{T}_\nu$ measures the fractional transmission that is due to the Earth's atmosphere, thereby connecting the observed intensity and the intensity on top of the Earth's atmosphere. This is, of course, irrelevant for satellite-based observations. (2) No realistic apparatus can be equally sensitive at all frequencies or be sensitive at only a given frequency ν_0. The factor $\mathcal{F}_\nu$ is the fractional sensitivity of the filter used in the telescope or other measuring apparatus at frequency ν. For practical purposes, the apparatus can be characterised by a mean frequency ν_0 at which it is most sensitive and a full width at half maximum (FWHM) that specifies the band of frequencies over which significant sensitivity exists. (3) $\mathcal{R}_\nu$ represents the efficiency of the detector and is the ratio between energy detected and energy incident upon the detector. The second equality in Eq. (1.12) gives the corresponding equation in terms of wavelength, with S_λ combining the effects of all the three factors. Among these factors, the filter response $\mathcal{F}_\nu$ is probably most important and is often used to characterise the intensity in different frequency bands such as the ultraviolet (U) band, blue (B) band, visible (V) band, red (R) band, etc. For the sake of standardisation, each of these bands is specified in terms of an effective wavelength (λ_{eff}) at which the band is centred and a FWHM.

For historical reasons, astronomical measurements (especially those in the optical band) are quoted in terms of another quantity, called *magnitude*, which is related to f in a logarithmic manner. This unit is not of any intrinsic value

Table 1.1. Filter characteristics of broadband photometric systems

Band	λ_{eff} nm	W_λ nm	$\frac{dE}{dt\,dA\,d\nu}$ (Jy)	$\frac{dE}{dt\,dA\,d\lambda}$*	$a^\dagger$	$b^\ddagger$	$c^\S$
U	365	66	1780	4.0×10^{-8}	22	150	9.9
B	445	94	4000	6.1×10^{-8}	23	100	9.4
V	551	88	3600	3.6×10^{-8}	22	170	15
R	658	138	3060	2.1×10^{-8}	21	250	35
I	806	149	2420	11.2×10^{-9}	18.5	1.5×10^3	223
J	1220	213	1570	3.07×10^{-9}	16	1.0×10^4	2.1×10^3
H	1630	307	1020	1.12×10^{-9}	13	5.6×10^4	1.7×10^4
K	2190	390	636	4.07×10^{-10}	12.5	4.4×10^4	1.8×10^3
L	3450	472	281	7.30×10^{-11}	5.5	8.0×10^6	3.8×10^6
M	4750	460	154	2.12×10^{-11}	2	1.0×10^8	4.6×10^7

*Flux density of a zero-magnitude star per unit wavelength [$f_\lambda(0)$/W m^{-2} μm^{-1}].
†Background intensity in magnitude arcsec^{-2}.
‡Background photon intensity per unit wave band [$I(\lambda)$/photons m^{-2} arcsec^{-2} s^{-1} μm^{-1}].
§Background photon intensity in standard wave band (I/ photons m^{-2} arcsec^{-2} s^{-1}), obtained as the product of the value given in b and W$_\lambda$.

and can be fairly confusing; however, as it is unlikely that optical astronomy will switch to a more rational and scientific unit of measurement in the near future, it is necessary for us to define and relate this archaic concept to the flux measured in physical units.

The total flux F_X in physical units integrated over the filter function with width W_X from an object with apparent magnitude m_X in the band X can be written as

$$F_X \equiv W_X f_X = (10^Q W_X) 10^{-0.4 m_X}, \quad (1.13)$$

where Q is defined as $\log f_X$ for a reference star with $m_X = 0$ and W_X is the FWHM for the band X; that is,

$$Q = \log\left(\frac{f_\lambda}{\text{ergs cm}^{-2} \text{ s}^{-1} \mu\text{m}^{-1}}\right), \quad (1.14)$$

where f_λ is the flux of a reference star with $m = 0$ and the width is conveniently measured in micrometers. For the U, B, V, R, I, J, and K bands, $Q = -4.37, -4.18, -4.42, -4.76, -5.08, -5.48$, and -6.40, respectively, by definition.

Table 1.1 gives the details for a commonly used photometric system. The fourth entry in the table (f_X) gives the flux density in the X band for an $m_X = 0$ reference star in units of 1 Jy $= 10^{-26}$ W m^{-2} Hz^{-1} $= 10^{-23}$ ergs cm^{-2} s^{-1} Hz^{-1} commonly used in radio astronomy. The fifth entry gives the flux density of a zero-magnitude star per unit wavelength [$f_\lambda(0)$/W m$^{-2}\mu$m^{-1}] $= 3 \times 10^{-6} f_X$(Jy)λ^{-2}(nm). It

1.3 Stellar Magnitudes and Colours

should be stressed that a wavelength interval, $\Delta\lambda$, corresponds to a frequency interval, $\Delta\nu$, with $|\Delta\nu| = (c/\lambda^2)\Delta\lambda$. (The entries a, b, and c will be described later on.)

Finally, we mention one more measure of brightness of celestial objects, called *apparent bolometric magnitude* m_{bol}, which is defined as

$$m_{\text{bol}} \equiv -2.5 \log\left(\int_0^\infty d\nu\, f_\nu\right) + C_{\text{bol}}, \qquad (1.15)$$

where the constant C_{bol} is decided by convention. This quantity is a measure of the total flux from the object, and we shall follow the convention in which the total flux F and m_{bol} are related by

$$\left(\frac{F}{\text{W m}^{-2}}\right) = (2.75 \times 10^{-8}) 10^{-0.4 m_{\text{bol}}}. \qquad (1.16)$$

Thus $m_{\text{bol}} = 0$ corresponds to a flux of $F_0 \simeq 2.75 \times 10^{-8}$ W m^{-2} = 2.75×10^{-5} ergs cm^{-2} s^{-1}; F and m are related approximately by $\log F \cong -8 - 0.4(m-1)$ when F is measured in watts per square meter.

As an example, consider a star such as the Sun with $R = 7 \times 10^{10}$ cm and a surface temperature of $T_s \approx 5500$ K; the Planckian radiation intensity f_λ corresponding to this temperature will peak at $\lambda \approx 5500$ Å (in the visible band). The total flux of radiation from such a star located at a distance of 10 pc will be approximately $F = 3 \times 10^{-10}$ W m^{-2}, showing that a sunlike star at 10 pc will have a bolometric magnitude of ~ 4.8. When the flux changes by 1 order of magnitude, the magnitude changes by 2.5.

The relation between visual magnitude m_V and the bolometric magnitude m_{bol} is given by $m_{\text{bol}} = m_V - \text{BC}$, where BC is called the *bolometric correction*. In general, the bolometric correction will depend on the spectrum and is usually tabulated for a blackbody spectrum as a function of temperature.

Starlight also contributes to the background light in the sky in the optical band. A solid angle $d\Omega$ will intercept a volume $(1/3)R^3 d\Omega$ of our galaxy if R is the radius of the galaxy. If the number density of bright stars with luminosity $L \simeq L_\odot$ is $n \simeq 0.1$ pc^{-3}, then the flux per steradian is

$$\mathcal{F} \simeq \frac{1}{3} R^3 (n L_\odot) \frac{1}{4\pi (R/2)^2} \simeq \frac{1}{3\pi} n L_\odot R \simeq 2.4 \times 10^{-5} \text{ W m}^{-2} \text{ rad}^{-2}, \qquad (1.17)$$

if $R = 10$ kpc. Using 1 rad$^2 \simeq 4 \times 10^{10}$ arcsec^2, we get a sky brightness that is due to integrated starlight of approximately 6×10^{-16} W m^{-2} arcsec^{-2}. Another (confusing) convention used in optical astronomy relates to expressing the flux per unit solid angle as equivalent magnitudes per unit solid angle. For example, a flux of "f_X ergs cm^{-2} s^{-1} μm^{-1} arcsec^{-2}" will be quoted as "m_X mag arcsec^{-2}," where f_X and m_X are related by Eq. (1.13). With this convention, the above sky background can be stated ~ 21 mag arcsec^{-2}.

We have seen above that a galaxy consists of $\sim 10^{11}$ stars and could be located at distances ranging from 1 to 4000 Mpc. At 10 Mpc, such a galaxy will subtend an angle of $\theta \simeq (2R/d) \simeq 200''$ and will have a flux of approximately 3×10^{-11} W m^{-2} if the size of the galaxy $R_{\rm gal} \simeq 10$ kpc. The surface brightness of the galaxy will be $\sim 10^{-16}$ W m^{-2} arcsec^{-2} or ~ 21 mag arcsec^{-2}. These galaxies also contribute to the background light in the optical band. Repeating the above analysis we did for stars with $L = 10^{11} L_\odot$, $R \simeq 4000$ Mpc, and $n \simeq 1$ Mpc^{-3}, we get a background of 2×10^{-17} W m^{-2} arcsec^{-2}, equivalent to 24.5 mag arcsec^{-2}.

The entries in Table 1.1 marked a, b, and c give the following quantities related to sky background: a, background intensity of the sky in magnitude arcsec^{-2}; b, background photon intensity per unit wave band $[I(\lambda)/\text{photons m}^{-2} \text{ arcsec}^{-2} \text{ s}^{-1} \mu\text{m}^{-1}]$; c, background photon intensity in standard wave band given in column three ($I/$ photons m^{-2} arcsec^{-2} s^{-1}).

Exercise 1.4

Practice with magnitudes: In a device that counts photons statistically, there will be a fluctuation around the mean number of photons observed around the average. Let us assume that when we expect to observe N photons on the average, the actual value can fluctuate by an amount $\pm\sqrt{N}$. (a) How many photons need to be collected if the magnitude of a star has to be measured to an accuracy of ± 0.02? How long an exposure will we require with a 1-m telescope to measure the B magnitude of an $M_B = 20$ star with the same accuracy? (b) Assume that the night sky has the same brightness as that produced by one 22.5-B-mag star per square arc second. Express the night-sky brightness in ergs cm^{-2} s^{-1} sr^{-1} Å^{-1}. (c) Consider an attempt to measure the magnitude of an $M_B = 24$ star within an accuracy of ± 0.02 mag. Assume that the starlight is spread out over 1 arcsec^2. How many photons are required to be collected to achieve this and what should be the exposure time of a 4-m telescope for this measurement? [Answer: (a) Because $F \propto 10^{-0.4m} \propto \exp -m(0.4 \ln 10) \propto \exp(-0.92\,m)$, it follows that $(\Delta F/F) = 0.92(\Delta m) = (\Delta N/N) = N^{-1/2}$. Therefore the number of photons needed to measure the magnitude with an accuracy Δm will be $N = (0.92\,\Delta m)^{-2}$. In this case, we get $N \approx 2.9 \times 10^3$ photons. To find out how long an exposure will give us this many photons, we first use Eq. (1.13) with $Q = -4.18$, $W_X = 0.094$ μm, and $m_X = 20$ to get $F = 6.2 \times 10^{-14}$ ergs cm^{-2} s^{-1}. The corresponding photon flux is $n = (F/h\nu) = 1.39 \times 10^{-2}$ photons cm^{-2} s^{-1}. The telescope with a 1-m aperture will have an area $A = (\pi/4)10^4$ cm^2 and will collect $nA = 1.1 \times 10^2$ photons s^{-1}. To collect $N \approx 2.9 \times 10^3$ photons, the exposure should be for ~ 26 s. (b) Application of Eq. (1.13) with $Q = -4.18$, $W_X = 0.094$ μm, and $m_X = 22.5$ now gives $F = 6.2 \times 10^{-15}$ ergs cm^{-2} s^{-1}. This is the flux received in a bandwidth of 94 nm, i.e., 940 Å and over 1 arcsec$^{-2} \approx 2.3 \times 10^{-11}$ rad^{-2}. Hence the flux is equivalent to $F = 3.5 \times 10^{-6}$ ergs cm^{-2} s^{-1} sr^{-1} Å^{-1}. (c) Because the magnitude of the star that is treated as being spread over 1 arcsec^2 is higher than that of the sky background, we will be limited by sky-background noise rather than photons from the source. The rest of the calculation proceeds as in part (a).]

1.3 Stellar Magnitudes and Colours

The amount of energy emitted by an object in different wavelength bands contains valuable information regarding the physical processes that occur in the object. Ideally, this would require the measurement of the entire spectrum (that is, the specific intensity at all wavelengths) of the object. In practical situations, we may often have to be content with the measurements of intensity in different bands such as U, B, V, etc. In such a case, some of the spectral properties of the object can still be usefully quantified in terms of a set of *colour indices* that describe the variation of f_λ with λ in a coarse-grained manner. If A and B denote two different filters, then the colour index corresponding to A and B is defined as

$$(\text{CI})_{AB} \equiv m_A - m_B = \text{constant} - 2.5 \log \frac{\int_0^\infty d\lambda \, S_\lambda(A) f_\lambda}{\int_0^\infty d\lambda \, S_\lambda(B) f_\lambda}. \qquad (1.18)$$

Quite often we simply denote the colour index by the notation B–V, U–B, etc. It is clear from Eq. (1.18) that the colour index measures the ratio of fluxes near the effective wavelengths of the two bands. Because it depends on only the ratio of the fluxes, it is an intrinsic property of the spectrum and is independent of the distance to the object.

The apparent magnitude defined above is based on the flux received by a detector – say, on Earth – and will vary as r^{-2}, where r is the distance to the source. The *absolute magnitude* M of an object is defined to be the apparent magnitude of that object if it is located at a standard distance r_0, which is conventionally taken to be 10 pc. It follows that

$$m - M = 5 \log r - 5, \qquad (1.19)$$

where r is measured in parsecs. The quantity $m - M$ is called the *distance modulus* of the object.

It should be noted that Eq. (1.19) incorporates only the inverse square falloff of the intensity. If the light from a source reaches us through, say, the ISM that causes absorption and scattering of photons, then the object will appear dimmer than it would in the absence of such physical processes. For a given object with an intrinsic absolute magnitude this will cause the apparent magnitude to be larger. It is again conventional to express the interstellar extinction in terms of magnitude and rewrite Eq. (1.19) as

$$m - M = 5 \log r - 5 + A, \qquad (1.20)$$

where the extinction has caused an increase by A magnitudes to the apparent brightness of the object. In general, A will depend strongly on the wave band.

Another complication that arises in the use of Eq. (1.19) has to do with the redshift of radiation. There are several physical processes (for example, cosmological expansion) that can cause the photons received in, say, the V band to have been actually emitted at shorter wavelengths. This effect will make m and M refer to different wavelength bands if Eq. (1.19) is used naively. To obtain the correct relation, it is necessary to apply another correction to this equation and write it in the form

$$m - M = 5 \log r - 5 + A + K, \tag{1.21}$$

where K is called the K correction. The actual form of this correction depends on the shape of the spectrum of the object.

1.4 Overview of Stellar Evolution

We next overview the key features of stellar evolution that will be elaborated on in Chaps. 3–7. Figure 1.2 summarises the entire evolutionary sequence of a

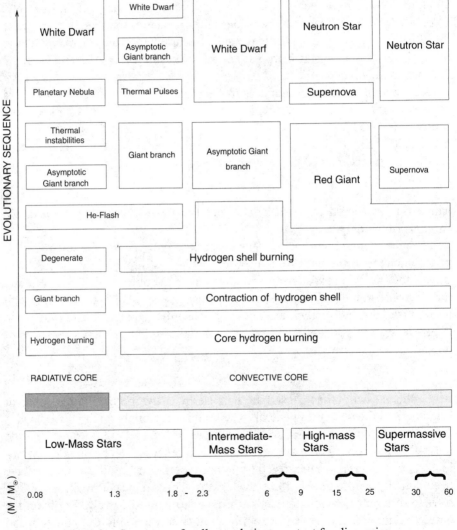

Fig. 1.2. Summary of stellar evolution; see text for discussion.

1.4 Overview of Stellar Evolution

star as a function of its initial mass. The following comments highlight the key points in the figure.

The bottommost line of Fig.1.2 gives the mass of the star in terms of solar mass. For $M < 0.08\,M_\odot$, nuclear reactions cannot be sustained in the contracting cloud for it to become a star. Such objects end up as *planets* or brown dwarfs (see Chap. 3, Subsection 3.3.4). Similarly, systems with masses higher than approximately (60–100) $M_\odot$ are unstable and cannot last for a significant period of time. This sets the upper and the lower limits in the bottommost line.

In the allowed mass range, stellar structure and evolution are characterised by some key values for the masses. Based on this fact, we have divided the stars into low-mass, intermediate-mass, high-mass, and supermassive stars. The division is based on some features in the evolution, which are described below. It should also be noted that stars with $M \lesssim 1.3\,M_\odot$ are structurally different from those with $M \gtrsim 1.3\,M_\odot$. We saw in Vol. I, Chap. 12, Section 12.4, that the hydrogen fusion leading to helium can occur either through p–p reactions or through a carbon–nitrogen–oxygen (CNO) cycle, depending on the temperature. Stars with $M \gtrsim 1.3\,M_\odot$ have the CNO cycle as the dominant reaction channel whereas stars with $M \lesssim 1.3\,M_\odot$ have p–p reactions as the dominant channel. Because the CNO cycle has a much greater temperature sensitivity, resulting in a larger temperature gradient in the core region, these stars have a convective core whereas stars with $M \lesssim 1.3\,M_\odot$ have a radiative core. This implies that hydrogen burning will occur in a convective environment and the ashes of burning will be uniformly distributed over the core region by convection in stars with $M \gtrsim 1.3\,M_\odot$. This is indicated by the shaded area in the figure.

(1) The first – and longest – phase of stellar evolution occurs when stars act as gravitationally bound systems in which nuclear reactions fusing hydrogen into helium are taking place in the centre. The process of combining four protons into a helium nuclei releases $\sim 0.03\,m_p c^2$ of energy, which is approximately a fraction, 0.007, of the original energy $4\,m_p c^2$. Assuming that a fraction $\epsilon \approx 0.01$ of the total rest-mass energy can be made available for this nuclear reaction, the lifetime of the nuclear burning phase of the star will be $t_{\text{star}} = \epsilon M/L \approx 3 \times 10^9$ yr $(\epsilon/0.01)(M/M_*)^{-2}$ if the opacity is due to Thomson scattering so that expression (1.9) is applicable. This defines the characteristic time scale in stellar evolution. In fact, t_{star} represents the longest stretch of time in a star's life spent in steady-state nuclear burning. Such stars are called *zero-age main-sequence* (ZAMS) stars. Most of the scaling relations obtained above are applicable to such stars, and they will form a band in the H-R diagram called the *main-sequence band*. Because any given star spends a maximum amount of time in the main sequence, it follows that most of the stars will be found along the main sequence (see Fig. 1.3).

After a characteristic time $t \gtrsim t_{\text{star}}$, the hydrogen will be exhausted in the core and the nuclear reaction will cease. The response of the star (and its further evolution) is now determined essentially by its mass.

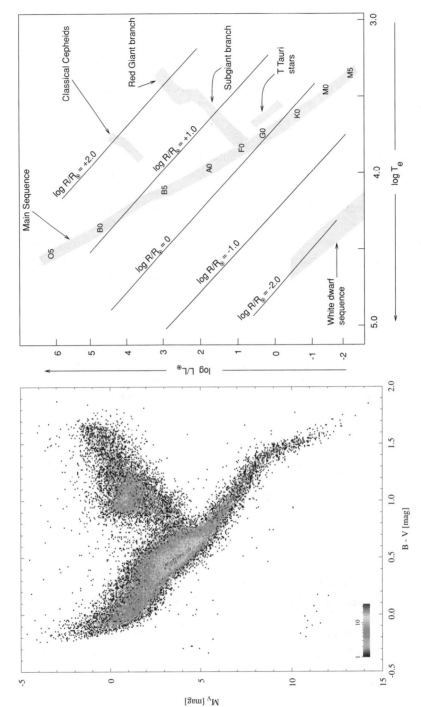

Fig. 1.3. The left panel shows the colour-luminosity diagram for a large number of stars in the Hipparcos database (figure based on the data made available at the Hipparcos website). The right panel is the theoretical version of the H-R diagram showing the distribution of stars in different branches.

(2) Let us first consider the evolution of low-mass stars. Once the hydrogen burning is over, the core contracts under its weight and the outer regions expand, causing the star to bloat up in size. (This epoch is called the *giant* or the *red giant* phase.) Eventually the core becomes sufficiently hot to trigger helium burning, but this occurs after the electrons in the core have already become degenerate. In such a case, the temperature increase does not affect the pressure significantly (as the pressure – provided by degenerate electrons – is mainly dependent on only the density) and the helium burning occurs as a runaway process (called *helium flash*) in a time scale of a few days. This part of the evolution is called the *asymptotic giant branch*. The low-mass stars do not become hot enough to ignite further nuclear reactions. During the later part of evolution, the outer regions of the star undergo violent instabilities, leading to an eventual ejection of a significant fraction of material in the form of a *planetary nebula*. The core region will become a *white dwarf* supported by the degeneracy pressure of electrons. To achieve this balance, the Fermi energy ϵ_F of the electrons must be larger than the gravitational potential energy $\epsilon_g \cong G m_p^2 N^{2/3} n^{1/3}$. When the particles are nonrelativistic, $\epsilon_F = (\hbar^2/2m_e)(3\pi^2)^{2/3} n^{2/3}$ and the condition $\epsilon_F \geq \epsilon_g$ can be satisfied (at equality) if

$$n^{1/3} = \frac{2}{(3\pi^2)^{2/3}} \left(\frac{G m_p^2 m_e}{\hbar^2} \right) N^{2/3}. \quad (1.22)$$

With $n = (3N/4\pi R^3)$ and $N = (M/m_p)$, this reduces to the following mass–radius relation:

$$RM^{1/3} \simeq \alpha_G^{-1} \lambda_e m_p^{1/3} \simeq 8.7 \times 10^{-3} \, R_\odot \, M_\odot^{1/3}. \quad (1.23)$$

Such structures supported by electron degeneracy pressure are called white dwarfs. A white dwarf with $M \simeq M_\odot$ will have $R \simeq 10^{-2} R_\odot$ and density $\rho \simeq 10^6 \rho_\odot$. The evolution of low-mass stars leads to a helium or a carbon core of the stars ending up as white dwarfs.

White dwarfs are fairly faint objects and are somewhat difficult to detect. For example, a 200-in. telescope can reach a V magnitude of 25. For $T_s = 10^4$ K, the bolometric correction is 0.3, making the corresponding bolometric limiting magnitude $25 - 0.3 = 24.7$. Consider a white dwarf with $T_s \approx 10^4$ K and radius $R \approx 6 \times 10^8$ cm located at a distance r from Earth. The flux received from it will be $F = \sigma T_s^4 (R/r)^2$. Using the fact that $m_{bol} = 0$ corresponds to $F_0 = 2.75 \times 10^{-5}$ ergs cm^{-2} s^{-1}, we can easily verify that the limiting bolometric magnitude is reached when $r \approx 2.5$ kpc. Because white dwarfs are significantly dimmer than the normal stars, they will lie to the left and below the main sequence in a H-R diagram. A typical carbon white dwarf could have a radius of $10^{-2} R_\odot$, a luminosity of $L \approx 0.03 L_\odot$, and a surface temperature of $T_s = 2.7 \times 10^4$ K.

(3) The evolution of intermediate-mass stars is similar, although many of the details vary with mass. For example, when the hydrogen in the core is exhausted

and the core contracts, the burning of hydrogen in a thin shell around the core occurs in these stars. Eventually these stars proceed to helium burning, but with a nondegenerate core. The final stage is again a white dwarf supported by degeneracy pressure. These stars also do not proceed beyond helium burning.

(4) High-mass stars with $M \gtrsim 8\,M_\odot$ proceed further in the nuclear cycle and ignite carbon burning in the core and – depending on the mass – proceed to even further stages in nuclear burning. A sufficiently high-mass star will eventually have an onion-ring structure of material, with iron, nickel, silicon, oxygen, neon, carbon, helium, and hydrogen occupying successive layers from the core to the surface. Such stars could also have different nuclear reactions, proceeding simultaneously in different regions, for example, in the core and in a shell.

(5) Several high-energy processes become important in the core region of the star at this stage. For example, neutrino interactions with nuclei and cooling of the core through neutrino emission can be significant even at the silicon burning phase. For a 20 $M_\odot$ star, the photon luminosity during this phase is approximately 4.4×10^{38} ergs s^{-1} whereas the neutrino luminosity can be as high as 3.1×10^{45} ergs s^{-1}. Neutrinos interact only very weakly with matter and are characterised by a very small scattering cross section σ. For example, the reaction $\bar{\nu}_e(p,n)e^+$ has a cross section $\sigma = 10^{-43}$ cm$^2(\epsilon_\nu/1\text{ MeV})^2$ whereas the reaction $\nu_e(e^-,e^-)\nu_e$ has a cross section $\sigma = 10^{-48}$ cm$^2(\epsilon_\nu/1\text{ MeV})$. The mean free path for neutrinos corresponding to the first reaction at $\epsilon_\nu \approx 1$ MeV is $\lambda = (n\sigma)^{-1} = (m_p/\rho\sigma) \simeq 10^{19}$ cm ρ^{-1}. This is $\sim 10^6\,R_\odot$ for typical stars and approximately $(10^3\text{--}10^4)\,R_{\rm wd}$ for white dwarfs. Therefore neutrinos can escape freely from stars and white dwarfs, carrying energy with them. The main drain of energy arises through the reactions

$$e^+ + (Z,A) \to (Z-1,A) + \nu \to (Z,A) + e + \bar{\nu} \qquad (1.24)$$

in which the kinetic energy of the electron is transferred to a ν–$\bar{\nu}$ pair. These reactions are significant at $T > 5 \times 10^9$ K. At these temperatures, e^+–e^- annihilations will also result in ν–$\bar{\nu}$ pair production in 1 out of 20 cases and γ production in the remaining 19 out of 20 cases. The energy production per volume in this process can be estimated as

$$\mathcal{E} \simeq (n_e n_{\bar{e}})\sigma c\,(k_B T) \simeq \left(\frac{aT^4}{k_B T}\right)^2 \sigma c\,(k_B T). \qquad (1.25)$$

Using $\sigma \approx 10^{-44}$ cm$^2\,(T/3 \times 10^9\text{ K})^2$, we get this energy-production rate as $\mathcal{E} \simeq 10^{20}$ ergs cm$^{-3}(T/3 \times 10^9\text{ K})^9$. The corresponding cooling time at this temperature $t_{\rm cool} \simeq (aT^4/\mathcal{E})$ is $\sim$2 h, showing that this can be a significant drain of the energy.

The photodisintegration of nuclei that is due to collisions with energetic photons present at high temperatures, as well as inverse beta decay, can reduce the pressure support of the core, triggering a rapid collapse. During the process, inverse beta decay converts a significant fraction of electrons and protons into

1.4 Overview of Stellar Evolution

neutrons, making the core neutron rich. As the core radius decreases and the density increases, the neutrinos from weak interactions will be trapped inside. An order-of-magnitude estimate for ν trapping, which illustrates the scalings involved, can be obtained as follows: When the collapse of the iron core begins, the physical parameters are $\rho_{\text{core}} \approx 10^{(9-10)}$ gm cm^{-3}, $t_{\text{collapse}} \approx (G\rho)^{-1/2} \approx 0.1$ s, and $\epsilon_F \approx 50$ MeV $(\rho_{12}/\mu_e)^{1/3}$. The neutrinos are trapped when their diffusive velocity v_{diff} becomes less than the in-fall velocity v_{collapse}. These two velocities scale as follows:

$$v_{\text{collapse}} \simeq \left(\frac{GM}{R}\right)^{1/2} \propto \sqrt{\rho}\, R \propto \rho^{1/6}, \tag{1.26}$$

as $R \propto \rho^{-1/3}$. Numerically, $v_{\text{collapse}} \approx 3 \times 10^9 \rho_{12}^{1/6}$ cm s^{-1}. On the other hand, $v_{\text{diff}} = (c/\tau)$, where τ, the optical depth, is given by

$$\tau \approx n\sigma R \propto \rho E_\nu^2 \rho^{-1/3} \propto \rho^{2/3} \epsilon_F^2, \tag{1.27}$$

as $E_\nu \approx \epsilon_F$. Using the fact that $\epsilon_F^2 \propto \rho^{2/3}$, we find that $\tau \propto \rho^{4/3}$, giving $v_{\text{diff}} \propto \rho^{-4/3}$. Numerically, $v_{\text{diff}} = 2 \times 10^8 \rho_{12}^{-4/3}$ cm s^{-1}. Equating $v_{\text{diff}} = v_{\text{collapse}}$ gives the critical density at which neutrinos are trapped, which is $\rho_{\text{trap}} \approx 2 \times 10^{11}$ gm cm^{-3}. When the nuclear reactions prevent further contraction of the core, a shock wave could be set up from the centre to the surface of the star, which is capable of throwing out the outer mantle of the star in a *supernova* explosion. The total kinetic energy of the outgoing shock is $\sim 10^{51}$ ergs, which is roughly 1% of the energy liberated in neutrinos. When the outer material expands to $\sim 10^{15}$ cm and becomes optically thin, an impressive optical display arises, releasing $\sim 10^{49}$ ergs of energy in photons with a peak luminosity of 10^{43} ergs s$^{-1} \simeq 10^9 L_\odot$. Such a supernova is one of the brightest phenomena in the universe in the optical band. The outgoing gas ploughs through the ISM with an ever-decreasing velocity and eventually will appear as a shell-like supernova remnant with a typical radius of ~ 40 pc. Different radiative processes operate in the supernova remnant during different stages of its evolution. Most important among them are the x-ray bremsstrahlung emission and radio synchrotron emission that are used to probe these remnants. Details of some of the dominant supernova remnants are given in Table 1.2. The remnant is a *neutron star* supported essentially by the nuclear forces at high densities.

The overall effect of supernova explosions on the evolution of the ISM and the galaxy is quite significant. To make a simple estimate, let us assume that the supernova explosion can be approximated as the instantaneous appearance of a remnant of radius $R = 100$ pc that survives for a time of $t \simeq 10^6$ yr. If the galaxy is modelled as a disk of 15 kpc radius and 200 pc thickness and if the supernova rate for our galaxy is $(1/30)$ yr^{-1} (occurring randomly in the disk), then the supernova rate in physical units is

$$\mathcal{R}_{\text{snr}} \approx 2.3 \times 10^{-13} \text{ pc}^{-3} \text{ yr}^{-1}. \tag{1.28}$$

Table 1.2. Some X-ray-emitting supernova remnants

Source	α 1950 δ 1950	Distance (kpc)	Diameter (pc)	Angular size (arc min)	Age (yr)	L_x (0.2–2 keV) (10^{35} ergs s^{-1})	Flux Density at 1 GHz (Jy)
Crab Nebula	$05^h 31^m$, $21°59'$	2	3	3.0×4.2	900	160	1000
Cas A	23 21, 58 33	3	3.5	4.0×3.8	300	30	3000
Cygnus Loop	20 49, 30 30	0.8	40	200×160	20000	8	180
Vela	8 32, −45 00	0.5	44	220×180	13000	4	1800
Tycho's SNR	00 22, 63 52	6	13	6.0×7.0	400	40	58
SN 1006	15 00, −41 45	1.2	10	30×22	970	0.2	25
PKS 1209-52	12 06, −52 10	2	40	86×75	20000	0.7	49

1.4 Overview of Stellar Evolution

The fraction of the volume filled by the supernova remnant (SNR) will be

$$f = 1 - \exp\left[-\left(\frac{4\pi}{3}\right) R^3 \mathcal{R}_{\text{snr}} t\right] \approx 0.55. \tag{1.29}$$

In other words, more than half the volume of the galaxy would have been run through a supernova remnant. Any given part of the disk will have this occurring in a period of $\sim 10^6$ yr on the average. Thus the entire ISM is affected by the occurrence of supernova remnants.

Equation (1.22) is still applicable to the remnant of a supernova explosion, with m_e replaced with m_n; correspondingly, the right-hand side of relation (1.23) is reduced by $(\lambda_n/\lambda_e) = (m_e/m_n) \simeq 10^{-3}$. Such objects – called neutron stars – will have a radius of $R \simeq 10^{-5} R_\odot$ and a density of $\rho \simeq 10^{15} \rho_\odot$ if $M \simeq M_\odot$. For such values, $(GM/c^2 R) \simeq 1$ and general relativistic effects are beginning to be important.

When the density of the core is still higher, the Fermi energy has to be supplied by relativistic neutrons and ϵ_F now becomes $\epsilon_F \simeq \hbar c n^{1/3}$, which scales as $\epsilon_F \propto n^{1/3}$, just like ϵ_g. Therefore the condition $\epsilon_F \geq \epsilon_g$ can be satisfied only if $\hbar c \geq G m_p^2 N^{2/3}$ or only if $N \leq \alpha_G^{-3/2} \simeq N_* \alpha^{-3/2}$, where N_* is defined in Eq. (1.4). The corresponding mass bound (called the Chandrasekhar limit) is $M \lesssim m_p \alpha_G^{-3/2} \simeq 1 M_\odot$. (A more precise calculation gives a slightly higher value.) If the mass of the stellar remnant is higher than $\alpha_G^{-3/2} m_p$, no physical process can provide support against the gravitational collapse. In such a case, the star will form a black hole and is likely to exert a very strong gravitational influence on its surroundings.

The tracks of the stars in the H-R diagram as they evolve will be different for different masses. This feature, along with the time scales spent in different phases, determines the density of stars in different regions of the actual H-R diagram. Figure 1.3 gives an actual H-R diagram of stars (on the left) based on the data of the Hipparcos – which is a High Precision Parallax Collecting Satellite that has determined the positions of more than a million stars with an accuracy of 0.001″ – and a theoretical schematic H-R diagram on the right. The main band running from left top to right bottom is the main sequence of stars and the branching towards the right top denotes the red giant branch. The actual data are given in the form of a colour-magnitude diagram, with the x axis giving the B-V magnitude. Large values of B-V will broadly indicate lower surface temperature compared with small values of B-V, and hence the x axis can be thought as measuring the surface temperature from right to left. The magnitudes on the y axis, of course, are a measure of the total luminosity. The right-hand figure makes this clear with the x and the y axes marked directly in terms of surface temperature T_s and luminosity $L \propto R^2 T^4$. Lines of constant R [corresponding to $(L^{1/2}/T^2) = $ constant] are also shown diagonally across the diagram. The main-sequence stars span a fairly wide range of temperature, luminosity, and radii. The subgiant and the red giant branches into which some main-sequence stars will evolve correspond to significantly higher radii (hence the name giants). On the other end are the white dwarfs, which form a distinct branch with much smaller radii and luminosities compared with those of the main-sequence stars. This component can also be seen faintly in the Hipparcos data.

Observationally the surface temperature of the stars are often measured by use of the properties of the stellar atmosphere (especially the strength of lines emitted by different elements), which we will discuss in detail in Chap. 2, Section 2.5. The basic idea is the following: Because the temperature decreases as we go from the centre to the surface of the star, the radiation emitted from the hot interior can be absorbed by the cooler gas in the stellar atmosphere, producing absorption lines characteristic of the local conditions prevailing in the outer regions. A given absorption line will be produced by a particular element, in a particular ionisation state, when an electron makes the transition between two energy levels. The probability for such an absorption by partially ionised elements in their excited states will depend sensitively on the temperature. At high temperatures, atoms tend to be completely ionised and hence will not possess discrete energy levels corresponding to bound states. At low temperatures there will be very few atoms in the excited state that can produce the given absorption line. It follows that absorption that is due to partially ionised atoms in their excited states will be maximum at some intermediate temperature – which can be estimated if the atomic properties are known. Further, absorption lines that are due to different elements will peak at different temperatures. Hence, by choosing different atomic line transitions, we can probe the stellar atmospheres effectively. Such a spectroscopic analysis has led to a classification of stars, usually denoted by the sequence of letters O, B, A, F, ..., etc., which are roughly in order of decreasing surface temperature. These letters are not of any particular significance, and the classification scheme is fairly archaic; nevertheless, it is widely used in astronomical literature and hence is marked along the main-sequence band in Fig. 1.3. Figure 1.4 shows the typical spectra of these classes of stars, indicating absorption by different elements in different bands.

Although most stars have a relatively stable luminosity during the significant part of their life, some stars show striking variability in their energy outputs. These intrinsically variable stars have their light output changing in a periodic or semiperiodic fashion because of internal processes. The simplest of the variations arises because of radial pulsations of the stars at some period P determined by the star's internal structure. (We will discuss this phenomenon in Chap. 3, Section 3.7; also see Ref. 3.15.) The period of oscillation is directly related to the mean density of the stars and hence can be connected to the luminosity and the surface temperature. A reasonable fit to the observations is given by

$$\log\left(\frac{L}{L_\odot}\right) = -17.1 + 1.49 \log\left(\frac{P}{1\ \text{day}}\right) + 5.15 \log T_{\text{eff}}. \qquad (1.30)$$

Because we can measure both T_{eff} and P without knowing the distance to the star, this relation allows us to determine the absolute luminosity of a variable

1.4 Overview of Stellar Evolution 23

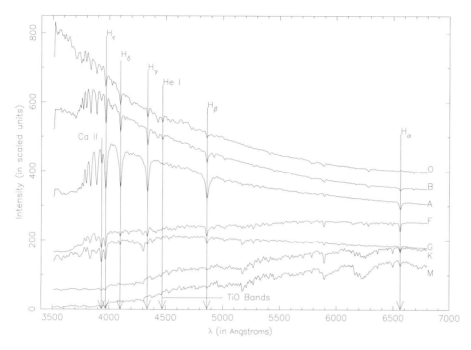

Fig. 1.4. Spectra of different types of stars indicating the characteristic features (figure courtesy of R. Gupta).

star without knowing its distance. By measuring the apparent luminosity and comparing it with absolute luminosity, we can estimate the distance to the star. In particular, if such a variable star is spotted in another galaxy, this technique will provide an estimate of the distance to the galaxy – which is of considerable significance in cosmology. This is one of the examples in which detailed modelling of a particular class of stars helps in the understanding of issues that are important in a wider context. The location of such a class of stars called classical cepheids is also shown in the right-hand panel of Fig. 1.3.

Stellar evolution also has the following significant feature. It synthesises a gamut of heavier nuclei starting from hydrogen and eventually distributes them in the ISM through the material that is ejected from the star in different stages. When the second-generation stars form from the enriched ISM they will contain a higher fraction of elements heavier than helium than the first generation of stars. It is conventional to call the first generation of stars, with mass fractions for hydrogen (X), helium (Y), and heavier elements (Z) of approximately $X = 0.76$, $Y = 0.24$, and $Z \approx 0$, *Population II* (Pop II) stars and the second generation of stars, with $X \cong 0.7$, $Y \cong 0.28$, and $Z \cong 0.02$, *Population I* (Pop I) stars. Because the equation of state as well as opacity depends on the metal fraction Z and the molecular weight, the evolution of Pop I and Pop II stars is measurably different. Although on the main sequence, Pop II stars are hotter and brighter; during the different evolutionary phases Pop II stars evolve

Table 1.3. Principle characteristics of stellar populations

Characteristics	Population I	Population II
Heavy-element content	2%–3%	<1%
Dominant spectral types	O, B, A	K, M
	Blue supergiants	Globular clusters, red giants
Variables	T Tauri	RR Lyrae
	δ Cepheids	Long-period variables
		Planetary nebulae
		Novae
Location and distribution	Galactic disk, primarily in spiral arms (patchy)	Galactic nucleus and halo (smooth)
Kinematics		
Height above disk	$\lesssim$200 pc	>400 pc
Average velocity	<10 km s^{-1}	$\gtrsim$20 km s^{-1}
Age	<1.5 × 10^9 yr	>1.5 × 10^9 yr

faster than Pop I stars. Some of the properties of these stars are described in Table 1.3.

Figure 1.5 summarises our knowledge of the contribution of stars to the mass and the luminosity of the galaxy.[7] The curves give (1) the cumulative number per 10^{-1} pc^{-3}, (2) the cumulative amount of mass (in units of 10^{-1} $M_\odot$ pc^{-3}), and (3) the cumulative amount of V-band luminosity (in units of 10^{-1} $L_\odot$ pc^{-3}) contributed by stars that are brighter than a given magnitude.

It is clear from the curves in Fig. 1.5 that most stars in the solar neighbourhood are faint, with the maximum number density being contributed by stars with $M = 15$. Nevertheless, most of the light comes from the bright stars, even though they are quite rare. In fact, stars fainter than approximately fourth magnitude contribute very little to the luminosity density. The average V-band luminosity that is due to stars is ∼0.053 $L_\odot$ pc^{-3}.

The situation is quite different as regards mass density, with all the stars in the magnitude range of ∼3–15 contributing nearly uniform amounts. The total mass density is ∼0.036 $M_\odot$ pc^{-3}. It should be stressed, however, that these data do not include the mass contributed by most of the white dwarfs, which have magnitudes $M = 13$–16 and masses in the range of (0.5–1) $M_\odot$. They contribute ∼0.003 $M_\odot$ pc^{-3} so that the total contribution to mass density is roughly 0.039 $M_\odot$ pc^{-3}.

It is also possible to investigate the mass density contributed by different subsets of stellar objects. The maximum mass density of ∼25 $M_\odot/10^3$ pc^3 is contributed by the dwarf M stars, with G and K dwarfs together giving ∼13 $M_\odot/10^3$ pc^3. The next major contributor is the white dwarf population, with

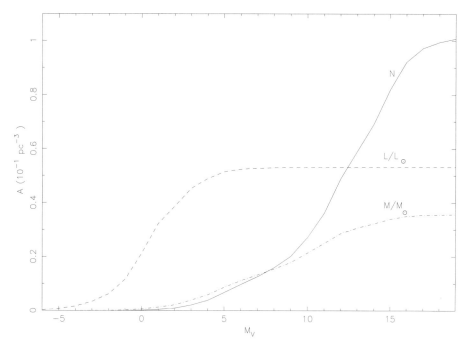

Fig. 1.5. Contribution of stars to number, luminosity, and mass densities in our galaxy.

20 $M_\odot/10^3$ pc^3. O, B, A, and F types give ~5 $M_\odot/10^3$ pc^3. The contribution from special kinds of stars like Cepheids (10^{-3} $M_\odot/10^3$ pc^3) and long-period variables are quite small. Galactic clusters give ~0.04 $M_\odot/10^3$ pc^3, and globular clusters give another 10^{-3} $M_\odot/10^3$ pc^3.

1.5 Pulsars

We have seen in the last section that stellar evolution can lead to remnants that are supported by the quantum degeneracy pressure of electrons (in white dwarfs) or by nuclear forces (in neutron stars). Because the original stellar core (from which the remnant is formed) is likely to have some angular momentum that is conserved during the collapse, the neutron star will be rotating fairly rapidly. Similarly, if the original star has some magnetic field, the flux of which is conserved during the contraction, the neutron star will end up with a high value of magnetic field. Such rotating, magnetic neutron stars emit a characteristic pattern of pulsed radiation with a remarkably stable period. These objects, called pulsars, are extensively studied and are used as probes of different aspects of astrophysics. (We will discuss pulsars in detail in Chaps. 6 and 7). Isolated pulsars allow the probing of the interstellar electron density and magnetic field by the processes of plasma dispersion and Faraday rotation. Pulsars in binary systems and clusters allow the study of a host of other phenomena related to the evolution of binary stars. They also act as clocks located in strong gravitational fields, thereby helping us to test specific predictions of general relativity.

The energy emitted by the pulsar spans a wide band of wavelengths – from radio to hard x rays – with most of the energy being emitted at higher frequencies. In spite of this fact, pulsars are most conveniently observed in radio frequencies, and over the years a significant amount of data has accumulated about radio pulsars. There are now ~1200 pulsars that are believed to be a galactic population of isolated neutron stars whose radiation is powered by rotation. The periods of rotation range from 1.5 ms to 8.5 s, with most of the pulsars being confined to periods of 0.2 to 2 s. As the rotational speed of the pulsar decreases, the period will increase, with the typical rate of change of period being $\dot{P} \approx 10^{-15}$. For a spherical neutron star, with $M \approx 1.4\ M_\odot$, $R \approx 12$ km, and $I \approx 1.4 \times 10^{45}$ gm cm^2, the total kinetic energy is $K = 2\pi^2(I/P^2) \simeq 2.5 \times 10^{49}$ ergs for $P \approx 0.03$ s. The Crab pulsar, for example, originated as a result of a supernova explosion that occurred in the year 1054 A.D. and hence is $T \cong 10^3$ yr old. Its period, $P = (2\pi/\Omega_0)$, is ~0.033 s. Determining $|\dot{\Omega}| = (\Omega_0/T) \approx 10^{-8}$ s^{-2}, we can estimate the energy-loss rate to be approximately $dE/dt = I\Omega\dot{\Omega} = 6.4 \times 10^{38}$ ergs s^{-1}. The magnetic moment of a neutron star is $m \approx (BR^3/2)$, and the magnetic dipole radiation from a rotating neutron star will be approximately

$$\frac{dE}{dt} = -\frac{2}{3c^3}|\ddot{\mathbf{m}}|^2 = -\frac{B^2 R^6 \Omega^4 \sin^2\alpha}{6c^3}, \quad (1.31)$$

where α is the angle between the magnetic dipole axis and the rotation axis (see Chap. 6 for a detailed discussion). If $\sin\alpha \approx 1$, then Eq. (1.31) allows us to determine the magnetic field as $B \approx 5.2 \times 10^{12}$ G. These are typical values for the pulsars.

Except for the slow decay, the periods of the pulsars happen to be remarkably stable, making them excellent clocks. The individual shapes of the pulses show a wide variety of microstructures with different numbers of peaks, subpulses, etc., which are not yet fully understood. Among the class of radio pulsars, 9 have also been observed in optical, 20 in x rays, and 7 in gamma rays. By and large, the luminosity at higher frequencies is ~4 orders of magnitude larger than the radio luminosity. The pulse peaks are also broader at higher frequencies and do not always show strict alignment across the frequency band. One of the most extensively studied pulsars is the Crab pulsar, mentioned above, in which radiation is observed over 60 octaves from 75 MHz to 500 GeV. It shows a double peak profile with a separation between the peaks of ~0.4 times the period at all energies. The spectral index varies widely from -2.7 in radio, -0.7 at x rays, and -1.1 in gamma rays; in between the index is positive, having a value of about ~2 at the IR band.

The top panel of Fig. 1.6 shows the distribution of pulsars in the sky in equatorial coordinate systems. They are mostly concentrated along the galactic plane that spans a particular curve in this coordinate system. Special kinds of pulsars

1.5 Pulsars

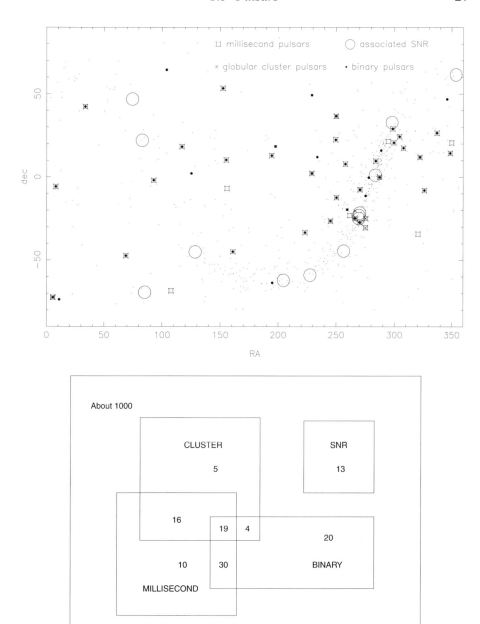

Fig. 1.6. The top panel shows the distribution of pulsars in the sky with specific symbols demarcating special class of pulsars. Pulsars are mostly confined to the galactic plane, which appears as a curved band in this figure. The lower panel summarises the statistics of $\sim 10^3$ pulsars, indicating the numbers available in different categories. The figure is based on the data made available on the Web by the Princeton Pulsar Group. (http://pulsar.princeton.edu/pulsar)

like those in a binary system or globular cluster, those associated with supernova remnants, and those which have very short periods (millisecond pulsars) are indicated by separate symbols. The bottom panel gives a rough Venn diagram of the pulsars, indicating the number of examples in each class.

1.6 Stellar Binaries

In the description of stellar evolution we have assumed that the star is an isolated system and is not affected by its surroundings, which is certainly not the case if the star is a part of a binary system. It is then possible for one of the binary stars to evolve into a compact remnant – white dwarf, neutron star, or even a black hole – and accrete matter out of the second star, leading to a wide variety of new physical phenomena. In an accretion process, the gravitational potential energy is converted into the kinetic energy of matter and dissipated as thermal radiation. Some of the very-high energy sources of radiation – both galactic and extragalactic – are generally believed to be powered by such accretion processes. In the galactic scale, accretion disks around stars can be a source of thermalised x-ray emission; in the extragalactic domain there are objects called *active galactic nuclei* and quasars (having a luminosity of about $\sim 10^{44}$ ergs s^{-1}) that are thought to be powered by accretion disks around very massive black holes.

The basic idea behind the process of accretion is the following: When a mass M falls from infinite distance to a radius R, in the gravitational field of a massive object with mass M_c, it gains the kinetic energy $E \simeq (GM_cM/R)$. If this kinetic energy is converted into radiation with efficiency ϵ, then the luminosity of the accreting system will be $L = \epsilon \, (dE/dt) = \epsilon \, (GM_c/R)(dM/dt)$. In stellar contexts, the compact object will have a mass in the range of $(0.6–1.4)\, M_\odot$ in the case of white dwarfs or neutron stars and somewhat higher in the case of black holes. For material accreting onto a neutron star of mass $1.4\, M_\odot$ and radius $R = 10^6$ cm, the accretion luminosity will be

$$L_{\rm acc} = \frac{GM_c \dot{M}}{R} = 2.95\, L_\odot \left(\frac{R}{10^6 \text{ cm}}\right)^{-1} \left(\frac{M_c}{1.4\, M_\odot}\right) \left(\frac{\dot{M}}{10^{-12}\, M_\odot \text{ yr}^{-1}}\right). \tag{1.32}$$

The accretion luminosity will therefore increase in proportion to $\dot{M}$.

There is, however, a natural upper bound to the luminosity generated by any accretion process onto a compact remnant of mass M_c (see Vol. I, Chap. 6, Section 6.8). The photons that are emitted by this process will be continuously interacting with the in-falling particles and will be exerting a force on the ionised gas. When this force is comparable with the gravitational force attracting the gas towards the central object, the accretion will effectively stop. The number density $n(r)$ of photons crossing a sphere of radius r, centred at the accreting object of luminosity L, is $(L/4\pi r^2)(\hbar\omega)^{-1}$, where ω is some average frequency. The rate of collisions between photons and the electrons in the ionised matter will be $n(r)\sigma_T$, and each collision will transfer a momentum $\hbar\omega/c$. Because electrons and ions are strongly coupled in a plasma, this force will be transferred to the protons. Hence the outward force on an in-falling proton at a distance r will be $f_{\rm rad} \cong (L\sigma_T/4\pi cr^2)$. This force will exceed the gravitational force attracting the proton, $f_g = (GM_c m_p/r^2)$, if $L > L_E$, where

$$L_E = \frac{4\pi G m_p c}{\sigma_T} M_c \simeq 1.3 \times 10^{46} \left(\frac{M_c}{10^8 \, M_\odot}\right) \text{ergs s}^{-1} = 10^{4.5} \left(\frac{M_c}{M_\odot}\right) L_\odot \quad (1.33)$$

is called the *Eddington luminosity*. The first form of the equation is applicable in the case of accretion onto supermassive black holes and thought to be relevant in the context of active galactic nuclei. The second form of the equality is relevant for galactic x-ray sources. The temperature of a system of size R radiating $L \simeq L_E$ will be determined by $(4\pi R^2)\sigma T^4 = L_E$; that is,

$$T \simeq 1.8 \times 10^8 \text{ K} \left(\frac{M_c}{M_\odot}\right)^{1/4} \left(\frac{R}{1 \text{ km}}\right)^{-1/2}. \quad (1.34)$$

For a solar-mass compact ($R \simeq 1$ km) stellar remnant, this radiation will peak in the x-ray band.

Equating the accretion luminosity in Eq. (1.32) to the maximum possible luminosity in Eq. (1.33), we obtain the maximum possible accretion rate as

$$\dot{M}_{\rm Edd} = (1.5 \times 10^{-8} \, M_\odot \text{ yr}^{-1}) \left(\frac{R}{10^6 \text{ cm}}\right). \quad (1.35)$$

For a typical neutron star with $M_c = 1.4 \, M_\odot$ and $R = 10^6$ cm, the maximum accretion rate that can be sustained in steady state is approximately $1.5 \times 10^{-8} \, M_\odot$ yr^{-1}. It is also clear from the above relations that an accretion rate of $(10^{-12}$–$10^{-8}) \, M_\odot$ yr^{-1} will produce luminosities in the range of $(3$–$10^4) \, L_\odot$.

These considerations lead to two interesting classes of accreting sources in the binary systems involving normal stars and a compact neutron star. In the first case, we have a neutron star orbiting around a high-mass normal star. High-mass stars have outer regions that are loosely bound, and hence there is a strong stellar wind from such stars. The orbiting neutron star can accrete a fraction of the material in the wind that is flowing past it, thereby leading to a steady $\dot{M}$. For typical stellar-wind losses, this accretion rate turns out to be

$$\dot{M}_{\rm acc} \approx 5 \times 10^{-20} \, M_\odot \text{ yr}^{-1} \left(\frac{M}{M_\odot}\right)^{5.67}, \quad (1.36)$$

where M is the mass of the star. (This will be discussed in more detail in Chaps. 3 and 7.) To have significant luminosity, we need $\dot{M}_{\rm acc}$ to be at least

$\sim 10^{-12}\ M_\odot\ \text{yr}^{-1}$. Equation (1.36) shows that the mass of the star has to be more than $\sim 20\ M_\odot$ for this to be feasible. These binary systems are called *high-mass x-ray binaries* (HMXBs).

Alternatively, one can consider a binary system with a low-mass star and a neutron star orbiting around each other. If, during the course of evolution, the radius of the low-mass star increases significantly, the strong gravitational field of the neutron star can remove the mass lying in the outer region of the companion star. This accretion can occur at two different time scales, depending on the mass of the star. If the mass of the star is more than that of the compact remnant (i.e, if $M \gtrsim 1.4\ M_\odot$), then the transfer can be unstable and occur at the thermal time scale of the star:

$$\tau_{\text{th}} = \frac{GM^2}{RL} \simeq 3 \times 10^7\ \text{yr}\ \frac{(M/M_\odot)^2}{(R/R_\odot)(L/L_\odot)}. \tag{1.37}$$

Using the scaling relations $L \propto M^3$, $R \propto M$ and assuming that most of the stellar mass (say, $\sim 0.8\ M_\odot$) is transferred at this time scale, we get the mass transfer rate of

$$\dot{M} \simeq \frac{M}{\tau_{\text{th}}} \simeq 3 \times 10^{-8} \left(\frac{M}{M_\odot}\right)^3 M_\odot\ \text{yr}^{-1}. \tag{1.38}$$

This is, however, *not* a feasible example because, for $M > 1.4\ M_\odot$, this will lead to an accretion rate higher than the maximum accretion rate allowed by Eq. (1.35). The second possibility is for stars with masses less than that of the compact remnant (i.e, if $M \lesssim 1.4\ M_\odot$). In this case, mass transfer takes place at a much slower (nuclear) time scale of the star:

$$\tau_{\text{nucl}} \approx 10^{10} \left(\frac{M}{M_\odot}\right) \left(\frac{L}{L_\odot}\right)^{-1}\ \text{yr}. \tag{1.39}$$

If $L \propto M^3$ we have $\dot{M} \approx 10^{-10}\ M_\odot\ \text{yr}^{-1}(M/M_\odot)^3$. This is a possible accretion scenario for $M \lesssim 1.4\ M_\odot$; for $M = 1.4\ M_\odot$ this will lead to a maximum accretion rate of approximately $3 \times 10^{-10}\ M_\odot\ \text{yr}^{-1}$, which will correspond to a luminosity of $L_{\text{acc}} \approx 10^3\ L_\odot$. Such systems are called *low-mass x-ray binaries* (LMXBs).

We note that feasible accretion-powered x-ray sources can exist only when the companion to the neutron star has a mass of less than $1.4\ M_\odot$ or greater than $\sim 20\ M_\odot$. These possibilities are illustrated schematically in Fig. 1.7, which gives the accretion rate as a function of the mass of the companion star. If $M < M_{\text{NS}} \approx 1.4\ M_\odot$ (marked by the vertical line), it is possible to have a LMXB with the accretion occurring at nuclear time scales (thick line at the lower left; the dashed line extending it to the right is not possible as it will require that $M > M_{\text{NS}}$, in which case the accretion rate will be at thermal time scales.) The dashed line on the left top of the diagram indicates the accretion rate if it occurs at a thermal time scale. This is, however, not possible as it requires that $M > M_{\text{NS}}$ when the accretion rate exceeds the Eddington rate. The thick line on the right half of the diagram

1.6 Stellar Binaries

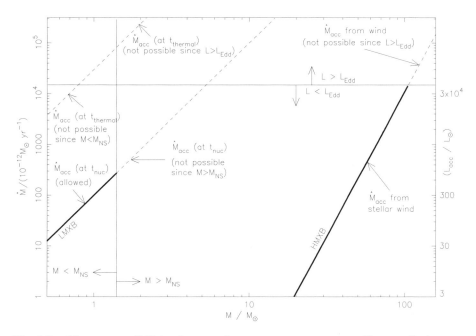

Fig. 1.7. The two possibilities for accretion onto a neutron star are illustrated schematically. The figure gives the accretion rate as a function of the mass of the companion star. If $M < M_{NS} \approx 1.4\, M_\odot$ (marked by the vertical line), it is possible to have a LMXB with the accretion occurring at nuclear time scales (thick line at the lower left; the dashed line extending it to the right is not possible as it will require that $M > M_{NS}$, in which case the accretion rate will be at thermal time scales.) The dashed line on the left top of the diagram indicates the accretion rate if it occurs at a thermal time scale. This case is, however, not feasible as it requires that $M > M_{NS}$ for which the accretion rate exceeds the Eddington rate. The thick line on the right half of the diagram illustrates the case for a HMXB in which the accretion occurs from the stellar wind.

illustrates the case for a HMXB in which the accretion occurs from the stellar wind. Obviously HMXBs can produce much higher luminosity than LMXBs.

A different class of phenomena occurs when the compact star in the binary is a white dwarf rather than a neutron star. Such configurations, made of a closed binary system with a white dwarf and a late-type main-sequence star or a red giant star, are called *cataclysmic variables*. These are characterised by abrupt increases in light output in the time scale of typically a day followed by a decline over a period lasting from weeks to a year. In some of these binary systems, mass transfer from the star to the compact remnant could trigger repetitive eruptions of light outbursts with the luminosity rapidly rising to a peak value of $\sim 10^{34}$ ergs s^{-1} with a slower decline followed by a quiescent period lasting from tens to hundreds of days. These systems are called *dwarf nova*. In yet another class of binary systems (called *polars*), the magnetic field of the white dwarfs plays a resistive role in the flow of matter from the main-sequence star.

Table 1.4. Types of X-ray binaries*

Type	Donor Star	Compact Object	Accretion Disk	Examples
HMXB	O-B	NS, BH	Small	Cen X-3; Cyg X-1
LMXB	K-M	NS, BH	Yes	Sco X-1
LMXB	A-F	NS, BH	Yes	Her X-1; Cyg X-2
LMXB	WD	NS	Yes	4U1820-30
CV (Dwarf nova)	K-M	WD	Yes	U Gem; SS Cyg
CV (Polar)	K-M	Magnetic WD	No	AM Her

*NS, neutron star; BH, black hole; WD, white dwarf; CV, cataclysmic variable.

Tables 1.4 and 1.5 summarise different types of such double-star systems, indicating the variety that is possible. These phenomena will be discussed in detail in Chap. 7.

Accretion to a white dwarf can lead to yet another class of phenomenon, of which the most important is a supernova explosion, called a type I supernova to distinguish it from the supernova occurring at the end stage of an isolated star, which is called type II. The type I supernova can be modelled as an explosion occurring in a accreting carbon–oxygen white dwarf that is a member of a closed binary system. When the mass of the white dwarf reaches about $\sim 1.3\ M_\odot$, carbon burning begins at the centre of the star in the degenerate core and moves towards the surface. This motion occurs subsonically (called deflagration front rather than detonation front), converting approximately one half of the white dwarf's mass into iron before the degeneracy is removed. Further expansion can

Table 1.5. Some galactic binaries

Source	α 1950 δ 1950	$L(X)$max (2–11) keV (ergs s^{-1})	V magnitude	$\dfrac{L(X)}{L_{\text{opt}}}$	Distance (kpc)
Vela X-1	$09^h 00^m 13^s.2$ $-40° 21' 25''$	1.4×10^{36}	6.9	3×10^{-3}	1.4
Cen X-3	11 19 01.9 $-60\ 20\ 57$	4×10^{37}	13.4	0.05	8
Sco X-1	16 17 04.5 $-15\ 31\ 15$	2×10^{37}	12.2–13.3	600	0.7
Her X-1	16 56 01.7 35 25 05	1.0×10^{37}	13.2	10	5
Cyg X-1	19 56 28.9 35 03 55.0	2×10^{37}	8.9	2×10^{-2}	2.5

1.7 Interstellar Medium

lead to a cooling that will eventually dampen the nuclear burning, leaving shells of intermediate-mass elements around a nickel–iron core. The energy released in the process can disrupt the star, resulting in the supernova explosion. This is consistent with the lack of hydrogen lines in the observed spectra of the type I supernova.

1.7 Interstellar Medium

The condensation of stars from galactic matter cannot be a totally efficient phenomenon; hence, we do expect a fair amount of matter to be distributed in the galaxy in different forms. This constitutes the *interstellar medium* (ISM) in which structures of very different densities and temperatures exist in pressure equilibrium. In our galaxy, the ISM contributes a mass of $\sim 10^9\ M_\odot$ and has a pressure of $P = nk_B T \simeq 10^{-12}$ dyn cm^{-2}. Table 1.6 and Fig. 1.8 provide a brief description of the components of the ISM.

There are several processes that couple the stars and the ISM strongly. To begin with, the formation of stars takes place inside dense molecular clouds in the ISM, thereby converting gas into stars. As the stars evolve, there is a feedback of matter into the ISM in different forms. Many giant stars have strong stellar winds involving a steady loss of outer layers of matter from the star to the ISM. Further, we saw in Section 1.4 that stars eject a large amount of matter in the form of planetary nebulas during their end stages. Finally, massive stars ending up in a supernova explosion feed back into the ISM a significant fraction of their mass.

Table 1.6. Physical characteristics of molecular regions in the ISM

Type of cloud	Density (cm^{-3})	T (K)	Mass $M_\odot$	Size (pc)	Velocity (km s^{-1})
Cirrus	10–50	30–150	$\sim$5000	10–30	0.5–5
Diffuse clouds	100–800	30–80	1–100	1–5	0.5–3
	50–260		2–260	1–5	
Translucent clouds	500–4000	13–35	3–100	0.15–1	0.5–3
Cold dark clouds					
Complex	10^2–10^3	≥ 10	10^3–10^4	6–20	1–3
Clouds	10^2–10^4	≥ 10	10–10^3	0.2–4	0.5–1.5
Cores/clumps	10^4–10^5	≈ 10	0.3–10	0.05–0.4	0.2–0.4
Giant molecular clouds					
Complex	100–300	15–20	10^3–3×10^5	20–80	6–15
Major fragments	10^2–10^4	≥ 20	10^3–10^5	3–20	3–12
Warm clouds	10^4–10^7	25–70	1–10^3	0.05–3	1–3
Hot cores	10^7–10^8	100–200	10–10^3	0.05–1	1–10

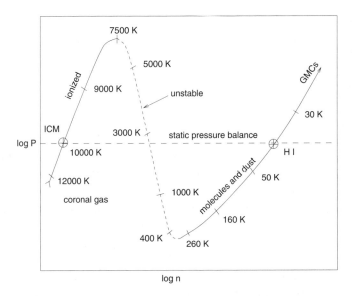

Fig. 1.8. Schematic description of pressure–density relationship for the ISM. A straight line of constant pressure intersects the curve at three points corresponding to three possible phases. In the middle phase, however, pressure increases with decreasing density and vice versa, which makes this phase unstable. Some of the components of the ISM are also marked in the diagram.

Because of these processes, the physical conditions that exist in the ISM vary dramatically from location to location. It is therefore often convenient to separate the study of ISM into (1) the study of reasonably well-defined substructures and (2) the study of the overall nature of the ISM. Among the substructures, the most important ones are planetary nebulas, supernova remnants, and HII regions; of these, planetary nebulas and supernova remnants arise from the matter ejected from the end stages of stellar evolution, as mentioned above. The HII regions, on the other hand, arise when a luminous young hot star ionises the region around it in a molecular cloud. Consider, for example, the region around a hot star with luminosity $L = 3.5 \times 10^{36}$ ergs s^{-1} and surface temperature $T_s \simeq 3 \times 10^4$ K. The rate of emission of ionising photons $\dot{N}_\gamma$ (with frequency $\nu > \nu_I$, where $h\nu_I = 13.6$ eV) by such a star can be estimated from the Planck spectrum and will be approximately 3×10^{48} s^{-1}. When both photoionisation and recombinations occur in a region, the equilibrium is described by the relation $n_e n_i \alpha_R \simeq \sigma_{\rm bf} n_H F$, where F is the flux of ionising photons with $\nu > \nu_I$, $\alpha_R \simeq 2 \times 10^{-14}(k_B T/10 \text{ eV})^{-1/2}$ cm^3 s^{-1} is the thermally averaged recombination coefficient, and $\sigma_{\rm bf}$ is the photoionisation cross section. (The expressions for α_R and $\sigma_{\rm bf}$ were derived in Vol. I, Chap. 6, Section. 6.12.) Taking $n_e = n_i = x n_0$ and $n_H = (1-x)n_0$, we can write this relation in the form

$$\frac{x^2}{(1-x)} \cong \left(\frac{\sigma_{\rm bf} c}{\alpha_R}\right)\left(\frac{F}{n_0 c}\right) \simeq 5 \times 10^4 \left(\frac{T}{10^4 \text{ K}}\right)^{1/2} \left(\frac{F}{n_0 c}\right), \quad (1.40)$$

which determines the ionisation fraction x in many astrophysical contexts. If a source of photons emitting $\dot{N}_\gamma$ ionising photons per second (with $\nu > \nu_I$) ionises a region of volume V around it, then the same argument gives $n_e n_i \alpha_R V \simeq \dot{N}_\gamma$. Taking $n_e = xn_0 \simeq n_0$, we get $V = (\dot{N}_\gamma/\alpha_R n_0^2)$. Using $n_0 \simeq 10$ cm^{-3}, we find that matter will be fully ionised for a region of radius $R = (3V/4\pi)^{1/3} \simeq 10$ pc. At a distance of 5 pc from the star, relation (1.40) gives $(1 - x) \simeq 10^{-3}$, indicating nearly total ionization. Such a local island of plasma in the ISM is called a *HII region*.

These clouds of HII regions also emit thermal bremsstrahlung radiation with a relatively flat spectrum. For example, the HII regions in Orion have $T \simeq 10^4$ K, $n_i \simeq 2 \times 10^3$ cm^{-3}, and an effective; line-of-sight thickness of 6×10^{-4} pc. The thermal bremsstrahlung emissivity; derived in Vol. I, Chap. 6, Subsection 6.9.3, is given by

$$j_\omega \equiv \left(\frac{d\mathcal{E}}{d\omega \, dt \, dV}\right) \simeq \left(\frac{q^6}{m_e^2 c^3}\right) \left(\frac{m_e}{k_B T}\right)^{1/2} n_e n_i \propto n^2 T^{-1/2}, \quad (\hbar\omega \lesssim k_B T). \tag{1.41}$$

From this equation we can estimate the flux to be approximately 3×10^{-21} ergs cm^{-2} s^{-1} Hz^{-1} rad^{-2} = 300 Jy, where 1 Jy = 10^{-23} ergs cm^{-2} s^{-1} sr^{-1} Hz^{-1}. This is typical of emission from HII regions. The total diffuse radio emission from the galaxy is in the range $(10^{29}$–$10^{33})$ W.

Figure 1.9 shows the distribution of supernova remnants and HII regions in our galaxy in the galactic coordinates, with the radius of the circle indicating the approximate size of the regions. It is clear that the ISM is permeated by these substructures. We will discuss the supernova remnants in Chap. 4, Section 4.9, and HII regions in Chap. 9, Section 9.3.

In proceeding from the study of individual objects to the overall structure of the ISM, we must take into account the balance among several processes. The first is the mechanical equilibrium, which requires the pressure in the ISM to be constant, which, in turn, requires high-temperature regions to have low densities and vice versa. The second is the balance among different atomic processes that are taking place in the ISM. The rate at which different processes populate a given atomic- or molecular-energy level should be balanced by the rate at which other processes depopulate the same level. Because of the rather low densities that prevail in the ISM, this condition can lead to situations that are not common in other systems. For example, some of the atomic transitions that are forbidden (that is, they occur with very low probability) in terrestrial conditions can play an important role in the cooling of the ISM. (We shall study some of these effects in Chap. 9, Section 9.6.) Finally, the net heating and cooling of the ISM should balance in order to conserve the energy. Because several different processes contribute to heating and cooling, it is not necessary that these conditions lead to a unique equilibrium solution or even to a unique, stable solution. In fact, it is possible for the ISM to be made of different phases – with significantly different

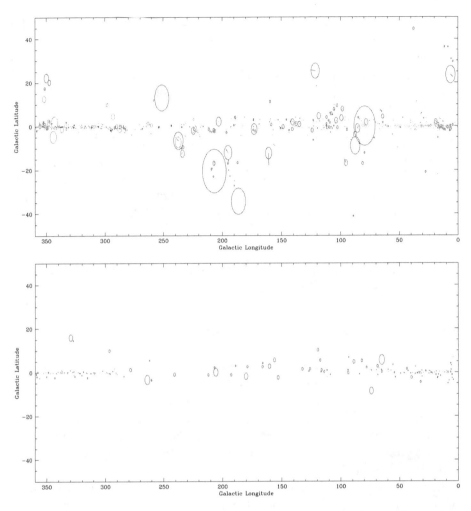

Fig. 1.9. The top frame gives the HII regions in our galaxy and the bottom frame gives the distribution of supernova remnants. These are plotted in galactic coordinates in which the disk of the galaxy is located as a thin region around the origin horizontally. (The data are adapted from the website http://www.nrao.edu/pbarnes/concord.)

temperatures and densities – in equilibrium with each other. This is illustrated in Fig. 1.8.

These complications also require us to use different techniques to probe different components of the ISM. Individual structures such as planetary nebulas, HII regions, and supernova remnants are probed by the continuum radiation (synchrotron, bremsstrahlung, etc.) as well as line radiation (hydrogen lines, radio recombination lines, forbidden fine-structure transitions, etc.) emitted by them. The distribution and the motion of bulk matter in the ISM is studied by the rotational transitions in the CO molecule and the 21-cm hyperfine transitions of the

1.8 Theoretical Constraints on Astronomical Observations

neutral hydrogen. Measurement of Faraday rotation and dispersion of the pulsar radiation give information about the ISM magnetic field and electron density.

1.8 Theoretical Constraints on Astronomical Observations

We shall now turn to a general discussion of astrophysical observations that have certain features which distinguish them from the experimental data used in other branches of applied physics. These differences arise from the fact that we often have far less control over the conditions (and quality) of the observations in astrophysics, compared with other branches of physics driven by laboratory observations, resulting in some limitations not shared by the laboratory observations. Some of these extraneous limitations are highlighted in this section, as they are vital in appreciating the nature of astronomical data.[8]

Because the primary source of information about distant astrophysical sources is the electromagnetic radiation detected from them, we shall concentrate on electromagnetic observations. There is a possibility that gravitational radiation and neutrino emission from astrophysical sources could possibly be used as a useful diagnostic at some future date. Except in very special cases, these observations do not provide a serious competition for observations based on electromagnetic radiation in the foreseeable future, and hence we shall not discuss these in any detail.

Electromagnetic radiation observed from cosmic sources ranges from very low-frequency ($\nu \approx 1$ MHz) radio waves to high-frequency ($\nu \approx 10^{24}$ Hz) gamma rays. Obviously, we need to use very different kind of techniques and instruments to study the radiation spanning such a large dynamic range of ~ 18 orders of magnitude. One key consideration that determines the nature of observations has to do with the manifestation of the quantum nature of radiation. This is most easily illustrated by the thermal radiation emitted by a body at temperature T. The mean number of photons per unit cell in phase space is then given by

$$n = \frac{1}{e^{h\nu/k_B T} - 1} \approx \begin{cases} k_B T / h\nu & (\text{for } h\nu \ll k_B T) \\ e^{-(h\nu/k_B T)} & (\text{for } h\nu \gg k_B T) \end{cases}. \quad (1.42)$$

It follows that, at long wavelengths (as in a radio band), $n \gg 1$, whereas at short wavelengths (such as x-ray and higher-frequency bands), $n \lesssim 1$. Therefore long-wavelength radiation can be thought of as arising out of a coherent superposition of several photons and can be described as a classical wave. Standard descriptions based on interference, diffraction, etc. are applicable in this context. At high energies, however, there is a paucity of photons per unit phase cell and the quantum nature of electromagnetic radiation becomes apparent. The description at these frequencies should be based on photon-counting techniques rather than wave optics. This feature can also be understood from the fluctuations in the number of photons in the unit phase cell that is given by

$(\Delta n)^2 = n + n^2$ (see Chap. 5, Section 5.11, of Vol. I). The first term dominates at high frequencies and leads to the Poissonian fluctuations $(\Delta n/n) \approx n^{-1/2}$ in the photon count. The second term dominates at low frequencies and gives $(\Delta n/n) \approx 1$, which is characteristic of classical waves. This slow transition from wave to particle nature of electromagnetic radiation plays a crucial role in observational astronomy as we proceed from radio wavelength to high-energy gamma rays.

One practical example of the above transition has to do with the detectors used for electromagnetic radiation. Classically, a dipole made of, say, two charges connected by a spring can serve as a detector for the electromagnetic wave. If the axis of the dipole is aligned along the direction of the electric field of the wave, it will be set into oscillation that – with suitable arrangements – can be converted into an oscillating electric current. This is basically the procedure used to detect long-wavelength radio waves with an antenna. However, for this method to be useful, the detecting component of the antenna should have dimensions comparable with the wavelength of the radiation, which is difficult to achieve at higher frequencies. In optical and shorter wavelengths, we essentially treat electromagnetic radiation as being made of photons, and we proceed to detect the effect of photons impinging on special materials like photographic emulsions. Similar issues arise when we try to focus the electromagnetic radiation on a detector. This is straightforward at radio bands and even at optical and UV bands. However, as we go to higher frequencies – such as x rays and gamma rays – changing the direction of propagation of the wave in order to focus it is not an easy task. This practical difference in the methods used for detection of the signals plays a vital role in all aspects of observational astronomy.

1.8.1 Limitations Due to Earth's Atmosphere

The energy levels of atoms and molecules have an important implication for observational astronomy because ground-based observations can detect only radiation that can penetrate through the Earth's atmosphere. The atoms of most elements have energy levels of the order of $E_0 \approx (1/2)\alpha^2 m_e c^2 \approx 10$ eV. Using the relation between photon energy and wavelength, $(E/1 \text{ eV}) \approx (\lambda/12345)^{-1}$ Å, we conclude that photons with $\lambda \lesssim 10^3$ Å will be absorbed by the atmosphere, leading to ionisation of the upper layers. Further, the rotational- and the vibrational-energy levels of molecules like H_2O and CO_2 (which exist in the atmosphere) fall in the IR band; this causes the IR radiation also to be absorbed by the atmosphere, although to somewhat lesser degree than the higher-energy radiation. Because of these effects, the ground-based observations are essentially limited to visible [$\lambda \approx (3000–6000)$ Å, $\nu \approx (10^{15}–5 \times 10^{14})$ Hz] and radio ($\lambda > 1$ cm, $\nu < 3 \times 10^{10}$ Hz) waves.

There is, however, another limitation arising from the fact that very long-wavelength radiation ($\lambda \gtrsim 100$ m) cannot propagate through the plasma in the

1.8 Theoretical Constraints on Astronomical Observations

ionosphere and is reflected back. Consider an electromagnetic wave that moves the electrons in a plasma (relative to ions) by a small distance δx along the x axis. This deposits a charge $Q \simeq e(nA\delta x)$ on a fictitious surface of area A perpendicular to the x axis. This charge density, in turn, will lead to an electric field $E_x \simeq 4\pi(Q/A) \simeq 4\pi e n \delta x$ that acts on the electrons in this small volume, pulling them back. Such a restoring force, proportional to displacement, gives electrons a characteristic frequency of oscillation (called *plasma frequency*):

$$\omega_p = \left(\frac{4\pi e^2 n}{m}\right)^{1/2} = 5.64 \times 10^4 \text{ Hz} \left(\frac{n}{1 \text{ cm}^{-3}}\right)^{1/2}. \quad (1.43)$$

Waves with frequencies lower than the plasma frequency cannot propagate through a plasma because the electrons can redistribute themselves sufficiently quickly to cancel the field of such an electromagnetic wave. We can estimate the number density n of electrons in the ionosphere by equating the ionisation rate that is due to solar radiation with the recombination rate. This gives an electron density of approximately 4×10^5 cm^{-3}; the corresponding plasma frequency is approximately $\nu_p = (\omega_p/2\pi) = 6$ MHz. Thus we cannot observe radio waves with frequencies lower than ~ 6 MHz, corresponding to wavelengths larger than ~ 50 m. (Interplanetary and interstellar space have electron densities of $n_e \simeq 1$ cm^{-3} and $n_e \simeq 0.03$ cm^{-3}. The corresponding frequencies are $\nu_p \simeq 9$ kHz and $\nu_p \simeq 1$ kHz, respectively.)

To obtain information about all other wavelength regimes, it is necessary to make observations from high altitudes, for example, from balloons, aircrafts, spacecrafts, satellites, etc. It is obvious that the kind of technology required for successfully operating a space-based observatory is quite different (and more difficult) compared with that for ground-based observations. As a result, the quality and quantity of data available in different wave bands can be significantly different.

Even ground-based observations can be of different qualities, depending on the observational site and the wave band in which the observations are carried out. For example, note that an isothermal atmosphere has the pressure variation $P(z) \propto \exp(-z/H)$ and the pressure scale height $H = (\mathcal{R}/\mu)(T/g)$, where $\mathcal{R} = 8.32$ J K^{-1} mol^{-1} is the gas constant, $\mu \simeq 30$ gm is the molecular weight for an atmosphere made of O_2 and N_2, T is the mean temperature, and g is the acceleration that is due to gravity. This gives $H \approx 8$ km for the pressure scale height. The water vapour, however, has a much smaller scale height of ~ 3 km. Hence observational sites at an altitude of ~ 3000 m or so will successfully reduce the absorption that is due to water vapour in the air.

The atmosphere not only absorbs radiation but also emits a background of photons, thereby adding to the noise in astronomical observations. To begin with, the atmosphere with a finite temperature will produce a thermal emission characteristic of this temperature. In the spectral bands L, M, N, and Q at wavelengths 3.4, 4.8, 10.2, and 21 μm, this thermal emission has intensities of 0.16, 22.5, 250, and 2100 Jansky arcsec^{-2}. (The corresponding magnitudes will be approximately 8.1, 2, -2.1, and -5.8 mag arcsec^{-2}.) It is necessary to subtract out this noise in the observations in IR wave bands. Second, the atmosphere also produces fluorescent emission from higher altitudes. This arises from the atoms created at excited states (which are due to, for example, the recombination of electrons with ions that have been produced during daytime photochemical dissociations) that are decaying radiatively rather than collisionally at the low densities that exist at high altitudes. For example, at 5500 Å the emission is approximately

$$s = 6.75 \times 10^{-17} \text{ W m}^{-2} \mu\text{m}^{-1} \text{ arcsec}^{-2}. \tag{1.44}$$

Because a monochromatic flux of 4×10^{-8} W m^{-2} μm^{-1} corresponds to zero magnitude in this wavelength band, this gives a sky background of $\sim$22 mag arcsec^{-2}. This sky background can be a major limiting factor in observing faint extended objects.

The effect of sky background is less for space-based observations, although not completely negligible because of the existence of zodiacal dust, geocoronal emission, interplanetary and interstellar emission, etc. The effect, of course, is less compared with that at ground level; in the wavelength band of (5000–9000) Å, the background is between 23 and 24 mag arcsec^{-2} for space-based observations.

Finally, the atmosphere also sets the limit for angular resolution because of the intrinsic patchiness. This effect was discussed in Vol. I, Chap. 3, Section 3.16, where it was shown that atmospheric turbulence sets the angular resolution to $\sim$1 arcsec for ground-based observations in the optical band.

1.8.2 Limitation on Resolution

In the astronomical observations there are three major resolving limits that are of interest. The first one is the angular resolution $\Delta\theta$ in the sky that specifies how well two separate directions can be distinguished. The second is the spectral resolution $\Delta\lambda/\lambda$ that decides the fractional accuracy with which features at two nearby wavelengths can be distinguished in a given spectroscopic technique. Finally, it is also important to have reasonable time resolution $\Delta t/t$ that sets the limit on our ability to monitor time-varying phenomena. We shall discuss each of these briefly.

The angular resolution of an optical system is typically of the order of $\theta \approx (1.22 \, \lambda/D)$, where λ is the operating wavelength and D is the effective aperture size. In direct observing techniques, D will be the physical diameter of, say, the optical telescope; in the case of interferometric techniques, D will be the effective baseline used for observation. This result has been obtained in Vol. I, Chap. 3, Section 3.13 and can be understood along the following lines. If a photon

1.8 Theoretical Constraints on Astronomical Observations

propagating along the positive x axis is confined by an aperture of transverse dimension D, it will acquire a transverse momentum of the order of $\Delta p_\perp = \hbar \Delta k_\perp \approx (\hbar/D)$. This will lead to an angular spreading by the amount $\Delta \theta = (\Delta k_\perp/k) \simeq (\lambda/D)$, which gives the correct order of magnitude. For ground-based observations in optical and nearby bands, the resolution is limited by atmospheric turbulence and is $\Delta \theta \simeq 1$ arcsec.

Although this is an intrinsic limitation, it could happen that there are other sources of limitations merely because of the technology used in the detection. This is particularly true in high-energy observations as in x-ray and gamma-ray bands in which focussing the radiation is more difficult. At present the limiting angular resolution is $\sim 10^3$ arcsec at energies higher than 100 keV, about ~ 0.1 arcsec for space telescopes, ~ 1 arcsec for large ground-based optical telescopes, and in the range from a few arcseconds to 100 arcsec for observations in 10 μm to a few millimeters. The resolution is significantly high and is $\sim 10^{-4}$ arcsec for radio interferometric arrays. This wide variation illustrates that the observing technique still plays an important role in determining the angular resolution. Figure 1.10 gives several characteristic stellar phenomena and the resolution required for observing them.

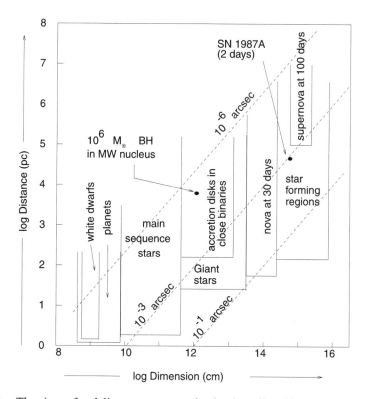

Fig. 1.10. The sizes of and distances to several galactic stellar objects. The dashed lines give the angular resolution required for resolving the object.

Whether a particular source is resolved or not by a telescope is important in deciding the relevant physical parameter that is being observed. If the resolution of the telescope is $\Delta\theta$ and the angle subtended by the source is $\theta \gg \Delta\theta$, then the telescope can measure the intensity of the radiation by using the energy passing through the aperture of the telescope and spread into a solid angle Ω_{min} corresponding to $\Delta\theta$. This quantity will be measured, for example, in units of ergs per second per square centimeter per steradian. Consider now the effect of moving the source further away from the telescope but still keeping it resolved. As it moves away, the energy received from each square centimeter on the surface of the source will decrease as the square of the distance. However, because the source is still resolved, the amount of area in the source that contributes to the solid angle Ω_{min} will increase as the square of the distance, thereby cancelling the previous effect. As a net result, the specific intensity of radiation received from a resolved source is independent of the distance to the source. The situation changes somewhat when the source is unresolved and $\Delta\theta \gtrsim \theta$. In that case the energy received from the entire source will be dispersed through the diffraction pattern of the telescope's aperture. Because the light, now arriving at the telescope, leaves the source at all angles, the detector is effectively integrating the specific intensity over the solid angle and is measuring the radiative flux from the source towards the telescope. If the distance to the source is now increased, the total amount of energy falling within the first diffraction pattern of the aperture will decrease as the square of the distance and hence there will be appreciable loss of signal to noise.

Although increasing the diameter of an optical telescope will help in achieving higher angular resolution, there are several practical difficulties in such an attempt. Even ignoring the cost and the technological challenges involved in fabricating large-aperture structures, we are still limited by the atmospheric fluctuations, leading to a limiting resolution of ~ 1 arcsec for ground-based optical astronomy. Large-aperture optical telescopes are therefore of use only when they are spaceborne. Although eventually technology will allow us to use large-diameter optical telescopes from space, at present there are several practical limitations to what can be achieved. The situation, however, is more favourable in radio wavelengths, in which the issue of resolution can be tackled by a synthesis interferometer. The principle behind this technique has already been discussed in Chap. 3, Section 3.14, of Vol. I, in which it was shown that if the degree of spatial coherence of a source [which is related to the average value, $\langle E(t, \mathbf{x})E(t, \mathbf{y})\rangle$, of the product of any one component of the electric field at two locations in space] is known, then the angular-intensity distribution of the source can be obtained by a Fourier transform. In principle, we can measure the quantity $\langle E(t, \mathbf{x})E(t, \mathbf{y})\rangle$ by keeping telescopes at $\mathbf{x}$ and $\mathbf{y}$ and correlating the signals. In practice, however, there are several difficulties. To begin with, it is difficult to fill all the spatial locations with telescopes, and hence we need to devise some other procedure either by moving the telescope or by making use of the natural motion of the telescope that is due to the rotation of the Earth. Second, in order to obtain such correlations precisely, it is important to make sure that any delays in the travel time of light to the different telescopes are precisely compensated for. This, in turn, would require adding or subtracting effective extra path lengths to the signal (typically of the order of the wavelength) before they are allowed to interfere. In radio bands, this would require path lengths of the order of centimeters to meters, which are easy to introduce; it is also possible to do this in IR bands by using laser heterodyne techniques that can introduce delays of the order of tens of microns; in optical wave bands, such delays have to be introduced by moving very stable mirrors over thousands of angstroms accurately and smoothly – which is far more difficult, but has been achieved. At higher wavelengths, the current technology does not allow such a technique to be used. Hence radio telescopes have achieved much higher angular resolutions than at other wave bands in spite of operating at very large wavelengths.

In principle, the correlation function of the square of the electric field $\langle E^2(t, \mathbf{x})E^2(t, \mathbf{y})\rangle$ also contains information about the angular-intensity pattern of the source. This correlation function is not limited by the requirement of suitable delays in order to give meaningful results. Unfortunately, this quantity has a much higher level of noise compared with $\langle E(t, \mathbf{x})E(t, \mathbf{y})\rangle$. Although intensity interferometry has been tried as an observational tool, it is limited in sensitivity compared with amplitude interferometry.

Let us next consider the spectral resolution that is predominantly determined by the actual technique available in the given wave band. The spectroscopic resolving power varies by $(\lambda/\Delta\lambda) \approx 10^2$ in the x-ray band and is significantly lower at energies above 10 MeV or so. Near $(3 \times 10^3$–$10^4)$Å in the optical and the IR bands, the resolving power is quite high and could be anywhere between 10^4 and 10^9, depending on the apparatus and the technique that are used. At wavelengths longer than 1 mm, the resolving power is $\sim 10^6$.

At optical wavelengths, low-resolution spectroscopy is usually done with different filters for U, B, V, etc., as described in Section 1.3. It is also possible to use a prism to resolve the light into different bands before the image formation in the telescope. For higher spectral resolution, we use beam splitters and interferometers such as the Fabry–Perot interferometer or the Michelson interferometer. This is also the preferred technique in radio wavelengths in which, thanks to the availability of high-speed digital electronics, it is actually possible to Fourier transform the signal in the time domain and obtain the spectral information.

It may be noted that, for radiation that is Doppler shifted, $\Delta\lambda/\lambda$ is of the order of v/c, where v is the speed causing the Doppler shift. Hence the spectral resolving power translates into a corresponding velocity resolution. To detect a velocity change of 1 m s^{-1} would require a resolution of approximately 3×10^8.

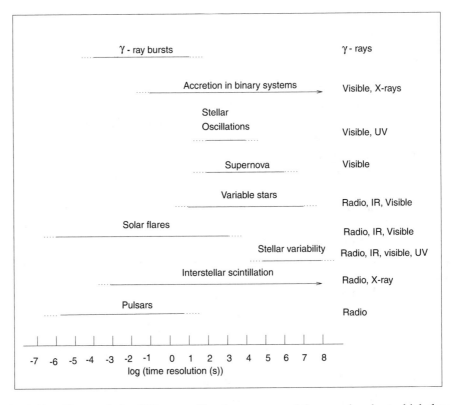

Fig. 1.11. Time scale for different stellar phenomena and the wave bands at which they take place.

The resolving power of $\sim 10^4$, available in optical and IR bands, is capable of measuring velocities of ~ 30 km s^{-1}.

The time resolution available in different wave bands is also limited by the observing technique and instrumentation. Typically, resolutions varying from 10^{-4} to 10^6 s is available in many wave bands. Figure 1.11 gives the characteristic time scales for different stellar phenomena and the relevant bands.

1.8.3 Sensitivity and Signal-to-Noise Ratio

Observational techniques in every wave band have their own intrinsic bounds on the sensitivity and the signal-to-noise ratio (SNR) that they can achieve. Once again, in some of the wave bands this limitation arises because of the particular techniques that are being adopted and by the currently available technology. In some other bands, the limitation is due to extraneous factors over which the astronomer has very little control. We shall try to summarise some of the basic limitations that are not based only on currently available technology.

1.8 Theoretical Constraints on Astronomical Observations

One of the basic limitations on the SNR in optical, UV, and higher frequencies is due to the random fluctuations in the photon counts in the detector. Consider an observation that attempts to detect a source producing $N_\nu = [F_{\text{source}}(\nu)/h\nu]$ photons per second per unit area of collecting device. Here $F_s(\nu)$ is the flux from the source in units of energy per second per unit area detectable in the entire bandwidth of observations, $\Delta \nu$. In an observation lasting for a time t, we would expect to detect a mean number of photons $\bar{N} = (F_s(\nu)/h\nu)A_{\text{eff}}\, t$, where A_{eff} is the effective collecting area of the device.

Even if there is no other source of photons contaminating the observations, there will be a random fluctuation of the order of $\pm\sqrt{\bar{N}}$ in the actual number of photons detected because of Poisson fluctuations. When other contaminating sources are present, the random fluctuation will be higher and the corresponding SNR will be lower. Usually two other sources of photon noise are relevant in such observations. The first is the source of background photons, say, those that are due to sky background or any other ambient source. If the background flux is $F_{\text{bg}}(\nu)$ ergs per second per unit area per steradian, then the mean number of background photons contaminating the observations will be $\bar{B} = [F_{\text{bg}}(\nu)/h\nu]A_{\text{eff}}\,\Omega t$, where Ω is the effective field of view of the telescope. Second, there could be a spurious counting of photons because of the detecting method itself. For example, in CCD detectors there could be a residual dark current that is equivalent to a steady detection of photons at some constant rate C; this adds a background of Ct photons over a time scale t. The total noise will therefore be $(\bar{N} + \bar{B} + Ct)^{1/2}$, so that the SNR becomes

$$\mathcal{R} = \left(\frac{S}{N}\right) = \frac{\bar{N}}{(\bar{N} + \bar{B} + Ct)^{1/2}}$$
$$= (A_{\text{eff}}\, t)^{1/2} \frac{(F_s/\nu)}{[(F_s/\nu) + (F_{\text{bg}}/\nu)\Omega + (C/A_{\text{eff}})]^{1/2}}. \quad (1.45)$$

This result shows that, other things remaining constant, the SNR increases as $\sqrt{t}$ in such observations. The same result can be stated in terms of the amount of integration time needed to achieve a given SNR: If the sensitivity is limited by the detector noise, then $t \propto A_{\text{eff}}^{-2}$ for a given SNR; if the sensitivity is limited by background noise, then $t \propto \Omega A_{\text{eff}}^{-1}$, whereas if only the fluctuations in the source dominate the noise, $t \propto A_{\text{eff}}^{-1}$.

The noise that is due to instruments (the term proportional to C) is quite small in modern CCDs and we will ignore it. In the case of observations limited by background, the minimum flux that can be detected at a SNR of $\mathcal{R}$ is given by

$$F_{\text{min}} \cong \mathcal{R}(F_{\text{bg}}\Omega \nu)^{1/2} (A_{\text{eff}}\, t)^{-1/2}. \quad (1.46)$$

As an example, consider an observation at 5500 Å with a telescope having a field of view of 2×2 arcsec^2 and a diameter of 3.6 m. At this wavelength, the sky background is ~22 mag arcsec^{-2}, corresponding to a flux of

$$f_{bg} = 4 \times 10^{-8} \text{ W m}^{-2} \mu\text{m}^{-1} \text{ arcsec}^{-2} \times 10^{-22/2.5}$$
$$= 6.2 \times 10^{-17} \text{ W m}^{-2} \mu\text{m}^{-1} \text{ arcsec}^{-2}. \qquad (1.47)$$

The bandwidth at 5500 Å is $\Delta\lambda = 0.088$ μm so that the total number of photons that are due to sky background at this wavelength is

$$N_\gamma = (f_{bg}\Delta\lambda)/(hc/\lambda) = 15 \text{ photons m}^{-2} \text{ arcsec}^{-2} \text{ s}^{-1}. \qquad (1.48)$$

The number of background photons collected in this observation will be

$$N_{bg} = N_\gamma \Omega A_{\text{eff}} t = \cong 2.2 \times 10^6 \text{ photons} \qquad (1.49)$$

for an integration time of 1 h. The fluctuation in this number will be approximately $\Delta N_{bg} = \sqrt{N_{bg}} \simeq 1.5 \times 10^3$. For a SNR of $\mathcal{R} = 3$, we should acquire a total of $N_s = 3\sqrt{N_{bg}} = 4.5 \times 10^3$ photons from the source. The corresponding flux of the source will be $F_s = (N_s h\nu)/(A_{\text{eff}} t) = 4.4 \times 10^{-20}$ W m^{-2}. Dividing by the bandwidth $\Delta\lambda = 0.089$ μm and $\Omega = 4$ arcsec^2, we get the flux per unit wavelength range as 1.23×10^{-19} W m^{-2} μm^{-1} arcsec^{-2}, which corresponds to a magnitude of ~28.5 mag arcsec^{-2}. Even if the integration time is extended to 1 month, the gain is only ~3.5 mag because of the $\sqrt{t}$ dependence.

Figure 1.12 shows the approximate limiting magnitude that can be reached in a telescope of diameters $D = 2$ and 10 m, a field of view of 2×2 arcsec^2, and a SNR of $\mathcal{R} = 5$. The x axis is proportional to $(\mathcal{R}/D)^2$; a higher SNR will move the curve horizontally to the right whereas a larger diameter will move the curves horizontally to the left. The change of slope in the curve around $m = 22$ is because the sky background starts acting as the limiting factor in attempts to detect fainter sources.

As we go to lower frequencies, the wave aspects of electromagnetic radiation dominate over the photon aspects. From the discussion in Vol. I, Chap. 5, Section 5.11, we know that the transition occurs for thermal radiation at $h\nu \approx k_B T$. Consequently, photon-counting fluctuations – which are important at optical and higher frequencies – are subdominant to thermal noise in the radio wavelength. The best way to characterise the noise in these long-wavelength bands is in terms of an equivalent *system temperature* T_s defined such that the noise power per unit frequency range is $k_B T_s$. If the beam size of the telescope is $\Omega_{\text{beam}} \approx (\lambda/D)^2$, where D is the effective diameter of the aperture of the telescope, then the noise will be equivalent to the intensity of

$$F_{\text{noise}} = B_\nu(T_s)\Omega_{\text{beam}} \approx \frac{2k_B T_s}{\lambda^2}\Omega_{\text{beam}}, \qquad (1.50)$$

where $B_\nu(T_s)$ is the blackbody intensity at the system temperature, and, in arriving at the second equality, we have used the Rayleigh–Jeans limit of the blackbody spectrum. To proceed below the noise limit, we must integrate for a

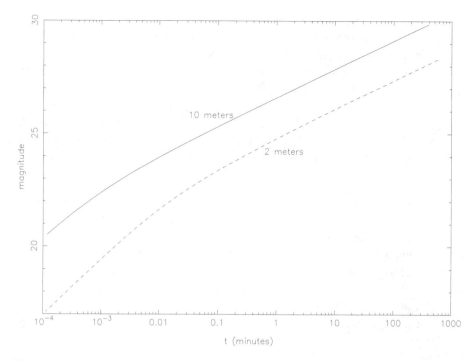

Fig. 1.12. The limiting magnitude reached by integration for a time t in a telescope of diameters $D = 2$ and 10 m, a field of view of 2×2 arcsec^2, and a SNR of $\mathcal{R} = 5$. The x-axis values scale in proportion to $(\mathcal{R}/D)^2$.

length of time t and use the basic fact that, while the signal increases as t, the noise increases only as $t^{1/2}$. If the bandwidth of observations is $\Delta \nu$, then the lowest time resolution available is $\Delta t \approx (1/\Delta \nu)$; so there are $\mathcal{N} = (t/\Delta t)$ independent observations in the observation lasting a time interval t. The minimum sensitivity reached will therefore be $F_{\min} \approx F_{\text{noise}}/\mathcal{N}^{1/2}$. Substituting for all the factors, we get an equivalent minimum rms flux as

$$F_{\min}(\nu) \simeq \frac{2 k_B T_s}{\lambda^2} \left(\frac{\lambda}{D^2}\right) \frac{1}{([\Delta \nu] t)^{1/2}}$$

$$\simeq 3 \left(\frac{T_s}{20\,\text{K}}\right) \left(\frac{\Delta \nu}{100\,\text{MHz}}\right)^{-1/2} \left(\frac{t}{10^4\,\text{s}}\right)^{-1/2} \left(\frac{D}{130\,\text{m}}\right)^{-2} \mu\text{Jy}. \quad (1.51)$$

This is the analogue of relation (1.46) in the optical (and higher-energy) observations; once again the sensitivity grows as $\sqrt{t}$.

Although this result is generally valid, more useful formulas can be obtained in special contexts, for example, in the case of detecting pulsar signals. We saw above in Section 1.5 that the signal from a pulsar has a fairly stable period P and has a particular shape and profile that differs from pulsar to pulsar. In the simplest context we can approximate the signal from a pulsar of period P as one of width W and height S. In this case, it is advantageous to fold the data repeatedly over stretches of length P in order to obtain an increased SNR. Finally, we will achieve a signal across the pulse period P, containing a number of bins N_b, so that the effective integration time will be t/N_b, where t is the total duration of observation, by use of which the folded profile is constructed. From expression (1.51), it follows that the minimum peak flux that can be detected scales as $S_{\text{min,peak}} \propto (t/N_b)^{-1/2} = Q(N_b/t)^{1/2}$, where Q stands for the rest of the quantities in expression (1.51). It is, however, customary to quote the detection limit in terms of the average flux $S_{\text{min,av}} \equiv (W/P)S_{\text{min,peak}}$ as

$$S_{\text{min,av}} = Q\sqrt{\frac{N_B}{t}}\left(\frac{W}{P}\right). \tag{1.52}$$

The best SNR is obtained if all the pulsar fluxes are collected into one bin, that is, if the bin width is W, giving $N_b = (P/W)$, provided that $W \ll P$. If $W \gtrsim (P/2)$, we have to take into account the off-pulse region, properly making $N_b = P/(P-W)$. It follows that $S_{\text{min,av}} \propto (W/P)^{1/2}$ for narrow pulses with $W \ll P$ and $S_{\text{min,av}} \propto W[P(P-W)]^{-1/2}$ for broad pulses with $W \gtrsim (P/2)$. In the literature, these two limits are usually handled by a simple interpolation formula of the form

$$S_{\text{min}} = \frac{Q}{\sqrt{t}}\left(\frac{W}{P-W}\right)^{1/2}. \tag{1.53}$$

This is equivalent to assuming that the entire on-pulse region is averaged to a bin of width W and the off-pulse region is averaged to a bin of width $P-W$ and suitably redefining the SNR. This formula also illustrates how the characteristics of the astronomical objects under study can influence the useful SNR achievable by a given instrument. Figure 1.13 summarises the sensitivity reached in different wave bands.

1.8.4 Dynamic Range of Observations

The fact that astrophysical phenomena span a wide dynamic range of ~ 17 decades in frequency – from $\sim 10^7$ Hz in radio to $\sim 10^{24}$ Hz in hard gamma rays – implies that complete information will be available only when all the bands are thoroughly surveyed. Because the degree of incompleteness of astronomical data is not often appreciated, we will illustrate it using a simple example.

Let us consider the task of detecting (hypothetical) sources emitting thermal radiation centred in different wave bands. For uniformly distributed sources, the number of sources is proportional to the volume of the space that is examined, whereas the flux observed from any given source will decrease as the inverse square of the distance from the source. Hence the number $N(>S)$ of sources with a flux greater than S scales as $N(>S) \propto S^{-3/2}$. If the minimum and

1.8 Theoretical Constraints on Astronomical Observations 49

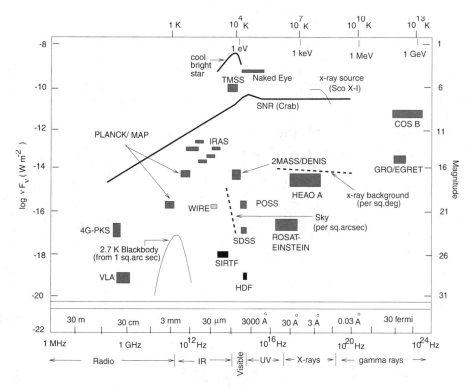

Fig. 1.13. Characteristic spectra of a class of stellar objects compared with the effective sensitivities reached in the surveys in different wave bands.

the maximum values of flux are fixed so that $N(>S_{max}) \approx 1$ and $N(>S_{min}) \approx 10^3$ – giving $N \approx 1000$ sources in a survey – then we require the observations to span a range of $S_{max}/S_{min} \approx 10^2$; that is, the survey should cover ~2 decades in S.

In the Planck spectrum, the quantity $\nu I(\nu)$ varies by 2 decades when the frequency ν varies by ~1.5 decades in the frequency around the peak. This suggests that there can be $(17/1.5) \approx 11$ distinct, nonoverlapping classes of thermally radiating sources in the universe, each having a population of ~1000. To do any meaningful analysis and relative calibration, surveys have to overlap to a certain extent. Assuming a conservative 50% overlap, we require ~20 surveys for detecting all these distinct classes of (hypothetical) thermal sources. At present, we possibly have approximately eight nonoverlapping sky surveys available in optical, IR, radio, x-ray, and gamma-ray bands with significant gaps around 300 μm, 0.1 μm, 5 mm, and 100 keV even in the range covered by these surveys. Tables 1.7 and 1.8 give some particulars about some of the currently available surveys. Although the argument given above is based on somewhat unrealistic (thermal) sources, it does illustrate the fact that the spectral information about the universe is possibly only 50% complete at the present stage.

Table 1.7. Radio to mid-IR surveys

Survey (Telescope)	Wave Band	Sky Coverage	Sensitivity	Angular Resolution
GB6 (Green Bank 91 m)	4850 MHz	$0 < d < +75$	18 mJ/beam	3.0'
PMN (Parkes 64 m)	4850 MHz	$-87 < d < +10$	30 mJ/beam	5.0'
HIPASS (Parkes 64 m)	1420 MHz	$d < 26$	40 mJ/beam	3.0'
ZOA (Parkes 64 m)	1420 MHz	$213 < l < 33$	20 mJ/beam	3.0'
Leiden (Dwingeloo 26 m)	1420 MHz	$d > -30$	0.2 K	0.5° grid
FIRST (NRAO VLA)	1400 MHz	NGP	1 mJ/beam	5.0''
NVSS (NRAO VLA)	1400 MHz	$d > -40$	2.5 mJ/beam	45''
SUMSS (Molonglo synthesis)	843 MHz	$d < -30$	1 mJ/beam	43''
WENSS (Westerbork synthesis)	326 MHz	$d > +30$	18 mJ/beam	54''
6C (Cambridge synthesis)	151 MHz	$d > +30$	120 mJ/beam	4.2'
7C (Cambridge synthesis)	151 MHz	$d > +20$	50 mJ/beam	70''
8C (Cambridge synthesis)	38 MHz	$d > +60$	1 J/beam	4.5'
UMSB (FCRAO 14 m)	12 CO ($J = 1$–0)	$+8 < l < +90$	0.4 K/channel	6', 18'
GRS (FCRAO 14 m)	13 CO, CS	$+18 < l < +52$, $-1 < b < +1$	0.2, 0.1K/channel	50''
ISS (IRAS IR array)	(12, 25, 60, 100)μ	All sky	0.5 Jy@12,25, 60;1.5 Jy@100	0.5'–2'
ISS (IRAS LRS)	8–22 μ	All sky	10 Jy	2'
ELAIS (ISO ISOCAM)	(6.7, 15)μ	HDF(N)	0.04, 0.2 mJy	
ISOGAL (ISO ISOCAM)	(7, 15)μ	$-10 < l < +30$, $-1 < b < +1$	10 mJy	6''

[GB: Green Bank; PMN: Parkes-Massachusetts Institute of Technology-National Radio Astronomy Observatory; HIPASS: HI Parkes All Sky-Surveys; ZOA: Zone Of Avoidance; FIRST: Faint Images of the Radio Sky at Twenty cm.; NGP: North Galactic Pole; NRAO: National Radio Astronomy Observatory; VLA: Very Large Array; NVSS: NRAO VLA Sky Survey; SUMSS: Sydney University Molonglo Sky Survey; WENSS: WEsterbork Northern Sky Survey; UMSB: University of Massachusetts-Stony Brook; FCRAO: Five College Radio Astronomy Observatory; GRS: Galactic Ring Survey; IRAS: InfraRed Astronomical Satellite; ISS: IRAS Sky Survey; LRS: Low Resolution Spectrograph; ISO: Infrared Space Observatory; ELAIS: European Large Area ISO Survey; ISOCAM: ISO CAMera; ISOGAL: ISO GALactic survey; HDF(N): Hubble Deep Field (North).]

Table 1.8. Near-IR to gamma-ray surveys

Survey (Telescope; Instrument)	Wave Band	Sky Coverage	Sensitivity	Angular Resolution
POSSII (Palomar; 48 in)	B, R, I	Northern hemisphere	22.5, 20.8, 19.5 mag	1″
DENIS (ESO Chile; 1 m IR CCD camera)	I, J, Kshort	Southern hemisphere	18,16,14 mag	2″
EIS (ESO NTT; EMMI)	U, G, R, I, K	Patches, HDF–S, etc.	23.5	—
HDF-N (HST; WFPC2, FOS)	UV, B, V, I	$12^h 36^m 49.4^s$ $+62° 12' 58''$	28 mag	—
HDF-S (HST; WFPC2, STIS, NIC)	UV–NearIR Four bands	$22^h 32^m 56.2^s$ $-60° 33' 3''$	30 mag	
SDSS (APO 2.5 m; CCD mosiac)	4/5 filters	$b > 30$ (north) 2×50 strip (south)	$R \sim 23(N), R \sim 25(S)$	3′
2MASS (CTIO 1.3 m; IR mosaic)	J, H, K	All sky	$R \sim 23(N), R \sim 25(S)$	1″
LASS (HEAO-1; A1)	0.25–25 keV	All sky	0.25 μJy	2°
RASS (ROSAT; PSPC, XIT)	0.25, 0.75, 1.5 keV	94% sky	$\sim 30 \times 10^{-6}$ cts s^{-1}arcmin^{-2}	3′
CXRB (ASCA; LSS, SIS)	2–10 keV	6° field Gal. plane	4×10^{-14} ergs cm^{-2} s^{-1}	20″
All sky (CGRO; EGRET)	$E > 100$ MeV	All sky	7×10^{-6} cm^{-2} s^{-1}	5′–10′
All sky (CGRO; COMPTEL)	1.8 MeV (26Al)	$-25 < b < +25$ All l	1.5×10^{-3} cm^{-2} s^{-1}	0.5°–1°
All sky (CGRO; BATSE)	0.02–100 MeV	Transients All sky	3×10^{-8} ergs cm^{-2} s^{-1}	3′

[POSS: Palomar Observatory Sky Survey; DENIS: DEep Near Infrared Survey of the Southern sky; CCD: Charge Coupled Device; ESO: European Southern Observatory; EIS: ESO Imaging Survey; NTT: New Technology Telescope; EMMI: ESO Multi Mode Instrument; HDF-N: Hubble Deep Field-North; HST: Hubble Space Telescope; WFPC: Wide-Field and Planetary Camera; FOS: Faint Object Spectrograph; HDF-S: Hubble Deep Field-South; STIS: Space Telescope Imaging Spectrograph; NIC: Near Infrared Camera; SDSS: Sloan Digital Sky Survey; APO: Appache Point Observatory; 2MASS: 2 Micron All Sky Survey; CTIO: Cerro Tololo Inter-American Observatory; LASS: Large Area Sky Survey; HEAO: High Energy Astrophysics Observatory; ROSAT: ROentgen SATellite; RASS: Rosat All Sky Survey; PSPC: Position Sensitive Proportional Counter; XIT: X-ray Imaging Telescope; CXRB: Cosmic X-Ray Background; ASCA: Advanced Satellite for Cosmology and Astrophysics; SIS: Solid state Imaging Spectrometer; CGRO: Compton Gamma Ray Observatory; EGRET: Energetic Gamma-Ray Experiment; COMPTEL: COMPton TELescope; BATSE: Burst And Transient Source Experiment.]

1.8.5 Luminosity Bias

In any survey that is sensitive down to a particular luminosity limit, we tend to pick up brighter objects preferentially. This implies that even when surveys are done with adequate resolution and dynamic range, it may not lead to a sample population that is representative of the original population. This phenomenon, usually called the Malquist bias, is important in any astronomical survey. We will now provide description of this effect.

Given a large collection of similar objects (such as stars in a galaxy), it is convenient to define a density function $\bar{\Phi}(M, \mathbf{x})$ such that

$$dN = \bar{\Phi}(M, \mathbf{x})\, dM\, d^3\mathbf{x} \tag{1.54}$$

represents the number of objects with absolute magnitudes between M and $M + dM$ in the spatial volume range $\mathbf{x}, \mathbf{x} + d^3\mathbf{x}$. Such a definition is quite general, and the conclusions we derive below are applicable not only to stars in our galaxy but also to galaxies, clusters of galaxies, and to any particular class of objects such as, say, pulsars, neutron stars etc. For the sake of definiteness, however, we shall describe the situation as regards stars. To make any progress regarding the structure of $\bar{\Phi}(M, \mathbf{x})$ it is necessary to assume that the spatial and luminosity distribution are uncorrelated so that we can write

$$dN = [\Phi(M)\, dM]\, [n(\mathbf{x})\, d^3\mathbf{x}], \tag{1.55}$$

where $\Phi(M)$ is called the luminosity function. The spatial distribution is decided by $n(\mathbf{x})$, which, in turn, will be determined by the dynamics and shall not concern us at present.

The luminosity function $\Phi(M)$ is of great importance in astronomy but is fairly difficult to determine accurately. What is usually determined observationally is the distribution of stars in apparent magnitude m. By straightforward counting, we can determine, for example, the quantity $A(m)$, where $dN = A(m)\, dm$ gives the number of stars in the apparent magnitude range $m, m + dm$. In any realistic situation, there will be a limiting magnitude m_1 so that we know only the functional form $A(m)$ for $m < m_1$. The question arises as to how best $\Phi(M)$ can be determined from the knowledge of $A(m)$ when the absolute and the apparent magnitudes are related by Eq. (1.20).

Given relation (1.20) and a luminosity function $\Phi(M)$, the number of stars with absolute magnitude in the range $M, M + dM$ *and* apparent magnitude in the range $m, m + dm$ can be calculated as

$$\frac{d^2N}{dm\, dM} = \Phi(M) \frac{d\mathcal{N}}{dr} \left(\frac{\partial r}{\partial m}\right)_M, \tag{1.56}$$

where $r(m, M)$ is determined by Eq. (1.20) and $\mathcal{N}(r)\, dr$ is the number of stars in the survey in the range $r, r + dr$; if the survey determining $A(m)$ has a solid angle coverage of Ω sr, $(d\mathcal{N}/dr) = \Omega r^2 n(r)$. Using this result and integrating

1.8 Theoretical Constraints on Astronomical Observations

over the absolute magnitude, we get

$$A(m) = \int_{-\infty}^{\infty} dM \frac{d^2 N}{dm\, dM} = \Omega \int_{-\infty}^{\infty} dM \Phi(M) \left(\frac{\partial r}{\partial m}\right)_M r^2 n(r)$$
$$= \Omega \int_{0}^{\infty} dr\, \Phi(M) r^2 n(r), \qquad (1.57)$$

where we have used the relation $(\partial M/\partial r)_m = -(\partial m/\partial r)_M$. Given the distribution $d^2 N/dM\, dm$, we can define averages of variables over M at a given observed value of m. For example, the mean absolute magnitude of stars in the sample that have an apparent magnitude m is given by

$$\langle M \rangle_m = \frac{\int_{-\infty}^{\infty} dM\, M \frac{d^2 N}{dm\, dM}}{\int_{-\infty}^{\infty} dM \frac{d^2 N}{dm\, dM}} = \frac{\int_{0}^{\infty} dr\, M \Phi(M) r^2 n(r)}{\int_{0}^{\infty} dr\, \Phi(M) r^2 n(r)}. \qquad (1.58)$$

It is now easy to derive the mean values of different derivatives of $\Phi(M)$ from Eq. (1.57). We get, for example,

$$\frac{1}{A}\frac{dA}{dm} = \left\langle \frac{1}{\Phi}\frac{d\Phi}{dM} \right\rangle_m, \qquad (1.59)$$

$$\frac{1}{A}\frac{d^2 A}{dm^2} = \left\langle \frac{1}{\Phi}\frac{d^2 \Phi}{dM^2} \right\rangle_m. \qquad (1.60)$$

These relations highlight an important point that can be illustrated with a specific example. Let us assume that $\Phi(M)$ is a Gaussian in M determined by two independent parameters, the mean M_0 and variance σ:

$$\Phi(M) = \frac{1}{(2\pi\sigma^2)^{1/2}} \exp\left[-\frac{(M-M_0)^2}{2\sigma^2}\right]. \qquad (1.61)$$

These are the means and the variances of the population of stars that would be obtained in a volume-limited sample that counts every star in some region of space. With Eqs. (1.59) and (1.60), it is straightforward to obtain the following relations:

$$\langle M \rangle_m - M_0 = -\sigma^2 \frac{d \ln A}{dm}, \quad \sigma_m^2 - \sigma^2 = \sigma^4 \frac{d^2 \ln A}{dm^2}, \qquad (1.62)$$

where

$$\sigma_m^2 \equiv \langle (M - \langle M \rangle_m)^2 \rangle_m = \langle M^2 \rangle_m - \langle M \rangle_m^2 \qquad (1.63)$$

is the variance of the measured absolute magnitudes. These relations show that, in general, the mean and the variance of the *magnitude-limited sample* will be different from that of the true values, i.e., the mean and the variance of the *volume-limited sample*. Because we usually have $dA/dm > 0$, Eqs. (1.62) show that the objects in the magnitude-limited survey of certain m will – on the average – be more luminous than the objects of the underlying population; this effect is called Malmquist bias. There is no general procedure available to correct for this bias, although particular situations can be handled by specific techniques.

Exercise 1.5
Examples of Malmquist bias: In the specific case in which $\Phi(M)$ is a Gaussian and $n(r) = $ constant, show that

$$A(m) \propto \exp[0.6 \ln 10(m - M_0)], \tag{1.64}$$

so that

$$\frac{d \log A}{dm} = \frac{d \ln A / \ln 10}{dm} = 0.6, \quad \frac{d^2 \ln A}{dm^2} = 0. \tag{1.65}$$

Repeat the calculation if $\Phi(M)$ is exponentially distributed and $n(r) = $ constant. How will the results change if $n(r)$ decreases with r?

One corollary of the luminosity bias is that we require adequate number of objects in a sample for maintaining acceptable accuracy at the faint end. For example, consider a population of sources such that the number of sources in the luminosity bin $L, L + dL$ is a power law: $N(L) \, dL \propto L^{-\alpha} \, dL$ or, equivalently,

$$\frac{dN}{d \ln L} \propto L^{1-\alpha}. \tag{1.66}$$

The maximum volume $V(L)$ in which a source of luminosity L could lie and still be detected in a flux-limited survey varies as $V \propto L^{3/2}$. It follows that the number of sources observed per logarithmic interval in the luminosity is

$$n(L) \propto V(L) \frac{dN}{d \ln L} \propto L^{5/2-\alpha}. \tag{1.67}$$

If we want to reach a sensitivity of $(L_{\min}/L_{\max}) = 10^{-y}$ with approximately $N_0 = 10^x$ objects in the faintest bin, it is necessary to have $N > 10^x 10^{y[(5/2)-\alpha]}$ objects in the survey. The random fluctuations in the faintest bin will give a fractional error of $N_0^{-1/2} = 10^{-x/2}$. To get 20% accuracy in the lowest (log L) bin, we need $0.2 \simeq N_0^{-1/2} \simeq 10^{-x/2}$, giving $x \simeq 1.5$. Then the constraint on N translates to ~ 2500 sources for $\alpha = 1.5, y = 2$, whereas we need approximately 2.5×10^5 sources for $\alpha = 1.5, y = 4$. If $\alpha = 1$, then we require 10^4 sources for $y = 2$ and 10^6 sources for $y = 4$. These numbers show that careful determination of the luminosity function of faint objects in order to make a census of the universe will require fairly elaborate surveys.

1.8.6 Confusion Limit

Surveys of the universe also suffer from a difficulty known as the *confusion limit*. To understand the origin of this limitation, let us consider a sky survey carried out by an instrument having a beam area of Ω sr and a flux limit of $F_{\min}$. If $\Omega_{\rm sky}$ sr denotes the entire solid angle that needs to be searched and if there are $N_{\rm tot}$ objects with the flux $F > F_{\min}$ in the universe, then the mean number of objects per beam expected in the survey will be $\bar{n}(\text{per beam}) = N_{\rm tot}(\Omega/\Omega_{\rm sky})$. Because the beam size sets the limit of resolution, we will end up misidentifying a source if more than one object occupies a beam. The probability that there will be n objects in a given beam is given by the Poisson distribution $P(n) = (\bar{n}^n/n!)\exp(-\bar{n})$. In practical situations, it is enough to consider the probability $P(2)$ to have, say, two objects in a beam. The number of beams in which such a mistaken identification of two objects as one will occur is $N_{\rm wrong} = P(2)(\Omega_{\rm sky}/\Omega)$. Hence the fraction of cases in which wrong identification will be made will be

$$\frac{N_{\rm wrong}}{N_{\rm tot}} = \frac{1}{2}\left(\frac{N_{\rm tot}\Omega}{\Omega_{\rm sky}}\right)\exp\left[-\left(\frac{N_{\rm tot}\Omega}{\Omega_{\rm sky}}\right)\right]. \quad (1.68)$$

To see the effect of this misidentification, consider a situation with $N_{\rm tot} = 1000$, $\Omega = 1 \text{ deg}^2$, and $\Omega_{\rm sky} = 4.1 \times 10^4 \text{ deg}^2$. Then $\bar{n} \approx 0.024$ and $P(2)$ is 3×10^{-4}, so that the number of beams in which two objects will be found is $N_{\rm wrong} \approx 12$. In other words, we have the correct identification in 99% of the cases. The situation, however, changes drastically when $N_{\rm tot}$ goes up to 10^4. Now, with $\bar{n} \approx 0.24$, nearly 10% of the beams will have wrong identification. Because $N_{\rm tot}$ does increase with increasing sensitivity of the instrument, this can be a real difficulty for highly sensitive surveys trying to identify a large number of objects.

It might seem that we could avoid the problem by narrowing the beam size Ω of the instrument so that $\bar{n}$ decreases. This, however, leads to two difficulties. First, this will be a waste of effort, with the instrument's pointing at empty fields a larger fraction of the time. Second, and more important, random fluctuations will lead to spurious detections of sources in such empty fields. To see this quantitatively, let us assume that a fluctuation in the intensity at a 3σ level, where σ is the noise limit of the instrument, will lead to an acceptable detection. Taking the instrument noise to be Gaussian, we find that the probability for a 3σ event is $G(3\sigma) \approx 0.0135$. Because the number of empty beams is $N_{\rm empty} \approx [(\Omega_{\rm sky}/\Omega) - N_{\rm tot}] = (\Omega_{\rm sky}/\Omega)[1 - (N_{\rm tot}\Omega/\Omega_{\rm sky})]$, the number of spurious detection will be $N_{\rm spu} = G(3\sigma)N_{\rm empty}$. Hence the fraction of cases in which there will be spurious detection confusing the survey will be

$$\frac{N_{\rm spu}}{N_{\rm tot}} = G(3\sigma)\left(\frac{\Omega_{\rm sky}}{N_{\rm tot}\Omega}\right)\left[1 - \left(\frac{N_{\rm tot}\Omega}{\Omega_{\rm sky}}\right)\right]$$

$$= G(3\sigma)\left[\left(\frac{\Omega_{\rm sky}}{N_{\rm tot}\Omega}\right) - 1\right] \approx G(3\sigma)\left(\frac{\Omega_{\rm sky}}{N_{\rm tot}\Omega}\right). \quad (1.69)$$

In the case of $N_{tot} = 1000$ and $\Omega = 1$ deg^2, there are approximately 4×10^4 empty beams, making the number of spurious sources $N_{spu} \approx 540$. In other words, nearly one third of the catalog based on the survey will be spurious, illustrating the danger of small beam size.

This issue can only be tackled by increasing the threshold of detection, say, to 5σ so that the probability for spurious detection drops to $G(5\sigma) \approx 3 \times 10^{-7}$. For a given σ this will, of course, reduce the number of objects that can be treated as detected, which is a trade-off we are forced to make. Of course, if the instrument sensitivity improves so that σ can be brought down, then the survey will become more efficient.

The above example illustrates the fine balance that needs to be maintained among the resolution Ω, the flux limit F_{min} (which indirectly determines N_{tot}), and the sensitivity limit σ in any given survey. This leads to yet another limitation on obtaining accurate information about the populations of objects in the universe.

2
Stellar Structure

2.1 Introduction

This chapter discusses the structure of stars that are in steady state. Concepts described in Vol. I, Chaps. 5–7, will be used extensively here. The models described here will be needed in several subsequent chapters dealing with stellar evolution, compact remnants, and binary stars.[1]

2.2 Equations of Stellar Structure

A self-gravitating body of mass M and radius R will have gravitational potential energy of $U \approx -(GM^2/R)$. If such a body is in equilibrium with the gas pressure balancing the gravity, the virial theorem implies that it will have temperature T such that $Nk_BT \approx (GM^2/R)$, that is, $T \approx (GMm_p/k_BR)$. For a sufficiently large value of M/R, this temperature can be high enough to ignite nuclear reactions at the centre of the body. The nuclear energy generated near the centre will be transported by radiation and convection towards the outer regions and will eventually escape from the body. This will establish a temperature gradient inside the body such that, in steady state, the energy produced by nuclear reactions is equal to the energy radiated away from the outer surface. Such a steady-state situation can last as long as the conditions in the body allow the generation of nuclear energy inside it. Observations suggest that the stars belong to such a category of self-gravitating bodies that are essentially powered by the nuclear reactions.

A complete study of such a star will require modelling of different stages of evolution. To begin with, we must understand how such isolated bodies of certain mass and size can form and determine their initial chemical composition. Next, to understand the structure of such a star in steady state, we need to model the mechanism for energy generation, the mechanism for energy transport, and the nature of hydrostatic equilibrium. Finally, we must take into account the fact that the nuclear reactions will lead to a gradual evolution of the star and will eventually lead to a phase in which the supply of nuclear energy is exhausted.

This chapter deals with the modelling of the stars in steady state, ignoring the questions related to the formation of the star and its evolution. (Stellar evolution will be discussed in the next chapter, and several special features relevant to the end stages of stellar evolution will be studied in Chaps. 4 and 5. Star formation will be covered in Chaps. 3 and 9.) We will start with a discussion in which effects of rotation, magnetic field, and other complications are ignored; these features are commented on at the end of this section.

Consider a spherically symmetric star in steady state in which all physical variables depend on only the radial coordinate r. The equation for hydrostatic equilibrium of such a star is given by

$$\frac{dP}{dr} = -G\frac{M(r)\rho(r)}{r^2}, \tag{2.1}$$

where $P(r)$ and $\rho(r)$ are the pressure and the density, respectively, at radius r and $M(r)$ is the mass contained within a sphere of radius r. (We use Newtonian gravity as it provides an excellent approximation to the processes described below. Modifications that are due to general relativity are important in compact stellar remnants and will be discussed in Chap. 5.) The latter two quantities are related by

$$\frac{dM(r)}{dr} = 4\pi r^2 \rho. \tag{2.2}$$

These *two* equations connect the *three* variables $P(r)$, $\rho(r)$, and $M(r)$; we need one more equation to close the system.

For barotropic fluids in which pressure can be expressed as a function of density, these equations can be integrated along with a barotropic equation of the form $P = P(\rho)$. (This was discussed in Vol. I, Chap. 8.) This can arise, for example, if the material is made of degenerate fermions. In that case, the relation between pressure and density can be expressed in parametric form (see Vol. I, Chap. 5) with $P \propto \rho^{5/3}$ or $P \propto \rho^{4/3}$ in the nonrelativistic and extreme relativistic limits, respectively. Such equations of state will be considered in Chap. 5 when we are modelling the end stages of stellar evolution. In general, however, the fluid will not be barotropic and the equation of state will be of the form $P = P(\rho, T)$, where $T(r)$ is the temperature. The equation of state will then depend on the relative importance of radiation and gas pressure as well as the composition and state of ionisation of the gas.

It is conventional to parameterise the composition of the star by giving the mass fraction of hydrogen (X), helium (Y), and that of all heavier elements (Z), with $X + Y + Z = 1$. The actual values of X, Y, and Z have to be determined either from observations or from theoretical considerations related to the formation of stars in the galaxy and the model for the galaxy at the time of star formation. Current theoretical ideas suggest that cosmological evolution leads to a composition with $X \approx 0.75$, $Y \approx 0.25$ at some very early stage in the formation of a galaxy. Very old stars that form initially in the galaxy will have this primordial composition at the moment of formation. We shall, however, see in the next chapter that stellar evolution can lead to the synthesis of heavier elements that could be redistributed in the galactic gas during the end stages of stellar evolution. The second and later generations of stars forming from the processed galactic material will have a higher initial value for Z. The primordial stars are called *Population II*, and the later generation of stars are called *Population I*. For population I stars, we can take the composition to be approximately $X \simeq 0.71$, $Y \simeq 0.27$, and $Z \simeq 0.02$.

In the case of a mixture of ideal gas and radiation, the equation of state is given by

$$P = \left(\frac{\rho}{\mu m_H}\right) k_B T + \frac{1}{3} a T^4 \equiv \left(\frac{\mathcal{R}}{\mu}\right) \rho T + \frac{1}{3} a T^4, \qquad (2.3)$$

where a is the Stefan–Boltzmann constant and μ is the mean molecular weight of the system. In the second form of the equation, we have introduced the gas constant $\mathcal{R} \equiv (k_B/m_H) = 8.31 \times 10^7$ ergs mol^{-1} K^{-1} (see Vol. I, Chap. 5, Section 5.6). For a fully ionised gas, we can write the molecular weight as

$$\mu = \left(\frac{1}{\mu_{\text{ion}}} + \frac{1}{\mu_{\text{ele}}}\right)^{-1} = \left(\sum_j \frac{X_j}{A_j} + \sum_j \frac{\mathcal{Z}_j}{A_j} X_j\right)^{-1}, \qquad (2.4)$$

where $\mathcal{Z}_j$ and A_j are the atomic number and the atomic weight of jth element and X_j is the mass fraction. If the mass fractions of hydrogen, helium, and heavier elements are denoted by X, Y, $Z = 1 - X - Y$, respectively, and if $Z \ll X$, then

$$\frac{1}{\mu_{\text{ion}}} = X + \frac{1}{4}Y + \left\langle\frac{1}{A}\right\rangle_Z Z \simeq \frac{4X+Y}{4}, \quad \frac{1}{\mu_{\text{elec}}} = X + \frac{1}{2}Y = \frac{2X+Y}{2}. \qquad (2.5)$$

Adding up and using $Y \simeq 1 - X$, we get

$$\mu \approx \frac{4}{3+5X} \simeq 0.5 \, X^{-0.57}. \qquad (2.6)$$

The second form is a power-law approximation that is useful for studying the X dependence of physical variables qualitatively. For stars with $X \approx 0.7$, we have $\mu \approx 0.6$.

Equations (2.1) and (2.2) and the equation of state $P = P(\rho, T)$ constitute *three* equations for the *four* variables $P(r)$, $\rho(r)$, $M(r)$, and $T(r)$. We therefore need another equation giving the variation of temperature with radial distance.

The form of such an equation depends crucially on the physical processes that are operating at a given region of a star. If the energy generated by nuclear reactions (or some other process) is transported outwards by radiation, then it is possible to relate the energy flux to the temperature gradient. Using the concepts developed Vol. I, Chap. 6, we can write

$$\mathcal{F}(r) = \frac{L(r)}{4\pi r^2} = -\frac{4acT^3}{3\rho\kappa}\left(\frac{dT}{dr}\right), \quad (2.7)$$

where $L(r)$ is the luminosity and $\mathcal{F}(r)$ is the energy flux. [Whenever convenient, we shall use the notation $L_r, \mathcal{F}_r, \ldots$, to denote $L(r), \mathcal{F}(r)$, etc.] The quantity κ here is the mean radiative opacity of the star and is, in general, a function of ρ and T; that is, $\kappa = \kappa(\rho, T)$. Relation (2.7) can be understood intuitively along the following lines. If the radiative-energy flux is $\mathcal{F}$, then the corresponding momentum flux is $\mathcal{F}/c$. If the number density of scatterers is n and the scattering cross section is σ, then the momentum scattered per second per unit volume will be $(n\sigma/c)\mathcal{F} = (\rho\kappa/c)\mathcal{F}$. This quantity will be the force per unit volume and hence should be equal to $-\nabla P_{\text{rad}} = -(4aT^3/3)\nabla T$. Equating the two expressions allows us to express the radiative flux $\mathcal{F}$ in terms of the temperature gradient as given by Eq. (2.7). Assuming that a suitable fitting function for opacity $\kappa(\rho, T)$ is available, Eq. (2.7) can be inverted to give the temperature variation that is due to radiative flux as

$$\frac{dT}{dr} = -\frac{3}{4ac}\frac{\kappa\rho}{T^3}\frac{L_r}{4\pi r^2} \quad \text{(radiation)}. \quad (2.8)$$

In Vol. I, Chap. 6, we have discussed several sources of opacity in the material medium. The actual form of κ depends on the composition of the medium and the dominant processes that are operating at the relevant temperatures and densities. It was mentioned above that observations and theory show that the stellar material is mostly made of hydrogen and helium with traces of heavier elements. For such a composition and for temperatures and densities that are prevalent in stars, the main source of opacities are the following:

(1) The scattering of radiation by free electrons contributes the *electron-scattering opacity* given by

$$\kappa_{\text{es}} = 0.2\,(1+X)\,\text{cm}^2\,\text{g}^{-1}, \quad (2.9)$$

where X is the mass fraction of hydrogen (see Vol. I, Chap. 6, Section 6.4).
(2) The ionisation of atoms leads to the *bound–free opacity*, which has the form

$$\kappa_{bf} \approx 4 \times 10^{25}\,\text{cm}^2\,\text{g}^{-1}\,Z(1+X)\rho T^{-3.5}, \quad (2.10)$$

where Z is the mass fraction of all elements heavier than helium. (See Vol. I, Chap. 6, Section 6.12.)
(3) The *free–free absorption* contributes an opacity given by (Vol. I, Chap. 6, Section 6.9)

$$\kappa_{ff} \approx 4 \times 10^{22}(X+Y)(1+X)\rho\,T^{-3.5}\,\text{cm}^2\,\text{g}^{-1}, \quad (2.11)$$

where Y is the mass fraction of helium. This form is usually called Kramer's opacity.

(4) Finally, it is important to note that the H$^-$ ion (discussed in Vol. I, Chap. 7) contributes significantly to the opacity when $3000\,\text{K} \lesssim T \lesssim 6000\,\text{K}$ and $10^{-10}\,\text{g cm}^{-3} \lesssim \rho \lesssim 10^{-5}\,\text{g cm}^{-3}$. If $X \approx 0.7$ and $0.001 \lesssim Z \lesssim 0.03$ then the opacity that is due to the hydrogen ion can be approximated by

$$\kappa_{\text{H}^-} \approx 2.5 \times 10^{-31}(Z/0.02)\,\rho^{1/2}\,T^9\;\text{cm}^2\,\text{g}^{-1}. \tag{2.12}$$

The above result of Eq. (2.8) for (dT/dr), of course, is valid only if the entire flux of energy $\mathcal{F}(r)$ is transferred through radiative processes. In general, convection could also provide a means of transferring the energy. We saw in Vol. I, Chap. 8, that a fluid will be unstable to convection if the temperature gradient is too steep. In the simplest context, the criterion for convective instability can be expressed as $\nabla > \nabla_{\text{ad}}$, where

$$\nabla \equiv \frac{d\ln T}{d\ln P}, \qquad \nabla_{\text{ad}} \equiv \frac{\gamma - 1}{\gamma}. \tag{2.13}$$

In the case of radiative transfer of energy, the gradient is given by

$$\nabla_{\text{rad}} \equiv \left(\frac{d\ln T}{d\ln P}\right)_{\text{rad}} = \frac{3\kappa}{16\pi ac}\left(\frac{P}{T^4}\right)\left(\frac{L_r}{GM_r}\right), \tag{2.14}$$

where Eqs. (2.1) and (2.8) have been used to eliminate the spatial derivatives of P and T. If the gradient ∇_{rad} arising out of radiative transfer is less than ∇_{ad}, then the fluid is stable against convection and we can use Eq. (2.8). On the other hand, if ∇_{rad} is greater than ∇_{ad}, we have to take into account the energy transported because of convection. This is a complicated problem and we do not yet have a fundamental theory describing convective-energy transfer. From the discussion in Vol. I, Chap. 8, Section 8.14, we may approximate the situation as one arising out of very efficient convection. In such a case, the gradient will tend to the adiabatic value and we may assume that $\nabla = \nabla_{\text{ad}}$, and write

$$\nabla = \frac{d\ln T}{d\ln P} = \left(\frac{P}{T}\right)\frac{(dT/dr)}{(dP/dr)} = -\frac{r^2}{GM_r}\left(\frac{P}{\rho T}\right)\frac{dT}{dr} = \nabla_{\text{ad}} = \left(1 - \frac{1}{\gamma}\right), \tag{2.15}$$

where we have used (2.1) to eliminate dP/dr. This gives

$$\frac{dT}{dr} = -\nabla_{\text{ad}}\left(\frac{GM_r}{r^2}\right)\left(\frac{\rho T}{P}\right) = -\left(1 - \frac{1}{\gamma}\right)\left(\frac{\rho T}{P}\right)\left(\frac{GM_r}{r^2}\right). \tag{2.16}$$

Putting together both convective and radiative transfers of energy, we can write the temperature gradient as

$$\frac{dT}{dr} = \begin{cases} -\dfrac{3\kappa}{4ac}\left(\dfrac{\rho}{T^3}\right)\left[\dfrac{L(r)}{4\pi r^2}\right] & [\text{if } \nabla_{\text{rad}}(r) \leq \nabla_{\text{ad}}] \\ -\nabla_{\text{ad}}\left(\dfrac{\rho T}{P}\right)\left(\dfrac{GM_r}{r^2}\right) & [\text{if } \nabla_{\text{rad}}(r) \geq \nabla_{\text{ad}}] \end{cases}. \quad (2.17)$$

This criterion should be applied locally at each r. In general, the value of ∇_{ad} may also change from place to place if ionisation processes are prevalent. The fact that the energy-transport mechanism can be convective or radiative, depending on the local conditions, introduces significant complications in the stellar modelling. For example, it allows for the possibility that the star may have a convective core and a radiative envelope or vice versa.

Equations (2.1), (2.2), and (2.17) form a set of *three* differential equations for the *five* variables $P(r)$, $M(r)$, $\rho(r)$, $T(r)$, and $L(r)$. Using the equation of state $P = P(\rho, T)$, we can eliminate one of these variables, reducing the set to three equations for four independent variables. To close the system, we need one more equation that relates the luminosity $L(r)$ to the energy-production rate in the star. Let $\epsilon(\rho, T)$ denote the amount of energy generated per unit mass of stellar material at density ρ and temperature T. In that case, we have, from the definition of luminosity, the relation

$$\frac{dL_r}{dr} = 4\pi r^2 \rho \epsilon. \quad (2.18)$$

This provides the last equation needed to close the set.

The form of $\epsilon(\rho, T)$ depends on the dominant nuclear reactions that are providing the energy in the star. These reactions were discussed in Vol. I, Chap. 12. If the energy is supplied by the proton–proton chain, then the energy-generation rate is given by

$$\epsilon_{pp}(\rho, T) \approx \frac{2.4 \times 10^4 \rho X^2}{T_9^{2/3}} e^{-3.380/T_9^{1/3}} \text{ ergs g}^{-1} \text{ s}^{-1}, \quad (2.19)$$

where $T_9 = (T/10^9 \text{ K})$, etc. On the other hand, for the CNO cycle, the energy-generation rate is

$$\epsilon_{\text{CNO}}(\rho, T) \approx \frac{4.4 \times 10^{25} \rho X Z}{T_9^{2/3}} e^{-15.228/T_9^{1/3}} \text{ ergs g}^{-1} \text{ s}^{-1}. \quad (2.20)$$

Both these reactions (and, in fact, more complicated helium burning, silicon burning, and oxygen burning reactions) can all be approximated by local power law

$$\epsilon = \epsilon_0 \rho^\lambda T^\nu, \quad (2.21)$$

where, for the p–p chain, $\lambda = 1$ and $\nu = 4$; for the CNO cycle, $\lambda = 1$ and $\nu = 15$–18; for helium burning $\lambda = 2$ and $\nu = 40$, etc. At high temperatures, the CNO cycle and other reactions dominate over the p–p chain as the energy-production mechanism. Because of the form of ϵ, it is possible for the star to have different energy-production mechanisms operating at different regions of the interior. This fact leads to additional complications in the study of stellar structure and plays a vital role in the stellar evolutionary processes.

Given the functions $P(\rho, T)$, $\kappa(\rho, T)$, and $\epsilon(\rho, T)$, we can integrate Eqs. (2.1), (2.2), (2.17), and (2.18) to determine the equilibrium structure of the stars, provided that we know the boundary conditions. From physical considerations, we demand that $M(r) \to 0$ and $L(r) \to 0$ as $r \to 0$. Further, we would expect P and ρ to vanish at the surface of the star, which we may take at some location $r = R$. It is often convenient to rewrite the equations of stellar structure with M_r as the independent variable (rather than r) and to treat $r = r(M_r)$ as a dependent function. We can easily obtain the corresponding equations by dividing Eqs. (2.1), (2.17), and (2.18) by Eq. (2.2). This gives the set

$$\frac{dP}{dM_r} = -\frac{GM_r}{4\pi r^4}, \tag{2.22}$$

$$\frac{dr}{dM_r} = \frac{1}{4\pi r^2 \rho}, \tag{2.23}$$

$$\frac{dT}{dM_r} = \begin{cases} -\frac{3\kappa}{64\pi^2 ac}\left(\frac{1}{T^3}\right)\left(\frac{L_r}{r^4}\right) & (\nabla_{\rm rad} \leq \nabla_{\rm ad}) \\ -\frac{\nabla_{\rm ad}}{4\pi}\left(\frac{T}{P}\right)\left(\frac{GM_r}{r^4}\right) & (\nabla_{\rm rad} \geq \nabla_{\rm ad}) \end{cases}, \tag{2.24}$$

$$\frac{dL_r}{dM_r} = \epsilon. \tag{2.25}$$

If the mass of the star is taken to be M, then the boundary conditions become

$$\text{at the centre} \quad (M_r = 0), \quad r = L_r = 0; \tag{2.26}$$
$$\text{at the surface} \quad (M_r = M), \quad \rho = T = 0. \tag{2.27}$$

The boundary condition at the origin is obvious, but the one at the surface should be thought of as an approximation. If the temperature at the stellar surface vanishes, no radiation will emerge from that surface. The outermost regions of the star need to be handled separately, and we will discuss this feature in Section 2.5. The key difficulty is that the outermost layers of any star are not in thermal equilibrium and the layers are not optically thick. Hence the radiative gradient does not govern the flow of energy in these layers. Accurate modelling of stellar atmospheres requires obtaining some semiempherical relationship between the temperature and the optical depth and matching it with the interior solutions in order to obtain the full structure. In fact, the outer regions of stars like the Sun contain a hot corona in which even hydrostatic equilibrium is not maintained. In spite of all these complications, the boundary conditions discussed above do provide a reasonable approximation as long as the resulting solution is used in only the interior. In other words, the actual structure of the *atmosphere* depends on the correct boundary conditions whereas the *interior* is reasonably independent of this boundary condition. A partial demonstration of this independence will be provided in Subsection 2.4.2.

Because we have four first-order ordinary differential equations and three auxiliary equations (in the form of ϵ, κ, and P) for the seven variables $r(M_r)$, ρ, T, L, P, κ, and ϵ, with four boundary conditions, the system is well determined. This implies that, given the total mass M of the star and the auxiliary functions, all other properties of the star should be well determined if the star has a homogeneous composition. Rigorously speaking, we need to know the composition profile of the matter inside the star. This, in turn, changes as nuclear reactions proceed in the star. Thus our description is strictly valid only for time scales that are small compared with the nuclear evolution time scale of the star. Conventionally such a star is called a *zero-age main-sequence* (ZAMS) star. Most of our discussion in this chapter regarding the structural properties will assume that we are considering a ZAMS star.

The detailed behaviour of solutions to stellar-structure equations depends fairly sensitively on the nature of the opacity, and hence it is important to determine κ accurately for different physical conditions that exist within the star. In this connection, the following points are crucial: (1) When multielectron ions or atoms are the cause of opacity, it is necessary to obtain their energy levels by some approximate method of solution to the Schroedinger equation. One popular technique is based on the variational principle, although it fails in some significant cases. For example, the lowest-order variational calculations do not predict the bound state for the H^- ion. (2) The opacity of helium and two electron ions also receives contributions from some other processes, in addition to those from the standard atomic processes. Normal atomic processes in the two electron systems will consist of one of the electrons being excited, with the other one remaining in the ground state. There are, however, doubly excited states for helium and two electron states that are discrete states embedded in a continuum because their energies are higher than their ionisation thresholds. Detailed tables of opacities, which take into account many of these effects, are now available for accurate modelling of stellar interiors.

Because of the rather complex nature of the equations, it is not easy to determine, in a mathematically rigorous form, whether the solution exists and is unique. It is clear, on physical grounds, that the solution may not exist for very low mass or for arbitrary composition. Explicit integration shows that solutions do exist for $M \gtrsim 0.1\ M_\odot$ with $X \approx 0.7$–1.0. As regards uniqueness, it is, in principle, possible to construct situations in which the solution is not unique. However, such situations usually turn out to be physically unacceptable. In physically realistic cases, the solution may be taken to be unique.

We note that two of the boundary conditions for the equations are provided at the origin whereas the other two are provided at the surface. Hence it is not easy to integrate these equations numerically in a straightforward manner, and special techniques have to be devised. The boundary conditions at the surface can also pose problems by making all the derivatives vanish there, thereby driving the equation towards the trivial solution. To avoid this, it is necessary to treat the region near the surface of star separately compared with the interior.

In principle, the above description is adequate to provide an understanding of a spherical star in steady state. In practice, however, several processes compete in different domains, thereby leading to a rich variety of possible structures. We shall therefore begin in the next section with a general description of stellar structure, based on the numerical solutions to the above set of equations. In the later sections we shall attempt to describe some of the features of this solution by using simplified models.

To avoid misunderstanding, we stress the following assumptions which have been made in arriving at the above description of a star[2]: (1) The star is assumed to be time independent; that is, the time scale over which any of the macroscopic physical properties of the star changes is considered to be significantly larger than the time scales involved in establishing and maintaining mechanical and thermal equilibrium. (2) As a corollary to the above assumption, we ignore any gradients in the composition of the interior and take it to be chemically homogeneous. (3) The star is taken to be spherically symmetric and all effects of rotation and magnetic field on the structure is ignored. This is equivalent to assuming that the kinetic-energy density that is due to rotation or magnetic-energy density is negligible compared with the thermal- and the gravitational-energy densities. (4) The convective transport of energy in the star is treated in a very simplified manner because of unavailability of a fundamental theory of convective transport. Because the star is assumed to be chemically homogeneous throughout, it is possible to use a fairly simplistic criterion for convection. We saw in Vol. I, Chap. 8, Section 8.14 that the criterion for convection $\nabla > \nabla_{ad}$ is true only if the star is homogeneous. (5) Simple boundary conditions are used near the surface of the star, and phenomena such as stellar winds are ignored. (6) The gravitational field is described by Newtonian theory and all general relativistic effects are ignored.

Exercise 2.1
Time-dependent stellar evolution: In the text, we discussed the equations of stellar structure, assuming that none of the physical variables change with time. Obtain the corresponding equations when the physical variables change with time.

Exercise 2.2

Relative speeds of convective and radiative transport: (a) Estimate the typical time taken by a photon to transport energy from the centre to the surface of a star, assuming that radiative diffusion is the process of energy transport. (b) Estimate the time scale for transport of energy from the centre to the surface in a fully convective star, making reasonable simplifying assumptions about convection. How do (a) and (b) compare?

2.3 Solutions to Equations of Stellar Structure

If the mass of the star M is specified, then the equations for stellar structure derived in the last section can be solved to determine various physical parameters as functions of the radius. These solutions also determine the total radius R, luminosity L, central temperature T_c, and central pressure P_c. We shall now discuss the properties of these solutions.[3]

Figures 2.1 and 2.2 give different properties of stellar structures obtained by numerical integration. To begin with, the four frames in these figures give the variation of the stellar properties as a function of the mass M, when the latter is varied by ~ 3 orders of magnitude from $M = 0.08\,M_\odot$ to $M = 60\,M_\odot$. All these stars were given a homogeneous composition, with $X = 0.74$ and $Y = 0.24$. The modelling is reasonably accurate for $M > M_\odot$ and gives qualitatively correct results for $M < M_\odot$. In Fig. 2.1(a) $R/R_\odot$ is plotted as a function of $M/M_\odot$ in logarithmic coordinates. The radius is an increasing function of mass with $R \propto M^{0.85}$ for $M \lesssim M_\odot$ and $R \propto M^{0.56}$ for $M \gtrsim M_\odot$. There is a distinct change around $M \approx M_\odot$. We shall see later that the internal structure of the star is quite different based on whether $M < M_\odot$ or $M > M_\odot$.

Figure 2.1(b) shows the variation of central temperature with the mass of the star. (The central temperature is measured in units of the central temperature of the Sun, $T_{c\odot} \approx 1.44 \times 10^7$ K.) This quantity does not vary much for $M \gtrsim M_\odot$ but increases with M for $M \lesssim M_\odot$. In general, T_c changes by only a factor of 3 or so when M varies by a factor of 60 for $M > M_\odot$. This curve also shows the general trend that the central temperature of the star increases with the mass of the star, albeit weakly. By and large, a similar increase of temperature is seen *throughout* the interior of the star when the total mass increases.

The actual value of the central temperature determines the dominant nuclear reaction process that is operating in the core of the star. For $T \gtrsim 1.4 \times 10^7$ K, the dominant process is the CNO cycle, whereas for $T \lesssim 1.4 \times 10^7$ K, it is the p–p chain [see Eqs. (2.19) and (2.20) as well as Vol. I, Chap. 12]. In other words, stars with $M \gtrsim M_\odot$ are powered by the CNO cycle, which has a very high temperature sensitivity, whereas stars with $M \lesssim M_\odot$ operate with the p–p chain, which has a weaker dependence on temperature. Strong temperature dependence of the nuclear reaction rate implies that there will be a steeper temperature gradient in the centre for stars with $M \gtrsim M_\odot$. This, in turn, will lead to

2.3 Solutions to Equations of Stellar Structure

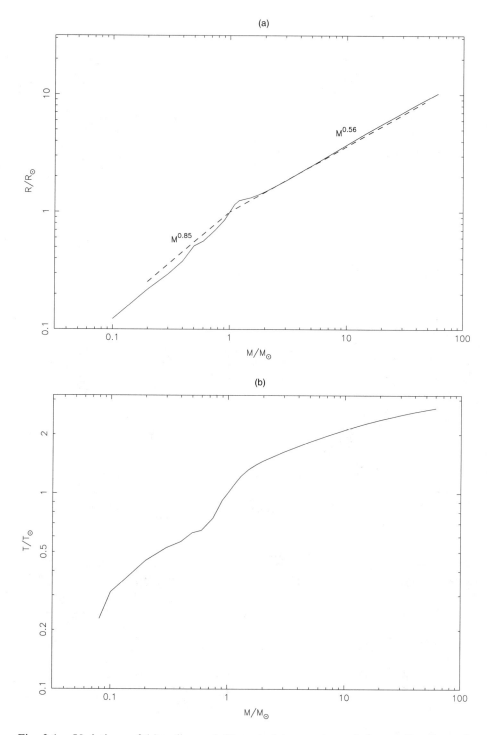

Fig. 2.1. Variations of (a) radius and (b) central temperature of stars as functions of the mass.

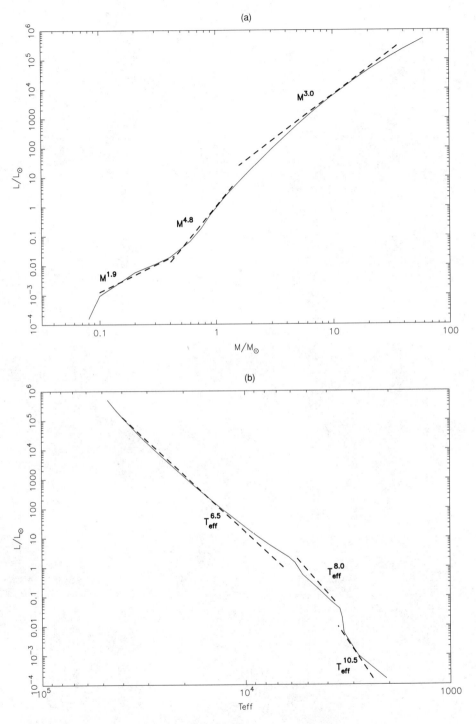

Fig. 2.2. (a) Variation of luminosity of stars as a function of the mass. (b) The main-sequence stars are plotted in the $T_{\text{eff}}-L$ plane as a H-R diagram.

2.3 Solutions to Equations of Stellar Structure

convective instabilities near the core. In fact, we shall see later (see page 79) that, for stars with $M \gtrsim 2\ M_\odot$, there exists a convective core near the origin containing more than $\sim 13\%$ of the total mass. On the other hand, subsolar-mass stars use radiative transport of energy near the core. This is one of the reasons why the physical properties in Fig. 2.1 show a strong 'break' around $M \approx M_\odot$.

From virial theorem, we expect $k_B T_c \propto (GMm_p/R) = |\phi|$. The ratio $(k_B T_c/|\phi|)$ varies between 0.2 and 0.8 when M varies by 3 orders of magnitude, confirming this expectation.

Figure 2.2(a) shows the variation of total luminosity of the star as a function of the mass of the star. Both the axes are logarithmic in this frame. The luminosity is clearly a strong function of the mass; for $M \gtrsim M_\odot$, the approximate scaling is given by

$$\frac{L}{L_\odot} \approx \left(\frac{M}{M_\odot}\right)^{3.5}, \qquad (2.28)$$

but a somewhat more accurate fitting function is

$$\frac{L}{L_\odot} \propto \begin{cases} 7^{1.75}(M/M_\odot)^3 & \text{(for } M \gtrsim 7\ M_\odot) \\ (M/M_\odot)^{4.8} & \text{(for } 0.4\ M_\odot \lesssim M \lesssim 7\ M_\odot) \\ 0.4^{2.85}(M/M_\odot)^{1.9} & \text{(for } M \lesssim 0.4\ M_\odot) \end{cases} \qquad (2.29)$$

The steep dependence in Eq. (2.28) arises essentially because of the strong dependence of the nuclear-energy-production rate on temperature for stars with $M \gtrsim M_\odot$.

This relation can be used to estimate the time scale over which the nuclear fuel can sustain the star. During hydrogen burning it is possible to release approximately $\epsilon = 6 \times 10^{18}$ ergs for every gram of hydrogen that is consumed. Assuming that a star ceases to be in steady state (and evolutionary effects become important) when, say, 10% of the original mass of the star undergoes conversion from hydrogen to helium, we can estimate the time scale for the existence of steady state to be

$$t_{\text{nuc}} \approx 0.1\,\epsilon\,\frac{XM}{L} \approx 10^{10} \left(\frac{M}{M_\odot}\right)\left(\frac{L}{L_\odot}\right)^{-1}\ \text{yr}, \qquad (2.30)$$

where we have used $X \approx 0.7$. Using Eq. (2.28), we get

$$t_{\text{nuc}} \approx 10^{10} \left(\frac{M}{M_\odot}\right)^{-2.5}\ \text{yr}. \qquad (2.31)$$

The stars that are in the process of converting hydrogen to helium are called main-sequence stars. From the above expression, we find that the main-sequence lifetime of a star with $M \simeq M_\odot$ is $\sim 10^{10}$ yr. More massive stars will spend less time in the main-sequence phase. This result, as we shall see in a later chapter, has important theoretical and observational consequences.

If a star is thought of as a blackbody at some temperature T_{eff} and radius R, then its luminosity will be given by $L = 4\pi R^2 \sigma T_{\text{eff}}^4$, where $\sigma = 5.67 \times 10^{-5}$ ergs cm^{-2} K^{-4} s^{-1}. Given L and R, this relation defines an effective temperature for the star. Because the continuum spectra of most stars can be approximated by a Planckian curve, this quantity T_{eff} is of direct observational significance. Numerically,

$$\frac{L}{L_\odot} = 8.973 \times 10^{-16} \left(\frac{R}{R_\odot}\right)^2 T_{\text{eff}}^4. \tag{2.32}$$

Using M–R scaling and Eq. (2.28), we find

$$T_{\text{eff}} \cong T_{\odot\text{eff}} \left(\frac{L}{L_\odot}\right)^\eta, \quad \eta = 0.12 - 0.17. \tag{2.33}$$

This relation is shown in Fig. 2.2(b) in logarithmic coordinates. (Note that the temperature *decreases* to the right along the x axis to conform to astronomical convention.) Because both T_{eff} and L are measurable, if the distance to the star is known, it is possible to plot the positions of all the stars in an L–T_{eff} plane. Such a diagram is called a H-R diagram and plays a vital role in the study of stellar evolution (see Chap. 3). Relation (2.33) gives the slope of the line in the H-R diagram for stars. As the mass increases, the main-sequence stars get bigger, brighter, and less dense.

Let us next consider the internal structure of the stars. The detailed profiles of various physical parameters of main-sequence stars are shown in Figs. 2.3–2.10. The masses of the seven stars chosen for display are $(M/M_\odot) = 60$ (thin solid curve), 20 (thin dashed curve), 5 (dotted–dashed curve), 1.3 (dotted curve), 1 (dashed–triple dotted curve), 0.6 (thick solid curve), and 0.4 (thick dashed curve).

Figure 2.3(a) denotes the fraction of the mass $M(r)/M$ contained within the fractional radius r/R, where M and R denote the total mass and the radius of the star, respectively. These curves show a systematic variation as the mass is changed. As we start from the high-mass end (thin solid curve) and decrease the mass, the curves fall on one side of the original thin solid curve up to $M = 1.3\,M_\odot$ (dotted curve). As the mass is lowered further, the curves start moving in the opposite direction and the curves for $M = (0.6, 0.4)\,M_\odot$ lie on the opposite side of the line for the 60 $M_\odot$ case.

The figure also shows that the outer regions of the star have very low density. Nearly 95% of the mass is contained within $\sim 65\%$ or less of the radius in stars with $M > M_\odot$. This fact is useful in modelling the outer regions of the star because most of the gravitational force in those regions is contributed by the core mass rather than by the self-gravity of the gas in the outer regions.

2.3 Solutions to Equations of Stellar Structure

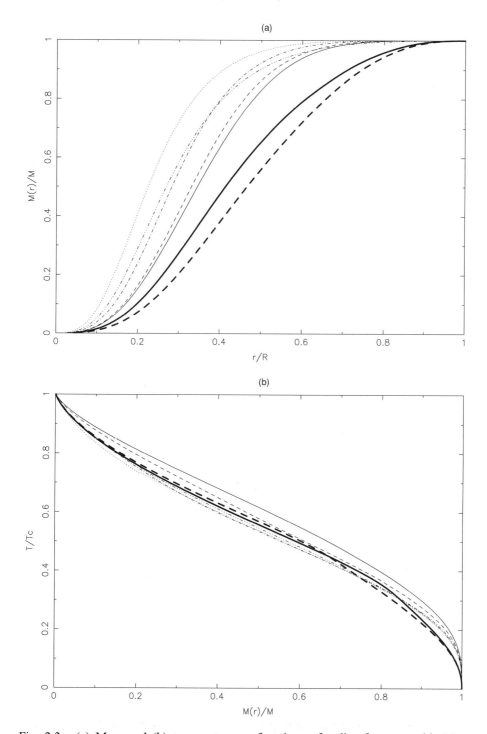

Fig. 2.3. (a) Mass and (b) temperature as functions of radius for stars with $M = (60, 20, 5, 1.3, 1, 0.6, 0.4)\, M_\odot$. The thin solid curve (—) corresponds to $M = 60\, M_\odot$, the thin dashed curve (- - -) is $M = 20\, M_\odot$, the dotted–dashed curve ($\cdot - \cdot - \cdot$) is $M = 5\, M_\odot$, the dotted curve ($\cdots$) is $M = 1.3\, M_\odot$, the dashed–triple dotted curve ($- \cdots - \cdots$) is $M = 1\, M_\odot$, the thick solid curve (—) is $M = 0.6\, M_\odot$, and the thick dashed curve (- - -) is $M = 0.4\, M_\odot$.

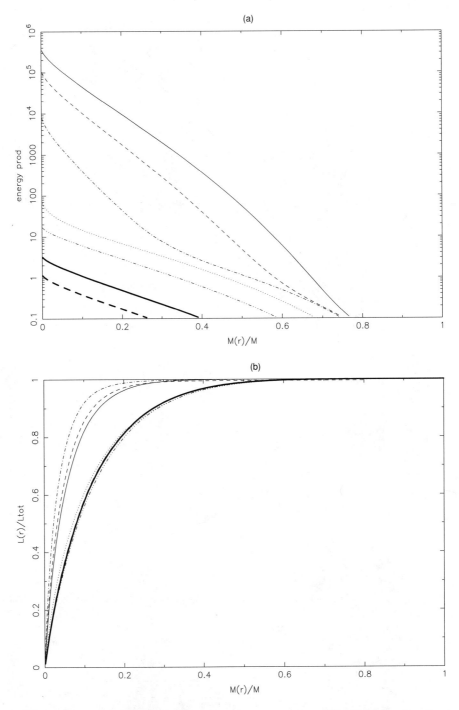

Fig. 2.4. (a) Energy production (in ergs per gram per second) and (b) luminosity as functions of radius for stars with $M = (60, 20, 5, 1.3, 1, 0.6, 0.4)\, M_\odot$. The curves are as given in Fig. 2.3.

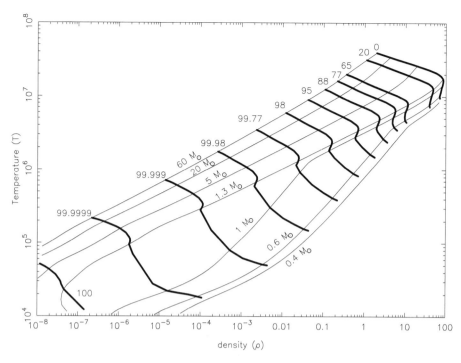

Fig. 2.5. The thick curves in this figure show the density and the temperature inside the stars at radii containing x percentage of mass, where $x = 0, 20, 65, 77, 88, 95, 98, 99.77$, 99.98, 99.999, 99.99999, and 100 for the 12 curves from top right to bottom left. (The last curve at the bottom left contains 100% mass within the accuracy of integration.) Stars with masses $M > 0.4\ M_\odot$ are used for generating the curves. The thin curves running across (from left bottom to right top) give ρ as a function of T for the seven stars with masses $M = (0.4, 0.6, 1, 1.3, 5, 20, 60)\ M_\odot$.

Figure 2.3(b) shows the variation of the temperature $T(r)$ with the fractional mass $M(r)/M$ contained inside a radius. The temperature is measured in terms of the central temperature. The temperature profiles, again, vary with stellar mass in a manner similar to that in Fig. 2.3(a). The behaviour is monotonic as the mass is decreased down to $M \simeq M_\odot$ and changes direction for lower mass.

Another aspect of the temperature variation is highlighted in Fig. 2.5. The thick curves in this figure show the density and temperature inside the stars at radii containing x percentage of mass, where $x = 0, 20, 65, 77, 88, 95, 98, 99.77, 99.98$, 99.999, and 99.99999 for the 11 curves from top right to bottom left. The last (12th) curve at the bottom left contains 100% mass within the accuracy of integration. Stars with masses $M > 0.4\ M_\odot$ are used for generating the curves. The thin curves running across (from left bottom to right top) give ρ as a function of T for the seven stars, varying in mass between 0.4 $M_\odot$ and 60 $M_\odot$. The following facts emerge from the examination of the actual numbers. The central temperature of these stars vary between 10^7 K and $10^{7.6}$ K. At the radius containing 50%

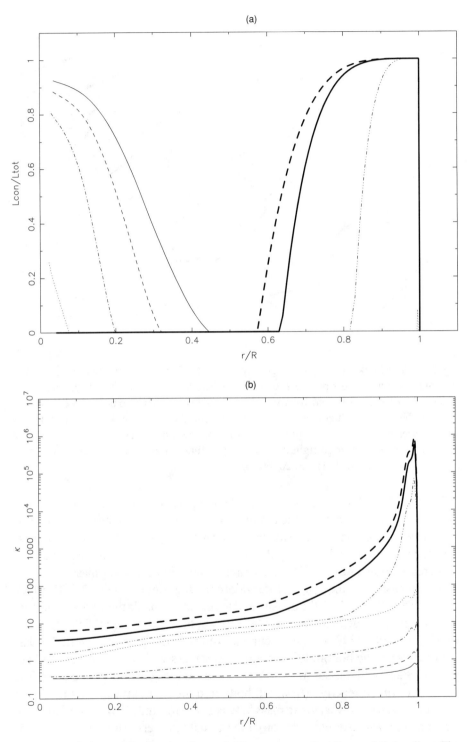

Fig. 2.6. (a) Convective luminosity and (b) opacity as functions of the radius. The curves are as given in Fig. 2.3.

2.3 Solutions to Equations of Stellar Structure

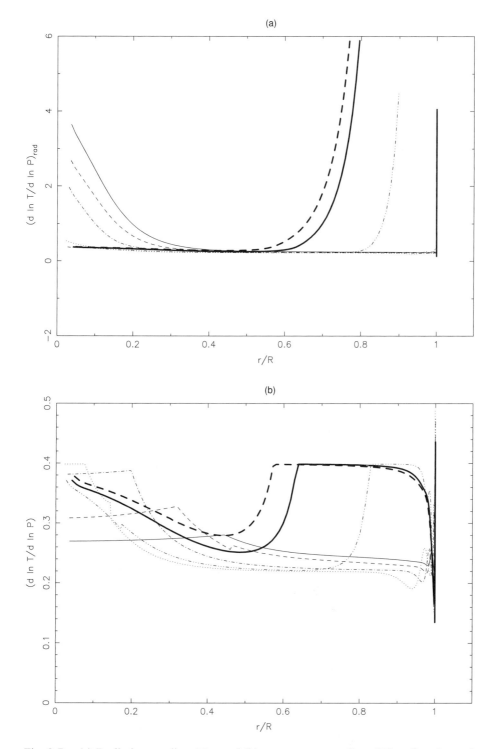

Fig. 2.7. (a) Radiation gradient ∇_{rad} and (b) temperature gradient (∇) as functions of the radius. The curves are as given in Fig. 2.3.

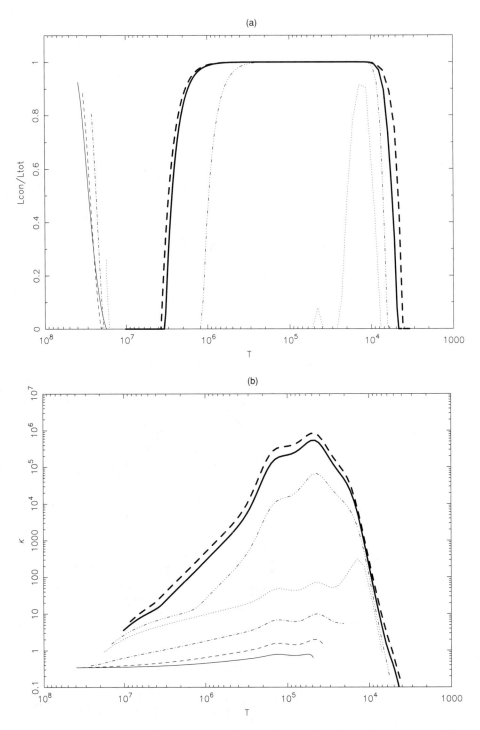

Fig. 2.8. (a) Convective luminosity and (b) opacity as functions of temperature. The curves are as given in Fig. 2.3.

2.3 Solutions to Equations of Stellar Structure

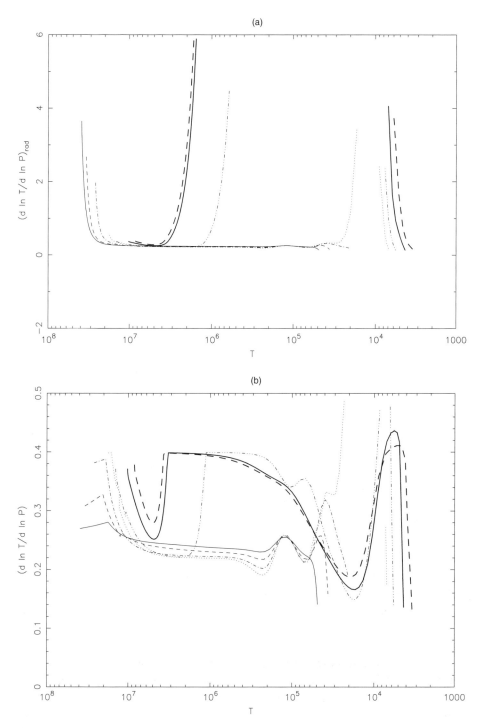

Fig. 2.9. (a) The radiation gradient and (b) the temperature gradient as functions of temperature. The curves are as given in Fig. 2.3.

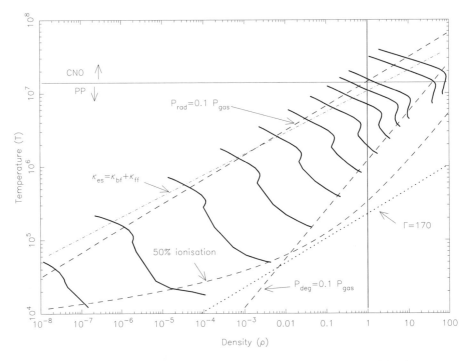

Fig. 2.10. The thick lines show the density and the temperature inside the stars at different radii containing a fixed percentage of mass. The parameterisation is same as that in Fig. 2.5. We obtain the topmost thick dashed line by equating radiation pressure to 10% of gas pressure. The middle thick dashed line demarcates the region above which hydrogen is ionised 50%. We obtain the lowest thick dashed line by equating nonrelativistic electron degeneracy pressure to 10% of the gas pressure. The vertical line at $\rho = 1$ gm cm^{-3} demarcates the region where pressure ionisation is important, and the horizontal line at $T = 1.4 \times 10^7$ K separates the region at which the CNO cycle dominates over the p–p cycle. The line $\Gamma = 170$ demarcates the region at which the Coulomb interaction becomes important.

of the total mass, the temperature varies between $10^{6.85}$ K and $10^{7.35}$ K; at the radius containing 99% of the mass, the temperature is still between $10^{5.8}$ K and $10^{6.4}$ K. At a radius containing 99.99999% of the mass, the temperature drops to a value between 10^4 K and 10^5 K. This shows that most of the mass in the star is in a fairly high-temperature region and the temperature decreases sharply in the outer low-density regime.

Figure 2.4(a) shows the energy-generation rate as a function of the fractional mass contained within a radius. It is clear that most of the energy generation occurs close to the centre and there is very little energy production beyond the central region, say, at $r \gtrsim 0.4\ R$. The energy generation goes up with the mass of the star, as to be expected. This, in turn, makes the luminosity of the star an increasing function of the stellar mass, as we have seen before. The same result is shown in a different manner in Fig. 2.4(b), in which the fraction of the luminosity $L(r)/L$ is plotted against the mass fraction. Clearly, most of the luminosity of the star is generated well within the central region containing less than 0.4 M.

The features shown above clearly indicate that there is a significant structural difference between the stars with $M \lesssim M_\odot$ and those with $M \gtrsim M_\odot$. The reason for such a difference is shown through Figs. 2.6–2.9.

Figure 2.6(a) plots the fraction of the total luminosity that is transported by convection in different regions of the star. The curves show that for stars with $M = (60, 20, 5, 1.3)\,M_\odot$ the central regions are convective and the rest of the star is radiative. The size of the central convective region increases monotonically with the mass of the star for $M \gtrsim 1.2\,M_\odot$. The fractions of the mass contained in the convective core in these models are 0.03, 0.23, 0.46, and 0.73 for $M/M_\odot = $ 1.3, 5, 20, and 60, respectively. On the other hand, for $M = (1.0, 0.6, 0.4)\,M_\odot$, the central region is radiative; that is, a negligible amount of energy is transported by convection near the origin. (The transition from convective core to radiative core occurs around $M \approx 1.2\,M_\odot$). The figure shows that these stars, however, transport most of the energy by convection in the outer envelope. The size of the outer convective region increases as the mass of the star is decreased. For $M \lesssim 0.3\,M_\odot$, the entire star becomes convective. Thus we see that stars with $M \gtrsim M_\odot$ have a convective core and radiative envelope whereas stars with $M \lesssim M_\odot$ have a radiative core and convective envelope. (Actually, the region very close to the surface of the star remains radiative even for $M \lesssim M_\odot$; we shall comment on this feature later in Section 2.4.)

The physical reasons for these two phenomena are different, although simple to understand. Let us consider stars with $M \gtrsim M_\odot$ first. From Fig. 2.5, it is clear that these stars have central temperatures higher than approximately 1.4×10^7 K and hence the CNO cycle will be the dominant energy-production mechanism near the core. Because this reaction is very sensitive to temperature, energy production will increase rapidly near the centre. The transport of large amounts of energy through a sphere of small surface area will lead to a large flux. If this flux is to be radiative, then Eq. (2.7) shows that we need a large, negative temperature gradient near the origin. Such a region with a large temperature gradient is unstable to convection, and the energy transport switches over to a convective mode near the centre. Efficient convection can transport large amounts of energy and keep the temperature gradient close to adiabatic value. At larger radii, at which the flux is lower and the required temperature gradient is not too large, this process gives way to radiative transport.

There are two more contributory factors to convective instability that are worth mentioning: (1) In the case of massive stars, when M increases, the central temperature goes up and the relative importance of $P_{\rm rad}$ increases. This causes the value of $\nabla_{\rm ad}$ to drop from the ideal gas value of 0.4. For example, in a 50-$M_\odot$ star, $P_{\rm rad} \approx 0.3\,P_{\rm total}$ near the centre and $\nabla_{\rm rad} \approx 0.27$. Because the condition for convective instability is $\nabla_{\rm rad} > \nabla_{\rm ad}$, the lowering of $\nabla_{\rm ad}$ helps convective instability. (2) The ionisation of hydrogen – as we move inwards from the cooler outer layers towards the hot interior – also plays an important role in lowering the adiabatic gradient, as we saw in Vol. I, Chap. 5, Section 5.6. In fact, the gradient can drop to nearly 0.11 in the hydrogen ionisation zone, because of which intermediate stars can have two convection zones in them where hydrogen and helium are being ionised.

As the mass of the star is reduced, the central temperature decreases and the relative importance of the CNO cycle decreases. This causes the size of the convective core to progressively shrink and finally disappear around $M \approx 1.2 \, M_\odot$. Around the same value of stellar mass, the outer regions begin to be convective. The reason for this is different from that given above. In the cooler outer region of these low-mass stars, the opacity is mainly due to free–free and bound–free transitions rather than electron scattering. This is shown in Fig. 2.10 by the dotted–dashed curve, which connects the density and the temperature at which $\kappa_{el} = \kappa_{bf} + \kappa_{ff}$. For regions above this line, electron scattering dominates whereas for regions below this line, atomic processes dominate. Now the bound–free and the free–free opacities become large when ionisation and the recombination of hydrogen take place significantly around 10^5 K. The radiative flux decreases with increasing opacity [see Eq. (2.7)] at a fixed temperature gradient; hence, to maintain the same outflow of energy with larger opacity, the temperature gradient again must increase in the outer region. This increase in the temperature gradient causes instabilities and convection takes over. As the mass of the star is lowered, the size of the outer convective region increases and, for $M \lesssim 0.3 \, M_\odot$, the entire star becomes convective. At such low temperatures and densities, the matter does not behave as an ideal gas and modelling is not easy.

These qualitative considerations are verified in Figs. 2.6(b), 2.7(a), and 2.7(b). Figure 2.6(b) shows the variation of the total opacity of the star as a function of the radius. (The y axis is logarithmic, and the x axis is linear.) For stars with $M \lesssim M_\odot$, the opacity shoots up to large values near the surface, as expected because of ionisation processes. This is the primary cause of convective instabilities in the outer regions of subsolar-mass stars. Figure 2.7(a) shows the variation of ∇_{rad} inside the star. For stars with $M \lesssim M_\odot$, this gradient is moderate in the central regions but shoots up near the surface; this is a direct consequence of the large increase in the opacity near the surface, seen in Fig. 2.6(b). Figure 2.7(a) also shows that ∇_{rad} is larger that ∇_{ad} near the origin in stars with $M \gtrsim M_\odot$. This is due to the temperature sensitivity of the CNO cycle, which dominates near the centre in these stars. This, as explained above, causes the convective core in the stars with $M \gtrsim M_\odot$. There is a clear one-to-one correspondence between the convective regions seen in Figs. 2.6(a) and 2.7(a). Finally, Fig. 2.7(b) shows the actual gradient ∇ as a function of the radius. In the convective regions, $\nabla = \nabla_{ad}$; this occurs near the centre for stars with $M \gtrsim M_\odot$ and near the surface for stars with $M \lesssim M_\odot$.

2.3 Solutions to Equations of Stellar Structure

The same results are shown again in Fig. 2.8 plotted against temperature. (The temperature decreases to the right along the x axis in order to keep relative orientations of the x axis the same in Figs. 2.8 and 2.9.) This has the advantage of stretching the outer layers. The increase in opacity [see Fig. 2.8(b)] around $T \approx 10^5$ K is due to the ionisation of helium, whereas the increase around $T \gtrsim 10^4$ K is due to the ionisation of hydrogen. The drop of ∇ (which is equal to ∇_{ad} in the convective region) around the ionisation regions is clear from Fig. 2.9(b) as well.

The statement that regions of high opacity have steep radiative temperature gradients and hence are apt to be convectively unstable is not quite precise. To be rigorous, it is the *rate of change* of opacity with depth that matters and not its absolute value. In most practical applications, however, a high value of opacity is also accompanied by a high value for its gradient and hence the less precise statement does not incur any error.

The discussion given above is based on the solutions to equations of stellar structure. These equations, however, do not ensure that the resulting solution is always stable. The analysis of stability of solutions to stellar structure equations is an extremely complex problem and is inherently time dependent. Some aspects of this issue will be discussed in the next chapter.

Exercise 2.3

Semiconvection: We saw in Vol. I, Chap. 8, Section 8.14 that a system can be stable against convection when $\nabla < [\nabla_{\text{ad}} + (\phi/\delta)\nabla_\mu]$, where

$$\phi = \left(\frac{\partial \ln \rho}{\partial \ln \mu}\right)_{T,p}, \quad \delta = -\left(\frac{\partial \ln \rho}{\partial \ln T}\right)_{p,\mu}, \quad \nabla_\mu = \left(\frac{\partial \ln \mu}{\partial \ln p}\right)_{\text{ext}}, \quad (2.34)$$

and μ is the chemical potential. The gradient of chemical composition contributes the extra term $(\phi/\delta)\nabla_\mu$. Now consider the case in which $\nabla_{\text{ad}} < \nabla < [\nabla_{\text{ad}} + (\phi/\mu)\nabla_\mu]$ and discuss what happens to a blob of gas when it is displaced. (a) Show, in particular, that the blob oscillates with increasing amplitude. (This process is called semiconvection.) (b) Estimate the relevant time scale. {Answers: If the blob of radius R and density ρ has temperature $T + \Delta T$ and the ambient temperature is T, the net radiation flux out of the blob is

$$L_{\text{blob}} \simeq \frac{4ac}{3\rho\kappa}\left(\frac{\partial \Delta T}{\partial x}\right)\pi R^2 \simeq \frac{4ac}{3\rho\kappa}\left(\frac{\Delta T}{R}\right)\pi R^2. \quad (2.35)$$

Equating this to $\rho V C_p(\partial \Delta T/\partial t)$, where C_p is specific heat per gram, we get $(\partial \Delta T/\partial t) \simeq (\Delta T/\tau)$, with $\tau \simeq (\kappa C_p/4ac)(\rho^2 R^2/T^3)$. At the bottom of convection zone of the Sun ($r \simeq 0.7 R_\odot$) we have $T \simeq 2 \times 10^6$ K, $\rho \simeq 0.2$ gm cm^{-3}, $\kappa \simeq 10$ gm cm^{-2}, and $C_p = (5/2)[k_B/(m_p/2)] = (5k_B/m_p) \simeq 5 \times 10^8$ ergs gm^{-1} K^{-1}. Taking $R \simeq 0.1 R_\odot$ gives $\tau \simeq 3 \times 10^4$ yr.}

2.4 Toy Stellar Models

The discussion of numerical solutions to the equations of stellar structure emphasises the fact that several complex phenomena play important roles in determining the overall structure of a star. This aspect is highlighted in Fig. 2.10, in which the ρ–T relations of several stars are replotted with different regions marked out:

(1) The horizontal line at $T = 1.4 \times 10^7$ K separates the region at which the CNO cycle begins to dominate over the p–p cycle. This divides the stars in terms of dominant energy-production mechanisms.
(2) The dotted–dashed line corresponds to $\kappa_{es} = \kappa_{bf} + \kappa_{ff}$; above this line, electron scattering is the dominant source of opacity, whereas, below this line, the bound–free transitions provide the opacity. The crossing of this line by the tracks of the stars shows that different stars are dominated by different radiative processes and, even in the same star, different mechanisms operate in different regimes.
(3) The other lines in the figure have a bearing on the equation of state. We obtain the topmost of the three thick dashed lines by equating the radiation pressure to 10% of gas pressure. Above this line, radiation pressure makes a reasonable contribution to total pressure. The fact that several star tracks cross this shows that we cannot ignore the effect of radiation pressure in massive stars. The middle thick dashed line gives the ρ–T relation at which hydrogen has 50% ionisation. The crossing of this line with outer regions of low-mass stars shows that the ionisation of hydrogen will play a crucial role in these stars. Finally, we obtain the third (lowermost) thick dashed line by equating the degeneracy pressure of a zero-temperature electron gas to 10% of gas pressure.

For most of the stars in the figure, degeneracy pressure does not play a vital role, although the low-mass stars are quite close to this point. Another aspect of the equation of state is brought out by the vertical line at $\rho = 1$ gm cm^{-3} above which pressure-ionisation effects are significant (see Exercise 2.4). This line is obtained by the condition $n^{-1/3} \simeq a_0$, where n is the number density of atoms (so that $n^{-1/3}$ is the interatomic spacing) and a_0 is the Bohr radius. This gives $n \approx 10^{24}$ cm^{-3}, corresponding to a density of $\rho \simeq m_H n \simeq 1$ gm cm^{-3}. The dotted line marked $\Gamma = 170$ demarcates the region where the potential energy of the nearest neighbours, because of the Coulomb interaction, becomes sufficiently important compared with the thermal energy leading to crystallisation. The quantity Γ is defined as the ratio between the Coulomb energy of the nearest neighbours and the thermal energy:

$$\Gamma \equiv \frac{(Ze)^2}{r_{ion} k_B T} = 2.7 \times 10^{-3} \left(\frac{Z^2 n_{ion}^{1/3}}{T} \right). \tag{2.36}$$

2.4 Toy Stellar Models

The Coulomb energy and the thermal energy are comparable when $\Gamma \approx 1$, but the effects of Coulomb interactions become really significant[4] when $\Gamma \approx 10^2$.

This figure illustrates clearly the complexity of stellar modelling. For this reason, it is not possible to obtain simple theoretical models that describe all stars in a unified manner. It is, however, possible to understand the qualitative behaviour of the numerical solutions by some simple models for stellar structure. In this section, we shall discuss some such models that are applicable in different contexts and with different levels of accuracy. It must be emphasised that these models make assumptions that are not quite realistic and reproduce only the qualitative features of stellar structure. For any serious work involving stellar structure, we should use the numerical solutions like those discussed in the last section.

Exercise 2.4

Ionisation in stellar cores: Compute the fractional ionisation at the centre of a star where the temperature is $T = 1.5 \times 10^7$ K and the number density is $n_H = 10^{26}$ cm^{-3} by using Saha's ionisation equation. How do you reconcile your answer with the fact that the solar interior is completely ionised? [Hint: If the hydrogen atoms are treated as systems with energy levels $\epsilon_j = -(13.6 \text{ eV}/j^2)$ then the sum defining the partition function $\sum_j \exp(-\beta \epsilon_j)$ will diverge. Argue that the modification of Coulomb potential to the form $\phi = (q/r) \exp(-r/L)$, where L is the Debye length, will cut off the sum at a maximum value $j_{\max}$, where $j_{\max}^2 = (L/2a_0)$ and $a_0 = (\hbar^2/m_e q^2)$ is the Bohr radius. What is $j_{\max}$ in the solar core?]

2.4.1 Homologous Stellar Models

Stellar-structure Eqs. (2.22) and (2.25) require three input functions, $P(\rho, T)$, $\kappa(\rho, T)$, and $\epsilon(\rho, T)$, for their integration. In general, these functions have fairly complicated forms because they incorporate several different effects. However, as a first approximation, we can think of all these functions as given by simple power laws like

$$P = P_0 P^a T^{-b}, \quad \kappa = \kappa_0 \rho^n T^{-s}, \quad \epsilon = \epsilon_0 \rho^\lambda T^\nu. \qquad (2.37)$$

Such an approximation for ρ and κ will be valid if any single process makes the dominant contribution to pressure and opacity. The approximation of ϵ by a power law is valid in limited ranges, depending on which nuclear reaction dominates the energy production. When the above approximation is valid, all the equations of stellar structure reduce to those involving powers of different variables on the right-hand side. This allows rescaling of the variables in the equations, reducing them to dimensionless form. To do this effectively, we replace the pressure P, temperature T, radius r, luminosity $L(r)$, and mass $M(r)$ with the dimensionless variables $p, t, x, l,$ and m through the scalings

$$P = P_c p, \quad T = T_c t, \quad r = Rx, \quad L_r = Ll, \quad M_r = Mm. \quad (2.38)$$

The quantities P_c and T_c are taken to be the central pressure and the central temperature of the star. The variables R, L, and M are chosen by the conditions

$$\frac{GM^2}{4\pi P_c R^4} = 1, \quad \frac{M}{4\pi \rho_0 R^3} \frac{T_c^b}{P_c^a} = L, \quad \frac{3\kappa_0 \rho_0^n}{64\pi^2 ac} \frac{P_c^{an} ML}{T_c^{4+s+nb} R^4} = 1. \quad (2.39)$$

These equations determine M, R, and L in terms of P_c and T_c. (In particular, note that they are *not* the radius, luminosity, and mass of the star.) Substituting Eq. (2.38) into the equations of stellar structure with radiative transport and using the above relations, we find that straightforward algebra leads to the following equations:

$$\frac{dp}{dm} = -\frac{m}{x^4}, \quad (2.40)$$

$$\frac{dx}{dm} = \frac{t^b}{x^2 p^a}, \quad (2.41)$$

$$\frac{dt}{dm} = -\frac{p^{an} l}{x^4 t^{3+s+bn}}, \quad (2.42)$$

$$\frac{dl}{dm} = A p^{a\lambda} t^{\nu - b\lambda}, \quad (2.43)$$

where A is a function of P_c and T_c, given by

$$A = \frac{3\kappa_0 \rho_0^n \epsilon_0}{16\pi Gac} \frac{P_c^{an+a\lambda+1}}{T_c^{4+s+nb+\lambda b - \nu}}. \quad (2.44)$$

In the case of convective transport of energy, the equation for dt/dm is replaced with the relation between pressure and temperature given by

$$P = KT^{5/2}, \quad (2.45)$$

which follows from setting $\nabla = \nabla_{\text{ad}} = 2/5$ in Eq. (2.15). In terms of dimensionless variables, this relation becomes

$$p = Bt^{5/2}, \quad B = 4\pi K G^{3/2} \left(\frac{\mu}{\mathcal{R}}\right)^{5/2} M^{1/2} R^{3/2}. \quad (2.46)$$

Using Eqs. (2.39), we can express B in terms of P_c and T_c, but we shall not need that expression. Equations (2.40), (2.41), (2.42), and (2.43) are four equations for the four variables $p(m)$, $x(m)$, $t(m)$, and $l(m)$, respectively, and require four boundary conditions for their solutions. By our definition, we must have $p(0) = 1$ and $t(0) = 1$. Further, regular behaviour at the origin requires that $x(0) = 0$ and $l(0) = 0$. Given these initial conditions near the origin, these equations can be integrated forward. It is clear from the signs on the right-hand side that p and t

2.4 Toy Stellar Models

are decreasing functions of m whereas x and l are increasing functions. It follows that p and t will go to zero at some values of m. However, there is no guarantee that p and t will vanish at the *same* value of m. Physically, we expect both the pressure and temperature to vanish at the surface of the star at which m takes some specified value, say, m_*. This condition will be satisfied only for a specific value of A or B, depending on whether the star is taken to be entirely radiative or convective. Given that particular value of A, say, we obtain a relation between P_c and T_c in the form

$$P_c^{a(n+\lambda)+1} \propto T_c^{4+s+b(n+\lambda)-\nu}. \tag{2.47}$$

It is now possible to express all other physical quantities such as M, R, and L in terms of T_c alone. Equivalently, we can trade off T_c in terms of M and determine scaling relations, expressing R, L, P_c, and T_c in terms of M. (A similar procedure can be used for a fully convective star as well, with B replacing A.) Omitting the straightforward algebra, we write the final result: If we take

$$R \propto M^{n_R}, \quad T \propto M^{n_T}, \quad L \propto M^{n_L}, \tag{2.48}$$

then, for a star with radiative transport, the indices are

$$n_R = \frac{1}{3}\left[1 - \frac{2}{D_r}\left(\frac{b}{a} + \nu - s - 4\right)\right], \tag{2.49}$$

$$n_T = -\frac{2}{D_r}\left(\frac{1}{a} + \lambda + n\right), \tag{2.50}$$

$$n_L = 1 + \frac{1}{D_r}\left[2\lambda\left(\frac{b}{a} + \nu - s - 4\right) - 2\nu\left(\frac{1}{a} + \lambda + n\right)\right], \tag{2.51}$$

where

$$D_r = \left(\frac{3}{a} - 4\right)(\nu - s - 4) - \frac{b}{a}(3\lambda + 3n + 4). \tag{2.52}$$

For a star with convective-energy transport, the corresponding results are

$$n_R = \frac{1}{3}\left(1 - \frac{2}{D_c}\right), \tag{2.53}$$

$$n_T = \frac{4}{3D_c}, \tag{2.54}$$

$$n_L = 1 + \frac{2}{D_c}\left(\lambda + \frac{2}{3}\nu\right), \tag{2.55}$$

with

$$D_c = \left(\frac{3}{a} - 4\right) + \frac{2b}{a} \tag{2.56}$$

Table 2.1(a). High-mass stars

	CNO Cycle, Ideal Gas, $a = b = 1$			
	$\lambda = 1, \nu = 15$		$\lambda = 1, \nu = 18$	
	Electron Scattering $(n = 0, s = 0)$	Convection	Electron Scattering $(n = 0, s = 0)$	Convection
n_R	0.78	−0.33	0.81	−0.33
n_T	0.22	1.33	0.19	1.33
n_L	3.0	23	3.0	27
n_{HR}	0.12	0.26	0.115	0.26

These relations also allow us to determine the slope of the curve in the L–T_{eff} plane. This index, n_{HR}, is given by

$$T_{\text{eff}} \propto \left(\frac{L}{R^2}\right)^{1/4} \propto L^{n_{HR}}, \quad n_{HR} = \frac{1}{4}\left(1 - \frac{2n_R}{n_L}\right). \quad (2.57)$$

The above relations, which express the physical parameters of the star in terms of any one chosen variable, say, M, are called *homology relations*. These relations arise because of our assumption that all input functions are power laws.

The numerical values of these indices are tabulated in Tables 2.1(a) and 2.1(b) for different physical models. For high-mass stars, with $M \gtrsim M_\odot$, we expect the CNO cycle to be the major energy-generating mechanism and the electron scattering to be the main source of opacity. For sufficiently high-mass stars, the radiation pressure could make a contribution to the total pressure. Further,

Table 2.1(b). Low-mass stars

	p–p Chain, Ideal Gas, $a = b = 1$					
	$\lambda = 1, \nu = 4$			$\lambda = 1, \nu = 5$		
	Kramer's $(n = 1, s = 3.5)$	Electron Scattering $(n = 0, s = 0)$	Convection	Kramer's $(n = 1, s = 3.5)$	Electron Scattering $(n = 0, s = 0)$	Convection
n_R	0.077	0.43	−0.33	0.2	0.5	−0.33
n_T	0.92	0.57	1.33	0.8	0.5	1.33
n_L	5.5	3.0	8.33	5.4	3.0	9.7
n_{HR}	0.24	0.18	0.27	0.23	0.17	0.27

2.4 Toy Stellar Models

we know that the central regions of these stars are convective. To take all these features into account, Table 2.1(a) gives the values of n_R, n_T, n_L, and n_{HR} for the following different cases: (1) ideal gas with electron scattering and (2) ideal gas with convective-energy transport. Each of these are evaluated for two indices of nuclear-power generation $\nu = 15$ and $\nu = 18$. Of these models, model (2) with convection is included essentially to demonstrate the effect of the convective core on the indices.

The examination of different indices show that the general trend of different physical parameters, obtained above by numerical integration, is reproduced by the homology relations; but the details do not match very well. For example, the numerical integration gives $n_R \approx 0.56$ in the high-mass end whereas the homology relation gives $n_R \approx 0.78$–0.81. However, note that the corresponding index for a purely convective star is negative. This suggests that the net effect of convection is to reduce the value of n_R. This effect will move n_R in Table 2.1(a) towards a more realistic, lower value. The homology models do somewhat better as regards n_L. The value in Table 2.1(a) is $n_L \approx 3.0$, which is to be compared with the observed relation of $n_L \approx 3.5$. Similarly, the homology predicts $n_{HR} \approx 0.12$, and the observed value is $n_{HR} \approx 0.17$. In the case of both n_{HR} and n_L the effect of the convective core is to increase the values of n_L and n_{HR}. Once again we see that this will move in the direction to reduce the discrepancy. Figures 2.1 and 2.2, based on numerical integration, show that the slopes of the curves are somewhat steeper around $M \approx M_\odot$. The predictions based on homology with $\nu = 15$–18 (CNO cycle), $a = 1$, $b = 1$ (ideal gas), and $n = 0$, $s = 0$ (electron scattering) are in better agreement with stars in the range $M = (1\text{–}10)\ M_\odot$.

Table 2.1(b) gives the corresponding values for low-mass stars, for which we expect the p–p chain to be the dominant process and the radiation pressure to be negligible. As regards energy transport, we consider all three cases: Kramer's opacity ($n = 1$, $s = 3.5$), electron scattering ($n = 0$, $s = 0$), and convective-energy transport. Once again we do not expect convection to play any significant role in deciding the overall properties of the star, as it is confined mainly to the outer regions; it is included only to show the effect of convection on the indices. The indices calculated in Table 2.1(b) fail to give good agreement with observations in the low-mass regime, except possibly for n_L and n_{HR}. Even here, the agreement is not as good as that for high-mass stars. It is obvious from the exact numerical integration (see Figs. 2.1 and 2.2) that these stars do not follow any simple homology relations. Hence it is not possible to model them along these lines in any realistic manner.

The above discussion also highlights the role played by convection and opacity in stellar modelling. The indices depend rather sensitively on the nature of opacity prevalent within the bulk of the star. Hence accurate stellar modelling requires a good understanding of opacity of matter under different conditions. As regards convection, the situation is different for high-mass and low-mass stars. For high-mass stars, the convective core does influence the various indices to some extent, but the bulk of the relationship still arises from the outer radiative envelope of the star. In the low-mass stars, convection does play a significant role and the accurate modelling of these systems is quite complicated. Even here, the extreme outer layers are still radiative, and this layer plays a crucial role in determining various indices. We shall have occasion to say more about this in the next section.

Similar homology relations can be used to study the dependence of stellar parameters on the mean atomic weight μ and the heavy-element abundance Z. The mean atomic weight μ enters the equations mainly through the ideal-gas law and somewhat more weakly through the opacity. Keeping the leading dependence that arises through the ideal-gas laws, we can easily show that the radius, temperature, and luminosity of a star of a given mass scale as

$$R \propto \mu^{m_R}, \quad T \propto \mu^{m_T}, \quad L \propto \mu^{m_L}, \tag{2.58}$$

where

$$m_R = \mathcal{B}, \quad m_T = (1 - \mathcal{B}), \quad m_L = (4 + n + s) + \mathcal{B}(3n - s), \tag{2.59}$$

with

$$\mathcal{B} = \frac{\nu - s - (4 + n)}{\nu - s + 3n + 4}. \tag{2.60}$$

For the CNO cycle ($\nu = 15\text{–}18$), electron-scattering opacity ($n = s = 0$), and the ideal-gas equation of state, we get

$$m_L = 4.0, \quad m_R = 0.58\text{–}0.63, \quad m_T = 0.42\text{–}0.36. \tag{2.61}$$

This shows that the radius and the temperature vary rather weakly with the mean atomic weight whereas the luminosity is a strongly increasing function. (For the p–p chain, with the ideal-gas equation of state and Kramer's opacity, the dependence is even steeper, and we get $m_L \approx 8$.) For the first case, the slope in the H-R diagram will be

$$m_{\text{HR}} = \frac{1}{4}\left(1 - \frac{2m_R}{m_L}\right) \approx 0.17, \tag{2.62}$$

which is higher than the corresponding value $n_{\text{HR}} \approx 0.12$. This implies that, as the molecular weight increases, say, because of the conversion of hydrogen into helium, the stars will have higher L but will evolve below the main-sequence line in the L–T_{eff} plane.

The above analysis of the dependence of stellar parameters on the mean atomic weight tacitly assumes that the star is well mixed and that the molecular weight changes uniformly everywhere in the star. This assumption is often violated in real stars and hence the homology relations are only suggestive of the direction of the dependence.

A similar analysis can be performed to determine the dependence of the parameters on the metal abundance of the stars, Z. Because Z is usually a small

2.4 Toy Stellar Models

number, the heavy-element abundance does not directly influence μ; however, the opacity has a dependence on Z. Using the power-law form for μ and κ given in Eqs. (2.6) and (2.10), for low-mass stars with a p–p chain reaction as the main source of energy and Kramer's opacity as the transport process, we can easily show that

$$R \propto Z^{0.15} X^{0.68} M^{0.077}, \quad L \propto Z^{-1.1} X^{-5.0} M^{5.46}, \quad T_{\text{eff}} \propto Z^{-0.35} X^{-1.6} M^{1.33}, \tag{2.63}$$

approximately. These combine to give

$$L \propto Z^{0.35} X^{1.55} T_{\text{eff}}^{4.12} \tag{2.64}$$

The actual Z dependence is weaker than indicated above because the electron-scattering opacity is not taken into account, but the relation shows the correct trend. Both luminosity and the effective temperature increase with decreasing Z. For a star of given mass but varying Z, we get the relation $T_{\text{eff}} \propto L^{0.33}$. In other words, the main sequence for metal-poor stars will lie below and to the left of the main sequence for metal-rich stars in the L–T_{eff} plane.

Exercise 2.5

Eddington's standard model: (a) Consider a star (of mass M and luminosity L) in thermal equilibrium with energy transport by radiative diffusion and pressure that is due to ideal gas and radiation. Let $\eta(r)$ denote the dimensionless ratio

$$\eta(r) = \frac{\langle \epsilon(r) \rangle}{\langle \epsilon(R) \rangle} = \frac{L_r/L}{M_r/M}, \tag{2.65}$$

and let $\beta(r)$ denote the ratio of ideal-gas pressure to the total pressure. Prove that

$$1 - \beta(r) = \frac{L}{4\pi c G M} \langle \kappa \eta(r) \rangle, \tag{2.66}$$

where the average on the right-hand side is defined by a pressure-weighted integral,

$$\langle \kappa \eta(r) \rangle = \frac{1}{P(r)} \int_0^{P(r)} \kappa \eta \, dP, \tag{2.67}$$

with $P(r)$ denoting the total pressure. (b) Argue that the product $\kappa \eta$ should not vary as strongly with position as either quantity individually. This suggests building a stellar model in which β is treated as a constant. (c) Show that such a model (with constant β) is equivalent to a polytrope of index $n = 3$. Using the numerical results available from the polytrope theory (e.g., Vol. I, Chap. 10, Section 10.3), show that β is related to the total mass of the star by

$$\frac{1-\beta}{\beta^4} = 0.002996 \, \mu^4 \left(\frac{M}{M_\odot}\right)^2 \tag{2.68}$$

and that the temperature can be related to the density by

$$T(r) = 4.62 \times 10^6 \beta \mu \left(\frac{M}{M_\odot}\right)^{2/3} \rho^{1/3}(r). \tag{2.69}$$

What does this result imply for the importance of radiative pressure in stars of different mass?

Exercise 2.6
Filling in the blanks: Derive the following relations, supplying the necessary intermediate steps: (2.40)–(2.43), (2.46), (2.47), (2.49)–(2.56), and (2.58).

Exercise 2.7
Analytic approximation for energy production: Consider a star in which the energy-generation rate per gram is given by the simple power law

$$\frac{\epsilon}{\epsilon_c} \approx \left(\frac{\rho}{\rho_c}\right)^\lambda \left(\frac{T}{T_c}\right)^\nu, \tag{2.70}$$

where the subscript c denotes central values. (a) When the ratio $\beta \equiv (P_{\rm rad}/P_{\rm tot})$ is approximately a constant throughout the star, show that the mean energy-generation rate, defined as

$$\langle \epsilon \rangle = \int_0^m \epsilon \, \frac{dm}{M}, \tag{2.71}$$

is given by

$$M\langle \epsilon \rangle = 4\pi \epsilon_c \rho_c a^3 \int_0^{\xi_1} \theta^{n(\lambda+1)+\nu} \xi^2 \, d\xi, \tag{2.72}$$

where $\theta = T/T_c$. (b) Explain why $\theta \simeq \exp(-\xi^2/6)$ is a fairly good approximation for the polytrope that we are considering. Using this, show that

$$\frac{\langle \epsilon \rangle}{\epsilon_c} = \frac{3.23}{(3\lambda + \nu + 1)^{3/2}} \quad \text{or} \quad \frac{\langle \epsilon \rangle}{\epsilon_c} = \frac{2.4}{(1.5\lambda + \nu + 1.5)^{3/2}} \tag{2.73}$$

for $n = 3$ and $n = 1.5$.

Exercise 2.8
Scaling relation and molecular weight: Using the scaling arguments, determine the dependence of the luminosity of a star on the mean molecular weight if the opacity is given by Kramer's formula; i.e., prove relations (2.63) and (2.64). Normalise this result by using the Sun, for which $X = 0.70$, $Y = 0.28$, and $Z = 0.02$. Hence determine how different the luminosity will be for a star with $X = 0.98$ and $Z = 0.02$ and no helium.

2.4.2 The Radiative Stellar Envelope

We have seen above that high-mass stars have a convective core and a radiative envelope, whereas low-mass stars have a radiative core and a convective envelope. However, even in the case of low-mass cool stars, there must exist a thin radiative layer near the surface. This follows from the fact that, in building the stellar model, we are imposing vanishing boundary conditions on temperature and pressure near the surface. Therefore, *all* stars are endowed with a surface

2.4 Toy Stellar Models

layer that is radiative. In the case of low-mass stars there is a radiative core near the centre, a convective region enveloping it, and a thinner radiative atmosphere near the surface; for high-mass stars, there is a convective core near the centre followed by a radiative envelope that may extend all the way to the surface. Many of the observed properties of a star, such as L, T_{eff}, etc., are decided by the nature of the radiative envelope near the surface in *all* stars. We shall now consider some properties of this envelope.[5]

Because most of the luminosity and the mass are contributed by the interior regions of a star, we can treat M and L as constants in the radiative envelope. The optical depth of this envelope is formally taken to be zero at the surface of the star and it increases rapidly as we go to the interior. The temperature of the star is also taken to be zero at the surface and increases rapidly towards the interior. We shall first determine the variation of temperature as a function of the optical depth τ and identify the optical depth at which $T = T_{\text{eff}}$.

Because the microscopic time scales are fairly short, it is reasonable to assume that the stellar material is in local thermodynamic equilibrium even at the envelope. In that case, the energy density of radiation is given by the Planck spectrum B_ν corresponding to the local temperature. The flux of radiation caused by the temperature gradient is given by

$$F_\nu = -\frac{4\pi}{3} \frac{1}{\kappa_\nu \rho} \frac{\partial B_\nu}{\partial r} = \frac{L_\nu}{4\pi r^2}, \qquad (2.74)$$

and the radiation pressure is given by

$$P_{\text{rad},\nu} = \frac{4\pi}{3c} B_\nu. \qquad (2.75)$$

[Relation (2.74) was derived in Vol. I, Chap. 6, Section 6.8, and is essentially the same as Eq. (2.7) applied individually at each frequency.] These equations can be combined, after integration over the frequency, to give

$$\frac{dP_{\text{rad}}}{dr} = -\frac{\kappa \rho L}{4\pi r^2 c}, \qquad (2.76)$$

where κ is defined by the relation

$$\kappa = \frac{1}{L} \int_0^\infty \kappa_\nu L_\nu \, d\nu. \qquad (2.77)$$

(This opacity is *not* the same as the Rosseland mean opacity defined in Vol. I, Chap. 6. However, we shall assume that these two expressions are comparable.) We now integrate Eq. (2.76) from the fiducial surface of the star (at which the optical depth $\tau = 0$) to some point in the interior of the stellar envelope. Because r varies very little, if we are interested in points close to the surface, although the optical depth increases more rapidly, we can evaluate the integral approximately as

$$P_{\text{rad}} = -\int_R^r \frac{L}{4\pi cr^2} \kappa\rho\, dr \cong \int_0^\tau \frac{L}{4\pi R^2 c}\, d\tau$$
$$\cong \frac{L}{4\pi R^2 c} \tau + P_{\text{rad}}(\tau = 0) = \frac{\sigma T_{\text{eff}}^4}{c}\tau + P_{\text{rad}}(\tau = 0), \quad (2.78)$$

where we have used the definition $\sigma T_{\text{eff}}^4 = (L/4\pi R^2)$. To determine $P_{\text{rad}}(\tau = 0)$ we proceed as follows. In general, the radiation pressure is related to the intensity $I(\theta)$,

$$P_{\text{rad}} = \frac{2\pi}{c} \int_0^\pi I(\theta) \cos^2\theta \sin\theta\, d\theta, \quad (2.79)$$

where $I(\theta) = (\sigma/\pi)T^4$ in radiative equilibrium. For such an isotropic intensity, it is clear that $P_{\text{rad}} = (4/3)(\sigma T^4/c)$. At the surface of the star from which the radiation is flowing out, it seems reasonable to assume that $I(\theta)$ is isotropic for all outgoing angles $0 < \theta < (\pi/2)$ but is zero for $(\pi/2) \leq \theta \leq \pi$. This is equivalent to assuming that no radiation enters the star from outside. In that case Eq. (2.79) can be integrated to give

$$P_{\text{rad}}(\tau = 0) = \frac{2\pi}{3c} I(\tau = 0). \quad (2.80)$$

We can compute the intensity $I(\tau = 0)$ by evaluating the net flux at the surface with $\tau = 0$ and equating it to L:

$$L = 4\pi R^2 2\pi \int_0^{\pi/2} I(\tau = 0)\cos\theta\sin\theta\, d\theta = 4\pi R^2 \pi I(\tau = 0). \quad (2.81)$$

Using this result to eliminate $I(\tau = 0)$, we get

$$P_{\text{rad}}(\tau = 0) = \frac{2}{3c}\frac{L}{4\pi R^2} = \frac{2}{3c}\sigma T_{\text{eff}}^4. \quad (2.82)$$

Substituting this back into Eq. (2.78), we find that

$$P_{\text{rad}}(\tau) = \frac{4}{3}\left(\frac{\sigma}{c}\right) T^4(\tau) = \frac{\sigma}{c}\left(\tau + \frac{2}{3}\right) T_{\text{eff}}^4, \quad (2.83)$$

or equivalently,

$$T^4(\tau) = \frac{1}{2} T_{\text{eff}}^4 \left(1 + \frac{3}{2}\tau\right). \quad (2.84)$$

This equation gives the change in the temperature as a function of the optical depth in the radiative stellar envelope. In particular, note that $T = T_{\text{eff}}$ at an optical depth of $\tau = (2/3)$. The radiation that is received from a star may be thought of as arising from a surface at which the optical depth has this particular value; this surface is usually called the photosphere.

2.4 Toy Stellar Models

We obtained the above results by making an ad hoc assumption (sometimes called the Eddington approximation) regarding the behaviour of intensity $I(\theta)$ at $\tau = 0$. This was necessary to get a closed solution without solving the full set of equations of radiative transfer. One consequence of this approximation is that the temperature at the $\tau = 0$ surface is not zero but has a value of $T_{\text{eff}}/2^{1/4}$. This, however, is not of major consequence because for all practical purposes the photosphere determines the surface of the star.

We shall next determine the variation of pressure and temperature in the radiative envelope. Ignoring convection, we can set $\nabla = \nabla_{\text{rad}}$, giving [see Eq. (2.14)]

$$\nabla = \frac{d \ln T}{d \ln P} = \frac{3\kappa}{16\pi acG} \frac{P}{T^4} \frac{L}{M}. \tag{2.85}$$

If $\kappa = \kappa_0 \rho^n T^{-s}$ and M and L are constant, this equation becomes

$$P^n dP = \frac{16\pi acGM}{3\kappa_1 L} T^{n+s+3} dT, \tag{2.86}$$

where

$$\kappa_1 \equiv \kappa_0 \left(\frac{\mu}{N_A k_B} \right)^n. \tag{2.87}$$

By integrating Eq. (2.86) from some value P_0, T_0 near the photosphere to some interior value $P(r)$, $T(r)$ in the stellar envelope [with $P(r) \geq P_0$, $T(r) \geq T_0$] and rearranging the result, we find

$$P^{n+1} = \frac{n+1}{n+s+4} \frac{16\pi acGM}{3\kappa_1 L} T^{n+s+4} \left[\frac{1 - (T_0/T)^{n+s+4}}{1 - (P_0/P)^{n+1}} \right]. \tag{2.88}$$

This gives the relation between pressure and temperature in the stellar envelope.

The behaviour of this system depends on the signs of the exponents $n + s + 4$ and $n + 1$. If $n + s + 4$ and $n + 1$ are both positive, then the quantity in the brackets will rapidly approach unity as we move towards the interior. The solution will then tend to

$$P^{n+1} \to \frac{n+1}{n+s+4} \frac{16\pi acGM}{3\kappa_1 L} T^{n+s+4}, \tag{2.89}$$

which is independent of the photospheric boundary conditions and is equivalent to the result that would have been obtained with the boundary conditions $P = 0$ at $T = 0$. This is clearly the case for Kramer's opacity ($n = 1$, $s = 3.5$) and electron-scattering opacity ($n = 0$, $s = 0$). For this reason, the standard boundary conditions used above continue to be valid even in the presence of a thin stellar envelope. (An important counterexample to this result is provided by the opacity that is due to H$^-$ ions, for which $n = 0.5$ and $s = -9$, giving $n + s + 4 = -4.5$. In this case, the boundary conditions do influence the interior solution and cannot be dropped. We shall discuss this case in Section 2.4.3.)

Assuming that Kramer's opacity or electron scattering dominates the stellar envelope, we conclude that the pressure–temperature relation fast approaches a polytropic law, with $\nabla(r)$ tending to

$$\nabla(r) \to \frac{n+1}{n+s+4} = \frac{1}{1+n_{\text{eff}}}, \quad n_{\text{eff}} = \frac{s+3}{n+1}, \qquad (2.90)$$

$$P = K' T^{1+n_{\text{eff}}}, \qquad (2.91)$$

where

$$K' = \left[\frac{1}{1+n_{\text{eff}}} \frac{16\pi a c G M}{3\kappa_0 L} \left(\frac{N_A k_B}{\mu} \right)^n \right]^{1/(n+1)}. \qquad (2.92)$$

For Kramer's opacity ($n = 1$, $s = 3.5$, $n_{\text{eff}} = 3.25$), the radiative gradient is $\nabla = 0.2353$. For electron-scattering opacity ($n = 0$, $s = 0$, $n_{\text{eff}} = 3$), the radiative gradient is $\nabla = 0.25$. If ionisation is ignored, then $\nabla_{\text{ad}} = (2/5) = 0.4$, showing that $\nabla < \nabla_{\text{ad}}$. In either of these cases, convection can be ignored and the envelope is radiative as originally assumed. The full solution for the stellar envelope regime can now be obtained from the standard solution for a polytrope with the index n_{eff}.

The variation of temperature in the stellar envelope can be determined approximately by the following arguments. If ∇ and M are treated as constants, then the equation for pressure equilibrium gives

$$\frac{dP}{dr} = \frac{P}{\nabla} \frac{1}{T} \frac{dT}{dr} = -\frac{GM}{r^2} \rho. \qquad (2.93)$$

With $\nabla = 1/(1+n_{\text{eff}})$ and the ideal-gas equation $P = \rho N_A k_B T/\mu$, this becomes

$$(n_{\text{eff}} + 1) \frac{dT}{dr} = -\frac{GM\mu}{N_A k_B} \frac{1}{r^2}, \qquad (2.94)$$

which may be integrated to yield

$$T(r) = \left(\frac{1}{1+n_{\text{eff}}} \right) \left(\frac{GM\mu}{N_A k_B} \right) \left(\frac{1}{r} - \frac{1}{R} \right)$$

$$= \frac{2.293 \times 10^7 \text{ K}}{1+n_{\text{eff}}} \mu \left(\frac{M}{M_\odot} \right) \left(\frac{R}{R_\odot} \right)^{-1} \left(\frac{R}{r} - 1 \right). \qquad (2.95)$$

This integration uses the explicit boundary condition $T = 0$ at $r = R$. For Kramer's opacity ($n_{\text{eff}} = 3.25$) and $\mu = 0.6$ [see Eq. (2.6) with $X = 0.7$], the temperature at $r = 0.99 R$ is $\sim 33{,}000$ K for the Sun, compared with T_{eff} of 5780 K. This shows that the temperature increases very rapidly inwards.

Knowing the variation of temperature and hence that of pressure, we can integrate all other equations for the envelope. In particular, we can verify that the mass contained in the envelope is negligibly small. It is also possible to calculate the total pressure of the system at the photosphere ($\tau_p = 2/3$) by integrating the

2.4 Toy Stellar Models

hydrostatic equation $(dP/dr) = -g\rho$, with g approximated by the surface value $g_s = (GM/R^2)$. Because $d\tau = \rho\kappa dr$ we get, for the photospheric pressure,

$$P(\tau_p) = g_s \int_0^{\tau_p} \frac{d\tau}{\kappa} + P(\tau = 0) \approx \frac{2}{3}\frac{g_s}{\kappa} + P(\tau = 0), \qquad (2.96)$$

provided that $\kappa \approx \kappa_p$ is treated as a constant. Substituting for $P(\tau = 0)$ from the first equality in Eq. (2.82), we get

$$P(\tau_p) = \frac{2}{3}\frac{g_s}{\kappa_p}\left(1 + \frac{\kappa_p L}{4\pi cGM}\right) \cong \frac{2}{3}\frac{g_s}{\kappa_p}. \qquad (2.97)$$

The second relation follows from the fact that the numerical value of the second term in the parentheses is small; this term,

$$\frac{\kappa_p L}{4\pi cGM} = 7.8 \times 10^{-5}\kappa_p \left(\frac{L}{L_\odot}\right)\left(\frac{M}{M_\odot}\right)^{-1}, \qquad (2.98)$$

can almost always be ignored unless the luminosity of the star is comparable with the *Eddington luminosity*, given by

$$L_{\text{Edd}} \equiv \frac{4\pi cGM}{\kappa_p}. \qquad (2.99)$$

Numerically, the Eddington luminosity is

$$\left(\frac{L_{\text{Edd}}}{L_\odot}\right) \approx 3.5 \times 10^4 \left(\frac{M}{M_\odot}\right) \qquad (2.100)$$

if $X = 0.7$ and $\kappa_p = \kappa_{\text{es}} = 0.34 \text{ cm}^2 \text{ g}^{-1}$. If the luminosity of the star is less than a few percent of the Eddington luminosity, then the second relation in Eq. (2.97) will be valid.

Exercise 2.9
Local thermodynamic equilibrium in stellar atmosphere: Estimate the ratio between the temperature scale height $h \equiv T(dT/dr)^{-1}$ and the mean free path of a hydrogen atom in the atmosphere of a star. Is local thermodynamic equilibrium a valid assumption in the stellar atmosphere? Where does it break down?

To complete the analysis, such an envelope solution should be matched smoothly to an interior solution at which energy generation is taking place. In high-mass stars, the core region is convective because of the temperature sensitivity of the energy generation. Then the core has to be matched to the outer radiative region at some intermediate point. Depending on the mass of the star, the radiative envelope may be rather large; if so, the above analysis will be applicable only near the surface and a more careful calculation is needed in the interior that takes into account the variation in $M(r)$ and $L(r)$. However, in principle, this can be done and the complete model can be generated. The observed properties of such high-mass stars are still determined essentially by the radiative envelope and the analysis reproduces results that are similar to those obtained by the homology arguments in the last subsection.

We shall provide a quick recap of the above results based on simple scalings, before proceeding further. The momentum flux associated with an energy flux of radiation F_{rad} is F_{rad}/c; if the scattering cross section is σ, then the force acting on a particle will be $\sigma(F_{\text{rad}}/c)$, so that the force acting per unit volume of material will be $n\sigma(F_{\text{rad}}/c) = \rho\kappa(F_{\text{rad}}/c)$, where n is the number density of particles and κ is the opacity. Equating this to the radiative pressure gradient, $-\nabla P_{\text{rad}} = -\nabla[(1/3)aT^4]$, we can express the radiative flux as $F_{\text{rad}} = -(ac/3\kappa\rho)\nabla T^4$. Using the relation $\nabla P = -g\rho$, we find that this becomes $F_{\text{rad}} = (gca/3\kappa)(dT^4/dP) \approx (gca/3\kappa)(T^4/P)$. Taking $g \propto (M/R^2)$, $\kappa \propto \rho T^{-3.5}$, and $\rho \propto (P/T)$, and using the fact that $F \propto (L/R^2)$ is a constant in the envelope, we get the relation $L/R^2 \propto (M/R^2)(T^{8.5}/P^2)$; i.e., $P \propto (M/L)^{1/2}T^{4.25}$ and $\rho \propto (M/L)^{1/2}T^{3.25}$. The radial dependence of temperature follows from the equation $(dP/dr) \approx (P/r) \approx -g\rho \propto -(M/r^2)(P/T)$; this gives $T \propto (1/r)$ in the envelope and fixes the variation of all relevant quantities in the stellar envelope.

The situation is different in the case of cool low-mass stars for which the radiative envelope is governed by opacity that is due to the H^- ion. In these stars, the radiative envelope is indeed thin but is preceded by a convective region and again a radiative core. This leads to some interesting features that will be discussed in the next subsection.

2.4.3 Fully Convective Stars with H^- Opacity

In the previous subsection, we concentrated on the envelope of stars in which the energy transport is dominated by radiative opacity. In the case of low-mass stars the radiative envelope is thin and is governed by H^- ion opacity. Below this envelope is a convective region. For very low mass, this convective region extends all the way down to the core. We shall now discuss some properties of such stars, called fully convective stars. Note that the existence of convection all the way to the core in these stars is due to the effect of opacity and not due to the temperature sensitivity of energy production.

We are primarily interested in modelling the radiative envelope and the convective interior of such a star. To do this rigorously, we must model the convective interior and the radiative envelope separately and match them at the point where $\nabla = \nabla_{\text{ad}}$. We shall, however, use a simpler approach to obtain the general scaling relation between T_{eff} and L for these stars. The key idea is to model the radiative envelope and the convective interior as two separate polytropes and match them at a suitable point with $T \approx T_{\text{eff}}$. Let us start with the radiative envelope for

2.4 Toy Stellar Models

which the P–T relationship was found above [see Eq. (2.88)]. Evaluated with $P_0 = P_p$ (pressure at photosphere) and $T_0 = T_{\text{eff}}$, Eq. (2.88) can be written as

$$P^{n+1} = P_p^{n+1} + \frac{A}{1+n_{\text{eff}}}\left(T^{n+s+4} - T_{\text{eff}}^{n+s+4}\right), \quad A = \frac{16\pi c G M}{3L\kappa_0}\left(\frac{N_A k_B}{\mu}\right)^n. \tag{2.101}$$

From Eq. (2.86), we get

$$\nabla = \frac{d\ln T}{d\ln P} = \frac{1}{A}\frac{P^{n+1}}{T^{n+s+4}}. \tag{2.102}$$

Using Eq. (2.101) for P^{n+1} and rearranging terms, we get the variation of ∇ as a function of temperature as

$$\nabla = \frac{1}{1+n_{\text{eff}}} + \left(\frac{T_{\text{eff}}}{T}\right)^{n+s+4}\left(\nabla_p - \frac{1}{1+n_{\text{eff}}}\right), \tag{2.103}$$

where $n_{\text{eff}} = (s+3)/(n+1)$ and ∇_p corresponds to the value of ∇ at the photosphere:

$$\nabla_p = \frac{3\kappa_0 L}{16\pi a c G M}\left(\frac{\mu}{N_A k_B}\right)^n \frac{P_p^{n+1}}{T_{\text{eff}}^{n+s+4}} = \frac{3L}{16\pi a c G M}\frac{P_p \kappa_p}{T_{\text{eff}}^4} = \frac{1}{8}. \tag{2.104}$$

The second equality follows from Eq. (2.85) evaluated at the photosphere and the numerical value follows from using $P_p \approx (2GM/3\kappa_p R^2)$, $L = 4\pi\sigma R^2 T_{\text{eff}}^4$, and $\sigma = (ac/4)$. If $n + s + 4$ is negative (as it is for H^- ion opacity), this relation shows that ∇ increases with depth. For the specific case with $n = 0.5$, $s = -9$, and $n_{\text{eff}} = -4$ we have

$$\nabla(r) = -\frac{1}{3} + \frac{11}{24}\left[\frac{T_{\text{eff}}}{T(r)}\right]^{-9/2}. \tag{2.105}$$

Because temperature increases with depth, so does ∇. Clearly ∇ will exceed the adiabatic value of $\nabla_{\text{ad}} = 2/5$ at some critical radius at which the temperature is T_{conv}. We can easily find the value of T_{conv} by setting $\nabla = 2/5$ in Eq. (2.103). This gives

$$\left(\frac{T_{\text{conv}}}{T_{\text{eff}}}\right) = \left[\frac{8}{5}\left(\frac{3-2n_{\text{eff}}}{7-n_{\text{eff}}}\right)\right]^{1/(n+s+4)}. \tag{2.106}$$

For H^- opacity, $T_{\text{conv}}/T_{\text{eff}} = (8/5)^{2/9} \approx 1.11$. Because T_{conv} is only 11% higher than T_{eff}, it follows that convection starts just below the photosphere. The corresponding value of pressure, P_{conv}, can be obtained from Eq. (2.88) in the form

$$\left(\frac{P}{P_p}\right)^{n+1} = 1 + \frac{1}{1+n_{\text{eff}}}\frac{1}{\nabla_p}\left[\left(\frac{T}{T_{\text{eff}}}\right)^{n+s+4} - 1\right]. \tag{2.107}$$

This gives

$$\frac{P_{\text{conv}}}{P_p} = \left\{1 + \frac{1}{1+n_{\text{eff}}} \frac{1}{\nabla_p} \left[\frac{11(1+n_{\text{eff}})}{5(7-n_{\text{eff}})}\right]\right\}^{1/n+1}. \quad (2.108)$$

In the case of the H$^-$ ion, $P_{\text{conv}} = 2^{2/3} P_p$.

The above analysis clearly shows that the matching of convective and radiative envelopes should take place at some radius at which the temperature is $T_{\text{conv}} \propto T_{\text{eff}}$ and $P_{\text{conv}} \propto P_p$. We shall now use this fact to obtain a simple scaling relation for these stars.

Let us start with the modelling of the interior. In the convective interior, we shall assume that $\nabla = \nabla_{\text{ad}} = 2/5$. This immediately leads to the relation between the pressure and the temperature in the form $P = KT^{5/2}$. For an ideal gas with $P \propto \rho T$, this gives $P \propto \rho^{5/3}$, which is a barotropic equation of state. (Models based on such an equation of state has been discussed in detail in Vol. I, Chap. 8. In this particular case, we are dealing with a polytrope of index 3/2). Further, for a convective star, Eq. (2.46) shows that $K \propto M^{-1/2} R^{-3/2}$, as B is a numerical constant for the model. Using the effective temperature, defined as $T_{\text{eff}} \propto (L/R^2)^{1/4}$, we can eliminate R in favour of the luminosity L and obtain $K \propto M^{-1/2} L^{-3/4} T_{\text{eff}}^3$. This shows that the pressure at the surface at which $T = T_{\text{conv}} \propto T_{\text{eff}}$ is given by

$$P_{\text{conv}} = KT_{\text{conv}}^{5/2} \propto KT_{\text{eff}}^{5/2} \propto M^{-1/2} L^{-3/4} T_{\text{eff}}^{11/2}. \quad (2.109)$$

We can obtain another relation among pressure, temperature, luminosity, and mass in the outer regions of the star by considering the radiative envelope. Because $P_p = (2GM/3\kappa_p R^2)$ and $\kappa_p = \kappa_0 \rho_p^n T_{\text{eff}}^{-s}$, we get

$$P_p = \frac{2}{3} \frac{GM}{R^2 \kappa_0} \left(\frac{T_{\text{eff}}^s}{\rho_p^n}\right). \quad (2.110)$$

Using the ideal-gas equation of state, $P = (N_A k/\mu)\rho T$, we can eliminate the density and express the pressure in terms of T_{eff} as

$$P_p = \left(\frac{2}{3}\frac{GM}{\kappa_0 R^2}\right)^{1/(n+1)} \left(\frac{\mu}{N_A k}\right)^{-n/(n+1)} T_{\text{eff}}^{(n+s)/(n+1)}. \quad (2.111)$$

Eliminating R in terms of L and T_{eff} and using the fact that $P_{\text{conv}} \propto P_p$, we get a second scaling relation

$$P_{\text{conv}} \propto \left(\frac{M}{L}\right)^{1/n+1} T^{(4+s+n)/(n+1)}. \quad (2.112)$$

We now have two expressions [Eq. (2.109) and relation (2.112)] for the pressure P_{conv}, and they have to match at the surface at which the transition from convective to radiative transport takes place. Setting the two expressions for P_{conv}

2.4 Toy Stellar Models

as proportional to each other, we can express T_{eff} in terms of M and L. We get

$$T_{\text{eff}} \propto M^{(n+3)/(9n+3-2s)} L^{(3n-1)/2(9n+3-2s)}. \tag{2.113}$$

This is the required relation that relates the effective temperature to the mass and the luminosity of a star that is fully convective in the interior and has a thin radiative atmosphere.

The explicit form of this scaling depends on the source of opacity. In several cool low-mass stars, the opacity at the surface layers is contributed by the negative hydrogen ion for which $n = 0.5$ and $s = -9$. This gives

$$T_{\text{eff}} \propto M^{7/51} L^{1/102}. \tag{2.114}$$

The very weak dependence on L shows that these stars will occupy a nearly vertical line in the T_{eff}–L plane, with small variations in T_{eff} leading to very large changes in L. We shall see in the next chapter that this result has important implications for different phases of stellar evolution.

Exercise 2.10
Uniformly rotating stars: Consider a chemically homogeneous star that is in hydrostatic and thermal balance. Let us further assume that it is rigidly rotating about some axis with an angular velocity Ω. The equation of state is that of an ideal gas and the energy transport is through radiative diffusion. Show that the above considerations require the nuclear-energy-production rate to satisfy the proportionality

$$\epsilon \propto \left(1 - \frac{\Omega^2}{2\pi G \rho}\right). \tag{2.115}$$

Because the nuclear-energy generation cannot be determined by the speed of stellar rotation, this result shows that a uniformly rotating star cannot be in radiative steady-state thermal equilibrium.

Exercise 2.11
Bounds on stellar variables: (a) Prove that in any star in which density ρ does not increase outwards we must have

$$\frac{1}{2}G\left(\frac{4\pi}{3}\right)^{1/3} \bar{\rho}^{4/3}(r) M^{2/3}(r) \leq (P_c - P) \leq \frac{1}{2}G\left(\frac{4\pi}{3}\right)^{1/3} \rho_c^{4/3} M^{2/3}(r), \tag{2.116}$$

where $\bar{\rho}(r) = [3M(r)/4\pi r^3]$ is the mean density inside a sphere of radius r and the subscript c refers to values of the variables at the centre. (b) As a corollary, prove that the bounds on the central pressure

$$\frac{1}{2}G\left(\frac{4\pi}{3}\right)^{1/3} \rho_c^{4/3} M^{2/3} \geq P_c \geq \frac{3}{8\pi} \frac{GM^2}{R^4}. \tag{2.117}$$

(c) Show that the magnitude of the gravitational potential energy is bounded by the inequalities

$$\frac{3}{5}\frac{GM^2}{r_c} \geq E_G \geq \frac{3}{5}\frac{GM^2}{R}, \qquad (2.118)$$

where r_c is defined through the relation $M = (4\pi \rho_c r_c^3/3)$. (d) For a star made of an ideal gas with homogeneous composition, show that the density-weighted average temperature, defined to be

$$\bar{T} = \frac{1}{M}\int_0^R T\, dM(r), \qquad (2.119)$$

satisfies the bound

$$\bar{T} > \frac{1}{6}\frac{\mu m_H}{k_B}\frac{GM}{R}. \qquad (2.120)$$

Exercise 2.12

Entropy gradient in a star: The first law of thermodynamics can be written in the form $TdS = dE - P(d\rho/\rho^2)$, where E is the energy per unit mass and S is the entropy per unit mass. (a) Treating $E = E(P, T)$ and $\rho = \rho(P, T)$, show that this equation can be reduced to the form

$$\frac{dS}{dr} = c_P(\nabla - \nabla_{ad})\frac{d\ln P}{dr}. \qquad (2.121)$$

What does this imply for a radiative and convectively unstable regions of a star? (b) For an ideal monoatomic gas we also know that

$$S(r) = \frac{3}{2}\frac{N_A k_B}{\mu}\ln\left[\frac{P(r)}{\rho^{5/3}(r)}\right]. \qquad (2.122)$$

Using the scalings $P \simeq (GM^2/R^4)$ and $\rho \simeq (M/R^3)$, convert this equation into the form

$$R \propto \exp\left(\frac{2\mu}{3 N_A k_B}\frac{S_{tot}}{M}\right). \qquad (2.123)$$

For a star in which the photosphere is dominated by H^- opacity, express the photospheric entropy in the form

$$S_p = \frac{N_A k_B}{\mu}\ln\left(\frac{3\, T_{eff}^{11.5}\rho_P^{0.5}}{2\, \kappa_0 g_s}\right). \qquad (2.124)$$

(c) Analyse what happens to the entropy as T_{eff} is reduced. Can T_{eff} fall arbitrarily low in the photosphere? This is another way of understanding why there is a forbidden region at the right top corner of the H-R diagram.

2.5 Observational Aspects of Stellar Atmospheres

We saw in the last section that the the radiation received from the stars originates from a surface with optical depth of approximately $\tau = 2/3$. The primary source of information about the stars is from the analysis of the spectrum of this radiation. We shall comment on several aspects of such an analysis that play a significant role in the observation and classification of stars.

Because local thermodynamic equilibrium is a reasonable assumption at the stellar envelope, the spectrum of a star will be expected to be nearly thermal, characterised by the temperature $T_{\rm eff}$, to the lowest order of approximation. This is indeed true for most stars. From Fig. 2.2(b), it is clear that $T_{\rm eff}$ varies from $\sim$2000 K for the low-mass stars to $\sim$50,000 K for the high-mass stars. The peak of νI_ν of the thermal radiation will vary from a wavelength of $\lambda_{\min} \approx 720$ Å (which is in the UV band) to $\lambda_{\max} = 18,000$ Å (which corresponds to IR) as the stellar temperatures vary in this range. It is conventional to associate a colour with a star based on the peak of the Planck spectrum. Many other aspects of the luminosity and the effective temperature of the stars mentioned in Chap. 1 follow directly from the above consideration.

The fact that an observer receives from a star photons that originate at the surface with $\tau = 2/3$ has some important observational consequences. To begin with, the above result leads to a phenomenon called *limb darkening*. The line of sight of an observer viewing the Sun, say, makes an increasingly large angle with the vertical near the edge, or limb, of the Sun. Hence, looking near the limb, the observer will not see as deeply into the star as along the centre. Thus the point at which $\tau = 2/3$ for a ray traversing near the limb will be at a lower temperature compared with a ray traversing towards the centre of the solar disk. As a result, the limb of the star appears darker than its centre.

Second, the definition of the optical depth τ,

$$\tau_\lambda = \int_0^s \kappa_\lambda \, \rho \, ds, \qquad (2.125)$$

shows that if the opacity κ_λ increases at some wavelength, then the physical distance along a ray to the point $\tau_\lambda = 2/3$ will decrease for that wavelength. (In arriving at this conclusion we have generalised the earlier result of $\tau = 2/3$ to each wavelength individually.) Because the temperature of the stellar atmosphere decreases outwards, the outer regions of the atmosphere will be cooler. Hence the intensity of radiation at $\tau_\lambda = 2/3$ will decline for those wavelengths at which opacity is largest. This wavelength dependence also shows that different parts of a spectral line are formed at different radius.

In such strict thermodynamic equilibrium, there will be no other form of line or continuum radiation. The spectral information, of course, is contained in the deviations of the spectrum from the blackbody spectrum corresponding to local thermodynamic equilibrium. We shall next discuss some general aspects of both continuum and line spectra of stars.

Exercise 2.13
Limb darkening in the Sun: The photographs of the Sun clearly show that the limb of the Sun appears darker than its centre. Model the atmosphere of the Sun along the lines given in Subsection 2.4.2 and show that the emergent intensity $I(\theta)$ at an angle θ is related to that at the centre of the star $I(0)$ by $I(\theta)/I(0) = (2/5) + (3/5)\cos\theta$ in this approximation. This gives a ratio of $2/5 = 0.4$ for the darkening of the limb ($\cos\theta = 0$) with respect to the centre ($\cos\theta = 1$); a more sophisticated analysis gives 0.344. {Hint: First show that the emergent intensity at an angle θ can be expressed in the form

$$I_\nu(\theta) = \int_0^\infty S(\tau_\nu) e^{-\tau_\nu \cos\theta} \sec\theta \, d\tau_\nu, \qquad (2.126)$$

where $S(\tau_\nu)$ is the source function at the optical depth τ_ν. Assume that $S = a + b\tau_\nu$ gives $I_\nu(\theta) = a + b\cos\theta$. Hence $[I_\nu(\theta)/I_\nu(0)] = (1-\beta) + \beta\cos\theta$, where $\beta = [1 + (a/b)]^{-1}$. Identify (b/a) for the approximation in the text, thereby completing the derivation.}

2.5.1 Continuum Radiation

The most important continuum absorption process that is of relevance in stellar atmospheres is the bound–free absorption; free–free absorption and electron scattering do not play really significant roles. The frequency-dependent bound–free absorption coefficient (derived in Vol. I, Chap. 6, Section 6.12) from a level with quantum number n is given by

$$\kappa_{bf} \cong \frac{2.815 \times 10^{29} Z^4}{n^5 \nu^3} N_n \text{ cm}^2 \text{ g}^{-1} \quad \left(\text{for } \nu > \nu_n = \frac{I_n}{h}\right), \qquad (2.127)$$

where Z is the atomic number, n is the principle quantum number of the bound state, I_n is the energy of the nth level, and N_n is the number of atoms or ions per unit mass populating the excited level n. In thermal equilibrium, N_n varies as $g_n \exp(-E_n/k_B T)$, where $g_n = 2n^2$ is the degeneracy of level n. (Near $\nu \approx \nu_n$ this formula ceases to be valid; further, we have ignored the Gaunt factor.) This coefficient increases with decreasing frequency up to ν_n. At any given frequency the main contribution to κ_{bf} will be from the lowest allowed value of n. The plot of κ_{bf} against frequency has a characteristic sawtooth shape; starting at the highest frequencies, κ_{bf} rises as ν^{-3} until the threshold for $n = 1$ is reached (see Fig. 2.11). In this regime it is essentially dominated by the $n = 1$ state. At lower frequencies, κ_{bf} is dominated by $n = 2$ so that there is an abrupt drop at the $n = 1$ threshold. It again picks up and rises as ν^{-3} until the $n = 2$ threshold is reached, etc.[6] (The value of κ_{bf} at the threshold $\nu_n \propto n^{-2}$ varies as $\kappa_{bf} \propto n^{-5} \nu_n^{-3} g_n \propto n^3$.)

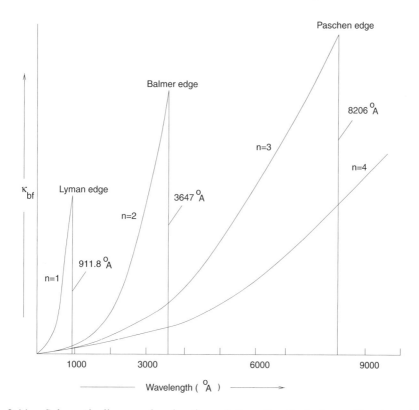

Fig. 2.11. Schematic diagram showing the variation of κ_{bf} for hydrogenlike atoms from each of the energy levels.

In the case of hydrogen, photoionisation from the ground state is possible for wavelengths of less than the Lyman limit $\lambda_{LL} = 912$ Å. Because most hydrogen atoms are in the ground state, this will contribute significantly to absorption for $\lambda < \lambda_{LL}$. The corresponding limit for photoionisation from the ground state of neutral helium is 504 Å, and this will make a contribution at still lower values. In massive stars with high surface temperatures, helium is ionised and HeII has a hydrogenlike spectrum with energies scaled up by $Z^2 = 4$. In this case, the HeII threshold occurs at 228 Å. Below this wavelength, HeII absorption is dominant and even outweighs the higher abundance of hydrogen.

As we cross $\lambda = \lambda_{LL}$ and move to higher wavelengths, there is an abrupt drop in the absorption coefficient. In these regimes, bound–free absorption is from the $n = 2$ level, and it increases up to the Balmer limit of $\lambda_{BL} = 3647$ Å. Once again the absorption coefficient drops, and this process goes on. (However, there will be only a small fraction of all hydrogen atoms at $n = 2, 3, \ldots$, levels, and this leads to a certain characteristic pattern of absorption as a function of gas temperature; we shall discuss this aspect in Section 2.5.2.) If hydrogen dominates the absorption, then the existence of ionisation edges will leave a signature in the continuum flux. As we cross λ_{BL} and proceed to longer wavelengths, the absorption coefficient drops and the stellar atmosphere becomes more transparent to radiation, allowing us to see the flux from deeper and hotter layers. This necessarily leads to an increase in the flux. The ratio of bound–free absorption coefficients across the Balmer discontinuity is given by

$$\frac{\kappa_{bf}(>3650\,\text{Å})}{\kappa_{bf}(<3650\,\text{Å})} = \frac{\kappa_{bf}(n=3)+\cdots}{\kappa_{bf}(n=2)+\kappa_{bf}(n=3)+\cdots} \simeq \frac{\kappa_{bf}(n=3)}{\kappa_{bf}(n=2)},$$

$$= \frac{8}{27} e^{-(E_3-E_2)/k_BT}, \qquad (2.128)$$

which has a value of 0.0037 at 5000 K and 0.033 at 10,000 K. (We have used $g_n = 2n^2$ to obtain the numerical coefficient.) This ratio decreases as we go to lower temperatures; that is, κ_{bf} at $\lambda > 3650$ Å will become progressively smaller compared with κ_{bf} at $\lambda < 3650$ Å as we go to lower temperatures, thereby increasing the discontinuity in the bound–free absorption coefficient at $\lambda = 365$ nm.

In cooler stars, the continuous absorption in the visible band arises from bound–free transitions of the H^- ions, which have one bound state with energy 1.2×10^{-12} ergs corresponding to a threshold of 16,450 Å. The H^- ion cannot be approximated as a single electron atom, and hence it does not have the sawtoothlike absorption pattern. The absorption cross section rises to a maximum of 4×10^{-9} cm^2 at 8500 Å and then falls smoothly to zero at the threshold of 16,500 Å. At solar temperatures, H^- absorption dominates at all wavelengths between λ_{BL} and the H^- threshold. Neutral hydrogen dominates in the visible band for temperatures greater than $\sim$7500 K.

It must be noted that the extra electron to form the H^- ion has to come from some other donor element in the cool stars. At 6000 K, elements like iron, silicon, and magnesium act as dominant donors whereas at still lower temperatures elements like sodium and potassium provide the electrons. (Because sodium and potassium are much less abundant than iron, silicon, and magnesium, their contribution is not important when the latter are ionized.) Other abundant elements have ionisation potentials comparable with or larger than hydrogen and hence cannot compete with hydrogen as a donor. In cool stars therefore the absorption coefficient depends on the donor and hence on the metallicity. A metal-poor star with a small line absorption coefficient for metal lines will also have a small continuous absorption coefficient.

Other elements can compete with hydrogen in spite of their lower abundance if the absorption in those elements takes place from the ground state; the corresponding absorption from the hydrogen has to come from an excited state. The most abundant elements after hydrogen have ionisation potentials that are larger than that of hydrogen (for example, helium and neon) or comparable with that of

2.5 Observational Aspects of Stellar Atmospheres

hydrogen (for example oxygen, carbon, and nitrogen). Silicon, magnesium, and iron have abundances that are $(2-4) \times 10^{-5}$ times that of hydrogen with ionisation edges at the wavelengths 1520 Å, 1620 Å, and 1570 Å, respectively. They can play an important role in cooler stars between these threshold values and the Lyman limit.

2.5.2 Lines

In hydrogenlike atoms, energy levels become closer as we go to higher n and the transitions with the shortest wavelengths are from the ground state. Because the transition probabilities and oscillator strengths decrease as we go to higher upper quantum numbers, Lyman-α lines from $n = 1$ to $n = 2$ will be the strongest. The lines produced by the transition between the ground state and the first excited state (which is permitted to decay radiatively to ground state) is called the *resonance line*; for the hydrogen atom, Lyman-α is the resonance line.

The energy-level structures of other atoms are more complex, but the strongest line is often the resonance line and stronger lines are usually at shorter wavelengths. Because the ionisation potential depends on the nuclear charge as well as on the shielding effect of the electrons, as we proceed to higher states of ionisation of a particular element the ionisation potentials, the excitation potentials, and the level separations will all increase.

The actual lines we observe in a particular spectrum depend vitally on the temperature and – to a lesser extent – on the pressure and abundances. Most of the observations made from Earth have an atmospheric cutoff around 3300 Å, although sensitive detectors can extend this up to 9000 Å. The most abundant elements (hydrogen, helium, oxygen, carbon, neon, and nitrogen) all have large ionisation potentials and resonance lines in the UV. At solar temperatures, helium and neon do not lead to observable visible lines, and there are only a few forbidden high-excitation lines of carbon, oxygen, and nitrogen. The key contribution therefore comes from hydrogen in the form of H_α, H_β, and H_γ lines of the Balmer series. The Balmer lines increase in strength with increasing temperature, reaching a maximum around 10^4 K and decreasing at higher temperatures. We shall now discuss, in some detail, how this comes about, as this phenomenon is quite generic and allows us to classify stars based on their spectra.

Balmer absorption lines of hydrogen arise when the electrons in hydrogen make an upward transition from the $n = 2$ energy level. The strength of such a line will therefore be proportional to the fraction N_2/N_H of hydrogen atoms in which the electron is found at the $n = 2$ energy level. This fraction, in turn, is the product of two factors: (1) the fraction of hydrogen atoms that are in the neutral state and (2) the fraction of neutral atoms in which the electron is at the excited $n = 2$ level. The first factor is determined by Saha's ionisation equation, and the second is determined by the Boltzmann formula. Combining these two together, we can find the expression for N_2/N_H that was obtained in Vol. I, Chap. 5:

$$\frac{N_2}{N_H} = \frac{g_2}{Z_0(T)} e^{-(\Delta E_{12}/k_B T)} \left[1 + \frac{2Z_1(T)}{n_e Z_0(T)} \left(\frac{m_e k_B T}{2\pi\hbar^2}\right)^{3/2} e^{-|E_0|/k_B T}\right]^{-1}. \tag{2.129}$$

Here, $\Delta E_{12} = 10.2$ eV is the energy difference between the $n = 1$ and the $n = 2$ levels of hydrogen; $|E_0| = 13.6$ eV is the ionisation potential; g_2 is the degeneracy factor for the $n = 2$ level of hydrogen (because the degeneracy factor is $g_n = 2n^2$, we have $g_1 = 1$, $g_2 = 2$); $Z_1(T)$ is the partition function for the ionised hydrogen that is unity because ionised hydrogen is just a proton; and $Z_0(T)$ is the partition function for the neutral hydrogen that may be approximated as

$$Z_0(T) \cong g_1 + g_2 \, e^{-(\Delta E_{12}/k_B T)} \cong 2 + 8 \, e^{-(\Delta E_{12}/k_B T)}, \tag{2.130}$$

with only the $n = 1$ and $n = 2$ states taken into account. Substituting all these values in Eq. (2.129) we get

$$\frac{N_2}{N_H} = \left(\frac{4 e^{-\beta \Delta E}}{1 + 4 e^{-\beta \Delta E}}\right) \left[1 + \frac{(m_e k_B T/2\pi\hbar^2)^{3/2}}{n_e(1 + 4 e^{-\beta \Delta E})} e^{-\beta E_0}\right]^{-1}. \tag{2.131}$$

It is clear from this equation that N_2/N_H is small for both low temperatures and high temperatures. The asymptotic forms are given by

$$\frac{N_2}{N_H} = \begin{cases} 4 e^{-\beta \Delta E} & (\beta \Delta E \gg 1) \\ 4 \left(\frac{P_e}{k_B T}\right) \left(\frac{2\pi\hbar^2}{m_e k_B T}\right)^{3/2} & (\beta \Delta E \ll 1) \end{cases}, \tag{2.132}$$

where we have expressed n_e in terms of the electron pressure by the relation $n_e = P_e/k_B T$. The vanishing of N_2/N_H at high temperatures is due to the nearly complete ionisation of hydrogen; because no neutral hydrogen is present, Balmer lines cannot be produced at high temperatures. The vanishing of the expression at low temperatures is due to the fact that most of the hydrogen will be in the ground state and not at the $n = 2$ state. It follows that Balmer lines will be produced predominantly when the temperature is in some intermediate regime.

Figure 2.12 shows a plot of (N_2/N_H) as a function of temperature, taking $P_e \approx 200$ dyn cm^{-2}. There is a distinct maximum around $T \approx 9900$ K that compares well with the observed fact that the Balmer lines in the stars reach maximum intensity at a temperature of $\sim$9520 K. A similar analysis can be performed for different elements and different ionisation states. It is clear from the above discussion that the line strengths arising from different physical processes will reach their maximum at some suitable temperature. This allows us to characterise and classify the stars based on the spectral features and connect it up with the T_{eff} of the star. Figure 2.13 shows the strength of spectral lines on the temperature and

2.5 *Observational Aspects of Stellar Atmospheres*   107

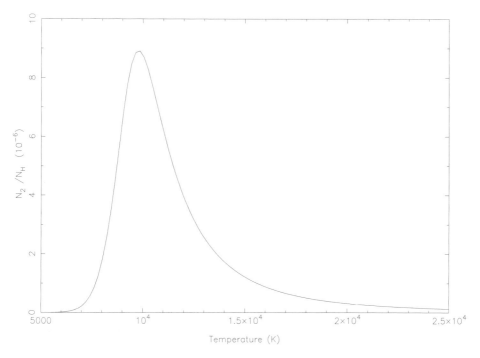

Fig. 2.12. The fraction of hydrogen atoms in the $n = 2$ state as a function of temperature in the stellar atmosphere. The distinct peak around $T = 9900$ K arises because of competition between the Boltzmann factor and the ionisation equilibrium.

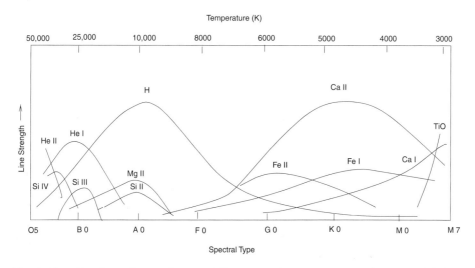

Fig. 2.13. The dependence of spectral line strength on temperature for stars with different effective surface temperatures. The lower x axis gives the standard classification of the star type.

Table 2.2. Spectral classification of stars

Type	Colour	Spectra
O	Blue–white	Strong HeII absorption/emission
B	Blue–white	HeI absorption
A	White	Balmer absorption lines
F	Yellow–white	CaII, FeI, CrI absorption lines
G	Yellow	Solar-type spectra; CaII, FeI becoming stronger
K	Orange	Spectra dominated by metal absorption lines
M	Red	Spectra dominated by molecular absorption bands, especially titanium oxide (TiO)

the corresponding characterisation of the star as belonging to type O, B, A, ..., etc. The standard descriptions used by astronomers are given in Table 2.2. Figure 1.4 in Chap. 1 illustrates the same point by use of realistic spectra of stars.

A similar analysis can be done to estimate the importance of H^- ion absorption in the stellar atmosphere. The ratio between the number density of hydrogen atoms at the $n = 3$ level (which are capable of producing Paschen continuum absorption in the visible band) to the number density of the H^- ion in the solar atmosphere is approximately 2×10^{-3}. (This can be computed along the lines similar to those given in the discussion above.) This shows that for continuum absorption in the visible band, H^- ions contribute significantly.

The second most abundant element, helium, has its energy levels split into singlets ($s = 0$) and triplets ($s = 1$), depending on the alignment of the electron spins. The ground state is a singlet with $1s^2$ but the lowest excited level is a triplet $1s\,2s$. The lowest excited singlet of $1s\,2s$ has a slightly higher energy level. Radiative transitions between singlet and triplet are weakly forbidden; so is the transition $1s^2 \to 1s\,2s$. Therefore the longest-wavelength line from the ground state of helium is from the transition $1s^2 \to 1s\,2p$ with a wavelength of 5840 Å. In the visible band, the transitions are from the first and the second levels of triplet systems: $1s\,2s\,^3S \to 1s\,3p\,^3P$ (3888 Å), $1s\,2p\,^3P \to 1s\,3d\,^3D$ (5875 Å), $1s\,2p\,^3P \to 1s\,3s\,^3S$ (7065 Å), and $1s\,2p\,^3P \to 1s\,4d\,^3D$ (4471 Å). There are also transitions arising from the first excited levels of singlet systems like $1s\,2s\,^1S \to 1s\,3p\,^1P$ (5015 Å). The high-excitation energy of the lower level in these transitions requires that the temperature be approximately 1.4×10^4 K. At still higher temperatures ($T \approx 2.5 \times 10^4$ K), helium becomes ionised. Because the energy-level structure of singly ionised helium is identical to that of hydrogen, except for multiplication by a factor $Z^2 = 4$, a family of lines of HeII coincides with that of hydrogen. As the temperature of the stellar atmosphere increases beyond 10^4 K, hydrogen lines weaken, HeI lines begin to appear, and later are replaced with HeII lines in still hotter stars.

2.5 Observational Aspects of Stellar Atmospheres

In the visible band, these lines dominate the spectra of O and B stars, although some of the lines from SiII, SiIII, and SiIV are present. In the UV, resonance lines of CIV (1548 Å) and SiIV (1393 Å) are very strong. As the temperature of the stellar atmosphere decreases, different sequences of lines come into play, as indicated in Table 2.2.

The ionisation of different elements in the stars also plays a crucial role in deciding the relative strength of different lines seen in the star. To illustrate this point, let us compare the relative strength of the absorption line that is due to hydrogen (the Balmer lines) and those that are due to, say, calcium (CaII K line). This calcium line has a wavelength of $\lambda = 3933$ Å and corresponds to the transition with energy difference $E_2 - E_1 = 3.12$ eV. The degeneracies for these two states are $g_1 = 2$ and $g_2 = 4$, and the partition functions for the calcium atom are $Z_I = 1.32$ and $Z_{II} = 2.30$. The ionisation energy of CaI is 6.11 eV, which is roughly half that of hydrogen.

Given these facts, we can compute the relative fraction of hydrogen and calcium atoms that are capable of producing the required absorption line. Consider the fraction of hydrogen atoms first; the ratio of ionised to neutral hydrogen is given by

$$\left[\frac{N_{II}}{N_I}\right]_H = \frac{2k_B T Z_{II}}{P_e Z_I}\left(\frac{2\pi m_e k_B T}{h^2}\right)^{3/2} e^{-\chi_1/k_B T} \cong 7.47 \times 10^{-5} \quad (2.133)$$

at the photospheric temperature $T = 5770$ K and the corresponding pressure $P_e = 15$ dyn cm^{-2} for the Sun. The result shows that only a small fraction of hydrogen is ionised at the Sun's surface. Using the Boltzmann equation, we can estimate the fraction of these atoms that will be in the first excited state. We find

$$\frac{N_2}{N_1} = \frac{g_2 e^{-E_2/k_B T}}{g_1 e^{-E_1/k_B T}} = \frac{g_2}{g_1} e^{-(E_2-E_1)/k_B T} \simeq 4.96 \times 10^{-9}. \quad (2.134)$$

Only a very small fraction of hydrogen atoms in the excited state are capable of producing Balmer lines:

$$\frac{N_2}{N_{\text{total}}} = \left(\frac{N_2}{N_1 + N_2}\right)\left(\frac{N_I}{N_{\text{total}}}\right) \cong \frac{N_2}{N_1} \simeq 4.96 \times 10^{-9}. \quad (2.135)$$

The first equality follows because most of the neutral hydrogen atoms are in state 1 or 2 so that $N_1 + N_2 \simeq N_I$; the second equality arises because $N_2 \ll N_1$ and $N_{II} \ll N_I$.

The situation is quite different as regards calcium. The ratio of ionised to neutral calcium on the solar surface is given by

$$\left[\frac{N_{II}}{N_I}\right]_{Ca} = \frac{2k_B T Z_{II}}{P_e Z_I}\left(\frac{2\pi m_e k_B T}{h^2}\right)^{3/2} e^{-\chi_1/k_B T} = 903, \quad (2.136)$$

where we have used the parameters mentioned above to perform the calculation. This shows that practically all the calcium atoms are ionised and are in the form of CaII. We can now use the Boltzmann equation to estimate the fraction of these calcium ions that are in the ground state capable of producing the CaII lines. The ratio of the number of CaII ions in the first excited state to those in ground state is

$$\left[\frac{N_2}{N_1}\right]_{\text{CaII}} = \frac{g_2}{g_1} e^{-(E_2-E_1)/k_B T} = 3.77 \times 10^{-3}. \tag{2.137}$$

This shows that almost all the calcium atoms are in the ground state and are capable of producing the CaII lines. Combining the results, we find that

$$\left[\frac{N_1}{N_{\text{total}}}\right]_{\text{CaII}} \simeq \left[\frac{N_1}{N_1+N_2}\right]_{\text{CaII}} \left[\frac{N_{\text{II}}}{N_{\text{total}}}\right]_{\text{Ca}} \cong 0.995, \tag{2.138}$$

which is nearly 100%. There are approximately 5×10^5 more hydrogen atoms than calcium atoms on the surface of Sun; but of these, only a fraction, 4.96×10^{-9}, are capable of producing the Balmer line. Multiplying these two factors, we get a value of $\sim 0.0025 = 1/400$; in other words, there are 400 times more CaII ions capable of producing CaII K absorption lines compared with those for the hydrogen Balmer lines.

Similar conclusions can be arrived at for the lines of different elements and different ionisation states. The temperature sensitivity of the states of excitation and ionisation allows us to use the observed line strengths for classifying the stars. The variation of some of the prominent lines as a function of temperature is shown in Fig. 2.13.

The line intensities of the different elements also allow us to determine the abundance of elements in a star. In general, this is done from the curve of growth discussed in Vol. I, Chap. 7, which gives the relation between the column density N_a and the equivalent width of the line. As an example, consider the two sodium lines that are produced when an electron makes an upward transition from the ground state of the neutral NaI atom. These lines have the wavelengths 3302.38 and 5889.97 Å. The observed equivalent widths W of these two lines in solar atmosphere are 0.088 and 0.073 Å, respectively. Given the curve of growth (which is essentially a plot of log W against log N; see Vol. I, Chap. 7, Section 7.3) and observed equivalent widths, we can obtain the abundance N. In this particular case, such an analysis gives log $N = 15.05$ and 14.95 for the two lines. Taking an average, we conclude that the column density of NaI atoms is $\sim 10^{15}$ cm^{-2}. This gives the number of absorbing sodium atoms in the ground state. From the excitation energy of sodium and the surface temperature of the Sun, we can easily show that the Boltzmann factor is approximately 5×10^{-4} and 10^{-2} for the two lines. Hence nearly all the neutral NaI atoms are in the ground state. To

2.5 Observational Aspects of Stellar Atmospheres

Table 2.3. Most abundant elements of the solar photosphere

Element	Atomic Number	Log Relative Abundance	Column Density (gm cm^{-2})
Hydrogen	1	12.00	1.1
Helium	2	10.99	4.3×10^{-1}
Oxygen	8	8.93	1.5×10^{-2}
Carbon	6	8.60	5.3×10^{-3}
Neon	10	8.09	2.7×10^{-3}
Nitrogen	7	8.00	1.5×10^{-3}
Iron	26	7.67	2.9×10^{-3}
Magnesium	12	7.58	1.0×10^{-3}
Silicon	14	7.55	1.1×10^{-3}
Sulphur	16	7.21	5.7×10^{-4}

determine the number of singly ionised sodium atoms N_{II}, we use the ionisation equation

$$\frac{N_{II}}{N_I} = \frac{2k_B T Z_{II}}{P_e Z_I} \left(\frac{2\pi m_e k_B T}{h^2}\right)^{3/2} e^{-\chi_I/k_B T}, \quad (2.139)$$

with $Z_I = 2.4$, $Z_{II} = 1$, and $\chi_I = 5.14$ eV. This gives the ratio $N_{II}/N_I = 2.43 \times 10^3$. Because the ionisation energy for NaII is ~ 47 eV, it is easy to see that the higher states of ionisation can be neglected. The column density of total number of sodium atoms in the solar photosphere is $N \approx 2.43 \times 10^3 N_I \approx 2.43 \times 10^{18}$ cm^{-2}. A similar analysis can be performed from the equivalent width and curves of growth for different elements. These lead to the results given in Table 2.3 for the relative abundance of some of the prominent elements in the solar photosphere.[7]

Exercise 2.14
Linewidths in solar atmosphere: Estimate the natural, Doppler, and collisional linewidths for hydrogen in the solar photosphere. [Answer: The lifetime of an excited state of hydrogen is approximately $\Delta t = 10^{-8}$ s. From the uncertainty principle, this leads to a natural linewidth $\Delta \lambda \approx (\lambda^2/2\pi c \Delta t) \approx 4.6 \times 10^{-4}$ Å for the H$_\alpha$ line with $\lambda = 6536$ Å. The Doppler broadening is given by $(\Delta \lambda/\lambda) \approx (v/c)$ with $v^2 = (2k_B T/m)$. At the Sun's photosphere with $T = 5770$ K, this leads to $\Delta \lambda \approx 0.43$ Å for the same line. The collisional broadening gives a linewidth $\Delta \lambda/\lambda \approx (\lambda/\pi c \Delta t_c)$, where $\Delta t_c \approx (n\sigma v)^{-1} \approx (n\sigma)^{-1}(2k_B T/m)^{-1/2}$. In the photosphere, with $n \approx 1.5 \times 10^{17}$ cm^{-3}, this will give $\Delta \lambda \approx 2.4 \times 10^{-4}$ Å.]

Exercise 2.15
Practice with hydrogen lines: The H^- ion has a binding energy $E = -0.75$ eV. (a) Use Saha's equation to show that in the Sun's photosphere $N(H^-)/N(H) \approx 10^{-7}$. (b) In the visible band, only the hydrogen atoms in $n = 3$ contribute to continuum absorption. Show that $N(H; n = 3)/N(H; n = 1) \approx 6 \times 10^{-10}$. This shows why H^- absorption is nearly 100 times stronger. (c) Show that $N(H; n = 2)/N(H; n = 1) \approx 10^{-8}$. Therefore $N(H; n = 2)/N(H^-) \approx 0.1$, and hence Balmer continuum absorption is not completely negligible for $\lambda < 3647$ Å.

3
Stellar Evolution

3.1 Introduction

This chapter deals with several time-dependent stellar phenomena and – in particular – with the time evolution of stellar structures.[1] It uses the results of the last chapter extensively and also draws on the material covered in Chaps. 5, 8, 10, and 12 of Vol. I.

In the last chapter we discussed the time-independent equilibrium configuration for stars, which were treated as self-gravitating bodies with ongoing nuclear reactions in the core. These stars have characteristic masses in the range $(0.1–60)\,M_\odot$ and central temperatures that are higher than $\sim 10^7$ K. Because nuclear reactions can fuel an object for only a finite period of time, of the order of $t_{\rm nuc} \approx 10^{10}(M/M_\odot)^{-2.5}$ yr [see Eq. (2.31) of Chap. 2], it is clear that any particular star must have formed at some *finite* time in the past. Similarly, the nuclear reactions will be able to provide a steady state for the star for only a finite period into the future. The structure of the star must evolve over time scales comparable with the nuclear-reaction time scale.

In studying such evolution, there are three phases that are best addressed individually. To begin with, we have to understand how the stars of different masses form out of gas in the interstellar medium (ISM). Second, we should follow the structural changes in the star as the nuclear reactions that power the star evolve in time. Finally, we need to address the end products of stellar evolution after they reach a steady state.

Of these three issues, the second and the third are better understood than the first. We do not currently have a comprehensive theory of star formation that allows us to understand the initial epochs in the history of the star. We shall discuss this topic briefly in the next section; some more issues related to star formation will be covered in Chap. 9 along with the discussion of the ISM. As regards the time-dependent changes that take place in a star as nuclear reactions evolve, there is a significant amount of data available both from numerical studies and from observations. The bulk of Sections 3.3–3.6 will concentrate on this issue.

These studies also reveal that the end products of stellar evolution can be of very different natures, depending on the mass of the star. Because of this diversity, it is necessary to study the late stages of evolution as well as the final stellar remnants separately. This will be done in Chaps. 4–6.

The evolution of a star can be significantly influenced if it is in close proximity to another star, as a part of a binary system. The evolution of such binary stars proceeds very differently from the evolution of single stars. This chapter concentrates on the evolution of isolated stars; binary-star evolution will be covered in Chap. 7.

3.2 Overview

The key physical processes that govern stellar evolution are fairly simple; however, these processes can occur in different contexts and in different combinations, leading to a wide variety of behaviour. Considering this complexity, it is important to have a general understanding of the overall trend in stellar evolution before discussing the specifics. We shall first provide such a general picture.

A system that is collapsing under self-gravity releases the gravitational potential energy as it contracts. If this energy can be trapped inside the body (because of some physical process) then the thermal energy of the body will increase. Depending on the equation of state of the body, this can lead to changes in the pressure, composition, etc. If the temperature increases sufficiently, then nuclear reactions can be triggered inside the body, thereby allowing a star to be formed out of a contracting mass of gas or to start a fresh set of nuclear reactions in the core. We have seen in the last chapter that stars are powered by the nuclear reactions that are taking place in their core region. When the nuclear reactions start in the core for the first time, the cloud will have a reasonably homogeneous chemical composition.

The nuclear reactions can sustain the star for a fairly long time in (approximate) steady state, as described in the last chapter. The next phase of evolution occurs when the nuclear reactions have progressed sufficiently, changing the structure of the stellar core. Let us assume that, during a particular phase of evolution, an element A is converted into an element B, providing the nuclear energy. When element A has been exhausted in the core region of the star, this production mechanism will cease to be operational; this causes the *local* luminosity $L(r)$, and hence the temperature gradient, to vanish. At this stage, we will be left with an isothermal core of material B covered by the rest of the stellar envelope made predominantly of A. In other words, nuclear reactions have introduced inhomogeneity in the distribution of elements inside the star.

The response of the core, which has no mechanism for energy production, will be to contract under its own weight, thereby reducing the core radius and increasing the core temperature. If the electrons in the core are not degenerate, then this process will be rapid; if electron degeneracy pressure is important, the core contraction could be slower or could even be halted. It is also possible that, at this stage, the temperature is still sufficiently high near the outer surface of the core region so that the nuclear reaction converting A to B continues in a thin shell around the core. This has the effect of adding material B to the core and accelerating core collapse.

Further evolution depends on the maximum temperature reached by the contracting core. If the core is degenerate and does not reach sufficiently high temperatures for element B to undergo further nuclear reactions, then the star has exhausted its energy supply and rapidly settles down to a final state made of the degenerate core. In this final configuration, degeneracy pressure balances gravity.

If the core reaches a sufficiently high temperature for element B to undergo further nuclear reactions, then the star is provided with a fresh energy source. If this happens, the process can now repeat with element B's converting into element C. (This will eventually lead to elements C, B, and A's existing in the star in successive layers from the centre to surface.) The response of the core to the fresh ignition of element B, in turn, will depend on the equation of state of the core material. If the core material is made of nearly ideal gas, then the increase in temperature will lead to an increase in pressure and consequent dynamical readjustment. If the electrons in the core were degenerate initially, then the nuclear burning will increase the temperature without significantly changing the pressure. In such a case, the nuclear burning will be a runaway process (usually called a flash), with the temperature shooting up rapidly in the core, until the degeneracy is removed.

This kind of evolutionary case can now repeat until the nuclear-energy supply is completely exhausted. Several other complications can arise in the intermediate stages (e.g., the element in a shell undergoing nuclear reaction, oscillations, instability), depending on the details.

The contraction of the core, by and large, has a tendency to make the envelope expand outwards, which is often a consequence of the virial theorem. Such an expansion has several observable consequences, as the star will now bloat up in size. The expansion will also cool the outer layers and will induce convective instability. As the outer convection zone moves towards the centre, it mixes up the composition and elements synthesised in the core will be redistributed such that even the photosphere will show traces of these elements.

The exhaustion of nuclear fuel at the core, contraction of the core, expansion of the envelope, degenerate or nondegenerate burning of fuel in the core and in the shells – these constitute the essential ingredients of stellar evolution. By and large, the evolutionary processes in a star are governed by the processes described above, with the mass of the star playing the crucial role in controlling the number of cycles of nuclear reactions that a star undergoes. We shall see specific examples of these phenomena in different combinations in the subsequent sections.

3.3 Pre-Main-Sequence Collapse

The discussion in the last chapter shows that stars can be modelled as self-gravitating clouds of gas fuelled by nuclear reactions. We shall see in Chap. 9 that the ISM contains large molecular clouds of masses in the range $(10^5–10^6)$ $M_\odot$, with temperatures of $(10–100)$ K and densities of $(10–10^4)$ cm^{-3}. To form stars by collapsing such gas clouds requires an increase in density by a factor of $\sim 10^{24}$ and an increase in temperature by a factor of $\sim 10^6$. This is an inherently complex process, and many of the details are still unclear. We shall provide below a simple discussion of a possible scenario and will comment on the complications towards the end of the section.[2]

3.3.1 Gravitational Instability and Mass Scales

If the effects of rotation and magnetic field are ignored and the collapse of the gas cloud is modelled as spherically symmetric, then the equation of motion for a shell of radius $r(m, t)$ enclosing a mass $m = m(r)$ is given by

$$\frac{\partial^2 r}{\partial t^2} = -\frac{Gm(r)}{r^2} - \frac{1}{\rho}\frac{\partial P}{\partial r} = -\frac{Gm}{r^2} - 4\pi r^2 \frac{\partial P}{\partial m(r)}. \tag{3.1}$$

The cloud will be in equilibrium if the two terms on the right-hand side balance each other; it will expand because of the pressure if the second term dominates in a characteristic time scale

$$t_s \approx (R/c_s) \propto (m/\rho)^{1/3} T^{-1/2} \propto m^{1/3} \rho^{-1/3} T^{-1/2}, \tag{3.2}$$

where $c_s \simeq (P/\rho)^{1/2}$ is the speed of sound in the gas. It will collapse under self-gravity if the first term dominates on the right-hand side in a time scale

$$t_{\rm ff} \approx \left(\frac{Gm}{r^3}\right)^{-1/2} \propto (G\rho)^{-1/2}. \tag{3.3}$$

This consideration suggests that clouds with large densities (which imply large m) will have $t_{\rm ff} \ll t_s$ and will be unstable for collapsing under self-gravity. The same result can be obtained in a different manner along the following lines: The gravitational potential energy of a cloud of mass M and radius R is $U = -(3/5)GM^2/R$ while the thermal kinetic energy is $(3/2)Nk_BT$, where $N = (M/\mu m_H)$ and μ is the mean molecular weight. It was shown in Vol. I, Chap. 8, that such a cloud is unstable to collapse if the magnitude of the gravitational potential energy is larger than (approximately) twice the kinetic energy. This gives the condition for collapse as

$$\frac{3Mk_BT}{\mu m_H} < \frac{3}{5}\frac{GM^2}{R}. \tag{3.4}$$

3.3 Pre-Main-Sequence Collapse

Eliminating the radius of the cloud in terms of the mean density $\rho_0 = (3M/4\pi R^3)$, we can express this criterion as $M > M_J$, where

$$M_J \simeq \left(\frac{5k_B T}{G\mu m_H}\right)^{3/2} \left(\frac{3}{4\pi \rho_0}\right)^{1/2}. \tag{3.5}$$

The quantity M_J is called the *Jeans mass*. Numerically, we find that

$$M_J \cong 1.2 \times 10^5 \, M_\odot \left(\frac{T}{100 \, K}\right)^{3/2} \left(\frac{\rho_0}{10^{-24} \, \text{gm cm}^{-3}}\right)^{-1/2} \mu^{-3/2}. \tag{3.6}$$

We shall see in Chap. 9 that the density $\rho \approx 10^{-24}$ gm cm^{-3} and temperature $T \approx 100$ K are reasonable for clouds in the ISM. In such a case, the above result shows that only objects with $M \gtrsim 10^5 \, M_\odot$ can collapse initially.

When such a large cloud starts collapsing under self-gravity, its physical parameters will change during the course of the collapse, thereby changing the Jeans mass. If the Jeans mass decreases as the cloud collapses, then subregions inside the original cloud can become locally unstable and start collapsing themselves. This process will continue with smaller and smaller inhomogeneities undergoing gravitational collapse as long as the Jeans mass continues to decrease. Therefore it is possible that lower-mass objects, which should act as pregenitors of stars, arise because of such a secondary phenomenon – namely, fragmentation inside the collapsing cloud.

The Jeans mass in relation (3.6) depends on the temperature and density as $M_J \propto T^{3/2} \rho^{-1/2}$. For a collapsing cloud, ρ keeps increasing; whether M_J increases or decreases will depend crucially on the evolution of the temperature T. As the cloud collapses, it releases gravitational potential energy that can either (1) increase the internal energy of the material and cause pressure readjustment or (2) be radiated away if the time scale for cooling t_{cool} is shorter than t_{ff}. If the cloud manages to radiate away the gravitational energy that is released, then it can contract almost isothermally. Expression (3.5) shows that, for isothermal contraction, M_J varies as $\rho^{-1/2}$. As the star contracts, ρ increases and the Jeans mass decreases, allowing smaller regions inside the gas cloud, containing fragments of smaller mass, to collapse on their own gravitationally. Under most of the interstellar conditions, the collapse time scale $t_{ff} \propto (G\rho)^{-1/2}$ is much larger than the cooling time scale of the gas (see Chap. 9 for a detailed discussion of heating and cooling of interstellar clouds). Hence it is reasonable to assume that the collapse of the cloud will be isothermal in the initial stages of the contraction.

This process will cease when the condition for isothermality is violated and the Jeans mass starts increasing. This will occur when, at some stage in the evolution, the cloud becomes optically thick to its own radiation. After this stage, the collapse will be adiabatic rather than isothermal. For an ideal, monatomic gas with $\nabla_{ad} = 2/5$, the temperature scales as $T \propto P^{2/5} \propto \rho^{2/3}$ during adiabatic compression. Then $M_J \propto T^{3/2} \rho^{-1/2} \propto \rho^{1/2}$, which means that the Jeans mass will grow with time.

This fact allows us to determine when the process of fragmentation will cease to be effective. The smallest mass scale that can form because of fragmentation will correspond to the Jeans mass at the moment when the cloud makes a transition from isothermal to adiabatic evolution. This lower limit can be estimated as follows. During the free-fall time of a fragment, $t_{\rm ff} \simeq (G\rho)^{-1/2}$, the total energy radiated away is of the order of $E \approx (GM^2/R)$. To maintain isothermality, the rate of radiation of energy has to be

$$A \approx \frac{E}{t_{\rm ff}} \simeq \frac{GM^2}{R}(G\rho)^{1/2} = \left(\frac{3}{4\pi}\right)^{1/2} \frac{G^{3/2} M^{5/2}}{R^{5/2}}. \tag{3.7}$$

However, we have seen in Vol. I, Chap. 6, that a body in thermal equilibrium at temperature T cannot radiate at a rate higher than that of a blackbody at the same temperature. (Because the microscopic time scales in a collapsing cloud are sufficiently small, the cloud may be assumed to be in local thermodynamic equilibrium). Hence the rate of radiation loss by a cloud fragment can be written as

$$B = (4\pi R^2)(\sigma T^4) f, \tag{3.8}$$

where f is a factor less than unity. For isothermal collapse, $B \gg A$, and a transition to adiabaticity occurs when $B \approx A$. This happens when

$$M^5 = \left[\frac{64\pi^3}{3}\right]\left[\frac{\sigma^2 f^2 T^8 R^9}{G^3}\right]. \tag{3.9}$$

The fragmentation would have reached its limit when the Jeans mass is equal to this value: $M_J = M$. Replacing M with M_J and expressing R in terms of the density, we can estimate the numerical value as

$$M_J = \left(\frac{3 \times 5^9}{64\pi^3}\right)^{1/4} \frac{1}{(\sigma G^3)^{1/2}} \left(\frac{k_B}{\mu m_{\rm H}}\right)^{9/4} f^{-1/2} T^{1/4} = 0.02\, M_\odot \frac{T^{1/4}}{f^{1/2} \mu^{9/4}}, \tag{3.10}$$

where T is in Kelvin. For $T \lesssim 10^3$ K and $f \approx 0.1$, we get $M \approx 0.36\, M_\odot$; this result does not change much if the parameters are varied within a reasonable range because of the weak dependence on T and f. Thus the collapse of a cloud can lead to fragments with masses of the order of the solar mass or above, but not significantly below.

3.3.2 Collapse of a Spherical Cloud

We shall now consider the spherical-collapse scenario in greater detail and describe the key physical processes that take place. Let us consider a region

3.3 Pre-Main-Sequence Collapse

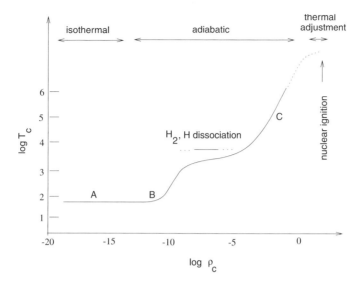

Fig. 3.1. Variation of central density and temperature in a collapsing protostellar cloud and the different physical processes taking place in it.

containing a mass of approximately $M = M_\odot$ that becomes unstable to gravitational collapse at some stage in the collapse of a larger cloud and contracts to form a virialised lump of matter with a mean temperature T_i. Given the mass and the initial temperature $T_i \approx 50$ K, we can estimate the initial radius R_i of the cloud from the condition for virial equilibrium, $(M_\odot v_{\rm rms}^2/2) \approx (3/5)(GM_\odot^2/R_i)$. For $T_i = 50$ K, $v_{i,\rm rms} \approx 10^5$ cm s^{-1}, we get $R_i \approx 2 \times 10^5 \, R_\odot$. The corresponding particle density is $n_i \approx (3 \, M_\odot/4\pi \mu m_{\rm H} R_i^3) \approx 10^8 \, \mu^{-1}$ cm^{-3} and the luminosity is $L_i = 4\pi \sigma R_i^2 T_i^4 \approx 10^2 \, L_\odot$. The initial position of the cloud is marked A in Fig. 3.1. The radiation from such a cloud will peak in the IR band. The central temperature will not be too different from the surface temperature at the initial stage.

Both observational and theoretical considerations suggest that the initial composition of the cloud will be mostly hydrogen (in molecular form, H$_2$), helium, and smaller quantities of higher elements. Depending on the initial conditions, the cloud will also contain traces of more complex molecules. Let us begin with the time evolution of such a collapsing cloud during the initial phase, in which the gravitational potential energy released because of contraction is efficiently radiated away. In this case, the cloud will contract isothermally, and the trajectory will be a horizontal line in Fig. 3.1 (and a vertical line in Fig. 3.2), marked AB. The temperature changes very little but R decreases, making $L \propto R^2 T_{\rm eff}^4$ decrease.

Because the pressure gradient per unit mass is $\rho^{-1}|(\partial P/\partial r)| \approx (P/\rho R) \propto (T/R)$ and the gravitational force per unit mass is GM/R^2, it is clear that gravity dominates as R decreases at constant M, T. Ignoring for simplicity the pressure

120   3 Stellar Evolution

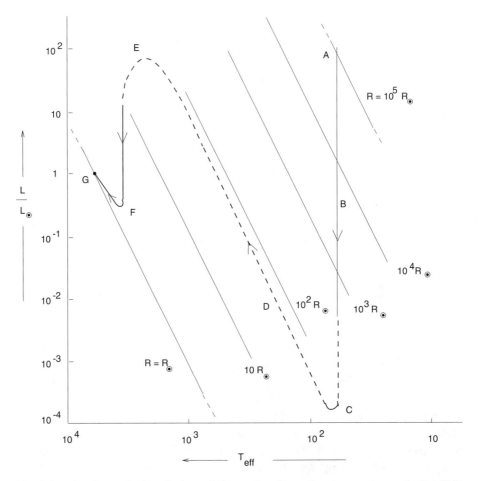

Fig. 3.2. A schematic description of the track of a collapsing protostar in the H-R diagram. See text for the description of different phases.

forces, we find that the evolution of the radius $r(t, m)$ of a sphere containing mass m is given by

$$\ddot{r} = -\frac{Gm}{r^2}. \tag{3.11}$$

The solution to this equation can be expressed in parametric form as

$$r = r_0 \cos^2 \zeta; \quad t = \left(\frac{8\pi G \rho_0}{3}\right)^{-1/2} \left(\zeta + \frac{1}{2} \sin 2\zeta\right), \tag{3.12}$$

where ρ_0 is the mean initial density within the sphere containing mass m and r_0 is the initial radius of this sphere. In such a case, all the mass contracts to a point

3.3 Pre-Main-Sequence Collapse

at $\zeta = \pi/2$ at $t = t_{\rm ff}$, where

$$t_{\rm ff} = \left(\frac{3\pi}{32\, G\rho_0}\right)^{1/2} \simeq 4.7 \times 10^3 \text{ yr} \left(\frac{\rho_0}{2 \times 10^{-16} \text{ gm cm}^{-3}}\right)^{-1/2}. \quad (3.13)$$

This is the time scale for isothermal evolution along AB in Fig. 3.1. For a dense molecular cloud with $T = 50$ K, $n = 10^8$ cm^{-3}, $\rho \simeq m_{\rm H} n_0 \simeq 2 \times 10^{-16}$ gm cm^{-3}, the free-fall time scale is $t_{\rm ff} \simeq 4700$ yr.

In any realistic case, the density of the cloud will not be a constant but will decrease outwards so that $t_{\rm ff}$ will be smaller for the denser, inner core. Then, as a cloud collapses, a central concentration will be formed at some moment; this concentration enhances itself because the local free-fall time ($t_{\rm ff} \propto \rho^{-1/2}$) is smaller near the denser central region than at the outer region. Because the radiation will find it harder to escape from the centre than from the surface, the central regions of the collapsing cloud will become opaque first and the collapse will be halted first in the central regions. The opacity is contributed mainly by the dust in the interstellar cloud (see Vol. I, Chaps. 9 and 6). If the opacity is $\kappa \approx 10^{-2}$ cm^2 gm^{-1}, then the central region will become opaque when $\kappa \rho R = 1$; using $R = (3\, M/4\pi \rho)^{1/3}$, we get $\rho_c \approx 10^{-13}$ gm cm^{-3}. The increase in pressure, when the cloud becomes opaque, substantially slows down the rate of collapse near the core. This leads to the formation of a core (usually called a protostar), with freely falling gas surrounding it. In this case, the problem reduces to that of accretion onto a central object, which was discussed in Vol. I, Chap. 10. The density profiles of the outer regions depend on whether the gravitational force is dominated by the central concentration or by the self-gravity of the cloud. When the self-gravity of the freely falling cloud can be ignored, the density profile of the outer regions can be determined by the following argument. In a steady radial in-fall of matter, the rate of mass flow $\dot{m} = 4\pi r^2 \rho v$ remains constant. If the velocity v is essentially generated by the core mass M, then $v = v_{\rm ff} \approx (2GM/r)^{1/2}$, giving

$$\rho(r) \propto r^{-3/2}. \quad (3.14)$$

In the other extreme limit, if we ignore the gravitational force of the central object and treat the rest of the cloud as a homologously collapsing isothermal sphere, then the density profile is given by

$$\rho(r, t) \propto r^{-2}. \quad (3.15)$$

(This is derived in Vol. I, Chap. 10.) During the initial phase, we expect the outer regions to follow a profile that is approximately r^{-2}, changing over to $r^{-3/2}$ gradually.

Once the cloud becomes optically thick, the gravitational potential energy, released during the collapse, is absorbed by the envelope and is radiated away as IR radiation. When the in-falling material hits the hydrostatic core, a shock wave will develop because the speed of the material exceeds the local sound speed. The kinetic energy of the in-falling material is lost at this shock front, which also goes to contribute to the cloud's luminosity. An estimate for the accretion rate can be obtained as

$$\dot{M} \approx \frac{M(r)}{t_{\text{ff}}(r)} \approx \frac{(v_{\text{esc}}^2 r/G)}{(r/v_{\text{esc}})} = \frac{v_{\text{esc}}^3(r)}{G}, \quad (3.16)$$

where $v_{\text{esc}}^2(r) \approx [GM(r)/r]$ is the escape velocity at radius r. The supersonic accretion occurs with $v_{\text{esc}} \approx c_s$, giving the accretion rate

$$\dot{M} \approx \frac{c_s^3}{G} \approx 2 \times 10^6 \left(\frac{T}{10\,\text{K}}\right)^{3/2} M_\odot\,\text{yr}^{-1}. \quad (3.17)$$

The mass and the density profiles in this case are given by $M(r) = (2rc_s^2/G)$ and $\rho(r) \approx (c_s^2/2\pi G r^2)$, respectively. With the above accretion rate, the time scale to build a protostar of mass M is given by $t_* \approx 5 \times 10^5 (M/M_\odot)(T/10\,\text{K})^{-3/2}$ yr.³

Several compositional changes occur in the core during the next stage of evolution. Initially, the gas consists (predominantly) of atomic and molecular hydrogen. When the central temperature reaches ~ 2000 K, the hydrogen molecules dissociate. During this phase, the energy supplied to the gas goes mainly into dissociation of the molecules, leading to a decrease of ∇_{ad}. For hydrogen molecules with $f = 5$ degrees of freedom, $\nabla_{\text{ad}} = (f+2)/f = 7/5 = 1.4$. This is quite close to the critical value of $4/3 \cong 1.33$ (see Vol. I, Chap. 8) for instability, and hence a slight reduction in ∇_{ad} can lead to an instability. The core becomes unstable, and the protostar starts to collapse again. When molecular dissociation in the core is complete, the temperature begins to increase again.

The evolution of central temperature and density of the cloud is shown schematically in Fig. 3.1. The initial phase is isothermal; for a $1 M_\odot$ cloud the central region becomes opaque when the central density is $\sim 10^{-13}$ gm cm^{-3}, the central temperature is $T \approx 50$ K, and the radius is 5 AU (AU stands for astronomical units). When the central core reaches hydrostatic equilibrium, it has a mass of $0.01\,M_\odot$, a central density of $\rho_c \simeq 2 \times 10^{-10}$ gm cm^{-3}, and a temperature of $T_c \simeq 170$ K. When the cloud becomes opaque, the temperature rises adiabatically with a slope of 0.4 corresponding to $\nabla_{\text{ad}} = 1.4$ for H_2; soon H_2 dissociates at nearly constant temperature followed by a phase in which the temperature increases adiabatically again with a slope of $2/3$ (corresponding to $\nabla_{\text{ad}} = 5/3$ for hydrogen gas). During the second core collapse, another shock front is established in the envelope. However, because the amount of mass available from the original cloud is substantially exhausted by now, the luminosity eventually decreases.

The ionisation process can be used to estimate the initial radius of the cloud (around C in Fig. 3.2) when the core is formed. We note that, during the quasi-static evolution, half the total gravitational energy released in the collapse goes

3.3 Pre-Main-Sequence Collapse

into the internal energy of the cloud and is used to dissociate H_2 molecules and ionise hydrogen and helium. The total energy needed for these processes can be estimated as (number of atoms of a given species/gm) × (mass fraction of the species) × (energy needed for dissociation or ionisation) for each species. The total will be

$$E_{\text{ion}} = (N_A X) E_H + \frac{1}{4}(N_A Y) E_{\text{He}} + \frac{1}{2}(N_A X) E_D = 1.9 \times 10^{13}(1-0.2X) \text{ ergs}, \quad (3.18)$$

where $E_H = 13.6$ eV is the ionisation energy of hydrogen, $E_D = 4.48$ eV is the dissociation energy of the hydrogen molecule, $E_{\text{He}} = E_{\text{HeI}} + E_{\text{HeII}} = 78.98$ eV is the total ionisation energy of helium, and N_A is Avogadro's number. The energy liberated when the cloud collapses from a large radius to R is approximately $(GM^2/2R)$. Equating the two, we find the maximum initial radius of the ionised cloud (at C in Fig. 3.2) to be

$$\frac{R}{R_\odot} = \frac{43.2(M/M_\odot)}{1-0.2X} \approx 50 \frac{M}{M_\odot}. \quad (3.19)$$

The corresponding mean temperature at C (which is mainly contributed by the central region) is $\bar{T} \simeq T_c \simeq (GM m_H / k_B R)$, so that

$$T_c \approx 3 \times 10^5 \mu (1-0.2 X) \text{ K} \approx 10^5 \text{ K}. \quad (3.20)$$

Thus, for $M = 1\, M_\odot$, the contraction along ABC (of Fig. 3.2) changes the radius from $10^5\, R_\odot$ to $\sim 50\, R_\odot$; the corresponding change in luminosity $L \propto T_{\text{eff}}^4 R^2$ at constant T_{eff} is by a factor $(10^5/50)^2 \approx 4 \times 10^6$, so that the luminosity at C is $\sim 10^{-4}\, L_\odot$.

Figure 3.2 shows the variation of T_{eff} and L for the contracting cloud schematically. T_{eff} does not change much during the phase ABC but changes drastically during the phase CDE for two reasons. The first reason has to do with the fact that, as the evolution proceeds, the photosphere moves inwards. We determine the effective temperature (as in the previous chapter) by computing the radius at which the optical depth τ is $\sim 2/3$. When the temperature is $\sim 10^3$ K, the dust begins to vaporise and the opacity drops. Consequently the radius of the photosphere (where $\tau = 2/3$) decreases substantially, approaching the surface of the hydrostatic core. Because the luminosity $L \propto R^2 T_{\text{eff}}^4$ remains constant during this phase, the effective temperature T_{eff} increases with decreasing R, reaching ~ 4000 K. (Because of these complications, the trajectory in the H-R diagram, corresponding to the part shown the dashed curve, is uncertain and model dependent.) During this part of evolution, a strong temperature gradient is established in the cloud, bringing convection into operation. This will also lead to higher values of T_{eff}. A change of T_{eff} by a factor of $4000/50 = 80$ will cause the luminosity to increase by $80^4 \approx 10^6$ from the initial value of $10^{-4}\, L_\odot$ (corresponding to point B) to a value of $10^2\, L_\odot$ at E. Hence we may estimate the values at point E to be approximately $(10\text{--}100)\,L_\odot$, $R \approx 50\,R_\odot$, and $T_{\text{eff}} \approx 4000$ K. The transition between C and E takes place very rapidly (in $\sim$300 days) because it is essentially caused by bulk gaseous motion.

As the collapse accelerates, both the luminosity and the temperature of the cloud increase. If the luminosity L is contributed by the release of gravitational energy $E_g \simeq (GM^2/R)$ in a time scale $t_g \simeq (GM/R^3)^{-1/2}$, then $L \simeq (E_g/t_g) \propto R^{-5/2}$. The effective temperature is $T_{\text{eff}} \propto (L/R^2)^{1/4} \propto R^{-9/8}$ so that

$$L \propto T_{\text{eff}}^{20/9} \propto T_{\text{eff}}^{2.2} \qquad (3.21)$$

during the free-fall collapse. This gives the initial part (CD in Fig. 3.2) of the curve in the H-R diagram. After the formation of a dense core, the luminosity is essentially decided by the nearly constant accretion rate $\dot{M}$ of the thinner outlying gas with $L \propto (GM\dot{M}/R)$. Initially, L increases because R decreases and $\dot{M}$ increases; later on, with the thinning of the outer regions, L remains constant (around E in Fig. 3.2).

We now have evolved a cloud of gas to somewhere near point E in Fig. 3.2. This cloud, which is fully convective, will move along EFG in order to reach the main sequence. The last two phases of evolution, along EF and FG, will be discussed in the next subsection.

3.3.3 Contraction onto the Main Sequence

With the steady increase of the temperature of the protostar, the opacity $\kappa \propto \rho^n T^{-s}$ of the outer layers will become dominated by the H$^-$ ion, with the extra electron being supplied by the partial ionisation of heavier elements in the original gas clouds. As in the case of regular stars, discussed in the last chapter, the high opacity causes the envelope to become convective. The depth of the convective zone depends on the mass of the cloud only very weakly. By and large the convective layer extends all the way down to the centre of the star, and we can model this phase of the evolution by using the results of Chap. 2, Subsection 2.4.3 for fully convective stars. It was shown there that such stars lie on a nearly vertical line in the H-R diagram determined by

$$T_{\text{eff}} \propto M^{(n+3)/(9n+3-2s)} L^{(3n-1)/2(9n+3-2s)}. \qquad (3.22)$$

Numerically, for H$^-$ opacity ($n = 0.5, s = -9$), we get

$$T_{\text{eff}} \propto \left(\frac{M}{M_\odot}\right)^{7/51} \left(\frac{L}{L_\odot}\right)^{1/102} \text{K}. \qquad (3.23)$$

As a star of given mass M contracts, its radius will decrease and it will move down in the H-R along the line $T_{\text{eff}} \propto L^{1/102}$. (For Kramer's opacity, with $n = 1$,

3.3 Pre-Main-Sequence Collapse

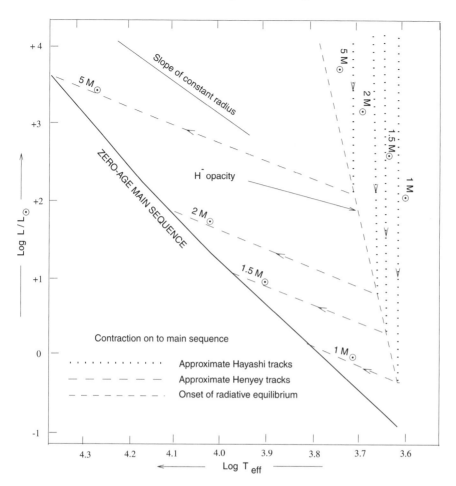

Fig. 3.3. The late stages in the collapse of a protostellar cloud towards the main sequence corresponding to the path EFG of Fig. 3.2. The nearly vertical dotted lines are the Hayashi tracks for different masses, and the sloping dashed lines are the Henyey tracks for the corresponding masses. The transition from the Hayashi to the Henyey track occurs along a line that is determined by the nature of the opacity. The line marked H^- opacity illustrates this transition.

$s = 3.5$, the corresponding result is $T_{\text{eff}} \propto M^{0.8} L^{0.2}$, which is still a weak dependence on L but not as weak as H^- opacity.) Such a line is called a Hayashi track and corresponds to the path EF in Fig. 3.2. The same line is shown in Fig. 3.3 for different values of mass.

Contraction along the Hayashi track will continue as long as the energy transport is due to convection. The evolutionary track will change when the temperature gradient favours radiative transport. This occurs at the point F in Fig. 3.2 and is also shown in Fig. 3.3 in greater detail. We shall now determine this transition point and the track along FG.

As the star moves down the Hayashi track, the internal temperature increases homologously as $T \propto (GM/R)$, and hence the actual temperature gradient will vary as $(\nabla T) \propto (GM/R^2)$. The radiative temperature gradient, for a given $L(r)$ and κ, varies as

$$\frac{dT}{dr} = -\frac{3\kappa \rho L(r)}{16\pi a c T^3 r^2} \sim \frac{\kappa(M/R^3)L}{(\mu M/R)^3 R^2} \sim \frac{\kappa L}{\mu^3 M^2 R^2}. \quad (3.24)$$

If $\kappa \propto \rho^n T^{-s}$, the ratio between the actual temperature gradient and the radiative gradient of the temperature will scale as

$$\mathcal{R} \sim \frac{M^{s-n+3}}{L R^{s-3n}}. \quad (3.25)$$

Depending on the dominant source of opacity, the radiative gradient can exceed the actual temperature gradient as the star contracts down the Hayashi track. Once this happens the convection will cease and radiative transfer of opacity will take over. Such a transition will first occur in the central region and the radiative core will propagate outwards as time proceeds. The location in the H-R diagram at which this transition occurs is given by $\mathcal{R} \approx 1$. Because T_{eff} is nearly constant along the Hayashi track, we can take $L \propto R^2$ in Eq. (3.25). Then the locus of points of constant $\mathcal{R}$ is given by

$$\frac{d \ln L}{d \ln M} = 2\left(\frac{s - n + 3}{s - 3n + 2}\right). \quad (3.26)$$

To get the slope in the L–T_{eff} plane, we need to know the derivative $(\partial \ln T_{\text{eff}}/\partial \ln M)$ at constant L. From expression (3.23) we know that this quantity is given by

$$\frac{\partial \ln T_{\text{eff}}}{\partial \ln M} = \frac{n + 3}{9n + 3 - 2s}. \quad (3.27)$$

Combining the relations (3.26) and (3.27) together we get

$$\frac{\partial \ln L}{\partial \ln T_{\text{eff}}} = \frac{2(9n + 3 - 2s)}{n + 3}\left(\frac{s - n + 3}{s - 3n + 2}\right) = \begin{cases} 39 & (\text{H}^- \text{ opacity}) \\ 5.5 & (\text{Kramer's opacity}) \end{cases}. \quad (3.28)$$

This line (also shown in Fig. 3.3) for H^- opacity marks the end of the Hayashi track and signals the onset of a radiative core.

During the convective regime, the H^- opacity of the envelope was the main barrier for radiation leakage. With the end of convection and formation of a radiative core it is the radiative opacity that determines the leakage of energy from the star. Further contraction and conversion of gravitational energy into radiation will proceed slowly but will lead to a steady increase in the temperature gradient and in the luminosity. The increased luminosity and the decreased radius will make the track in the H-R diagram move to the left and rise slightly

3.3 Pre-Main-Sequence Collapse

(FG in Fig. 3.2). In this part of the track (called the Henyey track) the luminosity of the star is mainly due to the release of gravitational potential energy; hence the analysis leading to expression (3.21) applies, and the star moves upwards and to the left on the H-R diagram with a slope of 2.2. This part of the trajectory is marked by the long-dashed lines in Fig. 3.3.

As the star evolves along the Henyey track, its core density increases. If the increase in the density leads to increasing temperature (as in the case of the ideal-gas equation of state) then the core temperature will eventually reach a value at which hydrogen fusion can take place. In such a case, the cloud reaches the appropriate point on the main sequence and is stabilised by the nuclear reactions.

The amount of time required for stars to collapse onto the main sequence is a decreasing function of the stellar mass; a $0.5 M_\odot$ star takes over 10^8 yr whereas a $15 M_\odot$ star requires only 6×10^4 yr. Figure 3.4 shows the amount of time spent by stars of different masses in the Hayashi track and in the Henyey track before reaching the main sequence. It is clear from this figure that lower-mass stars spend more time on the Hayashi track and less time on the Henyey track.[4]

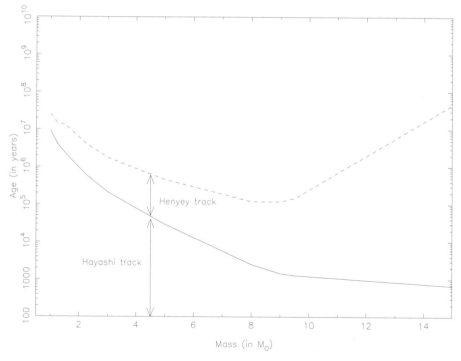

Fig. 3.4. Time spent by stars of different masses in the Hayashi and the Henyey tracks. High-mass stars spend less time on the Hayashi track, and more on the Henyey track, as is clear from Fig. 3.3 as well.

In general, lower-mass cloud fragments will be more abundant inside a given interstellar cloud. Hence the number of stars that form per unit volume per unit mass interval will be a strong function of the mass. Low-mass stars are formed in larger numbers and last longer, thereby exhibiting greater abundance. Because the process of star formation is extremely complex, it is not easy to obtain distribution functions for the number densities of stars of different mass ranges that are formed in the galaxy from first principles. One simple parameterisation (based on observations) that is extensively used is[5]

$$\psi_s(m)\,dm = 2 \times 10^{-12} m^{-2.35}\,dm \text{ pc}^{-3} \text{ yr}^{-1}, \quad m \equiv (M/M_\odot), \quad (3.29)$$

which gives the birth rate of stars of different masses in the range $0.4 \lesssim M/M_\odot \lesssim 10$.

The above discussion, as well as those in the last two subsections, suggests a possible mechanism for star formation arising from the collapse of interstellar clouds. This analysis, however, does not include effects of rotation and magnetic field – both of which have the effect of working against gravity and preventing the collapse. The existence of nonzero angular momentum for the original cloud will also cause the collapse to be axisymmetric rather than spherical and – in the extreme case – can lead to the formation of a protostellar disk of material accreting to the central core. The case described in this subsection implicitly assumes that the cloud has some efficient means of shedding the magnetic field and the angular momentum. The details of this process are, as yet, unclear.

The hot, massive stars of the O or the B spectral type formed inside a molecular cloud emit a significant amount of UV radiation that can ionise the neutral hydrogen gas in the ISM that surrounds the newborn star. In equilibrium, the rate of ionisation will equal the rate of recombination; each recombination, when it cascades through different levels of hydrogen, will lead to the emission of photons, many of which will be in the visible part of the spectrum. In particular, these regions around a newborn O or B star will be dominant sources of Balmer photons arising from the transition between $n = 3$ and $n = 2$ states. This feature is of some observational significance and will be discussed in greater detail in Chap. 9.

Exercise 3.1

Magnetic fields and transport of angular momentum: If the initial protostar has some angular momentum, then the collapse cannot be spherically symmetric. In general, the existence of angular momentum will work against gravity. Motivated by this, we would like to examine possible schemes by which a collapsing protostar can lose its angular momentum. One such possibility arises through the existence of magnetic fields, called magnetic breaking. (a) Consider a protostar that is rotating axisymmetrically with a velocity field $\mathbf{v} = r\Omega(r, z)\hat{\mathbf{e}}_\theta$. Let there be an axially symmetric magnetic field (which is independent of θ) threading through the protostar and the surrounding plasma. Show that the component of such a magnetic field can be written in the form $B_r = -(1/r)(\partial \Psi/\partial z)$, $B_z = (1/r)(\partial \Psi/\partial r)$ in terms of a scalar function $\Psi(r, z)$. In a steady-state magnetohydrodynamic (MHD) situation with highly conducting plasma, the magnetic field must satisfy the constraint $\nabla \times (\mathbf{v} \times \mathbf{B}) = 0$. Show that this is possible only if $\Omega = f(\Psi)$. Also argue that $\Psi(r, z)$ is constant along the magnetic-field lines, thereby concluding that the angular velocity should remain constant along the magnetic-field lines. (b) Consider a collapsing protostar surrounded by plasma. At some time t, let us assume that the plasma up to a distance $r = a$ is rotating with the angular velocity Ω of the collapsing cloud. Because magnetic disturbances propagate with Alfven speed, a shell of plasma between radii a and $a + v_A \delta t$ will be spun to the angular velocity Ω during an interval δt. Show that this process makes the collapsing protostar lose its angular velocity at the rate

$$\frac{2}{5} M a^2 \frac{d\Omega}{dt} = -\frac{8\pi}{3} a^4 \rho v_A \Omega. \tag{3.30}$$

Provide an approximate solution to this equation and estimate the final angular velocity for protostellar collapse.

If the core region becomes degenerate during the evolution, then the increase in density may not lead to an increase in temperature and the cloud may never make it to nuclear ignition. This effect of the equation of state on the evolution of the cloud is discussed in the next subsection.

3.3.4 Brown Dwarfs

During the contraction of a protostar, the central temperature and density rise and if they cross the threshold for nuclear reaction, then further contraction is halted in the protostar. It is, however, possible for the electrons to become degenerate at the centre of the protostar before the temperature becomes high enough to ignite nuclear reactions at a significant rate. Further contraction will now be prevented by the degeneracy pressure of the electrons, and such failed stars are called *brown dwarfs*. They will be fully convective from their centre to the photosphere.[6]

To study the structure of such stars, we first need to obtain a suitable equation of state for matter in brown dwarfs. This is a complicated problem because the matter is far from ideal at the densities we will be interested in. It is, however, possible to obtain an approximate equation of state that captures the essence of the physical processes and leads to numerical estimates that are fairly accurate. Such a model is described here, although it must be stressed that the numerical accuracy of the model is somewhat accidental.

In low-mass objects such as brown dwarfs, ions never become degenerate (unlike, for example, in neutron stars, as we will see in Chap. 5), nor do electrons become relativistic (unlike in massive white dwarfs; see Chap. 5). Therefore the pressure is provided essentially by nonrelativistic, degenerate electrons and an ideal gas of ions. For fully ionised hydrogen, treated as an ideal gas, the total pressure is $P = n k_B T = 2 \rho k_B T / m_p$ with the factor 2 coming from electrons and protons belonging to each hydrogen atom. Thus each component contributes

$P_{\text{ideal}} = (\rho k_B T/m_p) = 8.3 \times 10^7 \rho T$ in cgs units. The electron degeneracy pressure is given by Vol. I, Chap. 5, Section 5.9,

$$P_{e,\text{deg}} = (3\pi^2)^{2/3} \left(\frac{\hbar^2}{m_e}\right) \left(\frac{\rho}{m_p \mu_e}\right)^{5/3} \simeq 10^{13} \rho^{5/3}, \qquad (3.31)$$

in cgs units; μ_e is the electronic contribution to the molecular weight (defined in Vol. I, Chap. 5, Section 5.6),

$$\mu_e = \sum_j \left(\frac{Z_j}{A_j}\right) X_j, \qquad (3.32)$$

where the sum is over the species of particles contributing the free electrons. This expression assumes that electrons are fully degenerate, which is valid if $k_B T$ is negligible compared with ϵ_F; numerically, this is equivalent to the condition $T < 3 \times 10^5$ K $(\rho/1 \text{ gm cm}^{-3})^{2/3}$. We will now define a parameter

$$\xi = \frac{P_{\text{ideal}}}{P_{e,\text{deg}}} = 8 \times 10^{-6} T \rho^{-2/3} \propto \left(\frac{k_B T}{\epsilon_F}\right) \qquad (3.33)$$

in terms of which we can write the ion pressure as $P_{\text{ion}} = P_{\text{ideal}} = \xi P_{e,\text{deg}}$ and electron pressure as $P_{e,\text{deg}}$ (for $\xi \ll 1$; degenerate limit) or $P_{e,\text{deg}}\xi$ (for $\xi \gg 1$; nondegenerate limit). This suggests writing the total pressure in the form

$$P \simeq 10^{13} \text{ dyn cm}^{-2} \rho^{5/3} f(\xi) \equiv K \rho^{5/3}, \qquad (3.34)$$

where the function $f(\xi)$ has the behaviour

$$f(\xi) \to \begin{cases} 1 + \xi + \mathcal{O}(\xi^2) & (\text{for } \xi \ll 1) \\ 2\xi + \mathcal{O}(\xi^{-1}) & (\text{for } \xi \gg 1) \end{cases} \qquad (3.35)$$

in the two limits. This will lead to the correct limiting forms of pressure in the two extreme limits. The utility of this parameterisation stems from the fact that ξ is approximately constant throughout the brown dwarf. To see this, note that entropy per nucleon, s, for a nondegenerate ionised gas is given by

$$\frac{s}{k_B} = A \ln\left(\frac{T}{\rho^{2/3}}\right) - B, \qquad (3.36)$$

where A and B are constants. This result was obtained in Vol. I, Chap. 5, and can be easily understood by counting the number of microstates available for a particle. Because the typical momentum of a particle in a Boltzmann distribution is approximately $(2mk_B T)^{1/2}$, the number of quantum states available per particle is

$$\mathcal{N} \propto \frac{(2mk_B T)^{3/2} V}{N h^3} \propto \frac{T^{3/2}}{\rho}. \qquad (3.37)$$

Because $(s/k_B) = \ln \mathcal{N}$, it follows that the entropy scales as $(s/k_B) = (3/2)\ln(T\rho^{-2/3}) + $ constant, as claimed above. [In fact, for a degenerate plasma

3.3 Pre-Main-Sequence Collapse

with Coulomb interactions, numerical work suggests that the entropy per nucleon is given by $(s/k_B) = 2.2 \ln(T/\rho^{0.63}) - 11.6$; hence even in this limit, $(T/\rho^{2/3})$ is nearly constant if the entropy is constant.] Because the brown dwarf is fully convective, $s \approx$ constant, implying that ξ is a constant throughout the brown dwarf. This suggests that the equation of state is of the form $p \propto \rho^{5/3}$, making the brown dwarf an $n = 3/2$ polytrope. This is the crucial simplification that allows us to model the brown dwarfs.

Exercise 3.2

Review of ideal gases with constant specific heats: Consider an ideal gas with the equation of state $P = (N/V)k_B T = nk_B T = (\rho/m)k_B T$. The specific heats at constant volume and pressure are defined as $C_V = (\partial U/\partial T)_V = (\partial Q/\partial T)_V$ and $C_P = (\partial Q/\partial T)_P$.
(a) Show that, for any ideal gas,

$$C_P = C_V + Nk_B, \qquad (3.38)$$

$$dU = C_V\, dT = C_P \frac{d(PV)}{Nk_B} = \frac{d(PV)}{\gamma - 1}, \qquad (3.39)$$

where $\gamma = (C_P/C_V)$. (b) For an adiabatic process with $dQ = 0$, show that the combinations PV^γ, $P^{1-\gamma}T^\gamma$, and $TV^{\gamma-1}$ are all constants. Also show that the entropy of the ideal gas can be expressed in the form

$$S = Nk_B \ln\left[VT^{1/(\gamma-1)}\right] + \text{constant}. \qquad (3.40)$$

We can now use the standard results of the $n = 1.5$ polytrope to obtain the mass–radius relation for brown dwarfs. The numerical values given in Vol. I, Chap. 10, Subsection 10.3.1 show that, for $n = 1.5$, (1) the central density ρ_c and the mean density $\bar{\rho} = (3M/4\pi R^3)$ are related by $\rho_c = 5.991\bar{\rho}$ and (2) the radius and the central density are related by $R = 3.654 L_0$, with $L_0^2 = [(n+1)K/4\pi G\rho_c^{1/3}]$ and K defined by relation (3.34). Combining, we get

$$R \propto L_0 \propto K^{1/2}\rho_c^{-1/6} \propto K^{1/2}\bar{\rho}^{-1/6} \propto K^{1/2}M^{-1/6}R^{1/2}, \qquad (3.41)$$

or, equivalently, $R \propto KM^{-1/3}$. Inserting numerical factors and using relation (3.34), we get

$$R \equiv 2.8 \times 10^9 \text{ cm} \left(\frac{M_\odot}{M}\right)^{1/3} f(\xi) \equiv R_0 f(\xi). \qquad (3.42)$$

We next need to estimate ξ in terms of other parameters. Because ξ is approximately constant, it should be equal to its central value

$$\xi = \xi_c = 8 \times 10^{-6} T_c \rho_c^{-2/3} \propto T_c \bar{\rho}^{-2/3} \propto T_c M^{-2/3} R^2, \qquad (3.43)$$

where we have used the relation $\rho_c \propto \bar{\rho} \propto (M/R^3)$. Expressing R in terms of M and ξ by using Eq. (3.42) and evaluating the numerical quantities, we get the relation

$$\xi = 3 \times 10^{-9} T_c \left(\frac{M_\odot}{M}\right)^{4/3} f^2(\xi). \tag{3.44}$$

This can be solved to express the central temperature as

$$T_c = 3 \times 10^8 \text{ K} \left[\frac{\xi}{f^2(\xi)}\right] \left(\frac{M}{M_\odot}\right)^{4/3}. \tag{3.45}$$

Equations (3.42) and (3.45) allow us to study the variation of central temperature T_c with R for a fixed mass M, provided that the functional form of $f(\xi)$ is known. Varying R at fixed M is equivalent to varying ξ; from expression (3.35), it follows that the function $Q(\xi) = [\xi/f^2(\xi)]$ in Eq. (3.45) increases (as ξ) for small ξ but decreases (as $1/\xi$) for large ξ. Any physically sensible extrapolation formula for $f(\xi)$ will then lead to a distinct maximum for the function Q. We conclude that the temperature attained by the contracting (proto) brown dwarf has a distinct maximum.

To determine the maximum value, we need to obtain the form of $f(\xi)$, which is a difficult task. Writing $f(\xi) = 1 + \xi + f_1(\xi)$, we clearly see that $f_1(\xi)$ should vary as ξ^2 for small ξ and as ξ for large ξ. An infinite class of such functions exist, with one of the simplest choices being $f_1(\xi) = \xi^2(1+\xi)^{-1}$. We shall use this form for explicit computation. With this choice, the temperature becomes

$$T_c = 3 \times 10^8 \text{ K} \left(\frac{M}{M_\odot}\right)^{4/3} \left[\frac{\xi(1+\xi)^2}{(1+2\xi+2\xi^2)^2}\right] \equiv 3 \times 10^8 \text{ K} \left(\frac{M}{M_\odot}\right)^{4/3} Q(\xi). \tag{3.46}$$

The function $Q(\xi)$ is maximised at $\xi = 0.5514$, which corresponds to $R = 1.75 R_0$ from Eq. (3.42). Thus the central temperature is maximised at $R = 4.7 \times 10^9 (M_\odot/M)^{1/3}$ cm. For much larger R the star is nondegenerate ($\xi \gg 1$) and $T_c \propto (M/R)$. Figure 3.5 shows the variation of $T(R)$ for $M = (0.30, 0.20, 0.08, 0.04) M_\odot$ in a brown dwarf of pure hydrogen. The corresponding maximum value of the temperature is given by

$$T_{c,\text{max}} = 5.4 \times 10^7 \left(\frac{M}{M_\odot}\right)^{4/3} \text{ K} \tag{3.47}$$

if the system is made of pure hydrogen. In general, the pressure will scale with μ_e as $\mu_e^{-5/3}$ so that R_0 will also scale as $\mu_e^{-5/3}$. The ξ also has a μ dependence with $\xi \propto T(\mu_e/\rho)^{2/3}$. Taking all these into account, we find that T_c obtained above scales as $\mu_e^{8/3}$. If we take $\mu_e = 1.15$ corresponding to cosmic abundance, $T_{c,\text{max}}$ rises to $T_{c,\text{max}} = 8 \times 10^7 (M/M_\odot)^{4/3}$.

If the maximum temperature achieved during the contraction is high enough to cause significant nuclear burning, then the system would have become a star. To determine the mass at which the transition from brown dwarf to star takes place, it is better to compare two separate sources of luminosity. The first one is

3.3 Pre-Main-Sequence Collapse

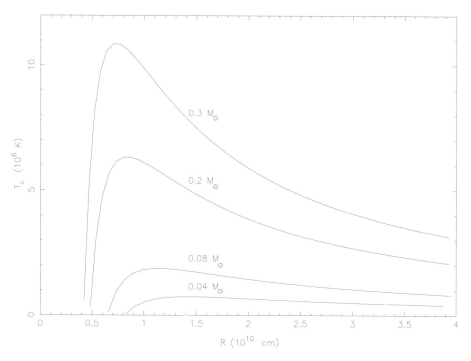

Fig. 3.5. The $T(R)$ relation for brown dwarfs. The curves are for masses $M = (0.3, 0.2, 0.08,$ and $0.04)$ $M_\odot$.

the standard luminosity L_{pp} due to the p–p reaction at the maximum temperature. The second one arises from the fact that when the cloud contracts to a radius $R(T_{max})$ in a time scale t, it would have released the gravitational potential energy $E_{grav} \propto [GM^2/R(T_{max})]$, thereby contributing a luminosity E_{grav}/t. To be considered as a star, it is necessary that $L_{p-p} > (E_{grav}/t)$ when the collapse occurs within the age of the universe $t \lesssim t_{univ} \approx 10^{10}$ yr. We shall now compute this criterion.

Let us begin with the luminosity of such a star arising from the p–p reaction. This luminosity is given approximately by $L_{pp} \approx \epsilon_{pp}(T_c) \rho_c R_c^3$, where $R_c \approx 0.38 R$ is the radius at which density drops to half the central value for the $n = 1.5$ polytrope, say. Expressing central density in terms of $\bar{\rho}$, we get $L_{pp} = 0.08 M \epsilon_{pp}(T_c)$. We can now estimate this by using the fitting formula for p–p energy-production rate described in Vol. I, Chap. 12:

$$\epsilon_{pp}(T_c) \cong 2.4 \times 10^6 \rho X^2 T_6^{-2/3} \exp(-33.8 T_6^{-1/3}) \text{ ergs gm}^{-1} \text{ s}^{-1}. \quad (3.48)$$

If we assume that the brown dwarf is made of pure hydrogen ($\mu_e = 1$, $X = 1$) and $T_c \cong T_{c,max}$, then we get

$$L_{pp,max} = 0.08 X^2 \rho_c M \mathcal{F}(T_c) = 4 \times 10^{36} \text{ ergs s}^{-1} \left(\frac{M}{M_\odot}\right)^3 \mathcal{F}(T_c), \quad (3.49)$$

where

$$\mathcal{F}(T_c) = 2.4 \times 10^6 T_6^{-2/3} \exp(-33.8 T_6^{-1/3}), \quad T_6 = \frac{T_c}{10^6 \text{ K}}. \quad (3.50)$$

In arriving at the second equality in Eq. (3.49), we have also expressed ρ_c in terms of M by using $\rho_c \propto (M/R^3)$ and $R \propto M^{-1/3}$.

On the other hand, the total energy radiated by a contracting star, when it reaches the radius $R(T_{c,\max})$, is the gravitational energy of an $n = 1.5$ polytrope (see Vol. I, Chap. 10, Subsection 10.3.1), given by

$$E = \frac{1}{2} \frac{3}{(5-1.5)} \frac{GM^2}{R(T_{c,\max})} = \frac{3}{7} \frac{GM^2}{R(T_{c,\max})} \simeq 2.1 \times 10^{49} \text{ ergs} \left(\frac{M}{M_\odot}\right)^{7/3}. \quad (3.51)$$

If this is to be radiated within the age of the universe, $t_u \approx 10^{10}$ yr, the luminosity must be at least

$$L_{\text{grav}} = \frac{E}{t_u} = 7 \times 10^{31} \left(\frac{M}{M_\odot}\right)^{7/3} \text{ ergs s}^{-1}. \quad (3.52)$$

Comparing L_{grav} with L_{pp} numerically (see Fig. 3.6), we see that only stars more massive than $\sim 0.085\,M_\odot$ could be sustained by fusion. This sets a limit of

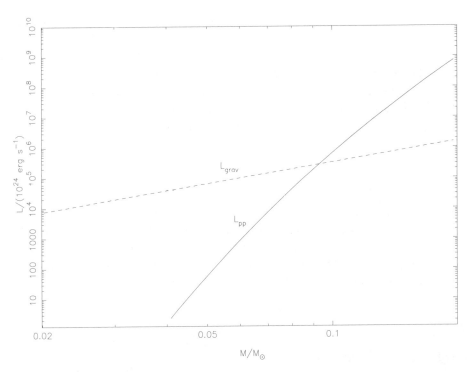

Fig. 3.6. The nuclear and gravitational luminosities of a brown dwarf in a simplified model.

$\sim 0.08\, M_\odot$ for stars reaching the main sequence. The actual numerical value obtained for the critical mass is close to the result from more sophisticated modelling of brown dwarfs. This agreement, of course, is to some extent due to the choice of the extrapolation function $f(\xi)$ and other numerical coincidences. The analysis, however, explains the physical origin of the cutoff and overall trend correctly.

It is also possible to obtain some general results about the cooling of brown dwarfs by modelling its photosphere. The outermost layers of the brown dwarf are cool enough that the hydrogen is molecular rather than atomic in this non-degenerate photosphere. The hydrostatic equilibrium for the photosphere can be expressed in the form $(dP/d\tau) = -g/\kappa$, where τ is the optical depth, g is the surface gravity, and κ is the opacity. This gives an estimate for photospheric pressure of $P_{\rm ph} \approx (2/3)(g/\kappa)$ [see Eq. (2.97), Chap. 2]. Equating this to the pressure of H_2 molecules given by $P_{H_2} = (\rho k_B T_{\rm eff}/2m_p)$ (where $T_{\rm eff}$ is the effective photospheric temperature), we get $\rho_{\rm ph} = (4/3)(GMm_p/k_B T_{\rm eff} R^2 \kappa)$. The entropy per nucleon for the classical H_2 gas is given by $S \propto \ln(TV^{\gamma-1})$, with $(\gamma - 1) = (1/C_V) = 2/5$ for a molecule with 5 degrees of freedom (see Exercise 3.2). This would give

$$\frac{S}{k_B} = A_1 \ln\left(\frac{T}{\rho^{n_1}}\right) - B_1, \tag{3.53}$$

with $A_1 = 2.5$ and $n_1 = 0.4$; a more precise calculation leads to $A_1 \approx 1.3$, $n_1 \approx 0.42$, and $B_1 \approx 3.0$. Substituting for $\rho_{\rm ph}$ and equating the entropy to its value at its centre, we can determine the photospheric entropy in terms of the other parameters in the problem. A straightforward calculation gives the photospheric entropy as

$$\frac{S_{\rm ph}}{k_B} = 1.8 \ln T_{\rm eff} + 1.08 \ln\left(\frac{R}{R_0}\right) + 0.54 \ln\left(\frac{\kappa}{10^{-2}}\right) - 0.82 \ln\left(\frac{M}{M_\odot}\right) - 4.6, \tag{3.54}$$

and the central entropy as

$$\frac{S_c}{k_B} = 2.2 \ln T_c - 2.78 \ln\left(\frac{M}{M_\odot}\right) + 4.17 \ln\left(\frac{R}{R_0}\right) - 28.0, \tag{3.55}$$

where we have used the fact that $\rho_c = 5.991(3M/4\pi R^3)$. Equating the two expressions and solving for $T_{\rm eff}$, we obtain

$$T_{\rm eff} = 2 \times 10^{-6} T_c^{1.22} \left(\frac{M}{M_\odot}\right)^{-1.05} \left(\frac{R}{R_0}\right)^{1.7} \left(\frac{\kappa}{10^{-2}}\right)^{-0.3}. \tag{3.56}$$

Because R is a function of M and T_c and $L = (4\pi R^2 \sigma T_{\rm eff}^4)$, it follows that the luminosity of the brown dwarf is entirely determined by its central temperature T_c and its photospheric opacity κ. The luminosity L decides the rate of loss of thermal energy of the ions in the central part, allowing us to estimate the cooling rate dT_c/dt in terms of the mass, central temperature, and photospheric opacity.

Integrating this equation, we get the cooling history $T_c(t)$ parameterised by the mass and the atmospheric opacity. Performing this straightforward but tedious calculation, we find that $\xi(t) \propto t^{-0.3}$ and $L \propto t^{-1.3}$ for $\xi \ll 1$, which describes the initial cooling of the brown dwarfs. Numerical integration of the relevant equations will provide a more exact evolutionary track.

3.3.5 General Discussion of Homologous Collapse

The analysis in the last subsection shows the importance of the equation of state in deciding the trajectory of the system in the ρ–T plane. If the core becomes degenerate during any stage of evolution, there arises a possibility that the temperature $T_c(\rho_c)$ has a distinct maximum, and hence further nuclear reactions may not take place. It is therefore important to obtain the condition under which this will happen. We shall now derive a general condition by considering some formal aspects of homologous collapse.

Consider a homologous contraction (or expansion) of a cloud in which a shell of radius r (containing certain mass) is displaced to the position $r + dr = r(1+x)$, where x is constant for the entire system. Under such a homologous change, the velocity of the shell $\dot{r}$ must be proportional to r with the same constant of proportionality for all shells. This constancy of $\dot{r}/r$ implies that

$$\frac{\partial}{\partial m}\left(\frac{\partial \ln r}{\partial t}\right) = 0. \tag{3.57}$$

Exchanging the order of the derivatives, we find that

$$\frac{\partial}{\partial t}\left(\frac{1}{r}\frac{\partial r}{\partial m}\right) = \frac{\partial}{\partial t}\left(\frac{1}{4\pi r^3 \rho}\right) = \frac{1}{4\pi r^3 \rho}\left(-3\frac{\dot{r}}{r} - \frac{\dot{\rho}}{\rho}\right) = 0, \tag{3.58}$$

leading to

$$\frac{\dot{\rho}}{\rho} = -3\frac{\dot{r}}{r}. \tag{3.59}$$

Consider next the change in pressure. From the equation for hydrostatic equilibrium, written in the integral form as

$$P = \int^M \frac{Gm}{4\pi r^4(m)}\, dm, \tag{3.60}$$

we get

$$\dot{P} = \int^M \frac{\partial}{\partial t}\left(\frac{1}{r^4}\right)\frac{Gm}{4\pi}\, dm = -4\frac{\dot{r}}{r}\int^M \frac{Gm}{4\pi r^4}\, dm, \tag{3.61}$$

where we have used the fact that $\dot{r}/r$ is a constant throughout the star. This gives

$$\frac{\dot{P}}{P} = -4\frac{\dot{r}}{r}. \tag{3.62}$$

3.3 Pre-Main-Sequence Collapse

Given the changes in density and pressure, we can determine the changes in the temperature. Defining

$$\frac{\dot\rho}{\rho} = \alpha\left(\frac{\dot P}{P}\right) - \delta\left(\frac{\dot T}{T}\right), \quad \alpha \equiv \left(\frac{\partial \ln \rho}{\partial \ln P}\right)_T, \quad \delta = -\left(\frac{\partial \ln \rho}{\partial \ln T}\right)_P, \quad (3.63)$$

we get

$$\frac{dT}{T} = -\left(\frac{4\alpha - 3}{\delta}\right)\frac{dr}{r} = \left(\frac{4\alpha - 3}{3\delta}\right)\frac{d\rho}{\rho} \quad (3.64)$$

for homologous collapse.

Equation (3.64) can be thought of as providing the direction of "flow" of system in the ρ–T plane. When the gas is ideal, $\alpha = \delta = 1$ and $dT/T = (1/3) d\rho/\rho$; so the temperature increases when the system contracts and ρ increases. The evolutionary track of such a system is a straight line with a slope of $1/3$ in the ρ–T plane (the lines marked A, B, and C all have this behaviour in the left part of Fig. 3.7), whereas the line separating degenerate and ideal behaviour has a slope of $2/3$ in the nonrelativistic region and $1/3$ in the relativistic region ($\rho \gtrsim 10^6$ gm cm^{-3}). This is shown by the thick line in Fig. 3.7.

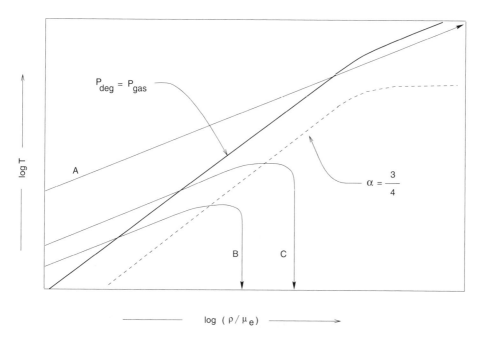

Fig. 3.7. The evolution of density and temperature during the collapse for degenerate matter. Degeneracy effects prevent the temperature increasing without bound in the collapse of nonrelativistic matter and lead to a maximum value for the possible temperature. The value of $T_{\max}$ depends on the mass of the cloud, and a star can form only if $T_{\max}$ is high enough to ignite nuclear reactions. There is no maximum temperature in the collapse of a relativistically degenerate system.

Clearly, the evolution takes the system towards this line. For a completely degenerate *nonrelativistic* system, $\alpha \to 3/5 = 0.6$ and $\delta \to 0$. As the system approaches complete degeneracy, α varies from 1 to 0.6 and δ changes from 1 to 0. Therefore α will pass through the value $3/4 = 0.75$ while δ is still nonzero and the sign of the right-hand side of Eq. (3.64) will change. Further collapse will lead to cooling and not heating. This reconfirms our earlier result that, for a partially degenerate system below a critical mass, collapse will lead to a maximum temperature followed by decrease in the temperature. Clouds with low mass reach a maximum temperature and cool (if the maximum temperature is not high enough to ignite the nuclear reactions). For clouds with high enough mass, the maximum temperature is high enough to ignite the nuclear reactions, thereby forming a star. The curve corresponding to $\alpha = 3/4$ (shown by the dashed curve in Fig. 3.7) demarcates the two regions in the ρ–T plane. This curve has a slope of $1/3$ initially (corresponding to $\alpha = 1 = \delta$ of an ideal gas); the slope keeps decreasing and flattens considerably for relativistic densities.

An explicit and useful model showing these effects is provided by a system in which the electrons are partially degenerate and provides the bulk of the pressure so that the equation of state can be approximated as

$$P \approx P_e = \frac{\mathcal{R}}{\mu_e}\rho T + K_\gamma \left(\frac{\rho}{\mu_e}\right)^\gamma. \tag{3.65}$$

The first term gives the ideal-gas pressure and the second term is the pressure of the degenerate gas of electrons. The index γ is *not* a constant and varies from $5/3$ for $\rho \ll 10^6$ gm cm^{-3} to $\gamma = 4/3$ for $\rho \gg 10^6$ gm cm^{-3}; the constant K_γ also varies smoothly with γ (see Chap. 5, Section 5.9) with γ; $K_{5/3} \approx 10^{13}$ and $K_{4/3} \approx 1.24 \times 10^{15}$ are in cgs units. The central pressure of the cloud in hydrostatic equilibrium is given approximately by

$$P_c \propto \frac{GM\rho}{R} = fGM^{2/3}\rho^{4/3}, \tag{3.66}$$

where f is a numerical constant. Using this relation to eliminate the pressure from Eq. (3.65), we find a relation between the temperature and the density:

$$\frac{\mathcal{R}}{\mu_e}T = fGM^{2/3}\rho^{1/3} - K_\gamma \rho^{\gamma-1}\mu_e^{-\gamma}. \tag{3.67}$$

In the initial stages, the first term on the right dominates and the temperature increases as $\rho^{1/3}$. If the system is nonrelativistic and $\gamma = 5/3$, then the temperature reaches a maximum value and then decreases, as in the case of brown dwarfs discussed in the last subsection. More interesting is the case in which the density is higher than $\sim 10^7$ gm cm^{-3} at the centre and the system is relativistically degenerate. Writing $\gamma \approx (4/3) + \chi$ (with $\chi \to 0$), the temperature–density relation now becomes

$$\frac{\mathcal{R}}{\mu_e} T = \rho^{1/3} \left[f G M^{2/3} - K_{(4/3+\chi)} \mu_e^{-(4/3+\chi)} \rho^\chi \right]. \qquad (3.68)$$

In this case, the temperature does not go to zero asymptotically but increases again as $T \propto \rho^{1/3}$, provided that

$$M > M_{\text{crit}} = \left(\frac{K_{4/3}}{fG} \right)^{3/2} \mu_e^{-2} \equiv \left(\frac{f}{0.58} \right)^{-3/2} \left[\frac{5.84}{\mu_e^2} M_\odot \right]. \qquad (3.69)$$

The quantity in brackets, $M_{\text{Ch}} \equiv 5.84 \, \mu_e^{-2} \, M_\odot$, called the *Chandrasekhar mass*, will play a crucial role in the evolution of stellar cores and stellar remnants. [We will derive the exact numerical value of M_{Ch} in Chap. 5; the scaling of f in Eq. (3.69) anticipates this result.] The behaviour of clouds of different masses is shown in Fig. 3.7. The curves marked B and C correspond to masses lower than the critical mass M_{Ch}; these curves reach a maximum temperature and decrease again. The curve marked A corresponds to $M > M_{\text{Ch}}$, and it does not show any maximum.

3.4 Evolution of High-Mass Stars

The first phase of stellar evolution, namely the formation of stars from interstellar clouds, was discussed in Section 3.3. Once the nuclear reaction has been ignited, the star remains in an approximate steady state for considerable period of time, viz. $t_{\text{nuc}} \cong 10^{10} (M/M_\odot)^{-2.5}$ yr. This phase was studied in Chap. 2. We shall now take up the discussion of stellar evolution during the later stages when the nuclear fuel in the core is beginning to be exhausted, i.e., for $t \gtrsim t_{\text{nuc}}$.

The post-main-sequence evolution depends strongly on the mass of the main-sequence star and somewhat weakly on the internal composition. Because the time-dependent evolutionary equations are extremely complex, most phases of stellar evolution have to be studied numerically. In what follows, the key physical features are presented that govern the stellar evolution, without the finer details that are model dependent.

We saw in the last chapter that low-mass stars have radiative cores whereas the high-mass stars have convective cores. This implies that the elements are fairly well mixed near the centre in the high-mass stars, leading to a fairly uniform distribution in the core. In contrast, the nuclear fuel has a stronger gradient (and is less uniform) in the cores of low-mass stars. Matter in the central regions of low-mass stars is also closer to degeneracy compared with that in high-mass stars. (This is clearly seen from the ρ_c–T_c tracks of Fig. 2.10 of Chap. 2. Low-mass stars have less central density and less central temperature, but because degeneracy depends more critically on temperature, the lowering of the central temperature makes the cores of the low-mass stars more degenerate than those of the high-mass stars.) These two features influence the further evolution of low- and high-mass stars significantly and make them evolve very differently. We shall begin with a discussion of high-mass stars; low-mass stars will be taken up in Section 3.5.

3.4.1 Initial Stages in Main Sequence

The main-sequence phase of a high-mass star (say, a $5M_\odot$ star) consists of converting the hydrogen in the core to helium through nuclear reactions (see Fig 3.8, phases 1–2).[7] As time goes on, this leads to a core region in the star essentially made of helium, surrounded by an envelope of hydrogen. This reaction increases the mean molecular weight of the core region and – according to the ideal-gas law – decreases the pressure. The core responds by contracting, thereby increasing the density and temperature. Half the potential energy, released by the contraction as well as the increased nuclear luminosity that is due to increased temperature, causes the overall luminosity of the star to increase. These two features affect the radius of the star in different directions: (1) An increase in the core temperature causes an increase in the mean temperature of the star and because virial equilibrium requires that $\langle T \rangle \propto (M/R)$, this leads to a decrease of radius, and (2) the release of core potential energy during contraction and an increase in luminosity cause the radius of outer shells to expand slightly. In this particular case, it arises as a consequence of the equation $(dT/dr) \propto (\kappa\rho/T^3)(L_r/r^2)$; the increase in L_r is accommodated by the increase in $r(m)$, with other quantities changing very little. (We shall see the core contraction and envelope expansion occurring together on several phases of stellar evolution. The reasons for this phenomenon can be quite different in different contexts.) Because the increase in luminosity is directly related core contraction that in turn is due to a change in molecular weight, we can take, to first approximation, $R \propto (\mu/T_c)$ with both μ and T_c increasing.

Which of these terms will dominate depends on the temperature sensitivity of the nuclear reactions. For stars with $M \gtrsim 3\,M_\odot$, powered by the highly temperature-dependent CNO cycle, the change in T_c is smaller and the radius increases with μ. This has the effect of reducing T_{eff}, and the star moves from 1 to 2 in Fig. 3.8. (For lower masses, the radius remains almost constant or can even decrease. The initial track of the star will be nearly along the main sequence; see the corresponding part marked 1–2 in Fig. 3.8.) Thus the radius, central temperature, and the luminosity of the star increase during the main-sequence phase itself.

3.4.2 Shell Burning of Hydrogen

At point 2 of Fig. 3.8, the hydrogen fuel in the core is nearly exhausted and the nuclear luminosity vanishes at the core. Because $L(r) \propto (dT/dr)$, the temperature gradient also vanishes and the core becomes isothermal. A core without temperature or density gradient cannot provide the pressure gradient to support the matter against gravity and there will be a slight overall contraction of the star.

3.4 Evolution of High-Mass Stars

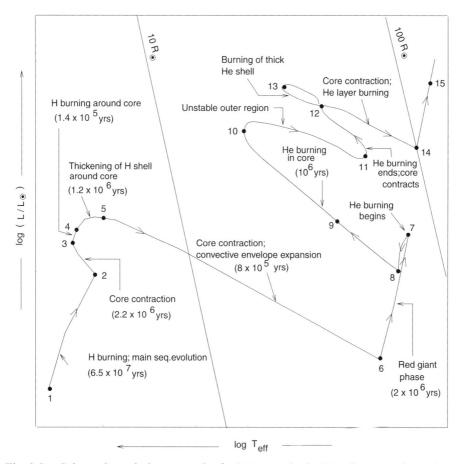

Fig. 3.8. Schematic evolutionary track of a 5-$M_\odot$ star in the H-R diagram. The various phases marked by the numbers 1–14 on the diagram are briefly described below. A more detailed description is given in the text. (a) 1–2, Evolution in main sequence with hydrogen burning (6.5×10^7 yr); (b) 2–3, contraction of the core (2.2×10^6 yr); (c) 3–4, beginning of hydrogen burning in a shell around the core (1.4×10^5 yr); (d) 4–5, thickening of the hydrogen burning shell (1.2×10^6 yr); (e) 5–6, rapid contraction of the core and expansion of the convective envelope (8×10^5 yr); (f) 6–7, red giant phase (2×10^6 yr) – helium burning begins around 7; (g) 7–10, helium burning in the core (10^7 yr); (h) 10–11, the outer regions of the star become unstable (10^6 yr); (i) 11–12, end of the helium burning in the core followed by contraction of the core; (j) 12–13, burning of helium in a thick shell; (k) 13–14, contraction of the core and helium burning in a thick layer.

This contraction has three effects: (1) Part of the gravitational potential energy goes to increasing the luminosity. The star moves along 2 to 3 in Fig. 3.8. By the time the star reaches point 3, the mass fraction of hydrogen in the core has been reduced to nearly 1%. This process takes approximately 2.2×10^6 yr for a $5M_\odot$ star. (2) A density gradient is established so as to provide the pressure gradient.

(3) The temperature of the edge of the core increases sufficiently for the hydrogen in a shell region around the core to undergo fusion. This shell burning is ignited fairly rapidly, causing the envelope to expand slightly. This moves the star from 3 to 5. At this stage, the internal structure of the star is as shown in Fig. 3.9(a).

For stars with $M \gtrsim 3\,M_\odot$ powered by the CNO cycle, the nuclear-reaction rate varies as a high power of T_c. The core has to collapse a fair bit for the temperature of the shell to increase sufficiently so that the CNO cycle can lead to significant energy production. The situation, as we shall see in the next section, is different for stars with $M \lesssim 3\,M_\odot$ powered by p–p reactions. Here the shell region is fairly close to the ignition temperature by the time the core helium is exhausted. Hence there is a smooth transition from the core burning to shell burning in the lower-mass stars.

Exercise 3.3

Aspects of shell burning: Consider a thin active shell source that separates the stellar core from the envelope. As the shell burns, its position tends to remain approximately constant. This feature, which arises in different stages of stellar evolution, can be understood as follows. (a) Suppose that there is an initial radial displacement $(\Delta R/R)_1$ in the position of the shell located at $r = R_1$. The star now makes a homologous change in its structure to regain hydrostatic equilibrium. If the energy-generation rate in the shell is given by $\epsilon = \epsilon_0 \rho T^\nu$, show that the fractional change in energy-generation is $(\Delta\epsilon/\epsilon) \approx -(3+\nu)(\Delta R/R)_1$. (b) To restore thermal equilibrium with the changed energy-generation rate, the temperature has to change. If the opacity is given by $\kappa = \kappa_0 P^a T^b$, show that the temperature changes by $(\Delta T/T) \simeq [(3a+\nu+3)/(4-b)](\Delta R/R)_1$. (c) This temperature change induces a pressure change $(\Delta P/P) = (\Delta T/T)$ at constant density, forcing the star out of hydrostatic equilibrium. The star now makes a second homologous change in R, restoring hydrostatic equilibrium. Show that

$$\left(\frac{\Delta R}{R}\right)_2 = -\left[\frac{3a+\nu+3}{4(4-b)}\right]\left(\frac{\Delta R}{R}\right)_1. \qquad (3.70)$$

Discuss the range of values for the coefficient on the right-hand side and argue why the shell is likely to remain stationary during the nuclear burning.

3.4.3 Core Contraction and Envelope Expansion

Further evolution depends on the structural changes that take place in the helium core (points 5–6 of Fig. 3.8). We stress the fact that the helium core is fairly homogeneous in a $5M_\odot$ star because of the mixing that is due to the original convective transport (the situation, as we shall see in the next section, is different in lower-mass stars). Further, it will be nearly isothermal because the vanishing of luminosity implies the vanishing of the temperature gradient. The equilibrium of such a star depends on the ability of an isothermal core (with mass $M_{ic} \equiv qM$)

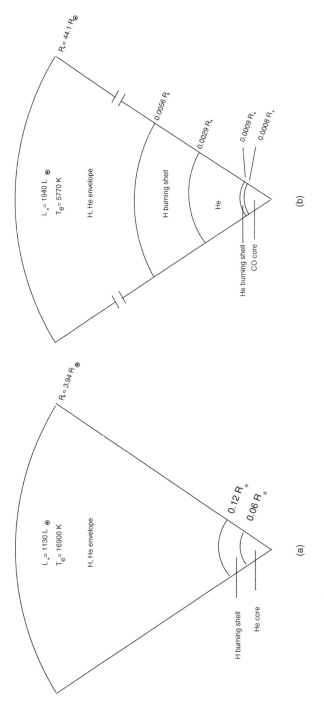

Fig. 3.9. (a) Internal structure of a $5M_\odot$ star with a helium core and a hydrogen burning shell, (b) internal structure of the same star approaching the asymptotic giant branch around point 11 in Fig. 3.8. Note that the scales of the outer shells have been enhanced for clarity.

to support the envelope of mass $(1-q)M$. It turns out that this is possible only if the fraction of the mass in the core is below a critical value called the *Chandrasekhar–Schoenberg* limit. The maximum fraction of a star's mass that can exist in the supporting isothermal core is given by

$$q_{\max} = \left(\frac{M_{\rm ic}}{M}\right)_{\rm CS} \simeq 0.37 \left(\frac{\mu_{\rm env}}{\mu_{\rm ic}}\right)^2, \tag{3.71}$$

where $\mu_{\rm env}$ and $\mu_{\rm ic}$ are the mean molecular weights in the envelope and the isothermal core, respectively. For example, if the initial composition of the star is $X = 0.708$, $Y = 0.272$, and $Z = 0.02$, then $\mu \approx 0.61$, assuming complete ionisation at the core–envelope boundary. If all the hydrogen has been converted into helium, $\mu_{\rm ic} \approx 1.33$ [see Eqs. (2.5) of Chap. 2]. In this case the maximum mass that can be accommodated by the core is $q_{\max} \approx 0.08$; the isothermal core will collapse if its mass exceeds 8% of the total mass.

The rigorous derivation of the Chandrasekhar–Schoenberg limit can be provided from the behaviour of polytrope solutions in the u–v plane. In Vol. I, Chap. 10, it was shown that the curve representing the isothermal sphere in the u–v plane lies below a maximum value $v_{\max} \approx 2.25$ (see Vol. I, Fig. 10.1 of Chap. 10). The radiative envelopes of the stars are characterised by a polytrope of $\gamma = 5/3$, $n = 3/2$. For a radiative envelope to match an isothermal core, it is necessary that these two curves should intersect in the u–v plane. This turns out to be impossible for $q \gtrsim 0.37 \, (\mu_{\rm env}/\mu_{\rm ic})^2$ because of the spiralling nature of the curve corresponding to the isothermal sphere in the u–v plane.

A more physical derivation of the same result can be provided along the following lines: For equilibrium, the pressure exerted by the envelope and by the isothermal core must match at the boundary between the core and the envelope. We will estimate these two pressures separately, starting with the pressure exerted by the envelope. From the equation for hydrostatic equilibrium, we can write

$$P_{\rm env} = \int_0^{P_{\rm env}} dP = -\int_M^{M_{\rm ic}} \frac{GM_r}{4\pi r^4} dM_r \simeq -\frac{G}{8\pi \langle r^4 \rangle} \left(M_{\rm ic}^2 - M^2\right), \tag{3.72}$$

where M is the total mass of the star and $\langle r^4 \rangle$ depends on the envelope profile. Assuming that $M_{\rm ic} \ll M$ and taking $\langle r^4 \rangle \approx R^4/2$, we get

$$P_{\rm env} \simeq \frac{G}{4\pi} \frac{M^2}{R^4}. \tag{3.73}$$

Using the ideal-gas law and estimating the density of the envelope by

$$T_{\rm ic} = \frac{P_{\rm env} \mu_{\rm env} m_{\rm H}}{\rho_{\rm env} k_B}, \quad \rho_{\rm env} \simeq \frac{M}{4\pi R^3/3}, \tag{3.74}$$

3.4 Evolution of High-Mass Stars

we can express the radius R in terms of T_{ic} as

$$R \simeq \frac{1}{3} \frac{GM}{T_{ic}} \frac{\mu_{env} m_H}{k_B}. \tag{3.75}$$

Substituting this back into relation (3.73), we find that the pressure of the envelope is given by

$$P_{env} \simeq \frac{81}{4\pi} \frac{1}{G^3 M^2} \left(\frac{k_B T_{ic}}{\mu_{env} m_H} \right)^4. \tag{3.76}$$

Note that this pressure depends on the *total* mass of the star and the temperature of the *core* but is independent of the mass of the core.

Let us next consider the pressure exerted by the isothermal core. The virial theorem applied to the isothermal core gives

$$4\pi R_{ic}^3 P_{ic} - 2K_{ic} = U_{ic}, \tag{3.77}$$

where K_{ic} and U_{ic} are the kinetic and the gravitational potential energies of the core given by

$$K_{ic} = \frac{3 M_{ic} k_B T_{ic}}{2 \mu_{ic} m_H}; \quad U_{ic} = -\frac{3}{5} \frac{GM_{ic}^2}{R_{ic}}. \tag{3.78}$$

Hence the pressure of isothermal core is

$$P_{ic} = \frac{3}{4\pi R_{ic}^3} \left(\frac{M_{ic} k_B T_{ic}}{\mu_{ic} m_H} - \frac{1}{5} \frac{GM_{ic}^2}{R_{ic}} \right). \tag{3.79}$$

[This is essentially the same result derived in Vol. I, Chap. 8, Section 8.5, with α of Eq. (8.70) set to 3/5.] Note that the thermal energy of the core tends to increase the pressure whereas the gravitational energy tends to decrease it. For some values of M_{ic} this pressure will be maximised; hence there is an upper limit on the pressure an isothermal core can exert in order to support the envelope. The maximum occurs when the core radius and mass are related by

$$R_{ic} = \frac{2}{5} \frac{GM_{ic} \mu_{ic} m_H}{k_B T_{ic}} \tag{3.80}$$

and the maximum value of the pressure is given by

$$P_{ic,max} = \frac{375}{64\pi} \frac{1}{G^3 M_{ic}^2} \left(\frac{k_B T_{ic}}{\mu_{ic} m_H} \right)^4. \tag{3.81}$$

As the core mass increases, the maximum pressure, decreases, and when the mass crosses a critical value, the core will be unable to support the envelope. Equating the maximum pressure to the pressure of the envelope, we get the upper

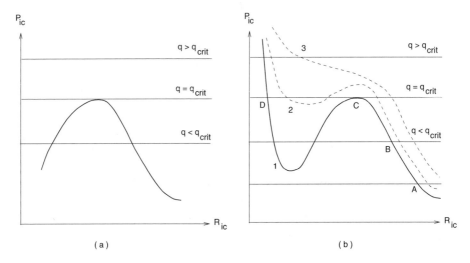

Fig. 3.10. The relation among pressures of isothermal cores as functions of the core radius (a) without degeneracy and (b) with degeneracy. The horizontal lines give the pressure of the envelope for three different values of $q \equiv (M_{\rm ic}/M)$. The equilibrium configuration is given by the intersection of these curves.

limit as

$$q_{\rm crit} = \left(\frac{M_{\rm ic}}{M}\right)_{\rm crit} \simeq 0.54 \left(\frac{\mu_{\rm en}}{\mu_{\rm ic}}\right)^2. \tag{3.82}$$

The thick curve in Fig. 3.10(a) gives a schematic plot of the core pressure as a function of the core radius based on Eq. (3.79) for a given value of $M_{\rm ic}$. The three horizontal lines correspond to the pressure of the envelope when q is greater than, equal to, or less than the critical value. (These lines can be thought of as parameterised by the value of total mass M; when M increases, q will decrease.) For $q < q_{\rm crit}$, these curves intersect at two points; the equilibrium at a larger value of R_c is stable whereas that corresponding to a smaller value of R_c is unstable. This is easily seen by the fact that if we slightly decrease the core radius around this equilibrium point, the core pressure decreases, thereby allowing the envelope to crush it further. The opposite behaviour occurs for the equilibrium at the larger radius.

The above analysis suggests that *no* equilibrium is possible for $q > q_{\rm crit}$. This result, however, arises because we have treated the core as nondegenerate and ideal. With future applications in mind, we shall mention the modification to the above results when degeneracy pressure is included.

To do this, it is best to start from the equation for hydrostatic equilibrium, written in the form

$$4\pi r^3 \frac{dP}{dM_r} = -\frac{GM_r}{r} = \frac{d(4\pi r^3 P)}{dM_r} - \frac{3P}{\rho}. \tag{3.83}$$

3.4 Evolution of High-Mass Stars

Then we integrate both sides over dM_r inside the core. This gives

$$\int_0^{M_{ic}} \frac{d(4\pi r^3 P)}{dM_r} dM_r - \int_0^{M_{ic}} \frac{3P}{\rho} dM_r = -\int_0^{M_{ic}} \frac{GM_r}{r} dM_r. \quad (3.84)$$

The first and the third terms lead to $4\pi R_{ic}^3 P_{ic}$ and U_{ic} of Eq. (3.77). The second term will give $2K_{ic}$ if we use the ideal-gas equation of state for the core. To take into account the degeneracy pressure we shall instead use an equation of state that extrapolates between an ideal gas and a nonrelativistic degenerate gas:

$$\frac{P}{\rho} = \frac{k_B T_{ic}}{\mu_{ic} m_H} + \frac{K_{5/3} \rho^{2/3}}{\mu_e^{5/3}}. \quad (3.85)$$

Then

$$\int_0^{M_{ic}} \frac{3P}{\rho} dM_r = 2K_{ic} + \frac{K_{5/3}}{\mu_e^{5/3}} \langle \rho^{2/3} \rangle M_{ic} = 2K_{ic} + \alpha \frac{K_{5/3}}{\mu_e^{5/3}} \frac{M_{ic}^{5/3}}{R_{ic}^2}, \quad (3.86)$$

where we have assumed that $\langle \rho^{2/3} \rangle \propto (M/R^3)^{2/3}$ and we have introduced a proportionality constant α. Using this expression, we find that the core pressure becomes

$$P_{ic} = \frac{3}{4\pi R_{ic}^3} \left(\frac{M_{ic} k T_{ic}}{\mu_{ic} m_H} - \frac{1}{5} \frac{GM_{ic}^2}{R_{ic}} + \frac{A}{3} \frac{M_{ic}^{5/3}}{R_{ic}^2} \right), \quad A \equiv \alpha \frac{K_{5/3}}{\mu_e^{5/3}} \quad (3.87)$$

instead of the one given by Eq. (3.79). The behaviour of P_{ic} as a function of R_{ic} is shown in Fig. 3.10(b). The turning points in the curve occur at

$$R_{ic} = \frac{2}{15} \mu_{ic} \frac{GM_{ic} m_H}{k_B T_{ic}} \left[1 \pm \left(1 - \frac{125 A}{4\mu_{ic} m_H G^2} \frac{k_B T_{ic}}{M_{ic}^{4/3}} \right)^{1/2} \right]. \quad (3.88)$$

If the core mass M_{ic} is larger than a critical value

$$M_{crit} = \left(\frac{125\alpha K_{5/3}}{4\mu_{ic}} \right)^{3/4} \left(\frac{\mathcal{R} T_{ic}}{G^2} \right)^{3/4} \frac{1}{\mu_e^{5/4}}, \quad (3.89)$$

where $\mathcal{R} = (k_B/m_H)$, both the roots are real and the curve will look like the one marked 1 in Fig. 3.10(b). As the core mass decreases, the turning points approach each other (see the curve marked 2) and for $M_{ic} = M_{crit}$ they coincide (see the curve marked 3). For lower masses, the turning points are imaginary and the curve is monotonic. For a core temperature of approximately 2×10^7 K the critical mass is $\sim 0.1 \, M_\odot$.

For stars with $M \gtrsim 3 \, M_\odot$, the core mass will be larger than this value and the curve will have two turning points. In that case, we arrive at the following sequence of equilibria as q increases: (1) When q is small, the intersection A is at some large radius and the system corresponds to an isothermal sphere that is supporting an envelope. (2) As q increases, the horizontal line in Fig. 3.10(b), marked $q < q_{\text{crit}}$, intersects the $P_{\text{ic}}(R_{\text{ic}})$ curve at three points. Of these three, the middle one is unstable, the lower one corresponds to a degenerate core supporting the envelope, and the one at largest radius corresponds to that of an ideal-gas core supporting the envelope. By continuity, the system will evolve to the equilibrium configuration corresponding to that with the largest radius. Thus, as q increases, the core radius moves along the path AB in Fig. 3.10(b). (3) This is possible only as long as $q \leq q_{\text{crit}}$. When this limit is exceeded, the only equilibrium is that with the degenerate core and the system has to make a rapid contraction from point C to point D in Fig. 3.10(b) [Note that this result is a direct consequence of the high mass of the star; for low-mass stars, curve 3 of Fig. 3.10(b) will be applicable and the transition can be continuous, as we shall see in the next section.]

Let us now consider the implication of the above result for a high-mass star. As the shell burning increases, the mass of the core q will exceed q_{crit} and the core will no longer be able to support the envelope. For a 5-$M_\odot$ star this occurs at 5 in Fig. 3.8, when the degeneracy effects in the core are negligible. Hence it will not be possible for the degeneracy effects to support the envelope. Because of the pressure imbalance, the core will star start collapsing and the evolution will proceed at a faster rate $t_{\text{KH}} \simeq (GM^2/RL)$, called the *Kelvin–Helmholtz* time scale; this is approximately 3×10^6 yr for a $5M_\odot$ star in this phase.

During this phase ($5 \to 6$ in Fig. 3.8), the core and the envelope regions behave in a very different way. The study of the trajectories of different mass shells inside the star as functions of time based on numerical integration of equations of stellar evolution shows that the core collapses while the envelope expands. The expansion of the envelope decreases T_{eff} and cools the outer layers of the star significantly; this causes the star to move from 5 to 6 in the H-R diagram.

In this context, the core contraction with envelope expansion arises for the following reason[8]: During this phase, the evolution occurs at the Kelvin–Helmholtz time scale. Therefore, for time scales shorter than the Kelvin–Helmholtz time scale but longer than the virialisation time scale, both energy conservation ($K + U = $ constant) and virial theorem ($2K + U = 0$) must hold. This requires U and K to be separately conserved. (Numerical integration shows that this is indeed true to ~10% accuracy). With $M_{\text{ic}} \gg M_{\text{en}}$, the magnitude of the potential energy can be written as

$$|U| \approx \frac{GM_{\text{ic}}^2}{R_c} + \frac{GM_{\text{ic}}M_{\text{en}}}{R_*} \approx \text{constant}, \qquad (3.90)$$

where R_* is the radius of the star. Further, we saw in Exercise 3.3 that the location of the burning shell does not change significantly, implying that both M_{ic} and

3.4 Evolution of High-Mass Stars

M_{en} are constant. Then

$$\frac{dR_*}{dR_c} \approx -\left(\frac{M_{\text{ic}}}{M_{\text{en}}}\right)\left(\frac{R_*}{R_c}\right)^2, \tag{3.91}$$

implying that the radius of the star increases as the core contracts.

This argument can be used to relate the slope of the trajectory in the H-R diagram during this phase to properties of core and envelope. The luminosity of the star arises essentially because of the gravitational contraction of the core that is due to the release of the energy (GM_{ic}^2/R_c) in a time scale $(G\rho_{\text{core}})^{-1/2}$. This gives

$$L \propto \frac{GM_{\text{ic}}^2}{R_c}\left(\frac{GM_{\text{ic}}}{R_c^3}\right)^{1/2} \propto R_c^{-5/2}, \quad T_{\text{eff}} \propto L^{1/4} R_*^{1/2} \propto R_c^{-5/8} R_*^{1/2}. \tag{3.92}$$

Taking logarithmic derivatives, we easily see that

$$\frac{d \ln L}{d \ln T_{\text{eff}}} = \frac{20}{5 + 4(R_* M_{\text{ic}}/R_c M_{\text{en}})}. \tag{3.93}$$

Writing the energy $|U|$ as

$$|U| = \frac{GM_{\text{ic}}^2}{R_c}\left(1 + \frac{R_c M_{\text{en}}}{R_* M_{\text{ic}}}\right), \tag{3.94}$$

we see that the quantity $(R_c M_{\text{en}}/R_* M_{\text{ic}})$ gives the fractional contribution of the envelope to the total energy. Numerical simulations show that this is $\sim 20\%$ so that $(R_* M_{\text{ic}}/R_c M_{\text{en}}) \approx 5$. Then we get a universal slope for the trajectory of the star in the H-R diagram that is $(d \ln L/d \ln T_{\text{eff}}) \approx (4/5) \approx 0.8$. The universality of the slope as well as its value close to 0.8 is seen in numerical evolutions for stars in the mass range of $\sim(4\text{--}8)\ M_\odot$.

The location of the nuclear burning shell remains stationary during this process; this thin shell is located at the zero of the velocity field, with matter at the lower radius contracting and matter at the larger radius expanding.

Because the contraction of the core, which causes the star to move from 5 to 6 in the H-R diagram of Fig. 3.8, is a very rapid process (a $5M_\odot$ star hardly spends $\sim 10^6$ yr in this phase), it is difficult to observe stars in this phase; the paucity of stars in this region of the H-R diagram is known as the *Hertzsprung gap*.

3.4.4 Red Giant Phase and Core Helium Ignition

With the expansion of the envelope, the opacity of the photosphere increases because of an additional contribution from H^- ions in the cool outer regions and a convection zone develops near the surface. The structure of such a star with a convective outer region is well approximated by the Hayashi track, and the star now evolves along the Hayashi track from point 6 to point 7 in Fig. 3.8. Because the convective layers have penetrated fairly deeply into the star by this time, some of the products of the nuclear reaction can be scooped up and distributed all the way to the stellar surface, and they make their presence known through spectral signatures from the photosphere. (This process is called *dredge-up*.)

In the initial stages of core collapse, the energy is supplied by gravitational contraction and the temperature gradient is established in the core. Because the massive stars start with a relatively low central density and remain nondegenerate during this phase, the core contraction leads to an increase in temperature. When the temperature reaches $\sim 10^8$ K, helium burning can be ignited (see Vol. I, Chap. 12) at the core (at point 7 of Fig. 3.8) and this new energy source stops the rapid contraction.

With the burning of helium in the core, the star again reaches thermal and hydrostatic equilibrium. The initial input of energy from the core helium's burning causes the core to expand slightly with the envelope's contracting. This is just the reverse of the process described above; as a result, the star moves from 7 to 8 in the H-R diagram.

3.4.5 Horizontal and Asymptotic Giant Branches

The evolution now is similar to the preceding main-sequence evolution except for the fact that helium has replaced hydrogen as the fuel. The helium burning increases both the luminosity and the temperature of the star, and it moves from 8 to 10 in Fig. 3.8. With the gradual depletion of fuel in the core, slow contraction sets in, and this phase is similar to the main-sequence evolution described in Subsection (3.4.1) above. The difference in molecular weight now is not as significant as in the case of hydrogen burning, which is reflected by the slope of the track. It must be stressed that the helium burning lasts only $\sim 10^7$ yr, roughly 20% of the duration of the main sequence; even during the helium burning, a fair share of the luminosity of the star is contributed by the hydrogen's burning in a thin shell around the core.

With the increase in temperature, the convective envelope retreats back towards the surface, while at the same time a convective core develops. The latter is again due to the high-temperature sensitivity of the triple-alpha reaction. When the star has reached point 10 of Fig. 3.8, the core helium burning gives way to shell burning around a carbon core. The evolution from 10 to 11 is now similar to the phase from 5 to 6; as the core contracts, the envelope expands, resulting in the drop in stellar luminosity. The part of the H-R diagram between points 8 and 11 is known as the *horizontal branch* and is populated by stars having a helium burning core and hydrogen burning shell. Because this evolution occurs with nuclear burning, it proceeds on a slower time scale and the stars in this region of the H-R diagram are observed. We shall see later in Section 3.7 that stars in this regime can also develop instabilities in their outer envelope that lead to pulsations that are observable.

The evolution from 11 to 14 is similar to evolution between points 8 and 11, with the shell helium burning replacing the core helium burning. A convective region develops near the surface and moves downwards into the star. Just as occurs around 6, the outer convective zone once again scoops the elements processed in the core and redistributes it throughout the star. The second upward evolution along the Hayashi track (14–15) by the star is referred to as the *asymptotic giant branch* (AGB). The internal structure of a star approaching the AGB is shown in Fig. 3.9(b). Note that there are two regions of shell burning and that the radial scale is exaggerated.

The evolution described above is qualitatively similar for all stars in which the core remains nondegenerate while helium burning is ignited. The number of left–right excursions made by the star in the H-R diagram and some other details will depend on the mass of the star, but the overall evolution up to this point proceeds as described above. The details of further evolution, however, depend strongly on the mass of the stars, and we will discuss some of these aspects in Section 3.6.

3.5 Evolution of Low-Mass Stars

The evolution of stars with masses below $\sim 2.3\ M_\odot$ is quite different from that of the high-mass stars. The differences arise because of the following facts: (1) Low-mass stars, contrary to high-mass stars, have no convective core; as a result, material is not mixed at the centre of the low-mass stars, and the transition between a core of helium and a shell of hydrogen is continuous in these stars. This point is illustrated in Fig. 3.11, which compares (schematically) the production of helium in the central regions in low- and high-mass stars. (2) Low-mass stars have central temperatures and densities closer to the degeneracy limit than high-mass stars. As the star evolves, the core quickly becomes degenerate, which leads to very different consequences compared with those of the high-mass stars. (3) Low-mass stars are closer to the Hayashi track in their main-sequence location compared with high-mass stars; this fact also plays a role in their later evolution.

3.5.1 Core and Shell Hydrogen Burning

The evolutionary track of the star in the H-R diagram is shown in the top frame of Fig. 3.12. The main-sequence evolution again consists of the gradual conversion of hydrogen to helium at the core along with slow expansion of the star, as described in Subsection 3.4.1 for high-mass stars. Because the *p*–*p* chain operating in the low-mass stars has lower temperature sensitivity, the radius changes more gradually compared with that of high-mass stars. This takes the star along 1–3 in Fig. 3.12, with the initial part (1–2) being along nearly constant radius on the main sequence. Around 3, the hydrogen is exhausted in the core. The core

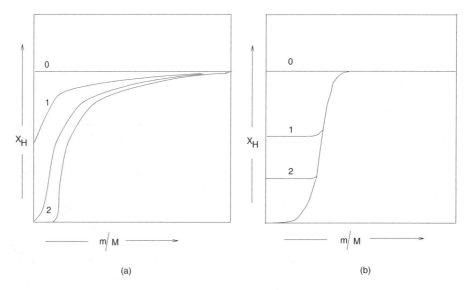

Fig. 3.11. The evolution of hydrogen fraction in the central regions of (a) a low-mass star and (b) a high-mass star. The curves marked 0, 1, and 2 correspond to different times $t_0 < t_1 < t_2$.

contracts, the envelope expands, decreasing $T_{\rm eff}$, and the hydrogen is ignited in the shell region during 4–5. The transition from core burning to shell burning is nearly continuous in low-mass stars, unlike the high-mass stars, in which the core has to contract in order to increase the shell temperature to the ignition point.

The effect of shell burning is, however, different in low-mass stars compared with what we have seen in high-mass stars. Because the cores are nearly degenerate, the Chandrasekhar–Schoenberg limit is fairly irrelevant for low-mass stars. As the burning shell causes the core mass to exceed $\sim 0.1\, M_\odot$, the core contraction would have produced sufficient degeneracy to circumvent the Chandrasekhar–Schoenberg constraint. At this stage, the core is made of degenerate, isothermal helium and no rapid core contraction occurs. Hence low-mass stars do not exhibit the Hertzsprung gap that arises when the star makes a rapid movement from left to right in the H-R diagram.

During the initial phase, the growth of the core mass is slow and the whole core settles to a temperature of the surrounding hydrogen burning shell. This temperature ($\sim 10^8$ K) is lower than the threshold for the ignition of helium. This is in contrast to what happens in high-mass stars in which the core contraction leads to helium ignition. The helium burning in low-mass stars, as we shall see, occurs at a later stage in a very different manner. The shell burning phase that occurs between the core hydrogen burning and the (later) core helium burning is a slow nuclear phase.

3.5 Evolution of Low-Mass Stars

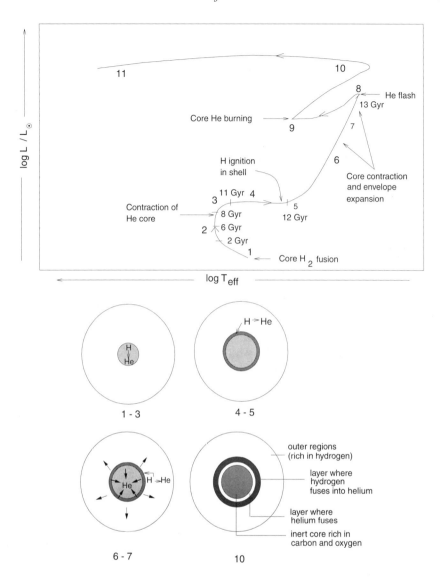

Fig. 3.12. The top frame shows the evolution of a star with $M = M_\odot$ in the H-R diagram and the bottom frames show the internal structure of the same star at different epochs: 1, hydrogen fusion in the centre of the star corresponding to the main-sequence phase; 2–3, the helium core begins to contract and the radius of the star increases slowly; 4–5, hydrogen is exhausted in the core and the contraction increases the temperature sufficiently to ignite helium in a shell; 6–8, the core contraction continues with the envelope expanding considerably – the star becomes a red giant and helium flash occurs at around 8; 9, helium combustion stabilises in the core, which expands while the envelope contracts. The internal structure of the star at selected phases is shown in the four frames at the bottom.

3.5.2 Red Giant Phase

The contraction of the core does lead to the expansion of the hydrogen-rich envelope outside the shell source and to a significant increase in the luminosity. This part of the evolution is essentially controlled by the core mass M_c and is independent of the mass $M - M_c$ of the envelope. An approximate analytic description of this phase can be provided along the following lines.

Let us model the star as being in equilibrium with a degenerate helium core of mass M_c and radius R_c surrounded by an extended envelope of hydrogen with some mass fraction X_H. As the shell burns, the core mass increases at the rate $\dot{M}_c = (L/X_H E_H)$, where E_H is the energy liberated per unit mass of hydrogen. As evolution proceeds, M_c increases and so does L, which is essentially governed by the core conditions. To determine this dependence, we shall assume that the opacity and the energy generation can be parameterised by

$$\kappa = \kappa_0 P^a T^b, \quad \epsilon = \epsilon_0 \rho^\lambda T^\nu. \tag{3.95}$$

The equation of state outside the core can be taken to be that of an ideal gas with $P = (\mathcal{R}/\mu)\rho T$. Under such conditions, homologous solutions will exist for the equations of stellar structure in which the core mass is taken to dominate over the mass of the envelope; that is, we can replace $M(r)$ with M_c in the equation for hydrostatic equilibrium. Straightforward analysis now leads to the scaling relations

$$T \propto \mu \frac{M_c}{R_c}, \quad L \propto \mu^{\sigma_3} M_c^{\sigma_1} R_c^{\sigma_2}, \tag{3.96}$$

where

$$\sigma_1 = \frac{\nu(1+a) - (\lambda+1)(a+b-4)}{2+\lambda+a},$$

$$\sigma_2 = \frac{(1+a)(3-\nu) + (\lambda+1)(a+b-3)}{2+\lambda+a}, \tag{3.97}$$

$$\sigma_3 = \frac{\nu(1+a) + (\lambda+1)(4-b)}{2+\lambda+a}.$$

As an illustration, consider the case of electron scattering ($a = b = 0$) and the CNO cycle ($\nu \approx 15; \lambda = 1$) for which

$$L \propto M_c^{7.7} R_c^{-6}, \quad T \propto \frac{M_c}{R_c}. \tag{3.98}$$

These relations, $T(M_c, R_c)$ and $L(M_c, R_c)$, can be used to determine the evolutionary track of the star in the H-R diagram if we can obtain an extra relation between M_c and R_c. For the *fully* degenerate nonrelativistic core, $M_c \propto R_c^{-3}$ (see Vol. I, Chap. 10, Section 10.3). In this particular case, the degeneracy varies from partial to total in a rather smooth manner and the relation $R_c(M_c)$ can be

3.5 Evolution of Low-Mass Stars

more complicated. However, relations (3.98) show that L is a strongly increasing function of M_c even if R_c is a constant; if R_c decreases with M_c, then this increase will be still steeper. Writing the scaling relations in the form

$$\frac{d \ln L}{d \ln M_c} = \sigma_1 + \sigma_2 \frac{d \ln R_c}{d \ln M_c} \qquad (3.99)$$

and using the approximate fitting function for $d \ln R_c/d \ln M_c$ [see Vol. I, Eq. (10.16)], we find that $d \ln L/d \ln M_c \approx 8\text{--}10$. The integrated form of this relation is given by the left-hand side of the solid curve in Fig. 3.13.

In the above discussion, we have ignored the radiation pressure that could be important in some contexts. For the sake of completeness, we shall now give the corresponding results with radiation pressure. In this case, the equation of state

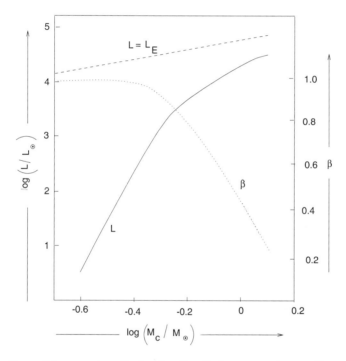

Fig. 3.13. The solid curve gives the luminosity of a low-mass star as a function of core mass obtained by the integration of Eq. (3.99). The left part of the figure is obtained without including the radiation component, which is negligible for low masses. The right top end of the solid curve shows the effect of radiation, which becomes important at high masses. The dotted curve gives the ratio of gas pressure to the total pressure obtained by the integration of Eq. (3.107) with the corresponding scale shown in the vertical axis on the right-hand side. It is clear that gas pressure becomes less and less important at large masses. The dashed curve gives the Eddington luminosity, which is the limiting form for $L(M)$.

is given by

$$P = P_{\text{gas}} + P_{\text{rad}} = \frac{\mathcal{R}}{\mu}\rho T + \frac{a}{3}T^4 \equiv \frac{1}{\beta}\frac{\mathcal{R}}{\mu}\rho T, \quad \beta \equiv \frac{P_{\text{gas}}}{P_{\text{gas}} + P_{\text{rad}}}. \quad (3.100)$$

Direct computation now shows that

$$\frac{\partial \ln P}{\partial \ln \rho} = \beta, \quad \frac{\partial \ln P}{\partial \ln T} = 4 - 3\beta, \quad (3.101)$$

allowing us to approximate the equation of state as

$$P \propto \rho^\beta T^{4-3\beta}. \quad (3.102)$$

[Another equivalent form of this equation of state is $\rho \propto P^\alpha T^{-\delta}$, with $\alpha = 1/\beta$ and $\delta = (4 - 3\beta)/\beta$.] Then the corresponding scaling relations turn out to be

$$T \propto M_c^{\psi_1} R_c^{\psi_2}, \quad L \propto M_c^{\sigma_1} R_c^{\sigma_2}, \quad \rho(r/R_c) \propto M_c^{\phi_1} R_c^{\phi_2}, \quad (3.103)$$

with

$$\psi_1 = \frac{\lambda + 2}{N}, \quad \psi_2 = \frac{2\beta - \lambda - 4}{N};$$

$$\sigma_1 = \frac{4(\lambda + 1) + \nu}{N}, \quad \sigma_2 = \frac{-\nu - 3\lambda}{N}\beta; \quad (3.104)$$

$$\phi_1 = \frac{4 - \nu}{N}, \quad \phi_2 = \frac{\nu - 12 + 6\beta}{N};$$

$$N = (4 - 3\beta)(2 + \lambda) + (1 - \beta)(\nu - 4). \quad (3.105)$$

When $\beta = 1$ and $a = b = 0$, these relations agree with previously derived results. With increasing core mass, β in the shell will decrease rapidly. Writing

$$\beta \propto \frac{\rho T}{P} \propto \rho^{1-\beta} T^{-3(1-\beta)} \quad (3.106)$$

and using the dependence of ρ and T on M_c and R_c, we easily derive the variation of β with M_c. We get

$$\frac{d \ln \beta}{d \ln M_c} = (1 - \beta)\left[(\phi_1 - 3\psi_1) + (\phi_2 - 3\psi_2)\frac{d \ln R_c}{d \ln M_c}\right]. \quad (3.107)$$

Given the fitting function for $R_c(M_c)$, this equation can be integrated to find $\beta(M_c)$. The results of such an integration for $\lambda = 1$ and $\nu = 16$ are shown in Fig. 3.13 by the dotted curve. Integration of Eq. (3.99) with the values of the parameters given by Eqs. (3.104) and $\beta = \beta(M_c)$ allows us to determine the luminosity including the effects of radiation pressure. This is shown by the solid curve. For small core masses, $\beta \approx 1$ and relation (3.98) holds, giving a steep increase of luminosity with core mass. For larger M_c, radiation pressure becomes important and β decreases, leading to a decrease in the slope of the

3.5 Evolution of Low-Mass Stars

$L(M_c)$ curve. In the limit of $\beta = 0$, we have $\sigma_1 = 1$ and $\sigma_2 = 0$, giving $L \propto M_c{}^9$. Numerical work confirms this result and gives a more exact fit of the form

$$\frac{L}{L_\odot} = 5.92 \times 10^4 \left(\frac{M_c}{M_\odot} - 0.52 \right). \tag{3.108}$$

As the star moves up and to the right (1 → 5 in Fig. 3.12), the energy production switches from core to shell burning and the core mass starts increasing with the ashes of shell burning being added to it. This causes the luminosity to increase by a large factor, and the star moves up very close to the Hayashi line along the ascending giant branch (5 → 7 in Fig. 3.12). As to be expected, the outer convective region develops and moves inwards, making more than 70% of the total mass convective. This fact also explains the closeness of the track to the Hayashi line. When the convective envelope reaches the region already contaminated by the products of nuclear evolution, the processed material gets well mixed and is also brought partially to the surface.

When the outer convective region and the hydrogen burning shell come into contact, the monotonic increase of the luminosity is disrupted. At this point, a discontinuity in molecular weight between the homogeneous hydrogen-rich outer layer and the helium-enriched layers below is produced by the mixing. When the shell reaches the discontinuity, the molecular weight of the shell becomes smaller. From the homology relations with $\nu = 16$, $\lambda = 1$, and $a = b = 0$, we see that $L \propto \mu^8$. The luminosity decreases with decreasing μ and the trajectory of the star in the H-R diagram actually drops back (not shown in Fig. 3.12), causing a transient reduction in L. After the shell source goes past the discontinuity, μ retains its lower value and the luminosity grows with increasing core mass.

Because the key properties are essentially determined by the core mass and are independent of the total mass, tracks of different stars come very close to each other while approaching the Hayashi line. The same convergence occurs for the central conditions; the tracks of different low-mass stars converge to the same region in the ρ_c–T_c plane (see Fig. 3.14).

3.5.3 Helium Flash

With growing core mass, the temperature of the core rises. When the core mass is $M_c \approx 0.45\, M_\odot$, the core temperature reaches $T_c \approx 10^8$ K, at which the helium is ignited (point marked 8 in Fig. 3.12). In the next phase of the star, the helium burns at the core explosively. We shall now discuss some of the details of this process, called the *helium flash*.

When the nuclear reactions are triggered in fully degenerate matter, the increase in temperature does not lead to a corresponding increase in pressure because the equation of state $P \propto \rho^{5/3}$ does not depend on temperature. The material therefore cannot expand and cool. The increase in temperature leads to

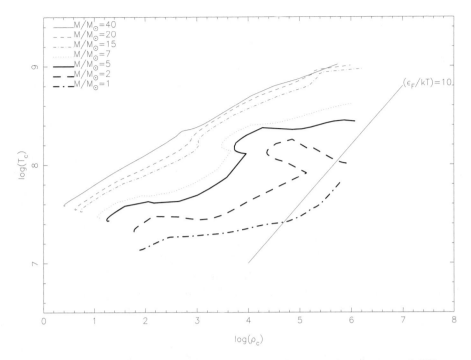

Fig. 3.14. The time evolution of central densities and temperatures in stars of different masses. The masses of the stars are shown in the legend. The straight line corresponds to $\epsilon_F = 10\, k_B T$ and shows that degeneracy pressure can be important for low-mass stars.

a further increase in the rate of energy production, quickly leading to a runaway situation.

The mathematical origin of helium flash is fairly straightforward. The rate of change of temperature is decided by

$$C_V \frac{dT}{dt} = C_V \frac{dE}{dt}\left(\frac{dE}{dT}\right)^{-1} = \epsilon, \tag{3.109}$$

where $C_V = C_{V,\text{ion}} + C_{V,\text{ele}}$. For fully ionised helium, $C_{V,\text{ion}} = (3/2)(k_B/4m_p)$, whereas the specific heat of degenerate electron gas [see Eq. (5.179) of Vol. I; we have divided by ρ to obtain specific heat for unit mass] is

$$C_{V,\text{ele}} \cong 1.35 \times 10^{14}\, T_9 x (1+x^2)^{1/2} \rho^{-1}, \quad x = \left(\frac{\rho}{9.74 \times 10^5 \mu_e \text{ gm cm}^{-3}}\right)^{1/3}. \tag{3.110}$$

The nuclear-energy-generation rate is given by the fitting formula

$$\epsilon = 5.1 \times 10^8 \rho^3 Y^2 T_9^{-3} \exp(-4.4027/T_9) \text{ ergs gm}^{-1}\, \text{s}^{-1}. \tag{3.111}$$

The equation for dT/dt can now be integrated with the initial values, say,

$T_9 = 0.15$, $\rho = 2 \times 10^5$ gm cm^{-3}, $\mu_e = 2$, and $Y = 1$. The integration has to be stopped when the electrons become nondegenerate, which will occur when $(\rho/\mu_e) = 6 \times 10^{-9} T^{3/2}$ in cgs units. The resulting temperature remains approximately constant up to about 5.7 days and then shoots up rapidly. This shows that the helium flash is nearly explosive with a time scale of ~6 days.

A general criterion describing such phenomena can be obtained along the following lines. From Eqs. (3.63) and (3.64), we can relate the fractional change of pressure to fractional change of temperature as

$$p_c \equiv \frac{dP_c}{P_c} = \frac{4\delta}{4\alpha - 3} \frac{dT_c}{T_c}. \tag{3.112}$$

From the first law of thermodynamics, we can find the heat dq added to the system as

$$dq = du + P dv = C_P T_c (\vartheta_c - \nabla_{\text{ad}} p_c) = c^* T_c \vartheta_c, \tag{3.113}$$

where $p_c \equiv dP_c/P_c$, $\vartheta_c \equiv (dT_c/T_c)$ and

$$c^* = C_P \left(1 - \nabla_{\text{ad}} \frac{4\delta}{4\alpha - 3}\right) \tag{3.114}$$

represents the gravitational specific heat per unit mass of material; $dT = dq/c^*$ gives the change in temperature of the core if the heat dq is added to it. This specific heat corresponds to the situation in which the gas pressure of the core balances the weight of the envelope. For an ideal monotonic gas, $\alpha = \delta = 1$ and $\nabla_{\text{ad}} = 2/5$, giving $c^* < 0$. This negative value for the gravitational specific heat is, in fact, necessary for stability. If the nuclear-energy generation increases slightly, leading to a perturbation with $dq > 0$, then the condition $c^* < 0$ ensures that $dT < 0$. The cooling of the region reduces the energy overproduction immediately. On the other hand, for a nonrelativistic degenerate gas, $\delta \to 0$ and $\alpha \to 3/5$, leading to $c^* > 0$. This clearly is unstable and the temperature of the region goes up rapidly until the degeneracy is removed.

The evolution of the star during the helium flash can be understood from Eq. (3.64), written in the form

$$\frac{d\rho_c}{\rho_c} = \frac{3\delta}{4\alpha - 3} \vartheta. \tag{3.115}$$

For $\alpha = 3/5$, $\delta = 0$, we have $d\rho_c = 0$; the central region evolves at constant density and is represented by a vertical line in the ρ_c–T_c plane. With increasing temperature at constant density, the degeneracy is finally removed when the vertical line cuts the $\alpha = 3/4$ line, giving the border between degenerate and ideal-gas. With a further increase in T, the core expands whereas the removal of degeneracy makes the gravitational specific heat negative. The helium burning then becomes stable, with the expansion stopping the increase in temperature. The star settles down for the values appropriate for quiet, stable helium burning.

3.5.4 Evolution after the Helium Flash

Until the onset of the helium burning, the total luminosity of the star (produced in the hydrogen burning shell), increases with increasing core mass; after the degeneracy is removed, the core expands and the R_c increases. During the short phase of the flash, M_c does not change significantly. Hence relation (3.98) predicts that the luminosity is to be appreciably reduced after the flash phase. This is indeed seen in parts 8–9 in Fig. 3.12. The helium burning now stabilises in the core and the evolution along 9–10 is similar to the initial evolution of the star along 2–3, with helium burning replacing the hydrogen burning.

The evolution of the star beyond point 9 involves several complications that have been only partly addressed in the literature. Broadly speaking, the star at this stage can be thought of as made of a helium burning core and a hydrogen envelope. Such a structure can be modelled with the standard equations of stellar structure with an extra parameter q_0 that gives the ratio of the mass in the helium (core) to the total mass. For $q_0 = 1$, the entire star is made of helium; such a (hypothetical) helium main-sequence star will lie to the left of the hydrogen main sequence in the H-R diagram because of the scaling with the molecular weight [see relations (2.58) of Chap. 2]. The situation, however, changes drastically if q_0 is reduced even slightly. For example, if $q_0 = 0.9$, the star moves fairly close to the hydrogen main sequence. This is to be expected because the position of the star in the H-R diagram depends crucially on the surface properties of the star, and even the addition of a small amount of hydrogen as an envelope will affect the position of the star in the H-R diagram significantly. The behaviour again changes completely when q_0 drops below a value between 0.8 and 0.7 (depending on the total mass M of the star). For q_0 below this critical value, the location moves further to the right and lies close to the Hayashi track. The closest approach to the Hayashi track occurs around $q_0 = 0.5$ and for smaller values of q_0 these tracks move back to the left.[10]

From this analysis, we would have expected the star with a helium burning core and a hydrogen envelope to move to the left in the H-R diagram towards the horizontal branch. The subsequent evolution parallels the original departure from zero-age main sequence (ZAMS), and the star moves back to the red giant branch along the track (9–10), usually called an AGB. This branch is close to the original giant branch but at a slightly higher effective temperature.

Before concluding this section, we shall comment briefly on the time scales involved in stellar evolution. The conversion of four hydrogen nuclei to one helium nucleus releases ~ 26.7 MeV, of which a small fraction is taken away by the neutrinos, leaving ~ 25 MeV trapped in the star. If a fraction f of the total mass of the star can be converted during the main-sequence phase, then the main-sequence lifetime will be $t_{\rm ms} \approx 0.007\, f(Mc^2/L)$. We saw that $\sim 10\%$ of the stellar mass gets converted in the initial phase, leading to $t_{\rm ms} \approx 10$ Gyr $(M/M_\odot)(L/L_\odot)^{-1}$.

All other phases of evolution last for much shorter period of time. For example, let us consider the horizontal branch. Stars with an initial mass less than approximately $(1.8-2.2)$ $M_\odot$ experience helium flash and have a luminosity of ~ 50 $L_\odot$ while on the horizontal branch. The helium core of these stars have masses of ~ 0.45 $M_\odot$. Approximately half of this helium is converted into carbon and the other half becomes oxygen, releasing $7.2 \times 10^{-4}(\Delta M)c^2$ of energy. Hence the time scale for the horizontal branch is approximately constant and is given by

$$t_{\text{hb}} \approx 7.2 \times 10^{-4} \frac{0.45 \, M_\odot c^2}{50 \, L_\odot} \approx 0.5 \text{ Gyr,} \qquad (3.116)$$

which is quite short compared with the main-sequence lifetime.

3.6 Late-Stage Evolution of Stars

In the last two sections, we discussed the key physical processes that govern the first part of stellar evolution. The evolution described so far takes the star to a fairly late stage in its life. We shall now consider what happens at the last phases of stellar evolution.[11] In this phase, the evolution becomes complicated because of two effects: (1) The hydrogen and the helium burning shells exhibit certain thermal instabilities that lead to periodic changes in the luminosity of the star, and (2) the envelope of the star sheds mass in the form of stellar wind at a fairly high rate and the star could lose a large fraction of its mass fairly rapidly. Let us now consider these effects:

(1) The thermal instability of the shell occurs for both low- and high-mass stars and hence can be studied in a general manner. To understand the origin of this instability, let us compare the core nuclear burning with shell nuclear burning. In the case of core burning, $m \propto \rho r^3$ and an expansion $dr > 0$ (without change of mass, $dm = 0$) requires that $(d\rho/\rho) = -3(dr/r)$. In the case of a shell source with thickness D, the mass scales as $m \propto \rho r^2 D$, where r is the radius of the base of the shell. If the shell expands (i.e., D becoming $D + dD$ with r = constant), then

$$\frac{d\rho}{\rho} = -\frac{dD}{D} = -\frac{r}{D}\frac{dr}{r}. \qquad (3.117)$$

The mass outside $r_0 + D$ may be assumed to expand or contract homologously, giving $(dP/P) = -4(dr/r)$. Repeating the original analysis that leads to Eq. (3.113) in the present case, we get, in place of Eq. (3.113), the result

$$c^* \frac{dT}{dt} = d\epsilon, \quad c^* = C_P\left(1 - \nabla_{\text{ad}} \frac{4\delta}{4\alpha - r/D}\right). \qquad (3.118)$$

The only change is the replacement of the factor 3 with r/D, the origin of which is obvious from Eq. (3.117).

The system is clearly unstable if c^* is positive. This occurs in the case of strong degeneracy with $\delta \approx 0$, which is the cause of, for example, the helium flash discussed above. (It is clear that such a flash can also occur in a shell of degenerate material and need not necessarily be in the core.) In addition, there arises a new instability even for an ideal monotonic gas with $\alpha = \delta = 1$ and $\nabla_{\rm ad} = 2/5$. If D/r is small enough (in the above case, if it is less than 0.25), then c^* is positive and the shell burning is unstable. The physical origin of this instability is the following. Suppose the shell tries to get rid of some excess energy by expansion. If $D/r \ll 1$, then a substantial fractional increase in dD/D leads to a corresponding substantial change in $d\rho/\rho$ but only a small change dr/r. The outer layers have moved negligibly and so their weight remains approximately constant, and hydrostatic equilibrium demands that dP/P should be negligible. Comparing the relation $dP/P = (4D/r)(d\rho/\rho)$ with the equation of state written in the form

$$\frac{d\rho}{\rho} = \alpha \frac{dP}{P} - \delta \frac{dT}{T} \tag{3.119}$$

shows that $d\rho/\rho < 0$ requires that $dT/T > 0$. In fact, when $dP/P \to 0$, we have $d\rho/\rho \approx -\delta(dT/T)$. In summary, the expansion of a thin nuclear burning shell does not stabilise it, but instead heats it up. The shell acts as though the equation of state is ρT = constant, which is clearly unstable.

The fact that outer layers of matter will be pushed out when the base of the envelope produces sufficiently high luminosity can be understood from fairly simple considerations. An element of matter with internal-energy density $U \approx (\rho k_B T/m_{\rm H})$ cannot move faster than the local speed of sound $v_s \approx (k_B T/\mu m_{\rm H})^{1/2}$ without forming a shock wave. Therefore the maximum energy that convective flux can carry is $U v_s$, giving the maximum convective luminosity as

$$L_{\rm max,c} \simeq (4\pi r^2) U v_s \simeq 4\pi r^2 \rho \mu \left(\frac{k_B T}{\mu m_{\rm H}}\right)^{3/2}. \tag{3.120}$$

The right-hand side is fixed by the matter in the envelope wheras the luminosity could be generated in a shell undergoing nuclear reaction at the base of the envelope. When $L > L_{\rm max,c}$, radiative transport takes over at the base of the envelope and a temperature gradient is set up. The corresponding pressure gradient, taking into account both the gas and radiation pressures, will be

$$\frac{dP}{dr} = \frac{dP_R}{dr} + \frac{dP_g}{dr} = \frac{4a}{3} T^3 \frac{dT}{dr} + \frac{k_B T}{\mu m_{\rm H}} \frac{d\rho}{dr} + \frac{k_B \rho}{\mu m_{\rm H}} \frac{dT}{dr}. \tag{3.121}$$

This can be converted by using the equation of motion and the expression for radiative luminosity in terms of the temperature gradient

$$L(r) = -\frac{16\pi ac}{3\kappa \rho} r^2 T^3 \frac{dT}{dr} \tag{3.122}$$

3.6 Late-Stage Evolution of Stars

to obtain

$$\frac{P_g}{\rho}\frac{d\rho}{dr} = \frac{1}{r^2}\left[\left(\frac{\kappa\rho L}{4\pi c}\right)\left(\frac{3P_g}{4P_R}+1\right) - mG\rho - \rho r^2\ddot{r}\right]. \quad (3.123)$$

In hydrostatic equilibrium, $\ddot{r}=0$ and $(d\rho/dr)$ is negative. This requires that

$$\frac{\kappa\rho L(r_1)}{4\pi G cm(r_1)}\left(\frac{3P_g}{4P_R}+1\right) \leq 1 \quad (3.124)$$

at the base of the envelope, $r=r_1$, say. If $L(r_1)$ becomes too large and violates this inequality, then the term $\rho r^2\ddot{r}$ must be retained with $\ddot{r}>0$ at the base of the envelope. When a shell flash is strong enough some of the envelope will be accelerated to escape velocities. Even when the shell flash is relatively weak, the expanding envelope will have a higher mean free path and the radiative-energy loss can lead to a lowering of pressure. The envelope will then collapse and will be recompressed as it reaches the core, sending out a compression wave. This wave can develop into a shock and eject the envelope. Numerical calculations indicate that such processes can eject red giant envelopes over a time scale of 10^2–10^3 yr.

Such thermal instabilities occur in different stars at different stages of evolution. For example, a star in the AGB, depending on its mass, may have more than one shell source of nuclear energy [see Fig. 3.9(b)]. In the AGB phase, the hydrogen shell reignites and the helium shell below it turns on and off periodically. The basic reason for this phenomenon is fairly simple. As the hydrogen burning shell dumps helium on the layer below, it can become slightly degenerate and undergo a helium shell flash. This drives the hydrogen burning shell outwards, causing it to cool and turn off for sometime. Eventually, the burning in the helium shell reduces, the hydrogen burning reoccurs, and the process repeats. The period between pulses depends on the mass of the star and varies from thousands of years (for stars with $M \sim 5\,M_\odot$) to hundreds of thousands of years (for stars with $M \sim 0.6\,M_\odot$).

(2) The rapid expansion of the star in the track 9–10 of Fig. 3.12 implies that the outer regions of the star are very loosely bound. Hence it is possible for matter to escape from the star in the form of a steady outflow, usually called a *stellar wind*. As a simple estimate of this effect, consider a point inside a star with temperature of approximately 5×10^4 K. The kinetic energy $(3/2)k_B T$ corresponding to this temperature is ~ 7 eV. In a $1\,M_\odot$ star, a hydrogen atom at a distance of $\sim 200\,R_\odot$ has approximately the same potential energy. This suggests that, if a $1\,M_\odot$ star expands beyond $200\,R_\odot$, such that a temperature of 5×10^4 K occurs around $200\,R_\odot$, the outer regions of the star will be rather loosely bound.

Modelling the resulting stellar wind from fundamental considerations is extremely difficult and no reliable theory exists at present. Semiempirical considerations based on the numerical simulations lead to the estimate of stellar-mass loss for which an approximate fit[12] is given by

$$\dot{M} \approx -2 \times 10^{-8} \left(\frac{L}{10^3 L_\odot}\right)^{3.7} \left(\frac{M}{M_\odot}\right)^{-3.1} \left(\frac{R}{10^2 R_\odot}\right) M_\odot \text{ yr}^{-1}. \quad (3.125)$$

If the mass and the radius of the star can be related to the luminosity by appropriate fitting functions, then this result can be expressed in a simpler form such as

$$\left(\frac{\dot{M}}{M_\odot}\right) \approx 10^{-14.97} \left(\frac{L}{L_\odot}\right)^{1.62} \text{ yr}^{-1}. \quad (3.126)$$

Using the mass–luminosity relation, $L \propto M^{3.5}$, we can convert this equation into a time scale for mass loss:

$$t_{\text{ml}} \approx 10^{14.97} \left(\frac{M}{M_\odot}\right)^{-4.67} \text{ yr}. \quad (3.127)$$

The ratio of this time scale to the nuclear-evolution time scale is

$$\frac{t_{\text{ml}}}{t_{\text{nuc}}} = 10^{5.0} \left(\frac{M}{M_\odot}\right)^{-2.17}. \quad (3.128)$$

A $60 M_\odot$ star has a time scale for mass loss that is $\sim$14 times larger than the evolution time scale. More accurate time scales give even shorter values for this number, showing that the mass loss is fairly significant for high-mass stars.

Exercise 3.4
Stellar feedback on the ISM: A massive O star has a typical luminosity of 3×10^{39} ergs s^{-1}, a lifetime of 3×10^6 yr, a stellar-wind velocity of 5000 km s^{-1}, and a mass-loss rate of 10^{-5} $M_\odot$ yr^{-1}. When it ends up as a supernova, $\sim$5 $M_\odot$ is ejected with a velocity of 5000 km s^{-1}. Estimate the contribution of these processes to the energy and the momentum of the ISM.

It is clear that the final stages of evolution will again be a sensitive function of the mass of the star. If the mass of the star is fairly low, nuclear evolution may never proceed beyond the conversion of hydrogen to helium. Such a star will slowly cool down after the nuclear-energy supply is switched off. For stars with initial masses below 8 $M_\odot$, evolution along the AGB leads to more and more conversion of helium into carbon and oxygen, as the shell burning adds mass to the core. This causes the core to contract slowly, causing the central density to increase. Depending on the star's mass, the emission of neutrinos will also tend to decrease the central temperature during this phase. The situation in such a star is similar to the development of an electron degenerate helium core in a low-mass star as the latter approaches the red giant branch. The further evolution of the star depends on whether the carbon–oxygen core will undergo nuclear burning.

3.6 Late-Stage Evolution of Stars

If the mass is less than $\sim 4\, M_\odot$, this does not occur and the basic nuclear evolution of the star comes to an end at this stage. For masses between 4 and 8 $M_\odot$, it is possible for the carbon–oxygen core to reach a sufficiently large mass such that it cannot remain in hydrostatic equilibrium even with degenerate electron pressure. If the core loses its ability for hydrostatic support, it undergoes a catastrophic collapse, with the temperature and density rising enormously. It is possible for such a carbon–oxygen core flash to be explosive, ejecting the outer mantle of the star. Such a conclusion necessarily assumes that during the evolution the mass loss from the star was not large enough for, say, an 8 $M_\odot$ star to become a 4 $M_\odot$ one. The evolutionary processes leading to such a stage are somewhat uncertain, but it is generally believed that large mass losses in the earlier phases prevent such an explosion from taking place.

In such a case, carbon ignition does not take place for $M \lesssim 8\, M_\odot$, and a more gradual ejection of material from the star in the form of shell flashes, winds, and envelope pulsations will lead to an expanding shell of gas around the core. This expanding shell of gas is called a *planetary nebula*. Once such an ejection takes place, the inner core of the star – made of degenerate carbon and oxygen surrounded by a thin layer of residual hydrogen and helium, with a surface temperature of $\sim 10^4$ K – will be visible. In the H-R diagram, the track now moves towards higher temperatures almost horizontally along 10–11 (see Fig. 3.12). The effective temperature increases but the radius decreases, keeping the luminosity constant. This broadly summarises the evolution of stars with masses below 8 $M_\odot$ or so.

Exercise 3.5

P-Cygni profiles: The process described above can lead to a situation in which a star is surrounded by a thin annular shell of matter with inner radius R_1 and outer radius R_2, with the material in the shell moving outwards with a constant velocity V, which is large compared with the thermal speed v_{th} of the gas. (a) Consider the radiation received from a point (r, θ) on the shell, where θ is the azimuthal angle with respect to the z axis, which is taken to be towards the observer. Show that the height of the column Δz that will contribute radiation at wavelength $\lambda = \lambda_0 + \Delta\lambda = \lambda_0[1 - (V/c)\cos\theta]$ is given by $\Delta z = (2 v_{\text{th}}/V)(r \csc^2 \theta)$. (b) Hence show that the total intensity at the wavelength $\lambda + \Delta\lambda$ is given by

$$I(\Delta\lambda) = \int_{R_1}^{R_2} \frac{\epsilon_0}{\alpha_0}(1 - e^{-\alpha_0 \Delta z})\, 2\pi\, \sin^2\theta\, r\, dr, \qquad (3.129)$$

where ϵ_0 is the emissivity and α_0 is the absorption coefficient. (c) Consider the optically thin and optically thick limits and determine the line profile, i.e., I as a function of $\Delta\lambda$ in the two cases. [Hint: (a) The height of the column is given by $\Delta z = 2(dz/d\Delta\lambda)\Delta\lambda_D$, where $\Delta\lambda_D = \lambda_0(v_{\text{th}}/c)$ is the Doppler width of the line. This gives $\Delta z = 2v_{\text{th}}(dz/dv_z)$. Using $(dv_z/dz) = -(V/r)\sin^2\theta$, we get the quoted result. (b) The integral for $I(\Delta\lambda)$ requires an integration over $dzdxdy$. The z integration is straightforward whereas $dxdy$

integration should be taken over a surface of constant $\Delta\lambda$; that is, a surface of constant θ. This converts $dx\,dy$ to $2\pi\sin^2\theta r\,dr$. (c) In the optically thin case, $I(\Delta\lambda)=$ constant; in the optically thick case, $I(\Delta\lambda)$ is quadratic in $\Delta\lambda$.]

When the star is in the AGB phase, mass loss is driven by large-amplitude radial pulsations and $\dot{M}$ increases with increasing luminosity, as shown by Eq. (3.125). While the envelope mass decreases because of this mass loss, the core mass is growing as a result of hydrogen and helium burning in the shells. If the stellar luminosity is provided essentially by hydrogen burning, then the rate of growth of core mass is proportional to the luminosity and inversely proportional to the efficiency of energy production: $\dot{M}_c = (L/0.007c^2)$ or

$$\dot{M}_c = 10^{-8}\left(\frac{L}{10^3 L_\odot}\right) M_\odot\,\mathrm{yr}^{-1}. \tag{3.130}$$

The time scale for the growth of the core is $\tau_c \equiv (M_c/\dot{M}_c)$. Because $M_c \geq 0.5\,M_\odot$ (which is the required minimum mass for core helium ignition during the earlier evolution) we must have $\tau_c \geq 5\times 10^7(L/10^3\,L_\odot)^{-1}$ yr. On the other hand, the time scale for the shrinkage of the envelope is given by $\tau_{\mathrm{env}} = (M_{\mathrm{env}}/\dot{M})$, where $\dot{M}$ is given by Eq. (3.125). When $\tau_c \gtrsim 10\tau_{\mathrm{env}}$, say, the core is effectively frozen and further envelope mass loss will expose the core, which will appear as the white dwarf. Writing $\tau_c = 10\tau_{\mathrm{env}}$ and using Eq. (3.125), we find that a white dwarf forms effectively at a luminosity at which

$$L = 2.35\times 10^3\, L_\odot \left(\frac{M}{M_\odot}\right)^{1.5}\left(\frac{R}{10^2\,R_\odot}\right)^{-0.37}. \tag{3.131}$$

The core mass at these luminosities can be obtained by fitting formula (3.108), which relates the luminosity of an AGB star with the mass of the degenerate core

$$\frac{L}{L_\odot} \approx 6\times 10^4 \left(\frac{M_c}{M_\odot} - 0.5\right). \tag{3.132}$$

This gives

$$\frac{M_c}{M_\odot} \approx 0.5 + \left[0.04\left(\frac{M}{M_\odot}\right)^{1.5}\left(\frac{R}{10^2\,R_\odot}\right)^{-0.37}\right]. \tag{3.133}$$

This shows that a change of a factor of 4 changes the final core mass only by $\sim 0.3\,M_\odot$. We would therefore expect white dwarfs to be narrowly clustered in mass around $0.6\,M_\odot$; this is indeed observed.

In a more massive star, further nuclear burning can take place with the production of carbon, oxygen, silicon, etc. In fact, as the heavier elements are built up near the centre, the new elements are likely to be evenly distributed in shells, giving an onion-skin model for the star. Figure 3.15 gives the mass of different elements synthesised in stars of different masses.[13] In such an onion-skin

3.6 Late-Stage Evolution of Stars

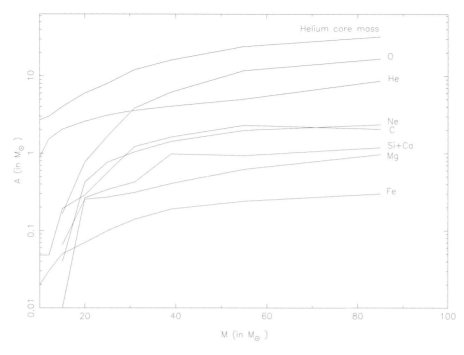

Fig. 3.15. Amount of different heavier elements synthesised in a star of a given total mass.

structure, nuclear reactions could be going on at different shell sources with periodic switching on and off. For stars with intermediate masses, the nuclear reactions will proceed up to some intermediate stage.

The behaviour of such a star in the late stages is easier to state in terms of the mass of the degenerate core M_c after helium burning (rather than in terms of the total mass) because of the uncertainty in estimating mass loss. The crucial factor that discriminates among different evolutionary cases is the ratio between the core mass M_c and the Chandrasekhar mass introduced in Eq. (3.69). If the core mass is larger than the Chandrasekhar mass, the core does not become degenerate, whereas if $M_c < M_{\text{Ch}}$ then degeneracy effects come to play. Broadly speaking, three different cases arise:

(1) $M_c \lesssim M_{\text{Ch}}$: In this case, the evolution depends on whether there exists a sufficiently massive envelope. If no massive envelope exists, then the core mass cannot approach the Chandrasekhar mass because of shell burning and the core temperature increases in the nondegenerate regime until a maximum is reached. Once the maximum is reached, the core becomes degenerate and starts to cool, and the star becomes a white dwarf.

If the envelope is sufficiently massive, then the shell burning can cause the core mass to grow up to the Chandrasekhar mass. The core becomes degenerate and cools after reaching a maximum temperature. However, the density increases with core mass and finally carbon burning is ignited. This starts in a highly degenerate state and could be explosive.

(2) $M_{Ch} \lesssim M_c \lesssim 40\, M_\odot$: In this case, the evolutionary track misses the non-relativistic degenerate region of the ρ–T plane, allowing the core to heat up, reaching successively higher nuclear reactions. Eventually the core will collapse and could possibly lead to a neutron star formation with the ejection of an envelope. This phenomenon, called a type II supernova, will be discussed in detail in Chap. 4. This core is likely to be made of heavier elements such as iron, with core densities of $\rho \gtrsim 10^{14}$ gm cm^{-3} and initial temperatures of $T \gtrsim 10^{10}$ K.

(3) $M_c > 40\, M_\odot$: In this case the core reaches the carbon burning stage while still remaining nondegenerate. However, further evolution can push the central temperature and density to a region of the ρ–T plane in which pair-creation (e^+e^-) effects are important. The pair creation can lead to an instability and core collapse until oxygen burning is triggered. Further evolution is somewhat uncertain, but is expected to be similar to that of case (2) above.

A summary of stellar evolution of different masses is given in Table 3.1 and in Fig. 3.16. (The table mentions two processes, called the r process and the s process, discussed in Vol. I, Chap. 12, Subsection 12.4.5, which will be described in more detail in the next chapter; see Section 4.11.) The total time taken for evolution from the main sequence to the planetary nebula stage for stars with $0.6 \lesssim M/M_\odot \lesssim 10$ is well approximated by the fitting function[14]

$$\log t_{evol} = 9.921 - 3.6648 \log\left(\frac{M}{M_\odot}\right) + 1.9697 \left[\log\left(\frac{M}{M_\odot}\right)\right]^2$$
$$- 0.9369 \left[\log\left(\frac{M}{M_\odot}\right)\right]^3, \tag{3.134}$$

where t_{evol} is measured in years. Similarly, the lifetime of a core helium burning stage on the horizontal branch is given by

$$\log t_{HB} \approx 7.74 - 2.2(M_{core} - 0.5)\ \text{yr}, \tag{3.135}$$

where M_{core} is the core mass of the helium at the onset of helium flash, given by

$$M_{core} \approx 0.476 - 0.221(Y - 0.3) - 0.009(3 + \log Z) - 0.023(M - 0.8). \tag{3.136}$$

The lowest characteristic mass relevant to stellar evolution is $M_{min} \approx 0.08\, M_\odot$. Objects with $M < M_{min}$ do not reach sufficiently high central temperatures to begin hydrogen ignition and end up as brown dwarfs. The next significant threshold mass is $M_{conv} \approx (1.1$–$1.3)\, M_\odot$. Stars with $M > M_{conv}$ have convective cores whereas stars with $M < M_{conv}$ are radiative in the core. The mass scale of $M_{HeF} \approx (1.8$–$2.2)\, M_\odot$ (depending on the composition) decides whether the helium flash occurs in the star. Stars with $M < M_{HeF}$ undergo helium flash

3.6 Late-Stage Evolution of Stars

Table 3.1. Summary of stellar evolution

Masses	Properties
$M \lesssim 0.08\ M_\odot$	No hydrogen burning; brown dwarfs belong to this class; locks up material and lives forever.
$0.08\ M_\odot \lesssim M \lesssim 0.5\ M_\odot$	Hydrogen burning occurs at the centre; helium core becomes degenerate at low temperatures so that helium is never ignited; end stage is a helium white dwarf.
$0.05\ M_\odot \lesssim M \lesssim M_{\mathrm{HeF}}$	Helium is ignited in a degenerate core as helium flash; end stage is a CO white dwarf; precise theoretical value of M_{HeF} depends on the treatment of convection and is in the range of $(1.85\text{–}2.2)\ M_\odot$; if $M \gtrsim 1\ M_\odot$, they contribute ^{4}He, ^{14}N to the ISM through stellar winds and planetary nebula; lifetime $(15\text{–}1)$ Gyr.
$M_{\mathrm{HeF}} \lesssim M \lesssim M_{\mathrm{CO-WD}}$	Helium is ignited in a nondegenerate core but the CO core is supported by electron degeneracy. End stage is CO white dwarf; contributes ^{4}He, ^{12}C, ^{13}C, ^{14}N, ^{17}O, and s-process elements to the ISM through stellar wind and planetary nebula; theoretical estimates give $M_{\mathrm{CO-WD}}$ approximately $8\ M_\odot$; lifetime is $(10^9\text{–}10^7)$ yr.
$M_{\mathrm{CO-WD}} \lesssim M \lesssim (10\text{–}12)\ M_\odot$	Carbon is ignited in a nondegenerate core; those that have helium cores in the mass range $(2.2\text{–}2.5)\ M_\odot$ ignite oxygen in a degenerate NeO core; in those with helium cores of mass $(2.5\text{–}3)\ M_\odot$, all the burning stages are nondegenerate and an Fe core is formed; lifetime is (several $\times\ 10^7$) years; end stage is a type II supernova; contributes to ISM mostly He because other elements are locked up in the collapsing cores.
$(10\text{–}12)\ M_\odot \lesssim M \lesssim M_{\mathrm{SNII}}$	Contributes bulk of O, Ne, Mg, Si, S, Ca, and possibly r-process elements while ending as a type II supernova; He, C, ^{22}Ne, and N are contributed to ISM through stellar wind; lifetime is $\sim 10^6$ years; calculations suggest that $M_{\mathrm{SNII}} \approx 40\ M_\odot$; those with $M \lesssim 20\ M_\odot$ leave a neutron star with $M_{\mathrm{NS}} \approx 1.4\ M_\odot$ as remnant; higher-mass stars are likely to leave a black hole, although the mass limit is uncertain.
$M \gtrsim M_{\mathrm{SNII}}$	At sufficiently high masses, several instabilities come into play; evolution is uncertain; lifetimes are less than 10^6 yr.

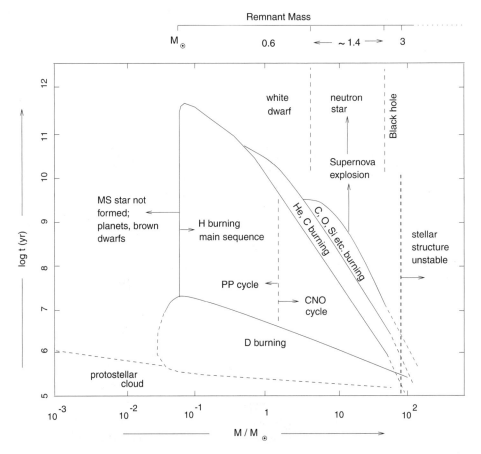

Fig. 3.16. Summary of stellar evolution for different initial masses as a function of time.

whereas those with higher masses do not. The mass scale $M_{car} \approx 8\,M_\odot$ determines the nature of carbon ignition. Stars with $M \gtrsim M_{car}$ undergo carbon ignition and, in general, nuclear burning proceeds all the way to iron. The largest characteristic mass that is relevant in stellar evolution is probably $M_{max} \approx 60\,M_\odot$. Stars with $M > M_{max}$ are fairly unstable and their evolution is quite uncertain.

These mass scales also determine the mass of the final remnant at the end of the stellar evolution. Stars with $M \gtrsim 0.8\,M_\odot$ could not have fully evolved within the age of the universe. Stars in the range $0.8\,M_\odot \lesssim M \lesssim M_{car}$ produce remnants in the mass range of $(0.5–1.5)\,M_\odot$, with the remnant mass monotonically increasing with the stellar mass. These remnants usually are white dwarfs. Stars in the range $M_{car} \lesssim M \lesssim M_{max}$ end up with a nearly constant mass remnant of $\sim 1.4\,M_\odot$, which will be a neutron star. The remnant mass increases linearly, with the stellar mass for $M > M_{max}$ reaching $\sim 3\,M_\odot$ for the stellar mass of $90\,M_\odot$. These remnants are usually black holes.

Stellar evolution also plays a vital role in connecting the currently observed distribution function of stars with the initial properties of the stars at the time of formation. If $dN = N_0 \xi(M)\,dM$ denotes the number of stars with masses in the range M–$(M+dM)$, when the stars are formed in a molecular cloud, then the quantity $\xi(M)$ is called the *initial mass function* (IMF). Observationally, however, what is (at best) determined is the luminosity function $\Phi(\mathcal{M})$, where $\mathcal{M}$ is the absolute luminosity of the star. To convert the current luminosity function to the initial luminosity function $\Phi_0(\mathcal{M})$, we need to make some assumptions regarding the star formation. If the populations of stars were all born at the same time, no correction would be required and $\Phi_0(\mathcal{M})$ would be equal to $\Phi(\mathcal{M})$; on the other hand, if the star-formation rate was roughly constant, then

$$\Phi_0(\mathcal{M}) = \Phi(\mathcal{M}) \times \begin{cases} t/\tau_{\rm ms}(\mathcal{M}) & (\text{for } \tau_{\rm ms}(\mathcal{M}) < t) \\ 1 & (\text{otherwise}) \end{cases}, \quad (3.137)$$

where t is the age of the population of stars under study and $\tau_{\rm ms}(\mathcal{M})$ is the main-sequence lifetime of a star with absolute magnitude $\mathcal{M}$. The extra factor corrects for the fact that only stars of magnitude $\mathcal{M}$ that we see are those that form in the last fraction τ/t of the population's lifetime. (Similar results can be obtained for any other assumption regarding the star-formation rate.) Given $\Phi_0(\mathcal{M})$, the initial mass function is, in principle, determined by

$$\xi(M) = \frac{d\mathcal{M}}{dM} \Phi_0[\mathcal{M}(M)], \quad (3.138)$$

where $\mathcal{M}(M)$ determines the relationship between mass and absolute magnitude of the stars. Given this relation, either from theory or from observation, and $\Phi_0(\mathcal{M})$ from the observations as well as assumptions regarding star-formation rate, we can obtain $\xi(M)$. There are large theoretical uncertainties for $M < 0.5\,M_\odot$, and observationally the slope of $\mathcal{M}(M)$ is hard to determine. In spite of these uncertainties, it is generally believed that $\xi(M)$ can be expressed as a power law $\xi \propto M^{-\alpha}$. The so-called Salpeter IMF uses $\alpha = 2.35$; a more sophisticated fit would be with $\alpha = 4.5, 2.2$, and 1.2 for stars with $M > 1\,M_\odot$, $0.5 < M < 1\,M_\odot$, and $M < 0.5\,M_\odot$, respectively.

3.7 Stellar Oscillations and Stability

The discussion of stellar structure in the previous chapter was based on static equilibrium solutions to the equations of hydrodynamics. In the study of stellar evolution in the previous sections we needed to consider time-dependent changes in the structure; however, the time scale over which these changes occur was very long. In fact, some of the features of stellar evolution discussed so far could have been obtained from the study of a sequence of instantaneous equilibrium models. Such an analysis, however, cannot answer a question that is important in the study of any dynamical system: Is the time-independent solution stable? A different kind of approach is needed to address the stability of equilibrium solutions.

The simplest approach to stability analysis proceeds as follows. We perturb the equilibrium solution by changing the physical variables by a small amount and examine how the deviations of the equilibrium evolve in time. If these deviations oscillate or decay, then the system may be thought of as stable. On the other hand, if the deviations grow with time, then the system is unstable and the static solution is of limited validity. This approach has the advantage of being mathematically tractable, as we need to study only perturbation equations that are linear in the small deviations. However, it may happen that in certain systems such a linear analysis is misleading. For example, it is possible to construct systems that are metastable; such systems are stable to small perturbations, but it is possible to destroy the equilibrium configuration by larger nonlinear perturbations. Similarly, it is possible to construct systems that are linearly unstable but could be stabilised by nonlinear effects. The study of nonlinear instability is highly system dependent and no general procedure can be devised. For the most part, our discussion shall be restricted to linear stability analysis.

The linear stability analysis of stellar models also provides another insight that is of importance in observations. Even a stable configuration can undergo oscillations in physical parameters when perturbed from the original equilibrium. Such oscillations of a star will lead to variability in its luminosity that can be observed. Thus the study of linear perturbations around equilibrium stellar configurations has a potential use in understanding the variability of stars.

The equilibrium configuration of a stellar system depends crucially on the equation of state. Similarly, the time evolution of the perturbation will depend on the relationship that is assumed between the perturbed values of pressure, temperature, and density. The simplest assumption regarding these perturbations will be to invoke an adiabatic relationship between perturbed pressure and density. We shall study adiabatic pulsations of stars in Subsection 3.7.1. Nonadiabatic perturbations are more difficult to model and will be taken up in Subsection 3.7.2.

In the study of stellar models we have used the assumption of a spherical symmetry to simplify the structure significantly. Even when the equilibrium configuration is spherically symmetric, there is no guarantee that the perturbations should possess this symmetry. We shall study the spherically symmetric perturbations, which are simpler to analyse mathematically, in Subsections 3.7.1 and 3.7.2. The more general case with angular dependence will be taken up in Subsection 3.7.3.

3.7.1 Adiabatic Spherical Pulsations

Consider a stellar configuration characterised by the functions $P_0(m)$, $r_0(m)$, and $\rho_0(m)$, where $m(r_0)$ is the mass contained within a sphere of radius r_0 and P_0 and ρ_0 denote the pressure and the density, respectively. We now perturb these

3.7 Stellar Oscillations and Stability

quantities by small amounts $P_1(m, t)$, $r_1(m, t)$, and $\rho_1(m, t)$. If the perturbed values of the physical variables are substituted into the equations of hydrodynamics and linearised in P_1, r_1, and ρ_1, we will obtain the equation governing the perturbed functions. Because the background solution is independent of time, we can assume, without any loss of generality, that the perturbed functions evolve with time as $\exp(i\omega t)$. (Because the equations are linear in the perturbed quantities, any general time-dependent solutions can be built by superposition of such modes.) Hence the total pressure, radius, and density of a shell containing a mass m can be taken to be

$$P(m, t) = P_0(m) + P_1(m, t) = P_0(m)[1 + p(m) e^{i\omega t}],$$
$$r(m, t) = r_0(m) + r_1(m, t) = r_0(m)[1 + x(m) e^{i\omega t}], \quad (3.139)$$
$$\rho(m, t) = \rho_0(m) + \rho_1(m, t) = \rho_0(m)[1 + d(m) e^{i\omega t}],$$

where p, x, and d are all assumed to be much smaller than unity. Substituting these into the equation of motion for the fluid shell,

$$\ddot{r} = -\frac{Gm}{r^2} - 4\pi r^2 \frac{\partial P}{\partial m}, \quad (3.140)$$

and linearising in the perturbation, we get

$$\frac{P_0}{\rho_0} \frac{\partial p}{\partial r_0} = \omega^2 r_0 x + g_0 (p + 4x), \quad (3.141)$$

where $g_0 = (Gm/r_0^2)$. Similarly, linearising the equation $(\partial r/\partial m) = (1/4\pi r^2 \rho)$, we get

$$r_0 \frac{\partial x}{\partial r_0} = -3x - d. \quad (3.142)$$

Equations (3.141) and (3.142) provide two partial differential equations for the three variables p, x, and d. To complete this set, we need an equation of state that connects the perturbed pressure to the perturbed density. As mentioned above, the simplest choice is to assume adiabaticity and set $p = \gamma_{\text{ad}} d$. (The simplicity of this relation arises from the use of m as the independent variable. This choice allows us to study the physics in a frame that is comoving with the fluid.) It is now possible to combine these relations to obtain a single relation that determines the evolution of $x(m)$. Writing Eq. (3.142) in the form

$$p = -3\gamma_{\text{ad}} x - \gamma_{\text{ad}} r_0 \frac{\partial x}{\partial r_0} \quad (3.143)$$

and differentiating both sides with respect to r_0, we can relate $(\partial p/\partial r_0)$ to the derivatives of $x(m)$. Substituting for $(\partial p/\partial r_0)$ from Eq. (3.141) and using Eq. (3.143) to eliminate p, we get the differential equation governing the perturbation $x(m)$ as

$$\frac{\partial}{\partial r_0}\left(\gamma_{\text{ad}}\frac{\partial x}{\partial r_0}\right) + \frac{4}{r_0}\frac{\partial}{\partial r_0}(\gamma_{\text{ad}}x) - \frac{\rho_0 g_0}{P_0}\gamma_{\text{ad}}\frac{\partial x}{\partial r_0} + \frac{\rho_0}{P_0}\left[\frac{g_0}{r_0}(4-3\gamma_{\text{ad}}) + \omega^2\right]x = 0.$$
(3.144)

This second-order differential equation requires two boundary conditions for its solution. To preserve radial symmetry, we must have $x = 0$ at $r_0 = 0$. At the surface of the star, the simplest boundary condition to assume will correspond to $p = 0$. Because the equation is linear, the solution can be scaled by any arbitrary constant. It is convenient to normalise the solution such that $x(M) = 1$.

With these conditions, Eq. (3.144) can be thought of as an eigenvalue equation for the quantity ω^2. If $\omega^2 > 0$, the solutions are oscillatory, and when $\omega^2 < 0$, the solutions contain an exponentially growing or decaying mode, indicating instability. Given the equilibrium structure of the star, all the coefficients appearing in Eq. (3.144) are known as functions of r_0, and this differential equation, in principle, can be solved to determine eigenvalues. Such a study will allow us to determine the regions of instability as well as the frequencies of oscillations of the star.

To analyse some of the general characteristics of this problem, it is convenient to rewrite Eq. (3.144) in a different form. Using the relation $(\partial P/\partial r) = -g\rho$, we can obtain, after some algebra, the following version of the same equation:

$$\hat{\mathcal{L}}x \equiv -\frac{1}{\rho_0 r_0^4}\frac{\partial}{\partial r_0}\left(\gamma_{\text{ad}}P_0 r_0^4 \frac{\partial x}{\partial r_0}\right) - \frac{1}{r_0\rho_0}\left\{\frac{\partial}{\partial r_0}[(3\gamma_{\text{ad}} - 4)P_0]\right\}x = \omega^2 x.$$
(3.145)

This equation defines the linear second-order differential operator $\mathcal{L}$, which has the eigenvalues ω^2. The form of the equation corresponds to the one called the *Sturm–Liouville* problem. From the general theory of eigenvalues of such operators, we can draw the following conclusions in this particular context: (1) The system will have countably infinite number of eigenvalues ω_n^2 labelled by an integer n. All these eigenvalues will be real and can be ordered as $\omega_0^2 < \omega_1^2 < \cdots$ with $\omega_n^2 \to \infty$ as $n \to \infty$. (2) The eigenfunction $x_0(r_0)$ corresponding to the lowest eigenvalue ω_0 will not have any node in the interval $0 < r_0 < R$. For $n > 0$, the eigenfunction x_n will have n nodes in the above interval. (3) The normalised eigenfunctions x_n form a complete set and can be taken to be orthonormal with the weightage function $\rho_0 r_0^4$ in the interval $[0, R]$.

These properties can be used to express the lowest eigenvalue in terms of the unperturbed solution. Integrating each of the terms in Eq. (3.145) between $[0, R]$, we get

$$-\left(\gamma_{\text{ad}}P_0 r_0^4 \frac{\partial x}{\partial r_0}\right)\Big|_0^R - \int_0^R r_0^3 \frac{\partial}{\partial r_0}[(3\gamma_{\text{ad}} - 4)P_0]x_0\, dr_0 = \omega_0^2 \int_0^R x_0 \rho_0 r_0^4\, dr_0.$$
(3.146)

3.7 Stellar Oscillations and Stability

The first term vanishes at both the limits, giving

$$\omega_0^2 = -\frac{\int_0^R r_0^3 \frac{\partial}{\partial r_0}[(3\gamma_{ad} - 4)P_0]x_0 \, dr_0}{\int_0^R x_0 \rho_0 r_0^4 \, dr_0}. \tag{3.147}$$

As a specific example, let us consider a star for which γ_{ad} is a constant throughout the star. Pulling $(3\gamma_{ad} - 4)$ outside the integral and using $(\partial P_0/\partial r_0) = -g_0\rho_0$, we get

$$\omega_0^2 = (3\gamma_{ad} - 4) \frac{\int_0^{R_0} r_0^3 \rho_0 g_0 x_0 \, dr_0}{\int_0^{R_0} r_0^4 \rho_0 x_0 \, dr_0}. \tag{3.148}$$

Because x_0 does not change sign in the interval, the sign of ω_0^2 is the same as the sign of $(3\gamma_{ad} - 4)$. Therefore, if $\gamma_{ad} > 4/3$, then $\omega_0^2 > 0$ and the equilibrium is stable. On the other hand, if $\gamma_{ad} < 4/3$, then the fundamental mode (as well as a finite number of overtones) can become dynamically unstable. When γ_{ad} is not a constant, we still have a similar result with a suitably defined average value for γ_{ad}.

An intuitive description of this result can be obtained as follows. We start with the equation

$$m\ddot{r} = 4\pi r^2 P - \frac{GMm}{r^2}. \tag{3.149}$$

In equilibrium, at $r = r_{eq}$, we have the condition $4\pi r_{eq}^2 P_{eq} = (GMm/r_{eq}^2)$. Perturbing the equation around this equilibrium value, we get

$$m\frac{d^2\delta r}{dt^2} = 4\pi r_{eq}^2 P_{eq}\left(\frac{2\delta r}{r} + \frac{\delta P}{P}\right) + \frac{GMm}{r_{eq}^2}\left(\frac{2\delta r}{r}\right), \tag{3.150}$$

which governs the radial displacement of a shell of matter confined by $r = r_{eq}$ originally. Writing $(\delta P/P) = \gamma(\delta\rho/\rho) = -3\gamma(\delta r/r)$ (where the last equality follows from the mass conservation $\rho r^3 = $ constant), we get

$$m\frac{d^2\delta r}{dt^2} = -\frac{GMm}{r_{eq}^3}(3\gamma - 4)\delta r. \tag{3.151}$$

This equation shows that δr is oscillatory with a local frequency $\omega^2 = (GM/r_{eq}^3)(3\gamma - 4) \propto \rho(3\gamma - 4)$. The frequency scales as $\sqrt{\rho}$, and the stability requires that $\gamma > (4/3)$.

One specific case in which the eigenvalue problem has been investigated in detail corresponds to the polytropic spheres with a constant γ_{ad}. In this case, it is more convenient to rewrite Eq. (3.144) in the form

$$x'' + \left(\frac{4}{r_0} - \frac{\rho_0 g_0}{P_0}\right)x' + \frac{\rho_0}{\gamma_{ad} P_0}\left[\omega^2 + (4 - 3\gamma_{ad})\frac{g_0}{r_0}\right]x = 0. \tag{3.152}$$

All the coefficients that appear in this equation can be expressed in terms of the Lane–Emden function $w(z)$, which satisfies the equation (see Vol. I, Chap. 10, Subsection 10.3.1)

$$\frac{1}{z^2}\frac{d}{dz}\left(z^2\frac{dw}{dz}\right) + w^n = 0. \tag{3.153}$$

In particular, if the equation of state is $P = K\rho^{(n+1)/n}$ and Φ is the gravitational potential, then we have

$$g_0 = \frac{\partial \Phi_0}{\partial r_0} = A\Phi_c \frac{dw}{dz}, \quad A^2 = \frac{4\pi G}{[(n+1)K]^n}(-\Phi_c)^{n-1},$$

$$\rho_0 = \left[\frac{-\Phi_c w}{(n+1)K}\right]^n, \quad \frac{\rho_0}{P_0} = \frac{1}{K}\rho^{-1/n} = -\frac{n+1}{\Phi_c w}, \tag{3.154}$$

where the subscript c denotes the central values of the unperturbed model. From these, we get

$$\frac{g_0\rho_0}{P_0} = -A\frac{n+1}{w}\frac{dw}{dz}, \quad \frac{g_0}{r_0} = \frac{\Phi_c A^2}{z}\frac{dw}{dz}. \tag{3.155}$$

Finally, changing the independent variable in Eq. (3.152) from r_0 to $z = Ar_0$, we get the perturbation equation as

$$\frac{d^2 x}{dz^2} + \left(\frac{4}{z} + \frac{n+1}{w}\frac{dw}{dz}\right)\frac{dx}{dz} + \left[\Omega^2 - \frac{(4-3\gamma_{\mathrm{ad}})(n+1)}{\gamma_{\mathrm{ad}}}\frac{1}{z}\frac{dw}{dz}\right]\frac{x}{w} = 0, \tag{3.156}$$

with Ω denoting a dimensionless frequency:

$$\Omega^2 = \frac{n+1}{\gamma_{\mathrm{ad}}(-\Phi_c)A^2}\omega^2. \tag{3.157}$$

Because Eq. (3.156) depends on only the Lane–Emden function, the dimensionless eigenvalue ω^2 depends on only n and γ_{ad} and not on other details of the polytropes. By using Eqs. (3.154) at the centre (where $w = 1$), we have

$$\omega^2 = \frac{\gamma_{\mathrm{ad}}(-\Phi_c)A^2}{n+1}\Omega^2 = \frac{4\pi G\gamma_{\mathrm{ad}}\rho_c}{n+1}\Omega^2. \tag{3.158}$$

For the polytropes of a given index n, the central density ρ_c and the mean density $\bar{\rho}$ differ by a constant factor that depends on only n; hence we find that $\omega^2 \propto \bar{\rho}$. The period of oscillation $\Pi = (2\pi/\omega)$ satisfies the relation

$$\Pi\sqrt{\bar{\rho}} = \left[\frac{(n+1)\pi}{\gamma_{\mathrm{ad}}G\Omega^2}\left(\frac{\bar{\rho}}{\rho_c}\right)_n\right]^{1/2}. \tag{3.159}$$

For a given mode, the right-hand side depends on only n and γ_{ad} and this relation

Table 3.2. Intrinsically variable stars

Kind of Variable	Period	Population	Spectral Type	M_V	Radial (R) or Nonradial (NR)
Long period, Miras	100–700 days	I, II	M,N,R,S	-2 to $+1$	R
RV Tauri	20–150 days	II	G–K	~ -3	R
Classical Cepheids	1–50 days	I	F6–K2	-6 to -0.5	R
RR Lyrae	1.5–24 h	II	A2–F2	$\langle M_V \rangle \approx 0.6$	R
The Sun	5–10 min	I	G2	$+4.83$	NR
White Dwarfs	100–1000 s	I, II	O,B2,A0	$+2,+7,+8$	NR

is called the period–density relation. For numerical estimates, this relation can be calibrated from the observed values of some variable star, the luminosity of which is known to oscillate with a certain period. For example, if we assume that a δ-Cephei star with $M = 7\,M_\odot$ and $R = 80\,R_\odot$ has the period $\Pi = 11$ days, then we get $\bar{\rho} \approx 2 \times 10^{-5}$ gm cm^{-3} and $\Pi\sqrt{\bar{\rho}} \approx 0.049$ in the given units. Using this constant, we find that a supergiant star with $\bar{\rho} = 5 \times 10^{-8}$ gm cm^{-3} will have a period of $\sim$220 days whereas a compact white dwarf with $\bar{\rho} = 10^6$ gm cm^{-3} will have a period of $\sim$4 s if γ_{ad} and r do not change drastically between the stars. These conclusions are broadly in agreement with observations. Table 3.2 gives a list of the main types of variable stars.

Given the structure of the star, we can relate the mean density to other physical parameters by using homology-type relations. In that case, we can relate the period Π to other physical characteristics of the star. One such analysis, in which numerical fits to theoretical stellar-structure results are used, gives the following empirical relation[15]:

$$\log \Pi \approx -0.340 + 0.825(\log L - 1.7) - 3.34(\log T_{\text{eff}} - 3.85) \\ - 0.63(\log M + 0.19), \qquad (3.160)$$

where Π is in days and mass and luminosity are in solar units. Because Π, L, and T_{eff} are all directly observable, this result can be used to determine the mass of the star.

More detailed modelling of the pulsation is possible if the underlying properties of the star is given. In all practical cases, the frequency spectrum has to be determined numerically for low-order modes. It is, however, possible to make some general comments regarding the high-frequency oscillations by use of the asymptotic form of the perturbation equation. To do this, we begin by eliminating the first derivative term in Eq. (3.145) by changing the variable to

$$q(r_0) = r_0^2(\gamma_{ad} P_0)^{1/2} x(r_0). \tag{3.161}$$

Straightforward algebra now gives the equation

$$\frac{d^2q}{dr_0^2} + \left[\frac{\omega^2}{v_s^2} - \phi(r_0)\right] q = 0, \tag{3.162}$$

where $v_s^2 \equiv \gamma_{ad}(P_0/\rho_0)$ is the square of the local speed of sound and $\phi(r)$ is the function

$$\phi(r) = \frac{2}{r^2} + \frac{2}{\gamma_{ad} P_0 r} \frac{d(\gamma_{ad} P_0)}{dr} - \left[\frac{1}{2\gamma_{ad} P_0} \frac{d(\gamma_{ad} P_0)}{dr}\right]^2$$

$$+ \frac{1}{2\gamma_{ad} P_0} \frac{d^2(\gamma_{ad} P_0)}{dr^2} - \frac{1}{\gamma_{ad} P_0 r} \frac{d}{dr}[(3\gamma_{ad} - 4)P_0]. \tag{3.163}$$

We can perform the asymptotic analysis of such an equation by assuming that the solutions are of the form

$$q(r_0) \propto \exp\left[i \int k(r_0) dr_0\right], \quad k^2(r_0) = \frac{\omega^2}{v_s^2(r_0) - \phi(r_0)}. \tag{3.164}$$

In this limit, the eigenfrequencies are determined by the condition

$$\int_a^b k_r \, dr = (n+1)\pi, \tag{3.165}$$

provided that the left-hand side is sufficiently large (and real), which implies that $(\omega^2/v_s^2) \gg \phi$. Ignoring $\phi(r)$, we can recast the above relation in the form

$$\omega = (n+1)\pi \left(\int_a^b \frac{dr}{v_s}\right)^{-1} = (n+1)\omega_0. \tag{3.166}$$

In this limit the eigenfrequencies are equally spaced. The quantity $dt = (dr/v_s)$ gives the time taken by sound to travel a distance dr, and it is clear that the eigenfrequencies given above are the inverses of this time scale. For this reason these modes are called radial-acoustic modes or radial-pressure modes.

3.7.2 Nonadiabatic Effects in Radial Pulsations

The analysis in the previous subsection shows that a star is capable of pulsating at different modes with eigenfrequencies ω_n. This result, obtained from linear perturbation analysis and the assumption of adiabaticity, requires modification in realistic situations. In general, such a pulsating star will not be strictly adiabatic,

3.7 Stellar Oscillations and Stability

and, over a period of time, the oscillations cannot be sustained unless there is a mechanism that powers the standing waves. That is, we need a physical mechanism that will work on the pulsating modes of the star and will supply sufficient energy to maintain the oscillations. Such an effect is clearly nonadiabatic and is at least of second order in the perturbations.

The net amount of work done by each layer of the star during one cycle of oscillation is the difference between the heat flowing into the gas and the heat leaving the gas. To drive the oscillations, heat must enter the layer during the high-temperature part of the cycle and exit during the low-temperature part. Different layers of the star may have different phase relations as regards such a process, and whether the oscillations will be sustained or not will depend on the net effect.

Favourable circumstances for sustained oscillations occur if a layer of a star becomes more opaque on compression (thereby storing the energy flowing towards the surface) and uses it to push the surface layers upwards. When the expanding layer becomes more transparent, the trapped heat can escape and the layer will fall back to start a new cycle. For this mechanism to work, the opacity must increase when the layer is compressed. Under normal circumstances, opacity actually decreases with compression. For example, consider the Kramer's opacity $\kappa \propto (\rho/T^{3.5})$. During the compression of a layer, both the density and the temperature will increase but because the opacity is more sensitive to temperature, κ actually decreases. An exception occurs in the layers of the star that are partially ionised. During the compression of such a layer, part of the work done goes to ionising the system further rather than raising the temperature; with a smaller temperature rise, the density increase leads to corresponding increase in Kramer's opacity. The converse is true during expansion. Thus a partially ionised zone of a star can absorb heat during compression and release the heat during expansion, thereby sustaining the oscillation. This mechanism is called the κ *mechanism*.

Two ionisation zones exists in most of the stars. The first one is a broad region corresponding to the layers with the temperature range $(1–1.5) \times 10^4$ K, where the ionisation of hydrogen (HI $\to$ HII) as well as the single ionisation of helium (HeI $\to$ HeII) takes place. The second ionisation layer occurs around a temperature of 4×10^4 K and corresponds to the ionisation of helium HeII $\to$ HeIII. It is usual to call the first layer the *hydrogen ionisation zone* and the second the *helium ionisation zone*. The actual location of these zones will depend on the effective temperature of the star. In a hot star with $T_{\text{eff}} \approx 7500$ K, the zones will be located very near the surface and hence there will not be enough mass available to drive the oscillations effectively. As T_{eff} is lowered, the ionisation zones move deeper into the star, allowing the overtones and eventually the fundamental mode of pulsation to be driven. The latter effect occurs in stars with $T_{\text{eff}} \approx 5500$ K. If the surface temperature becomes still lower, then the efficient convection in the outer layers will work against systematic oscillations and will dampen them.

The above arguments suggest that sustained nonadiabatic oscillations can take place in a particular band of stars in the H-R diagram. We can estimate the characteristic region along the following lines. We begin by noting that whether a layer of star pulsates adiabatically or not depends on the ratio between the thermal time scale t_{th} and the dynamical time scale t_{dyn}. The latter may be estimated by the period of oscillation Π of the star. We can find the thermal time scale for a shell of mass δM by dividing its heat content $\delta M\, C_V T$ by the luminosity L that passes through it. The ratio between these two is given by

$$\mathcal{R}(\Delta M) \equiv \frac{t_{\text{th}}}{t_{\text{dyn}}} = \frac{C_V T \delta M}{\Pi L}. \tag{3.167}$$

When $\mathcal{R}$ is large, the motion will be nearly adiabatic and the nonadiabaticity will set in at $\mathcal{R} \approx 1$. For the κ mechanism to work, we expect $\mathcal{R}$ to become unity at some temperature $T = T_{\text{tr}} \approx 4 \times 10^4$ K, corresponding to the helium ionisation zone. In this case, $\delta M = M_{\text{env}}$ will correspond to the mass of the outer envelope. Setting $\mathcal{R} = $ constant and treating C_V and T also as constants, we find that

$$L \propto \frac{M_{\text{env}}}{\Pi} \propto M_{\text{env}}\, M^{1/2} R^{-3/2}. \tag{3.168}$$

This is the relation that has to be satisfied by the stars for which the κ mechanism can lead to an instability. We shall now convert this into a relation between L and T_{eff} for these stars. We first relate M_{env} to other variables by equating the pressure of the envelope in the form $P = (GM M_{\text{env}}/R^4)$ to the expression for envelope pressure found in Chap. 2, Subsection 2.4.2 [see Eq. (2.91)]. This gives

$$\frac{GM M_{\text{env}}}{R^4} = K' T_{\text{tr}}^{1+n_{\text{eff}}} \propto \left(\frac{M}{L}\right)^{\frac{1}{n+1}} T_{\text{tr}}^{1+n_{\text{eff}}}, \quad n_{\text{eff}} = \frac{s+3}{n+1}, \tag{3.169}$$

where we have used the scaling for $K' \propto (M/L)^{1/(n+1)}$. The transition temperature T_{tr} is a constant as far as this discussion is concerned, which allows us to write

$$M_{\text{env}} \propto \frac{R^4}{M} \left(\frac{M}{L}\right)^{\frac{1}{n+1}}. \tag{3.170}$$

Substituting relation (3.170) into relation (3.168), we find the luminosity of these stars in terms of their radius and mass:

$$L \propto R^{\frac{5}{2}\left(\frac{n+1}{n+2}\right)} M^{-\frac{1}{2}\left(\frac{n-1}{n+2}\right)}. \tag{3.171}$$

From the definition of T_{eff}, we can eliminate R in favour of T_{eff} to obtain

$$L \propto T_{\text{eff}}^{\frac{4(5n+11)}{n+3}} M^{\frac{2(n-1)}{n+3}}. \tag{3.172}$$

3.7 Stellar Oscillations and Stability

This relation shows that L is a steeply increasing function of T_{eff} for any realistic values of n; its dependence on M on the other hand is fairly weak. In other words, the region of instability in the H-R diagram will be a nearly vertical line with a high value of slope. If we assume, for example, that the stars have $L \propto M^4$, we get

$$L \propto T_{\text{eff}}^{\frac{8(5n+11)}{n+7}}. \qquad (3.173)$$

For $n = 1$, $L \propto T_{\text{eff}}^{16}$, and for $n = 2$, $L \propto T_{\text{eff}}^{19}$. This result shows that the stars with sustained oscillations and instabilities will occur in a nearly vertical strip with a slope of ~16–19 and effective temperatures in the range of $T_{\text{eff}} = $ (5500–7500) K. This result agrees reasonably well with observations (except for the sign of the slope, which is not correctly reproduced by the above analysis).

3.7.3 Adiabatic Nonradial Oscillations

The study of nonradial oscillations, in principle, proceeds exactly as before. Because the displacement field can now have three independent components, the structure of the linearised equations becomes complicated and analytically intractable. We shall consider these equations under certain simplifying assumptions in order to gain physical insight into the problem.

In this case, it proves to be more convenient to start from the hydrodynamic equations and perturb them rather than to start from the equations of stellar structure. Ignoring all dissipative effects, we can write the equations of hydrodynamics as

$$\nabla^2 \Phi = 4\pi G \rho, \qquad (3.174)$$

$$\frac{\partial \rho}{\partial t} + \nabla \cdot (\rho \mathbf{v}) = 0, \qquad (3.175)$$

$$\rho \left(\frac{\partial}{\partial t} + \mathbf{v} \cdot \nabla \right) \mathbf{v} = -\nabla P - \rho \nabla \Phi. \qquad (3.176)$$

The equilibrium solutions to these equations will be given by the functions $P(r)$, $\rho(r)$, $\Phi(r)$, and $\mathbf{v}(r)$, which depend on only the magnitude of the radius vector r. Let us imagine that each fluid element in the star is displaced from its equilibrium position by the amount $\mathbf{x}(\mathbf{r}, t)$. The perturbed quantities in the Eulerian framework will be denoted by a prime and the corresponding Lagrangian perturbations will be denoted by the addition of the prefix δ to the quantity. For example, the Eulerian and Lagrangian density perturbations are connected by the relation

$$\delta\rho = \rho' + \mathbf{x} \cdot \nabla \rho. \qquad (3.177)$$

The corresponding time derivatives in Eulerian and Lagrangian frameworks are as usual related by

$$\frac{d}{dt} = \frac{\partial}{\partial t} + \mathbf{v} \cdot \nabla. \tag{3.178}$$

Using these relations, we can obtain the linearised equations for the perturbed quantities. To obtain a closed set, we also need to relate the perturbed pressure and perturbed density. These Lagrangian perturbations will be taken to be connected by

$$\frac{\delta P}{P} = \Gamma_1 \frac{\delta \rho}{\rho}, \tag{3.179}$$

where Γ_1 is the adiabatic index. The perturbed equation for the displacement is given by

$$\rho \frac{\partial^2 \mathbf{x}}{\partial t^2} = -\nabla P' - \rho \nabla \Phi' - \rho' \nabla \Phi, \tag{3.180}$$

and the perturbation in the density is related to the expansion of the vector field $\mathbf{x}$ by

$$\frac{\delta \rho}{\rho} = -\nabla \cdot \mathbf{x}. \tag{3.181}$$

In general, these equations have to be coupled to the perturbed gravitational potential through the Poisson equations. However, to simplify matters, we shall ignore the variations in the perturbed gravitational potential. Then we can separate the time dependence of the modes by taking $\mathbf{x}(\mathbf{r}, t) = \mathbf{x}(\mathbf{r}) \exp i\sigma t$. Taking the components of Eq. (3.180) and manipulating the terms, we can reduce the equations to the form

$$\sigma^2 x_r = \frac{\partial}{\partial r}\left(\frac{P'}{\rho}\right) - A \frac{\Gamma_1 P}{\rho} \nabla \cdot \mathbf{x}, \tag{3.182}$$

$$\sigma^2 x_\theta(r, \theta, \phi) = \frac{\partial}{\partial \theta}\left[\frac{1}{r}\frac{P'}{\rho}(\mathbf{r})\right], \tag{3.183}$$

$$\sigma^2 x_\phi(r, \theta, \phi) = \frac{1}{\sin\theta} \frac{\partial}{\partial \phi}\left[\frac{1}{r}\frac{P'}{\rho}(\mathbf{r})\right], \tag{3.184}$$

where

$$A(r) = \frac{d \ln \rho}{dr} - \frac{1}{\Gamma_1}\frac{d \ln P}{dr}. \tag{3.185}$$

It may appear that these constitute three equations for the four unknowns P'/ρ, $\mathbf{x}(t, \mathbf{r})$; however, there are only two independent components in $\mathbf{x}$ because the background is spherically symmetric. This is most easily seen if we write the

3.7 Stellar Oscillations and Stability

variables in terms of the spherical harmonics $Y_{lm}(\theta, \phi)$ in the form

$$\mathbf{x}(r, \theta, \phi) = x_r(r, \theta, \phi)\mathbf{e_r} + x_\theta(r, \theta, \phi)\mathbf{e_\theta} + x_\phi(r, \theta, \phi)\mathbf{e_\phi}$$
$$= \left[x_r(r)\mathbf{e_r} + x_t(r)\mathbf{e_\theta}\frac{\partial}{\partial \theta} + x_t(r)\mathbf{e_\phi}\frac{1}{\sin\theta}\frac{\partial}{\partial \phi}\right]Y_{lm}(\theta, \phi), \quad (3.186)$$

$$\frac{P'(\mathbf{r})}{\rho} = \frac{P'(r)}{\rho}Y_{lm}(\theta, \phi), \quad (3.187)$$

where the transverse component $x_t(r)$ is related to P'/ρ by

$$x_t(r) = \frac{1}{\sigma^2}\frac{1}{r}\frac{P'(r)}{\rho}. \quad (3.188)$$

With this choice, the θ and ϕ components of the equations are identically satisfied and the nontrivial dynamics is contained in only the radial component. Using the standard equation satisfied by the spherical harmonics,

$$\frac{1}{\sin\theta}\frac{\partial}{\partial\theta}\left(\sin\theta\frac{\partial Y_{lm}}{\partial\theta}\right) + \frac{1}{\sin^2\theta}\frac{\partial^2 Y_{lm}}{\partial\phi^2} + l(l+1)Y_{lm} = 0, \quad (3.189)$$

and the expansion

$$\nabla \cdot \mathbf{x} = \frac{1}{r^2}\frac{d}{dr}(r^2 x_r) + \frac{1}{r\sin\theta}\frac{\partial}{\partial\theta}(x_\theta \sin\theta) + \frac{1}{r\sin\theta}\frac{\partial}{\partial\phi}(x_\phi), \quad (3.190)$$

we can easily show that

$$\nabla \cdot \mathbf{x} = \frac{1}{r^2}\frac{d}{dr}(r^2 x_r)Y_{lm} - \frac{l(l+1)}{r}x_t Y_{lm}. \quad (3.191)$$

On the other hand, the density perturbation can be written in the form

$$\frac{\delta\rho}{\rho} = \frac{\rho}{\Gamma_1 P}(\sigma^2 r x_t - g x_r), \quad (3.192)$$

where the pressure gradient has been replaced with $-g\rho$ by use of the equation for hydrostatic equilibrium. Equating $\nabla \cdot \mathbf{x}$ and $-\delta\rho/\rho$ gives one differential equation whereas using the definition of x_t from Eq. (3.188) and the expression for $\delta\rho/\rho$ gives another. These equations are

$$r\frac{dx_r}{dr} = \left(\frac{k_t^2 gr}{S_l^2} - 2\right)x_r + r^2 k_t^2\left(1 - \frac{\sigma^2}{S_l^2}\right)x_t, \quad (3.193)$$

$$r\frac{dx_t}{dr} = \left(1 - \frac{N^2}{\sigma^2}\right)x_r + \left(\frac{r}{g}N^2 - 1\right)x_t, \quad (3.194)$$

where the following quantities have been defined:

$$N^2 \equiv -Ag = -g\left(\frac{d\ln\rho}{dr} - \frac{1}{\Gamma_1}\frac{d\ln P}{dr}\right), \quad (3.195)$$

$$S_l^2 \equiv \frac{l(l+1)}{r^2}\frac{\Gamma_1 P}{\rho} = \frac{l(l+1)}{r^2}v_s^2, \quad (3.196)$$

$$k_l^2 = \frac{l(l+1)}{r^2} = \frac{S_l^2}{v_s^2}. \quad (3.197)$$

Both these frequencies N and S_l have direct physical interpretations: The quantity S_l (called the Lamb frequency) is related to the transverse wave number k_t by $k_t v_s = S_l$, showing that S_l^{-1} is the time taken for a sound wave to travel the wavelength corresponding to this transverse wave number; N gives the frequency (called the Brunt–Vaisala frequency) of oscillation for a fluid mass that is perturbed from its original position in a convectively stable star. To see this, consider a blob with density $\rho_{\text{blob}}(r)$ that is displaced vertically upwards by a small amount q inside a star. The ambient density at $r+q$ is given by

$$\rho_*(r+q) \simeq \rho_*(r) + \left(\frac{d\rho}{dr}\right)_* q, \quad (3.198)$$

where the subscript $*$ denotes quantities pertaining to the star. If the blob is displaced adiabatically, then its density at the new location is given by

$$\rho_{\text{blob}}(r+q) \simeq \rho_{\text{blob}}(r) + \left(\frac{\partial\rho}{\partial r}\right)_S q = \rho_*(r) + \left(\frac{\partial\rho}{\partial r}\right)_S q, \quad (3.199)$$

where the second equality follows from the fact that originally $\rho_{\text{blob}}(r) = \rho_*(r)$. The net difference in the density,

$$\Delta\rho = \rho_{\text{blob}}(r+q) - \rho_*(r+q) = \left[\left(\frac{\partial\rho}{\partial r}\right)_S - \left(\frac{d\rho}{dr}\right)_*\right]q, \quad (3.200)$$

will cause a buoyancy force $g\Delta\rho$, causing the blob to move according to the equation $\rho\ddot{q} = g\Delta\rho$ or

$$\ddot{q} = -\frac{g}{\rho}\Delta\rho = \frac{g}{\rho}\left[\left(\frac{d\rho}{dr}\right)_* - \left(\frac{\partial\rho}{\partial r}\right)_S\right]q$$

$$= g\left[\left(\frac{d\ln\rho}{dr}\right)_* - \frac{1}{\Gamma_1}\left(\frac{d\ln P}{dr}\right)\right]q = -N^2 q, \quad (3.201)$$

where $\Gamma_1 = (\partial\ln P/\partial\ln\rho)_S$. This shows that N^2 determines the oscillation frequency of a fluid elemet. For future reference, note that N^2 can also be written as

$$N^2 = -g\left(\frac{P}{\rho}\right)\left(\frac{d\ln P}{dr}\right)\left[\left(\frac{d\rho}{dP}\right)_* - \left(\frac{\partial\rho}{\partial P}\right)_S\right] = g\left[\left(\frac{d\rho}{dP}\right)_* - \left(\frac{\partial\rho}{\partial P}\right)_S\right], \quad (3.202)$$

where the second equality follows from the equation for pressure support $(dP/dr) = -g\rho$.

To solve these differential equations we need the boundary conditions for the physical variables. The actual details of the boundary conditions near the origin depend on how A and P vary with the radius near the centre. In general, the regular solutions go to zero as r^{l-1} near the origin and the radial and the transverse components are related by

$$x_r(r) = l x_t(r) \quad (as\ r \to 0). \quad (3.203)$$

The boundary conditions at the surface depend on the details of the photosphere assumed for the star. The simplest condition corresponds to taking $\delta P = 0$ at the surface. From Eq. (3.192), rewritten as

$$\frac{\delta P}{P} = \frac{\rho}{P}(\sigma^2 r x_t - g x_r), \quad (3.204)$$

we can determine the conditions on the displacement field. The quantity ρ/P tends to infinity near the surface; hence if $\delta P/P$ has to vanish, we must have the condition

$$x_t(R) = \frac{g_s x_r(R)}{\sigma^2 R} \quad (3.205)$$

satisfied at the surface (g_s is the surface gravity). Once again, this is equivalent to complete reflection of the waves at the surface. In addition to these, we shall impose the normalisation conditions that $x_r = 1$ at the surface. With these conditions we have obtained an eigenvalue problem for σ^2.

There is a significant mathematical difference between the differential equations for radial and nonradial oscillations. In the case of the radial oscillations, we had a well-defined Sturm–Liouville problem that permitted the eigenvalues to be ordered with a definite lower bound. In the nonradial case, we do not have a well-defined Sturm–Liouville problem and hence there is no guarantee that the eigenvalues will be ordered in any form or even that σ^2 will have a lower bound. All that can be verified from the analysis of the differential equation is that σ^2 will be real and the eigenfunctions will be orthogonal. We will now discuss some of the elementary properties of the solutions.

From the standard analysis of the equation governing the spherical harmonics, we can ascertain the following facts: (1) The solutions are characterised by two integers l and m, where l must be zero or a positive integer and $m = -l, -l+1, \ldots, 0, \ldots, l-1, l$. Thus for a given value of l there are $(2l+1)$ permitted values for m. (2) $l = 0$ corresponds to the spherically symmetric situation and in this case the problem reduces to the one studied earlier in Subsection 3.7.1. For nonradial oscillations we need to consider only l greater than or equal to unity.

The modes with $m = 0$ are called *zonal modes*, and those with $|m| = l$ are called *sectoral modes*. The in-between values are called tesseral modes. As a simple example, consider the sectoral mode with $l = m = 1$. The phase factor representing such a mode will vary as $\exp i(\sigma t + m\phi)$, and thus the lines of constant phase satisfy the relation $(d\phi/dt) = -\sigma/m$. For positive values of m, this node moves in the retrograde azimuthal direction as the fluid sloshes back and forth periodically.

We can obtain some insights into the radial nature of the solutions through a WKB analysis of the differential equations by assuming that the radial variation is of the form $\exp[ik_r(r)r]$ in the limit when $k_r r \gg 1$. In this limit, the dispersion relation for the radial wave number is given by

$$k_r^2 = \frac{k_t^2}{\sigma^2 S_l^2}(\sigma^2 - N^2)(\sigma^2 - S_l^2). \tag{3.206}$$

For solutions that are oscillatory in time, σ^2 is positive. If σ^2 is greater or less than both N^2 and S_l^2, then k_r^2 is positive and we have spatially oscillatory solutions. On the other hand, if σ^2 lies between N^2 and S_l^2, the solutions have a transient character in the radial direction. In the case of propagating waves, we therefore have to consider two separate kinds of modes: The first type has σ^2 that is much larger than both N^2 and S_l^2 and the second case has σ^2 that is much less than both N^2 and S_l^2.

In the first case, it often happens that N^2 is less than S_l^2 and the solution to Eq. (3.206) is given by

$$\sigma_P^2 \approx \frac{K^2}{k_t^2} S_l^2 = (k_r^2 + k_t^2) v_s^2 \quad (\sigma^2 \gg N^2, S_l^2), \tag{3.207}$$

with $K = \sqrt{(k_r^2 + k_t^2)}$ representing something analogous to the total length of the wave vector. These modes are called P modes or pressure modes. They are very similar to the radial modes obtained in Subsection 3.7.1 and reduce to those modes when $k_t = 0$.

In the second case, the solution to Eq. (3.206) can be approximated by

$$\sigma_p^2 \approx \frac{k_t^2}{(k_r^2 + k_t^2)} N^2 \quad (\sigma^2 \ll N^2, S_l^2). \tag{3.208}$$

These are called g modes or gravity modes because buoyancy acts as a restoring force in this case. Further, when N^2 is negative (representing a convectively unstable situation), σ_g becomes pure imaginary and the perturbation is unstable or transient.

3.7 Stellar Oscillations and Stability

The corresponding eigenfunctions can be estimated by standard WKB analysis. The ratio of radial and tangential eigenfunctions for the two cases is given by

$$\left|\frac{x_r}{x_t}\right| \sim \begin{cases} rk_r & (P \text{ modes}) \\ l(l+1)/rk_r & (g \text{ modes}) \end{cases}. \tag{3.209}$$

For large wave numbers for which WKB analysis is applicable, the P modes correspond to radial fluid motions whereas the g modes correspond to transverse fluid motions. In the same approximation we can also obtain an explicit expression for the frequencies σ_P and σ_g in terms of the number n describing the eigenfunctions. In general, the number of nodes can be approximated as

$$n \approx 2 \int_0^R \frac{dr}{\lambda(r)} \approx \frac{1}{\pi} \int_0^R k_r(r)\, dr. \tag{3.210}$$

Ignoring the transverse contributions for simplicity, we get for the P and g modes the relations

$$\sigma_P \approx n\pi \left(\int_0^R \frac{dr}{v_s}\right)^{-1}, \tag{3.211}$$

$$\Pi_g = \frac{2\pi}{\sigma_g} \approx n \frac{2\pi^2}{[l(l+1)]^{1/2}} \left(\int_0^R \frac{N}{r}\, dr\right)^{-1}. \tag{3.212}$$

As in the previous case, the P mode frequencies are equally spaced for large n. The spacing depends essentially on the speed of sound, which in turn depends on the temperature. Thus in stars like the Sun, P modes provide a useful probe of the temperature structure of the star. On the other hand, for g modes, it is not the frequencies but the periods Π_g that are equally spaced. These periods depend sensitively on l and increase with N, which is in direct contrast with the P modes. The structure of these modes and the distribution of eigenfrequencies will be useful in the study of solar physics and oscillations of white dwarfs.

4
Supernova (Type II)

4.1 Introduction

This chapter deals with supernova explosions that occur at the final stages of evolution of isolated massive stars, conventionally called *type II supernova*.[1] It uses the material developed in the past chapters as well as that of Chaps. 5, 6, 8–10, and 12 of Vol. I. There exists another class of supernova (called *type I*), which are believed to arise because of the accretion of mass by one of the stars in a binary system. This topic will be covered in Chap. 7.

4.2 Overview

A sufficiently massive star will have different nuclear reactions taking place inside it at different radii during the late stages of its evolution. For example, in a star with $M > 8\, M_\odot$, carbon burning will be ignited at the core, and a variety of heavier nuclei like $^{16}_{8}O$, $^{20}_{10}Ne$, $^{23}_{11}Na$, $^{23}_{12}Mg$, and $^{24}_{12}Mg$ will undergo nuclear burning in different shells with the exact details depending in a complex manner on the mass of the star. Eventually, oxygen and silicon burning will take place in the core, leading to the synthesis of a host of nuclei centred around the $^{56}_{26}Fe$ peak of the nuclear-binding-energy curve. The most abundant of these nuclei are probably $^{54}_{26}Fe$, $^{56}_{26}Fe$, and $^{56}_{28}Ni$; together, these products are christened as forming the iron core. Once an iron core is formed at the centre of a sufficiently massive star, the future evolution may be roughly summarised along the following sequence of events:

(1) Continued silicon burning and other shell burning processes add to the core mass and eventually drive it towards the limiting value for mass $M_{ch}(\simeq 1\, M_\odot)$, called the Chandrasekhar mass. At the central conditions prevailing during this stage, neutrino interactions with the nuclei play a significant role, and neutrino processes can act as the main cooling mechanism for the core. For example, during the silicon burning phase, the photon luminosity of a $20 M_\odot$ star is approximately 4.4×10^{38} ergs s^{-1} and the neutrino luminosity is approximately 3.1×10^{45} ergs s^{-1}.

4.2 Overview

(2) At the very high temperature that exists in the core, energetic photons can disintegrate the iron nuclei into α particles and protons. In a way, this has the effect of undoing what the star has been trying to achieve over its entire lifetime. The process of breaking up the heavy nuclei is endothermic and hence takes away the energy from the photon gas and consequently lowers the pressure support at the core.

(3) With these central conditions, it also becomes possible for inverse beta decay to produce neutrons (and neutrinos) by combining protons and electrons. Each of these reactions leads to the emission of a neutrino that can escape from the star under the prevailing conditions. The vanishing of the electron population again leads to a rapid decrease in the degeneracy pressure that they supply at the core.

(4) Both the processes described above – the photodisintegration of the iron core and the neutronisation – have the effect of reducing the pressure support at the core of the star, triggering a rapid collapse of the core. In the inner part of the core, the collapse is homologous with the speed of in-fall's being proportional to the distance from the centre of the star. This necessarily would lead to supersonic velocities at some radius beyond which the collapse ceases to be homologous. The inner core collapses subsonically and is decoupled from the outer mantle, which collapses supersonically almost in free fall. This process is very fast, and the speeds in the outer core can be as high as 70,000 km s^{-1}.

(5) The homologous collapse of the inner core continues until the density is approximately 8×10^{14} gm cm^{-3}, which is roughly three times the nuclear density. Models for nuclear structure suggest that nuclear interactions can produce an effective repulsive force at high densities. The collapse is halted around this density because the packed nucleons in the inner core feel the short-distance repulsive part of the nuclear force. Consequently, the inner core steepens, leading to a rebounding of the inner region, sending a pressure wave outwards through the in-falling matter.

(6) The propagation of the shock wave outwards is not easy to analyse, and the final picture from the numerical simulations is not completely clear. By and large, two possible cases can be envisaged.

(a) As the shock propagates outwards through the inner core, it will break down the iron nuclei through photodisintegration, which is a costly process in terms of energy: It takes approximately 1.7×10^{51} ergs of energy to disintegrate 0.1 $M_\odot$ of iron. If the iron core is not too massive, the shock can emerge at the outer region without too much loss of energy. When the shock energy is deposited on the outer material, an explosion can result, spewing out the mantle. This process is called a prompt hydrodynamic explosion.
(b) On the other hand, if the iron core is fairly massive, the shock can lose a fair amount of energy and its propagation can stall. When the shock becomes stationary, the accreting matter on the core will produce an accretion shock

as in the case of a protostar collapse. However, now the matter is sufficiently dense and even neutrinos cannot escape freely from it. This leads to the formation of a neutrino sphere (analogous to the ordinary photosphere for radiation in stars), and a small fraction (~0.05) of the neutrino energy is deposited in the matter behind the shock. This additional energy can heat up the matter and allow the shock to continue forward, resulting in an explosion. This case, which relies heavily on the results of numerical simulations for its justification, is called a *delayed-explosion mechanism*.

(7) The total kinetic energy of the outgoing shock is $\sim 10^{51}$ ergs, which is roughly 1% of the energy liberated in neutrinos. When the outer material expands to $\sim 10^{15}$ cm and becomes optically thin, an impressive optical display arises, releasing $\sim 10^{49}$ ergs of energy in photons with a peak luminosity of $\sim 10^{43}$ ergs s$^{-1} \simeq 10^9 L_\odot$.

(8) The nature of the remnant depends on its mass, which in turn depends on the original mass of the star. If the original mass was less than $\sim 25 M_\odot$, the inner core, mostly made of neutrons, can be stabilised by the degeneracy pressure of the neutrons, and the resulting structure is called a neutron star. If the mass of the original star was greater than $25 M_\odot$, the remnant mass is larger than the Chandrasekhar mass for the neutron sphere, and the remnant will collapse to form a black hole.

(9) One clear signature of the core collapse followed by the formation of a compact object is the emission of approximately 3×10^{53} ergs of energy in the form of neutrinos. Hence the detection of neutrinos from a supernova can be of significant help in checking the theoretical models.

We shall now consider each of these features in greater detail.

4.3 Formation of Iron Cores and Neutrino Cooling

We can understand the major qualitative difference between the evolution of massive stars and low-mass stars by applying the virial theorem to a degenerate electron gas. In this particular case, the virial theorem implies that

$$E_G + 3 \int P \, dV = 0, \qquad (4.1)$$

where $E_G \approx -(GM^2/R)$ is the gravitational potential energy of the star. The exact expression for the pressure P is complicated because the electrons (which contribute significantly to the pressure) could be either degenerate or nondegenerate and also relativistic or nonrelativistic. The region that separates conditions in which electrons at the core are degenerate from the ones in which they are nondegenerate is determined by the ratio between Fermi energy and thermal energy:

$$\frac{\epsilon_F}{k_B T} = \begin{cases} 2.97 \times 10^5 T^{-1} (\rho Y_e)^{2/3} & \text{(nonrelativistic)} \\ 5.93 \times 10^7 T^{-1} (\rho Y_e)^{1/3} & \text{(relativistic)} \end{cases}, \qquad (4.2)$$

4.3 Formation of Iron Cores and Neutrino Cooling

where Y_e is the number of electrons per baryon and ρ and T are measured in cgs units. (Note that $Y_e = \mu_e^{-1}$; it is conventional to use Y_e rather than μ_e in discussing supernova.) Because, for degenerate electrons the density is related to Fermi momentum by

$$\rho = 0.97 \times 10^6 (\text{gm cm}^{-3}) \frac{A}{Z} \left(\frac{p_F}{m_e c}\right)^3, \qquad (4.3)$$

they become relativistic (with $p_F \to m_e c$) for $\rho \gtrsim 10^6$ gm cm^{-3}. Although an exact analysis is complicated, we can capture the essential physics and all the limiting values by adding the zero-temperature degeneracy pressure of the electrons $P_e(T=0)$ to the ideal-gas pressure by taking $P \cong P_e(T=0) + nk_BT$. For $P_e(T=0)$ we have to further extrapolate between the relativistic case [$P_R = (1/3)\epsilon_0$, $\gamma = 4/3$, where ϵ_0 is the energy density] and the nonrelativistic case [$P_{NR} = (2/3)\epsilon_0$, $\gamma = 5/3$]. We achieve this by writing $P_e(T=0) \cong (\gamma - 1)\epsilon_0$. Finally, the energy density ϵ_0 can be approximated as $\epsilon_0 \cong n_e[(p_F^2 c^2 + m_e^2 c^4)^{1/2} - m_e c^2]$, where p_F is the Fermi momentum and n_e is the number density of electrons. Putting all these together, we have

$$P \cong P_e(T=0) + nk_BT = (\gamma - 1)\epsilon_0 + nk_BT$$
$$= (\gamma - 1)n_e m_e c^2[(1+x^2)^{1/2} - 1] + nk_BT, \qquad (4.4)$$

with

$$x \equiv \frac{p_F}{m_e c} \simeq \frac{h}{m_e l_{ee} c} \simeq \frac{n_e^{1/3} h}{m_e c} \simeq \lambda_e \frac{N_e^{1/3}}{R}, \qquad (4.5)$$

where l_{ee} is the mean separation between the electrons, N_e is the total number of electrons, $\lambda_e \equiv (\hbar/m_e c)$ is the Compton wavelength of the electrons, and R is the radius of the star. If $N = (N_e/Y_e \mu)$ is the total number of particles of all kinds, with Y_e and μ denoting the number of electrons per baryon and molecular weight, respectively, then virial theorem (4.1) becomes

$$-\frac{GM^2}{R} + 3Nk_B\bar{T} + 3(\gamma - 1)N_e m_e(\sqrt{x^2+1} - 1) \simeq 0. \qquad (4.6)$$

Defining

$$N_0 \equiv \frac{m_{\text{Pl}}^3}{m_N^3}, \quad N_{e0}^{1/3} \equiv N_0^{1/3} Y_e = \frac{m_{\text{Pl}}}{m_N} Y_e, \qquad (4.7)$$

where $m_{\text{Pl}} \equiv (\hbar c/G)^{1/2}$ is the Planck mass and m_N is the mass of the nucleon and writing the first term in relation (4.6) as $(-GM^2/R) = -GM^2(x/\lambda_e N_e^{1/3})$, we obtain a relation between the temperature and the density (parameterised x) in the form

$$\frac{k_B\bar{T}}{m_e c^2} \simeq \frac{1}{3}\mu Y_e \left[\left(\frac{N_e}{N_{e0}}\right)^{2/3} x + 3(\gamma - 1)(1 - \sqrt{1+x^2})\right]. \qquad (4.8)$$

This equation, which provides an approximate description of the variation of central parameters of a star, allows us to draw several key conclusions: (1) The factor $3(\gamma - 1)$ varies between 1 and 2 and can be approximated for our purposes as unity. With increasing density ρ, the Fermi momentum p_F increases [see Eq. (4.3)] and hence $x \propto p_F$ increases monotonically; but $T(x)$ in relation (4.8) reaches a maximum and then decreases monotonically if $N_e \lesssim N_{e0}$. On the other hand, if $N_e \gtrsim N_{e0}$, the temperature increases without limit scaling asymptotically as $T \propto x \propto p_F \propto \rho^{1/3}$. The mass corresponding to the critical number N_{e0} is $M_{ch} \simeq (N_{e0}/Y_e)m_N = (m_{Pl}^3/m_N^2)Y_e^2$. Except for a numerical factor, this is same as the limiting mass called the Chandrasekhar mass (see Chap. 1, Section 1.4, and Chap. 3, Section 3.3; we will discuss this limiting mass in detail in Chap. 5). The rigorous definition of the Chandrasekhar mass, which we shall obtain in the next chapter, is

$$M_{Ch} = 3.1 \, Y_e^2 \, \frac{m_{Pl}^3}{m_N^2} = 5.84 \, Y_e^2 \, M_\odot \qquad (4.9)$$

(2) If $M < M_{Ch}$ and $x \ll 1$, the maximum temperature scales with the mass as $(k_B T / m_e c^2)_{\max} \simeq (M/M_{Ch})^{4/3}$.

These considerations show that the mass scale for the core evolution is set by the Chandrasekhar mass. As long as $M_{core} \gtrsim M_{Ch}$, the central temperature and the density of sufficiently massive stars ($M > 10 \, M_\odot$) increase almost monotonically during their evolution (see Fig. 4.1). Such a star possesses a core that contracts continually except during nuclear-energy generation by the burning of hydrogen, helium, carbon, oxygen, and finally silicon. The stars with M greater than $\sim 10 \, M_\odot$ go through the entire sequence, thereby having a onion-skin-like structure at late stages. Silicon burning occurs at strongly degenerate conditions when the core mass M_{core} is greater than the Chandrasekhar mass, leading to the formation of an iron core in the centre of the star.[2]

Calculations also suggest that M_{core} is driven towards M_{Ch} around the time of silicon burning. Silicon burning requires a temperature of approximately $k_B T \gtrsim 0.6 m_e c^2$, which in turn requires an ignition mass of $M_{\min} \approx M_{Ch}$. If $M_{core} < M_{Ch}$, then the ignition of silicon is delayed until the shell burning of lighter nuclei increases the core mass M_{core}. If, on the other hand, $M_{core} > M_{\min}$ at some earlier epoch in the evolution of the star, core contraction will lead to rather high central temperatures during the evolution. As we shall see below, cooling by neutrinos, which varies as a high power of T, will then inhibit the growth of the convective core at each intermediate stage of nuclear burning. This has the effect of shrinking the size and the mass of the core, thereby again driving M_{core} to M_{Ch}. Detailed numerical simulations show that a $15 M_\odot$ star, for example, will have $Y_e = 0.42$, $\rho_c = 3.7 \times 10^9$ gm cm^{-3}, $T_c = 8 \times 10^9$ K $= 0.69$ MeV, and a core mass of $1.5 \, M_\odot \gtrsim M_{Ch} \, (Y_e = 0.42)$ at the onset of silicon burning.

At these temperatures and densities, the emission of neutrinos can be a major source of energy loss for the star. We shall now briefly describe different neutrino-emission processes that play a role in evolution.

4.3 Formation of Iron Cores and Neutrino Cooling

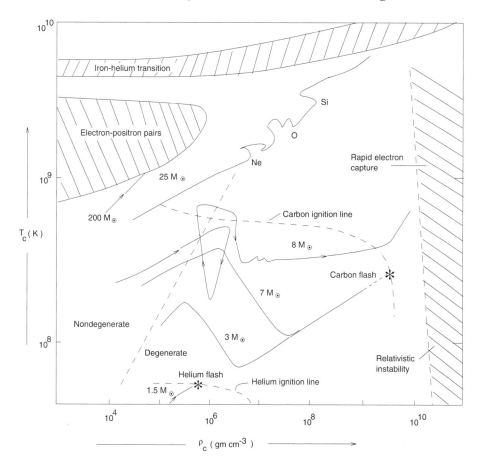

Fig. 4.1. Evolutionary track of stars of different mass in terms of central temperature and density. The tracks for low-mass stars show that carbon ignition occurs under degenerate conditions whereas the evolution of a $25 M_\odot$ star shows that the carbon ignition occurs under nondegenerate conditions for high-mass stars.

Neutrino emission can be broadly classified as nuclear and leptonic. The first set contains processes in which neutrino emission accompanies a group of nuclear reactions. Many such examples were discussed in the study of nuclear reactions in Vol. I, Chap. 12, Section 12.13. In addition to these, there also exists a phenomenon called the Urca process, which involves the following two reactions in sequence[3]:

$$(Z, A) + e^- \to (Z-1, A) + \nu,$$
$$(Z-1, A) \to (Z, A) + e^- + \bar{\nu}. \quad (4.10)$$

Because the original nucleus (Z, A) is regained, the net effect of the Urca process is a loss of thermal energy of the stellar matter into a $\nu\bar{\nu}$ pair. The details of the process depend on the availability of a pair of isobaric nuclei, with $(Z-1, A)$ having a slightly higher energy and being unstable to beta decay.

When appropriate nuclei are present, the energy loss increases montonically with ρ and T.

The leptonic processes, on the other hand, occur without nuclear reactions and arise because (according to the theory of weak interactions) there exists a direct coupling between electrons and neutrinos, just as in the theory of electromagnetic interactions there exists a direct coupling between electrons and photons. The most important among these processes are the following:

$$
\begin{aligned}
\text{pair annihilation:} \quad & e^- + e^+ \to \nu + \bar{\nu}, \\
\text{photoneutrino process:} \quad & \gamma + e^- \to e^- + \nu + \bar{\nu}, \\
\text{plasmon neutrino process:} \quad & \text{plasmon} \to \nu + 2\bar{\nu}.
\end{aligned} \tag{4.11}
$$

The first process arises because there is a small probability for electron–positron annihilation to emit two neutrinos rather than two photons. The second process is analogous to Compton scattering in which the incoming photon is absorbed by the electron but – instead of being reemitted as a photon – a $\nu\bar{\nu}$ pair is emitted. The last one arises from the fact that the quanta of any plasma wave (like the ones discussed Vol. I, Chap. 9, Section 9.4) can be thought of as a quasiparticle with a definite energy and momentum. For example, the simplest type of wave has the dispersion relation $\omega^2 = (k^2 c^2 + \omega_{pl}^2)$, where ω_{pl} is the plasma frequency. This is equivalent to a quasiparticle with a rest-mass energy $\hbar\omega_{pl}$. Such a quasiparticle corresponding to the collective mode in plasma can lose its energy in the form of neutrinos. In addition to these, it is also possible to emit neutrinos during the deceleration of an electron in the Coulomb field of a nucleus in a manner analogous to normal bremsstrahlung.

It is possible to compute the energy loss per unit time per unit volume that is due to all these processes from standard weak-interaction theory. For example, the energy-loss rate (ergs cm^{-3} s^{-1}) in the case of pair production is given by

$$
\epsilon_{\text{pair}} \simeq \frac{4}{9\pi^5}\left[\frac{1}{2}(C_V^2 + C_A^2) + (C_V - 1)^2 + (C_A - 1)^2\right]\frac{1}{\hbar^4 c^3}
$$
$$
\times \frac{G_F^2}{(\hbar c)^6}(m_e c^2)^9 \left(\frac{k_B T}{m_e c^2}\right)^9 F(\epsilon_F/k_B T), \tag{4.12}
$$

with

$$
F(y) \equiv \int_0^\infty \frac{u^4\, du}{1 + e^{u-y}} \int_0^\infty \frac{v^3\, dv}{1 + e^{v+y}} + \int_0^\infty \frac{u^3\, du}{1 + e^{u-y}} \int_0^\infty \frac{v^4\, dv}{1 + e^{v+y}}, \tag{4.13}
$$

where C_V and C_A are the weak-interaction coupling constants and G_F is the Fermi coupling constant. Numerically $C_V \simeq 0.96$, $C_A \simeq 0.5$, and $[G_F/(\hbar c)^3] = 1.166 \times 10^{-11}$ MeV^{-2}. The factor within the brackets in relation (4.12) is ~ 0.84 in standard electroweak theory. In the limit in which electrons are degenerate,

4.3 Formation of Iron Cores and Neutrino Cooling

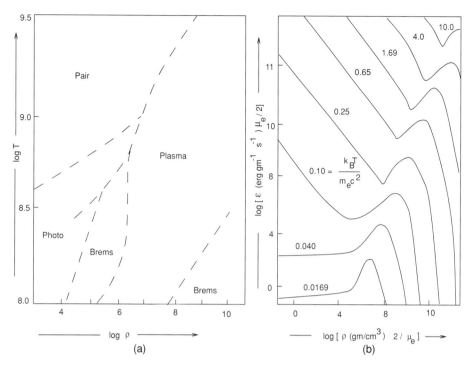

Fig. 4.2. Effects of neutrino cooling: (a) various regions in which different ν processes are operating; (b) the neutrino luminosity.

$F(y)$ can be approximated by

$$F(y) \simeq \frac{6}{5}\left(y^5 + \frac{5}{y}\right)e^{-y}, \tag{4.14}$$

and the corresponding energy loss, in the case of extreme relativistic and degenerate electrons, will be

$$\epsilon_{\text{pair}} \simeq 1.2 \times 10^{15} \rho Y_e \left(\frac{\epsilon_F}{m_e c^2}\right)^2 T_{10}^4 e^{-\epsilon_F/k_B T} \text{ ergs cm}^{-3} \text{ s}^{-1}. \tag{4.15}$$

If $\rho \approx 10^9$ gm cm^{-3} and $k_B T \approx m_e c^2$, then the energy loss is $(\epsilon_{\text{pair}}/\rho) \simeq 6 \times 10^{11}$ ergs gm^{-1} s^{-1}. For a 1-$M_\odot$ core, this corresponds to a luminosity of $\sim 10^{11} L_\odot$. A comparison with the radiation-energy density $aT^4 = 7.6 \times 10^{17} T_{10}^4$ ergs cm^{-3} shows that the cooling time $(aT^4/\epsilon) = 10^{-3}$ s is extremely short.

The energy loss that is due to all other neutrino processes can be computed along similar lines. Figure 4.2 summarises the various effects of neutrino cooling. Figure 4.2(a) indicates the dominant neutrino process that operates at different regions of the T–ρ plane, and Fig. 4.2(b) gives the total ν luminosity as a function of the density of the material parameterised by the temperature.[4]

4.4 Photodisintegration and Neutronisation

At the temperatures relevant to our discussion, the photons have sufficient energy to dissociate the iron nuclei into α particles by means of the reaction

$$\gamma + {}^{56}_{26}\text{Fe} \rightleftarrows 13\alpha + 4n. \tag{4.16}$$

The threshold energy needed for this process is

$$Q = (13m_\alpha + 4m_n - m_{\text{Fe}})c^2 = 124.4 \text{ MeV}. \tag{4.17}$$

When this reaction is in equilibrium, the resulting composition of the core can be obtained from the condition for nuclear statistical equilibrium (described in Vol. I, Chap. 5, Subsection 5.12.3). Using the degrees of freedom $g_\alpha = 1$, $g_n = 2$, and $g_{\text{Fe}} = 1.4$, where the last result arises from taking into account some of the low-lying excited states of the nucleus, we get

$$\frac{n_\alpha^{13} n_n^4}{n_{\text{Fe}}} = \frac{g_\alpha^{13} g_n^4}{g_{\text{Fe}}} \left[\left(\frac{k_B T}{2\pi \hbar^2} \right)^{3/2} \right]^{13+4-1} \left(\frac{m_\alpha^{13} m_n^4}{m_{\text{Fe}}} \right)^{3/2} e^{-Q/k_B T}. \tag{4.18}$$

Replacing mass m_A of a nucleus of atomic weight A with $A m_N$, where m_N is the nucleon mass, we find that this relation becomes

$$\frac{n_\alpha^{13} n_n^4}{n_{\text{Fe}}} = \frac{2^{43}}{(56)^{3/2}(1.4)} \left(\frac{m_N k_B T}{2\pi \hbar^2} \right)^{24} e^{-Q/k_B T} \tag{4.19}$$

for equilibrium between α particles, neutrons, and iron. Further, if we assume that ^{56}Fe is the most abundant heavy nucleus, then reaction (4.16) implies that

$$n_n = \frac{4}{13} n_\alpha. \tag{4.20}$$

Relations (4.19) and (4.20) allow us to calculate the degree of dissociation of the iron core at any given ρ and T. In particular, half the mass of the system will be in the form of ^{56}Fe at a temperature and density related by

$$\log \rho = 11.62 + 1.5 \log T_9 - \frac{39.17}{T_9}. \tag{4.21}$$

For $\rho \simeq 3.7 \times 10^9$ gm cm^{-3}, the characteristic temperature at which the disintegration of iron becomes important is $T_9 \approx 11$, which is not very different from the initial core temperature. (In the above calculation we have approximated the mass of a nuclei with atomic number A as $A m_H$; this is clearly imprecise as it ignores the binding energy of the nuclei. The final relation obtained, however, is not changed significantly by this approximation.)

At somewhat higher temperatures (at the same densities), the α particles will dissociate into nucleons by means of the reaction

$$\gamma + {}^4\text{He} \rightleftarrows 2p + 2n. \tag{4.22}$$

4.4 Photodisintegration and Neutronisation

This process has a threshold of $Q' = 28.3$ MeV. The corresponding equation in this case will be

$$\frac{n_p^2 n_n^2}{n_\alpha} = 2\left(\frac{m_N k_B T}{2\pi \hbar^2}\right)^{9/2} \exp\left(-\frac{Q'}{k_B T}\right). \qquad (4.23)$$

If, for example, $\rho = 10^9$ gm cm^{-3}, then the temperature at which half the mass will be ^{56}Fe and the other half will be ^{4}He and free neutrons will be $T_9 \approx 9.6$. However, the temperature at which ^{4}He is also split into nucleons is somewhat higher and is given by $T_9 \approx 15.2$. For more realistic estimates, we need to take into account the whole chain of nuclear reactions; however, the above analysis clearly shows that the photodisintegration of iron group elements occurs at these temperatures.

The dissociation of iron has the effect of reducing the adiabatic index Γ_1 below $4/3$, thereby initiating a catastrophic collapse of the core. To estimate this effect we can proceed as follows. Let $\mathcal{E}(P, v)$ be the specific internal energy of the system, where P is the pressure and v is the specific volume. For adiabatic changes, $d\mathcal{E} = -P dv = \mathcal{E}_{,v} dv + \mathcal{E}_{,p} dp$, giving

$$\Gamma_1 = \left(\frac{\partial \ln P}{\partial \ln \rho}\right)_{ad} = -\frac{v}{P}\left(\frac{\partial P}{\partial v}\right)_{ad} = \frac{Pv + v\mathcal{E}_{,v}}{P\mathcal{E}_{,P}}. \qquad (4.24)$$

If we assume that $\mathcal{E}$ is a function only of Pv (as in most cases), then we obtain

$$\Gamma_1 = 1 + \frac{1}{\mathcal{E}'}, \qquad (4.25)$$

where the prime denotes the derivative with respect to the argument Pv. For the iron–helium transition, the changes in $\mathcal{E}$ and Pv can be easily estimated. Because reaction (4.16) allows the release of $Q \simeq 124$ MeV per $56\, m_N$ of mass, $\Delta \mathcal{E}$ is

$$\Delta \mathcal{E} = 2 \times 10^{18} \text{ ergs gm}^{-1}. \qquad (4.26)$$

On the other hand, $P = (N/V)k_B T$ and $v = (V/N\bar{m})$ (where $\bar{m}$ is the mean mass per particle), giving $Pv = (k_B T/\bar{m})$. Initially $\bar{m} = 56\, m_N$; finally, there are 13 alpha particles and 4 neutrons, giving $\bar{m} = [56/(13+4)]m_N$. This changes $1/\bar{m}$ by $(2/7)\, m_N^{-1}$ so that

$$\Delta(Pv) = \left(\frac{2}{7}\right)\frac{k_B T}{m_N} = 2.4 \times 10^{17} \text{ ergs gm}^{-1} \quad (\text{for } T \simeq 10^{10} \text{ K}). \qquad (4.27)$$

Therefore

$$\Gamma_1 \simeq 1 + \frac{\Delta(Pv)}{\Delta \mathcal{E}} \simeq 1.1 < 4/3, \qquad (4.28)$$

indicating an instability. The above estimate ignores electrons, which is justifiable because they have an adiabatic index close to $4/3$.

The second key process that takes place in the core as it starts to collapse is the neutronisation, which has the effect of decreasing the number of free electrons and protons and increasing the number density of neutrons. Initially, the Fermi energy $\epsilon_{F,e}$ of the electron is typically 6 MeV. It rises rapidly as the core collapses and the density increases by means of the relation

$$\frac{\epsilon_{F,e}}{m_e c^2} \simeq 100\, Y_e^{1/3} \rho_{12}^{1/3}. \tag{4.29}$$

The dominant neutronisation reaction is $e^- + p \to n + \nu_e$, which is essentially an inverse beta decay. Although, in principle, capture can occur with protons that are bound in the nuclei, this is a subdominant effect that is due to blocking from the nuclear-shell effects. Hence the process is essentially limited by the abundance of free protons. The total capture rate per proton $\mathcal{R}$ in the extreme relativistic limit can be estimated from the standard weak-interaction theory, and the result is given by

$$\mathcal{R} = 1.3 \times 10^{-4} \left(\frac{\epsilon_F}{m_e c^2}\right)^5 \text{s}^{-1}\, \text{proton}^{-1} = 1.3 \times 10^6\, Y_e^{5/3} \rho_{12}^{5/3}\, \text{s}^{-1}\, \text{proton}^{-1}, \tag{4.30}$$

provided that the neutrinos escape freely from the core region. The neutronisation also leads to a further decrease of Γ_1 below $4/3$. At high densities, taking the system to be described by a zero-temperature Fermi gas, we have

$$\Gamma_1 = \frac{d \ln P}{d \ln n_e} \frac{d \ln n_e}{d \ln \rho} = \Gamma_1^{(0)}(1 - \delta), \tag{4.31}$$

where n_e is the electron number density and Γ_1^0 is the value of Γ_1 without loss of electrons through inverse beta decay. For relativistic electrons, the first factor is given by

$$\Gamma_1^{(0)} = \frac{4}{3} + \frac{2}{3(p_F/m_e c)^2} + \cdots, \tag{4.32}$$

which is quite close to $4/3$ in the regime under discussion. For inverse beta decay, $\delta > 0$ and hence Γ_1 drops below $4/3$.

At sufficiently high densities the neutrino mean free path decreases and they get trapped in the core. Once this happens, further electron capture by means of inverse beta decay is disfavoured because of the Pauli principle, which does not allow the ν emitted in the reaction to occupy the states that are already blocked out by the trapped neutrinos. We shall see in the next section that the density at which this trapping occurs is $\rho_{\text{trap}} \approx 3 \times 10^{11}$ gm cm^{-3}.

4.5 Neutrino Opacity and Trapping

The above analysis assumed that the neutrinos could escape from the stellar core essentially at the speed of light; i.e., the diffusive time scale of the neutrino is

4.5 Neutrino Opacity and Trapping

much smaller than the characteristic time scale for collapse. As the density of material increases, this is no longer true, and the mean free path of the neutrinos becomes progressively smaller because of its interaction with stellar matter. At a critical density, the star becomes optically thick for the neutrinos and the neutrinos are trapped inside and could diffuse out only like the photons from the photosphere of normal stars.

There are several processes that contribute to the coupling between neutrinos and matter of, which the most important ones are the following: (1) The first one is the scattering between free nucleons and neutrinos given by reactions of the kind $\nu + n \to \nu + n$ and $\nu + p \to \nu + p$. (2) The same process can also proceed through coherent scattering of neutrinos with several of the nucleons in heavy nuclei by means of $\nu + (Z, A) \to \nu + (Z, A)$. The cross section for this process scales as the square of the atomic mass A^2 rather than as A, which would have been the result if the nucleons acted incoherently and independently. The persistence of heavy nuclei at high densities makes coherent scattering a dominant effect. (These two processes exist in the weak-interaction theory because one of the mediating vector bosons of the weak interaction – the Z boson – is neutral; that is, it has no electric charge. Hence these two reactions are said to be mediated by means of a neutral current.) (3) Neutrinos also participate in nucleon absorption reactions like inverse beta decay ($\nu_e + n \to p + e^-$). For low-energy neutrinos ($E_\nu \ll m_N c^2$) the scattering from nucleons and nuclei is elastic and the neutrino energy changes very little. (4) Finally, there is also a scattering of neutrinos by electrons; $e^- + \nu \to e^- + \nu$. The ν_e scattering, in contrast to nucleon scattering, is nonconservative in the sense that it can lead to appreciable neutrino-energy loss during the scattering. This process is thus quite important in thermalising the neutrinos.

The scattering cross section for all these effects can be computed from the standard weak-interaction theory. The result for coherent neutrino-nucleus scattering can be expressed by the differential cross section

$$\frac{d\sigma}{d\Omega} = \sigma_0 \left(\frac{A^2}{64\pi}\right) \left(\frac{E_\nu}{m_e c^2}\right)^2 (1 + \cos\theta), \tag{4.33}$$

where

$$\sigma_0 = \frac{4}{\pi} \frac{G_F^2 m_e^2}{h^4} = 1.7 \times 10^{-44} \text{ cm}^2 \tag{4.34}$$

and G_F is the Fermi coupling constant. We can obtain the total cross section for momentum transfer by multiplying the angular-scattering cross section $(d\sigma/d\Omega)$ by $(1 - \cos\theta)$ and integrating over all solid angles. This gives the momentum-transfer cross section as

$$\Sigma_{\nu A} = \sigma_0 \left(\frac{A^2}{24}\right) \left(\frac{E_\nu}{m_e c^2}\right)^2. \tag{4.35}$$

The corresponding result for scattering by neutrons and protons is given by

$$\frac{d\sigma_{vn}}{d\Omega} = \frac{G_F^2}{(2\pi)^2} \frac{1}{4} E_v^2 \left[(1+\cos\theta) + 3g^2 \left(1 - \frac{1}{3}\cos\theta\right) \right], \quad (4.36)$$

where $g \approx 1.25$ is a constant occurring in the weak-interaction theory. The cross sections for momentum transfer in the vn and vp scatterings are almost equal and are given by

$$\Sigma_v \equiv \Sigma_{vp} \approx \Sigma_{vn} = \left(\frac{\sigma_0}{24}\right) \left(\frac{E_v}{m_e c^2}\right)^2 (1 + 5g^2). \quad (4.37)$$

Given the scattering cross section for the momentum transfer, we can obtain the mean free path that is due to neutrino scattering on the nucleons and nuclei. A simple calculation gives

$$\lambda_v = \frac{m_N}{\rho \Sigma_v} \left(X_N + \frac{\bar{N}^2}{A(1+5g^2)} X_A \right)^{-1}, \quad (4.38)$$

where X_A and X_N are the mass fractions of heavy nuclei and free nucleons, respectively, and $\bar{N}$ is the average number of neutrons in each heavy nucleus. Numerically,

$$\lambda_v \simeq (1.0 \times 10^6 \text{ cm}) \rho_{12}^{-1} \left(\frac{\langle E_v \rangle}{10 \text{ MeV}}\right)^{-2} \left(X_N + \frac{\bar{N}^2}{A(1+5 \times 1.25^2)} X_A \right)^{-1}. \quad (4.39)$$

Let us consider a case with $\langle E_v \rangle \approx \epsilon_{F,e}$. Using expression (4.29) with $Y_e \approx 0.4$, we get $\langle E_v \rangle \approx 40 \rho_{12}^{1/3}$ MeV. [Once the collapse is underway, the most significant production mechanism for neutrinos is by means of electron capture. In that case, we can compute the mean energy of the neutrino by averaging the neutrino energy over the cross section for electron capture. Such a calculation changes the result slightly and gives $\langle E_v \rangle = (5/6)\epsilon_{F,e}$. This has the effect of reducing $\langle E_v \rangle$ by a factor of $5/6$ compared with the naive estimate.] Taking the nuclear composition to be characterised by $X_N = 0, (Z, A) = {}^{56}\text{Fe}, X_A = 1$, we find that the mean free path becomes

$$\lambda_v \simeq 0.3 \times 10^5 \rho_{12}^{-5/3} \text{ cm}. \quad (4.40)$$

If the size of the region is R and the number of scatterings is N_{scat} before the neutrino escapes, then the random-walk argument gives $R \simeq N_{\text{scat}}^{1/2} \lambda_v$ and the diffusion time scales as

$$\tau_{\text{diff}} \simeq N_{\text{scat}} \left(\frac{\lambda_v}{c}\right) \simeq \left(\frac{R}{\lambda_v}\right)^2 \left(\frac{\lambda_v}{c}\right) \simeq \frac{R^2}{c\lambda_v} \propto \frac{\rho^{-2/3}}{\rho^{-5/3}} \propto \rho. \quad (4.41)$$

4.5 Neutrino Opacity and Trapping

Taking into account the spherical geometry (by solving the diffusion equation in spherical coordinates) changes the formula by a factor of $3/\pi^2$ and we get

$$\tau_{\text{diff}} \simeq \frac{R^2(\rho)}{\pi^2 c \lambda_\nu(\rho)/3} \simeq 0.02 \, \rho_{12} \text{ s}. \tag{4.42}$$

This becomes larger than the free-fall collapse time,

$$t_{\text{coll}} = \left(\frac{8\pi G\rho}{3}\right)^{-1/2} = 1.3 \times 10^{-3} \rho_{12}^{-1/2} \text{s}, \tag{4.43}$$

when the density of matter is higher than ρ_{trap}, where

$$\rho_{\text{trap}} \simeq 1.5 \times 10^{11} \text{ gm cm}^{-3}. \tag{4.44}$$

At higher densities, neutrinos are trapped inside the star for the duration of the collapse. Most of the neutrinos from electron capture remain in the star and hence the lepton number per baryon, Y_l, does not change. The neutrino density increases very fast, and the neutrinos soon become degenerate. In fact, the entire physical state of the system made of baryons, leptons, and photons can now be uniquely specified by three quantities: T, ρ, and Y_e, or, equivalently, by s, ρ, and Y_e. Collapse proceeds further until nuclear densities $\rho_{\text{nucl}} \approx [3\, m_N/4\pi(1.2 \text{ fm})^3] \approx 10^{14}$ gm cm^{-3} are reached.

The neutrino trapping also has the effect of reducing their luminosity significantly. When the star collapses, from some radius R to a radius R_{nuc} at which the density is of the order of nuclear densities, the amount of gravitational potential energy released is of the order of (GM^2/R_{nuc}) because $R_{\text{nuc}} \ll R$. Writing $\rho_{\text{nuc}} \approx (3M/4\pi R_{\text{nuc}}^3)$, we get $R_{\text{nuc}} \approx 12$ km for $M \approx M_\odot$. If this energy was released in the form of neutrinos, within a *collapse* time scale, then the luminosity can be easily estimated to be

$$L_{\nu,\text{max}} \simeq \frac{(GM^2/R_{\text{nuc}})}{t_{\text{coll}}} \simeq 10^{57} \text{ ergs s}^{-1}. \tag{4.45}$$

On the other hand, neutrino trapping leads to the same energy's being liberated over the *diffusive* time scale with $\tau_{\text{diff}} \gg t_{\text{coll}}$ when $\rho \approx \rho_{\text{nuc}}$. As a result, the actual neutrino luminosity turns out to be more like

$$L_\nu \simeq \frac{(GM^2/R_{\text{nuc}})}{\tau_{\text{diff}}} \simeq 10^{52} \text{ ergs s}^{-1}. \tag{4.46}$$

Hence the bulk of the liberated gravitational energy is actually transformed internally to other forms of energy such as thermal energy, excitation energy of the nuclear state, etc., rather than being lost to the system. This suggests that after the neutrino trapping occurs, the collapse proceeds almost adiabatically in the system.

Exercise 4.1

Neutrinos from supernova: Assume that the neutrinos emitted from the supernova are in the form of a thermal radiation from the surface of a neutron star of mass $M = 1.4 \, M_\odot$ and radius $R \approx 10$ km. All three types of neutrinos and their antineutrinos are emitted in equal numbers. Compute the effective temperature and the mean energy of the neutrino radiation. [Answer: Treating neutrino species as zero-mass fermions with $\epsilon_F = 0$ and spin degeneracy $g_s = 1$, the energy density obtained by integrating over the Fermi–Dirac distribution will be $(7/16)aT^4$ per species. The total energy density for three species each of ν and $\bar{\nu}$ will be six times $U_\nu = (21/8)aT^4$. The corresponding flux will be $F_\nu = (21/8)\sigma T_{\text{eff}}^4$. Equating it to $(L_\nu/4\pi R^2)$, we get $T_{\text{eff}} = (2L_\nu/21\pi\sigma R^2)^{1/4}$. The mean energy of a Fermi distribution is given by $\langle E_\nu \rangle \approx 3.15 \, k_B T_{\text{eff}}$. Taking $L_\nu \approx (GM^2/Rt_{\text{diff}}) \approx 10^{52}$ ergs s^{-1}, we get $T_{\text{eff}} \approx 5 \times 10^{10}$ K and the average value of energy to be ~ 13 MeV.]

Exercise 4.2

Bound on the neutrino mass: Neutrinos with energies of (7.5–35) MeV were detected over a time interval of ~ 12.4 s from supernova 1987A, which occurred at a distance of ~ 50 kpc from Earth. Assuming that all these neutrinos were emitted simultaneously and the delay in arrival time is due to neutrinos of different energies' having different speeds, put an upper bound on the possible mass of the neutrino. [Answer: $m_\nu c^2 \lesssim 17$ eV.]

4.6 Core Collapse

The processes described in the last section show that the core of the star will undergo a collapse; we shall now discuss some aspects of this collapse. The key feature of the collapse of the iron core is that the entropy per baryon of the system remains fairly low. Hence the entire collapse case can be modelled as an adiabatic process with $P \propto \rho^\gamma$, with $\gamma \approx 4/3$. The low entropy per baryon (an initial value of $s_i \approx 1$, increasing slightly by $\delta s \lesssim 0.5$ before neutrino trapping, remaining constant at $s_f \approx 1$–2 during the later stages) together with the high lepton number arising from neutrino trapping preserves the heavy nuclei right up to nuclear densities. As we shall see in next chapter, this is in marked contrast to situations in cold matter at high densities. In the latter case, a process called *neutron drip* is already underway at $\rho = \rho_{\text{drip}} \approx 4.3 \times 10^{11}$ gm cm$^{-3} \ll \rho_{\text{nuc}}$ and free neutrons dominate the composition for $\rho \gtrsim \rho_{\text{drip}}$.

The entropy per electron for the relativistic electrons can be computed with the relations

$$S = \frac{1}{T}(PV + U - \mu N), \quad PV = \frac{1}{3}U \quad (4.47)$$

and the Fermi–Dirac distribution, leading to

$$s_e = \frac{4}{3}\frac{F_3(\mu_e/k_B T)}{F_2(\mu_e/k_B T)} - \frac{\mu_e}{k_B T}, \quad (4.48)$$

4.6 Core Collapse

where

$$F_n(\eta) = \int_0^\infty \frac{x^n}{e^{x-\eta}+1} dx \qquad (4.49)$$

(see Vol. I, Chap. 5, Section 5.9). In the strongly degenerate case, we have the standard approximation

$$\int_0^\infty \frac{\phi'(x)}{e^{x-\eta}+1} dx = \phi(\eta) + 2\left[\frac{\pi^2}{12}\phi''(\eta) + \frac{7\pi^4}{720}\phi^{(iv)}(\eta) + \cdots\right], \qquad (4.50)$$

which gives the entropy per electron as

$$s_e = \pi^2 \frac{k_B T}{\epsilon_{F,\text{ele}}} \quad \text{(relativistic; per electron)}. \qquad (4.51)$$

The entropy per nucleus arising from normal translational motion is given by (Vol. I, Chap. 5, Section 5.6)

$$(s_{\text{Fe}})_{\text{per nucleus}} = \frac{5}{2} + \ln\left[\frac{1}{n_{\text{Fe}}}\left(\frac{m_{\text{Fe}} k_B T}{2\pi \hbar^2}\right)^{3/2}\right], \qquad (4.52)$$

where n_{Fe} and m_{Fe} are the number density and the mass of the iron nuclei, respectively. Taking $\rho = 3.7 \times 10^9$ gm cm^{-3}, $T_c = 8 \times 10^9$ K, $Y_e = (26/56) = 0.46$, and using relation (4.29) to estimate $\epsilon_{F,\text{el}}$, we find $s_e \simeq 1.1$ and $(s_{\text{Fe}})_{\text{per nucleus}} \simeq 16.7$. Because each iron nucleus has 56 nucleons and 26 electrons, the entropy per nucleon is

$$\left(s_{\text{Fe}}^{\text{trans}}\right)_{\text{nucleon}} + (s_e)_{\text{nucleon}} = \frac{1}{56}[16.7 + (26 \times 1.1)] = 0.81. \qquad (4.53)$$

For the nuclei, we should also include the contribution of excited states to the entropy that we can obtain by assuming, to the lowest order of approximation, that the nucleus behaves as a nonrelativistic Fermi gas. In that case, the entropy per nucleon that is due to the excited states will be $s_{\text{ex}} \approx (\pi^2/2)(k_B T/\epsilon_F)$, where the Fermi energy of the nucleus is

$$\epsilon_{F,\text{nuc}} = \frac{(3\pi^2 \rho)^{2/3}}{(2m_N)^{5/3}} = 39 \text{ MeV} \qquad (4.54)$$

evaluated for $\rho = \rho_{\text{nuc}} \approx 2.8 \times 10^{14}$ gm cm^{-3}. At $T = 8 \times 10^9$ K, this gives a contribution of $s \simeq 0.1$, and hence we conclude that the initial entropy per nucleon is $s_i \approx 0.9$.

This entropy does not increase much until the neutrinos are trapped. An elementary estimate shows that the lepton number changes very little during the initial stages of collapse. To estimate the net change in the electron number ΔY_e, we integrate the equation $\dot{Y}_e = -\mathcal{R} Y_p = -\mathcal{R} Y_e$, with $\mathcal{R}$ given by Eq. (4.30), from the initial state until $\rho \approx \rho_{\text{trap}}$. For an order-of-magnitude estimate, we can take

ΔY_e to be $\dot{Y}_e(\rho_{\text{trap}})$ multiplied by the free-fall time scale $\tau_{\text{ff}}(\rho_{\text{trap}})$ corresponding to the density $\rho = \rho_{\text{trap}}$, where

$$\tau_{\text{ff}} = \left(\frac{8\pi}{3}G\rho\right)^{-1/2} = 1.3 \times 10^{-3}\rho_{12}^{-1/2} \text{ s} \qquad (4.55)$$

is the free-fall time for the collapse at density ρ. This gives

$$Y_{e,f} - Y_{e,i} = \int \dot{Y}_e \, dt \simeq [\dot{Y}_e(\rho)\tau_{\text{ff}}(\rho)]|_{\rho=\rho_{\text{trap}}} \simeq -2 \times 10^{-3} Y_{e,f}^{8/3} \rho_{12,\text{trap}}^{7/6}, \qquad (4.56)$$

from which we find that $Y_{e,f} \approx 0.1$. A more careful analysis shows that this is an underestimate and the actual lepton number is $Y_l \approx 0.35$ and that the net change in the entropy is ~ 0.5 for a wide class of physical situations.

Given that the entropy changes very little before neutrino trapping and remains constant after that, the collapse essentially proceeds along a low-entropy adiabat. We obtain the equation of state for such a system by setting the entropy to a constant. To the lowest order of approximation, the total entropy per baryon – as a function of ρ, Y_e, and T – is

$$s = \frac{1}{\bar{A}}\left\{\frac{5}{2} + \ln\left[\frac{\bar{A}m_u}{\rho}\left(\frac{\bar{A}m_u k_B T}{2\pi\hbar^2}\right)^{3/2}\right]\right\} + \frac{\pi^2 k_B T Y_e}{\epsilon_{F,\text{ele}}} + \frac{\pi^2 k_B T}{2\epsilon'_{F,\text{nuc}}} = \text{constant}. \qquad (4.57)$$

This expression ignores the contribution from free baryons and alpha particles and treats all nucleons as bound in a nucleus of mean atomic weight $\bar{A}$. At the onset of the collapse, the first two terms dominate the entropy; of these, the first one is logarithmic, and hence the initial part of the $T - \rho$ trajectory will be governed by the constancy of the second term. Because $\epsilon_F \propto \rho^{1/3}$, we expect the adiabat to follow $T \propto \rho^{1/3}$ when Y_e remains constant. Figure 4.3 shows the adiabat for a low value of entropy from which it is clear that the nuclei survive all the way to nuclear densities. From the Fermi gas model, we see that $s = 1.5$, say, corresponds to a temperature of ~ 10 MeV, which is fairly low compared with $\epsilon_{F,\text{nuc}}$; as a result, thermal dissociation does not take place until the nuclei fuse together to form effectively a stellar-sized nucleus at nuclear densities.[5] These considerations show that the equation of state can be well approximated by a pressure dominated by relativistic degenerate electrons all the way up to nuclear densities with $P \propto Y_e^{4/3} \rho^{4/3}$.

When the equation of state is of the form $P = \kappa\rho^\gamma$, it is possible to solve the time-dependent, spherically symmetric, gravitational-collapse problem by use of self-similarity. (This method has been used earlier in connection with the collapse of an isothermal sphere in Vol. I, Chap. 10, Section 10.3.1.) Such solutions arise because there are only two-dimensional parameters (κ and G) in the problem. The relevant similarity variable on which all quantities depend can

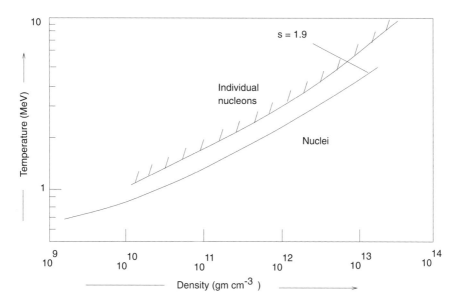

Fig. 4.3. Typical trajectories in the T–ρ plane during core collapse.

be chosen to be

$$\chi = r(-t)^{\gamma-2}\kappa^{-1/2}G^{(\gamma-1)/2}, \tag{4.58}$$

where $t = 0$ is chosen to be the time at which the central density becomes infinite; the precollapse stage corresponds to $t < 0$. Taking the density, velocity, and the mass enclosed within a sphere of radius r to be of the form

$$\begin{aligned}
\rho &= G^{-1}(-t)^{-2}D(\chi), \\
v &= \kappa^{1/2}G^{(1-\gamma)/2}(-t)^{1-\gamma}V(\chi), \\
m &= \kappa^{3/2}G^{(1-3\gamma)/2}(-t)^{4-3\gamma}M(\chi),
\end{aligned} \tag{4.59}$$

we can reduce the collapse equations to ordinary differential equations connecting D, V, and M. One of the relations between the variables is, of course,

$$M(\chi) = 4\pi \int_0^\chi x^2 D(x)\,dx, \tag{4.60}$$

and the continuity and the Euler equations lead to

$$\begin{aligned}
{[V - (\gamma - 2)\chi]}\frac{D'}{D} + V' &= -2 - \frac{2}{\chi}V, \\
\gamma D^{\gamma-2}D' + [V + (2-\gamma)\chi]V' &= -\frac{M}{\chi^2}(\gamma - 2)V.
\end{aligned} \tag{4.61}$$

For a general value of γ, these equations have to be integrated numerically. Figure 4.4 shows the resulting dimensionless in-fall velocity $-V$ as a function

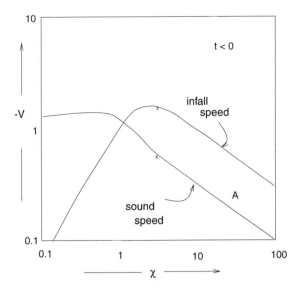

Fig. 4.4. In-fall velocity and speed of sound for adiabatic collapse. The solution is obtained by the integration of Eqs. (4.61) for $\gamma = 1.3$.

χ for $\gamma = 1.30$. For comparison, the speed of sound is also plotted on the same graph.

The general behaviour of the solution can be understood from the asymptotic limits of Eqs. (4.61). It is easy to see that the asymptotic limits of the solution are

$$V(\chi) \simeq -2\chi/3 \quad (\chi \ll 1),$$
$$V(\chi) \propto \chi^{(1-\gamma)/(2-\gamma)} \quad (\chi \gg 1). \quad (4.62)$$

Most of the inner homologous core is subsonic, with the sonic point occurring around $\chi \approx 1.4$. The maximal in-fall velocity is at $\chi = 2.96$, where the Mach number is ~2.1. The outer core has almost constant Mach number with an asymptotic value of ~2.94. These results are in agreement with more detailed numerical calculations and show that the inner core does collapse homologously and adiabatically. Figure 4.5 shows the density profile of such a collapsing structure, which is typical of stars having total masses in the range of 15–25 $M_\odot$. The heavy bottom curve shows the structure at the end of silicon burning with $Y_e \approx 0.48$. The other curve corresponds to end of hydrostatic electron capture with $Y_e \approx 0.42$. The outer regions are not seriously affected by the collapse.

4.7 Bounce and Possible Shock Formation

The low-entropy increase and small lepton loss will allow the core collapse to continue all the way to nuclear densities. Because the strong nuclear forces have a repulsive component at short distances, the compression can be resisted when

4.7 Bounce and Possible Shock Formation

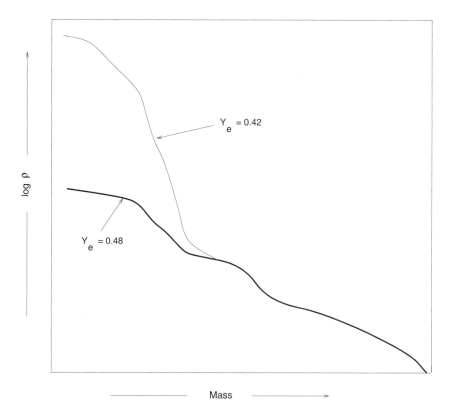

Fig. 4.5. Schematic diagram of the density profile during the core collapse of a star.

the material reaches nuclear densities with all the nucleons merging to form one enormous nucleus, described as a Fermi liquid. As the density rises a little higher, nuclear forces become repulsive, making the central matter stiffen up at a short time scale. This sends out a shock wave through the outer regions. The crucial question is how strong this shock will be by the time it has traversed the inner core and reaches the outer mantle. In particular, we would like to know whether the shock will have sufficient strength – on reaching the outer mantle – to eject the stellar material in the form of an explosive event.

Such a complex issue needs to be addressed by detailed numerical simulations of the system. Although it is generally believed that there exists a range of values for the parameters for which ejection of the outer mantle indeed occurs, thereby providing an explanation for supernova explosion, the detailed results of the simulations are at present still inconclusive. We shall now discuss several facets of these investigations in order to provide an overall picture.

The broad class of conditions under which the bounce and shock can occur can be conveniently bracketed by two limiting cases. In the first case, we can invoke the complete trapping of neutrinos with minimum neutrino transport and a hard equation of state at transnuclear densities. This will give a strong bounce shock. On the other extreme, we can think of neutrino transport as essentially decided by flux-limited diffusion that incorporates all known scattering and absorption processes of neutrinos. The equation of state can also be taken to be relatively softer in order to have a more moderate initiation of the shock.

Until the epoch of core bounce, these two models will lead to similar evolutionary sequences except that the second case will have a smaller homologous core, for two reasons: (1) Neutrino transport allows some neutrinos to escape, leading to smaller pressure differences, and (2) neutrino transport has the effect of rearranging the pressure distribution to slow the outer parts of the homologous core. In $\sim 10^{-3}$ s after the trapping of neutrinos, the effects of the equation of state become evident and the evolutionary sequence differs. In both the cases a bounce shock does form and starts moving out through the core towards the outer region. However, in a large class of models, the explosive shock has very little outward speed by the time it has reached the radius containing $\sim 1.1\ M_\odot$. This radius is well within the silicon and the oxygen burning shells and so we must conclude that the explosion has failed.

The qualitative reason is the following. At the point of bounce, the energy available to the shock is the kinetic energy, which – in the first model – is approximately 1.2×10^{52} ergs. As the shock moves, the matter behind is hot and in hydrostatic equilibrium. It is the excess energy that drives the shock with a small amount of matter moving outwards. The heating that is due to the shock leads to nuclear dissociation, requiring ~ 8 MeV per nucleon or 1.6×10^{52} ergs per $M_\odot$. In most stellar models, the shock has to move through $\sim 0.9\ M_\odot$ of matter before it can even reach the region where oxygen shell burning is occurring. This requires approximately 1.44×10^{52} ergs, which is more than what was available. Hence the shock barely reaches the region of silicon burning covering $\sim 0.5\ M_\odot$ of material. The above discussion shows that, in some sense, the numbers are evenly matched and small effects can change the conclusion in either direction; to that extent, the order-of-magnitude description is *not* trustworthy. However, it should be stressed that the above description was most optimally biased towards an explosive shock and, in fact, grossly underestimates important energy losses. The second model mentioned in the last paragraph is more realistic because it takes neutrino-transport properties into account better. In that case, the conclusion is still the same.

Matter outside the homologous core, which was originally in free fall, will lead to an accretion shock (to be distinguished from the bounce shock) in a manner similar to the case of protostellar collapse. As the evolution proceeds, more matter will undergo the accretion shock by the expansion of surrounding matter inwards to the newly formed core. The rate at which mass is being fed into the accretion shock is $\dot{M} \approx (4\pi r^2 \rho c_s)$, with $c_s^2 = (\gamma P/\rho) \approx (4P/3\rho) \propto \rho^{1/3}$. To a very good approximation, $\rho \propto r^{-3}$, giving $\dot{M} \propto r^{-3/2} \propto \rho^{1/2}$. Thus the rate of mass infall decreases as accretion continues. The energy reserves that were originally present in the bounced shock, added to the dynamical evolution triggered by the accretion shock, may be able to lead to the ejection of the outer mantle in some cases. Most of these cases involve energy that is being supplied, not from the outburst of neutrinos in the bounce shock, but from the longer-term loss from the core. Neutrinos that emerge from the neutrinosphere have a distribution function that is similar to that of the electrons from which they are produced and with which they interact. The typical temperature is approximately (1.5–2) MeV, and the chemical potential is ∼8 MeV. The temperature of matter above the neutrinosphere will be lower so that the flow of neutrinos tend to heat the matter through various interactions. The heating rate per unit volume will scale as ρ as it depends on the amount of matter available to interact with neutrinos. The cooling, rate of gas, however, will increase as ρ^2. It follows that, at low densities, heating will dominate over cooling, causing the temperature to rise. If energy can be transferred to this region quickly, pressure can build up to cause an explosion. This effect is also contributed to by possible convective transport in the outer regions. However, it is not clear how generic this situation is.

The discussion in this section was based on several implicit assumptions, of which the most important are the following: (1) The supernova explosion is spherically symmetric or nearly so. (2) The energy E involved in the explosion and the mass M that is ejected are such that $\eta \equiv (E/Mc^2) < 1$. This makes the explosion nonrelativistic in the sense that the ejecta move with nonrelativistic velocities. (3) The major component of energy involved in the explosion rests in the kinetic energy of the baryons that are ejected. Given the complexity of the underlying process, it is important to bear in mind that totally different cases are possible if the above assumptions are modified. One extreme variant of a supernova explosion – assumed to be relevant for the phenomena of gamma-ray bursts – involves energy that is transported out almost as pure radiation ("fireball") with very little baryonic contamination. Such a radiation fireball will initially expand from some radius R_i because of its own radiation pressure $P = (\rho/3)$. Initially, the energy E will be high enough to produce, a copious number of e^+–e^- pairs that will be present in equilibrium with the photons of the fireball. As the fireball expands and cools, e^+–e^- annihilation will reduce the pairs to a sufficiently small number and the fireball will become transparent to the photons, which can escape to infinity. A small amount of baryons that could be present in the fireball will be accelerated, thereby converting part of the radiation energy into bulk kinetic energy. Such a fireball can be characterised by two parameters, $\eta \equiv (E/Mc^2)$ and

$$\eta_b \equiv \left(\frac{3\sigma_T E}{8\pi m_p c^2 R_i^2}\right)^{1/3} \simeq 10^5 \left(\frac{E}{10^{52}\text{ ergs}}\right)^{1/3} \left(\frac{R_i}{10^7\text{ cm}}\right)^{-2/3}. \quad (4.63)$$

The situation described above will arise if $1 \ll \eta \ll \eta_b$; such a fireball will become matter dominated before it becomes optically thin. Most of the initial energy will eventually be converted to bulk kinetic energy of the baryons with a final Lorentz factor of $\Gamma_f \simeq \eta$. The ejected mass in this case has to be less than 5×10^{-3} $M_\odot$ if $E \simeq 10^{52}$ ergs. This is quite different from a conventional supernova explosion in which nearly 1 $M_\odot$ of material is ejected. The further evolution and radiation pattern of such an explosion will be quite different from what was described above.[6]

Exercise 4.3
Meaning of η_b: Explain the physical meaning of the parameter η_b and the condition $1 < \eta < \eta_b$.

4.8 Supernova Luminosity and Light Curves

For the progenitor of the supernova, in hydrostatic equilibrium, $c_s^2 \approx (P/\rho) \approx (GM/R)$, which is the gravitational binding energy per unit mass of the star. Because the explosive energy per unit mass, $\mathcal{E} = (v^2/2)$, of the supernova is much larger than the gravitational binding energy per unit mass, it follows that the explosion velocities are supersonic with $v \gg c_s$. When matter is ejected with such high speeds and expands, different physical processes take place in it, eventually leading to the optical emission that characterises the supernova. One of the main characteristics of a supernova is the actual light curve that is the plot of the supernova luminosity as a function of time. We shall next consider the question of relating the light curve to the explosion model.

In general, the evolution of such a system can be expressed in terms of the law of conservation of energy expressed as the first law of thermodynamics:

$$\dot{E} + P\dot{V} = \epsilon - \frac{\partial L}{\partial m}, \tag{4.64}$$

where E is the energy per unit mass and $V = (1/\rho)$ is the volume per unit mass on the left-hand side. On the right-hand side, ϵ denotes the heating term; the second term $(\partial L/\partial m)$ denotes the radiative losses through diffusion, which can be approximated as

$$\frac{\partial L}{\partial m} \approx \frac{L}{M} = \frac{E}{\tau_{\text{diff}}}, \tag{4.65}$$

where M is the total mass and $\tau_{\text{diff}} = f(\kappa M/cR)$ is the time scale for radiative diffusion ($f \approx 0.07$ is a numerical constant that depends on the actual geometry). Taking $\kappa \approx 0.4$ cm^2 g^{-1}, we find that

$$\tau_{\text{diff}} \approx 2 \times 10^7 \text{ s} \left(\frac{M}{M_\odot}\right)\left(\frac{R}{10^{14} \text{ cm}}\right)^{-1}. \tag{4.66}$$

The hydrodynamic time scale, $\tau_h \simeq (R/v)$, on the other hand, is

$$\tau_h \equiv 10^5 \text{ s} \left(\frac{R}{10^{14} \text{ cm}}\right)\left(\frac{v}{10^9 \text{ cm s}^{-1}}\right)^{-1}. \tag{4.67}$$

4.8 Supernova Luminosity and Light Curves

For reasonable parameters, $\tau_{\text{diff}} \gg \tau_h$; but note that, although τ_h increases with time, τ_{diff} decreases and these two become equal around $R \approx 1.4 \times 10^{15}$ cm $(mv_9)^{1/2}$, where $m = (M/M_\odot)$. The high-luminosity phase is bound by this value in the simple model for a supernova.

Let us now consider the solution to Eq. (4.64) within different contexts. The simplest one will correspond to ignoring the right-hand side and treating the expansion of the radiation-dominated gas as adiabatic. In this case, $\dot{E} + P\dot{V} \approx 0$, giving $(\dot{T}/T) \approx -(\dot{R}/R)$ and then $T \propto R^{-1}$ and $E \propto T^4 R^3 \propto R^{-1}$. In that case, the adiabatic expansion cools the radiation gas possibly even before an optical light curve is generated.

To the next order of approximation, we can ignore the heating sources ($\epsilon = 0$), set $P = (E/3V)$, and combine relations (4.64) and (4.65) into

$$\frac{\dot{E}}{E} + \frac{1}{3}\frac{\dot{V}}{V} = \frac{d}{dt}(\ln ER) = \frac{d}{dt}(\ln T^4 R^4) \cong -\frac{1}{\tau_{\text{diff}}}. \tag{4.68}$$

Next we will approximate $\tau_{\text{diff}} \propto R^{-1}$ as

$$\frac{1}{\tau_{\text{diff}}(t)} = \frac{R(t)}{R_0 \tau_{\text{diff}}(0)}. \tag{4.69}$$

If the explosion causes the material to move with speed v_a, then $R(t) = R_0 + v_a t$ and

$$\frac{1}{\tau_{\text{diff}}(t)} = \frac{R_0 + v_a t}{R_0 \tau_d(0)} \equiv \frac{\tau_h + t}{\tau_h \tau_d(0)}, \quad \tau_h \equiv \frac{R_0}{v_a}, \tag{4.70}$$

so that

$$\frac{d \ln(T^4 R^4)}{dt} = -\frac{(\tau_h + t)}{\tau_h \tau_d(0)}. \tag{4.71}$$

Here, note that $\tau_h \equiv (R_0/v_a)$ is the hydrodynamical time scale, with v_a denoting the asymptotic fluid velocity corresponding to the energy of the supernova explosion and $\tau_d(0) \equiv f(\kappa M/cR_0)$. Taking the opacity to be a constant, we can integrate this equation to give

$$\left(\frac{TR}{T_0^4 R_0^4}\right)^4 = \exp\left\{-\left(\frac{\tau_h t + t^2/2}{\tau_h \tau_d(0)}\right)\right\}, \tag{4.72}$$

so that the luminosity is

$$L(t) = \frac{ME}{\tau_d} = \left(\frac{4\pi ac}{3f}\right)\frac{T^4 R^4}{\kappa M} = L_0 \exp\left\{-\left(\frac{\tau_h t + t^2/2}{\tau_h \tau_d(0)}\right)\right\}, \tag{4.73}$$

with $L_0 = (4\pi ac/3f)(T_0^4 R_0^4/\kappa M)$. [Note that ME is the total energy of the system, which is equal to $(4\pi/3)R^3(aT^4)$.] For small τ_h, the luminosity has a

Gaussian dependence in t with a time constant determined by the product

$$\tau_{\text{lum}} = [2\tau_h \tau_d(0)]^{1/2} \approx 2 \times 10^6 \text{ s} \left(\frac{M}{M_\odot}\right)^{1/2} \left(\frac{v}{10^9 \text{ cm s}^{-1}}\right)^{-1/2}. \quad (4.74)$$

This is the typical time scale characterising the supernova. It is clear that objects that have low mass and low opacity and that expand faster have the light curves, which also evolve faster. For time scales that are short compared with this characteristic time, the luminosity is approximately constant. If the initial radiation energy $\mathcal{E}_{\text{rad}} = (4\pi/3)a R_0^3 T_0^4$ is taken to be half of the supernova explosion energy $\mathcal{E}_{\text{sn}}$, then the constant luminosity can be estimated as

$$L_0 = \frac{1}{2f}\left(\frac{\mathcal{E}_{\text{sn}}}{M}\right) R_0 \frac{c}{\kappa} \approx 2.5 \times 10^{43} \text{ ergs s}^{-1} \left(\frac{\mathcal{E}}{10^{51}}\right)\left(\frac{R}{10^{14} \text{ cm}}\right)\left(\frac{M}{M_\odot}\right)^{-1}. \quad (4.75)$$

For $(\mathcal{E}_{51}/m) \approx 0.1$ and $R_{14} \approx 1$, where $\mathcal{E}_{51}$ is the explosion energy in units of 10^{51} ergs etc., the luminosity is $\sim 3 \times 10^{42}$ ergs s^{-1}.

This simplified analysis contains two features that are generic. (1) Lower opacity, implying a shorter time scale for radiative diffusion, leads to higher luminosity even if the opacity is not constant in time or space, and (2) the luminosity scales in proportion to energy per unit mass $\mathcal{E}_{\text{sn}}$ so that, for the same $\mathcal{E}_{\text{sn}}$, less-massive stars are brighter. Tenuous large stars will be brighter than denser brighter stars, other things being equal. This is because denser stars have to expand more before the radiation can escape and in the process suffer more adiabatic losses.

The above analysis ignored the heating sources in a supernova. In general, the most important heating arises from the radioactive decay of iron group elements, especially from the decays of ^{56}Ni to ^{56}Co and ^{56}Co to ^{56}Ni. The energy of the radioactive decay, of course, is immune to processes like adiabatic cooling, and whether the energy released by the decay will contribute to the luminosity will depend on whether the photons can readily escape, thereby influencing the observed light curve. In general, radioactive decay leads to the production of gamma rays and positrons that have to be thermalised by different processes and downgraded in energy in order to affect the UV and the optical parts of the light curve. The decay of ^{56}Ni to ^{56}Co (in terrestrial conditions) is by means of electron capture from a K-shell electron and has a half-life of ~ 6.10 days. The decay product is in the 1.72-MeV excited state of ^{56}Co that in turn decays to the ground state through a series of gamma-ray transitions. The interaction of gamma rays with electrons has been discussed in Vol. I, Chap. 6, Section 6.4. At low energies, with $E_\gamma \lesssim 2m_e c^2$, the dominant process is Klein–Nishina scattering with the scattering cross section $\sigma_{\text{KN}} \approx 0.3\sigma_T$. At high energies, $E_\gamma \gtrsim 2m_e c^2$, the dominant process is the production of e^+e^- pairs. The scattering transfers energy from gamma rays to electrons and when – after several scatterings – the gamma-ray energy is reduced to that of K-shell electrons, photoelectronic absorption on Ni, Co, or Fe becomes dominant. Once this happens, thermalisation is fairly easy. Therefore the key factor deciding thermalisation is whether a sufficient number of scatterings of the gamma ray occur, degrading their energy. The ratio of the radius R to the Klein–Nishina mean free path $\lambda = (\rho \kappa)^{-1}$ is

$$\frac{R}{\lambda} = \frac{3\kappa_{\rm KN} M}{4\pi R^2} \approx 32 \left(\frac{M}{M_\odot}\right)\left(\frac{R}{10^{15}\,{\rm cm}}\right)^{-2}. \tag{4.76}$$

When $(R/\lambda)^2 \approx 1$, the gamma rays start escaping freely, which occurs at $R_{15} \approx 5.6\sqrt{m}$. Dividing the radius by the velocity of expansion will give the characteristic time scale for the emission of gamma rays. For $v \simeq 10^9$ cm s^{-1}, the time scale is $t_\gamma \simeq 10^7$ s $\simeq 120$ days. For larger masses and slower explosion velocities the epoch of gamma-ray emission occurs later. Supernovas that are of low mass and expanding rapidly with the ejection of a large amount of radioactive matter will have high level of gamma-ray emission. In the case of SN 1987A, for example, x rays were detected starting from ~ 130 days after the collapse; these x-rays are believed to be downgraded photons emitted as gamma rays in the radioactive decay.

The second decay of ^{56}Co to ^{56}Fe is accompanied by positron emission 19% of the time. The positron scatters readily with electrons leading to thermalisation and, eventually, when the positron energies are sufficiently low, annihilations of electrons and positrons occur, leading to secondary gamma rays with 0.51-MeV energy. Whether these gamma rays will escape or not can be estimated along similar lines and depends on the model parameters. Even if they are not thermalised, the original positron kinetic energy *is* thermalised, thereby leading to a minimum amount of radioactive yield.

When such heating and cooling are present, the full equation (4.64) needs to be solved. Assuming spherical symmetry and radiative diffusion as the dominant source of transfer of photon energy, we can write

$$\frac{L}{4\pi r^2} = -\left(\frac{\lambda c}{3}\right)\frac{\partial a T^4}{\partial r} = -\left(\frac{ac}{3\rho\kappa}\right)\frac{\partial T^4}{\partial r} \tag{4.77}$$

so as to get

$$4T^4\left(\frac{\dot T}{T} + \frac{\dot V}{3V}\right) = \frac{1}{r^2}\frac{\partial}{\partial r}\left(\frac{c}{3\kappa\rho}r^2\frac{\partial T^4}{\partial r}\right) + \epsilon. \tag{4.78}$$

[Because the variables in this equation refer to unit mass, $V = (1/\rho)$.] In general, this is a fairly complex problem; but it is possible to obtain a simplified solution if we assume that the evolution is self-similar. In such a case, we assume that all variables can be expressed in the form $f_1(t) f_2[r/R(t)]$, where $R(t)$ is a specified function characterising the expansion and f_1 and f_2 are different for different physical variables such as temperature, density, etc. In other words, the spatial dependence is assumed to be through the variable $x \equiv [r/R(t)]$. In the same spirit, we shall also assume that the t and the x dependences can be separated in the equations. The temperature and the density are taken to be of the form

$$T(r,t)^4 \equiv F_1(x)F_2(t)T(0,0)^4 R(0)^4/R(t)^4, \tag{4.79}$$

$$\rho(r,t) = \rho(0,0)F_3(x)[R(0)/R(t)]^3, \tag{4.80}$$

respectively. To obtain a consistent solution, it is also necessary to assume that $\kappa = \kappa(x)$ and $\epsilon = \epsilon_0 F_4(x)F_5(t)$. Substituting such an ansatz into our equation and separating the variables, we easily obtain the two ordinary differential equations of the form

$$\alpha = -\frac{1}{x^2 F_1(x)} \frac{\partial}{\partial x}\left[\frac{x^2}{F_3(x)}\frac{\kappa(0)}{\kappa(x)}\frac{\partial F_1}{\partial x}\right], \tag{4.81}$$

$$\frac{1}{F_2}\frac{\partial F_2}{\partial t} = -\frac{R(t)}{R(0)\tau_0} + \frac{\epsilon}{aT^4 V}, \tag{4.82}$$

where α is a separation constant and

$$\tau_0 \equiv 3R(0)^2 \rho(0,0)\kappa(0)/ac \tag{4.83}$$

is the diffusion time scale at $t = 0$. For these equations to be consistent with our assumptions, the quantity $(\epsilon/aT^4 V) = (\epsilon\rho/aT^4)$ must depend on only t, that is, in

$$\frac{\epsilon}{aT^4 V} = \left[\frac{\epsilon_0}{aT^4(0,0)V(0,0)}\right]\left[\frac{F_4(x)F_3(x)}{F_1(x)}\right]F_5(t), \tag{4.84}$$

with $V(0,0) = \rho(0,0)^{-1}$, the quantity in the second set of brackets $\beta \equiv [F_4(x)F_3(x)/F_1(x)]$, must be a constant.

Treating just the Ni decay, for which $\epsilon_0 \simeq 4.78 \times 10^{10}$ ergs gm^{-1} s^{-1} and $F_5(t) = e^{-t/\tau_{\text{Ni}}}$ with $\tau_{\text{Ni}} \simeq 7.6 \times 10^5$ s, we find that Eq. (4.82) becomes

$$\dot{F}_2 + \frac{F_2}{\tau_0}\frac{R(t)}{R(0)} = \bar{\epsilon}\frac{R(t)}{R(0)}e^{-t/\tau_{\text{Ni}}}, \tag{4.85}$$

where

$$\bar{\epsilon} \equiv \beta\epsilon_0/aT(0,0)^4 V(0,0) \tag{4.86}$$

is a constant. For the expansion radius we will take $R(t) = R(0) + v_{\text{sc}}t$. It is also convenient to introduce the hydrodynamical time scale $\tau_h \equiv R(0)/v_{\text{sc}}$. In that case, the solution to Eq. (4.85) can be expressed in the form

$$F_2(t) = \{e^{-u(\bar{x})} + [\epsilon_0 M^0_{\text{Ni}}\tau_0/E_{\text{Th}}(0)]F_6(\bar{x}, y, w)\}F_2(0), \tag{4.87}$$

4.8 Supernova Luminosity and Light Curves

where $\bar{x} \equiv (t/\tau_m) \equiv (t/\sqrt{2\tau_0\tau_h})$, $y \equiv (\tau_m/2\tau_{Ni}) = (2f\kappa M/cv_{sc})^{1/2}/2\tau_{Ni}$, $w \equiv (2\tau_h/\tau_0)^{1/2}$, $u(\bar{x}) = w\bar{x} + \bar{x}^2$,

$$F_6(\bar{x}, y, w) \equiv e^{-u(\bar{x})} \int_0^{\bar{x}} (w + 2z) e^{-2yz + u(z)} \, dz, \qquad (4.88)$$

and we have taken $F_2(0) = 1$ without any loss of generality. M_{Ni}^0 is the mass of nickel originally present, given by

$$M_{Ni}^0 = 4\pi R(0)^3 V(0, 0)^{-1} \int_0^1 F_4(x) F_3(x) x^2 \, dx. \qquad (4.89)$$

$E_{Th}(0)$ is the total thermal energy present at $t = 0$, given by

$$E_{Th} = \frac{4\pi R(0)^3 aT(0,0)^4 I_{Th} F_2(t) R(0)}{R(t)}, \qquad I_{Th} \equiv \int_0^1 F_1(x) x^2 \, dx. \qquad (4.90)$$

This provides the complete solution to the problem.

If we ignore the radioactive heating, then $\epsilon_0 \to 0$ and the solution reduces to $F_2(t) = \exp[-u(\bar{x})]$, which was obtained above. A somewhat less drastic approximation involved ignoring $R(0)$ and treating $R(t) \approx v_{sc}t$. This will correspond to the limit $w \to 0$, leading to the solution

$$F_2(t) = \frac{\epsilon_0 M_{Ni}^0 \tau_0}{E_{Th}(0)} F_7(\bar{x}, y) + e^{-\bar{x}^2} \qquad (4.91)$$

with

$$F_7(\bar{x}, y) \equiv e^{-\bar{x}^2} \int_0^1 e^{-2zy + z^2} 2z \, dz. \qquad (4.92)$$

The parameter y is small for $\kappa M \ll v_{sc}$. In this limit of $y \ll \bar{x}$, the function F_7 is given by the simple form

$$F_7(\bar{x}, y) \approx e^{-t/\tau_{Ni}} - e^{-(t/2y\tau_{Ni})^2}. \qquad (4.93)$$

This form is essentially determined by the radioactive decay rate, $\exp(-t/\tau_{Ni})$, of ^{56}Ni. In the other limit of $y \gg \bar{x}$, which is appropriate for the rising part of the light curve, F_7 is given by

$$F_7(\bar{x}, y) \approx [1 - e^{2\bar{x}y}(1 + 2\bar{x}y)]/2. \qquad (4.94)$$

Because $(t/\tau_{Ni}) = 2\bar{x}y$, this gives a rapidly rising curve that reaches saturation in a few decay times.

Exercise 4.4

Radioactive decay and light curves: Recall from the text that ^{56}Ni decays to ^{56}Co with a half-life of 6.1 days, which, in turn, decays ^{56}Fe with a half-life of 77.1 days. Estimate the form of the luminosity of a supernova as a function of time when the cobalt decay is dominant. For supernova 1987A, take $L \approx 10^{42}$ ergs s^{-1} at $t = 0$ for the purpose of cobalt decay extrapolation. What is the total amount of energy released and the mass of nickel synthesised? [Answer: Because the decay proceeds as $2^{-t/\tau}$, where $\tau = 77.1$ days, the luminosity varies with time as $\log L = -(\log 2/\tau)t+$ constant $= -3.9 \times 10^{-3} t_{\text{day}} +$ constant. The constant is 42 from the data given in the problem so that $\log L(t) = 42 - 4.5 \times 10^{-8} t$ (in cgs units). Integrating L over t we can estimate the total energy release to be $E \approx 10^{49}$ ergs. The ^{56}Co to ^{56}Fe releases $Q = 6.4 \times 10^{16}$ ergs gm^{-1}. Hence the mass involved is $M = E/Q \approx 0.08\ M_\odot$. Note that this form of $L(t)$ is valid in the IR to UV band only if the gamma rays released in the decay are efficiently thermalised. For $t \gtrsim 500$ days, the material becomes optically thin, and hence the decay will be faster.]

4.9 Evolution of Supernova Remnants

When a supernova explosion takes place and stellar material is ejected, the local region of the ISM is significantly disturbed. The ejected material moves out and will eventually be the source of electromagnetic radiation in different wave bands. The study of such objects, called supernova remnants, helps us in understanding several evolutionary processes in the ISM.

Consider a typical supernova explosion in which an energy of $E \approx 10^{51}$ ergs is released at a time scale that is short compared with any other relevant time scale in the problem. The velocity of the ejected matter v_{eject} is related to the energy E by

$$v_{\text{eject}} \approx 10^4 \text{ km s}^{-1} E_{51}^{1/2} m^{-1/2} \approx 10^{-2} \text{ pc yr}^{-1} E_{51}^{1/2} m^{-1/2}, \quad (4.95)$$

where $m \equiv (M/M_\odot)$ and we have used the convenient conversion 10^6 km s^{-1} ≈ 1 pc yr^{-1}. If this entire energy is thermalised, the temperature of the shocked ISM will rise to $\sim 10^9$ K. The first question to decide will be whether fluid approximation leading to standard shock-wave theory (developed in Vol. I, Chap. 8, Section 8.11) will be applicable. An ejection velocity of 10^4 km s^{-1} corresponds to an individual kinetic energy of ~ 2 MeV for, say, each proton. When such protons hit the hydrogen atom, ionisation will occur; the cross section for this process is $\sigma_{\text{ion}} \approx a_0^2 \simeq 10^{-17}$ cm^2, and the energy lost per ionisation is ~ 50 eV. If the medium has a density of ~ 1 hydrogen atom per cubic centimeter, the stopping length l for a 2-MeV proton is

$$l \simeq \left(\frac{2 \text{ MeV}}{50 \text{ eV}}\right) \frac{1}{n_l \sigma_{\text{ion}}} \simeq 10^3 \text{ pc}. \quad (4.96)$$

This is larger than any macroscopic-length scale of interest in this case, and hence we might think that fluid approximation is inapplicable. The situation here is, however, more complex because the ISM contains magnetic fields that affect the motion of charged particles. A magnetic-field strength of, say, $B \approx 3 \times 10^{-6}$ G would result in a Larmor radius of $R_L \approx 2 \times 10^{10}$ cm $\approx 10^{-8}$ pc.

4.9 Evolution of Supernova Remnants

Even though the energy density of such a magnetic field is not of much direct dynamical consequence, it certainly will lead to the formation of a collisionless hydromagnetic shock, with R_L playing the role of the effective mean free path. (We shall encounter a similar situation in the study of solar wind in the vicinity of the Earth's magnetic field, in Chap. 8.) The magnetic field itself is swept up by this process and is collected along with the gas by the moving shock front. The shock scenario here is certainly more complicated than the normal collisional shock and we expect part of the shock-deposited energy to appear as the relativistic kinetic energy of a fraction of the charged particles. In spite of these differences, the Larmor radius acts as the effective mean free path for the system and fluid approximation is at least approximately valid.

Hence, in the simplest picture for the supersonic expansion of a hot, gaseous sphere, there will be an abrupt discontinuity between the expanding gas and the swept-up material in front. The situation is very similar to the case of the supersonic piston studied in Vol. I, Chap. 8, Section 8.11. As in that case, a shock wave will run ahead of the contact discontinuity, and the region between the piston and the shock will be heated to a high temperature. In the case of strong shocks with high Mach number M_1, the ratio of densities on either side of shock wave is $\rho_2/\rho_1 = (\gamma+1)/(\gamma-1) = 4$ if $\gamma = 5/3$. The temperature ratio, $T_2/T_1 = 2\gamma(\gamma-1)M_1^2/(\gamma+1)^2 = (5/16)M_1^2$ for $\gamma = 5/3$, can be very large for strong shocks and we expects the shocked gases of supernova remnants to be intense x-ray emitters.

In the expansion of such remnants, we can distinguish several different phases of evolution. In the first phase, the material moves with uniform velocity as a blast wave with $r \propto t$. This phase ends at a radius $r = r_f$ when the mass of the ISM swept up by the blast wave, $M_{\text{sweptup}} \approx (4\pi/3)\rho_{\text{ISM}}r_f^3$, is comparable with the ejected mass M_{ej}. This gives

$$r_f \simeq 2 \text{ pc} \left(\frac{M_{\text{ej}}}{M_\odot}\right)^{1/3} \left(\frac{\rho_{\text{ISM}}}{2 \times 10^{-24} \text{ gm cm}^{-3}}\right)^{-1/3}. \tag{4.97}$$

The corresponding time scale is

$$t_f \simeq \frac{r_f}{v_{\text{ej}}} \simeq 200 \text{ yr} \left(\frac{M_{\text{ej}}}{M_\odot}\right)^{5/6} E_{51}^{-1/2} \rho_{24}^{-1/3}. \tag{4.98}$$

For comparison, it may be noted that the electron–electron and electron–proton collision time scales for a plasma are given by (see Vol. I, Chap. 9, Section 9.3)

$$t_{ee} \simeq \frac{300}{n} T_9^{3/2} \text{ yr}, \quad t_{ep} \simeq \frac{5 \times 10^5}{n} T_9^{3/2} \text{ yr}. \tag{4.99}$$

(The discussion above assumes that the density of the ISM around the supernova is constant. If the density varies as a power law, it is possible to generalise these results in a fairly simple manner.)

The next phase is well described by the Sedov solution studied in Vol. I, Chap. 8, Section 8.12, in which it was found that $r \propto t^{2/5}$. This phase could also be interpreted as the one in which the energy is constant. Note that the total energy of the system can be expressed as

$$E \approx \frac{1}{2}\left(\frac{4\pi}{3}\right)\rho r^3 v^2 + \text{(constant)} (\rho v^2)\left(\frac{4\pi}{3}\right) r^3, \tag{4.100}$$

where the first term is the kinetic energy of bulk motion and the second term is the internal energy of the system. The constancy of E implies that $r^3 v^2 = $ constant. Using $v = \dot{r}$ and integrating, we obtain $r \propto (E/\rho)^{1/5} t^{2/5}$. We have seen in Vol. I, Chap. 8, Section 8.12 that, for $\gamma = 5/3$, the constant of proportionality is of the order of unity. In that case, the Sedov phase is characterised by the scalings

$$r \cong \left(\frac{E}{\rho}\right)^{1/5} t^{2/5} \simeq 0.3 \text{ pc } E_{51}^{1/5} n_H^{-1/5} t_{yr}^{2/5},$$

$$v = \dot{r} \simeq 5000 \text{ km s}^{-1} \left(\frac{r}{2 \text{ pc}}\right)^{-3/2} E_{51}^{1/2} n_H^{-1/2},$$

$$T \simeq 6 \times 10^8 \text{ K} \left(\frac{r}{2 \text{ pc}}\right)^{-3} E_{51} n_H^{-1} \simeq (50 \text{ keV}) \left(\frac{r}{2 \text{ pc}}\right)^{-3} E_{51} n_H^{-1} \tag{4.101}$$

$$\simeq 10^6 \text{ K } E_{51}^{2/5} n_H^{-2/5} \left(\frac{t}{3 \times 10^4 \text{ yr}}\right)^{-6/5}$$

$$\simeq (100 \text{ eV}) E_{51}^{2/5} n_H^{-2/5} \left(\frac{t}{3 \times 10^4 \text{ yr}}\right)^{-6/5}.$$

These scalings show that temperatures are approximately 3×10^6 K for a supernova remnant with an age of $\sim 10^4$ yr. Such a system should radiate strongly in x rays. Also note that Eqs. (4.101) define a characteristic time scale

$$t_{\text{sedov}} \simeq 3 \times 10^4 \, T_6^{-5/6} E_{51}^{1/3} n_H^{-1/3} \tag{4.102}$$

as a function of the temperature. For example, the supernova remnant Cygnus Loop has reached the end of phase II right now with $R_{\text{now}} \approx 20$ pc and $v_{\text{now}} \approx 115$ km s^{-1}. The lifetime obtained from this relation is $t \simeq 65,000$ yr.

As the deceleration of the expanding sphere becomes significant, several other effects come into play. The most important is the fact that the outer shells of the expanding spheres are decelerated first, allowing the material in the inner regions to catch up with the material in the outer regions. In fact, as the deceleration continues, the flow of gas into the outer layers itself becomes supersonic (relative to the sound speed inside the sphere) so that a shock wave can form on the inner edge, thereby strongly heating the matter in the outer shell. In the initial phase, the kinetic energy of the expansion of the gas was communicated to the swept-up interstellar gas, whereas in the deceleration phase the kinetic energy was fed back into the ejected material itself, making it a strong soft-x-ray emitter.

The shell region is also subject to Rayleigh–Taylor instability during this phase. The deceleration acts as effective gravity in this case, and we have the situation of a lighter shocked interstellar gas resting on top of a denser shell of gas. In these instabilities, seed magnetic fields are magnified by turbulent motion and we would expect this region to have strong clumps of magnetic fields. The presence of charged particles in a magnetic field will lead to synchrotron radiation from this region, which we shall discuss towards the end of this section.

The Sedov phase, which conserves the energy, ends when the energy loss that is due to radiative cooling becomes significant. As the temperature falls below 10^6 K, the abundant ions of C, N, and O will acquire electrons and form atomic systems, and cascading down to the ground state through the emission of photons will cause cooling of the gas. The rate of cooling is given by $n_H^2 \Lambda(T) \approx 10^{-22}$ ergs cm^3 s^{-1} $n_H^2 T_6^{-1/2}$ (see Vol. I, Chap. 6). The third phase will be the one in which the cooling losses are important. The transition to this phase can be estimated by the condition $t_{cool} \lesssim t_{sedov}$, where

$$t_{cool} \simeq \frac{n k_B T}{n^2 \Lambda(T)} \simeq 4 \times 10^4 \frac{T_6^{3/2}}{n_H} \text{ yr,} \qquad (4.103)$$

and t_{sedov} is defined in relation (4.102). The condition $t_{cool} < t_{sedov}$ leads to $T_6 < E_{51}^{1/7} n^{2/7}$ or, equivalently,

$$\dot{r} \propto T^{1/2} < 200 \text{ km s}^{-1} \left(E_{51} n_H^2\right)^{1/14}. \qquad (4.104)$$

This relation shows that the velocity at the end of the Sedov phase is ~ 200 km s^{-1} and is only weakly dependent on E and n_H.

At this stage, the shock becomes isothermal, as described in Vol. I, Chap. 8, Section 8.11. There is a hot interior region [with radius $R(t)$] surrounded by a cool dense shell. The shell moves with approximately constant radial momentum, piling up interstellar gas like a snow plough piles up snow. [Because the shell is thin, we can describe it by a single radius $R(t)$.] The cooling takes away most of the shock energy, and the motion of the shell forward is described approximately by the equation for momentum conservation:

$$\frac{d}{dt}[(4\pi/3)\rho R^3 \dot{R}] \approx 0. \qquad (4.105)$$

If the thin shell is formed at a time t_0 with radius $R = R_0$ and velocity v_0, then the first integral to Eq. (4.105) gives

$$(4\pi/3)\rho R^3 \dot{R} = (4\pi/3)\rho R_0^3 v_0. \qquad (4.106)$$

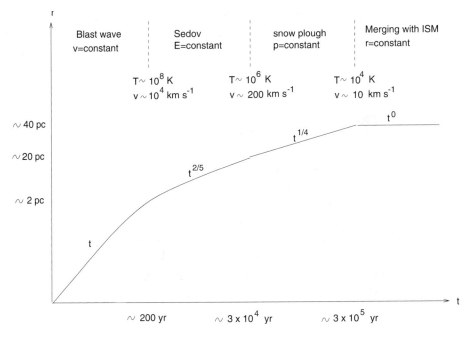

Fig. 4.6. The radius of the supernova shell as a function of time during the different phases.

This integrates to give

$$R = R_0\left[1 + 4\frac{v_0}{R_0}(t - t_0)\right]^{1/4}, \quad \dot{R} = v_0\left[1 + 4\frac{v_0}{R_0}(t - t_0)\right]^{-3/4}. \quad (4.107)$$

For large t, $R \propto t^{1/4}$ and

$$\dot{R} \propto t^{-3/4} \simeq 200 \text{ km s}^{-1} \, (t/3 \times 10^4 \text{ yr})^{-3/4}. \quad (4.108)$$

The time constant in relation (4.108) is fixed by equating the Sedov phase velocity of Eq. (4.101) to 200 km s^{-1}.

In the final phase, the speed of the shell drops below the sound velocity of the ISM, which is approximately (10–100) km s^{-1} in a time scale of $t \approx (1\text{–}5) \times 10^5$ yr. Around this time scale, the remnant loses its identity, and it is dispersed by random motions in the ISM. The evolution is shown schematically in Fig. 4.6.

It should be noted that supernova explosions and their eventual dispersion of ejected material have the effect of enriching the ISM with the material processed in stellar interiors. In particular, the heavy elements synthesised inside a star reach the ISM through this process. Because massive stars evolve at shorter time scales and also are more likely to end up as supernovas, the evolution of the first generation of massive stars changes the character of the ISM. Second and later generations of stars condense out of this enriched ISM and will have a higher proportion of heavier elements.

4.9 Evolution of Supernova Remnants

A supernova explosion can also trigger optical phenomena of different kinds from the surrounding ISM. The ISM surrounding the original star might contain a gaseous nebulalike region centred on the star arising from the stellar wind of the star in the presupernova phase. The intense radiation from the supernova can heat and ionise such a region, leading to the emission of radiation. For example, an expanding luminous ring (with a speed of 10 km s^{-1}) emitting at 5007 Å from OIII was detected around supernova SN 1987A at a distance of $\sim$0.2 pc from the centre of the explosion, possibly because of the above phenomena. Supernovas also lead to light echoes from the dust in the ISM because of the phenomena discussed in Vol. I, Chap. 3, Exercise 3.5. In the case of SN 1987A, two light echoes were detected at distances of 6.8 and 11.4 pc from the supernova approximately 1 yr after the explosion.

A supernova emits x rays profusely during the first two phases from the hot material behind the shock. The key emission process is thermal bremsstrahlung from the plasma at a temperature of 10^6 K or higher. The bright optical filaments are formed during phase 3, and their emission comes in the form of line radiation in the material with a temperature of 10^4 K. The emission is clearly characteristic of the radiating atoms. In addition to these x-ray and optical emissions, supernova remnants are also strong sources of radio waves because of electrons that are spiraling in the magnetic fields, leading to synchrotron radiation. We saw in Vol. I, Chap. 6, Section 6.11, that if the differential number density of relativistic electrons per unit volume is taken to be

$$n(E)\,dE = KE^{-p}\,dE, \qquad (4.109)$$

then the total flux of an optically thin source at a distance d at frequency ν can be expressed as

$$S_\nu = \frac{G}{d^2} V K B^{(1+p)/2} \nu^{-(p-1)/2}, \quad G = b(p)\frac{e^3}{mc^2}\left(\frac{3e}{4\pi m^3 c^5}\right)^{(p-1)/2}, \qquad (4.110)$$

where V is the volume of the source, B is the average magnetic field, and $b(p)$ is a numerical factor. In the case of a supernova remnant, the gas at $T > 10^4$ is strongly ionised during the Sedov phase and hence the magnetic field will be frozen to the plasma fluid. It follows that the field will decay as

$$B(r) = B_0 \left(\frac{r_0}{r}\right)^2. \qquad (4.111)$$

If the energy of individual relativistic electrons decays because of the adiabatic expansion of the volume, the energy density ϵ varies as $d(\epsilon V) = -P dV$. With the pressure of relativistic electrons taken as $P \simeq (1/3)\epsilon$, this equation integrates to give $\epsilon \propto r^{-4}$. The total energy of the electrons $E = \epsilon V \propto r^{-1}$, allowing us to write

$$E(r) = E_0 \left(\frac{r_0}{r}\right). \tag{4.112}$$

This, in turn, implies that the condition for the conservation of total number of electrons,

$$V_0 \int_{E_1}^{E_2} K_0 E^{-p} \, dE = V_0 \left(\frac{r}{r_0}\right)^3 \int_{E_1 r_0/r}^{E_2 r_0/r} K(r) E^{-p} \, dE, \tag{4.113}$$

can be satisfied only if

$$\frac{K(r)}{K_0} = \left(\frac{r}{r_0}\right)^{-(2+p)}. \tag{4.114}$$

Using Eqs. (4.114) and (4.111), we get

$$S_\nu(r) = S_\nu(r_0) \left(\frac{r}{r_0}\right)^{-2p} = S_\nu(t_0) \left(\frac{t}{t_0}\right)^{-4p/5}. \tag{4.115}$$

The surface brightness $\Sigma_\nu = (S_\nu/\pi r^2)$ scales as

$$\Sigma_\nu(r) = \Sigma_\nu(r_0) \left(\frac{r}{r_0}\right)^{-2(p+1)}, \tag{4.116}$$

which relates observable quantities to the index p for electron distribution.

The actual estimates of the synchrotron emission, etc., would require an estimate of the magnetic field in the remnant. Although direct measurements are difficult, it is possible to show that the total energy content of the system (emitting a certain amount of synchrotron radiation) has a minimum as a function of the magnetic-field strength. This comes about as follows. Consider a source of volume V from which synchrotron radiation is detected. Assume that the radiation is emitted by relativistic electrons with a power-law distribution in energy: $n(E) \, dE \propto E^{-p} \, dE$. The total energy content of the system will be the sum of the magnetic energy and the energy of the relativistic particles. Let the energy content in protons be β times that of relativistic electrons. The magnetic energy of the source, on the other hand, is $U_B = (B^2 V/8\pi)$, and the kinetic energy of the particles (protons and electrons) can be taken to be $(1 + \beta) U_{\text{elec}}$. Hence the total energy density is

$$\frac{U_{\text{tot}}}{V} = (1 + \beta) \int_{E_1}^{E_2} K E n(E) \, dE + \frac{B^2}{8\pi}. \tag{4.117}$$

The typical frequency ν at which an electron of energy E will be emitting the radiation is given by $\nu = C B E^2$, where C is a constant [see Eq. (6.249), Vol. I, Chap. 6, Section 6.10]. Thus we can set $E_1 = (\nu_1/CB)^{1/2}$ and $E_2 = (\nu_2/CB)^{1/2}$ if the radiation is detected in the range of frequencies $\nu_1 < \nu < \nu_2$. The energy

4.9 Evolution of Supernova Remnants

content in the particles now becomes

$$\frac{U_{\text{part}}}{V} = (1+\beta) \int_{E_1}^{E_2} KE^{1-p}\, dE = \frac{(1+\beta)K}{(p-2)}(CB)^{(p-2)/2}\left[v_1^{(2-p)/2} - v_2^{(2-p)/2}\right]. \tag{4.118}$$

Substituting for K in terms of S_ν and B from Eqs. (4.110), we get

$$\frac{U_{\text{part}}}{V} = \frac{(1+\beta)}{(p-2)}\left[\frac{S_\nu}{AVB^{(1+p)/2}\nu^{(1-p)/2}}\right](CB)^{(p-2)/2}\left[v_1^{(2-p)/2} - v_2^{(2-p)/2}\right]$$
$$\equiv Q(1+\beta)S_\nu B^{-3/2}, \tag{4.119}$$

where $A \equiv (G/d^2)$ and we have shown explicitly only the dependence on β, B, and S_ν. The total energy density now becomes

$$\frac{U_{\text{tot}}}{V} = \frac{B^2}{8\pi} + Q(1+\beta)S_\nu B^{-3/2}. \tag{4.120}$$

This expression is the sum of two terms, of which one increases monotonically with B while the other decreases. If we detect a certain luminosity of synchrotron radiation from a source and all other parameters are fixed, then there is a critical value of the magnetic field that will minimise the total energy content. This occurs when the magnetic field has the value

$$B_{\min} = [6\pi Q(1+\beta)S_\nu]^{2/7}. \tag{4.121}$$

Thus, for producing a particular amount of synchrotron luminosity by using high-energy particles, we require a magnetic field of this minimum strength. It is easy to see that, when $B = B_{\min}$, the particle energy U_{part} and the magnetic energy U_B are comparable and $U_B = (3/4)U_{\text{part}}$.

If the system indeed has minimum energy subject to other constraints, then the above argument allows us to estimate the parameters of the system from observed quantities. It should, however, be noted that there is no physical reason for the system to have reached equipartition or a minimum-energy configuration. The calculations also assume that both the magnetic field and the charged particles fill the source volume uniformly; if not, the volume has to be corrected by a filling factor.

Exercise 4.5
Supernova in an ISM bubble: It is possible that the strong stellar winds preceding the supernova explosion of a star could have led to a density profile in the local ISM, which has an approximate form $\rho_{\text{ISM}} \propto r^{-\beta}$. Repeat the analysis of supernova-blast wave propagation in such a medium and obtain the scaling relations. In particular, show that the Sedov phase now has $R \propto t^{2/(5-\beta)}$. This indicates that the shock wave accelerates for $\beta > 3$ and decelerates for $\beta < 3$. Give a physical reason for this behaviour.

4.10 Shock Acceleration in Supernova Remnants

One of the key assumptions that went into modelling the synchrotron radiation from supernova remnants was the existence of high-energy particles with a power-law spectrum. Maintaining such a spectrum of particles is not an easy task, as a relativistic particle loses its energy because of adiabatic expansion as $E \propto r^{-1}$. To keep a constant supply of such energetically charged particles, it is necessary to accelerate the charged particles to high energies in the system. It is generally believed that this could be done by the particle-acceleration mechanisms taking place in the strong shocks. Because the issue of particle acceleration arises in many astrophysical phenomena, we will first consider a specific process that is of relevance in the case of supernova remnants and then provide a general description.

Consider a supernova shock moving through the ISM with a flux of high-energy particles existing both in front of and behind the shock. The particles are assumed to be of very high energy so that their speeds are considerably higher than the speed of shock propagation. The gyroradius of such high-energy particles in the magnetic field near the vicinity of the shock will also be much larger than the thickness of the shock. As the particles cross the shock front and are scattered by the irregularities either behind it or in front of it, their velocity distribution becomes isotropic quite rapidly on either side of the shock front *in the frames of reference in which the fluid is at rest locally*. As we shall show, this allows the particle to gain energy when it crosses the shock front in either direction.

Let the shock front move with velocity U from left to right. The unshocked material on the right-hand side is characterised by P_1, T_1, ρ_1, and v_1, and the corresponding variables behind the shock on the left-hand side are P_2, T_2, ρ_2, and v_2. For a gas with $\gamma = 5/3$, we have $(\rho_2/\rho_1) = 4$ and $(v_2/v_1) = 1/4$. In the frame in which the shock front is at rest, the unshocked upstream material will be flowing from right to left into the shock with velocity $v_1 = |U|$. The velocity of material behind the shock (on the left-hand side) will be $v_2 = (1/4)|U|$. (The flow everywhere is from right to left.) The situation is interestingly different in the frames of reference in which the unshocked or the shocked material is at rest.

Consider first the frame in which the unshocked upstream matter is at rest. In this frame, the shocked material will be moving from left to right with a speed $|v_2 - v_1| = (3/4)U$. A high-energy particle on the right-hand side can cross the shock front, which is moving towards the particle with a velocity $V = (3/4)U$. Performing a Lorentz transformation, we can compute the energy of the particle when it crosses into the downstream region as $E' = \gamma(E + p_x V)$. If the shock is nonrelativistic but particles are ultrarelativistic, we can set $V \ll c$, $\gamma \simeq 1$, $E \simeq pc$, and $p_x \simeq (E/c)\cos\theta$. In that case

$$\frac{\Delta E}{E} = \frac{V}{c}\cos\theta. \tag{4.122}$$

We now need the probability that the particles that cross the plane of the shock

4.10 Shock Acceleration in Supernova Remnants

hit it at an angle θ. The number of particles incident between the angles θ and $(\theta + d\theta)$ is proportional to $\sin\theta \, d\theta$; but the rate at which they approach the shock front is proportional to the x component of their velocities, $c \cos\theta$. Thus the total probability should be proportional to $\sin\theta \cos\theta \, d\theta$. Averaging the expression $(\Delta E/E)$ using properly normalised distribution, $P(\theta) = 2\sin\theta \cos\theta \, d\theta$, we find that

$$\left\langle \frac{\Delta E}{E} \right\rangle = \frac{V}{c} \int_0^{\pi/2} 2\cos^2\theta \sin\theta \, d\theta = \frac{2}{3}\frac{V}{c}. \tag{4.123}$$

The velocity vector of the particle will be randomised by scattering in the downstream region without any loss of energy. Thus particles in the vicinity of a strong shock gains energy while crossing the shock front from the upstream direction.

Consider next a situation in a frame at which the material behind the shock is at rest. In this frame the upstream gas will be flowing into the shock with the speed $|v_1 - v_2| = (3/4)U$. It is clear that exactly the same argument as that given above is applicable in this case as well, leading to the same fractional energy increase when the particle crosses the shock front from downstream to upstream. Thus high-energy particles gain energy when they cross the shock front in either direction. Our analysis makes clear the reason for this seemingly counterintuitive behaviour: There exists randomising processes that make the velocity isotropic *with respect to the local rest frame* on either side of the shock front.

Given any physical process that is repeated randomly and leads to a fixed fractional gain of energy, we can work out the resulting spectrum of particles along the following lines. Let an acceleration process increase the energy of the particle by a factor of β each time, the particle undergoes the basic process and let the mean time between the two occurrences of the process be τ so that the rate of occurrence is τ^{-1}. There is also a probability $P(t) = e^{(-t/T)}$ for a particle to remain in the accelerating region for time t, where T is the time scale characterising the escape process. Then $P_0 = \exp(-\tau/T)$ is the probability that the particle remains within the accelerating region after undergoing the basic process once. The number of particles that survive in the accelerating region for the time interval $(t, t+dt)$ is

$$n(t)\,dt = n_0 e^{-t/T} \left(\frac{dt}{T} \right). \tag{4.124}$$

Because the mean number of occurrences of the process in time t is (t/τ), the typical energy of these particles will be

$$E(t) = E_0 \beta^{t/\tau} = E_0 \exp\left[\left(\frac{\ln\beta}{\tau}\right)t\right]. \tag{4.125}$$

It follows that $t = (\tau/\ln\beta)\ln(E/E_0)$; so

$$n(E)\,dE = n(t)\left(\frac{dt}{dE}\right)dE \propto E^{-1}\exp\left[-\left(\frac{\tau}{T\ln\beta}\right)(\ln E)\right]dE$$
$$\propto E^{-[1+\tau/(T\ln\beta)]}\,dE. \tag{4.126}$$

Because the probability for a particle to remain in the accelerating region after the occurrence of the process once is $P_0 = \exp(-\tau/T)$, we have $(\tau/T) = -\ln P_0$. Therefore the energy spectrum can be written as

$$n(E) \propto E^{-p}\,dE, \quad p = 1 + \frac{\tau}{T\ln\beta} = 1 - \frac{\ln P_0}{\ln\beta}, \tag{4.127}$$

which is a power law. Given a specific process, we will know P_0 and β and hence we can compute p.

In our case we need to estimate these two quantities when the basic process is the crossing and the recrossing of the shock front by a particle. Because the same amount of fractional gain occurs when the particle crosses from downstream to upstream, $\beta = [1 + (4V/3c)] = (1 + U/c)$ for this process. To determine the escape probability we may proceed as follows: The number of particles crossing the shock is $nc/4$, where n is the number density of the particles. This is the average number of particles crossing the shock in either direction. Downstream, however, the particles are swept away from the shock front by the bulk velocity of the flow at the rate $nV = (nU/4)$. Thus the fraction of particles lost per unit time away from the shock front is $[(1/4)nU/(1/4)nc] = (U/c)$. Because $U \ll c$ we may take the escape probability to be $P_0 \cong (1 - U/c)$. We therefore find that

$$\ln P_0 = \ln\left(1 - \frac{U}{c}\right) \approx -\frac{U}{c}, \quad \ln\beta = \ln\left(1 + \frac{U}{c}\right) \approx \frac{U}{c}, \tag{4.128}$$

giving $n(E)\,dE \propto E^{-2}\,dE$. It follows that the crisscrossing of the supernova shock by high-energy particles (which are scattered by different physical processes on either side of the shock front) can lead to a power-law spectrum with a specific index (-2). This mechanism has the advantage in that the acceleration occurs in situ near the shock front and hence is immune to adiabatic losses.

It is possible to derive the above result in a somewhat more formal and general setting, which we shall now briefly describe. Let f denote the distribution function for a bunch of relativistic electrons that are interacting with a plasma moving through a shock front. Conventionally, we would have taken f to be a function of the position variable $\mathbf{x}$ and the momentum variable $\mathbf{p}$, with both vectors evaluated in the lab frame in which the plasma is moving. This, however, is inconvenient when the electron distribution is isotropic in the rest frame of the plasma rather than in the lab frame. To incorporate this feature, it is convenient to treat f as a function of $\mathbf{x}, \mathbf{p}$, where $\mathbf{x}$ is measured in the lab frame but $\mathbf{p}$ is

4.10 Shock Acceleration in Supernova Remnants

measured in the local rest frame of the background plasma. In the lab frame, the plasma is moving with a velocity **u**, which need not be a constant. As long as this plasma velocity is small (even though the electrons themselves are relativistic), the momentum in the lab frames, **p**′ and **p** are connected by

$$\mathbf{p} = \mathbf{p}' - m\mathbf{u}, \tag{4.129}$$

where m is the electron mass. Further, the spatial derivatives in the two frames transform as

$$\left(\frac{\partial}{\partial x_i'}\right)_{\mathbf{p}'} = \left(\frac{\partial}{\partial x_i'}\right)_{\mathbf{p}} + \left(\frac{\partial p_j}{\partial x_i'}\right)_{\mathbf{p}'}\left(\frac{\partial}{\partial p_j}\right), \tag{4.130}$$

with a similar transformation for the time derivative. The Boltzmann equation for the distribution function involving $(df/dt) \equiv (\partial f/\partial t) + \mathbf{v} \cdot \nabla f$ now becomes

$$\begin{aligned}\frac{df}{dt} &= \frac{\partial f}{\partial t} - m\frac{\partial u_i}{\partial t}\frac{\partial f}{\partial p_i} + \left(\frac{p_i}{m} + u_i\right)\left[\frac{\partial f}{\partial x_i'} - m\frac{\partial u_j}{\partial x_i'}\frac{\partial f}{\partial p_j}\right] \\ &= \frac{\partial f}{\partial t} + u_i\frac{\partial f}{\partial x_i'} - \frac{\partial u_j}{\partial x_i'}p_i\frac{\partial f}{\partial p_j} \\ &\quad - m\frac{\partial u_i}{\partial t}\frac{\partial f}{\partial p_i} - mu_i\frac{\partial u_j}{\partial x_i'}\frac{\partial f}{\partial p_j} + \frac{p_i}{m}\frac{\partial f}{\partial x_i'}.\end{aligned} \tag{4.131}$$

This equation simplifies significantly when the scattering is strong in the local rest frame so that we can average the terms over the direction of the momentum. Odd powers of momenta average to zero, leading to a simpler equation

$$\frac{df}{dt} = \frac{\partial f}{\partial t} + \mathbf{u} \cdot \nabla f - \frac{1}{3}(\nabla \cdot \mathbf{u})\, p\, \frac{\partial f}{\partial p}. \tag{4.132}$$

When the diffusive processes of the shock are taken into account by a tensor diffusion coefficient D_{ij}, the Boltzmann equation can be written in the form

$$\frac{df}{dt} = \frac{\partial}{\partial x_i}\left(D_{ij}\frac{\partial f}{\partial x_j}\right). \tag{4.133}$$

The analysis simplifies when we study the one-dimensional flow of the plasma with the diffusion coefficient D_{ij} replaced with a scalar quantity $D = (1/3)v\lambda_{\text{mfp}}$, where λ_{mfp} is the mean free path. Let us first consider the equilibrium solution describing the stationary transport of the fluid in the diffusion approximation with a constant velocity u. In the frame in which the shock is stationary and located at $x = x_s = 0$, the spatial evolution of the distribution function for the particles can be described by an equation of the form

$$\frac{\partial f}{\partial t} + u\frac{\partial f}{\partial x} = u\frac{\partial f}{\partial x} = \frac{\partial}{\partial x}\left(D\frac{\partial f}{\partial x}\right), \tag{4.134}$$

where we have set the time derivative to zero in the steady state. Upstream of the shock, $(x < 0)$, the velocity is given by $u = u_s$, whereas downstream $(x > 0)$ we take $u = (u_s/\mathcal{R})$, where $\mathcal{R}$ is the compression ratio of the shock. We can easily integrate this equation by converting it into a differential equation for df/dx. On the upstream side, the time-independent solution to the Boltzmann equation is given by

$$f(x, p) = f_u(p) + [f(0) - f_u(p)] \exp\left[-\int_x^0 dx' \frac{u}{D(x')}\right]. \quad (4.135)$$

With this solution, $f \to f_u$ and the gradient of f vanishes as $x \to -\infty$. Downstream of the shock, both diffusion and the flow tend to take the particles towards $x = +\infty$ so that the only possible solution is $f(x, p) = f_d$, where f_d is the downstream limit of f. These two solutions must be now smoothly joined at the shock front, with the particle number flux kept continuous across the front. This flux is given by

$$\mathcal{F} = -D\frac{\partial f}{\partial x} - u\left(\frac{1}{3} p \frac{\partial f}{\partial p}\right), \quad (4.136)$$

where the first term arises from diffusion and the second term arises from our using the momentum variable in the rest frame and spatial variables in the lab frame. The nature of the second term is obvious from Eq. (4.132). Physically, this is the lowest-order correction to the diffusive flux that arises because of the combination of two facts. The particle distribution is more nearly isotropic in the fluid frame than in the shock frame, and hence the distribution function will depend on only the magnitude of the momentum in the fluid frame. However, the particles having momentum p in the shock frame have a spread of the order of $(u/c)p$ in the fluid frame. This leads to the second term in the flux. Evaluating the continuity equation at $x = 0$ gives

$$\left(D\frac{\partial f}{\partial x} + \frac{u}{3} p \frac{\partial f}{\partial p}\right)_{\text{upstream}} = \left(D\frac{\partial f}{\partial x} + \frac{u}{3} p \frac{\partial f}{\partial p}\right)_{\text{downstream}}. \quad (4.137)$$

Using $u_{\text{up}} = u_s$, $u_{\text{down}} = u_s/\mathcal{R}$, $D(\partial f/\partial x)_{\text{down}} = 0$, and $D(\partial f/\partial x)_{\text{up}} = u_s[f(0, p) - f_u(p)]$ at $x = 0$, we get a differential equation for $f(0, p) \equiv f(p)$ at $x = 0$:

$$p\frac{\partial f}{\partial p} = \frac{3\mathcal{R}}{\mathcal{R} - 1}[f_u(p) - f(p)], \quad (4.138)$$

which has the solution

$$f(p) = Ap^{-q} + qp^{-q} \int_0^p dp'\, f_u(p')\, p'^{q-1}, \quad q = \frac{3\mathcal{R}}{\mathcal{R} - 1}, \quad (4.139)$$

where the first term with some constant A denotes the homogeneous part of the solution. The second part of the solution, which depends on $f_u(p')$, shows that particles with a delta-function distribution of energies will acquire a power-law distribution with index q on passing through the shock front. For a strong adiabatic nonrelativistic shock, $\mathcal{R} = 4$, giving $q = 4$. Because $E \propto p^2$, this leads to an energy distribution of the form E^{-2}, as derived above. Note that the power-law index is entirely determined by the compression ratio of the shock and is independent of the diffusion coefficient D. If $f_u(p') \propto p'^{-n}$ is a power law, then the resulting solution at high momentum is also a power law, with the index dominated by the smaller of n and q. Even when n is smaller than q, the spectrum is amplified by the factor $q(q-n)^{-1}$.

In more general contexts, the evolution of the particle spectrum undergoing stochastic random acceleration has to be studied with a diffusion-type equation. If $N(E, t)$ represents the spectrum in energy space at time t, its evolution can, in general, be described by an evolution of the kind

$$\frac{\partial N}{\partial t} = \frac{\partial}{\partial E}[b(E) N(E)] + \frac{1}{2}\frac{\partial^2}{\partial E^2}[d(E) N(E)] - \frac{N}{\tau_{\rm esc}} + Q(E), \quad (4.140)$$

where the various terms have the following meaning. The first two terms on the right-hand side describe the change in the mean value of the energy and the rms value of the energy that is due to the stochastic acceleration process. The quantity $b(E) \equiv -(dE/dt)$ represents the change of energy in the individual acceleration event; because the mean-square fluctuation in the energy ΔE^2 will grow linearly with time, $d(E)$ can be interpreted as $d(E) \equiv d\langle(\Delta E)^2\rangle/dt$. (These two terms have been derived earlier in the case of Kompaneets equation in Vol. I, Chap. 6, Section 6.7. The analysis for a general case is identical.) The last two terms describe the gain and the loss of particles that are due to external processes; $Q(E)$ represents the rate of injection of particles at energy E and $N/\tau_{\rm esc}$ describes the random-escape probability from the region.

There are several physical processes in which $d(E)$ and $b(E)$ are such that the steady-state solutions to this equation, in the absence of particle injection, are power laws. For example, in the case considered above, $\Delta E = \beta E$ and $(\Delta E)^2 = (9/8)\beta^2 E^2$ per collision, implying that $d(E)/b(E) = -(9/8)\beta E$. In general, if we take $b(E) = -\beta E$ and $d(E)/b(E) = -\mu\beta E$, where μ is a numerical coefficient, then the above equation [with $Q = 0$ and $(\partial N/\partial t) = 0$] admits power-law solutions of the form $N \propto E^{-x}$, where x satisfies the quadratic equation

$$x^2 + x\left(\frac{2}{\mu\beta} - 3\right) - 2\left(\frac{1}{\mu\beta} + \frac{1}{\tau_{\rm esc}\mu\beta^2} - 1\right) = 0. \quad (4.141)$$

Given the physical process, the above relation determines the power-law index.

Exercise 4.6

First- and second-order Fermi processes: There is an alternative way of understanding the acceleration of particles in a wide class of situations that is quite useful. This exercise develops this idea in a fairly general context. Consider a cloud containing magnetic inhomogeneties that is moving with a nonrelativistic velocity v with respect to the lab frame. A charged particle with energy E_1 and momentum $\mathbf{p}_1$ enters this cloud. It undergoes interactions with the magnetic inhomogeneities inside the cloud and eventually leaves the cloud with energy E_2 and momentum $\mathbf{p}_2$. Assume that the charged particle is ultrarelativistic whereas the cloud is moving with nonrelativistic velocity. (a) In the rest frame of the cloud, there will be no change in the energy for the particle because the magnetic inhomogeneities cannot do any work on the particle. Show that the situation is different in the lab frame. In particular, prove that the fractional energy change of the particle $(\Delta E/E) \equiv (E_2 - E_1)/E_1$ is given by

$$\frac{\Delta E}{E} = (1-\beta^2)^{-1}(1 - \beta\cos\theta_1 + \beta\cos\theta_2' - \beta^2\cos\theta_1\cos\theta_2') - 1, \quad (4.142)$$

where $\beta = (v/c)$, θ_1 is the angle between $\mathbf{p}_1$ and $\mathbf{v}$, θ_2 is the angle between $\mathbf{p}_2$ and $\mathbf{v}$, and the prime denotes quantities measured in the rest frame of the cloud. (b) Argue that when averaging over the angles is performed we have $\langle\cos\theta_2'\rangle = 0$ whereas $\langle\cos\theta_1\rangle = -\beta/3$. Hence show that

$$\left\langle\frac{\Delta E}{E}\right\rangle = \frac{4}{3}\frac{\beta^2}{1-\beta^2} \approx \frac{4}{3}\beta^2. \quad (4.143)$$

Explain physically why the result is quadratic in β.

Consider now a situation in which a relativistic particle crosses a shock front from the upstream towards the downstream, enters a cloud of magnetic inhomogeneties on the downstream side, leaves the cloud, and crosses back to the upstream region. Show that Eq. (4.142) is still applicable, with the angles now being defined as between the momenta of the particle and the direction of flow of the shock front. A crucial difference arises when the expression is averaged over all angles. Show that in the present case we have $\langle\cos\theta_2'\rangle = (2/3)$ whereas $\langle\cos\theta_1\rangle = -(2/3)$. Now the average energy gained will be

$$\left\langle\frac{\Delta E}{E}\right\rangle \approx \frac{4}{3}\beta + \mathcal{O}(\beta^2). \quad (4.144)$$

Explain physically why the result is now first order in β.

4.11 Stellar Nucleosynthesis

The chemical elements and their isotopes are seen to be distributed with widely different abundances among different astrophysical systems in the universe such as comets, meteors, Earth, Sun, ISM, etc. It is necessary to seek a broad picture of the variation of the abundances and an explanation for each.[7]

As we shall see in Vol. III, the element helium could be synthesised with an abundance of $y \simeq 0.24$ during the very early phases of evolution of the universe, long before any of the astrophysical structures have formed. The cosmological nucleosynthesis can also produce small quantities of many other elements but not at a level comparable with that of helium. When the galaxy and

4.11 Stellar Nucleosynthesis

Table 4.1. Most abundant nuclei

Number	Element	Z	A	Abundance	Generating Process
1	H	1	1	7.057×10^{-1}	Big bang
2	He	2	4	2.752×10^{-1}	Big bang, CNO, p–p
3	O	8	16	9.592×10^{-3}	Helium burning
4	C	6	12	3.032×10^{-3}	Helium burning
5	Ne	10	20	1.548×10^{-3}	Carbon burning
6	Fe	26	56	1.169×10^{-3}	Explosive synthesis
7	N	7	14	1.105×10^{-3}	CNO
8	Si	14	28	6.530×10^{-4}	Oxygen burning
9	Mg	12	24	5.130×10^{-4}	Carbon burning
10	S	16	32	3.958×10^{-4}	Oxygen burning
11	Ne	10	22	2.076×10^{-4}	Helium burning
12	Mg	12	26	7.892×10^{-5}	Carbon burning
13	Ar	18	36	7.740×10^{-5}	Silicon, oxygen burning
14	Fe	26	54	7.158×10^{-5}	Explosive synthesis, silicon burning
15	Mg	12	25	6.893×10^{-5}	Carbon burning

first-generation stars form, the initial composition will be determined by this cosmological abundance. The first-generation stars, however, can synthesise heavier elements through a series of nuclear reactions that we have considered in Vol. I, Chap. 12, Section 12.4 and in Chaps. 2 and 3. When these stars explode as supernovas, the ISM in enriched by these elements and the second- and later-generation stars as well as planetary systems, etc., imbibe a signature of the abundance determined essentially by the stellar-nucleosynthesis processes. Some amount of nucleosynthesis takes place in the stars during slow hydrostatic evolution by means of the burning of different elements. In addition to this, elements are also synthesised during the explosive phase of the supernova.

To fix the ideas, let us consider the abundance of different elements in the solar system. The first 15 of the most abundant elements are listed in decreasing order of abundance in Table 4.1. These abundances, which arise essentially from stellar nuclear reactions, can be understood from considerations of nuclear stability. The first 14 in the list have even A, and all of them, except ^{14}N, are doubly even. We have seen in Vol. I, Chap. 12, Section 12.2 that such configurations are preferred in nuclear structures. Also note that, except for ^{56}Fe, most of the abundant doubly even nuclei also have $Z = N$ and hence are called alpha-particle nuclei. These are easy to generate through alpha-capture reactions. High abundance also results for nuclei with doubly magic numbers: ^{4}He with $Z = N = 2$, ^{16}O with $Z = N = 8$, ^{40}Ca with $Z = N = 20$, and ^{56}Ni with $Z = N = 28$, with the decay sequence Ni → Co → Fe, leading to the iron abundance. The overall trend in the low – A

elements can therefore be understood by the above considerations. We now give a brief description of the abundances, in order of increasing atomic number.

(1) Hydrogen is the most abundant element, followed by helium. Most of the helium (0.24 by mass) was produced in cosmological nucleosynthesis, whereas the subsequent hydrogen burning in stars could have changed the abundance to a higher value of about ~ 0.28. By measuring the helium abundance of the oldest and highly metal-poor stars, we would, in principle, be able to estimate the primordial helium abundance. Unfortunately the old stars that are hot enough to excite the spectral lines of helium will also be evolved stars that would have synthesised additional helium during the evolution. Because it is necessary to measure Y of reasonably unevolved stars, the following procedure is often adopted. Both Y and the metal fraction Z are measured in a population of hot young stars in a set of metal-poor galaxies and then the function $Y(Z)$ is extrapolated back to $Z = 0$. (Here Z is the metallicity and not the atomic number.) Such analysis leads to the value $Y = 0.235 \pm 0.005$, which is used in the standard stellar-evolution model.

(2) The fragile nuclei of lithium, beryllium, and boron are very scarce, and they are invariably destroyed in stellar interiors. Approximately 10% of ^{7}Li can also come from cosmological nucleosynthesis. Deuterium is more abundant than the above three nuclei even though it is more fragile. Because deuterium is completely destroyed in stellar evolution and because there is no plausible mechanism for creating in the stars, it must be primordially generated in the cosmological nucleosynthesis. In fact, the observed deuterium abundance sets a lower limit to the primordial abundance and – as we shall see in Vol. III – serves as a sensitive test in cosmological model building.

(3) Nuclei from carbon to calcium show an overall decrease in abundance except for the effects that can be accounted for by the odd–even nature and shell structure of the nuclei. All these arise from successive stages of nuclear evolution.

(4) The iron group elements show an abundance that is approximately in agreement with the result of nuclear statistical equilibrium at a temperature of $\sim 2 \times 10^9$ K. These elements originate in explosive nucleosynthesis that can occur in the shock that emerges from the supernova.

The nuclear reactions discussed so far, which are capable of synthesising progressively heavier elements starting from light elements, cannot produce elements significantly heavier than iron because the binding energy per nucleon decreases for such elements. There are other reactions that are capable of generating heavier elements starting from elements that are to the left of the iron peak (i.e., with $Z < 56$). One set of such reactions relies on the capture of neutrons by nuclei – a process that is not affected by Coulomb repulsion. We shall briefly describe this process, which can take place in the supernova.

Consider a collection of nuclei that experience a flux of neutrons. Absorption of a neutron by an element (Z, A) leads to the reaction

$$(Z, A) + n \rightarrow (Z, A+1) + \gamma, \tag{4.145}$$

4.11 Stellar Nucleosynthesis

thereby forming a nucleus $(Z, A + 1)$. If this nucleus is stable, then the process can continue with the absorption of additional neutrons leading to $(Z, A + 2)$, $(Z, A + 3), \ldots$, etc. At some stage in this sequence, possibly even at the first step, we may reach an element that is unstable because of beta decay:

$$(Z, A + 1) \to (Z + 1, A + 1) + \bar{\nu} + e^-. \tag{4.146}$$

If this happens then the further evolution will depend on stability or otherwise of the elements $(Z + 1, A + 1)$. The crucial parameter that determines the different channels through which such reactions proceed is the neutron flux.

The mean time between the capture of two neutrons is of the order of $\tau_{\text{cap}} \approx (\sigma_n \Phi_n)^{-1} = (n_n \langle \sigma_n v \rangle)^{-1}$, where Φ_n is the neutron flux, n_n is the number density of the neutrons, v the speed of the neutrons, and σ_n is the neutron-absorption cross section. For $\sigma_n \approx 10^{-25}$ cm^2 and $v \approx 3 \times 10^8$ cm s^{-1}, we have $\tau_{\text{cap}} \approx 10^9$ yr n_n^{-1}. If τ_{cap} is smaller than the lifetime τ_β of the nucleus against beta decay, then a further addition of neutrons is very likely. On the other hand, if $\tau_{\text{cap}} > \tau_\beta$, the nucleus is likely to undergo beta decay before forming the next isotope. The τ_β generally ranges from a fraction of a second to a few years and is obviously independent of Φ_n; τ_{cap}, on the other hand, varies inversely as the neutron flux Φ_n and can be larger than τ_β for low fluxes. In such a situation, beta decay occurs before the second neutron is captured and the nuclei formed by this process lie along the continuous path of stable nuclei in a nuclear chart. Such elements are called *s-process* elements, with *s* signifying slow neutron capture. On the other hand, if the neutron flux is large, heavier and heavier isotopes can be synthesised by the addition of neutrons before eventually leading to an unstable nucleus in which the (γ, n) reaction occurs before the (n, γ) reaction occurs. This leads to a sequence of *r-process* elements (with *r* standing for rapid) that lie along the neutron-rich side of stable elements. For $n_n \lesssim 10^5$ cm^{-3}, $\tau_{\text{cap}} \gtrsim 10^4$ yr is larger than any beta-decay lifetime; such low neutron fluxes correspond to *s* process. When $n_n \gtrsim 10^{22}$ cm^{-3}, $\tau_n \lesssim 10^{-6}$ s; such a neutron-rich environment will be dominated by the *r* process.

To provide a simple illustration of these processes in the context of nucleosynthesis in a neutron-rich environment, we study a section of the nuclear chart shown in Fig. 4.7. The x axis of this figure gives the number of neutrons, and the y axis has number of protons. The horizontal rows therefore represent isotopes of different elements, and the vertical columns are nuclei that have the same neutron number. Isobars (which are nuclei of equal atomic weight) lie along a diagonal from the upper left to the lower right. The neutron absorption will take us to the right, by one column, along the same horizontal line. The beta decay, on the other hand, will take diagonally to the left and above; i.e., it decreases N by 1 and increases Z by 1. In the chart is shown an *s*-process path that moves through cadmium, indium, tin, and antimony. The isotopic composition of each element is indicated in the figure by the percentage of total element abundance. Also given in the figure is the lifetime against beta decay for the unstable elements.

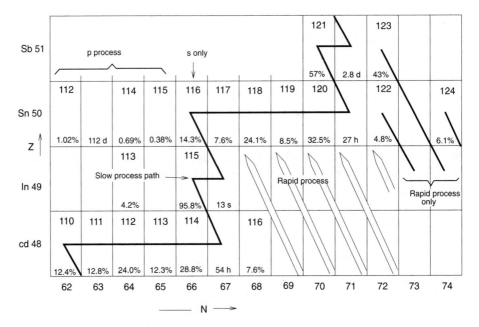

Fig. 4.7. The *r* and the *s* processes for the synthesis of heavy elements.

Starting from $^{48}_{110}$Cd, slow neutron absorption takes us through to $^{48}_{114}$Cd. The addition of an extra neutron leads to the unstable element $^{48}_{115}$Cd, which undergoes beta decay with a lifetime of 54 h, thereby taking us diagonally to $^{49}_{115}$In. The capture of another neutron will lead to the unstable element $^{49}_{116}$In with a half-life for beta decay of 13 s. This decay takes us to $^{50}_{116}$Sn and by neutron capture we can proceed up to $^{50}_{120}$Sn. The addition of a neutron now will lead to an unstable element with $^{50}_{121}$Sn, followed by beta decay, etc.

The *r*-process elements can be identified by a similar procedure. Now the neutron-rich nuclei will continue to beta decay until they reach a stable configuration. Hence, for an *r* process, we move along a beta-decay track (from the lower right to the upper left diagonally) until we reach a stable element. Some of these are also marked in the figure.

Such an exercise can be done for any *s*- or *r*-process case. If the evolutionary time scale is t_{evol} (which is typically 10^6 yr for an *s* process) then we treat all nuclei with a half-life of less than t_{evol} as unstable and the others as stable. The temperature and the neutron density will determine whether an *r* or an *s* process dominates. In the case of the *s* process, we will also expect the equilibrium abundance of isotopes to be determined by the relation $\sigma_A n_A =$ constant, where σ_A is the neutron-capture cross section and n_A is the relative abundance. Observations show that the quantity $\sigma_A n_A$ is indeed reasonably constant for the *s*-process isotopes, for example (^{122}Te, ^{123}Te, and ^{124}Te), (^{134}Ba and ^{136}Ba), and (^{148}Sm, ^{149}Sm, and ^{150}Sm).

4.11 Stellar Nucleosynthesis

The r- and the s-process reactions involving neutron absorption can also take place in red giants as a consequence of helium shell flashes. During peak burning phases, the temperatures within helium burning shells can be approximately 3×10^8 K and neutrons can be formed by means of reactions such as ^{22}Ne $(\alpha, n)^{25}$Mg and ^{13}C $(\alpha, n)^{16}$O.

Although the theory of stellar nucleosynthesis leads to predictions that are generally in agreement with observations, there does exist some noteworthy anomalies. Some of the heavy elements are underproduced even after s and r processes are taken into account. [It is believed that species on the proton-rich side arise predominantly from the (γ, n) photodisintegration of the neighbouring nuclei.] Similarly, lithium, beryllium, and boron, found in cosmic rays, are in excess of the abundances predicted by these models. It is generally believed that the latter difficulty can be circumvented by cosmic-ray spallation reactions between abundant elements, such as carbon, nitrogen, and oxygen, and protons.

Exercise 4.7

Practice with r and s processes: A good way to familiarise yourself regarding the vagaries of r and s processes is to work out a sequence of nuclei synthesised in a neutron-rich environment of a star by using a chart of nuclides and a table of neutron-capture cross sections. Assume that the star has a temperature of approximately 3×10^8 K, ambient neutron density $n_n = 10^7$ cm^{-3}, and an evolutionary time scale of $\sim 10^6$ yr. Start with any nuclei in the atomic number range $Z > 75$ and obtain the sequence of nuclei synthesised in such a neutron-rich environment. Which of these nuclides will be produced in the s process and approximately by what factor will the abundance be enhanced? Next, repeat the same procedure in the environment of a supernova with $T = 3 \times 10^9$ K, $n_N = 10^{21}$ cm^{-3} and an evolutionary time scale of few minutes. (Hint: When the evolutionary time scale is $\sim 10^6$ yr, we can assume that elements with a half-life less than 10^6 yr are unstable whereas those with a half-life greater than 10^6 yr are stable. In equilibrium we must have $\sigma n = $ constant for the stable nuclei, where n is the abundance and σ is the relevant neutron-capture cross section. As regards the r process, the key elements will be the first stable ones along the β-decay tracks with constant A.)

5

White Dwarfs, Neutron Stars, and Black Holes

5.1 Introduction

This chapter deals with three possible stellar remnants: white dwarfs, neutron stars, and black holes.[1] It relies heavily on the previous two chapters as well as on Chaps. 3, 5, and 9–12 of Vol. I. The material covered here will be needed in Chap. 6 (pulsars), Chap. 7 (binary stars), and in the study of active galactic nuclei in Vol. III.

Another closely related class of remnants, called pulsars, are known to be rotating neutron stars and will be discussed separately in Chap. 6. An entirely new class of physical phenomena arises when a compact object forms a constituent of a binary system. The role of stellar remnants in binary systems will be studied separately in Chap. 7.

5.2 Structure of White Dwarfs

It was seen in Chap. 3 that the end point of stellar evolution can lead to self-gravitating objects supported by degeneracy pressure. Such astrophysical objects are usually termed compact because, as we shall see, their sizes are significantly smaller than main-sequence stars of similar mass. Depending on the average density, these compact objects can have different internal structures:

(1) For densities in the range of 10^5 gm cm$^{-3} \lesssim \rho \lesssim 10^9$ gm cm^{-3}, the remnant is made of an ideal nondegenerate gas of ions and a degenerate gas of electrons. A typical example could be a carbon white dwarf with a temperature of $T_{\text{eff}} = 2.7 \times 10^4$ K, a luminosity of $L = 0.03\ L_\odot$, and mass $M = 1\ M_\odot$. Its radius, estimated as $R = (L/4\pi\sigma T_{\text{eff}}^4)^{1/2}$, is ~ 5000 km, giving a mean density of $\rho \approx (3\ M_\odot/4\pi R^3) \approx 4 \times 10^6$ gm cm^{-3}. For this density, we can estimate the mean distance between the carbon ions as $d_{ii} \approx (\rho/m_C)^{-1/3} \approx 0.02$ Å, where $m_C \approx 12 m_H$ is the mass of the carbon nucleus. This is smaller than the radius of the carbon atom, $r_C \approx (a_0/Z) \approx (a_0/6) \approx 0.08$ Å. Hence the electrons will be stripped off the nuclei and will exist as a separate fermionic gas. Further,

the mean distance between the electrons will be $d_{ee} \approx (Z\rho/m_C)^{-1/3} \approx 10^{-2}$ Å and the thermal wavelength of the electron is $\lambda_e = (h^2/m_e k_B T)^{1/2} \approx 10$ Å. Because $\lambda_e \gtrsim d_{ee}$, quantum-mechanical treatment is required. In general, the system will be degenerate if $T < T_c$ and nondegenerate if $T > T_c$, where $T_c \approx 3 \times 10^9$ K $(\rho/\rho_c)^{2/3}$ (see Vol. I, Chap. 5, Section 5.9).

To decide whether the electrons are relativistic or not, we should compare $m_e c$ with the Fermi momentum $p_F = (3\pi^2)^{1/3} \hbar n_e^{1/3} = (3\pi^2)^{1/3} \hbar (\rho/\mu_e)^{1/3}$, where $\mu_e = (\rho/n_e m_p) = 2(1+X)^{-1}$ is the mass per electron. The Fermi momentum of the electron will be equal to $m_e c$ at the density $\rho = \rho_c$, where

$$\rho_c \equiv \frac{8\pi}{3} m_p \mu_e \frac{m_e c^3}{h} \approx 10^6 \mu_e \text{ gm cm}^{-3}. \quad (5.1)$$

In our case the actual density is marginally in the relativistic domain.

(2) At lower densities, ($\rho \ll 10^5$ gm cm^{-3}), ions form a crystal lattice structure and the electron density is nonuniform over the scale of the lattice. The Coulomb interactions become important in deciding the structure of the matter and the equation of state becomes stiffer; eventually, the electrostatic forces lead to the occurrence of nonzero density at approximately zero pressure, as in the case of normal solids made of nondegenerate matter. The Coulomb interaction between electrons is characterised by the energy $E_c = (Ze^2/d_{ee}) \approx (Ze^2 n_e^{1/3})$, where n_e is the number density of the electrons. The ratio of Coulomb energy to Fermi energy is $(E_c/\epsilon_F) \approx Z^{2/3} \alpha (\rho_c/\rho)^{1/3}$, where α is the fine-structure constant. When $E_c \ll \epsilon_F$, we can treat the gas as ideal. This approximation breaks down at low densities ($\rho \lesssim \rho_c Z^2 \alpha^3 \simeq Z^2 \mu_e$) and – as we shall see – Coulomb interactions lead to the formation of a crystal structure with finite density at zero pressure.

(3) For densities $\rho \gtrsim 10^9$ gm cm^{-3}, the electrons are capable of combining with the protons in the nuclei and thus changing the composition of matter. The equation of state at higher densities should be determined by a study of the the minimum-energy configuration of matter, allowing for nuclear reactions.

We shall begin our discussion with *white dwarfs*, which essentially correspond to the situation in case (1) and are formed from the evolution of stars with an initial mass of approximately (1–8) $M_\odot$. The high-density regime, described by case (3), is relevant in the study of *neutron stars* formed from stars with an initial mass of approximately (8–50) $M_\odot$, which we shall take up in Sections 5.5–5.7. (These mass ranges are very approximate.) Finally, in Sections 5.8 and 5.9 we will study black holes, which are the predicted end stage for a self-gravitating system in which the gravitational force cannot be balanced by the degeneracy pressure of matter.

The formation rate of any stellar remnant can be easily estimated if the initial mass function for stars is known. If we use the empirically determined initial mass function of the form

$$\psi_s dm = 2 \times 10^{-12} m^{-2.35} dm \text{ stars pc}^{-3} \text{ yr}^{-1}, \quad m \equiv \left(\frac{M}{M_\odot}\right), \quad (5.2)$$

then the birthrate $\mathcal{B}$ of stars in the range of masses M_1, M_2 in the galaxy is given by

$$\mathcal{B} = V_{\text{disk}} \int_{(M_1/M_\odot)}^{(M_2/M_\odot)} \psi_s dm = 0.19 \, m^{-1.35} \Big|_{(M_2/M_\odot)}^{(M_1/M_\odot)} \text{yr}^{-1}, \quad (5.3)$$

where

$$V_{\text{disk}} = \pi r^2 (2H) = 1.3 \times 10^{11} \, \text{pc}^3 \quad (5.4)$$

is the volume of the disk of the galaxy with vertical height $H = 90$ pc and radius $r = 15$ kpc, which is appropriate for stars more massive than 1 $M_\odot$. The number density of compact objects formed from these parent stars will be

$$n = T_0 \int_{(M_1/M_\odot)}^{(M_2/M_\odot)} \psi_s dm = 0.018 \, m^{-1.35} \Big|_{(M_2/M_\odot)}^{(M_1/M_\odot)} \text{pc}^{-3}, \quad (5.5)$$

where $T_0 \approx 12 \times 10^9$ yr is the age of the galaxy. For white dwarfs, taking the mass range of the parent star to be (1–8) $M_\odot$, we get an integrated galactic birthrate of 0.18 per year with $n_{\text{wd}} \simeq 1.7 \times 10^{-2}$ pc^{-3}. Neutron stars, on the other hand, originate from stars with an initial mass of (8–50) $M_\odot$, and we get an integrated galactic birthrate of 0.01 per year with $n_{\text{ns}} \simeq 10^{-3}$ pc^{-3}. The mean distance ($n^{-1/3}$) between the white dwarfs should be ~ 3.8 pc whereas that between neutron stars should be ~ 10 pc.

To determine the mass contributed by these objects, we have to take into account the fact that much of the initial mass will be lost during the evolutionary process. White dwarfs and neutron stars have average masses of approximately $\langle M \rangle_{\text{wd}} = 0.65 \, M_\odot$ and $\langle M \rangle_{\text{ns}} = 1.4 \, M_\odot$ so that

$$\frac{\rho}{\rho_T} = \frac{n \langle M \rangle}{\rho_T}, \quad (5.6)$$

where $\rho_T \approx 0.14 \, M_\odot$ pc^{-3} is the mass density of the galaxy. This gives $\rho/\rho_T \simeq 0.08$ for white dwarfs and $\rho/\rho_T \simeq 0.01$ for neutron stars.

5.2.1 White Dwarfs Supported by Degenerate Electron Pressure

The end point in the evolution of main-sequence stars with an initial mass of $M \lesssim 8 \, M_\odot$ consists of a degenerate core after the ejection of significant amount of matter in the form of a planetary nebula. The core will predominantly be made of helium or carbon with very little hydrogen. In such a remnant, the pressure support in the core is provided by an ideal (noninteracting) gas of degenerate electrons, whereas most of the mass density is contributed by a nondegenerate gas of carbon or helium ions. As we shall see, the Fermi energy of the electrons in such a system is higher than the kinetic temperature of the system (except near the surface of the white dwarf, which we shall discuss separately) and hence the electrons can be taken to be a zero-temperature Fermi gas.

5.2 Structure of White Dwarfs

For such a system, the equation of state $P_e(\rho)$ is given by an implicit relation [see Vol. I, Chap. 5, Section 5.9, Eq. (5.156)]. The number of electrons per unit volume, with momentum $p < p_F$, is $n_e = 2(4\pi/3)(p_F/h)^3$. If $\mu_e \equiv [(\rho/m_p)/n_e] = 2(1+X)^{-1}$ is the electronic molecular weight, then $(\rho/\mu_e) = m_p n_e = (8\pi/3) m_p (p_F/h)^3$. The pressure of degenerate electrons is given by $P_e = (1/3)\langle n\mathbf{p}\cdot\mathbf{v}\rangle$ which – for a zero-temperature Fermi gas – becomes

$$P_e = \frac{1}{3}\int_0^{p_F} pv(p)\left[\frac{2d^3p}{h^3}\right] = \frac{8\pi c^2}{3h^3}\int_0^{p_F}\frac{p^4 dp}{(p^2 c^2 + m_e^2 c^4)^{1/2}}, \quad (5.7)$$

where we have used the relations $v = (pc^2/\epsilon)$ and $\epsilon^2 = (p^2 c^2 + m_e^2 c^4)$. Introducing a parameter $x \equiv (p_F/m_e c)$, we can express the density and pressure as $\rho = \rho(x)$ and $P = P_e(x)$, respectively, thereby providing the equation of state $P(\rho)$ in parameterised form:

$$\frac{\rho}{\mu_e} = \frac{8\pi m_p}{3}\left(\frac{h}{m_e c}\right)^{-3} x^3 \equiv Bx^3, \quad (5.8)$$

and

$$P \approx P_e = \frac{8\pi\, m_e^4 c^5}{3\, h^3}\int_0^x \frac{y^4 dy}{(1+y^2)^{1/2}} \equiv Af(x), \quad (5.9)$$

where

$$f(x) = x(2x^2 - 3)(1+x^2)^{1/2} + 3\sinh^{-1} x, \quad (5.10)$$

and

$$A = \frac{\pi}{3}\left(\frac{h}{m_e c}\right)^{-3} m_e c^2 = 6.002\times 10^{22}\text{ dyn cm}^{-2}, \quad B \approx 9.74\times 10^5\text{ gm cm}^{-3}. \quad (5.11)$$

In equilibrium, the gravitational force is balanced by the gradient of the degeneracy pressure; i.e., $\rho^{-1}\nabla P_e = -\nabla\phi$. Taking the divergence of this equation and using $\nabla^2\phi = 4\pi G\rho$, we get, in a spherically symmetric case,

$$\frac{1}{r^2}\frac{d}{dr}\left(\frac{r^2}{\rho}\frac{dP_e}{dr}\right) = -4\pi G\rho. \quad (5.12)$$

To cast this equation into a more convenient form, we introduce a variable $z^2 \equiv (x^2+1)$ [which is essentially $(\epsilon/m_e c^2)$, as $x = (p/m_e c)$] and note that

$$\frac{1}{\rho}\frac{dP_e}{dr} = \frac{1}{\rho}\frac{dP_e}{dx}\frac{dx}{dz}\frac{dz}{dr} = \left(\frac{1}{B\mu_e x^3}\right)\left[\frac{8Ax^4}{(1+x^2)^{1/2}}\right]\left(\frac{z}{x}\right)\frac{dz}{dr} = \frac{8A}{B\mu_e}\left(\frac{dz}{dr}\right). \quad (5.13)$$

Substituting Eq. (5.13) into Eq. (5.12), we get

$$\frac{1}{r^2}\frac{d}{dr}\left(r^2\frac{dz}{dr}\right) = -\frac{\pi}{2}\frac{GB^2\mu_e^2}{A}(z^2-1)^{3/2}. \tag{5.14}$$

Let z_c be the value of z at the centre $r=0$. Rescaling the variables to

$$Q \equiv \frac{z}{z_c}, \quad \zeta \equiv \frac{r}{\alpha}, \quad \alpha \equiv \sqrt{\frac{2A}{\pi G}}\left(\frac{1}{B\mu_e z_c}\right), \tag{5.15}$$

we find that Eq. (5.14) becomes, in terms of Q and ζ,

$$\frac{d^2Q}{d\zeta^2} + \frac{2}{\zeta}\frac{dQ}{d\zeta} + \left(Q^2 - \frac{1}{z_c^2}\right)^{3/2} = 0. \tag{5.16}$$

The boundary conditions at the origin for this equation is

$$Q = 1, \quad Q' = 0 \quad (\text{at } \zeta = 0). \tag{5.17}$$

The first condition follows from the definition of Q whereas the second condition is needed to ensure the vanishing of the pressure gradient at the origin. Given any value of z_c greater than unity, this equation can be integrated outwards (numerically) from $\zeta = 0$ by the above boundary condition. The density at any given location is given by

$$\rho = B\mu_e x^3 = B\mu_e(z^2-1)^{3/2} = B\mu_e z_c^3\left(Q^2 - \frac{1}{z_c^2}\right)^{3/2}. \tag{5.18}$$

We can determine the radius of the object by noting that $\rho = 0$ at $r = R$, corresponding to some value $\zeta = \zeta_1$. Denoting the value of the variables at the surface $r = R$ by the subscript 1, we have, at the surface,

$$x_1 = 0, \quad z_1 = 1, \quad Q_1 = 1/z_c \quad (\text{at } \zeta = \zeta_1), \tag{5.19}$$

$$R = \alpha\zeta_1 = \sqrt{\frac{2A}{\pi G}}\frac{1}{B\mu_e z_c}\zeta_1. \tag{5.20}$$

The total mass of the system is given by

$$M = \int_0^R 4\pi r^2 \rho \, dr = 4\pi\alpha^3 B\mu_e z_c^3 \int_0^{Q_1} \zeta^2\left(Q^2 - \frac{1}{z_c^2}\right)^{3/2} d\zeta$$

$$= 4\pi\alpha^3 B\mu_e z_c^3\left(-\zeta^2\frac{dQ}{d\zeta}\right)_1 = \frac{4\pi}{(B\mu_e)^2}\left(\frac{2A}{\pi G}\right)^{3/2}\left(-\zeta^2\frac{dQ}{d\zeta}\right)_1. \tag{5.21}$$

In proceeding from the first line to the second, we have used Eq. (5.16) to complete the integration. The results of the numerical integration of Eq. (5.16) for several choices of z_c are given in Table 5.1. The following features may be noted as regards these results:

5.2 Structure of White Dwarfs

Table 5.1. *Properties of a simple model for white dwarfs*

$1/z_c$	x_c	ζ_1	$(-\zeta^2 dQ/d\zeta)_1$	ρ_c/μ_e (gm cm^{-3})	$\mu_e^2 \dfrac{M}{M_\odot}$	$\left(\mu_e \dfrac{R}{R_\odot}\right) \times 10^2$
0	∞	6.8968	2.0182	∞	5.84	0
0.01	9.95	5.3571	1.9321	9.48×10^8	5.60	0.48
0.02	7	4.9857	1.8652	3.31×10^8	5.41	0.67
0.05	4.36	4.4601	1.7096	7.98×10^7	4.95	1.02
0.1	3	4.0690	1.5186	2.59×10^7	4.40	1.35
0.2	2	3.7271	1.2430	7.70×10^6	3.60	1.79
0.3	1.53	3.5803	1.0337	3.43×10^6	2.99	2.17
0.5	1	3.5330	0.7070	9.63×10^5	2.04	2.73
0.8	0.5	4.0446	0.3091	1.21×10^5	0.89	3.97
1.0	0	∞	0	0	0	∞

(1) As z_c varies from very large values to unity, x_c varies from very large values (corresponding to a fully relativistic degenerate gas) to $x_c = 0$ (corresponding to a nonrelativistic degenerate gas). (2) The central density is fairly high for these models (compared with that of ordinary stars) and decreases with decreasing x_c; the relativistic compact objects are denser compared with those of the nonrelativistic ones, as to be expected. (3) The mass of the system is bounded from above and cannot exceed the value of approximately

$$M_{\text{ch}} = \frac{5.84}{\mu_e^2} M_\odot = \left(\frac{2}{\mu_e}\right)^2 \times 1.459 \, M_\odot, \tag{5.22}$$

which is the Chandrasekhar mass introduced in different contexts in earlier chapters; this equation provides its precise definition. [If the stellar core, from which the white dwarf forms, is made of helium or heavier elements with $X \approx 0$, then $\mu_e \approx 2$ – which explains the scalings used in Eq. (5.22).] Therefore there exists a maximum mass that can be supported by the degeneracy pressure of electrons. (4) The radius of the system decreases with increasing x_c; that is, massive white dwarfs are smaller in size. The relativistic objects are very compact compared with the nonrelativistic ones. A reasonable fit to $R(M)$ is given by (Vol. I, Chap. 10, Subsection 10.3.2)

$$R(M) \simeq 0.022 \mu_e^{-1} R_\odot \left(\frac{M}{M_{\text{ch}}}\right)^{-1/3} \left[1 - \left(\frac{M}{M_{\text{ch}}}\right)^{4/3}\right]^{1/2}. \tag{5.23}$$

In the nonrelativistic range, somewhat simpler fits are

$$\frac{\mu_e R(M)}{R_\odot} \simeq 4.06 \times 10^{-2} \left(\frac{M \mu_e^2}{M_\odot}\right)^{-1/3}, \quad \rho_c \simeq 7.8 \times 10^6 \text{ gm cm}^{-3} \mu_e^5 \left(\frac{M}{M_\odot}\right)^2. \tag{5.24}$$

The behaviour of $R(M)$ in the nonrelativistic and extreme relativistic limits is a simple consequence of the limiting forms of the equation of state. We have already seen in Vol. I, Chap. 10 that the relativistic degenerate system corresponds to a Lane–Emden polytrope with the index $n = 3$ whereas a nonrelativistic degenerate system has the index $n = 3/2$. From the general mass–radius relation for polytropes, $M \propto R^{(3-n)/(1-n)}$, it follows that $M \propto R^{-3}$ for the nonrelativistic case and $M = $ constant for the relativistic case. It can be easily seen that Eq. (5.16) reduces to the corresponding Lane–Emden equations in these two limiting cases, thereby explaining the limiting behaviour.

The same result can be interpreted in terms of the pressure balance along the following lines. Pressure balance against gravity requires the ratio $\mathcal{R}$ between (GM^2/R^4) and P to be a constant. With $P \propto \rho^\gamma \propto (M/R^3)^\gamma$, this ratio becomes

$$\mathcal{R} = (GM^2/PR^4) \propto M^{2-\gamma} R^{3\gamma-4}. \tag{5.25}$$

For a nonrelativistic system with $\gamma = 5/3$, we have $\mathcal{R} \propto M^{1/3} R$. In equilibrium, $\mathcal{R}$ will have some definite numerical value; if M is now increased, equilibrium can be restored by R decreasing so as to maintain the constancy of $M^{1/3} R$. Hence a system supported by nonrelativistic degeneracy pressure will contract if more mass is added. On the other hand, $\mathcal{R} \propto M^{2/3}$ for the relativistic systems with $\gamma = 4/3$. If $M < M_{\text{ch}}$, pressure exceeds gravity and the star will expand, thereby making the system eventually nonrelativistic; an equilibrium solution can be found at a suitable value of M and R. However, if $M > M_{\text{ch}}$, then the gravitational force exceeds the pressure and the system will contract. However, because $\mathcal{R}$ is independent of R, the contraction will *not* lead to a new equilibrium configuration and the system should continue to collapse under self-gravity. Figure 5.1 gives the fitting function to the M–R relation obtained above.

Although the above analysis gives the broad picture of the structure of a white dwarf, it is flawed as regards several details. To begin with, the equation of state is incorrect at low densities (because of Coulomb interactions) and at high densities (because of nuclear interactions). Further, for a compact high-density configuration, we have to take into account the general relativistic effects. We shall now discuss the relevant modifications.

Exercise 5.1

Model equation of state for degenerate fermions: A simple equation of state, which can extrapolate between the nonrelativistic and the extreme relativistic limits of a degenerate Fermi gas, is given by

$$P_e = \frac{8\pi}{3h^3} \frac{cp_F^4}{(4+q)} \left(\frac{p_F}{m_e c}\right)^q, \tag{5.26}$$

5.2 Structure of White Dwarfs

where $q = 1$ and $q = 0$ correspond to the two limits. Given the standard relation between Fermi momentum and the density, this provides a convenient equation of state parameterised by the constant q. Determine the variation of the central density and the radius of a white dwarf on the mass for different values of q. In particular, consider $q = 1, 0.5, 0.2, 0.1, 0.01$. (Use simple extrapolations from the tabulated values of Lane–Emden functions to obtain the numerical coefficients.)

Exercise 5.2
Temperature correction to the Chandrasekhar mass: The expression for Chandrasekhar mass, viz., $M_{Ch} = 5.84 Y_e^2 \, M_\odot$, is computed with the zero-temperature limit of a Fermi gas. Show that the lowest-order finite-temperature corrections increase it by a factor of $[1 + (\pi k_B T / \epsilon_F)^2]$.

5.2.2 Coulomb Corrections at Low Densities

According to the above analysis, the radius of the system goes to infinity as the mean density goes to zero. This is clearly incorrect and arises because of our using the equation of state of degenerate matter at low densities where it is not

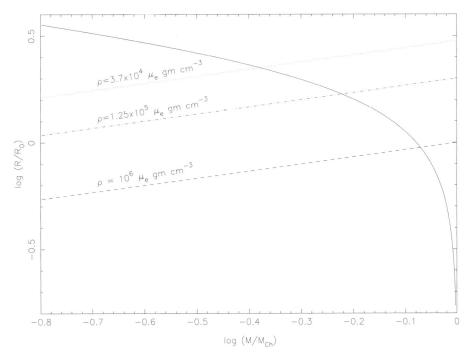

Fig. 5.1. The mass–radius relation for a fully degenerate nonrelativistic fermionic sphere in logarithmic coordinates with $R_0 = 10^{-2} \mu^{-1} R_\odot$. The constant-density systems with $M = (4\pi/3)\rho R^3$ are marked by sloping lines.

valid. Proper analysis should reveal that the system makes a smooth transition to planetlike structure at low densities.

At low densities, we cannot treat ions as an ideal gas and must take into account the fact that they will form a solid lattice structure. Such a transition occurs essentially because of the Coulomb interaction, which we shall now take into account. To compute the dominant correction, which arises from the Coloumb energy of the lattice, let us divide the crystal lattice into spheres of radius r_0 such that $(4\pi r_0^3/3)n_e = Z$, where n_e is the number density of electrons. Each of these spheres contains one ion with charge $+Z$ at the centre and Z electrons that are distributed with uniform density so that each sphere has zero total net charge. The Coloumb energy per electron is the sum of the self-energy E_{ee} of a negatively charged sphere of radius r_0 and the electrostatic interaction energy E_{ei} between the nucleus and the electron cloud, both of which are easy to compute. Because $E_{ee} = (3/5)(Z^2 e^2 / r_0)$ and $E_{ei} = -(3/2)(Z^2 e^2 / r_0)$, the total Coloumb energy of a single cell of radius r_0 is given by

$$E_c = E_{ee} + E_{ei} = -\frac{9}{10} \frac{Z^2 e^2}{r_0}. \tag{5.27}$$

The interaction between different cells is negligible because, by construction, each of the cells is charge neutral. Taking $n_e = (3Z/4\pi r_0^3)$, we find that the Coloumb energy per electron is given by

$$\epsilon(n_e) \equiv \frac{E_c}{Z} = -\frac{9}{10} \left(\frac{4\pi}{3}\right)^{1/3} Z^{2/3} e^2 n_e^{1/3}. \tag{5.28}$$

The corresponding pressure, obtained by $dE/dV = dE/d(N/n) = d\epsilon/d(1/n)$, where $\epsilon = (E/N)$ is the energy per particle, will be

$$P_c = \frac{\partial \epsilon}{\partial(1/n_e)} = n_e^2 \frac{d(E_c/Z)}{dn_e} = -\frac{3}{10} \left(\frac{4\pi}{3}\right)^{1/3} Z^{2/3} e^2 n_e^{4/3}. \tag{5.29}$$

This pressure P_c arising because of electrostatic interactions should be added to the pressure P_0 of the ideal Fermi gas to get the total pressure.

Let us consider the correction in the two limits: In the extreme relativistic limit, $P_0 \simeq [(3\pi^2)^{1/3}/4] \hbar c n_e^{4/3}$ and the Coulomb correction changes the pressure by the ratio

$$\frac{P}{P_0} \equiv \frac{P_0 + P_c}{P_0} = 1 - \frac{2^{5/3}}{5} \left(\frac{3}{\pi}\right)^{1/3} \alpha Z^{2/3}, \tag{5.30}$$

where $\alpha = (e^2/\hbar c) \simeq 10^{-2}$. This correction is small in the cases of astrophysical interest. More significant is the correction in the nonrelativistic limit, where

5.2 Structure of White Dwarfs

$P_0 \simeq [(3\pi^2)^{2/3}/5]\hbar^2 m_e^{-1} n_e^{5/3}$ and the corresponding ratio is

$$\frac{P}{P_0} = 1 - \frac{Z^{2/3}}{2^{1/3}\pi a_0 n_e^{1/3}}, \qquad (5.31)$$

where a_0 is the Bohr radius. Such a correction predicts that the pressure P will vanish at a finite density corresponding to $n_e = n_{e0} \equiv (Z^2/2\pi^3 a_0^3)$ and $\rho_0 = \mu_e m_p n_e \approx 0.4 Z^2 \mu_e$ gm cm^{-3}; i.e., a solid structure with finite density and zero pressure can arise because of electrostatic correction. Because of the lowering of pressure for a given density, the mass–radius relation turns around at low values of mass. The total pressure now becomes

$$P = P_0 \left[1 - \left(\frac{\rho_0}{\rho}\right)^{1/3}\right] \simeq K\rho^{5/3}\left[1 - \left(\frac{\rho_0}{\rho}\right)^{1/3}\right],$$

$$K = \frac{(3\pi^2)^{2/3}}{5}\frac{\hbar^2}{m_e}\left(\frac{1}{m_p \mu_e}\right)^{5/3}, \quad \rho_0 \simeq 0.4 Z^2 \mu_e \text{ gm cm}^{-3}. \qquad (5.32)$$

Equating this to (GM^2/R^4) and expressing ρ as $\rho = (3M/4\pi R^3)$ will lead to a mass–radius relation that incorporates the Coulomb corrections. Straightforward algebra gives

$$\frac{R}{R_\odot} = \left[\left(\frac{M}{M_\odot}\right)^{1/3} + \left(\frac{M_c}{M}\right)^{1/3}\right]^{-1}, \qquad (5.33)$$

with

$$M_c = 2.8 \times 10^{-7} Z^2 \mu_e M_\odot. \qquad (5.34)$$

The radius varies as $R \propto M^{1/3}$ (constant-density solid) for $M \lesssim M_{\text{max}}$ and as $R \propto M^{-1/3}$ (nonrelativistic white dwarf) for $M \gg M_{\text{max}}$, where the turnaround occurs at

$$M_{\text{max}} \equiv (M_0 M_c)^{1/2} \cong 1.1 \times 10^{-3} Z \left(\frac{\mu}{2}\right)^3 M_\odot,$$

$$R_{\text{max}} \simeq 8.3 \times 10^{-2} R_\odot \times \left(\frac{\mu}{2}\right)^{2/3}\left(\frac{1}{z}\right)^{1/3}, \qquad (5.35)$$

which are close to planetary masses. Figure 5.2 shows the R–M relation for $Z = 1, \mu = 1$ and can be rescaled for other values.

Exercise 5.3

More on Coulomb corrections: A better description of the equation of state, taking into account the nonuniform distribution of electrons within each cell, can be obtained as follows. Within each cell, we will assume that the electrons move in a spherically symmetric potential $V(r)$ and use a local version of Fermi–Dirac theory. At any given radius r, the Fermi momentum will be $p_F(r)$ and the Fermi energy will be $\epsilon_F = -eV(r) + (p_F^2(r)/2m_e)$. In an equilibrium situation, ϵ_F (which is also the chemical potential of the system) must not have any spatial gradient. The number density of the electrons is therefore given by

$$n_e = \frac{8\pi}{3h^3} p_F^3 = \frac{8\pi}{3h^3} \{2m_e[\epsilon_F + eV(r)]\}^{3/2}, \tag{5.36}$$

where ϵ_F is a constant and $V(r)$ has to be determined self-consistently through the Poisson equation

$$\nabla^2 V = 4\pi e n_e + 4\pi Z e \delta_D(\mathbf{r}), \tag{5.37}$$

where the second term is due to the nuclear charge concentrated at the origin. Solve this equation approximately and show that the Coulomb interactions contribute a pressure $P(\rho)$ that varies asymptotically as $P \propto \rho^{10/3}$. How is the M–R relation modified in this case?

5.2.3 Corrections at High Densities

Let us next consider the high-density limit, in which the naive theory predicts infinitely high central densities as the mass approaches the Chandrasekhar mass. In reality the equation of state ceases to be valid at high densities because of several other effects. The most important among them are the following:

Consider a system made of nuclei with atomic weight and atomic number A and Z, respectively, surrounded by a cloud of electrons. For densities below $\rho_c \cong 10^6$ gm cm^{-3}, the system may be described as a solid lattice of ions with nonrelativistic degenerate gas of electrons. At a density of $\rho_c \cong 10^6$ gm cm^{-3}, electrons become relativistic and the system can again be described as consisting of a lattice of ions and a relativistic degenerate pressure of electrons. When the density increases still further, the Fermi energy of electrons will become high enough to induce the inverse beta-decay reaction $e^-(p, n)\nu$ by combining with the protons in the nuclei. (The direct decay $n \to p + e + \bar{\nu}$ is blocked at high enough densities because of the lack of available energy states for electrons.)

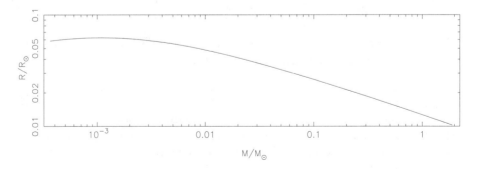

Fig. 5.2. R–M relation for white dwarfs, including the Coulomb corrections.

5.2 Structure of White Dwarfs

Table 5.2. Neutronisation thresholds

Reaction	Neutronisation Threshold (MeV)	ρ_0 (gm cm^{-3})
$^1_1H \to n$	0.782	1.22×10^7
$^4_2He \to ^3_1H + n \to 4n$	20.596	1.37×10^{11}
$^{12}_6C \to ^{12}_5B \to ^{12}_4Be$	13.370	3.90×10^{10}
$^{16}_8O \to ^{16}_7N \to ^{16}_6C$	10.419	1.90×10^{10}

Because this reduces the number of electrons – and thus the electron pressure – the system becomes unstable. (This process, called *neutronisation*, has already been encountered in the last chapter.)

The threshold density for neutronisation can be estimated as follows: The Fermi energy of a relativistic, degenerate electron is $m_e c^2 (1 + x_e^2)^{1/2}$, with $x_e = (p_F/m_e c)$. If this energy is larger than the difference Q in the binding energy of the two nuclei (A, Z) and $(A, Z - 1)$, then neutronisation will be favoured. The threshold value of the Fermi momentum at which this occurs can be computed from the relation $m_e c^2 (1 + x_e^2)^{1/2} = Q$, and the corresponding number density is

$$n = \frac{1}{3\pi^2 \lambda_e^3} \left[\left(\frac{Q}{m_e c^2}\right)^2 - 1 \right]^{3/2}, \quad \lambda_e = \frac{\hbar}{m_e c}. \tag{5.38}$$

The neutronisation threshold for different reactions and the corresponding densities at which they occur are given in Table 5.2. The most relevant ones for white dwarfs are the carbon and helium thresholds at 3.9×10^{10} gm cm^{-3} and 1.37×10^{11} gm cm^{-3}. Above these densities, our description clearly fails and the white dwarfs become unstable as a result of a rapid disappearance of electron pressure.

The second process that affects the equation of state at high densities is the possibility that nuclear reaction can take place even at zero temperature in a crystal lattice at sufficiently high densities. This process, called a *pycnonuclear reaction*, arises because of the zero-point oscillations of the nuclei which allow them to tunnel through the potential barrier to the neighbouring site and induce nuclear reactions. We shall now provide a derivation of this effect.

Let us approximate the potential felt by an ion in a crystal lattice by that for an ion located between two fixed identical ions separated by a distance $2R_0$ along any one dimension. This potential is given by

$$V(x) = \frac{Z^2 e^2}{R_0 - x} + \frac{Z^2 e^2}{R_0 + x} - \frac{2Z^2 e^2}{R_0} = \frac{2Z^2 e^2}{R_0} \frac{1}{[(R_0/x)^2 - 1]} \quad (|x| < R_0 - R_n), \tag{5.39}$$

where R_n is the range of the nuclear force. Close to the nucleus we will smoothly join this potential to the attractive nuclear potential of width R_n, as shown in

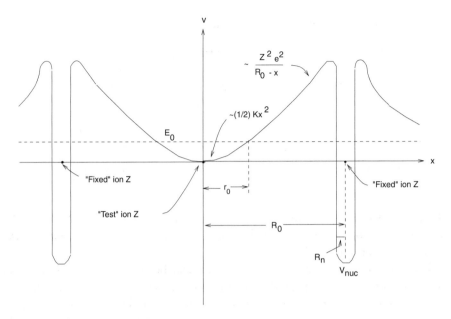

Fig. 5.3. Lattice potential felt by an ion undergoing pycnonuclear reaction.

Fig. 5.3. For $x \ll R_0$, the potential may be approximated as a harmonic oscillator with

$$V(x) \cong \frac{1}{2} M\Omega_0^2 x^2, \quad \Omega_0^2 = \frac{4Z^2 e^2}{MR_0^3}. \tag{5.40}$$

The ground state of the ion of mass M has the energy $E_0 \cong (1/2)\hbar\Omega_0$ and the particle is confined to region of size $r_0 \equiv (2\hbar/M\Omega_0)^{1/2}$. The ground-state wave function can be approximated as $|\psi|^2 \cong (\pi^{3/2} r_0^3)^{-1}$ near the origin.

The zero-point oscillations of the central ion allow it to have a finite probability of tunnelling through the potential $\mathcal{T}$ to the adjacent sides. The tunnelling probability for an ion with energy E_0, computed along the lines of the discussion in Chap. 12 in Vol. I, is given by

$$\mathcal{T} = \exp\left[-2 \int_a^b k(x) \, dx\right] \equiv \exp\left\{-\frac{2}{\hbar} \int_a^b dx [2M(V(x) - E_0)]^{1/2}\right\}, \tag{5.41}$$

where a and b are the limits between which tunnelling occurs. Taking $a = r_0$ and $b = R_0 - R_n$ and using $V(x)$ in Eq. (5.39), we get

$$\mathcal{T} = \exp\left[-2\left(\frac{4MZ^2 e^2 R_0}{\hbar^2}\right)^{1/2} \int_{r_0/R_0}^{1-R_n/R_0} \left(\frac{u^2}{1-u^2} - \alpha\right)^{1/2} du\right], \tag{5.42}$$

5.2 Structure of White Dwarfs

where $u = (x/R_0)$ and

$$\alpha = \frac{E_0 R_0}{2Z^2 e^2}, \quad r_0 = \left(\frac{h}{2Ze}\right)^{1/2} \left(\frac{R_0^3}{M}\right)^{1/4}. \tag{5.43}$$

In the limit of $R_n \to 0$ and $(r_0/R_0) \ll 1$, this integral can be simplified to give

$$T \cong \frac{R_0}{r_0} \exp\left(-2\frac{R_0^2}{r_0^2}\right). \tag{5.44}$$

The reaction rate per ion pair is given by

$$W = v|\psi_{\text{inc}}|^2 \frac{T S(E)}{E}, \tag{5.45}$$

where $v = \sqrt{2E/M}$ is the velocity of the ion and $|\psi_{\text{inc}}|^2 \approx (r_0^3 \pi^{3/2})^{-1}$ is the incident-wave function for the ground state of the harmonic oscillator localised within a volume r_0^3; $S(E)$ is the standard reaction cross section for nuclear reactions described in Vol. I, Chap. 12. Putting all these together, we get

$$W = 4 \left(\frac{2}{\pi^3}\right)^{1/2} S \frac{(Z^2 e^2 M)^{3/4}}{(h^2 R_0)^{5/4}} \exp\left[-4Ze \frac{(MR_0)^{1/2}}{h}\right]. \tag{5.46}$$

If n_A is the number density of ions evaluated when one ion is assigned to each sphere of radius $R_0/2$, then the number of reactions per unit volume per second is given by

$$\mathcal{R} = n_A W = \left(\frac{\rho}{A}\right) A^2 Z^4 S \gamma \lambda^{5/4} \exp(-\epsilon \lambda^{-1/2}), \tag{5.47}$$

where A is the atomic weight, $\gamma \simeq 1.1 \times 10^{44}$, $\epsilon \simeq 2.85$, and

$$\lambda \equiv \frac{h^2}{2MZ^2 e^2} \left(\frac{n_A}{2}\right)^{1/3} = \frac{h^2}{2MZ^2 e^2} \left(\frac{3}{\pi}\right)^{1/3} \frac{1}{R_0}. \tag{5.48}$$

To determine the critical densities at which these reactions become important, we can define a time scale t by writing $\mathcal{R}t = n_A$. For $t = 10^5$ yr, we can solve for ρ if we know the reaction rates $S(E)$. For carbon, $S_{\text{cc}} = 8.83 \times 10^{16}$ MeV b, giving $\rho = 6 \times 10^9$ gm cm^{-3}. This process converts H to ^{4}He, ^{4}He to ^{12}C, and ^{12}C to ^{24}Mg at a time scale of $\sim 10^5$ yr at densities above 5×10^4, 8×10^8, and 6×10^9 gm cm^{-3}, respectively. This shows that pycnonuclear reactions can be major sources of instability at white dwarf densities.

5.2.4 General Relativistic Corrections

The discussion of white dwarfs so far has been based on Newtonian gravity. At high densities, however, it is necessary to take into account general relativistic effects on the structure of white dwarfs, which could also – as we shall see – lead to an instability.

The physical reason for this instability is the following: In the Newtonian theory, we obtain an equilibrium configuration by balancing the gravitational field of the mass density by the pressure. In principle, we can arrange the equation of state such that the pressure is arbitrarily high for any given density, thereby balancing any given gravitational force. In general relativity, however, pressure also contributes an effective mass (see discussion in Vol. I, Section 11.8), and hence increasing the pressure will also increase the gravitational force. Therefore we cannot ensure that the system is stable by increasing the pressure arbitrarily.

To analyse the stability, we can proceed as follows: We have seen in Section 5.2.1 that compact objects with barotropic equations of state form a one-parameter family of solutions for any given equation of state. It is now convenient to take this independent parameter to be the central density ρ_c. Let the energy of the configuration be some function $E(\rho_c)$. At equilibrium, $(dE/d\rho_c) = 0$ and the stability of the system is clearly related to the sign of $(d^2E/d\rho_c^2)$. If $(d^2E/d\rho_c^2) > 0$, the system is in a minimum-energy configuration and is stable, whereas if $(d^2E/d\rho_c^2) < 0$, then the system is in a local maxima and is unstable. The marginal stability can be determined by the condition $(d^2E/d\rho_c^2) = 0$. We shall now determine the form of the function $E(\rho_c)$ for a polytrope of index n, including the lowest-order effects of general relativity, and use this condition to determine the stability.

The energy E of a *Newtonian* polytrope with the equation of state $P = K\rho^\Gamma$ can be expressed as the sum of the internal energy $E_{\text{int}} \propto PV$ and the gravitational potential energy $U_{\text{grav}} \propto -(GM^2/R)$. Denoting various constants of proportionality by $k_0, k_1, \ldots$, etc., we get

$$E = k_0 PV - k_1 \frac{GM^2}{R} = k_0 MP\rho^{-1} - k_1 \frac{GM^2}{R} = k_2 M\rho_c^{\Gamma-1} - k_3 GM^{5/3}\rho_c^{1/3}, \tag{5.49}$$

where the subscript c denotes the central values. The equilibrium condition is given by $(\partial E/\partial\rho_c) = 0$, which leads to the relation $M \propto \rho_c^{(3/2)(\Gamma-4/3)}$. For stability, an increase in mass must lead to increase in central density; that is, we need $(d \ln M/d \ln \rho_c) = (\Gamma - 4/3) > 0$, which is the standard result. [The second derivative $(\partial^2 E/\partial\rho_c^2)$ at the turning point is proportional to $GM^{5/3}\rho^{-5/3}(\Gamma - 4/3)$, so we get the same result from the condition $(\partial^2 E/\partial\rho_c^2) > 0$, as to be expected.] To proceed from Newtonian theory to general relativity, we need the corresponding expression for $E(\rho_c)$ in general relativity. The lowest-order general relativistic correction can be obtained as follows: The gravitational energy in Newtonian theory is given by $E_1 = -\alpha_1(GM^2/R)$, where α_1 is some numerical constant. In relativity, this will be equivalent to the mass $m_{\text{eff}} = (|E_1|/c^2)$. The coupling of this mass with M will provide the lowest-order corrections to the energy, which will be $E_{\text{corr}} = -\alpha_2(GMm_{\text{eff}}/R)$, where α_2 is another numerical

5.2 Structure of White Dwarfs

constant. Using the expression for m_{eff}, we get $E_{\text{corr}} = -\alpha_1\alpha_2(G^2M^3/c^2R^2)$. Writing $\rho_c = \alpha_3^3(M/R^3)$ and eliminating R gives

$$E_{\text{corr}} = -\left(\frac{\alpha_1\alpha_2}{\alpha_3^2}\right)\left(\frac{G}{c}\right)^2 M^{7/3}\rho_c^{2/3}. \tag{5.50}$$

The numerical constant $k_4 \equiv (\alpha_1\alpha_2/\alpha_3^2)$ depends on the actual distribution of matter and cannot be estimated by this argument. Including this correction from general relativity, the expression for energy becomes

$$E = k_2 M\rho_c^{\Gamma-1} - k_3 G M^{5/3}\rho_c^{1/3} - k_4 \frac{G^2}{c^2} M^{7/3}\rho_c^{2/3}. \tag{5.51}$$

Repeating the stability analysis by computing $(\partial^2 E/\partial\rho_c^2)$ and writing $\rho_c^{1/3} = \alpha_3(M^{1/3}/R)$, we find that the condition on Γ becomes

$$\Gamma > \frac{4}{3} + \frac{2}{3}\left(\frac{k_4\alpha_3}{k_3}\right)\left(\frac{GM}{c^2R}\right). \tag{5.52}$$

This shows that general relativity changes the critical Γ to a larger value, thereby increasing the domain of instability. [For an $n = 3$ polytrope, the correction term has the numerical value of $2.25(GM/c^2R)$.]

In determining the actual Γ, we now need to take into account corrections from the value for a relativistically degenerate electron gas $\Gamma_{\text{deg,R}} = 4/3$. In the general expression for pressure,

$$P_e = \frac{8\pi}{3h^3}\int_0^{p_F} pv(p)p^2 dp, \tag{5.53}$$

we expand $v(p)$ in a Taylor series in $(m_e c/p)$, retaining the first nontrivial term:

$$v = c\left(1 + \frac{m_e^2 c^2}{p^2}\right)^{-1/2} \cong c\left(1 - \frac{m_e^2 c^2}{2p^2}\right). \tag{5.54}$$

Using this in Eq. (5.53) and integrating, we get

$$P_e = \frac{2\pi c}{3h^3} p_F^4\left(1 - \frac{m_e^2 c^2}{p_F^2}\right). \tag{5.55}$$

Because the density is related to the Fermi momentum by $\rho \propto n_e \propto p_F^3$, we can write $p_F = K\rho^{1/3}$. Taking the logarithm of Eq. (5.55), we get

$$\ln P_e \cong 4\ln p_F - \frac{m_e^2 c^2}{p_F^2} + \text{constant} = \frac{4}{3}\ln\rho - \frac{m_e^2 c^2}{K^2}\rho^{-2/3} + \text{constant}. \tag{5.56}$$

Hence

$$\Gamma = \frac{d \ln P_e}{d \ln \rho} = \frac{4}{3} + \frac{2}{3} \frac{m_e^2 c^2}{K^2} \rho^{-2/3} = \frac{4}{3} + \frac{2}{3} \left(\frac{m_e c}{p_F}\right)^2. \tag{5.57}$$

The condition for stability now becomes, omitting factors of the order of unity, $(GM/Rc^2) < (m_e c/p_F)^2$. Expressing p_F in terms of ρ, writing $\rho \approx (M/R^3)$, and again omitting factors of order unity, we find that this condition becomes $(GM^{5/3}/R^3 c^2) < (m_e c/h)^2 (\mu_e m_p)^{2/3}$ or, numerically,

$$R > 10^8 \left(\frac{M}{M_\odot}\right)^{5/9} \mu_e^{-2/9} \text{ cm}. \tag{5.58}$$

On the other hand, for a relativistic white dwarf, at the Chandrasekhar mass limit (with $\mu_e = 2$, $M = 1.4 \, M_\odot$ and $\rho \gtrsim \rho_{\text{rel}} = 10^6 \mu_e$ gm cm^{-3}), the radius scales as $R \approx 10^9 (\rho_{\text{rel}}/\rho)^{1/3}$ cm. This radius will fall below the bound obtained in inequality (5.58) if $\rho > 600 \rho_{\text{rel}}$; that is, the white dwarf will be unstable because of general relativity effects if $\rho > 1.2 \times 10^9$ gm cm^{-3}. This is the key new feature arising from general relativistic effects.

We shall now provide a more detailed and rigorous evaluation of these results. The total energy of a general relativistic, spherical mass distribution with total mass M and baryon number N, excluding the rest energy, is $E = Mc^2 - m_B c^2 N$, where

$$M = \int_0^R \rho \, 4\pi r^2 \, dr = \int_0^R \rho_0 (1 + u) \, 4\pi r^2 \, dr = m_B \int_0^R n(1 + u) \, 4\pi r^2 \, dr, \tag{5.59}$$

$$N = \int_0^R n \sqrt{^3 g} \, d^3 x \equiv \int_0^R n \, dV = \int_0^R n \left[1 - \frac{2m(r)}{r}\right]^{-1/2} 4\pi r^2 \, dr. \tag{5.60}$$

Here m_B is the mass of the individual baryon, ρ_0 is the rest-mass density, u is the internal-energy density (both of which contribute to mass in general relativity), n is the baryon number density, and $dV \equiv \sqrt{^3 g} d^3 x$ is the proper volume element of the 3-space. The metric for a spherically symmetric space time is taken to be of the form

$$ds^2 = -e^{2\Phi} dt^2 + e^{2\lambda} dr^2 + r^2 d\Omega^2, \tag{5.61}$$

and we have defined a new function $m(r)$ by the relation

$$e^{2\lambda} \equiv \left(1 - \frac{2m}{r}\right)^{-1} \tag{5.62}$$

5.2 Structure of White Dwarfs

(see Vol. I, Chap. 11). So the total energy can be written as

$$E = \int_0^R \left[\rho\left(1 - \frac{2m}{r}\right)^{1/2} - \rho_0 \right] d\mathcal{V} \approx \int_0^R \rho_0 \left[u - \frac{m}{r} - u\frac{m}{r} - \frac{1}{2}\left(\frac{m}{r}\right)^2 \right] d\mathcal{V}. \tag{5.63}$$

In arriving at the second approximation, we have used a Taylor series expansion in the small quantities u, $m(r)/r$. Note that $\rho_0 d\mathcal{V}$ can be treated as an invariant and hence need not be expanded. The corresponding expression in the Newtonian theory is

$$E_{\text{Newt}} = \int_0^R \rho_0 u \, d\mathcal{V} - \int_0^M \frac{m'}{r'} \, dm', \tag{5.64}$$

where $dm' = \rho_0 d\mathcal{V}$ and $r' \equiv (3\mathcal{V}/4\pi)^{1/3}$. The correspondence with Newtonian theory is made by a comparison of systems for which same number of baryons exists in same proper volume – i.e., the functional form of $\rho_0(\mathcal{V})$ is the same in general relativity and Newtonian theory. The correction to the Newtonian expression that is due to general relativity is therefore

$$\Delta E_{\text{GTR}} = \int_0^R \rho_0 \, d\mathcal{V} \left[-u\frac{m}{r} - \frac{1}{2}\left(\frac{m}{r}\right)^2 + \frac{m'}{r'} - \frac{m}{r} \right]. \tag{5.65}$$

From these definitions it follows that, to the lowest order,

$$\mathcal{V} = \frac{4\pi r^3}{3}\left(1 + \frac{3}{r^3}\int_0^r mr \, dr\right), \tag{5.66}$$

$$r' - r = \frac{1}{r^2}\int_0^r mr \, dr, \tag{5.67}$$

$$m'(\mathcal{V}) - m(\mathcal{V}) = \int_0^{\mathcal{V}} d\mathcal{V}\left[\rho_0 - \rho\left(1 - \frac{2m}{r}\right)^{1/2}\right] = -\int_0^{\mathcal{V}} d\mathcal{V}\left(u - \frac{m}{r}\right). \tag{5.68}$$

Substituting these expressions into Eq. (5.65) and writing

$$\frac{m'}{r'} - \frac{m}{r} = \frac{m' - m}{r'} - \frac{m(r' - r)}{rr'}, \tag{5.69}$$

we can evaluate the integrals with the Newtonian relations for ρ_0, r, etc., because the lowest-order corrections are explicitly taken care of. The calculation is now straightforward and leads to an expression of the form

$$\Delta E_{\text{GTR}} = I_1 + I_2 + I_3 + I_4 + I_5, \tag{5.70}$$

where

$$I_1 = -\int_0^M u\frac{m}{r}dm, \quad I_2 = -\frac{1}{2}\int_0^M \left(\frac{m}{r}\right)^2 dm,$$

$$I_3 = -\int_0^M \frac{dm}{r}\int_0^m u\,dm, \quad (5.71)$$

$$I_4 = \int_0^M \frac{dm}{r}\int_0^m \frac{m}{r}dm, \quad I_5 = -\int_0^M \frac{m\,dm}{r^4}\int_0^r mr\,dr. \quad (5.72)$$

This result is true for any given equation a state; given a Newtonian theory, the general relativistic corrections are given by these terms correct to quadratic order in the small parameters $u, m/r$.

In the case of a polytropic equation of state, we can simplify these expressions further by using the results

$$u = n\frac{P}{\rho_0}, \quad \frac{1}{\rho_0}\frac{dP}{dr} = -\frac{m}{r^2}. \quad (5.73)$$

Using these and the standard expressions from polytropic theory, we can relate the I_j's to each other,

$$I_5 = \frac{1}{n}I_1, \quad I_4 = 2I_2 - \frac{2}{n}I_1 - \frac{3}{n}I_3, \quad I_3 = I_1 - \frac{2n}{n+1}(I_2 + I_4), \quad (5.74)$$

and we can express the correction ΔE_{GTR} as

$$\Delta E_{\text{GTR}} = \frac{5 + 2n - n^2}{n(5-n)}2I_1 + \frac{n-1}{5-n}3I_2 = -kM^{7/3}\rho_c^{2/3}, \quad (5.75)$$

where k is numerical constant expressible in terms of the Lane–Emden function $\theta(\xi)$:

$$k = \frac{(4\pi)^{2/3}}{(5-n)[\xi_1^2|\theta'(\xi_1)|]^{7/3}}\left[-\frac{5+2n-n^2}{(n+1)}2\int_0^{\xi_1}\xi^3\theta'\theta^{n+1}d\xi\right.$$

$$\left. + \frac{3}{2}(n-1)\int_0^{\xi_1}\xi^4\theta'^2\theta^n\,d\xi\right]. \quad (5.76)$$

To convert to conventional units, the expression should be multiplied by $(G/c)^2$. The function $\theta(\xi)$ satisfies the equation

$$\frac{1}{\xi^2}\frac{d}{d\xi}\left(\xi^2\frac{d\theta}{d\xi}\right) = -\theta^n \quad (5.77)$$

with the boundary conditions $\theta(0) = 1$ and $\theta'(0) = 0$. Note that the correction term has the same form as determined above by simple arguments: see Eq. (5.51).

We shall now apply this result for a relativistic white dwarf corresponding to an $n = 3$ polytrope. The total energy for a polytrope with $P = K\rho^{(n+1)/n}$ is

5.2 Structure of White Dwarfs

given by

$$E = E_{\text{int}} + E_{\text{grav}} + \Delta E_{\text{int}} + \Delta E_{\text{GTR}}, \qquad (5.78)$$

where the internal thermal energy E_{int} and the Newtonian gravitational potential energy E_{grav} are

$$E_{\text{int}} = \int u\, dm = \int \frac{nP}{\rho} dm = K\rho_c^{1/n} M \frac{n}{|\xi_1^2 \theta'|} \int_0^{\xi_1} \xi^2 \theta^{n+1} d\xi = k_1 K \rho_c^{1/n} M, \qquad (5.79)$$

$$E_{\text{grav}} = -G \int \frac{m}{r} dm = (4\pi \rho_c)^{1/3} \frac{GM^{5/3}}{|\xi_1^2 \theta'|^{5/3}} \int_0^{\xi_1} \xi^3 \theta' \theta^n d\xi = -k_2 G \rho_c^{1/3} M^{5/3}, \qquad (5.80)$$

with

$$k_1 = \frac{n(n+1)}{5-n} \frac{|\xi_1^2 \theta'|}{\xi_1} = 1.75579, \quad k_2 = \frac{3}{5-n} \frac{|4\pi \xi_1^2 \theta'|^{1/3}}{\xi_1} = 0.639001, \qquad (5.81)$$

$$K = \frac{3^{1/3} \pi^{2/3}}{4} \frac{\hbar c}{m_p^{4/3} \mu_e^{4/3}} = \frac{1.2435 \times 10^{15}}{\mu_e^{4/3}} \text{ cgs} \qquad (5.82)$$

for $n = 3$. The coefficient of general relativistic correction has the value $k = 0.918294$ for this polytrope so that

$$\Delta E_{\text{GTR}} = -k_4 \frac{G^2}{c^2} M^{7/3} \rho_c^{2/3}, \quad k_4 = 0.918294. \qquad (5.83)$$

The correction term ΔE_{int} represents the contribution of the electrons to the internal energy arising from the departure from the extreme relativistic limit of $n = 3$, which is of comparable order with general relativistic corrections and hence should be retained. The internal-energy density of a relativistic gas can be expanded in a Taylor series in $x \equiv (p_F/m_e c)$ by use of the following relations (see Vol. I, Chap. 5, Section 5.9):

$$\frac{E}{V} = \frac{1}{24\pi^2} \left(\frac{m_e c^2}{\lambda_e^3} \right) G(x) = \frac{1}{24\pi^2} \left(\frac{m_e c^2}{\lambda_e^3} \right)(6x^4 - 8x^3 + 6x^2 \cdots),$$

$$n = \frac{1}{3\pi^2} \frac{1}{\lambda_e^3} x^3. \qquad (5.84)$$

This gives, for energy per particle,

$$\frac{E}{nV} = \frac{3}{4} m_e c^2 \left(x - \frac{4}{3} + \frac{1}{x} \cdots \right). \qquad (5.85)$$

5 White Dwarfs, Neutron Stars, and Black Holes

Correspondingly, the energy per unit mass will be

$$u = \frac{3}{4}\frac{m_e c^2}{\mu_e m_p}\left(x - \frac{4}{3} + \frac{1}{x} + \cdots\right). \tag{5.86}$$

The first term is the one that gives E_{int}; the second term is the negative of the total rest energy $-m_e c^2 N$ and can be dropped with the understanding that E now measures the total energy of the system, including the rest energy; so the first nontrivial correction term is the third one, which contributes an amount

$$\Delta E_{\text{int}} = \frac{3}{4}\frac{m_e c^2}{\mu_e m_p}\int\frac{1}{x}dm, \quad x = \left(\frac{3\pi^2 \rho \lambda_e^3}{\mu_e m_p}\right)^{1/3}. \tag{5.87}$$

This gives

$$\Delta E_{\text{int}} = k_3 \frac{m_e^2 c^3}{\hbar(\mu_e m_p)^{2/3}} M \rho_c^{-1/3}, \tag{5.88}$$

with

$$k_3 = \frac{3}{4}\frac{1}{(3\pi^2)^{1/3}}\frac{1}{|\xi_1^2 \theta'|}\int_0^{\xi_1} \xi^2 \theta^2 d\xi = 0.519723. \tag{5.89}$$

Putting together all the contributions to energy for an $n = 3$ polytrope, we get

$$E = (AM - BM^{5/3})\rho_c^{1/3} + CM\rho_c^{-1/3} - DM^{7/3}\rho_c^{2/3}, \tag{5.90}$$

with

$$A = k_1 K, \quad B = k_2 G, \quad C = k_3 \frac{m_e^2 c^3}{\hbar(\mu_e m_p)^{2/3}}, \quad D = k_4 \frac{G^2}{c^2}. \tag{5.91}$$

The equilibrium configuration is determined by the condition $(dE/d\rho_c = 0)$, leading to

$$(AM - BM^{5/3})\frac{1}{3}\rho_c^{-2/3} - \frac{1}{3}CM\rho_c^{-4/3} - \frac{2}{3}DM^{7/3}\rho_c^{-1/3} = 0. \tag{5.92}$$

To the lowest order, ignoring the terms proportional to C and D, we get

$$M = \left(\frac{A}{B}\right)^{3/2} = 1.457\left(\frac{\mu_e}{2}\right)^{-2} M_\odot, \tag{5.93}$$

which shows that equilibrium for relativistically degenerate Fermions is possible at *only* Chandrasekhar mass. The C, D terms provide small corrections to this expression. The stability limit is determined by $(d^2 E/d\rho_c^2) = 0$, which gives

$$-\frac{2}{9}(AM - BM^{5/3})\rho_c^{-5/3} + \frac{4}{9}CM\rho_c^{-7/3} + \frac{2}{9}DM^{7/3}\rho_c^{-4/3} = 0. \tag{5.94}$$

Substituting for $(AM - BM^{5/3})$ from Eq. (5.92), all the terms will be of same order and M can be replaced with $(A/B)^{3/2}$. Then we get the critical density as

$$\rho_c = \frac{CB^2}{DA^2} = \frac{16k_3k_2^2}{(3\pi^2)^{2/3}k_4k_1^2} \frac{m_p^2\mu_e^2}{\lambda_e^3 m_e} = 2.9 \times 10^{10}\left(\frac{\mu_e}{2}\right)^2 \text{gm cm}^{-3}. \quad (5.95)$$

Note that the existence of this density is a purely relativistic effect and will not occur if ΔE_{GTR} (proportional to D) is not present. However, the numerical value of the critical density is independent of G. For ^{56}Fe with $\mu_e = 2.154$, we have $\rho_c = 3.07 \times 10^{10}$ gm cm^{-3}, which is higher than the neutronisation threshold of 1.14×10^9 gm cm^{-3}, making this instability irrelevant; for helium and carbon, however, $\rho_c = 2.65 \times 10^{10}$ gm cm^{-3}, which is lower than the neutronisation thresholds of 1.37×10^{11} and 3.9×10^{10} gm cm^{-3}, respectively. Thus in helium–carbon white dwarfs, the general relativistic effects provide the limit on the stability.

The ratio between Newtonian and general relativistic contributions at this critical value is $\mathcal{R} \equiv (\Delta E_{\text{GTR}}/E_{\text{grav}}) = (k_4/k_2)(G/c^2)M^{2/3}\rho_c^{1/3} \approx 6.57 \times 10^{-3}$, and hence the perturbative treatment used above is valid.

The realistic effects described in Subsections 5.2.2–5.2.4 show that the limiting configuration for white dwarfs is set by different physical processes rather than by the Chandrasekhar mass described in Subsection 5.21 above. The latter limit is of only theoretical significance.

5.2.5 Effect of Rotation and Magnetic Field

The maximum mass found above for white dwarfs was for a spherically symmetric configuration without any magnetic field. Because the existence of the maximum mass is an important theoretical concept, it is necessary to investigate the effect of other physical phenomena (such as rotation or the magnetic field) on this result.

To provide an exact theory of a magnetic (or rotating) white dwarf, we will have to solve the full equations of magnetohydrodynamics (MHD) or hydrodynamics for a degenerate configuration. This is an extremely complex problem, but – fortunately – it is not necessary to adopt this procedure. The physical effects of magnetic fields and rotation can be easily determined from the virial theorem derived in Vol. I, Chap. 9, Section 9.6.

Let us consider the effect of the magnetic field first. The virial theorem in the presence of the magnetic field can be stated in the form

$$W + 3M\left\langle\frac{P}{\rho}\right\rangle + \left\langle\frac{B^2}{8\pi}\right\rangle\frac{4}{3}\pi R^3 = 0, \quad (5.96)$$

where W denotes the gravitational potential energy and the angle brackets denote averaging over the star. In the limit of high conductivity, the magnetic flux,

$\Phi_M \propto R^2 \langle B \rangle$, is conserved, allowing us to write $\langle B^2 \rangle \propto (\Phi_M^2/R^4)$. We will take the equation of state to be $P \propto \rho^{5/3}$ in the nonrelativistic limit and $P \propto \rho^{4/3}$ in the relativistic limit. In that case, the virial theorem for the two limits reduces to

$$0 = -\alpha_{3/2}\frac{GM^2}{R} + \beta_{3/2}\frac{M^{5/3}}{R^2} + \gamma_{3/2}\frac{\Phi_M^2}{R} \quad \text{(nonrelativistic)}, \quad (5.97)$$

$$0 = -\alpha_3 \frac{GM^2}{R} + \beta_3 \frac{M^{4/3}}{R^2} + \gamma_3 \frac{\Phi_M^2}{R} \quad \text{(extreme relativistic)}, \quad (5.98)$$

where the subscripts refer to the polytrope index $n = 3/2$ and $n = 3$ and the constants of proportionality are calculable, in principle, given the internal structure. It is clear that the net effect of the magnetic field is to reduce the effective value of G, thereby expanding the star at any given mass. In the nonrelativistic case, the equilibrium radius is given by

$$R = \frac{R_0}{1 - \mathcal{M}/|W|}, \quad \mathcal{M} \equiv \left\langle \frac{B^2}{8\pi} \right\rangle \frac{4}{3}\pi R^3, \quad (5.99)$$

where R_0 is the radius in the absence of the magnetic field. This effect is not very important for the nonrelativistic configuration.

More significant is the effect of the magnetic field in the relativistic limit. Here the mass changes by only a small amount,

$$M^{2/3} = \frac{\beta_3}{\alpha_3 G}\left(1 + \frac{\gamma_3 \Phi_M^2}{\beta_3 M^{4/3}}\right) \simeq \frac{\beta_3}{\alpha_3 G}\left(1 + \frac{\gamma_3 \Phi_M^2}{\alpha_3 GM^2}\right), \quad (5.100)$$

if $\delta \equiv (\mathcal{M}/|W|)$ is small. However, the radius of the configuration can change significantly even for small δ. Because $E/|W| = -(3-n)/3$, even a small change in E can lead to a large change in R when $n \approx 3$. More precisely, for a polytrope,

$$\frac{\Delta R}{R} = \frac{3}{3-n}\frac{\Delta \mathcal{M}}{|W|}, \quad (5.101)$$

which may be integrated at constant n to give

$$R = R_0 \exp\left(\frac{3}{3-n}\frac{\mathcal{M}}{|W|}\right). \quad (5.102)$$

Therefore the radius can increase significantly even with a small $(\mathcal{M}/|W|)$. Although this result is of some theoretical significance, we do not have evidence for strong magnetic fields (such that magnetic-energy density is comparable with gravitational potential energy) in white dwarfs.

Let us next consider the effect of rotation on the structure of white dwarfs. Taking the rotational kinetic energy as $T \propto M\Omega^2 R^2 \propto (J^2/MR^2)$, where J is the conserved angular momentum, the virial equation becomes

$$0 = -\alpha_{3/2}\frac{GM^2}{R} + \kappa_{3/2}\frac{J^2}{MR^2} + \beta_{3/2}\frac{M^{5/3}}{R^2} \quad \text{(nonrelativistic)}, \quad (5.103)$$

$$0 = -\alpha_3\frac{GM^2}{R} + \kappa_3\frac{J^2}{MR^2} + \beta_3\frac{M^{4/3}}{R} \quad \text{(extreme relativistic)}. \quad (5.104)$$

The nonrelativistic limit is again not of much significance because it provides only small corrections to the mass–radius relationship. The situation, however, is different in the relativistic case. In the absence of rotation ($J = 0$), Eq. (5.104) can be satisfied for only a specific value of M because the remaining terms have the same dependence on R. However, because the rotational term ($\propto R^{-2}$) has a steeper dependence on the radius than the other two terms, it is now possible to obtain an equilibrium model for *any* mass by decreasing the radius sufficiently. This is in sharp contrast to the nonrotating case of a relativistic degenerate gas, which led to the Chandrasekhar limit. In reality, however, very small radii, corresponding to very high densities, will be inadmissible because of the physical instabilities discussed in the subsections above. However, this argument shows that the mass bound is increased by rotation. Writing the virial equation in terms of the total kinetic energy and potential energy,

$$0 = -\alpha_3\frac{GM^2}{R}\left(1 - \frac{2T}{|W|}\right) + \beta_3\frac{M^{4/3}}{R}, \quad (5.105)$$

and solving for M, we get

$$M = \left[\frac{\beta_3}{\alpha_3 G(1 - 2T/|W|)}\right]^{3/2} = \frac{M_0}{(1 - 2T/|W|)^{3/2}}. \quad (5.106)$$

The bound on M now arises from the maximum value allowed for the ratio ($T/|W|$) for different rotational configurations. The actual bound depends on the detailed assumptions regarding the shape of the configuration, and some of these aspects have been discussed in Vol. I, Chap. 10. In general, a range of values around ($T/|W|$) ≈ 0.14–0.26 is obtained as maximum bounds in different cases. If this ratio is taken to be 0.2, then the maximum mass is increased by a factor of ~ 2. Thus the maximum mass of a rapidly rotating white dwarf can be nearly twice as large as the nonrotating one.

5.3 Surface Structure and Thermal Evolution of White Dwarfs

Because the white dwarfs actually have a nonzero temperature, they will have a finite luminosity that is due to the radiation of heat energy. To study the evolution of the white dwarf, it is necessary to model its cooling and the change of the temperature. The luminosity will again be determined by the nature of the physical processes taking place near the surface of the white dwarf. It should be noted, however, that the equation of state for degenerate matter will be inapplicable near the surface of a white dwarf (whatever may be the internal structure) because we expect $\rho \to 0$ near the surface. The surface layer of the white dwarf needs to be treated separately.

The internal density of a white dwarf decreases from some central value ρ_c at $r = 0$ to $\rho = 0$ at the surface, $r = R$. Let us assume that the equation of state ceases to be that of a degenerate gas at some intermediate point $r = r_0$; that is, for $r < r_0$, we shall assume that the equation of state is that of a degenerate gas whereas at $r > r_0$, the equation of state is that of an ideal gas. At $r = r_0$, the Fermi energy $\epsilon_F = (p_F^2/2m_e)$ of the system will be equal to the thermal energy $k_B T$. Writing $p_F \equiv (m_e c)x$ and $\epsilon_F = (1/2)m_e c^2 x^2$ and using Eq. (5.8) to express x in terms of ρ, we find that the condition $\epsilon_F = k_B T$ becomes

$$\rho = \frac{m_p \mu_e}{3\pi^2} \left(\frac{\hbar}{m_e c}\right)^{-3} \left(\frac{2k_B T}{m_e c^2}\right)^{3/2}. \tag{5.107}$$

This gives the relation between the density ρ_0 and temperature T_0 at the transition point, which numerically, in cgs units, is

$$\rho_0 \approx 6 \times 10^{-9} \mu_e T_0^{3/2}. \tag{5.108}$$

Further, we can easily show that the temperature is a constant (at $T = T_0$) for all $r < r_0$ because of the large thermal conductivity of the degenerate electron gas. We saw in Vol. I, Chap. 8, Section 8.14 that the thermal conductivity of a degenerate electron gas is quite high. The diffusion constant $\chi_{\text{th}} = (1/3)v\lambda$ [where $\lambda = (n_i \sigma)^{-1}$ is the mean free path and $\sigma \simeq (2Ze^2/m_e v^2)^2$] governing thermal conduction is $\chi_{\text{th}} \cong (1/12\pi)(m_e/Ze^2)^2(v^5/n_i)$. If $\rho \simeq 10^6$ gm cm^{-3} and $v \simeq v_F \simeq c$, then $\chi_{\text{th}} \simeq 200$ cm^2 s^{-1}; the corresponding diffusion time scale in the white dwarf interior is $t_{\text{cond}} \simeq (R^2/\chi) \simeq 10^6$ yr. This shows that thermal conduction can quickly equalise the temperature at $r < r_0$, so we may take $T = T_0$ for all $r < r_0$. It follows that it is the thin outer layer of nondegenerate matter that prevents efficient cooling of white dwarfs and determines the luminosity.

The scaling relation, relating the luminosity of the white dwarf to other parameters, can be obtained fairly easily. We saw in Chap. 2, Subsection 2.4.2 that in a radiative envelope $\rho \propto (M/L)^{1/2} T^{3.25}$. At the point of matching ($r = r_0$) between the degenerate core and the envelope, we must have $T = T_{\text{crit}} \propto \rho^{2/3}$; so $\rho_0 \propto T_0^{3/2}$. Combining the two relations, we get $(M/L)^{1/2} T_0^{3.25} \propto T_0^{3/2}$ or $L \propto MT_0^{7/2} \propto MT_{\text{core}}^{7/2}$, where the last proportionality follows from the fact that the core is isothermal. The characteristic cooling time for such a system is $t_{\text{cool}} \approx (U/L) \propto (MT_{\text{core}}/MT_{\text{core}}^{7/2}) \propto T_{\text{core}}^{-5/2}$. This leads to the scaling law $L \propto MT_{\text{core}}^{7/2} \propto Mt_{\text{cool}}^{-7/5}$. We shall now provide a rigorous derivation of this result.

If the region at $r > r_0$ is described by a radiative polytrope with zero boundary conditions and the opacity law $\kappa = \kappa_0 \rho^n T^{-s}$, then the pressure and the temperature of the envelope are related by $P = K'T^{1+n_{\text{eff}}}$, where $n_{\text{eff}} = (s+3)/(n+1)$

5.3 Surface Structure and Thermal Evolution of White Dwarfs

and

$$K' = \left[\frac{1}{1+n_{\text{eff}}} \frac{16\pi ac GM}{3\kappa_0 L} \left(\frac{N_A k_B}{\mu} \right)^n \right]^{1/(n+1)} \quad (5.109)$$

[see Chap. 2, Eq. (2.91)]. Equating the pressure from the ideal-gas law, $P_0 = (N_A k_B/\mu)\rho_0 T_0$, at $r = r_0$ to the pressure obtained from the envelope solution, we get

$$P_0 = K' T_0^{1+n_{\text{eff}}} = \rho_0 \frac{N_A k_B}{\mu} T_0 = 6 \times 10^{-9} \mu_e \frac{N_A k_B}{\mu} T_0^{5/2}, \quad (5.110)$$

where we have used Eq. (5.108) to express ρ_0 in terms of T. From the models for stellar evolution, we know that stellar remnants could be made of degenerate cores of carbon and oxygen with $\mu_e \approx 2$. In such a case, the standard form of bound–free opacity with $\kappa \approx 4 \times 10^{25} \rho T^{-3.5}$ cm^2 gm^{-1} will provide a reasonable description, giving $n = 1, s = 3.5$, and

$$K' \approx 8.1 \times 10^{-15} \mu^{-1/2} \left(\frac{M/M_\odot}{L/L_\odot} \right)^{1/2}. \quad (5.111)$$

Taking $\mu_e = 2$ and using the above relation in Eq. (5.110), we get

$$\frac{L}{L_\odot} \approx 6.6 \times 10^{-29} \mu \frac{M}{M_\odot} T_0^{7/2}, \quad (5.112)$$

or, equivalently,

$$T_0 = 4.6 \times 10^6 \text{ K} \left(\frac{L}{10^{-4} L_\odot} \right)^{2/7} \left(\frac{M}{0.6 M_\odot} \right)^{-2/7} \left(\frac{\mu}{12} \right)^{-2/7}, \quad (5.113)$$

which relates the core temperature to the mass and the luminosity. Note that the high conductivity of the core implies that $T = T_0$ for all $r < r_0$. If the luminosity is increased to $100 \, L_\odot$, the temperature rises to only 2.8×10^8 K because of the rather weak dependence. These high temperatures effectively rule out the presence of hydrogen or helium in the core.

To find the actual thickness of the envelope, we can use the temperature profile of the envelope obtained in Chap. 2, Eq. (2.95), namely,

$$T(r) = \frac{2.293 \times 10^7}{1+n_{\text{eff}}} \mu \left(\frac{M}{M_\odot} \right) \left(\frac{R}{R_\odot} \right)^{-1} \left(\frac{R}{r} - 1 \right) \text{ K}. \quad (5.114)$$

From this we get

$$\frac{R - r_0}{R} \cong 10^{-3} \left(\frac{\mu}{12} \right)^{-9/7} \left(\frac{M}{0.6 M_\odot} \right)^{-9/7} \left(\frac{L}{10^{-4} L_\odot} \right)^{2/7} \left(\frac{R}{10^{-2} R_\odot} \right), \quad (5.115)$$

showing that the envelope is really thin.

The nonzero luminosity of the white dwarf causes it to lose its thermal energy and cool over a period of time. If C_V is the specific heat per unit mass of the white dwarf, the rate of loss of thermal energy will be $(-C_V M \dot{T})$, which should match the luminosity determined above. Integrating this equation, we can find the temperature of the white dwarf $T(t)$ as a function of time. To do this we need to know the specific heat C_V of ions and degenerate electrons. The latter was determined in Vol. I, Chap. 5, and is given by (with $x = p_F/m_e c^2$)

$$C_V^{\text{el}} = \frac{\pi^2 k_B^2}{m_e c^2} \frac{Z}{A m_p} \frac{\sqrt{1+x^2}}{x^2} T$$

$$\approx \frac{\pi^2 k_B}{2} \frac{Z}{A m_p} \frac{k_B T}{\epsilon_F}, \quad \text{(for } x \ll 1\text{)}, \quad (5.116)$$

where A is the atomic weight of the ion. [The expression in Vol. I, Chap. 5, Section 5.9, Eq. (5.179) is the same as the one given above if the number of electrons per unit mass is taken to be $(Z/A m_p)$.] The specific heat of ions is more complicated. At high temperatures, ions will behave as an ideal gas and the specific heat per unit mass will be given by

$$C_V^{\text{ion}} = \frac{3}{2} \frac{k_B}{A m_p}, \quad (5.117)$$

so that the ratio between the two specific heats is $(C_V^{\text{el}}/C_V^{\text{ion}}) \approx (\pi^2 Z/3)(k_B T/\epsilon_F)$ for $(k_B T/\epsilon_F) \ll 1$. In this limit, the ion specific heat dominates and we can set $C_V = C_V^{\text{ion}}$. Combining the relations $L \propto \dot{T}$ and $L \propto T^{7/2}$, we find that $\dot{T} \propto T^{7/2}$ or, equivalently, $\dot{L} \propto -L^{12/7}$. Including all the numerical factors and integrating, we find that the luminosity evolution is

$$t = 6.3 \times 10^6 \text{ yr} \left(\frac{A}{12}\right)^{-1} \left(\frac{M}{M_\odot}\right)^{5/7} \left(\frac{\mu}{2}\right)^{-2/7} \left[\left(\frac{L}{L_\odot}\right)^{-5/7} - \left(\frac{L_0}{L_\odot}\right)^{-5/7}\right]. \quad (5.118)$$

This expression relates the time t to the luminosity $L(t)$ if the initial luminosity (at $t = 0$) is L_0. Ignoring the L_0 term at late times, we get the cooling time as

$$t_{\text{cool}} \approx 3.9 \times 10^6 \text{ yr} \left(\frac{A}{12}\right)^{-1} \left(\frac{M}{M_\odot}\right)^{5/7} \left(\frac{L}{L_\odot}\right)^{-5/7}. \quad (5.119)$$

If $N(L)$ denotes the number density of white dwarfs in the range of luminosities $(L, L + dL)$, then the rate of change of this number density is given by

$$\dot{N} = \dot{L}\frac{dN}{dL} = \left(\frac{\dot{L}}{L}\right)\left(\frac{dN}{d\ln L}\right) = \left[\frac{1}{t_{\text{cool}}(L)}\right]\left(\frac{dN}{d\ln L}\right) \propto \left(\frac{L}{M}\right)^{5/7}\left(\frac{dN}{d\ln L}\right). \quad (5.120)$$

5.3 Surface Structure and Thermal Evolution of White Dwarfs

If the white dwarfs are produced at an approximately constant rate, then we expect $\dot{N} \approx$ constant. Then

$$\frac{dN}{d\ln L} \propto \left(\frac{L}{M}\right)^{-5/7}. \tag{5.121}$$

Thus the luminosity function, giving the number density of white dwarfs per logarithmic interval of luminosity, is a power law with a slope of $(-5/7)$. We do expect this power law to be truncated at the extremes. At low luminosities and temperatures, crystallisation will modify the structure whereas at very high luminosities and temperatures, we must take into account the effects that are due to neutrino cooling. In the intermediate range, we expect the above power law to hold.

The coolest white dwarfs with $L \approx 10^{-4.5} L_\odot$ have a cooling time of $\sim 10^{10}$ yr, which is comparable with the age of the galaxy. Figure 5.4 shows the fitting curve for the observed luminosity function for the white dwarfs. The dramatic drop in the population for $L \simeq 10^{-4.5} L_\odot$ is obvious and is consistent with an age of $(9 \pm 1.8) \times 10^9$ yr. Adding the time spent in the pre-white-dwarf stage in the stellar evolution would suggest that star formation in the disk of the galaxy began $(9.3 \pm 2) \times 10^9$ yr ago. This is $\sim 6 \times 10^9$ yr shorter than the age of the globular clusters and has implications for the formation scenarios of the disk and globular clusters.[2]

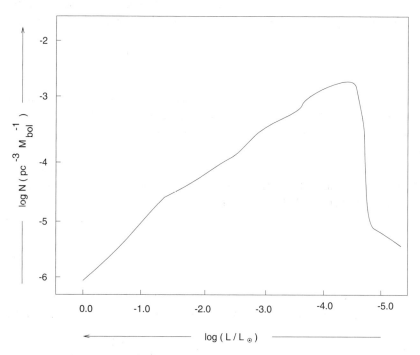

Fig. 5.4. Fitting curve for the observed luminosity function of white dwarfs.

Exercise 5.4

Numbers for white dwarfs: For a $1 M_\odot$ carbon white dwarf with a surface temperature of $T_{\text{eff}} = 10^4$ K, estimate the following: (a) escape velocity from the surface, (b) central pressure, (c) Fermi energy for the electron at the centre, (d) central temperature, (e) age of the white dwarf, (f) maximum possible frequency of rotation. [Answers: From the results of Table 5.1, we see that for $\mu_e \approx 2$, $M = M_\odot$, the radius is $R \cong 11.5 \mu_e^{-1}$ km $\simeq 5.8$ km and the central density is $\rho_c \simeq 3 \times 10^7$ gm cm^{-3}. (a) Given M and R, we get the escape velocity as $(2GM/R)^{1/2} \cong 0.02c$. (b) Given ρ_c, we get $x \approx 2.5$, giving $P_c \approx 4 \times 10^{24}$ dyn cm^{-2}. (c) $p_F = x m_e c$, which determines $\epsilon_F = (1+x^2)^{1/2} m_e c^2 \simeq 1.4$ MeV. (d) Given R and T_{eff}, we can determine $L = (4\pi R^2) \sigma T_{\text{eff}}^4$ and, through Eq. (5.113), the core temperature $T_{\text{core}} = T_0 \simeq 10^7$ K. (e) Approximation (5.119) gives the age $t \simeq 10^9$ yr. (f) $\Omega_{\max} \approx (GM/r^3)^{1/2} \approx 0.4$ Hz.]

Because $\dot{T} \propto (1/C_V)$, larger values of C_V lead to larger cooling times, t_{cool}. The expression for C_V used above needs to be corrected for different physical processes in order to estimate the cooling time more accurately. To begin with, for smaller M and larger T and Z, we cannot ignore the electronic contribution; for example, $C_V^{\text{elec}} \approx 0.25 C_V^{\text{ion}}$ for a carbon–oxygen mixture with $T = 10^7$ K and $M = 0.5 M_\odot$. We can incorporate this effect by taking the total specific heat of the system and numerically integrating the resulting equation.

Further, we must note that a plasma will cool, crystallise, and release the latent heat of crystallisation when the Coulomb interaction energy is comparable with the thermal energy. We define the ratio between these two by using a parameter

$$\Gamma_c \equiv \frac{(Ze)^2}{r_{\text{ion}} k_B T} = 2.7 \times 10^{-3} \frac{Z^2 n_{\text{ion}}^{1/3}}{T}. \qquad (5.122)$$

A cooling plasma of ions will crystallise when $\Gamma_c \approx 10^2$. When such a crystallisation occurs, the latent heat of ions will be released, thereby contributing additionally to the luminosity. The ideal-gas approximation used above is valid for only $\Gamma_c \lesssim 1$.

Once the ionic lattice is formed, we have to take into consideration the appearance of an additional degree of freedom in the form of ionic oscillations. The lattice vibrations of the solid will have a characteristic Debye temperature Θ (see Vol. I, Chap. 5, Exercise 5.15) related to the plasma frequency Ω_p by

$$k_B \Theta = \hbar \Omega_p, \quad \Omega_p = \frac{2e}{m_p} \left(\frac{\pi Z \rho}{A} \right)^{1/2}. \qquad (5.123)$$

(Because the system has a nonnegligible Coulomb interaction, it is the plasma frequency that provides the cutoff for normal modes of vibration.) Numerically,

$$\Theta = \frac{he}{k_B m_p \sqrt{\pi}} \frac{Z^{1/2}}{A} \rho^{1/2} \approx 7.8 \times 10^3 \text{ K} \left(\frac{Z}{A} \right)^{1/2} \rho^{1/2}, \qquad (5.124)$$

5.3 Surface Structure and Thermal Evolution of White Dwarfs

in cgs units. Hence $k_B\Theta$ is the characteristic energy of lattice oscillations that *cannot* be excited for $(T/\Theta) \lesssim 1$. At lower temperatures, ionic specific heat decreases as T^3 (see Vol. I, Chap. 5, Section 5.11). For typical white dwarfs composed of carbon and oxygen, the Debye temperature is lower than the crystallisation temperature. As the temperature is lowered still further, the specific heat decreases because fewer oscillation modes are excited, and it goes to zero as $C_V \propto T^3$ near $T = 0$.

Taking all these into account, we can roughly summarise the thermal energy of a white dwarf as follows: At any given temperature T, only the fraction k_BT/ϵ_F of all electrons can be thermally excited, with each carrying a typical energy of k_BT. Hence the thermal energy per electron is of the order of $(k_BT)^2/\epsilon_F$. The thermal energy per ion is approximately k_BT if $T \gg \Theta$ and is approximately $k_BT(T/\Theta)^3$ if $T \lesssim \Theta$. At high temperatures, the ratio of electronic energy to ionic energy is $[Z(k_BT)^2/\epsilon_F]/(k_BT) \approx Z(k_BT/\epsilon_F)$; therefore ions dominate at higher temperatures. The situation is different for $T \lesssim \Theta$. The electronic and the ionic contributions become comparable when $Z(k_BT/\epsilon_F)(\Theta/T)^3 \approx 1$. Hence at $(T/\Theta) < (kZ\Theta/\epsilon_F)^{1/2}$ the electronic energy dominates. For a $1M_\odot$ carbon white dwarf, $\Theta \approx 2 \times 10^7$ K, and this condition becomes $(T/\Theta) \lesssim 0.09$; that is, electronic energy dominates for $T \lesssim 2 \times 10^6$ K. The large variations in specific heat – an increase by a factor of 2 followed by a rapid drop to zero – influence the cooling times somewhat. These effects are to be taken into account by actual numerical calculations, as simple analytical models are not available to describe these processes.

Given the time dependence of luminosity and the fact that the radius $R \approx$ constant (being determined by M), we can follow the evolution of white dwarfs in the H-R diagram. Most of them lie in a narrow strip roughly parallel and below the main sequence with surface temperatures between 5000 and 80,000 K. Because any star with a helium core mass higher than 0.5 $M_\odot$ will undergo fusion, most of the white dwarfs consist of ionised carbon and oxygen nuclei. The masses are peaked at 0.56 $M_\odot$ with 80% of white dwarfs lying between 0.42 and 0.70 $M_\odot$.

Even though white dwarfs do not have a significant amount of hydrogen, they do contain small traces of it from earlier phases of stellar evolution. These hydrogen atoms, being lighter than the other ions, diffuse to the surface of the white dwarf and produce pressure-broadened hydrogen lines. Such white dwarfs are called type DA. They constitute the bulk of white dwarfs. Other conventions used in the spectral classification of white dwarfs are given in Table 5.3.

White dwarfs with surface temperatures of $\sim$12,000 K fall within the instability strip of the H-R diagram. Their pulsation period is between 10^2 and 10^3 s, and they are called ZZ Ceti variables or DAV stars. The estimate of p−mode frequencies in Chap. 3 shows that, for a mean density of 10^6 gm cm^{-3}, the period should be less than $\sim$4 s. Hence the observed pulsations cannot be p modes and

Table 5.3. White dwarf spectroscopic classification scheme

Spectral Type	Characteristics
DA	Balmer lines only: no HeI or metals present
DB	HeI lines: no H or metals present
DC	Continuous spectrum with no readily apparent lines
DO	HeII strong: HeI or H present
DZ	Metal lines only: no H or He
DQ	Carbon features of any kind

must arise from the g modes. The latter depends on the Brunt–Vaisala frequency N^2, which needs to be determined numerically for a given structure of the white dwarf. We recall from Chap. 3 that the Brunt–Vaisala frequency can be expressed in the form

$$N^2 = g^2 \left[\left.\frac{d\rho}{dp}\right|_* - \left.\frac{\partial\rho}{\partial p}\right|_s \right], \qquad (5.125)$$

where the first derivative is for the unperturbed star and the second is the thermodynamic derivative at constant entropy. For a strictly zero-temperature (degenerate) Fermi gas, it is easy to see (from the definition) that $N^2 = 0$. This follows from the fact that, in a degenerate $T=0$ electron gas, each electron has only one possible state and hence the configuration is isoentropic. Therefore $(dp/d\rho)_* = (dp/d\rho)_s$. To evaluate N^2, we therefore need the form of $P(\rho)$ for a finite-temperature Fermi gas. When $k_B T \ll \epsilon_F$, the electron distribution function is modified in a range only of the order of $k_B T$ around ϵ_F. In other words, a fraction $k_B T/\epsilon_F$ of electrons will increase their energy by a factor of $k_B T/\epsilon_F$. Because the energy density will be proportional to pressure in both the nonrelativistic and the relativistic limits, it follows that the finite-temperature corrections to the pressure are of the form

$$P(T) - P(0) \approx P(0)\, a \left(\frac{k_B T}{\epsilon_F}\right)^2, \qquad (5.126)$$

where a is a coefficient of the order of unity. A similar correction exists for density as well and hence we can write, for a finite-temperature Fermi gas,

$$P(T) = P(0)\left[1 + a\left(\frac{k_B T}{\epsilon_F}\right)^2\right], \quad \rho(T) = \rho(0)\left[1 + b\left(\frac{k_B T}{\epsilon_F}\right)^2\right], \qquad (5.127)$$

where $b = \mathcal{O}(1)$ is another numerical factor. Evaluating $(d\rho/dP)_* = (d\rho/dr)/$

(dP/dr), we get

$$\left.\frac{d\rho}{dP}\right|_* = \left(\frac{dP}{dr}\right)^{-1} \left\{\frac{d\rho_0}{dr}\left[1 + b\left(\frac{k_BT}{\epsilon_F}\right)^2\right] + \rho_0 b \frac{d}{dr}\left(\frac{k_BT}{\epsilon_F}\right)^2\right\}$$

$$= \left.\frac{d\rho_0}{dP}\right|_*\left[1 + b\left(\frac{k_BT}{\epsilon_F}\right)^2\right] + \rho_0 b \left(\frac{dP}{dr}\right)^{-1}\frac{d}{dr}\left(\frac{k_BT}{\epsilon_F}\right)^2. \quad (5.128)$$

The entropy of the Fermi gas at a finite temperature is $(s/Nk_B) = (\pi^2/2)(k_BT/\epsilon_F)$; therefore derivatives at constant s must be taken at constant values of k_BT/ϵ_F. Hence

$$\left.\frac{\partial \rho}{\partial P}\right|_s = \left(\frac{\partial P}{\partial r}\right)^{-1}\left.\frac{\partial \rho}{\partial r}\right|_s = \left.\frac{d\rho_0}{dP}\right|_s\left[1 + b\left(\frac{k_BT}{\epsilon_F}\right)^2\right]. \quad (5.129)$$

A simple calculation now gives the Brunt–Vaisala frequency as

$$N^2 \approx -gb\frac{\partial}{\partial r}\left(\frac{k_BT}{\epsilon_F}\right)^2. \quad (5.130)$$

In general, $b < 0$ and (k_BT/ϵ_F) increases outwards, making $N^2 > 0$.

The contribution to N^2 can arise from the ionic pressure and from the deviation of electron gas from complete degeneracy. The ionic contribution to the pressure is proportional to k_BT, and hence we will now have

$$P(T) = P(0)\left[1 + a\left(\frac{k_BT}{\epsilon_F}\right)\right]. \quad (5.131)$$

Repeating the analysis, we will now find that

$$N^2 \approx -gb\frac{\partial}{\partial r}\left(\frac{k_BT}{\epsilon_F}\right). \quad (5.132)$$

Because the outer regions are less degenerate compared with the interior, k_BT/ϵ_F increases, with r suggesting that $N^2 > 0$ in the outer regions, and vanishes inside the core. The Lamb frequency S_l, on the other hand, will be large in the interior but becomes small in the envelope. From the discussion of conditions for wave propagation in Chap. 3, Section 3.7, it is clear that g modes propagate in the envelope region whereas p modes (if they exist) will propagate in the deep interior. It may be noted that this behaviour is opposite to that of normal stars like the Sun.

5.4 Equation of State at Higher Densities

A different class of remnants is formed in the case of stars with masses $M \gtrsim 8\,M_\odot$ when the star is likely to undergo a supernova explosion, leaving behind a core. This core is likely to be made of heavier elements such as iron, with core densities of $\rho \gtrsim 10^{13}$ gm cm^{-3} and initial temperatures of $T \gtrsim 10^{10}$ K. In the initial stages, this remnant will cool rapidly through several processes involving neutrinos, which were described in Chap. 4, Section 4.3. The resultant structure is determined by the equation of state for matter at high densities. We shall now consider the physical processes that determine the equation of state of matter at high densities.

We have seen before in Section 5.2 that, for densities well below 10^6 gm cm^{-3}, the system may be described by a solid lattice of ions with a nonrelativistic degenerate gas of electrons. At a density of $\sim 10^6$ gm cm^{-3}, electrons become relativistic and the system can again be described as consisting of a lattice of ions and a relativistic degenerate pressure of electrons. When the density increases still further, the Fermi energy of electrons will become high enough to induce neutronisation through the inverse beta-decay reaction. Because such a process involves transformation of the nuclei, we must find the equation of state by first determining the minimum-energy configuration for the system with specific conservation laws. In other words, we have to find the lowest energy state of a system with $A \approx 10^{57}$ baryons possibly made of separate nuclei in beta equilibrium with a relativistic electron gas.

From standard nuclear physics, we know that, for systems with $A \lesssim 90$, the lowest energy state consists of the iron nucleus that has the tightest binding. For $A \gtrsim 90$, the lowest energy state will again be made of several such iron nuclei. This situation, however, gets modified when electrons are capable of combining with protons in the nuclei. To see this, we must recall that the tight binding for $A = 56$ arises because the balance between Coloumb repulsion of protons and the attractive nuclear forces. When electrons combine with protons to form neutrons, the Coloumb repulsion of the nuclei decreases and it is possible for larger nuclei (with $A \gg 56$) to be preferred energetically. We shall see below that the critical ratio for n/p is reached when the density is approximately 4×10^{11} gm cm^{-3}. Any further increase in density leads to a two-phase system in which electrons, nuclei, and *free neutrons* coexist. When the density grows above this critical value, we are led to higher (n/p) ratios and more and more free neutrons. Finally, when the density exceeds approximately 4×10^{12} gm cm^{-3}, the free neutron gas provides most of the pressure rather than electrons.

The energy density of a mixture of nuclei, free electrons, and free neutrons can be written in the form

$$\epsilon = n_N M(A, Z) + \epsilon'_e(n_e) + \epsilon_n(n_n), \qquad (5.133)$$

where the subscript N refers to nucleons, e to electrons, n to neutrons, etc., and

5.4 Equation of State at Higher Densities

$M(A, Z)$ is the energy of the nucleus characterised by A, Z:

$$M(Z, A) = [(A - Z)m_n c^2 + Z(m_p + m_e)c^2 - A\bar{E}_b]$$
$$= m_N c^2 \left[b_1 A + b_2 A^{2/3} - b_3 Z + b_4 A \left(\frac{1}{2} - \frac{Z}{A} \right)^2 + \frac{b_5 Z^2}{A^{1/3}} \right],$$
(5.134)

with

$$b_1 = 0.991749, \quad b_2 = 0.01911, \quad b_3 = 0.000840, \quad b_4 = 0.10175,$$
$$b_5 = 0.000763 \quad (5.135)$$

(see Vol. I, Chap. 12). It is conventional in nuclear physics to include rest energy of nucleons and electrons in $M(A, Z)$, and hence ϵ'_e does not have the rest energy. Denoting the conserved baryon number density by $n = n_N A + n_n$ and the neutron fraction as $Y_n \equiv n_n/n$, we can reexpress the energy as

$$\epsilon = n(1 - Y_n) \frac{M(A, Z)}{A} + \epsilon'_e(n_e) + \epsilon_n(n_n), \quad n_e = n(1 - Y) \frac{Z}{A}, \quad n_n = n Y_n.$$
(5.136)

We have to minimise the above energy $\epsilon(n, A, Z, Y_n)$ with respect to A, Z, and Y_n at constant n to find the equilibrium configuration. This is straightforward and leads to the following equations:

$$\frac{\partial M}{\partial Z} = -(\epsilon_{F,e} - m_e c^2), \quad A^2 \frac{\partial}{\partial A} \left(\frac{M}{A} \right) = Z(\epsilon_{F,e} - m_e c^2), \quad \frac{\partial M}{\partial A} = \epsilon_{F,e}$$
(5.137)

or

$$b_3 + b_4 \left(1 - \frac{2Z}{A} \right) - 2b_5 \frac{Z}{A^{1/3}} = \left[(1 + x_e^2)^{1/2} - 1 \right] \frac{m_e}{m_N}, \quad (5.138)$$

$$Z = \left(\frac{b_2}{2b_5} \right)^{1/2} A^{1/2} = 3.54 \, A^{1/2}, \quad (5.139)$$

$$b_1 + \frac{2b_2 A^{-1/3}}{3} + b_4 \left(\frac{1}{4} - \frac{Z^2}{A^2} \right) - \frac{b_5 Z^2}{3 A^{4/3}} = (1 + x_n^2)^{1/2} \frac{m_n}{m_N}, \quad (5.140)$$

where $x = (p_F/mc)$ for each species. Note that Z increases with A whereas the ratio Z/A decreases as $A^{-1/2}$ at high densities.

These equations can be used to determine the pressure–density relation (i.e., the equation of state) along the following lines: (1) For a given value of $A > 56$, Eq. (5.139) gives a value of Z. (2) for this value, we should check whether x_n^2 is greater than zero or not; if $x_n > 0$, then free neutrons exist in the system (neutron drip has occurred) and their energy density and pressure can be computed by standard formulas for degenerate neutron gas. If $x_n^2 < 0$, free neutrons are not present and the energy and pressure of free neutrons can be set to zero. (3) Equation (5.138) gives x_e and hence we can compute ϵ'_e, P_e, n_e. Given these, the equation of state, $P(\rho)$ is implicitly given by

$$\rho = \frac{\epsilon}{c^2} = \frac{n_e M(A,Z)/Z + \epsilon'_e + \epsilon_n}{c^2}, \quad P = P_e + P_n, \quad n = n_e \frac{A}{Z} + n_n. \tag{5.141}$$

It is clear that a similar analysis can be performed for any given form of energy $\epsilon(n, A, Z, Y_n)$ arising from some theory for nuclear interactions. Such models have been constructed with different input physics regarding the nuclear interactions, leading to different forms of equations of state. The broad features of such analysis is plotted as a P–ρ relation in Fig. 5.5.

This figure shows that the deviations from an ideal degenerate electron gas is apparent for $\rho > 10^7$ gm cm^{-3}. The neutron drip occurs around $\rho_{\text{drip}} \approx 10^{12}$

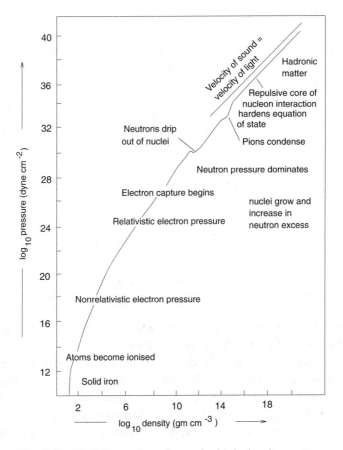

Fig. 5.5. Model equation of state for high-density matter.

gm cm^{-3} when the maximum corresponds to $A = 122$, $Z = 39$, and $E_{F_e} \approx$ 23.6 MeV. Above this density, free neutrons contribute more and more of the pressure. Above the density of 4.54×10^{12} gm cm^{-3} neutrons provide most of the pressure and the equation of state can be approximated by that of a cold gas of electrons, protons, and neutrons in beta equilibrium. This description is valid for $\rho_{\text{drip}} \lesssim \rho \lesssim \rho_{\text{nuc}}$, where $\rho_{\text{nuc}} \approx 2.8 \times 10^{14}$ gm cm^{-3} is the nuclear density. For $\rho \gtrsim \rho_{\text{nuc}}$, individual nuclei begin to interact strongly and dissolve together; the equation of state above these densities is uncertain and reliable conclusions are difficult to arrive at. Some of the possibilities will be mentioned at the end of the next section.

5.5 Neutron Star Models

The simplest model for a neutron star will be based on an ideal, degenerate equation of state for nonrelativistic neutrons and Newtonian gravity. This is identical to that of an $n = 3/2$ polytrope, and we have the standard $MR^3 =$ constant relation for neutrons. Numerically,

$$M = 1.102 \left(\frac{\rho_c}{10^{15} \text{ gm cm}^{-3}}\right)^{1/2} M_\odot = \left(\frac{15.12 \text{ km}}{R}\right)^3 M_\odot, \quad (5.142)$$

$$R = 14.64 \left(\frac{\rho_c}{10^{15} \text{ gm cm}^{-3}}\right)^{-1/6} \text{ km}. \quad (5.143)$$

Such a description is valid only when neutrons are stable against beta decay and contribute most of the pressure. From the discussion of the equation of state in the last section, we know that this is possible only for $\rho \gtrsim \rho_{\min} \equiv 4.5 \times 10^{12}$ gm cm^{-3}. From the above formula we get $R \approx 36$ km and $M \approx 0.1 \, M_\odot$ for $\rho = \rho_{\min}$. These values are not quite accurate because the equation of state at this density is *not* well approximated by an ideal neutron gas. Nevertheless, it illustrates the important fact that there is a minimum central density above which stable neutron stars can exist. The maximum central density permitted for a white dwarf, it may be recalled, is $\sim 10^{10}$ gm cm^{-3}. In the gap 10^{10} gm cm$^{-3} \lesssim \rho \lesssim 10^{12}$ gm cm^{-3} no stable compact objects can exist. It should also be noted that the minimum mass for a neutron star (corresponding to $\rho_c = \rho_{\min}$) is lower than the maximum stable mass for the white dwarf. Hence there exist completely different stable structures for the in-between mass range. For example, we can have a white dwarf *or* a neutron star with a total mass of $0.4 \, M_\odot$, depending on the internal composition. The neutron star, of course, will be significantly denser and more compact than a white dwarf.

To avoid misunderstanding, we stress the following fact: White dwarfs originate as remnants of stars of a certain mass range in which the nuclear reactions have not proceeded to the maximum possible limit. For example, most of the white dwarfs are made of carbon and helium in which further nuclear reactions have not taken place. In this sense, they do not constitute the lowest possible energy state for an object with a certain number of baryons. They are, however, quite stable even in cosmological time scales, as there is no possibility of further nuclear reactions taking place in the low-density isolated white dwarfs. Neutron stars, on the other hand, arise as the stellar remnants in which nuclear burning could have produced the ^{56}Fe nuclei in the core, which represent the lowest energy state at moderate densities. In computing the structure of the neutron star at higher densities, we will use the equation of state that we compute by taking into account explicitly the requirement of lowest energy (as in the last section). Because of this reason, neutron stars can be thought of as true minimum-energy configurations. It is therefore clear that no stable structures can exist in the intermediate regime that corresponds, naively, to the transition from the white dwarf to the neutron star structure.

For compact neutron stars, GM/Rc^2 is not significantly small compared with unity and we need to take into account the general relativistic effects. Given the equation of state $P(\rho)$, the structure of the star and the M–R relation must be determined by integration of the Oppenheimer–Volkoff equations of general relativity (see Vol. I, Chap. 11, Section 11.8). Such a numerical integration, which uses the equation of state like those described in the last section, leads to the M–R relation, with the minimum and maximum masses for the neutron stars given by

$$M_{\min} \simeq 0.18\, M_\odot, \quad R \simeq 300 \text{ km}, \quad \rho_c \simeq 2.6 \times 10^{13} \text{ gm cm}^{-3},$$
(5.144)

$$M_{\max} \simeq 0.72\, M_\odot, \quad R \simeq 8.8 \text{ km}, \quad \rho_c \simeq 5.8 \times 10^{15} \text{ gm cm}^{-3}.$$
(5.145)

The existence of the maximum mass can be understood from the general relativistic stability considerations, as in the case of white dwarfs. Consider now an $n = 3/2$ polytrope made of nonrelativistic neutrons. The contributions to the total energy from internal energy, Newtonian gravitational energy, and general relativistic corrections for this polytrope are

$$E_{\text{int}} = k_1 K \rho_c^{2/3} M, \quad k_1 = 0.795873,$$
(5.146)

$$E_{\text{grav}} = -k_2 G \rho_c^{1/3} M^{5/3}, \quad k_2 = 0.760777,$$
(5.147)

$$\Delta E_{\text{GTR}} = -k_4 \frac{G^2}{c^2} M^{7/3} \rho_c^{2/3}; \quad k_4 = 0.6807.$$
(5.148)

To compute ΔE_{int} we now use the nonrelativistic expansion [obtained in a manner similar to obtaining Eq. (5.86)]

$$u = c^2 \left(\frac{3}{10} x^2 - \frac{3}{56} x^4 \right).$$
(5.149)

5.5 Neutron Star Models

The first term gives E_{int}, and the second term gives the correction

$$\Delta E_{\text{int}} = -\frac{3}{56}c^2 \int x^4 \, dm. \qquad (5.150)$$

Using the relation

$$\rho_0 = m_n n_n = \frac{m_n x^3}{3\pi^2 \lambda_n^3} \qquad (5.151)$$

we can evaluate this as

$$\Delta E_{\text{int}} = -k_3 \frac{\hbar^4}{m_n^{16/3} c^2} M \rho_c^{4/3}, \qquad (5.152)$$

where

$$k_3 = \frac{3}{56}(3\pi^2)^{4/3} \frac{1}{|\xi_1^2 \theta'(\xi_1)|} \int_0^{\xi_1} \theta^{3.5} \xi^2 d\xi = 1.1651. \qquad (5.153)$$

Adding all the contributions together, we obtain the total energy as

$$E = AM\rho_c^{2/3} - BM^{5/3}\rho_c^{1/3} - CM\rho_c^{4/3} - DM^{7/3}\rho_c^{2/3}, \qquad (5.154)$$

with

$$A = k_1 K, \quad B = k_2 G, \quad C = \frac{k_3 \hbar^4}{m_n^{16/3} c^2}, \quad D = k_4 \frac{G^2}{c^2}. \qquad (5.155)$$

The condition for equilibrium $(\partial E / \partial \rho_c) = 0$ gives

$$2A\rho_c^{-1/3} - BM^{2/3}\rho_c^{-2/3} - 4C\rho_c^{1/3} - 2DM^{4/3}\rho_c^{-1/3} = 0. \qquad (5.156)$$

The first two terms give the standard result for the $n = 3/2$ polytrope, but keeping all the terms leads to a better M–ρ_c relation. The stability condition $(\partial^2 E / \partial \rho_c^2) = 0$ gives

$$-2A\rho_c^{-1/3} + 2BM^{2/3}\rho_c^{-2/3} - 4C\rho_c^{1/3} + 2DM^{4/3}\rho_c^{-1/3} = 0. \qquad (5.157)$$

Changing to the variable $y = M^{4/9}$, we can easily manipulate these two equations to give the cubic

$$2A - 3B^{2/3}C^{1/3}y - 2Dy^3 = 0, \quad \rho_c = \frac{BM^{2/3}}{8C}. \qquad (5.158)$$

The positive root is at $y = 6.605 \times 10^{14}$ in cgs units corresponding to

$$M = 1.11 \, M_\odot, \quad \rho_c = 7.43 \times 10^{15} \, \text{gm cm}^{-3}. \qquad (5.159)$$

For these values the total energy is given by $E = -0.08 \, M_\odot c^2$. The total mass of the system is $M_{\text{tot}} = M - E/c^2 = 1.03 \, M_\odot$. This maximum mass and the corresponding maximum density are comparable with the numerical estimate quoted above.

The ratio between Newtonian and general relativistic contributions at this critical value is $\mathcal{R} \equiv (E_{\text{GTR}}/E_{\text{grav}}) = (k_4/k_2)(G/c^2)M^{2/3}\rho_c^{1/3} \approx 0.12$, and hence we do not expect more than a few tens of percent accuracy in the above estimate. In fact, if the same calculation is repeated with an $n = 3$ polytrope index for system, the corresponding ratio is $\mathcal{R} \approx 0.61$ and the total mass M_{tot} turns out to be negative. This clearly shows that the perturbative treatment is of only limited validity and has to be used with caution.

5.6 Mass Bounds for Neutron Stars

It is clear from our discussion that neutron stars are the most compact configurations of matter that can possibly withstand gravitational force. Hence the maximum mass of a neutron star provides a bound on the stable, lowest-energy configuration of matter in its most compact form. For this reason, it is important to estimate, as accurately as possible, the value of this mass.

Given the equation of state of matter, the maximum mass can be estimated, in principle, in a straightforward manner. The integration of the Oppenheimer–Volkoff equation with a specific equation of state will lead to an $M(\rho_c)$ curve (like the one shown in Fig. 5.6) from which the maximum mass can be determined. Unfortunately, the equation of state for matter at supernuclear densities is highly

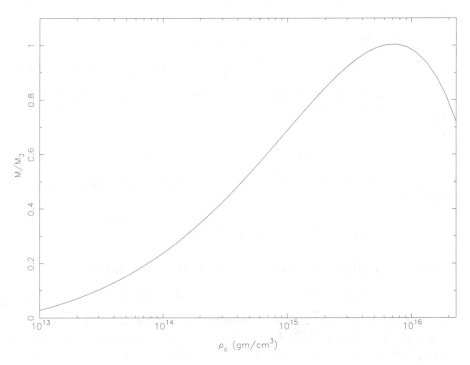

Fig. 5.6. M–R relation, based on a simplified model, for neutron stars.

5.6 Mass Bounds for Neutron Stars

uncertain. Hence the question of maximum mass has to be addressed in an indirect manner, as follows.

Let us assume that the equation of state is known to be given by some specific function $P(\rho)$ up to some density $\rho = \rho_0$. For $\rho > \rho_0$, we would like to continue this equation of state in such a manner as to provide maximum possible pressure at a given density. Because we expect the velocity of sound to be less than or equal to the velocity of light, any equation of state must obey the constraint $(dp/d\rho) \leq c^2$. Hence maximum pressure can be provided at a given density $\rho > \rho_0$ if the equation of state is taken to be

$$P = P_0 + (\rho - \rho_c)c^2, \quad \rho \geq \rho_0. \tag{5.160}$$

Given any reasonable equation of state for $\rho < \rho_0$ and the above functional form $\rho > \rho_0$, we can integrate the Oppenheimer–Volkoff equation to obtain the maximum mass. Such an analysis, with the equation of state developed in Section 5.4 with $\rho_0 = 4.6 \times 10^{14}$ gm cm^{-3}, gives the maximum mass as $\sim 3.2\ M_\odot$. In fact, the maximum mass scales with the chosen value of ρ_0 approximately as

$$M_{\max} \simeq 3.2 \left(\frac{\rho_0}{4.6 \times 10^{14} \text{ gm cm}^{-3}} \right)^{-1/2} M_\odot. \tag{5.161}$$

We shall now provide a heuristic derivation of the above result, using a constant-density solution to Oppenheimer–Volkoff equation.

Consider matter with an equation of state defined by $\rho(n)$, where ρ is the energy density and n is the baryon number density. The pressure for such a relativistic system is determined by

$$\frac{d\rho}{dn} = \frac{\rho + (P/c^2)}{n}. \tag{5.162}$$

These two relations implicitly determine the functional form $P(\rho)$. Let us now construct a constant-density star of radius R, mass M, and total baryon number A by using the material that satisfies the above equation of state. In units with $c = G = 1$, the total mass and the baryon number of such a configuration are given by

$$M = 4\pi \int_0^R \rho r^2 dr = \frac{4}{3}\pi R^3 \rho, \tag{5.163}$$

$$A = 4\pi \int_0^R \frac{nr^2\, dr}{[1 - 2m(r)/r]^{1/2}} = 2\pi n \left(\frac{3}{8\pi\rho} \right)^{3/2} (\chi - \sin\chi \cos\chi), \tag{5.164}$$

where we have defined an angle χ by the relation

$$\sin\chi = \left(\frac{8\pi\rho}{3} \right)^{1/2} R \tag{5.165}$$

so that $\sin^2 \chi = (2M/R)$. The equilibrium configuration for such a system must correspond to the minimum energy $E = M$ for a fixed value of A. With χ treated as an independent variable, this condition reduces to $(\partial M/\partial \chi)_A = 0$. With the above expressions, a straightforward calculation reduces this condition to the form $(P/\rho) = \zeta(\chi)$, where

$$\zeta(\chi) \equiv \frac{6 \cos \chi}{9 \cos \chi - 2 \sin^3 \chi/(\chi - \sin \chi \cos \chi)} - 1. \quad (5.166)$$

Given an equation of state $\rho(n)$, P is determined by Eq. (5.162); the value of χ is determined through the condition $P/\rho = \zeta(\chi)$; R, M, and A are determined through relations (5.163)–(5.165). These relations completely determine our model.

The stability condition, given by $(\partial^2 M/\partial \chi^2)_A > 0$, can be reduced, after some straightforward but tedious algebra, to the form $\Gamma > \Gamma_c(\chi)$, where

$$\Gamma \equiv \frac{\partial \ln P}{\partial \ln n} = \left(1 + \frac{\rho}{P}\right) \frac{dP}{d\rho}, \quad (5.167)$$

and

$$\Gamma_c(\chi) = (\zeta + 1) \left\{ 1 + \frac{(3\zeta + 1)}{2} \left[\frac{(\zeta + 1)}{6\zeta} \tan^2 \chi - 1 \right] \right\}. \quad (5.168)$$

The condition on the pressure

$$P \leq P_0 + v^2(\rho - \rho_0) \quad \text{(for } \rho \geq \rho_0\text{)} \quad (5.169)$$

reaches the critical limit of equality when P/ρ reaches a critical value ζ_c corresponding to $\Gamma = \Gamma_c(\chi_c)$; that is, we need

$$\frac{\zeta_c + 1}{\zeta_c} v^2 = \Gamma_c. \quad (5.170)$$

The condition on the pressure now becomes

$$\rho_c \geq \frac{\rho_0 - (P_0/v^2)}{1 - (\zeta + 1)/\Gamma_c} \quad (5.171)$$

or, equivalently,

$$M \leq \frac{1}{2} \left(\frac{3}{8\pi}\right)^{1/2} \left(\frac{1 - (\zeta + 1)/\Gamma_c}{\rho_0 - (P_0/v^2)}\right)^{1/2} \sin^3 \chi_c. \quad (5.172)$$

We can now use any reliable equation of state to set a bound on $[\rho_0 - (P_0/v^2)]$. Taking $\rho = 5 \times 10^{14}$ gm cm^{-3} and the equation of state discussed in Section 5.4 gives the maximum mass for $v = 1$ as $\sim 3.6 \, M_\odot$.

It is actually possible to obtain a closely related bound along the following lines, even without using the condition of causality. Note that when $v \to \infty$, so does Γ_c. From Eq. (168), we see that $\Gamma_c \propto \zeta^2$ for large values of ζ and $\chi \to 0.4\pi$ as $\zeta \to \infty$. This gives a maximum mass of

$$M_{\max} \simeq \left(\frac{3}{8\pi}\right)^{1/2} \frac{1}{2}\left(\frac{1}{\rho_0}\right)^{1/2} \sin^3 \chi_c = 6.05 \left(\frac{4.6 \times 10^{14} \text{ gm cm}^{-3}}{\rho_0}\right)^{1/2} M_\odot. \tag{5.173}$$

These results suggest that there exists a rigorous upper bound for the mass of the neutron star that is in the range (3–5) $M_\odot$. Its precise value depends on the nature of the assumptions that we are prepared to make.

In the case of white dwarfs, we saw that rotation can increase the maximum mass by even as much as a factor of 2. The corresponding analysis in general relativity is considerably more difficult as rapidly rotating configurations and their stability criteria are not known. Virial theorem analysis can, of course, be repeated, taking into account the general relativistic correction as a perturbation. In that case, it is straightforward to show that the maximum mass increases by only ~20%.

The above analysis shows that there exists a bound on the maximum mass that can be supported against gravity by a neutron star structure. If the actual mass of the star is larger than this bound, then the system will continue to contract under self-gravity and will eventually form a black hole. This is a purely general relativistic phenomenon, which we shall study in Sections 5.8 and 5.9.

5.7 Internal Structure of Neutron Stars

In the discussion so far, we have treated the neutron star as a homogeneous entity with a given total mass and radius. Because the density of the neutron star varies with radial distance, different physical phenomena occur at different layers, giving a distinct internal structure to the neutron star. The details of this structure, of course, depend on the equation of state, which, unfortunately, is known only rather poorly. Nevertheless, some general comments can be made regarding the different layers.

5.7.1 Layers Inside a Neutron Star

Right at the top of the neutron star lies the surface layer that has densities of less than ~10^6 gm cm^{-3}. At these densities, matter will be made of the most stable nuclei such as ^{56}Fe, and will form a close-packed solid. The surface magnetic field influences the structure of the atoms significantly (We shall say more about this in Chap. 6, Section 6.4), and the solid will have high conductivity parallel to the magnetic field and zero conductivity transverse to it.

In the next layer, containing matter with a density of 10^6 gm cm$^{-3} \lesssim \rho \lesssim 4.3 \times 10^{11}$ gm cm^{-3}, the structure is similar to that of a white dwarf. This is a solid region, with the heavy nuclei forming a Coloumb lattice embedded in a relativistic degenerate gas of electrons. In this density range, the energy of the electron can be high enough to induce inverse beta decay, producing ^{62}Ni (at 3×10^8 gm cm^{-3}), ^{80}Zn (at 5×10^{10} gm cm^{-3}), ^{118}Kr (at 4×10^{11} gm cm^{-3}), etc. This region is usually called the *outer crust*.

For densities ranging between 4.3×10^{11} gm cm^{-3} and 2×10^{14} gm cm^{-3}, the lattice begins to give way to neutron-rich nuclei, free degenerate neutrons, and degenerate relativistic electron gas. (This layer is called the *inner crust*). As the density increases, nuclei with specific identity begin to dissolve and the neutron fluid provides most of the pressure. At densities greater than approximately 2×10^{14} gm cm^{-3}, the matter is in the form of neutron liquid with a very small concentration of protons and electrons.

The situation is unclear about the *core region* with densities higher than 3×10^{15} gm cm^{-3}. The existence of this phase is not generic but depends on the equation of state for bulk matter at high energies and densities. Many models for stable neutron stars do not possess such a core region, but, given the uncertainties of nuclear many-body theories, it is not possible to rule out phases involving more exotic phases such as quark matter or pion condensates.

There is one more complication that needs to be kept in mind in determining the internal structure of neutron stars. A collection of fermions (like protons or neutrons) can, under certain circumstances, become a superconductor or a superfluid. In a superconducting state, charged fermions will exhibit zero resistance and certain other peculiar properties; a superfluid will similarly exhibit zero viscosity for its flow. We shall now describe briefly the salient features of superfluidity that may be of relevance in understanding neutron stars.

5.7.2 Superfluidity in Neutron Stars

The most dramatic feature of a superfluid is its ability to flow without any viscosity. We shall first provide a simple criterion that must be satisfied by a system if dissipitative processes are not to be effective; then we shall discuss a microscopic model in which such a criterion is realised.

Let us consider a fluid moving along a tube with constant velocity **v**. The friction against the wall of the tube as well as the internal friction will cause the kinetic energy of the fluid to be dissipated over a period of time, with the flow gradually becoming slower. It is more convenient to discuss this flow in the frame in which the fluid is at rest, but the tube is moving with a velocity $-\mathbf{v}$. The effect of viscosity now will be to make the fluid begin to move, eventually achieving the same speed as the tube; when this happens the flow has ceased in the original laboratory frame. Microscopically, the movement of the liquid arises because of the gradual excitation of the internal degrees of freedom of the fluid. Quantum mechanically, these elementary excitations can be thought of as some kind of pseudoparticles, each having a momentum **p** and energy $\epsilon(p)$. When a single elementary excitation appears in the liquid, the energy E_0 and momentum

5.7 Internal Structure of Neutron Stars

$\mathbf{P}_0$ of the liquid (in the coordinate system in which it was originally at rest) will be equal to those of the elementary excitation; that is, $E_0 = \epsilon(p)$ and $\mathbf{P}_0 = \mathbf{p}$. Transforming back to the original frame in which the tube was at rest we obtain for the energy E and momentum $\mathbf{P}$ of the liquid

$$E = E_0 + \mathbf{P}_0 \cdot \mathbf{v} + \frac{1}{2}Mv^2, \quad \mathbf{P} = \mathbf{P}_0 + M\mathbf{v}, \tag{5.174}$$

where M is the mass of the liquid. Substituting $E_0 = \epsilon$ and $\mathbf{P}_0 = \mathbf{p}$ we get

$$E = \epsilon + \mathbf{p} \cdot \mathbf{v} + \frac{1}{2}Mv^2. \tag{5.175}$$

In this frame the liquid originally had a kinetic energy of $(1/2)Mv^2$; hence the change in energy that is due to the appearance of an excitation must be $\epsilon + \mathbf{p} \cdot \mathbf{v}$. Because the energy of the liquid should decrease because of viscous dissipation, this change must be negative and we must have $\epsilon + \mathbf{p} \cdot \mathbf{v} < 0$. The minimum value of the left-hand side occurs when $\mathbf{p}$ and $\mathbf{v}$ are antiparallel. In that case, we need to satisfy the condition

$$v > \frac{\epsilon(p)}{p} \tag{5.176}$$

for some value of p.

Simple excitations of the kind $\epsilon(p) = (p^2/2m_{\text{eff}})$ will lead to the condition $v > (p/2m_{\text{eff}})$, which can always be satisfied for sufficiently small values of p. In contrast, consider an excitation that has the dispersion relation

$$\epsilon(p) = \left[\left(\frac{p^2}{2m}\right)^2 + \Delta^2\right]^{1/2}. \tag{5.177}$$

In this case the right-hand side of condition (5.176), given by

$$\frac{\epsilon(p)}{p} = \left(\frac{p^2}{4m^2} + \frac{\Delta^2}{p^2}\right)^{1/2}, \tag{5.178}$$

has a distinct minimum at $p_{\min} = \sqrt{2m\Delta}$. Hence, unless the velocity is larger than $(2\Delta/m)^{1/2}$, condition (5.176) cannot be satisfied and the elementary excitations cannot be produced. In other words, no dissipation can take place in such liquids, for flow velocities below a critical value.

It is clear from the above analysis that whether a system will exhibit superfluidity (flow without dissipation) crucially depends on the dispersion relation for elementary excitations $\epsilon(p)$. The dispersion relations of the form of Eq. (5.177) are said to have a bandgap or energy gap; that is, we need to supply an energy Δ even at zero momentum to produce this excitation. Whether the elementary excitations of a system will possess a bandgap or not depends sensitively on the dynamics. Systems exhibiting superfluidity and superconductivity (which arise for similar reasons) possess interactions that lead to a bandgap.

Inside a neutron star, the attractive nuclear interactions among the neutrons make them superfluids at all densities above the neutron drip ($\rho \gtrsim 4 \times 10^{11}$ gm cm^{-3}). The protons are superconducting throughout the core. These features, however, do not affect the equation of state because the condensate energy E_c is far less than the Fermi energy. The normal matter in the neutron star consists essentially of (1) the outer crust, (2) the lattice nuclei in the inner crust made of all protons and some neutrons in bound state, and (3) the electrons throughout the star. It is the electrons that provide the coupling between the normal and the superstates through dissipative processes between the crust and the core of the neutron star.

The existence of an energy gap Δ between the ground state and the first excited state has several consequences, which we shall now elaborate. To begin with, it implies that the ground state acts as a coherent condensate with the condensation energy per particle being of the order of $E_c \approx (\Delta^2/\epsilon_F)$. (This is similar to the formation of a Bose–Einstein condensate in which a finite fraction of the particles occupy the zero-momentum state, as described in Vol. I, Chap. 5, Section 5.10.) At any finite temperature we can describe such a system by superfluid components of particles that are in the ground state as having a density n_0 and a normal component of excitations, with the density proportional to $n_{\text{ex}} \propto \exp(-\Delta/k_B T)$. The latter component is negligible in the neutron star interior.

Because of the coherence forced on the system by the energy gap, the particles can now be described by a single wave function $\psi \propto n_0^{1/2} \exp(i\phi)$, where $|\psi|^2$ is proportional to the density of particles and the phase ϕ is related to the macroscopic velocity of the particles by $\mathbf{v} \propto \nabla\phi$. The existence of a single wave function for all particles with a well-defined ϕ has interesting implications when any superfluid is rotated. Rigid-body rotation with angular velocity $\mathbf{\Omega}$ requires that $\mathbf{v} = \mathbf{\Omega} \times \mathbf{r}$ so that $\nabla \times \mathbf{v} = 2\mathbf{\Omega} \neq 0$. On the other hand, if the macroscopic velocity field is proportional to the gradient of the wave function describing all the particles, $\mathbf{v} \propto \nabla\phi$, then we would have naively expected $\nabla \times \mathbf{v} = 0$. This suggests that a superfluid cannot maintain rigid-body rotation in the manner envisaged above. When a superfluid is rotated, it breaks into a series of microscopic vortex lines, each carrying a certain amount of angular momentum. The total angular velocity is provided by a large number of such vortices rather than by rigid rotation of all the particles. In this case, the velocity field $\mathbf{v}$ becomes singular on the vortex, allowing the $\nabla \times \mathbf{v}$ to vanish everywhere except at the origin.

To see this explicitly, consider a single vortex line treated as infinitely long along the z axis. Suppose that the microscopic velocity field along this vortex is given by

$$\mathbf{v} = \frac{\hbar}{2m_n}\nabla\phi = \frac{\hbar}{2m_n}\frac{\hat{\mathbf{e}}_\phi}{r}, \qquad (5.179)$$

where m_n is the mass of the neutron and r is the radial distance from the vortex

5.7 Internal Structure of Neutron Stars

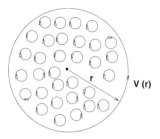

Fig. 5.7. Mimicking the rigid-body rotation of a superfluid by a large number of vortex lines.

line in the standard cylindrical (r, ϕ, z) coordinates. Because there is a singularity at $r = 0$, the phase of the wave function can, in general, change by an amount $2\pi n$ ($n = 1, 2, 3, \ldots$, etc.), as we go around the singularity. Therefore the line integral of $\mathbf{v}$ around a circle gives

$$\oint \mathbf{v} \cdot d\mathbf{l} = \left(\frac{\pi h}{m_n}\right) n \equiv nK, \quad K \equiv \frac{\pi h}{m_n} = \frac{h}{2m_n} \simeq 2 \times 10^{-3} \text{ cm}^2 \text{ s}^{-1}. \tag{5.180}$$

In differential form, this is equivalent to the relation

$$\nabla \times \mathbf{v} = nK\delta_D^{(2)}(\mathbf{r}), \tag{5.181}$$

where $\delta_D^{(2)}(\mathbf{r})$ is the two-dimensional Dirac delta function. A number of such vortices can mimic the macroscopic rigid-body rotation along the lines shown in Fig. 5.7. In this case, we have

$$\oint \mathbf{v}(\mathbf{r}) \cdot d\mathbf{l} = \int (\nabla \times \mathbf{v}) \cdot d\mathbf{S} = \sum_i K \int \delta_D^{(2)}(\mathbf{r} - \mathbf{r}_i) dS$$

$$= K \int_0^r n(r') 2\pi r' dr' \equiv KN(r), \tag{5.182}$$

where $n(r)$ is the number of vortices per unit area and $N(r)$ is the number of vortices enclosed by a circle of radius r. When $\mathbf{v}(\mathbf{r})$ is along the $\mathbf{e}_\phi$ direction, this equation simplifies to

$$2\pi r v(r) = 2\pi r^2 \Omega(r) = KN(r), \quad \frac{1}{r}\frac{\partial}{\partial r}(r^2 \Omega) = Kn(r). \tag{5.183}$$

In the case of rigid-body rotation, $(\partial \Omega/\partial r) = 0$ and the number density of vortices is given by $n = (2\Omega/K)$. Thus a uniform vortex density that is proportional to rotational velocity Ω is set up in a superfluid when it is rotating rigidly. The average spacing between the vortices is given by

$$\rho_{\text{vortex}} \simeq n^{-1/2} \simeq 3 \times 10^{-3} \left(\frac{\Omega}{100 \text{ rad s}^{-1}}\right)^{-1/2} \text{ cm.} \quad (5.184)$$

The situation is more complicated as regards protons that are charged particles. In this case, the canonical momentum is given by (see Vol. I, Chap. 3, Section 3.7) $\mathbf{p} = m_p \mathbf{v} + (e/c)\mathbf{A}$. Therefore, for protons, the gradient of the phase of the wave function (which we denote by χ) is related to the velocity by

$$\mathbf{v} = \frac{\hbar}{2m_p} \nabla \chi - \frac{e}{m_p c} \mathbf{A}. \quad (5.185)$$

If we attempt to keep the phase χ nonsingular, then, taking the curl of this equation, we get

$$2\Omega = \nabla \times \mathbf{v} = -\frac{e}{m_p c} \nabla \times \mathbf{A} = -\frac{e}{m_p c} \mathbf{B}. \quad (5.186)$$

This is, however, unrealistic for the neutron star as it requires an angular velocity of $\Omega = (eB/2m_p c) \approx 3 \times 10^{15} B_{12}$ rad s^{-1}, which is enormous. It is therefore necessary to again take the option of singular phases. It is now possible, however, to minimise the kinetic energy by taking $\mathbf{v} = 0$ and setting

$$\frac{\hbar}{2m_p} \nabla \chi = \frac{e}{m_p c} \mathbf{A} \quad \text{(except at singularities).} \quad (5.187)$$

Because the curl of the right-hand side is nonzero, we must again have a singular behaviour for $\nabla \chi$. Evaluating the line integral of both sides over a circle and using the fact that when $\nabla \chi$ is integrated around a singularity it can change by $(2\pi n)$, $(n = 1, 2, 3, \ldots,)$, we get

$$\Phi = \oint \mathbf{A} \cdot d\mathbf{l} = \frac{hc}{2e} \oint \nabla \chi \cdot d\mathbf{l} = \frac{hc}{2e} \cdot 2\pi n = \frac{hc}{2e} n. \quad (5.188)$$

This shows that the magnetic flux must be quantised in units of $\Phi_0 \equiv (hc/2e) \approx 2 \times 10^{-7}$ G cm^2. The quantity $\mathbf{B}/2$ is related to the flux-tube lines in exactly the same way Ω is related to the distribution of vortex lines. If $n_f(r)$ is the density of flux tubes, then the equation analogous to Eq. (5.183) is

$$\frac{1}{r} \frac{\partial}{\partial r} \left(\frac{1}{2} B r^2\right) = \Phi_0 n_f(r). \quad (5.189)$$

The spacing between the flux lines is

$$l_f \simeq n_f^{-1/2} \cong 5 \times 10^{-10} B_{12}^{-1/2} \text{ cm} \quad (5.190)$$

for a uniform magnetic field.

The singular phases obtained in the above description are, of course, a mathematical simplification. A vortex line is not physically singular at the microscopic

5.7 Internal Structure of Neutron Stars

level. When the kinetic energy $(1/2)m_n v^2$ arising from the velocity field $v = (K/2\pi r)$ of the single vortex line is comparable with the condensation energy per particle, $E_c \approx (\Delta^2/\epsilon_F)$, the particles will get out of the condensed phase. Equating the kinetic energy to E_c defines a coherence length ξ through the equation

$$E_c \simeq \frac{\Delta^2}{\epsilon_F} = \frac{1}{2} m_n \frac{K^2}{4\pi \xi^2} \tag{5.191}$$

or

$$\xi \simeq \left(\frac{m_n K^2}{8\pi \Delta^2} \epsilon_F \right)^{1/2} \simeq \left(\frac{\pi \hbar^2}{8 m_n} \frac{\epsilon_F}{\Delta^2} \right)^{1/2}. \tag{5.192}$$

Numerically, $\xi \approx (10\text{--}30)$ fm for a neutron star.

5.7.3 Microscopic Origin of a Bandgap

The results obtained in the previous subsections depended on the existence of a bandgap in the dispersion relation for the system. For the sake of completeness, we shall now provide a microscopic quantum-mechanical model illustrating how such dispersion relations arise.

Consider a set of particles such as neutrons in a neutron star or electrons in a solid. The behaviour of such a system will be governed by a fairly complex many-body interaction between large numbers of particles, and the system will usually try to occupy a state of minimum free energy, subject to external constraints. Under certain circumstances, there could arise a weak, attractive interaction between pairs of particles making up the system. In the case of electrons in a solid, this attraction could be envisaged as follows: An electron disturbs the lattice in a particular way; the deformed lattice influences another electron as though the net effect is like a weak attractive force between the two electrons. Such forces are of fairly long range in real space and hence will correspond to *nearly* zero-momentum states in the Fourier space. In the case of neutrons in a neutron star, a similar effect can arise because of nuclear forces.

When such a small attractive force exists between the fermions in the system, the ground state of the system can exhibit certain peculiarities, which we shall now discuss. We begin by writing the Hamiltonian for such a system in the Fourier space by adding to the Hamiltonian for noninteracting particles the Hamiltonian that is due to the weak attraction. If there are $n_\mathbf{k}$ particles each of energy $\epsilon_\mathbf{k}$ and $\hbar \mathbf{k}$ is the momentum of the particle, then a *noninteracting* system of particles will be described by the Hamiltonian

$$H_0 = \sum_\mathbf{k} \epsilon_\mathbf{k} n_\mathbf{k} = \sum_\mathbf{k} \epsilon_\mathbf{k} a_\mathbf{k}^\dagger a_\mathbf{k}, \tag{5.193}$$

where $a_\mathbf{k}^\dagger$ and $a_\mathbf{k}$ are the creation and the annihilation operators, respectively, for particles with momentum $\hbar \mathbf{k}$ and energy $\epsilon_\mathbf{k}$. (This is the standard Hamiltonian for a system of noninteracting particles; see, for example, Vol. I, Chap. 4, Section 4.5.) When there is a small attractive interaction between pairs of particles we should add to H_0 the interaction term

$$H_{\text{int}} = -G \sum_{\mathbf{k},\mathbf{k}'>0} a_\mathbf{k}^\dagger a_{-\mathbf{k}}^\dagger a_{-\mathbf{k}'} a_{\mathbf{k}'}. \tag{5.194}$$

Such an interaction term couples particles of momentum $\mathbf{k}$ with particles of momentum $-\mathbf{k}$ with a constant strength $-G$. The combination $a_{-\mathbf{k}'} a_{\mathbf{k}'}$ annihilates a pair with momenta $(\mathbf{k}', -\mathbf{k}')$ and the combination $a_\mathbf{k}^\dagger a_{-\mathbf{k}}^\dagger$ creates a pair with momenta $(\mathbf{k}, -\mathbf{k})$; this could be thought of as scattering of a pair from state $\mathbf{k}'$ to state $\mathbf{k}$. Although such a coupling is very simple (and somewhat unrealistic), it does capture – as we shall see – the essential features of the system.

The original ground state of the system $|0>$ is annihilated by all the annihilation operators $a_\mathbf{k}$, and the expectation value of H_0 in this state is zero. To find the new ground state in the presence of interaction, we can try to minimise the expectation value of the total Hamiltonian $H = H_0 + H_{\text{int}}$ subject to the boundary condition. A trial quantum state (called the *Bardeen–Cooper–Schrieffer* state in superconductivity) is given by the form

$$|\psi\rangle = \prod_{\mathbf{k}>0} \left(u_\mathbf{k} + v_\mathbf{k} a_\mathbf{k}^\dagger a_{-\mathbf{k}}^\dagger \right) |0\rangle, \tag{5.195}$$

where $u_\mathbf{k}$ and $v_\mathbf{k}$ are related by the normalisation condition

$$u_\mathbf{k}^2 + v_\mathbf{k}^2 = 1. \tag{5.196}$$

Because of this condition, the quantum state is essentially parameterised by one set of numbers, say, $v_\mathbf{k}$. We now need to minimise $\langle \psi | H | \psi \rangle$ subject to the constraint that the total number of particles is conserved; that is, we must keep

$$\left\langle \psi \left| \sum_\mathbf{k} a_\mathbf{k}^\dagger a_\mathbf{k} \right| \psi \right\rangle = N \tag{5.197}$$

while minimising the energy. Imposing this condition by means of a Lagrange multiplier and simplifying the variational equation

$$\frac{\delta}{\delta v_\mathbf{k}} \left\langle \psi \left| H - \lambda \sum_\mathbf{k} a_\mathbf{k}^\dagger a_\mathbf{k} \right| \psi \right\rangle = 0 \tag{5.198}$$

leads to the result

$$v_\mathbf{k}^2 = \frac{1}{2}\{1 - (\epsilon_\mathbf{k} - \lambda)[(\epsilon_\mathbf{k} - \lambda)^2 + \Delta^2]^{-1/2}\}, \tag{5.199}$$

5.7 Internal Structure of Neutron Stars

where Δ is given by

$$\Delta = G \sum_k u_k v_k. \tag{5.200}$$

Because Lagrange multiplier λ corresponds to the conservation of the total number of particles, it can be interpreted as the chemical potential or – equivalently – the Fermi energy ϵ_F. Similarly, because $v_k^2 = \langle \psi | a_k^\dagger a_k | \psi \rangle$, we can think of v_k^2 as the probability of occupation of a given state. Determining u_k^2 from Eq. (5.196) and substituting into Eq. (5.200), we obtain the dispersion relation for the *energy gap* Δ:

$$\Delta = \frac{1}{2} G \Delta \sum_k [(\epsilon_k - \lambda)^2 + \Delta^2]^{-1/2} = \frac{1}{2} G \Delta \sum_k [(\epsilon_k - \epsilon_F)^2 + \Delta^2]^{-1/2}. \tag{5.201}$$

In arriving at the last equality, we have also used the identification $\lambda = \epsilon_F$. The trivial solution is $\Delta = 0$, which corresponds to a normal, zero-temperature Fermi distribution; on the other hand, there exists a nontrivial solution of the form

$$\frac{2}{G} = \sum_k \{(\epsilon_k - \epsilon_F)^2 + \Delta^2\}^{-1/2}, \tag{5.202}$$

essentially because $G \neq 0$. Assuming that $\epsilon_k = (k^2/2m)$ and assuming a constant density of states, we can convert the summation to an integral, leading to

$$\frac{G}{2} \int_a^b \frac{4\pi k^2 dk}{(2\pi \hbar)^3} \frac{1}{[(k^2/2m - \epsilon_F)^2 + \Delta^2]^{1/2}} = 1. \tag{5.203}$$

To determine the nature of the integral we note that we can write

$$\left(\frac{k^2}{2m} - \epsilon_F\right)^2 + \Delta^2 = \left(\frac{1}{2m}\right)^2 (k - p_F)^2 (k + p_F)^2 + \Delta^2. \tag{5.204}$$

If our integral has dominant contributions around $k \approx p_F$, this expression can be approximated as $[(k - p_F)^2 v_F^2 + \Delta^2]$, where $v_F = p_F/m$. Choosing the limits of the integral such that $\Delta \ll v_F |p_F - k| \ll \epsilon_F$ leads to an approximate evaluation of the integral in Eq. (5.203):

$$\frac{G}{2} \frac{4\pi}{(2\pi \hbar)^3} \int \frac{k^2 dk}{[v_F^2 (p_F - k)^2 + \Delta^2]^{1/2}} \simeq \frac{G}{4\pi^2 \hbar^3} \frac{p_F^2}{v_F} \int_\Delta^{\epsilon_F} \frac{dq}{(\Delta^2 + q^2)^{1/2}}$$

$$\simeq \frac{G}{4\pi^2 \hbar^3} \left(\frac{2p_F^2}{v_F}\right) \ln\left(\frac{\epsilon_F}{\Delta}\right). \tag{5.205}$$

[This integral is formally divergent if evaluated in the range $(0, \infty)$. The procedure we have adopted to get the finite result, however, can be justified by more formal regularisation techniques.] Condition (5.203) can now be used to determine Δ as

$$\Delta = \epsilon_F \exp\left(-\frac{2\pi^2 \hbar^3}{G m p_F}\right) \equiv \epsilon_F \exp\left(-\frac{1}{G\eta}\right), \quad \eta \equiv \frac{m p_F}{2\pi^2 \hbar^3}. \quad (5.206)$$

Note that Δ is a nonanalytic function of G, and hence this result could *not* have been obtained by a perturbation series in G. The effect of paired correlations is to lower the ground-state energy by an amount

$$\Delta E = \sum_{\epsilon_k < \epsilon_F} \epsilon_k - \sum_k (\epsilon_k - \epsilon_F) v_k^2 + G\left(\sum_k v_k u_k\right)^2 \simeq \eta \Delta^2. \quad (5.207)$$

The physical meaning of the lower-energy ground state can be understood along the following lines. Consider a transformation from the original creation and annihilation operators $(a_k^\dagger, a_k)$ to a new set $(\alpha_k^\dagger, \alpha_k)$, where

$$\alpha_k^\dagger = v_k a_k^\dagger + u_k a_k, \quad (5.208)$$

etc. These quantities satisfy the same commutation relation as the original set and hence can be thought of as creation and annihilation operators of some well-defined *quasiparticles*. The original vacuum state $|0>$ is, of course, not annihilated by the new annihilation operator α_k, but the *total* Hamiltonian H can be expressed in almost diagonal form in terms of the new creation and annihilation operators, with

$$H = \sum_k E_k \alpha_k^\dagger \alpha_k + \text{(terms with vanishing ground-state expectation value)}, \quad (5.209)$$

where the new quasiparticle energies are

$$E_k = [(\epsilon_k - \epsilon_F)^2 + \Delta^2]^{1/2}. \quad (5.210)$$

Note that the first excited state for the quasiparticles has a bandgap Δ with respect to the ground state. In contrast, ordinary quantum particles of a free-field theory, say, can have a continuous set of energies arbitrarily close to the rest energy mc^2 with no bandgap. The existence of the bandgap implies that any dissipative process with a characteristic energy less than Δ cannot excite the quasiparticles and change the system from the ground state to an excited state. All the consequences discussed in the last Subsection 5.7.2 follow from this dispersion relation.

It is more difficult to evaluate the bandgap (and related properties) at finite temperatures because we must perform the integral in Eq. (5.202) with finite-temperature Fermi distribution. Numerical calculations show that the gap vanishes at some critical temperature T_c, where $k_B T_c \approx 0.57\Delta$. For $k_B T_c \ll \Delta$ the bandgap is given by[3]

$$\Delta = \Delta_0 \left[1 - \left(\frac{2\pi k_B T}{\Delta_0} \right)^{1/2} \exp\left(-\frac{\Delta_0}{k_B T} \right) \right], \quad \Delta_0 \equiv \Delta(T=0), \tag{5.211}$$

whereas near $T \approx T_c$ it is given by

$$\Delta = 3.06\, k_B T_c \sqrt{\left(1 - \frac{T}{T_c}\right)}. \tag{5.212}$$

In general, at any finite temperature, the system is made of a superfluid component with density ρ_s and a normal component with density ρ_n, with

$$\frac{\rho_n}{\rho} = \left(\frac{2\pi \Delta_0}{k_B T} \right)^{1/2} \exp\left(-\frac{\Delta_0}{k_B T} \right) \quad \text{(for } k_B T \ll \Delta\text{)}$$

$$= \left(\frac{2T}{T_c} - 1 \right) \quad \text{(for } T \simeq T_c\text{)}. \tag{5.213}$$

Although there is no attractive force between free neutrons, such a force arises in the context of a nuclear many-body problem. (This is similar to the lattice providing an attractive residual force between the electrons in a solid.) The typical energy is, based on the nuclear-reaction scale, $\sim$3 MeV. In the inner-crust and neutron-liquid regions of a neutron star, the temperatures [$k_B T \approx (1-10)$ keV] are significantly lower than the pairing energy that is due to the residual force, and hence we expect the neutrons in these regions to be superfluid. By the same reasoning, the protons in the inner region will be superconducting. The electrons will be a normal Fermi fluid as they do not feel the nuclear force.

5.8 Gravitational Collapse and Black Holes

If the mass of the compact stellar remnant is higher than the maximum mass allowed for the neutron star, then such a configuration of matter does not posses a static equilibrium solution and will continue to collapse. As the radius of the configuration becomes sufficiently small so that the ratio $(GM/c^2 R)$ approaches unity, general relativistic effects will become important in determining the evolution. The gravitational field around such a configuration is described by the Schwarzschild metric derived in Vol. I, Chap. 11, Section 11.8, and the collapse of matter leads to a configuration called a black hole with several peculiar features. We shall now discuss these phenomena.

The Schwarzschild metric that is due to a mass M has a characteristic length scale given by $r_g \equiv (2GM/c^2)$. The g_{00} component vanishes and g_{rr} diverges at $r = r_g$, showing that this radius must play a key role in the description of physics in Schwarzschild metric. It must, however, be remembered that the Schwarzschild metric describes the *empty* region outside a spherical source; in the nonempty region occupied by the source, the metric will be quite different. [If the source is static then the inside metric is given by Eq. (5.61).] If the radius of the source is L, then the Schwarzschild metric is not valid for $r < L$, and if $r_g < L$, then the singular behaviour of the metric coefficients at $r = r_g$ is not of any concern. However, if the size of the source is such that $L < r_g$, then several peculiar features come into play that need to be discussed.

Such a situation can arise, for example, in the gravitational collapse of matter. Consider, for simplicity, a sphere of dust with $p = 0$, $\rho \neq 0$ that is occupying a spherical region of radius $r = L_1$ at time time $t = t_1$. Let the total mass of the dust sphere be such that $L_1 > r_g$. At $t = t_1$, the metric, at $r > L_1$ is given by the Schwarzschild metric, which is everywhere well behaved because of our assumption that $L_1 > r_g$. This initial configuration, however, cannot be static because there is no pressure gradient to balance the gravitational attraction. We expect the dust sphere to collapse under its own gravitational force, thereby decreasing its size. When its surface contracts to a radius of less than r_g, the metric outside becomes ill behaved at $r = r_g$ and the question rises as to what the new features are that such a collapse introduces.

To understand this situation, let us study the radial trajectory of a particle of dust located at the surface of the dust sphere. (We will use units with $c = 1$.) In the absence of pressure, this particle follows a radial geodesic in the Schwarzschild metric. The position r of this particle at any given time t can be determined from standard Hamilton–Jacobi theory. The trajectory for a particle with angular momentum L and energy $\mathcal{E}$ is given implicitly by (see Vol. I, Chap. 11, Section 11.9)

$$t = \frac{\mathcal{E}}{m} \int dr \left(1 - \frac{r_g}{r}\right)^{-1} \left[\left(\frac{\mathcal{E}}{m}\right)^2 - \left(1 + \frac{L^2}{m^2 r^2}\right)\left(1 - \frac{r_g}{r}\right)\right]^{-1/2}, \tag{5.214}$$

and we obtain the radial trajectory by setting $L = 0$. The integral is elementary (although involved) and gives

$$r = \frac{R}{2}(1 + \cos \eta), \tag{5.215}$$

$$t = \left[\left(\frac{R}{2} + r_g\right)\left(\frac{R}{r_g} - 1\right)^{1/2}\right]\eta + \frac{R}{2}\left(\frac{R}{r_g} - 1\right)^{1/2} \sin \eta$$

$$+ r_g \ln \left|\frac{(R/r_g - 1)^{1/2} + \tan(\eta/2)}{(R/r_g - 1)^{1/2} - \tan(\eta/2)}\right|, \tag{5.216}$$

where R is the initial radius at which the particle has zero velocity. These two equations give the trajectory of the particle $r(t)$ in terms of the time coordinate

t of an observer located at a large distance from the origin in the Schwarzschild metric. The proper time τ, shown by a clock moving along with the dust particle, is related to the coordinate time t by $(dt/d\tau) = (\mathcal{E}/m)[1 - (r_g/r)]^{-1}$. Therefore the relation between the proper time and the radial coordinate is given by

$$\tau = \int d\tau = \int \frac{dr}{[r_g/r - r_g/R]^{1/2}}. \tag{5.217}$$

Integrating this expression, we get

$$\tau = \frac{R}{2}\left(\frac{R}{r_g}\right)^{1/2}(\eta + \sin\eta). \tag{5.218}$$

Equations (5.215), (5.216), and (5.218), which describe the trajectory of the dust particle in terms of the coordinate time and the proper time carried by its own clock, lead to several important conclusions. To begin with, we see that the total *proper time* for the dust particle to fall from $r = R$ to $r = 0$ is given by

$$\tau = \frac{\pi}{2}R\left(\frac{R}{r_g}\right)^{1/2}, \tag{5.219}$$

and is finite. Because the dust particle is at the surface of a sphere, this shows that the entire body will collapse to a point in finite proper time, as measured by clocks located at the surface of the body. As the surface crosses $r = r_g$, no peculiar effects will be noted by an observer located on the surface. The trajectory $r(\tau)$ is smooth throughout the motion.

The situation is quite different when viewed by an outside observer using the time coordinate t. From expressions (5.215) and (5.216) it is easy to see that $t \to \infty$ as $r \to r_g$; that is, it takes infinite *coordinate time* for the particle at the surface of the dust sphere to reach the Schwarzschild radius $r = r_g$. An outside observer can conclude that it takes infinite time for the body to collapse to the size $r = r_g$, even though it collapses to $r = 0$ within a finite proper time as determined by the clocks on the surface. The asymptotic form of the trajectory $r(t)$ as r approaches r_g is easy to determine. We find that

$$r - r_g = \text{constant } e^{-(t/r_g)}. \tag{5.220}$$

The behaviour is shown in Fig. 5.8. This is an extreme example of the gravitational field's affecting the clock rate.

As the body collapses, any light emitted from its surface will get progressively redshifted. Consider an electromagnetic wave with a wave vector k^a propagating radially from a radius r to infinity. An observer who is stationary at r will have the four velocity $u^a = g_{00}^{(-1/2)}(1, 0, 0, 0)$. (The g_{00} factor is needed to ensure the normalisation $u^a u_a = 1$.) The frequency of the electromagnetic wave as measured by this observer will be $\omega(r) = k^a u_a = k_0 u^0(r)$. The ratio of frequencies measured by the observer at infinity and at r will be

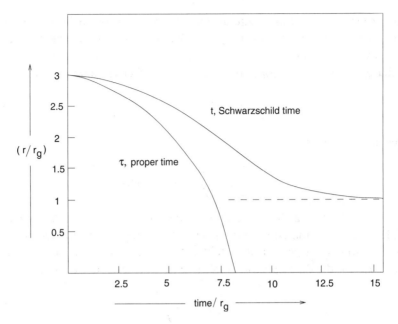

Fig. 5.8. The variation of coordinate time and proper time as a particle moves in a radial trajectory.

$[\omega(\infty)/\omega(r)] = [u^0(\infty)/u^0(r)] = \sqrt{g_{00}(r)}$. Therefore

$$\omega(\infty) = \omega(r)\left(1 - \frac{r_g}{r}\right)^{1/2}. \qquad (5.221)$$

As the dust sphere approaches r_g, $\omega(\infty) \to 0$ and the wave suffers infinite redshift.

The collapse also affects the communication between an observer located on the surface of the dust sphere and and a distant, stationary observer. Let us suppose that the former is sending light signals to latter at periodic intervals. Because the time of propagation for the radial light signals is given by $dt = dr/(1 - r_g/r)$, the time taken by the light signal to propagate from some point r to $r_0 > r$ is given by

$$\Delta t = \int_r^{r_0} \frac{dr}{(1 - r_g/r)} = r_0 - r + r_g \ln \frac{r_0 - r_g}{r - r_g}, \qquad (5.222)$$

which diverges as $r \to r_g$. Hence the signals take a progressively longer time to reach the distant observer and the signal sent when the dust sphere crosses $r = r_g$ (as indicated by the proper time on its clock) reaches the distant observer only after an infinite duration of time. The signals sent by the observer on the collapsing sphere after it has contracted through $r = r_g$ do not reach the outside observer at all. It follows that the region inside $r = r_g$ cannot communicate or

5.8 Gravitational Collapse and Black Holes

influence the outside region. For this reason, the surface $r = r_g$ is called the event horizon, and the body that has collapsed through its event horizon is called a black hole. The fact that part of the space–time region gets cut off from the rest is a special feature of Einstein's gravity.

To see this effect more clearly and also to understand some of the more general features related to the collapse, we rewrite the Schwarzschild metric by using a different set of coordinates called *Kruskal–Szekeres coordinates*. These coordinates (v, u, θ, ϕ) are related to the Schwarzschild coordinates (t, r, θ, ϕ) by

$$\left. \begin{array}{l} u = (r/r_g - 1)^{1/2} e^{r/2r_g} \cosh(t/2r_g) \\ v = (r/r_g - 1)^{1/2} e^{r/2r_g} \sinh(t/2r_g) \end{array} \right\} \quad r > r_g, \quad (5.223)$$

$$\left. \begin{array}{l} u = (1 - r/r_g)^{1/2} e^{r/2r_g} \sinh(t/2r_g) \\ v = (1 - r/r_g)^{1/2} e^{r/2r_g} \cosh(t/2r_g) \end{array} \right\} \quad r < r_g. \quad (5.224)$$

In these coordinates the Schwarzschild line element is given by

$$ds^2 = \frac{4r_g^3}{r} e^{-r/r_g} (dv^2 - du^2) - r^2(d\theta^2 + \sin^2\theta \, d\phi^2), \quad (5.225)$$

with r being given as an implicit function of u and v:

$$\left(\frac{r}{r_g} - 1\right) e^{r/r_g} = u^2 - v^2. \quad (5.226)$$

The major advantage of these coordinates is that radial ($d\theta = d\phi = 0$) light rays obeying $ds^2 = 0$ are given by $dv = \pm du$. Thus light cones are made of $45°$ lines in the u–v coordinates just as in flat space–time. Figure 5.9 shows a collapsing spherical region in the u–v coordinates and compares it with the corresponding region in the Schwarzschild coordinate. Curves of constant r become hyperbolas in Kruskal coordinates and curves of constant t become straight lines passing through the origin. The line $u = v$ corresponds to the event horizon $r = r_g$ and $t = \infty$. Because light cones are made of $45°$ lines, it is obvious that the region inside $r = r_g$ cannot send signals to the region with $r > r_g$.

The Kruskal coordinates also make clear two other facts: (1) The space–time metric is perfectly regular at $r = r_g$ and hence the pathological behaviour seen in Schwarzschild coordinates is an artificiality arising out of a bad choice for the coordinate system. (2) Our earlier discussion used the collapse of pressureless dust. However, even if the material has pressure, the same conclusions should still hold. Any spherical mass distribution that collapses to $r < r_g$ will not be able to communicate with the region at $r > r_g$. This is obvious from the nature of light cones in Kruskal coordinates. In the case of more complicated collapse examples, the trajectories of the collapsing material may be quite complex, but once the surface contracts below r_g, no light signal can escape to infinity.

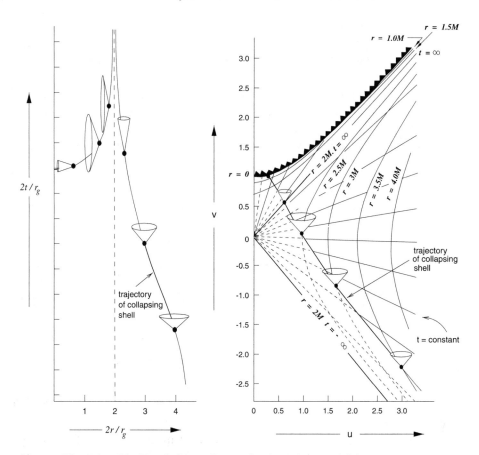

Fig. 5.9. The Kruskal coordinates for the Schwarzschild geometry.

The question arises as to what happens to the material that collapses in the above-mentioned manner. It can be easily verified that the curvature of the Schwarzschild metric becomes infinitely large at $r = 0$. In other words, $r = 0$ is a genuine singularity in space–time with an infinite gravitational field in contrast to the event horizon at $r = r_g$ at which all geometrical quantities are finite. According to our considerations, the collapsing matter should hit the singularity at $r = 0$ in finite proper time. It is very likely that Einstein's equations become invalid near $r = 0$ where the curvature is arbitrarily strong, possibly because of quantum gravitational effects. No theory is currently available to describe this situation, and hence the issue of final state of matter that collapses to form a black hole is, at present, unsettled.

In the above discussion we have assumed strict spherical symmetry for the collapsing matter. It turns out, however, that our conclusions have much more general validity. Calculations show that small deviations from spherical symmetry are wiped out during the late stages of gravitational collapse and the final state of the black hole is described by its mass alone. There are only two exceptions to this result. If the initial configuration possesses an electric charge or angular momentum, then the final state is described by a more complicated metric, called the *Kerr–Newman metric*. This metric depends on the mass, charge, and angular momentum of the source but again is completely independent of any other characteristics.

Exercise 5.5
Capture of particles by a black hole: We have seen in Vol. I, Chap. 11, Section 11.9, that the condition for the capture of a nonrelativistic particle of mass m by a black hole of mass M can be stated as $(E/m)^2 > V_{\max}$, where $V_{\max}$ is the maximum value of the effective potential. (a) For nonrelativistic particles with $(E/m) \simeq 1$, show that this condition translates to a constraint on the angular momentum per unit mass:

$$J_{\min}(E) = \frac{4GM}{c} \quad \text{(nonrelativistic particles)}. \tag{5.227}$$

(b) Consider now a swarm of particles described by a distribution function f and located around a black hole. Because of the existance of a loss cone $J < J_{\min}(E)$, it is convenient to express the particle density as a function of E and J. Show that the velocity element d^3v becomes

$$d^3v = 2\pi v_t dv_t dv_r = \frac{4\pi J\, dJ\, dE}{r^2 |v_r|}, \tag{5.228}$$

where the extra factor 2 arises from the fact that for a given E the radial velocity v_r can be positive or negative. Hence show that the number of particles captured per unit time is given by

$$\dot{N}_{\text{tot}} = 8\pi^2 \int_{\Phi(r)}^{\infty} dE \int_{J=0}^{J_{\min}(E)} dJ\, Jf. \tag{5.229}$$

(c) Assume that there is an infinite bath of nonrelativistic particles with a monoenergetic distribution function

$$f = f(E) = n_\infty \frac{\delta(E - R_\infty)}{4\pi (2E_\infty)^{1/2}}, \tag{5.230}$$

where E is the energy per unit mass. Estimate the mass accretion rate and show that

$$\dot{M}(E > 0) = m\dot{N}(E > 0) = 16\pi \left(\frac{GM}{c}\right)^2 \frac{\rho_\infty}{v_\infty}, \tag{5.231}$$

where v_∞ and ρ_∞ are the velocity and the density of the particles, respectively. (d) Show that the corresponding results for a Newtonian star of radius R is

$$\dot{M}(E > 0) = m\dot{N}(E > 0) = 2\pi GMR\frac{\rho_\infty}{v_\infty}\left(1 + \frac{v_\infty^2 R}{2GM}\right). \tag{5.232}$$

5.9 Rotating Black Holes

In astrophysical contexts, it is very unlikely that a significant amount of residual charge will exist in a collapsing structure. However, it is possible that such a structure will possess a nonzero angular momentum. Hence the metric representing a rotating black hole with nonzero mass M and angular momentum J is of particular interest. This metric, called the *Kerr metric*, can be expressed in the form[4] (in units which $c = 1$)

$$ds^2 = \left(1 - \frac{2Mr}{\Sigma}\right) dt^2 + \frac{4aMr\sin^2\theta}{\Sigma} dt d\phi - \frac{\Sigma}{\Delta} dr^2$$
$$- \Sigma d\theta^2 - \left(r^2 + a^2 + \frac{2Mra^2\sin^2\theta}{\Sigma}\right) \sin^2\theta d\phi^2, \quad (5.233)$$

where the black hole is taken to rotate in the ϕ direction and

$$M = \frac{r_g}{2}, \quad a \equiv \frac{J}{M}, \quad \Delta \equiv r^2 - 2Mr + a^2, \quad \Sigma \equiv r^2 + a^2 \cos^2\theta. \quad (5.234)$$

This metrics is clearly axisymmetric and is independent of t. At large distances, the asymptotic form of the metric agrees with that for an object of mass M and angular momentum J (located at origin) derived in Vol. I, Chap. 11, Section 11.9. It can also be verified by fairly tedious computation that this metric is a solution to Einstien's equation in vacuum. For these reasons, it is generally believed that the above metric could represent the outside region of a rotating black hole. Unfortunately, however, there is no rigorous proof available for this claim. No suitable interior solution corresponding to a rotating body that has collapsed to form a black hole is available. This is unlike the situation for the Schwarzschild metric, for which a collapsing interior solution is available and can be matched to the Schwarzschild exterior.

The event horizon for the Kerr metric occurs at the location where Δ vanishes. Taking the larger root of the quadratic equation $\Delta = 0$, we get

$$r_+ = M + (M^2 - a^2)^{1/2}. \quad (5.235)$$

The event horizon exists only if $a < M$, which is the case in the astrophysical context. (It is at present unknown whether the condition $a < M$ can be violated in general relativity for realistic, collapsing bodies.) As in the case of the Schwarzschild metric, the event horizon forms the boundary of a region from the inside of which no signals can reach spatial infinity.

In the case of the Schwarzschild metric, no static observers could exist inside the event horizon; any observer will be moving radially inwards towards the singularity. The situation is more complicated in the case of a rotating black hole: It turns out that the angular velocity of any observer has to be between a minimum and a maximum value at any given raidus, and – close to the black hole – it is not possible to have observers who are nonrotating. To obtain this result, consider a class of observers at fixed r and θ but rotating with constant

5.9 Rotating Black Holes

angular velocity

$$\Omega = \frac{d\phi}{dt} = \frac{u^\phi}{u^t}. \tag{5.236}$$

The condition $u^a u_a = 1$ translates to

$$1 = (u^t)^2 [g_{tt} + 2\Omega g_{t\phi} + \Omega^2 g_{\phi\phi}]. \tag{5.237}$$

The quantity in the brackets has to be positive; this is possible only if Ω lies between the root of the quadratic equation obtained when the bracketed expression is set to zero. Hence $\Omega_{\min} < \Omega < \Omega_{\max}$, where

$$\Omega_{\min \atop \max} = \frac{-g_{t\phi} \pm (g_{t\phi}^2 - g_{tt}g_{\phi\phi})^{1/2}}{g_{\phi\phi}}. \tag{5.238}$$

At large distances from the body, this condition merely implies that $|\Omega r \sin\theta| < c$, which is physically reasonable. As we approach the black hole, $\Omega_{\min}$ increases and we reach the limit $\Omega_{\min} = 0$ on the surface at which $g_{tt} = 0$. This surface, called the *ergoshere*, is given by

$$r_0 = M + (M^2 - a^2 \cos^2\theta)^{1/2}. \tag{5.239}$$

Nonrotating observers with $\Omega = 0$ can exist for only $r > r_0$. In the range $r_+ < r < r_0$, we must have $\Omega > 0$; that is, all observers must corotate with the black hole. It is possible for observers in this range r to send signals to large spatial distances. Hence the causality limit and the limit on stationary observers occur at different surfaces for a rotating black hole.

The motion of particles in the Kerr metric can be studied aling the lines similar to those of the Schwarzschild metric (see Vol. I, Chap. 11). Of particular interest is the radius of the smallest stable circular orbit, which is given by

$$r_{\text{ms}} = M\{3 + Z_2 \mp [(3 - Z_1)(3 + Z_1 + 2Z_2)]^{1/2}\},$$

$$Z_1 \equiv 1 + \left(1 - \frac{a^2}{M^2}\right)^{1/3} \left[\left(1 + \frac{a}{M}\right)^{1/3} + \left(1 - \frac{a}{M}\right)^{1/3}\right], \tag{5.240}$$

$$Z_2 \equiv \left(3\frac{a^2}{M^2} + Z_1^2\right)^{1/2}.$$

The binding energy of the marginally stable circular orbit is related to the parameter a/M by

$$\frac{a}{M} = \mp \frac{4\sqrt{2}(1 - \bar{E}^2)^{1/2} - 2\bar{E}}{3\sqrt{3}(1 - \bar{E})}. \tag{5.241}$$

It is easy to show from this expression that the matter moving through successive circular orbits of progressively smaller radius can release up to 42% of the rest-mass energy in the process of falling into the black hole.

6
Pulsars

6.1 Introduction

This chapter deals with the physics of isolated pulsars and draws heavily from the last two chapters. It also uses concepts developed in Vol. I, Chaps. 3, 5, 6, 9, and 11. There are several astrophysical phenomena that involve pulsars that are not isolated but exist as a part of, say, a binary system.[1] These phenomena will be discussed in Chap. 7.

6.2 Overview

We saw in the last two chapters that one of the possible end stages of stellar evolution could be a highly compact, dense distribution of matter in the form of a neutron star. When the neutron star forms as a result of a collapse, it conserves its angular momentum to a large degree. If the initial angular momentum of the precollapse core is $J_{in} \cong M\Omega_{in}R_{in}^2$, then the conservation of angular momentum $J_{in} = J_{final}$ requires that $\Omega_{final} \approx \Omega_{in}(R_{in}/R_{final})^2$. If the ratio between the initial and the final radii is $\sim 10^5$ and $\Omega_{in} = (2\pi/20 \text{ days})$, this leads to a fairly rapid rotation for the neutron star with $\Omega_{final} \approx (2\pi/10^{-4} \text{ s})$.

As the core collapses to form a neutron star, its electrical conductivity becomes very high. In the limit of infinite conductivity, the magnetic flux through the star, $\Phi \propto BR^2$, will also be conserved (see Vol. I, Chap. 9, Section 9.6). This requires that the final magnetic field be $B_{final} = B_{in}(R_{in}/R_{final})^2$. If $B_{in} \approx 100$ G and $(R_{final}/R_{in}) \approx 10^5$, we get $B_{final} \approx 10^{12}$ G. In other words, the neutron star will be threaded by a very strong magnetic field.

It is therefore clear that we are led to stellar remnants made of highly conducting objects threaded by a very high magnetic field and in a state of rapid rotation. The pulsars belong to such a class of objects and give rise to a host of complex astronomical phenomena, not all of which are yet completely understood. We shall first describe in qualitative terms the physical processes that can arise in such a situation and then discuss them in detail in the remaining sections.

6.2 Overview

Rapidly rotating neutron stars with strong magnetic fields will lead to the emission of radiation by means of different processes. Because the primary probe of such structures is through the electromagnetic radiation received from them, we shall begin by discussing these processes. The rotation and the magnetic field define (at least) two different axes in a pulsar, even in the simplest case in which the interior magnetic field is taken to be characterised by a single axis. In general, these two axes (the rotation axis and the magnetic-field axis) need not be aligned with each other. If they are misaligned, then the pulsar acts as a rotating magnetic dipole that will emit electromagnetic radiation essentially at the frequency of rotation, with a characteristic dipole pattern. If this radiation can propagate away from the pulsar, we will be able to detect it as a periodic signal. There are, however, other effects that could prevent this from happening – or at least make it subdominant. The model of a magnetic dipole rotator assumes that the region around the pulsar is a vacuum. As we shall see in Subsection 6.3.1, a rotating magnetic dipole will create a strong electric field near the surface of the pulsar that may be capable of pulling out charged particles from the surface. If this happens, there is a distinct possibility of the pulsar's being surrounded by a plasma that will also have a very high conductivity. We are thus faced with a problem of a rotating compact object surrounded by a corotating plasma, and the net radiation emitted by the system should take into account all the processes operating in this magnetosphere of the plasma. For realistic values for the plasma density, the dipole radiation will not be able to propagate through the plasma and hence may not be detectable outside. The plasma in the ISM could also block the radiation.

The magnetosphere, however, can lead to several different radiative processes of its own. To begin with, the charged particles spiralling around the magnetic field lines will emit curvature radiation, as discussed in Vol. I, Chap. 6, Section 6.11. Further, any high-energy gamma-ray photon propagating in the magnetosphere of the pulsar could lead to electron–positron pair production, thereby modifying the radiation pattern. Last, we should note that the plasma cannot be strictly corotating with the neutron star at an arbitrarily large radius away from the surface. This is because the translational velocity of the charged particle in the corotating plasma, located at a distance r from the rotation axis, will be $v = \Omega r$. This will exceed the speed of light at a distance

$$R_L \equiv \frac{c}{\Omega} = 4.77 \times 10^9 \text{ cm} \left(\frac{P}{1 \text{ s}}\right), \tag{6.1}$$

called the radius of the light cylinder. Obviously the corotation of the plasma ceases at $r \lesssim R$. In fact, the fluid approximation used in the MHD description of the plasma breaks down around $r \approx R_L$ and we are led to a situation similar to that of a collisionless shock. The net effect of the existence of a light cylinder is to reconfigure the electromagnetic fields in such a way that the magnetospheric plasma emits radiation to large distances. The resulting radiation (arising primarily from the charged particles moving along the magnetic-field lines that originate near the poles of the pulsar) can be detected on Earth periodically when the radiation cone sweeps past us at the rotational period of the pulsar.

It should be stressed that the magnetospheric radiation depends crucially on the existence of a plasma around the pulsar but is independent of the details of alignment between the magnetic and the rotation axes. If the magnetic axis is perfectly aligned with the rotation axis, we will not get magnetic dipole radiation because of a rotating dipole; but even in that case (called the *aligned rotator*), the large electric fields near the surface can lead to a magnetosphere containing plasma and radiation can originate from the plasma. Because the aligned rotator is easier to handle mathematically, it is often adopted as a simple model for the pulsar.

The details regarding the generation and the sustenance of the plasma around the pulsar can be quite complicated and are ill understood. In particular, the generation of plasma could depend on complex solid-state phenomena's taking place near the surface of the neutron star, which is not easy to handle theoretically. The situation is made more difficult by the fact that general relativistic effects may not be completely ignorable in modelling the pulsar, although for obtaining an analytically tractable solution it is often necessary to ignore complications from general relativity.

The phenomena described above, however, lead to two elementary conclusions that are reasonably robust and independent of the detailed nature of the emission mechanism. The first is that pulsars will emit radiation with well-defined periodicity. In fact, most of the astrophysical phenomena involving (and using) pulsars rely heavily on the stability of such pulsed radiation. Second, the pulsar must be constantly losing energy because of a variety of radiation processes. Because this energy has to be supplied from the rotational kinetic energy of the pulsar, the angular velocity of the pulsars must be decreasing over time; equivalently, the pulsar periods must be increasing. Because the period P can be measured with a high level of accuracy, we can estimate $\dot{P} \equiv (dP/dt)$ and consequently test several of the model predictions. (The above discussion assumed that the only energy source for the radiation is the rotational energy of the pulsar. This is not true if the pulsar is a part of binary system, as we shall see in Chap. 7.)

The extraordinary stability of the pulsar period makes pulsars a useful probe of several other physical phenomena. For example, the radiation from a pulsar received on Earth propagates through the ISM that contains plasma and magnetic fields. By estimating the Faraday rotation and plasma dispersion for the radiation (Vol. I, Chap. 9), we can estimate the interstellar magnetic fields and interstellar electron density. In a more general setting, even weak perturbations that cause variations in the period of the pulsar signal can be measured accurately and thus can be used to model or probe the source of perturbation. We shall now discuss these phenomena in detail.

Exercise 6.1

Incoherent emission: A classic example of incoherent emission of radiation is by particles in thermal equilibrium. We know that, in this case, the maximum emissivity is when the spectrum is that of a blackbody with a temperature T. Hence estimating the temperature T for a radiating system can help us to see whether incoherent emission can be responsible for the observed radiation. (a) The radio emission from a pulsar of radius $R \approx 10$ km located at a distance $D \simeq 1$ kpc is found to be $F \approx 10^{-25}$ W m^{-2} Hz^{-1} at a frequency $\nu \approx 100$ MHz. What is the brightness temperature in radio? (b) The same pulsar produces an x-ray flux of 10^{-30} W m^{-2} Hz^{-1}. What is the brightness temperature in x rays? [Answers: (a) The intensity I_ν is given by $I_\nu = (F/\Omega) = F(D/R)^2$, where Ω is the solid angle. This gives $I_\nu = 10^6$ W m^{-2} Hz^{-1} sr^{-1}. The brightness temperature is $T = (I_\nu c^2/2k_B \nu^2) \simeq 3 \times 10^{29}$ K. Radio emission cannot be from an incoherent process. (b) The corresponding x-ray intensity is $I_\nu \approx 10$ W m^{-2} Hz^{-1} sr^{-1}. The brightness temperature now becomes $T \simeq 3 \times 10^4$ K, which can be due to incoherent processes.]

6.3 Electromagnetic Field Around the Pulsar

Because the radiation from the pulsar can arise from different processes, it is convenient to discuss different possible cases for the pulsar-emission mechanism separately. In the lowest order of approximation, we can think of pulsars as rotating magnetic dipole moments emitting magnetic dipole radiation. Although this model is too simplistic, it leads to the correct order-of-magnitude estimates for radiation and hence is worth mentioning briefly.

Consider a rotating neutron star with a magnetic dipole moment **m**, radius R, and angular velocity Ω. Let the direction of **m** be inclined at an angle α to the rotation axis. In that case, the magnetic moment **m** varies with time as

$$\mathbf{m} = \frac{1}{2}BR^3(\mathbf{e}_\parallel \cos\alpha + \mathbf{e}_\perp \sin\alpha \cos\Omega t + \mathbf{e}'_\perp \sin\alpha \sin\Omega t), \quad (6.2)$$

where $\mathbf{e}_\parallel$ is the unit vector along the rotation axis and $\mathbf{e}_\perp$, $\mathbf{e}'_\perp$ are (fixed) mutually orthogonal unit vectors perpendicular to $\mathbf{e}_\parallel$. Note that the magnitude of the magnetic moment at the pole, $\alpha = 0$, must be $|\mathbf{m}| = (BR^3/2)$, which fixes the overall scaling in the above equation. The time-varying magnetic dipole moment emits radiation at a rate

$$\frac{dE}{dt} = -\frac{2}{3c^3}|\ddot{\mathbf{m}}|^2 = -\frac{B^2 R^6 \Omega^4 \sin^2\alpha}{6c^3}. \quad (6.3)$$

[This is completely analogous to the electric dipole radiation discussed in Chap. 4, Subsection 4.3.1. Equation (6.3) is the magnetic counterpart of Eq. (4.39) of Vol. I.] This radiation is emitted at the frequency of rotation Ω of the pulsar and vanishes in the case of an aligned rotator with $\alpha = 0$. Assuming that the energy is supplied by the kinetic energy of rotation, $K = (1/2)I\Omega^2$, it follows that

$$\frac{dE}{dt} = I\Omega\dot{\Omega} = -\frac{B^2 R^6 \Omega^4 \sin^2\alpha}{6c^3}. \tag{6.4}$$

Using the expression for (dE/dt), we can define a characteristic time scale T by

$$T \equiv -\left(\frac{\Omega}{\dot{\Omega}}\right)_0 = \frac{6Ic^3}{B^2 R^6 \Omega_0^2 \sin^2\alpha}, \tag{6.5}$$

where the subscript 0 denotes the current value. Integrating Eq. (6.4) and fixing the initial conditions correctly, we get

$$\Omega = \Omega_i \left(1 + \frac{2\Omega_i^2}{\Omega_0^2}\frac{t}{T}\right)^{-1/2}, \tag{6.6}$$

where Ω_i is the initial angular velocity at $t = 0$. This equation gives the angular velocity of the pulsar as a decreasing function of time; clearly, the pulsar is spinning down because of energy loss. Setting $\Omega = \Omega_0$ gives the present age of the pulsar as

$$t = \frac{T}{2}\left(1 - \frac{\Omega_0^2}{\Omega_i^2}\right) \simeq \frac{T}{2} \quad \text{(for } \Omega_0 \ll \Omega_i\text{)}. \tag{6.7}$$

These formulas can be used to make some simple estimates regarding the pulsar emission. For a spherical neutron star, with $M \approx 1.4\,M_\odot$, $R \approx 12$ km, and $I \approx 1.4 \times 10^{45}$ gm cm^2, the total kinetic energy is $K = 2\pi^2(I/P^2) \simeq 2.5 \times 10^{49}$ ergs for $P \approx 0.03$ s. The Crab pulsar, for example, originated as the result of a supernova explosion that occurred in 1054 A.D. and hence is $T \cong 10^3$ yr old. Its period, $P = (2\pi/\Omega_0)$, is ~0.033 s. Determining $|\dot{\Omega}| = (\Omega_0/T) \approx 10^{-8}$ s^{-2}, we can estimate the energy-loss rate to be approximately $(dE/dt) = 6.4 \times 10^{38}$ ergs s^{-1}. If $\sin\alpha \approx 1$, then Eq. (6.5) allows us to determine the magnetic field as $B \approx 5.2 \times 10^{12}$ G. These values appear to be reasonable. Figure 6.1 gives the P and the $\dot{P}$ values for over 700 pulsars with lines of constant age marked on it. (The line marked "spin-up limit" will be discussed in Chap. 7, Section 7.6.)

The dipole radiation model, described above, leads to a relation of the form $\dot{\Omega} \propto -\Omega^n$ with $n = 3$. Motivated by this form, it is possible to define an index n, in general, as $n \equiv (\Omega\ddot{\Omega}/\dot{\Omega}^2)$. The quantities on the right-hand side of this equation can be determined observationally – in principle – thereby leading to an estimate of n that can be compared with theory. Unfortunately, observations are not accurate enough (in many cases) to constrain the theory effectively.

For a general n, the equation $\Omega\ddot{\Omega} = n\dot{\Omega}^2$ has the solution

$$\Omega(t) = \Omega_0\left(1 + \frac{t}{\tau}\right)^{-1/(n-1)} \equiv \Omega_0\left(1 + \frac{t}{\tau}\right)^{-\alpha}, \tag{6.8}$$

where $\alpha = 1/(n-1)$ and Ω_0, τ are constants. From Eq. (6.4), it follows that

6.3 Electromagnetic Field Around the Pulsar

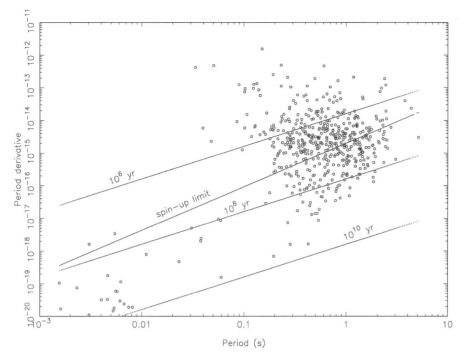

Fig. 6.1. Period and period derivative for a sample of over 700 pulsars. This figure is based on the catalogue maintained by the Princeton Pulsar Group.

$B^2(t) \propto \dot{\Omega}\Omega^{-3}$ varies as

$$B(t) = B_0\left(1 + \frac{t}{\tau}\right)^{-\frac{1}{2}\left(\frac{n-3}{n-1}\right)} \equiv B_0\left(1 + \frac{t}{\tau}\right)^{\alpha-(1/2)}. \quad (6.9)$$

When $n > 3$ [so that $\alpha < (1/2)$], such a model could represent a pulsar whose magnetic field is decaying in time because of some process. If, for example, $n = 4$, then

$$\Omega = \Omega_0\left(1 + \frac{t}{\tau}\right)^{-1/3}, \quad B = B_0\left(1 + \frac{t}{\tau}\right)^{-1/6}, \quad B_0^2 \equiv \left(\frac{2I}{R^6}\right)\frac{c^3}{\Omega_0^2 \tau}\frac{1}{\sin^2\alpha}. \quad (6.10)$$

Assuming that a pulsar is born with $P_0 \approx 1$ ms, $B_0 \approx 10^{12}$ G, $I \approx 10^{45}$ gm cm^2, and $R \approx 10$ km, then we can estimate τ to be $\tau \approx (4/\sin^2\alpha)$ yr. If we take $\alpha \approx (\pi/4)$ as a typical value, giving $\sin^2\alpha \approx 0.5$, then the decay time scale is $\tau \approx 8$ yr.

Exercise 6.2

Gravitational radiation from a deformed neutron star: We have seen in Vol. I, Chap. 11, Section 11.12 that a system with a time-varying quadrupole moment will emit gravitational radiation. If the neutron star is slightly deformed so that it can be described as a homogeneous ellipsoid with moment of inertia I and ellipticity ϵ (which is defined as the ratio between the difference in the equatorial radius and mean equatorial radius), then it will emit gravitational radiation as it spins. (a) Show that the energy loss that is due to gravitational radiation is given by

$$\dot{E}_{GW} = -\frac{32}{5}\frac{G}{c^5}I^2\epsilon^2\Omega^6 = 1.4 \times 10^{38} \text{ ergs s}^{-1} \left(\frac{I}{1.4 \times 10^{45} \text{ gm cm}^{-3}}\right)^2$$

$$\times \left(\frac{P}{0.0331 \text{ s}}\right)^{-6}\left(\frac{\epsilon}{3 \times 10^{-4}}\right)^2. \tag{6.11}$$

(b) When both the magnetic dipole *and* gravitational radiation are present as sources of energy loss, the spin-down rate is determined by an equation of the type

$$I\Omega\dot{\Omega} = -\beta\Omega^4 - \gamma\Omega^6. \tag{6.12}$$

Integrate this equation to find the age t as

$$(1+\lambda)\left(1 - \mu + \lambda \log \frac{\lambda + \mu}{\lambda + 1}\right) = \frac{2t}{T}, \tag{6.13}$$

where

$$\frac{1}{T} \equiv -\frac{\dot{\Omega}}{\Omega}\bigg|_0 = \frac{\beta\Omega_0^2 + \gamma\Omega_0^4}{I}, \quad \lambda \equiv \frac{\gamma\Omega_0^2}{\beta}, \quad \mu \equiv \frac{\Omega_0^2}{\Omega_i^2}, \tag{6.14}$$

and Ω_i is the initial angular velocity at $t = 0$. (c) For the Crab pulsar, using $t \simeq 900$ yr and $T \simeq 2500$ yr, show that $\lambda \simeq 0.27$ and is insensitive to μ for $\mu \lesssim 0.01$. Hence show that $\epsilon \simeq 3 \times 10^{-4}$ and $B \sin\alpha \simeq 4.6 \times 10^{12}$ G. What fraction of the energy is radiated as gravitational waves at the present epoch? [Hint: By using the quadrupole formula for radiation derived in Vol. I, Chap. 11 (see Eq. 11.232), it is easy to show that, for a rotating ellipsoid with $\phi = \Omega t$,

$$\frac{dE}{dt} = -\frac{1}{5}\frac{G}{c^5}\langle \dddot{I}_{xx}^2 + 2\dddot{I}_{xy}^2 + \dddot{I}_{yy}^2 \rangle$$

$$= -\frac{1}{5}\frac{G}{c^5}\frac{1}{4}(2\Omega)^6(I_1 - I_2)^2 \langle \cos^2 2\phi + 2\sin^2 2\phi + \cos^2 2\phi \rangle$$

$$= -\frac{32}{5}\frac{G}{c^2}(I_1 - I_2)^2\Omega^6, \tag{6.15}$$

where the averaging is over one period and

$$I_1 = \frac{1}{5}M(b^2 + c^2), \quad I_2 = \frac{1}{5}M(a^2 + c^2), \quad I_3 = \frac{1}{5}M(a^2 + b^2), \tag{6.16}$$

with a, b, and c denoting the semiaxes of the homogeneous ellipsoid. When $a \simeq b$ and $\epsilon \simeq 2(a-b)/(a+b) \ll 1$, this reduces to the result quoted in the problem.]

Given the above parameters, we can also estimate several other useful quantities. To begin with, consider the radio luminosity of the pulsar that may be taken to be a fixed fraction $f \approx 10^{-5}$, say, of the total loss of rotational energy. From

6.3 Electromagnetic Field Around the Pulsar

the above results,

$$L_{\text{radio}} = f|\dot{E}| = \left(\frac{fI\omega_0^2}{3\tau}\right)\left(1 + \frac{t}{\tau}\right)^{-(2\alpha+1)}$$

$$\approx 4 \times 10^{38} \text{ ergs s}^{-1} \left(1 + \frac{t}{\tau}\right)^{-(2\alpha+1)} \left(\frac{f}{10^{-5}}\right). \quad (6.17)$$

The radio flux from such a pulsar located at a distance R along the galactic disk is given by $F = (L/4\pi R^2)$. Assuming that we can detect radio fluxes above a particular threshold value $F_{\min}$, we will be able to see all the pulsars up to a distance $R_{\max}$, where $R_{\max}$ and $F_{\min}$ are related by the equation $R_{\max}^2 \propto (L/F_{\min}) \propto F_{\min}^{-1} P^{-(2\alpha+1)/\alpha}$. For $\alpha = (1/3)$, this gives

$$R_{\max}^2 = \frac{L}{4\pi F_{\min}} = \left(\frac{\pi}{3}\right)\left(\frac{fIP_0^3}{\tau}\right)\left(\frac{1}{F_{\min}P^5}\right). \quad (6.18)$$

Because this relationship involves the period P, there is a selection effect in any flux-limited sample for different periods of the pulsar. To make this quantitative, let us assume that the spatial number density of pulsars with a period in the range $(P, P + dP)$ is given by $n(P)\,dP$. If the pulsars are born with negligible spread in the initial period with a birthrate b, it follows that $b\,dt = n(P)\,dP$, where $P(t) = P_0[1 + (t/\tau)]^\alpha$. Then

$$n(P) = b\frac{dt}{dP} = \frac{b\tau}{\alpha P_0}\left(\frac{P}{P_0}\right)^{(1/\alpha)-1}. \quad (6.19)$$

The total number of pulsars with period P that can be observed in a flux-limited sample will be $N(P) = n(P)[\pi R_{\max}^2 h]$, where h is the thickness of the galactic disk. Using the expression for $R_{\max}^2$, we get

$$N(P) \propto n(P)R_{\max}^2 \propto P^{[(1/\alpha)-1]-[2+(1/\alpha)]} \propto P^{-3}. \quad (6.20)$$

We thus arrive at the interesting result that the P-dependence of the distribution of $N(P)$ is independent of the breaking index n or, equivalently, α. For the case of $n = 4$, $\alpha = 1/3$, we get

$$N(P) = \left(\frac{\pi^2}{3}\right)\left(\frac{fIhP_0^3}{\tau F_{\min}}\right)\left[\frac{n(P)}{P^5}\right] = \left(\frac{\pi^2 fIhb}{F_{\min}}\right)\left(\frac{1}{P^3}\right). \quad (6.21)$$

The cumulative number of pulsars with period greater than P that we will observe scales as

$$N(>P) = \int_P^\infty N(P')\,dP' = \frac{\pi^2 fIhb}{2}\left(\frac{1}{F_{\min}P^2}\right) \propto (F_{\min}P^2)^{-1}, \quad (6.22)$$

showing that we would preferentially detect lower-period pulsars in a flux-limited survey. To make a numerical estimate of the total number of pulsars in a given survey, consider an example with $F_{\min} \approx 10^{-18}$ ergs s^{-1} cm^{-2}, $f = 10^{-5}$, $\tau = 8$ yr,

$P_0 = 10^{-3}$ s, and $I = 10^{45}$ gm cm². Let us also assume that the pulsar birthrate is comparable with the galactic supernova rate of about 1 in 50 yr per galactic volume. Because the radius of the region surveyed has to be less than the size of the galaxy, we expect $R_{\max} < R_{\text{gal}} \simeq 10$ kpc. Equation (6.18) then shows that pulsars with periods larger than the minimum value of $P_{\min} \approx 0.13$ s can be detected wherever they are in the galaxy whereas those with periods in the range $(P_0, P_{\min})$ will be detected only if they are at $R < R_{\max}(P)$. Hence the total number of pulsars is given by

$$N_{\text{tot}} = \int_{P_0}^{P_{\min}} n(P) \pi R_{\text{gal}}^2 h \, dP + \int_{P_{\min}}^{\infty} n(P) \pi R_{\max}^2(P) h \, dP. \quad (6.23)$$

Substituting the numbers, we find that $N(>0.13 \text{ s}) \approx 6.51 \times 10^5$. The contribution from pulsars in the period $P_0 \approx 10^{-3}$ s to $P_{\min} \approx 0.13$ s is approximately 4.36×10^5. The total number of pulsars we would expect to find in a galaxy with a radius of ~ 10 kpc is $\sim 10^6$.

In the above analysis we have tacitly assumed that (1) pulsars are born with constant parameter values in the galactic disk, (2) their motion away from the galactic disk is ignorable, (3) they have a breaking index $n = 4$, and (4) they radiate isotropically. Each of these assumptions can be relaxed for a more sophisticated analysis.

6.3.1 The Aligned Rotator

The model for the pulsar described in the last section is based on several approximations: (1) We have assumed that the region surrounding the pulsar is a vacuum and that the radiation escapes freely, (2) we have also ignored all general relativistic effects near the neutron star, and (3) we have not bothered to solve for the electromagnetic fields both inside and outside the neutron star and to match it correctly at the surface. We shall now tackle some of these issues, starting from the simplest possible modification.

We can obtain useful model for the pulsar and the magnetosphere by treating the neutron star as a conducting, rotating sphere of mass M, radius R, angular velocity Ω, and with a magnetic field B inside. If we take the outside region to be a vacuum, then we are required to solve Maxwell's equations both inside a rotating spherical conductor and on the outside and to match the solutions at the surface. Because the surface gravity of a neutron star is $g = (GM/R^2) \approx 1.9 \times 10^{14}$ cm s^{-2}, the scale height of the hydrogen atmosphere at a temperature T is $H = (k_B T/m_H g) \approx 0.4(T/10^6 \text{ K})$ cm. This is smaller than the other relevant length scales in the problem and hence the surface between the conductor and vacuum can be thought of as a sharp discontinuity.

The simplest internal magnetic-field configuration corresponds to either (1) a uniformly magnetised sphere with $\mathbf{B} = B_0 \mathbf{e}_z$ or (2) a point dipole at the origin

6.3 Electromagnetic Field Around the Pulsar

with its magnetic moment along the z axis. Because the rotation is about the z axis, it is clear that there will be no radiation field arising from time-varying multipole moments. We shall discuss case (1) in detail and quote the results for case (2).

The velocity at any point $\mathbf{r}$ in the conductor, corresponding to the uniform rotation, is given by $\mathbf{v} = \mathbf{\Omega} \times \mathbf{r} = (\Omega r \sin\theta)\mathbf{e}_\phi$, where we are using the standard spherical coordinates (r, θ, ϕ). When the neutron star rotates, a charge q in the interior feels a Lorentz force $(q/c)(\mathbf{v} \times \mathbf{B})$ and moves until an electric field $\mathbf{E}$ is generated, which can compensate for the magnetic force. The vanishing of the net force implies that

$$\mathbf{E} = -\frac{\mathbf{v} \times \mathbf{B}}{c} = -\frac{\mathbf{\Omega} \times \mathbf{r}}{c} \times \mathbf{B} = -\frac{\Omega B_0 r \sin\theta}{c}(\sin\theta\, \mathbf{e}_r + \cos\theta\, \mathbf{e}_\theta). \quad (6.24)$$

This electric field satisfies the condition $\nabla \times \mathbf{E} = 0$ and hence can be expressed as $\mathbf{E} = -\nabla \phi_{\text{in}}(r, \theta)$. Integrating along $l \equiv r\sin\theta$ from the origin to the surface, we can determine $\phi_{\text{in}}(r, \theta)$ as

$$\phi_{\text{in}} = \left(\frac{\Omega B_0}{2c}\right) r^2 \sin^2\theta + \text{constant} = -\left(\frac{\Omega B_0 r^2}{3c}\right)[P_2(\cos\theta) - 1] + \phi_0, \quad (6.25)$$

where $P_2(\cos\theta)$ is the Legendre polynomial and ϕ_0 is a constant. Because $\mathbf{E} \cdot \mathbf{B} = 0$ inside the star, the magnetic-field lines are equipotentials labelled by the voltage, which in turn is determined by the location at which the particular field line emerges on the star's surface. Outside the star, the electric field is given by $\mathbf{E} = -\nabla \phi_{\text{out}}(r, \theta)$, where $\phi_{\text{out}}(r, \theta)$ satisfies the Laplace equation $\nabla^2 \phi_{\text{out}} = 0$. Taking the general solution to this equation in the form

$$\phi_{\text{out}}(r, \theta) = \sum_{l=1}^{\infty} \frac{a_l}{r^{l+1}} P_l(\cos\theta) \quad (6.26)$$

and matching the potential at the surface of the star $r = R$, we easily see that the outside solution must be

$$\phi_{\text{out}}(r, \theta) = -\frac{\Omega B_0 R^5}{6cr^3}(3\cos^2\theta - 1) \quad (6.27)$$

with the constant ϕ_0 in Eq. (6.25) set to $\phi_0 = -(\Omega B_0 R^2 / 3)$. The corresponding electric field on the outside is given by

$$\mathbf{E}(r, \theta) = -\frac{\Omega B_0 R^5}{2cr^4}(3\cos^2\theta - 1)\mathbf{e}_r - \frac{\Omega B_0 R^5}{cr^4}\sin\theta\cos\theta\, \mathbf{e}_\theta. \quad (6.28)$$

The magnetic field outside is, of course, that of a dipole with moment $m = B_0 R^3$; i.e.,

$$\mathbf{B} = \frac{B_0 R^3 \cos\theta}{r^3}\mathbf{e}_r + \frac{B_0 R^3 \sin\theta}{2r^3}\mathbf{e}_\theta. \quad (6.29)$$

These solutions also ensure that the radial component B_r of the magnetic field and the tangential component E_θ of the electric field are continuous at the surface of the star. The equation for the field lines corresponding to such a magnetic dipole will be of interest later on. These lines $r(\theta)$ satisfy the equation $(B_r/B_\theta) = (dr/rd\theta) = (2\cos\theta/\sin\theta)$, which can be integrated to give

$$r = K \sin^2\theta, \tag{6.30}$$

where K is a constant identifying the field line.

Before proceeding further, it is important that we determine the charge and the current distribution that give rise to these fields by using the equations $\rho = (\nabla \cdot \mathbf{E}/4\pi)$ and $\mathbf{J} = (c/4\pi)\nabla \times \mathbf{B}$. Inside the star these relations give

$$\rho_{\text{in}} = \left(\frac{\nabla \cdot \mathbf{E}}{4\pi}\right) = \left(\frac{\Omega B_0}{2\pi c}\right), \quad \mathbf{J}_{\text{in}} = 0 \quad \text{(interior)}. \tag{6.31}$$

On the outside, both charge and current vanish by explicit construction:

$$\rho_{\text{out}} = \left(\frac{\nabla \cdot \mathbf{E}}{4\pi}\right) = 0, \quad \mathbf{J}_{\text{out}} = 0 \quad \text{(exterior)}. \tag{6.32}$$

To determine the charge and the current on the surface of the star, we have to calculate the discontinuities in the electric and the magnetic fields. The charge density on the surface is given by $\rho_{\text{sur}} = [E_r]/4\pi$, where $[E_r]$ denotes the discontinuity in the radial component of the electric field across the surface. Using this relation we get

$$\rho_{\text{sur}} = \frac{1}{4\pi}(-\phi_{,r}^{\text{out}} + \phi_{,r}^{\text{in}}) = \frac{\Omega B_0 R}{12\pi}[2 - 5P_2(\cos\theta)]. \tag{6.33}$$

Similarly, the surface current is given by

$$\mathbf{J}_{\text{sur}} = \frac{c}{4\pi}(\mathbf{B}^{\text{out}} - \mathbf{B}^{\text{in}}) = \frac{cB_0}{8\pi}\sin\theta\, \mathbf{e}_\theta. \tag{6.34}$$

These results show that an aligned rotator will require (1) a charge density on the inside as well as on the surface and (2) a current flowing on its surface. Given these conditions, we could also check whether the conditions for the validity of the MHD approximation are satisfied. It can be directly verified that the charge density averaged over the star is $\bar{\rho} \approx (\Omega B_0/c)$ and the average current density is $\bar{J} \approx (cB_0/R)$. To satisfy the condition $\bar{\rho}c \ll \bar{J}$, we need $\Omega R \ll c$. This condition also ensures that $v \ll c$ and $E \approx (\Omega R/c)B_0 \ll B_0$, thereby allowing MHD approximations to be used. For typical values of a neutron star, $(\Omega R/c) \approx 10^{-2}$, and hence this approximation can be used with some faith. The same analysis, however, indicates that difficulties will arise, *vis a vis*, the MHD approximation at distances of the order of $r \gtrsim R_L = (c/\Omega)$. We shall see later (see page 313) that this is indeed the case.

We shall now show that, for realistic values of neutron star parameters, the region outside the star is unlikely to be a vacuum as assumed above. This has to do with the high values for the electric field and voltages near the surfaces of the star. Using the scalings obtained above, we clearly see that the voltage and the electric field near the star's surface are

$$\phi_{\text{sur}} \approx \frac{\Omega B_0 R^2}{2c} \approx 3 \times 10^{16} \text{ V} \left(\frac{\Omega/2\pi}{30 \text{ s}^{-1}}\right)\left(\frac{B_0}{10^{12} \text{ G}}\right)\left(\frac{R}{10 \text{ km}}\right)^2, \quad (6.35)$$

and

$$E \approx \frac{\Omega R}{c} B_0 \approx 2 \times 10^8 \left(\frac{B_0}{10^{12} \text{ G}}\right)\left(\frac{\Omega/2\pi}{1 \text{ s}^{-1}}\right) \text{ esu.} \quad (6.36)$$

The electric force that is due to this field on a charged particle is very much stronger than the gravitational force (by a factor of $\sim 10^8$ for a proton), and hence charged particles *can* be pulled out of the neutron star surface to create a magnetosphere around the star. Whether this *actually* occurs or not depends on the details of solid-state physics near the surface of the neutron star, and we shall discuss some aspects of this complex phenomenon in Section 6.4. Assuming for a moment that this does happen, we can ask how the structure of the magnetosphere and the nature of our solution change because of this process.

The charged particle pulled out from the surface will spiral around the magnetic-field lines and will drift along it. The rapid spiralling motion will cause the charge to radiate away the kinetic energy corresponding to transverse motion so that the main residual motion of the charged particle will be along the magnetic-field line. It should also be noted that the particles moving along the magnetic-field lines will be accelerated by the electric-field component $E_\parallel = (\mathbf{E} \cdot \mathbf{B}/|\mathbf{B}|)$ in the direction of the magnetic field. Near the surface, $E_\parallel$ varies as

$$E_\parallel = \frac{\mathbf{E} \cdot \mathbf{B}}{|\mathbf{B}|} = -\frac{\Omega R}{c} B_0 \cos^3 \theta. \quad (6.37)$$

The global motion of the charged particle will depend on whether the magnetic field line to which they are attached closes back before reaching the light cylinder or not. For a dipole field, these lines are described by $(\sin^2 \theta / r) = \text{constant}$ [see Eq. (6.30)]. Field lines starting out within an angular region $\theta < \theta_P$ near the polar cap will cross the light cylinder whereas those that originate at $\theta > \theta_P$ will loop back before reaching the light cylinder. The critical field line, starting at $r = R, \theta = \theta_P$, should have $r = R_L$ at $\theta = (\pi/2)$. The constancy of $(\sin^2 \theta / r)$ implies that $\sin \theta_P = (R/R_L)^{1/2}$. The corresponding radius R_P of the polar cap region is given by

$$R_P \simeq R \sin \theta_P = R\left(\frac{R}{R_L}\right)^{1/2} = 1.4 \times 10^4 \left(\frac{R}{10^6 \text{ cm}}\right)^{3/2}\left(\frac{P}{1 \text{ s}}\right)^{-1/2} \text{ cm.} \quad (6.38)$$

Charges that are pulled out from outside the polar cap region can, in principle, redistribute themselves around the star, forming a corotating magnetosphere. If

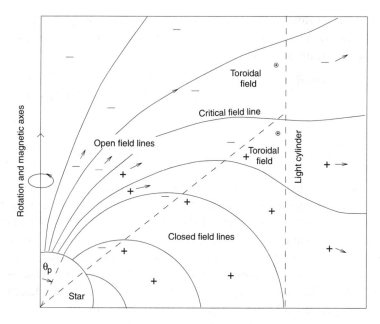

Fig. 6.2. Magnetosphere of a pulsar.

we neglect the inertia of the particles we conclude that the charges will rearrange themselves so that no net electromagnetic force acts on them; that is, $\mathbf{E} + (\mathbf{v}/c) \times \mathbf{B} = 0$, where $\mathbf{v} = \mathbf{\Omega} \times \mathbf{r}$. From the resulting electric field $\mathbf{E}$, we can determine the charge density ρ through the relation $\nabla \cdot \mathbf{E} = (4\pi\rho)$. This gives

$$4\pi c\rho = \nabla \cdot [\mathbf{B} \times (\mathbf{\Omega} \times \mathbf{r})] = -2\mathbf{\Omega} \cdot \mathbf{B}, \qquad (6.39)$$

showing that regions with $\mathbf{\Omega} \cdot \mathbf{B} = \Omega B_z > 0$ will have negative ρ and regions with $B_z < 0$ will have positive ρ. The number density of charges is approximately

$$n_e = 7 \times 10^{-2} B_z \left(\frac{P}{1\,\text{s}}\right)^{-1} \text{cm}^{-3}. \qquad (6.40)$$

The dashed line in Fig. 6.2 (corresponding to $B_z = 0$) separates regions of positive and negative space charge.

Charged particles that are pulled out from within the polar cap region will move along the magnetic field lines towards the light cylinder at $r = R_L$, where the relativistic effects will lead to a breakdown of our MHD approximations and prevent the plasma from corotating with the star. This is also obvious from the fact that a corotating plasma near the light cylinder will acquire speeds close to the speed of light. The structure of field lines as well as the charge density in the magnetosphere are modified drastically near $r = R_L$. The field lines are expected to be swept back near the light cylinder and stream off to infinity outside the light cylinder and are called open field lines. If they exist, then plasma will flow away from the pulsar along the open field lines, which – in the above simplified analysis – arise from a region near the polar cap. The potential difference $\Delta\phi$ between the centre and the edge of the polar cap can be estimated from our solution. If $\theta_P \ll 1$, then

$$\Delta\phi = \frac{\Omega B_0 R^2}{2c} \cdot \frac{R}{R_L} = 6 \times 10^{12} \left(\frac{B}{10^{12}\,\text{G}}\right)\left(\frac{P}{1\,\text{s}}\right)^{-2} \text{V}. \qquad (6.41)$$

The potential difference along the magnetic-field lines over a distance of the order of R, given by $\Delta\phi \approx (\mathbf{E}\cdot\mathbf{B}/B)R$, will also be of the same order as that given above. Charged particles, flowing along open field lines, will be accelerated by a voltage of this order somewhere along their path. This corresponds to energies of 6×10^{12} eV (B_{12}/P^2) and can make electrons highly relativistic with a gamma factor of $\gamma \approx 10^7(B_{12}/P^2)$, where P is measured in seconds.

Because the configuration studied here is axisymmetric, we may believe that there cannot be any energy-loss mechanism. This, however, is not true because of the existence of the light cylinder and open field lines. There is a flow of electromagnetic energy through the light cylinder (and a consequent torque on the pulsar) that can be estimated as follows. The magnetic field strength at the light cylinder is approximately $(B_0 R^3/2R_L^3)$. Because of the twisting of the field lines near the light cylinder we would expect $B_r \approx B_\phi \approx (B_0 R^3/2\sqrt{2}R_L^3)$. The moment of the $r\phi$ component of the magnetic stress tensor, $T_{r\phi} = (B_r B_\phi/4\pi)$, will lead to the torque:

$$I\frac{d\Omega}{dt} = \frac{B_r B_\phi}{4\pi}\left(4\pi R_L^2\right)R_L \simeq -\frac{K}{8c^3}(B_0 R^3)^2 \Omega^3, \qquad (6.42)$$

where K is a numerical factor of the order of unity. Apart from a constant numerical factor, this will give the same spin-down rate for the pulsar as for the rotating dipole model discussed above. The reason for this similarity lies in the fact that the only dimensional combination for the torque involving the correct monotonicity is the expression given above. This result also shows that the estimate based on the simple rotating dipole model has a somewhat more general domain of validity, except for numerical constants of the order of unity. A schematic representation of the magnetic field lines is given in Fig. 6.3 for the case in which the magnetic axis is orthogonal to the rotation axis.

It is possible to provide a more general discussion of the axisymmetric rotating magnetosphere around a pulsar if we assume that the plasma is force free; that is, it satisfies, in the MHD limit, the equation $\mathbf{J}\times\mathbf{B} = 0$, which ensures that the bulk force on the plasma from the magnetic field vanishes. In this case, we must have $\mathbf{J} = \mu\mathbf{B}$ (where μ is a scalar function of spatial coordinates) with an electric current flowing along the direction of the magnetic field. Because the configuration is assumed to be stationary in time, conservation of the electric

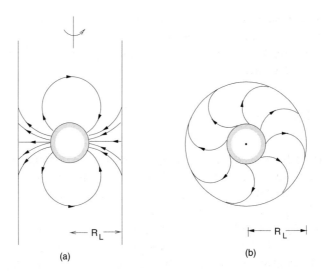

Fig. 6.3. The twisting of magnetic-field lines near the light cylinder.

charge requires that

$$0 = \nabla \cdot \mathbf{J} = \nabla \cdot \mu \mathbf{B} = (\mathbf{B} \cdot \nabla)\mu, \tag{6.43}$$

where we have used the fact that $\nabla \cdot \mathbf{B} = 0$. This equation shows that $\mu =$ constant along each of the field lines. In the axisymmetric case, each magnetic field line will sweep out a magnetic surface when rotated around the symmetry axis of the pulsar (see Fig. 6.4). Because no field line can cross this surface, it can be labelled by the total magnetic flux ψ contained inside it:

$$\psi = \int \mathbf{B} \cdot d\mathbf{A} = \frac{1}{\mu} \int \mathbf{J} \cdot d\mathbf{A} = -\frac{I}{\mu}, \tag{6.44}$$

where I is the current defined by the surface. (The minus sign is conventional and indicates the flow of current *into* the polar region of the star.) The fact that both $I(\mathbf{x})$ and $\psi(\mathbf{x})$ are defined by the same surface implies that we can express them in terms of each other, say, by a functional relation of the form $I = I(\psi)$. From their definitions, it also follows that $\nabla I = -\mu \nabla \psi$.

To proceed further, it is convenient to separate the magnetic field into a poloidal part (which has components in the r, θ directions) and a toroidal part (which is in the ϕ direction). Formally, these two components are defined as

$$\mathbf{B}^T = (\mathbf{B} \cdot \mathbf{e}_\phi) \mathbf{e}_\phi, \quad \mathbf{B}^P = \mathbf{B} - \mathbf{B}^T. \tag{6.45}$$

Because $\psi(\mathbf{x})$ is axisymmetric and does not change along the poloidal magnetic field, it is clear that $\mathbf{B}^P$ is perpendicular to both $\nabla \psi$ and $\mathbf{e}_\phi$. Let us now consider the magnetic flux through an annulus of width dl and an area orthogonal to $\mathbf{B}^P$. If $\mathcal{R} \equiv r \sin \theta$ denotes the radial coordinate in the cylindrical $(\mathcal{R}, \phi, z)$ coordinate

6.3 Electromagnetic Field Around the Pulsar

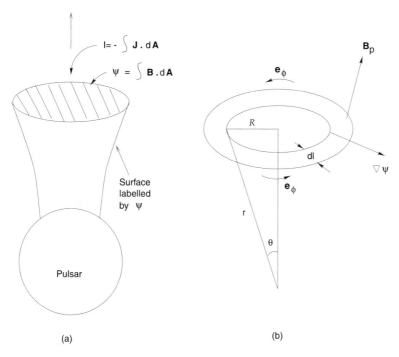

Fig. 6.4. Geometry showing the axisymmetric field and current configuration. (a) The surface generated by the rotation of a magnetic-field line around the axis of symmetry of the pulsar is characterised by the magnetic flux ψ and the current I traversing it. (b) The annular ring of width dl and area defined by a normal parallel to the poloidal magnetic field. The direction of the other vectors and $\nabla \psi$ are shown schematically.

system, then this flux is given by

$$2\pi \mathcal{R} B^P dl = \psi(x + dl) - \psi(x) = d\mathbf{l} \cdot \nabla \psi, \qquad (6.46)$$

which can be expressed as a vector equation:

$$\mathbf{B}^P = -\frac{1}{2\pi \mathcal{R}}(\mathbf{e}_\phi \times \nabla \psi). \qquad (6.47)$$

Thus the poloidal part of the magnetic field is directly related to $\psi(\mathbf{x})$. By repeating the same analysis using I and $\mathbf{J}$, we will arrive at the relation

$$\mathbf{J}^P = +\frac{1}{2\pi \mathcal{R}}(\mathbf{e}_\phi \times \nabla I). \qquad (6.48)$$

The toroidal part of the magnetic field can be determined from Ampere's law, expressed in the form

$$\oint \mathbf{B} \cdot d\mathbf{l} = \frac{4\pi}{c} \int \mathbf{J} \cdot d\mathbf{A} = -\frac{4\pi}{c} I \qquad (6.49)$$

312   6 *Pulsars*

and applied to a circle of constant $\mathcal{R}$ and z, leading to

$$\mathbf{B}^T = -\frac{2I}{c\mathcal{R}} \mathbf{e}_\phi. \tag{6.50}$$

The corresponding result for the toroidal part of the current will be

$$\mathbf{J}^T = -\frac{2\mu I}{c\mathcal{R}} \mathbf{e}_\phi = \frac{2I}{c\mathcal{R}}\left(\frac{dI}{d\psi}\right)\mathbf{e}_\phi, \tag{6.51}$$

where we have used the relation $(dI/d\psi) = -(J/B) = -\mu$ to eliminate μ. The electric field $\mathbf{E}$ can be determined from the expressions for the magnetic field because, in the force-free case, $\mathbf{E} = -(\mathbf{v}/c) \times \mathbf{B}$ with $\mathbf{v} = \mathbf{\Omega} \times \mathbf{r} = \Omega\mathcal{R}\mathbf{e}_\phi$. Clearly the electric field is poloidal and is determined entirely by the poloidal part of the magnetic field:

$$\mathbf{E} = \mathbf{E}^P = -\frac{\mathbf{v}}{c} \times \mathbf{B} = -\left(\frac{\Omega\mathcal{R}}{c}\right)(\mathbf{e}_\phi \times \mathbf{B}^P). \tag{6.52}$$

We have thus determined all the fields in terms of the functions $I(\psi)$ and $\psi(\mathbf{x})$. To determine a relation between these two functions we can proceed as follows. Using Maxwell's equation $\mathbf{J} = (c/4\pi)\nabla \times \mathbf{B}$, we get an expression for the toroidal part of current as

$$\mathbf{J}^T = \frac{c}{4\pi}(\nabla \times \mathbf{B}^P) = -\frac{c}{8\pi^2}\nabla \times \left(\frac{\mathbf{e}_\phi \times \nabla\psi}{\mathcal{R}}\right), \tag{6.53}$$

where we have used Eq. (6.47). Equating this expression to Eq. (6.51), we get

$$\nabla \times \left(\frac{\mathbf{e}_\phi \times \nabla\psi}{\mathcal{R}}\right) + \left(\frac{16\pi^2}{c^2}\right)\left(\frac{I}{\mathcal{R}}\right)\left(\frac{dI}{d\psi}\right)\mathbf{e}_\phi = 0. \tag{6.54}$$

To simplify this equation, we expand the first term, obtaining

$$\nabla \times \left(\frac{\mathbf{e}_\phi \times \nabla\psi}{\mathcal{R}}\right) = \mathcal{R}\mathbf{e}_\phi \nabla \cdot \left(\frac{\nabla\psi}{\mathcal{R}^2}\right) + \left(\frac{\nabla\psi}{\mathcal{R}^2} \cdot \nabla\right)\mathcal{R}\mathbf{e}_\phi - \mathcal{R}(\mathbf{e}_\phi \cdot \nabla)\left(\frac{\nabla\psi}{\mathcal{R}^2}\right), \tag{6.55}$$

where we have used the fact that $\nabla(\mathcal{R}\mathbf{e}_\phi) = 0$. Further, because $\mathbf{e}_\phi$ and $\nabla\psi$ are orthogonal and $\nabla\mathcal{R} = \mathbf{e}_\mathcal{R}$ and $(\partial\mathbf{e}_\mathcal{R}/\partial\phi) = \mathbf{e}_\phi$, we get

$$\left(\frac{\nabla\psi}{\mathcal{R}^2} \cdot \nabla\right)\mathcal{R}\mathbf{e}_\phi = \left(\frac{\nabla\psi}{\mathcal{R}^2}\right)_\mathcal{R} \mathbf{e}_\phi;$$

$$\mathcal{R}(\mathbf{e}_\phi \cdot \nabla)\left(\frac{\nabla\psi}{\mathcal{R}^2}\right) = \frac{\partial}{\partial\phi}\left(\frac{\nabla\psi}{\mathcal{R}^2}\right) = \left(\frac{\nabla\psi}{\mathcal{R}^2}\right)_\mathcal{R} \mathbf{e}_\phi, \tag{6.56}$$

showing that the last two terms on the right-hand side of Eq. (6.55) cancel each other. Substituting the surviving first term into Eq. (6.54), we get the final equation

$$\nabla \cdot \left(\frac{\nabla \psi}{\mathcal{R}^2}\right) + \frac{16\pi^2}{c^2} \frac{I}{\mathcal{R}^2} \frac{dI}{d\psi} = 0, \quad (6.57)$$

which relates the function $\psi(\mathbf{x})$ to $I(\psi)$. Once the latter is specified, this equation can be solved to give $\psi(\mathbf{x})$ and thus determine all other fields.

In a self-consistent model for the pulsar, the function $I(\psi)$ is determined by the behaviour of the fields and plasma near and outside the light cylinder as a boundary condition. In this region, the MHD approximation breaks down and charged particles are not tied down to move along the field lines. Because the particles can move across the magnetic field, there is a possibility that currents can be closed outside the light cylinder, leading to a consistent picture. In the absence of such a detailed physical model, we can still try different possible choices for $I(\psi)$ and explore the consequences; the above equation can serve as a good phenomenological tool for such model building.

The relation $\mathbf{J} = \mu \mathbf{B}$ shows that the magnetic-field lines act as current-carrying transmission lines. In such a case, the associated electromagnetic field will lead to a poloidal component of the Poynting flux, given by

$$\left(T^{0j}\mathbf{e}_j\right)^P = \frac{c}{4\pi}(\mathbf{E} \times \mathbf{B})^P = \frac{\Omega I}{2\pi c} \mathbf{B}^P. \quad (6.58)$$

In arriving at the last step, we have used the explicit forms obtained for the electric and the magnetic fields and have simplified the resulting expression by using elementary vector identities. There is also an accompanying flux of the z component of the angular momentum; the poloidal part of this angular-momentum flux is given by

$$\mathbf{L}^P = -\frac{\mathcal{R}}{4\pi} B_\phi \mathbf{B}^P = \frac{I}{2\pi c} \mathbf{B}^P = \frac{1}{\Omega}\left(T^{0j}\mathbf{e}_j\right)^P. \quad (6.59)$$

The loss of energy and angular momentum (and the resulting spin-down of the pulsar) are of the same order of magnitude as in the simpler case discussed above.

The aligned rotator, however, does not pulse. However, we can handle this by modifying it as an oblique rotator in which the magnetic field and the rotation axis are misaligned. Many of the qualitative features do not change when the misalignment is introduced.

There are, however, other fundamental problems with the above model: (1) To begin with, some of the open-field-line regions have a space charge of one sign but the current is carried by the motion of charges of opposite sign; (2) it is also not clear whether the current closes back properly in this model; (3) finally, we need to demonstrate that charged particles can indeed be pulled out from the surface of the pulsar in the manner assumed above to make the discussion complete; this is also not an easy task.

6.3.2 Pair Production in a Pulsar Atmosphere

Even if charged particles could not be pulled out copiously, a magnetosphere made of electron–positron pairs can still arise around the pulsar for the following reason: Any stray electron in the magnetosphere, spiralling along a magnetic-field line, will be accelerated by the electric-field component $E_\| = (\mathbf{E} \cdot \mathbf{B}/|\mathbf{B}|)$ parallel to the magnetic field. Near the surface, $E_\|$, given in Eq. (6.37), is capable of accelerating electrons to high energies. As we see from Eq. (6.41), this leads to highly relativistic electrons with a gamma factor of $\gamma \approx 10^7 (B_{12}/P^2)$. These electrons spiral around the magnetic field lines with a mean drift motion along the direction of the field lines. The kinetic energy that is due to the spiralling motion, transverse to the local direction of magnetic field, is quickly lost through synchrotron emission. We saw in Vol. I, Chap. 6, Section 6.10, that the lifetime for such synchrotron radiation is given by

$$\tau \simeq 1.5 \times 10^{-11} B_{12}^{-3/2} \left(\frac{\nu_c}{1 \text{ GHz}}\right)^{-1/2} \text{ s}, \tag{6.60}$$

where $\nu_c \equiv (c/2\pi\rho)\gamma^3$ and ρ is the radius of curvature of the orbit. This time scale τ is quite short for the pulsars. Parallel to the magnetic field, however, the $E_\|$ component of the electric field accelerates the particles to a high value of γ. Because the magnetic-field lines are curved, the forward motion is also on a curved track with a radius of curvature of the order of $\rho \approx (rc/\Omega)^{1/2}$. Such a relativistic particle, with charge e, mass m, and energy $\mathcal{E} = \gamma m c^2$, moving in an orbit with a radius of curvature ρ, will radiate a synchrotron spectrum of the form

$$I(\nu) \approx \frac{e^2}{2\pi c} \frac{c}{\rho} \left(\frac{\nu}{\nu_c}\right)^{1/3} \gamma \quad (\text{for } \nu \lesssim \nu_c), \tag{6.61}$$

with

$$\nu_c = \frac{c}{2\pi\rho}\gamma^3 = \frac{\gamma^3}{2\pi}\left(\frac{c\Omega}{r}\right)^{1/2} \approx 10^{23}\left(\frac{\gamma}{10^7}\right)^3 \left(\frac{\rho}{10^8 \text{ cm}}\right)^{-1} \text{ Hz}$$

$$\approx 10^8 \left(\frac{\gamma}{10^7}\right)^3 \left(\frac{\rho}{10^8 \text{ cm}}\right)^{-1} \text{ (eV/h)}. \tag{6.62}$$

The primary charges around the polar region will radiate gamma rays tangentially, and these rays, if they are energetic enough, will give rise to electron–positron pair production. These secondary charges will be accelerated in opposite directions and will, in turn, give rise to more electron–positron pairs, thereby establishing a cascade (see Fig. 6.5). We will now discuss some aspects of quantum-field-theoretical processes taking place in such a context.

A photon in vacuum cannot convert itself into an electron–positron pair without violating the conservation of energy and momentum. The photon, however, can convert itself into a virtual pair of particles of mass m and charge q for a

6.3 Electromagnetic Field Around the Pulsar

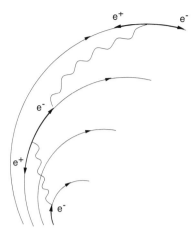

Fig. 6.5. Curvature radiation and pair cascading.

short period of time $\Delta t \approx (\hbar/mc^2)$, after which the charged particles will recombine. The situation is different when a strong electric field is present, as the electric field can now accelerate the virtual particles sufficiently to allow the virtual pair production to become a real one. This effect becomes important when the work done by the electric field in moving the charged particle over a distance $\hbar/mc$ is comparable with the rest-mass energy of the particles mc^2. This occurs for electric fields with a strength $E > E_c$, where $qE_c(\hbar/mc) = mc^2$ or $E_c = (m^2c^3/q\hbar)$. Because magnetic fields cannot work on a charged particle, such an effect normally does not take place in a pure magnetic field; that is, this effect requires that the Lorentz invariant condition $(E^2 - B^2) > 0$ be satisfied in any reference frame.

Although a pure magnetic field or a pure gamma-ray photon cannot, by itself, lead to pair production, the existence of both together can lead to this effect. Qualitatively, this arises because the total electric field now is due to that of the gamma-ray photon (treated as electromagnetic radiation) whereas the total magnetic field is the sum of the pulsar magnetic field and that of the electromagnetic radiation. It is possible that $(E_{\text{tot}}^2 - B_{\text{tot}}^2) > 0$, thereby leading to pair production. The energy $\hbar\omega$ of the initial photon will now go into producing a pair of particles with a γ factor $\gamma \approx (\hbar\omega/2mc^2)$.

The mean free path $l = \kappa^{-1}$ for photons with energy $\hbar\omega > 2m_ec^2$ moving at an angle θ to the magnetic field can be determined from theory to be[2]

$$l = \left(\frac{4.4}{e^2/\hbar c}\right)\left(\frac{\hbar}{m_e c}\right)\left(\frac{B_q}{B \sin\theta}\right) \exp(4/3\chi), \tag{6.63}$$

where

$$\chi = \frac{\hbar\omega}{m_e c^2} \frac{B \sin\theta}{B_q}, \quad B_q = \frac{m_e^2 c^3}{e\hbar} \approx 4.4 \times 10^{13} \text{ G}, \tag{6.64}$$

and the expressions are valid for $\chi \ll 1$. For a magnetic field of $B \approx 10^{12}$ G, we have $l^{-1} = \kappa \approx 2 \times 10^6$ cm^{-1} for typical values of the parameters. The exponential sensitivity shows that a tiny change in χ will lead to large changes in l; thus the magnetosphere acts as a surface of an opaque solid with attenuation varying from zero to unity over a short distance. We can estimate the location r of this surface by ignoring the spatial variations of B and θ and setting $\kappa r = 1$. For a specific photon energy, viewing angle, and magnetic-field variation $B(r)$, this equation defines a three-dimensional surface surrounding a pulsar. A photon of energy γ produced inside this surface would have degraded into $e^+ - e^-$ pairs, leading to a cascade.

In this secondary pair-creation case, it is vital that the electrons accelerated across the potential drop be able to create energetic photons by curvature radiation, which can produce e^+e^- pairs off the magnetic field. It is possible to derive an interesting inequality in the B–P plane that needs to be satisfied if the pulsar magnetosphere is produced by this model. The typical potential drop across the open field lines is of the order of $\Delta V \approx (B\Omega^2 R^3/2c^2)$ [see Eq. (6.41)]. Equating the energy $e\Delta V$ acquired by an electron in this potential drop to $\gamma m_e c^2$, we find that the typical γ factor of the accelerated electron will be

$$\gamma = \left(\frac{2\pi^2 e}{m_e c^4}\right) \frac{BR^3}{P^2} \approx 1.3 \times 10^7 \left(\frac{B}{10^{12}\ \text{G}}\right)\left(\frac{P}{1\ \text{s}}\right)^{-2}. \tag{6.65}$$

The typical energy of the photons radiated because of curvature radiation will be $\hbar\omega \approx \gamma^3(\hbar c/\mathcal{R})$, where $\mathcal{R}$ is the curvature of the magnetic-field line [see Eq. (6.62)]. Assuming that $\mathcal{R} \approx R \approx 10^6$ cm, we find that $\hbar\omega \approx 0.1 B_{12}^3 P^{-6}$ ergs. To induce pair production, we need to satisfy the condition $(\hbar\omega/2m_e c^2) > 1$. This gives

$$B_{12} P^{-2} \geq 0.025. \tag{6.66}$$

This condition is valid at reasonably strong fields.

At somewhat lower values of the field, a more stringent condition has to be met, viz., the mean free path of photons produced by curvature radiation must be less than the typical curvature scale length R. This mean free path $l = \kappa^{-1}$ for photons with energy $\hbar\omega > 2m_e c^2$ moving at an angle θ to the magnetic field is determined from Eq. (6.63). Using the condition $l < R$, and substituting for $\hbar\omega$ from the previous analysis, we find that a straightforward calculation gives the condition

$$P^6 < \frac{9\pi^6}{2} \frac{m_e^2 R^8}{\hbar^2 c^4} \left(\frac{B}{B_q}\right)^4 \ln\left[\frac{Rm_e e^2}{4.4\hbar^2} \frac{B}{B_q}\right]. \tag{6.67}$$

Substituting the numerical values for different terms, we find that the logarithmic

6.3 Electromagnetic Field Around the Pulsar 317

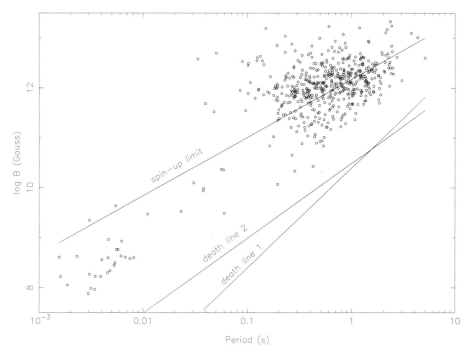

Fig. 6.6. Positions of over 700 pulsars in the B–P plane. The two death lines are based on $B \propto P^2$ and $B \propto P^{3/2}$ derived in Section 6.3. (Millisecond pulsars, if they were spun up by magnetic torque, must lie below the line marked "spin-up limit"; this will be discussed in Chap. 7.) The figure is produced from the data maintained by the Princeton Pulsar Group (http://pulsar.princeton.edu/pulsar).

dependence in B is subdominant and the constraint can be expressed in the form

$$\left(\frac{B}{10^9 \text{ G}}\right)\left(\frac{P}{1 \text{ s}}\right)^{-3/2} > 31. \tag{6.68}$$

This condition is stronger at low magnetic-field strengths. The two constraints (6.66) and (6.68) lead to a line in the B–P plane such that if the condition is violated (for example, because of the decay of the magnetic field) then the coherent radio emission by secondary pair creation cannot take place. It is conventional to call this the *death line* for the pulsars. Figure 6.6 shows over 700 pulsars in the B–P plane with the death-line constraints marked on it. (The line marked "spin-up limit" will be discussed in Chap. 7.)

The situation is different if the primary charges are protons. Because of their greater mass, they will be accelerated less and their energy might never reach the radiation-reaction limit. The pair cascade will have to be established at a longer distance from the surface of the star where the magnetic-field strength is weaker. This has the effect that the synchrotron self-absorption peak will move to lower frequencies, that is, into the radio region, leaving the optical and the x-ray pulses to escape. The corresponding upper limit on the period will decrease by a factor $(m_p/m_e)^{2/3}$ and will be ~ 0.05 s. In some of the models for pulsar emission, a preferential ejection of electrons could be achieved if the magnetic field is oriented antiparallel to the rotation axis so that the electric field in the polar cap points inwards.

6.4 Atomic Structures in Strong Magnetic Fields

The original model for populating the magnetosphere required the parallel component of the electric field to pull out charged particles from the surface of the pulsar. To understand this process in complete detail, it is necessary to study the surface properties of the pulsar and, in particular, to estimate the field required for inducing the emission of charged particles into the magnetosphere. This problem turns out to be quite intractable, and the final answer is, at present, unknown. To illustrate the complexities involved, we shall describe some simple aspects of a solid permeated by ultrastrong magnetic fields.

Near the surface of the neutron star, the scale height for atomic atmosphere is ~ 0.4 cm and matter is basically atomic in nature with densities in the range of $(10^4 – 10^5)$ gm cm^{-3}. However, the presence of a strong magnetic field changes the atomic structure drastically compared with what is known in the laboratory for weak field strengths. This is most easily seen by an estimate of the energies involved. For a hydrogen atom in a uniform magnetic field $B \approx 10^{12}$ G, the interaction energy of the electron spin with the magnetic field is $E_{\text{spin}-B} \approx \mu_B B \approx 10$ keV which is much larger than the electron–proton Coulomb energy $E_{\text{Coul}} \approx 10$ eV. For comparison, note that the thermal energy per particle is $E_{\text{thermal}} \approx 90$ eV at $T \approx 10^6$ K and the proton-spin interaction energy is $E_{\text{proton}-B} \approx \mu_N B \approx 5$ eV. It follows that the dominant feature is the interaction of the electron spin with the magnetic field with all other interactions treated as perturbations. This is in contrast to standard laboratory atomic physics in which the Coulomb interaction is the main force and the magnetic field is treated as a perturbation.

At $T \approx 10^6$ K, the electron spin will be totally polarised whereas the proton spins will be still randomly oriented. The energy estimates also show that a nonrelativistic treatment of electrons in a magnetic field is valid; then the energy eigenvalue equation for an electron in a uniform magnetic field can be written as

$$\frac{[\mathbf{p} - (e/c)\mathbf{A}]^2}{2m_e} \psi_n = E_n \psi_n, \quad (6.69)$$

where the vector potential $\mathbf{A}$ for a magnetic field B along the z axis is

$$\mathbf{A} = \frac{1}{2} B(-y, x, 0), \quad \nabla \cdot \mathbf{A} = 0. \quad (6.70)$$

6.4 Atomic Structures in Strong Magnetic Fields

Expanding the Schroedinger equation, we obtain

$$-\frac{\hbar^2}{2m_e}\nabla^2\psi_n(r) + \frac{e\hbar B}{4m_e c}i\left(y\frac{\partial}{\partial x} - x\frac{\partial}{\partial y}\right)\psi_n(r)$$
$$+ \frac{e^2 B^2}{8m_e c^2}(x^2 + y^2)\psi_n(r) = E_n\psi_n(r). \tag{6.71}$$

The z component of the motion is separable, with a dependence $\exp(ikz)$ contributing an amount $(\hbar^2 k^2/2m_e)$ to the total energy. The second term on the left-hand side is proportional to the angular momentum that will be conserved because of cylindrical symmetry. Using a cylindrical coordinate system (ρ, ϕ, z) and taking the angular part as $\exp(im\phi)$ [which contributes $(-e\hbar/2m_e c)Bm$ to the total energy], we can reduce the Schroedinger equation to the one for the radial part:

$$-\frac{\hbar^2}{2m_e}\left\{\frac{1}{\rho}\frac{\partial}{\partial\rho}\rho\frac{\partial}{\partial\rho} - \frac{m^2}{\rho^2} + k^2\right\}R_n(\rho) - \left(\frac{eB}{2m_e c}\right)(m\hbar)R_n(\rho)$$
$$+ \frac{e^2 B^2}{8m_e c^2}\rho^2 R_n(\rho) = E_n R_n(\rho). \tag{6.72}$$

We can solve this equation by transforming it to a two-dimensional harmonic-oscillator equation [which should be obvious from Eq. (6.71)]. The eigenfunctions vary as

$$R \propto e^{-\rho^2/2b^2}\left(\frac{\rho}{b}\right)^{|m|}\left(\frac{\rho}{b}\right)^{|m|/2} F[n, |m|+1, (\rho/b)^2], \tag{6.73}$$

where $b = (2\hbar c/eB)^{1/2}$ and F is the confluent hypergeometric function. The corresponding energy levels are given by

$$E_n = 2\left(n + \frac{1}{2}|m| + \frac{1}{2}\right)\hbar\omega, \quad \hbar\omega = \mu_B B = \left(\frac{e\hbar}{2m_e c}\right)B, \tag{6.74}$$

and the total energy of the state specified by the integers n, m is

$$E_{nm}(k) = 2\left(n + \frac{1}{2}|m| - \frac{1}{2}m + \frac{1}{2}\right)\hbar\omega + \frac{\hbar^2 k^2}{2m_e}. \tag{6.75}$$

In the pulsar magnetic field of $B \approx 10^{12}$ G, $\hbar\omega \approx 10$ keV; hence exciting the radial degree of freedom is difficult and we need consider only the oscillator ground state with $n = 0$. This ground state is infinitely degenerate with m, taking values of $0, 1, 2, \ldots$. For motion centred on the z axis, the allowed spiral orbits have the quantised radii

$$\rho_m = (2m + 1)^{1/2} b. \tag{6.76}$$

In the ground state, $\rho_0 = b \approx 10^{-10}$ cm, which is ~ 50 times smaller than the radius of the normal hydrogen atom.

Let us next consider the effect of Coloumb interaction as a perturbation on these levels. Because of the weakness of the perturbation, only the z dependence of the wave function will change appreciably. Assuming that higher oscillator states are not excited, the Coloumb potential can be estimated as

$$\frac{e^2}{r} \equiv \frac{e^2}{\sqrt{(\rho^2 + z^2)}} \simeq \frac{e^2}{\sqrt{(\rho_0^2 + z^2)}} = \frac{e^2}{\sqrt{(b^2 + z^2)}}, \qquad (6.77)$$

and the corresponding z dependence of the wave function is governed by

$$\left[-\frac{\hbar^2}{2m_e} \frac{\partial^2}{\partial z^2} - \frac{e^2}{\sqrt{(b^2 + z^2)}} \right] \phi_j(z) = E_j \phi_j(z). \qquad (6.78)$$

This equation needs to be solved numerically, but some of the qualitative features can be understood from studying the asymptotic behaviour. We have

$$\frac{e^2}{\sqrt{(b^2 + z^2)}} \simeq \begin{cases} \dfrac{e^2}{z}\left(1 - \dfrac{1}{2}\dfrac{b^2}{z^2} + \cdots\right) & (z \gg b) \\[2mm] \dfrac{e^2}{b}\left(1 - \dfrac{1}{2}\dfrac{z^2}{b^2} + \cdots\right) & (z \ll b) \end{cases}. \qquad (6.79)$$

At long distances there arises a one-dimensional Coloumb potential that confines the electron in the z direction, whereas at short distances the potential is that of an oscillator well with depth $(e^2/b) \approx 1$ keV and ground-state energy $\hbar\omega_z \approx [e\hbar/(m_e b^3)^{1/2}] \approx 10$ keV. Because the zero-point energy is much bigger than the depth of the potential, this cannot lead to any bound state. Thus the essential modification arises from only the long-distance behaviour, which gives the asymptotic form of the Schroedinger equation as

$$\left(-\frac{\hbar^2}{2m_e} \frac{\partial^2}{\partial z^2} - \frac{e^2}{z} \right) \phi_j(z) = E_j \phi_j(z). \qquad (6.80)$$

Because the Bohr radius a_0 corresponding to this equation is far greater than b, the asymptotic form is appropriate. The characteristic length scale in the z direction will be of the order of a_0 and the binding energies will be about 10 eV. As B increases, b decreases and the approximation gets better.

Because the potential now extends over the range $-\infty \leq z \leq +\infty$ (unlike the radial coordinate in the hydrogen atom, for which the range is $0 \leq r \leq \infty$), the ground state cannot have any node in the z direction. The standard ground state of the hydrogen atom, which vanishes at the origin, is actually the first excited state of this system. To find the ground state for our problem, we can assume that the electronic distribution is confined to a cylindrical region of length l

6.4 Atomic Structures in Strong Magnetic Fields

and radius b with an approximate energy

$$\epsilon_0 \simeq \frac{\hbar^2}{m_e l^2} - \frac{e^2}{l} \ln\left(\frac{l}{b}\right). \tag{6.81}$$

Minimising this expression with respect to l, we can determine the ground-state energy. For large B and $b \ll l$, we get

$$l \simeq \frac{2a_0}{\ln(a_0/b)} \simeq \frac{a_0}{2}, \quad \epsilon_0 \simeq -\frac{\hbar^2}{4ma_0^2}\left[\ln\left(\frac{a_0}{b}\right)\right]^2 \simeq -100 \text{ eV}. \tag{6.82}$$

In the absence of Coloumb perturbation along the z direction, each of the energy states is infinitely degenerate, corresponding to different z components of the angular momentum labelled by $m = 0, 1, 2, \ldots$. A state with angular momentum $m\hbar$ has an effective radius of approximately $\rho_m = (2m+1)^{1/2}b$. In the presence of the Coloumb perturbation, the states with different m have slightly different values for the energy, thereby removing the degeneracy. These energy levels are separated by the amount $(e^2/a_0)(b/a_0) \approx 0.2$ eV. (Levels with different m are usually called Landau orbitals.)

To summarise, we now have the following structure of energy levels for the atom in a strong magnetic field. The ground state of the atom has an energy of $\epsilon_0 \approx -100$ eV. This ground state corresponds to $n = j = m = 0$. The lowest excited states with $n = j = 0, m = 1, 2, 3, \ldots$, have excitation energies of the order of a fraction of electron volts. The excited states of the atom corresponding to different E_j's of Eq. (6.80) are similar to the ground and the excited states of normal hydrogen atoms and hence lead to binding energies of the order of $E_j \approx -10$ eV; that is, the first excited state of the system described by Eq. (6.80) is ~ 90 eV above the ground state. These levels correspond to excitations of the z component with $n = 0, j = 1, 2, 3, \ldots, m = 0, 1, 2, 3, \ldots$; changing m, keeping other quantum numbers fixed, leads to small splittings of these levels. We must remember that all these states correspond to $n = 0$ as far as the excitation of the radial harmonic oscillator is concerned. The state with $n = 1$ is ~ 10 keV higher than the states with $n = 0$.

The conventional structure of the periodic table is now totally lost; electrons in a multielectron atom will now fill successive orbitals, with all spins being antiparallel to the magnetic field. The length of the atoms is essentially independent of the atomic number Z whereas the radius varies as $(2Z+1)^{1/2}$ because the effective radius of the orbital varies as $(2m+1)^{1/2}b$. When the interaction between the electrons is taken into account by approximation techniques, we find that the ionisation energy varies weakly as $\log Z$. The filling of the orbitals with same value of n and j and different values of m will cease when it is energetically more favourable for an electron to go to the $j = 1, m = 0$ orbital rather than to the $j = 0, m = Z$ orbital. To determine when this happens, it is convenient to define a ratio between the Bohr radius a_0/Z and the radii of the orbitals $(2Z+1)^{1/2}b$:

$$\eta = \frac{a_0}{Z(2Z+1)^{1/2}b} \simeq 10\left(\frac{B_{12}}{Z^3}\right)^{1/2}, \qquad (6.83)$$

where B_{12} is measured in 10^{12} G and we have taken $Z \gg 1$. When $\eta \gg 1$, Landau orbitals are filled and we get cylindrical atoms, whereas for $\eta \ll 1$ we get spherical atoms. When $\eta \approx 1$, we have a spherical core that shields the outer electrons, reducing the effective Z, and the valance electrons once again populate the Landau orbitals, giving rise to the cylindrical shape. In the context of the pulsar, we need to consider ^{56}Fe with $Z = 26$ and ^{4}He with $Z = 2$ in magnetic fields $B \approx 10^{12}$ G. We have $\eta \approx 0.1$ for iron and $\eta \approx 3.5$ for helium.

The situation becomes more complicated when we consider a *lattice* of atoms bound together. In conventional solids, the binding is relatively weak and is due to valence electrons of the atoms. However, in a strong magnetic field, the atoms are magnetically distorted into cylindrical shapes, giving rise to enormous electric quadrupole moments. The quadrupole–quadrupole coupling is extremely strong and such atoms will be bound along the cylindrical axis, forming a polymeric chain. Two distorted atoms initially in their ground states, approaching each other, would have their electrons in the $m = 0$ Landau orbitals and the quadrupole–quadrupole interaction would prefer bonding along the axis of the cylinder. As the atoms overlap, the Pauli exclusion principle requires that one of the electrons in any pair of atoms that are bound together be at the $m = 1$ orbital; but this excitation energy is only approximately a fraction of electron volts compared with the electron binding energy of $\sim$100 eV. It is therefore possible to bind a series of distorted atoms with each alternate atom being in the $m = 1$ state along the cylindrical axis. In such a configuration, it can be shown that the binding energy per atom in an infinitely long polymeric chain is approximately $Z^3(e^2/a_0)\eta^{4/5}$ compared with the energy of an isolated atom $Z^3(e^2/a_0)[\ln \eta]^2$. Therefore for $\eta \gg 1$ the chain is more stable than isolated atoms. When $\eta \approx 1$, the atoms will have a spherical core region with a cylinder sheath of valence electrons that contribute to the bonding. In this case, the bonding is reduced compared with the case with $\eta \gg 1$. The resulting structure is shown schematically in Fig. 6.7.

At the next level, polymeric chains can bind to each other, again because of quadrupole–quadrupole coupling. The equilibrium configuration of such a solid will be a tetragonal space lattice with a Young's modulus of the order of $Y \approx (E/lb^2) \approx 10^{19}$ dyn cm^{-2}, which is 10^6 times larger than terrestrial steel. Because the size of the atom is $\sim lb^2$, the close-packed configuration will have a density $\sim 10^4$ times that of terrestrial solids. Further, because the binding energy is $\sim$10 keV, such a structure will be stable up to temperatures of 10^8–10^9 K. In other words, the neutron star will not have a gaseous atmosphere

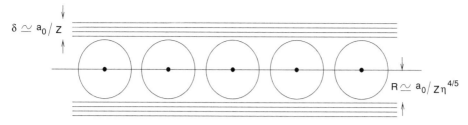

Fig. 6.7. Polymeric chain of atoms in a strong magnetic field.

near the surface but will be made of a magnetic polymeric solid with the density rising to 10^4 gm cm^{-3} within a Bohr radius. Then there will be a depth of ~ 1 cm in which the density rises relatively slowly to 10^5 gm cm^{-3}. Below this is the crust, with densities rising to above 10^7 gm cm^{-3}. The surface polymeric solid behaves as a one-dimensional metal with excellent electrical conductivity parallel to the magnetic field and nearly no conductivity perpendicular to it. When $\eta \approx 1$, only the valence electrons contribute to conduction and the electron work function for the solid is expected to be ~ 100 eV. Only the surface electric fields parallel to magnetic fields will be effective in inducing emission of electrons from the surface into the magnetosphere. Because of the high anisotropy of the structure, it is difficult to estimate the work function of electrons and other properties of the solids forming the surface layers of the pulsar.

6.5 Glitches in Pulsars

The pulsar periods are remarkably stable once the systematic increase that is due to radiation loss is taken into account. Because of this, any observed random change in the period is of considerable importance and has been studied extensively. These changes occur in two varieties. The first is the continuous and random changes of period of small magnitude, known as timing noise. The second is the sudden random change of much larger magnitude, known as glitches, which occur, for example, once in a few years in Crab and Vela pulsars. One possible explanation for the glitches is the following.

We saw in Chap. 5, Section 5.7 that the interior of the neutron star is likely to contain neutrons in the superfluid state and protons in the superconducting state, which cannot sustain rigid-body rotation. The existence of vortex lines and flux tubes causes several new features in a rotating neutron star, especially when the loss of energy makes the rotational velocity decrease. In a normal fluid, the angular velocity of rotation can decrease continuously while maintaining rigid rotation. In a superfluid the angular velocity is carried by a large number of vortices with quantised circulation. The spin-down of such a superfluid can occur only through the outward motion of the vortices. From the time derivative of the line integral of the velocity field over a circle of radius R and using Eq. (5.180), we have

$$\frac{\partial}{\partial t} \oint_R \mathbf{V} \cdot d\mathbf{l} = \frac{\partial}{\partial t} \int (\nabla \times \mathbf{V}) \cdot d\mathbf{S} = \frac{\partial}{\partial t} \int n_{\text{vortex}} K \, dS$$

$$= -K \int (\nabla \cdot \mathbf{j}_{\text{vortex}}) \, dS, \qquad (6.84)$$

where $\mathbf{j}_{\text{vortex}}$ is the current of the vortices. Converting the final integral to a line integral by using the Gauss theorem in two dimensions and writing the radial component of $\mathbf{j}$ as nV_r, we get

$$(2\pi R) R \frac{\partial \Omega}{\partial t} = -K(nV_r)(2\pi R), \qquad (6.85)$$

giving

$$\dot{\Omega} = -\frac{K}{R}(nV_r). \qquad (6.86)$$

The dynamics is now entirely determined by the equation of motion for the vortex line through the neutron star interior, which will determine V_r. This is a fairly complex problem to work out from first principles, and it is usually assumed that the force per unit length of the vortex line can be expressed in the form

$$\mathbf{f} = \rho \mathbf{K} \times (\mathbf{V}_s - \mathbf{V}_{\text{vortex}}), \qquad (6.87)$$

where ρ is the density of superfluid, $\mathbf{K}$ is the vorticity of the line, and $\mathbf{V}_s$ and $\mathbf{V}_{\text{vortex}}$ are the velocities of the superfluid and the vortex line, respectively. It is essentially this drag force that governs the spin-down of the superfluid in the neutron star.

As the crust slows down, the pinning of the vortices will lead to a lag between the superfluid rotation rate and the rotation rate of the crust. Because the pinned vortices are rotating with the crust, a relative velocity will develop between the vortex line and the superfluid, causing a frictional drag. This drag will increase and will eventually lead to unpinning of the vortices, thereby resulting in the slowdown of the superfluid. If the unpinning is sudden and catastrophic, it could lead to the glitch phenomena observed in the pulsars.

There are several details about this model that are still uncertain. It is, however, possible to develop a purely phenomenological model for the glitch based on the following assumptions: (1) The crust spin-up is communicated rapidly to the charged particles in the interior by the strong magnetic fields in a time scale of $\sim 10^2$ s. This time scale is decided by the time taken for the Alfven waves travelling with speeds $v_A \simeq (B^2/8\pi\rho)^{1/2}$ to cross the radius of the neutron star (Vol. I, Chap. 9, Subsection 9.7.1). For realistic parameters, this is $\sim 10^2$ s. (2) the response of the neutron superfluid to the spin-up of crust and charged particles is considerably slower because of the weak frictional coupling between the normal and the superfluid components.

The weak coupling between the neutron superfluid and the normal component leads the former to be spun up over some time scale τ_c. Let us assume that the interior of the neutron star is made of two components: a superfluid neutron component and a normal component. The moments of inertia of these components are taken to be I_{sf} and I_{norm}, respectively. After the glitch takes place, the superfluid component is spun up at a rate determined by the weak coupling between the superfluid and the normal components. In that case the system can be modelled purely phenomenologically by the equations

$$I_{norm}\dot{\Omega} = -\alpha - \frac{I_{norm}(\Omega - \Omega_{sf})}{\tau_c}, \quad I_{sf}\dot{\Omega}_{sf} = \frac{I_{norm}(\Omega - \Omega_{sf})}{\tau_c}. \quad (6.88)$$

Here α determines the steady change in the rotational speed that is due to the loss of rotational energy by dipole radiation, etc., and $1/\tau_c$ determines the strength of coupling between normal and superfluid components, where τ_c is the characteristic time scale. These equations can be integrated to give a solution

$$\Omega = -\frac{\alpha}{I}t + \frac{I_{sf}}{I}\Omega_1 e^{-t/\tau} + \Omega_2, \quad \Omega_{sf} = \Omega - \Omega_1 e^{-t/\tau} + \frac{\alpha\tau}{I_{norm}}, \quad (6.89)$$

where Ω_1 and Ω_2 are constants of integration, $I = I_{sf} + I_{norm}$, and $\tau = \tau_c I_{sf}/I$. Asymptotically (as $t/\tau \to \infty$) the solution tends to

$$\Omega_{sf} - \Omega \cong \frac{\alpha\tau}{I_{norm}} = \left(\frac{I_{sf}}{I_{norm}}\right)\left(\frac{\tau_c}{T}\right)\Omega, \quad T \equiv \frac{\alpha}{I\Omega}. \quad (6.90)$$

It is conventional to write solution (6.89) in the form

$$\Omega(t) = \Omega_0(t) + \Delta\Omega_0[Q\exp(-t/\tau) + 1 - Q], \quad \Omega_0(t) = \frac{\Omega_0 - \alpha t}{I}. \quad (6.91)$$

The behaviour of the solution is plotted in Fig. 6.8 and does model the glitches fairly well, at least in the case of the Crab pulsar. For the Vela pulsar, τ is of the order of months whereas for Crab it is of the order of weeks. The model predicts that τ_c and Q will be the same for all glitches occurring in a given pulsar.

If the nucleon fluid were a normal Fermi fluid, then the coupling would be strong enough for the relaxation time to be 10^{-4} s. When there is a significant superfluid component, whose circulation is confined to the vortex cores, the spin-up of the superfluid requires that angular momentum be transferred to these cores. Because the proton abundance is small, the direct electrostatic coupling between the protons and the electrons is as effective as the magnetic interaction of electrons with the neutrons. The total coupling is thus proportional to the number of nucleons in the vortex cores that, in turn, depends on the size of the vortices, their density, and the probability of occurrence of a nucleon quasiparticle excitation. The latter probability is proportional to $\exp[-(\Delta E/k_B T)]$, where

$$\Delta E \simeq \frac{\hbar^2}{2M\xi^2} \equiv \frac{\pi^2\Delta^2}{4\epsilon_F}, \quad \xi = \frac{\hbar v_F}{\pi\Delta}, \quad (6.92)$$

with M denoting nucleon mass, and ϵ_F, v_F, and ξ being the nucleon Fermi energy, the Fermi velocity, and the superfluid coherence length, respectively. At

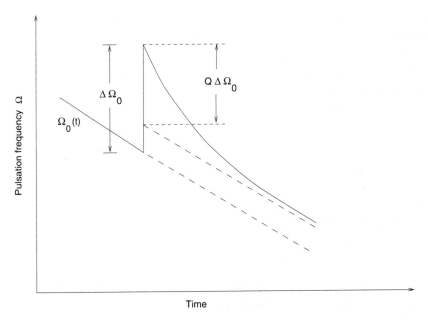

Fig. 6.8. Angular velocity Ω as a function of time for a pulsar undergoing a glitch.

$\rho = 10^{14}$ gm cm^{-3} the superfluid energy gap is $\Delta \simeq (1-2)$ MeV and $\epsilon_F \gtrsim 50$ MeV. Taking the temperature to be ~ 0.01 MeV, we find that the exponential factor is approximately $\exp[-(\pi^2 \Delta^2/4\epsilon_F k_B T)] \simeq e^{-5}$, which is partly the reason for the weak coupling.

6.6 Pulsar Timing

The pulses of radiation emitted by the rotating neutron star through any of the processes described in Section 6.3 are remarkably stable and act as a very precise clock. The study of arrival time of the pulses from the pulsar over a long duration of time allows us to determine the period P and its time derivative $\dot{P}$ with considerable accuracy. In some cases the precision in the measurement of P is comparable with that obtainable in terrestrial atomic clock cycles. Timing residuals of $(10-100)$ μs per observation are possible and – for periodic quantities – observations averaged over N cycles can improve the accuracy by a factor of $N^{-1/2}$. Secular effects can be measured to an even more startling precession. If the phase ϕ of a pulse is defined in the interval $[0,1]$ covering the whole pulse period, then

$$\phi = \int_0^t \frac{dt'}{P(t')} = \frac{t}{P(0)} - \frac{1}{2}\dot{P}(0)\left[\frac{t}{P(0)}\right]^2$$
$$+ \frac{1}{6}[2\dot{P}(0)^2 - P(0)\ddot{P}(0)]\left[\frac{t}{P(0)}\right]^3 + \cdots \qquad (6.93)$$

if the pulsar is monitored for a time t. If the phase ϕ is measured N times in this interval, with a single measurement residual of $\sim \epsilon P$ (with $\epsilon \approx 10^{-2}$) then the various terms in Eq. (6.93) can be fitted to an accuracy $(Q\epsilon/\sqrt{N})$, where $Q \approx 20$ is a factor that corrects for the fact that several parameters are being fitted simultaneously. For a stable pulsar and long observation time, N can be quite high.

In particular, the timing measurements can be used to determine the pulsar positions in the sky. Because the Earth orbits around the Sun, the radiation from the pulsar has to travel different distances during different times of the year. Assuming for simplicity that the orbit of the Earth is circular, the time delay t_c is given by

$$t_c = A \cos(\omega t - \lambda) \cos \beta, \qquad (6.94)$$

where A is the light travel time from Sun to Earth, ω is the angular velocity of the Earth in its orbit, and λ and β are the ecliptic longitude and latitude, respectively, of the pulsar. An error in the assumed coordinates $(\delta\lambda, \delta\beta)$ will give rise to periodic timing errors:

$$\delta t_c = A\delta\lambda \sin(\omega t - \lambda) \cos \beta - A\delta\beta \cos(\omega t - \lambda) \sin \beta. \qquad (6.95)$$

Typically, an observation of a pulsar with period $P \approx 1$ s can give an accuracy of 0.2 ms in the timing curve. With several such observations through 1 yr, the position of the pulsar can be determined to an accuracy much better than 0.1 arcsec.

However, to apply this method, it is necessary to determine and correct for several other known effects: (1) The pulsar period lengthens as the pulsar slows down, modifying the sinusoidal variation in Eq. (6.94). This introduces the parameter $\dot{P}(0)$, which also needs to be determined by pulsar timing. (2) The rotation of the Earth introduces a variable time delay that is of the order of 21 ms, corresponding to the light transit time over the radius of the Earth. The fact that the Earth's orbit is elliptical and not circular introduces a small delay. (3) The centre of mass of the solar system would have been at the centre of Sun if the Sun were infinitely massive. In reality, the existence of other planets, especially Jupiter, causes this centre of mass, known as the solar system barycentre, to be just outside the surface of the Sun. The barycentre moves uniformly through space but the motion of the telescope with respect to the barycentre needs to be corrected for. (4) The rate of clocks is affected by the depth of gravitational potential in which the clock is situated (see Vol. I, Chap. 11, Section 11.2). As the Earth moves around the Sun in an elliptical orbit, it samples a slightly different gravitational potential due to Sun, and hence the Earth-bound clock rate will show a small annual variation compared with a clock in a circular orbit around the Sun. The pulses are also affected by the gravitational potential well of the Sun, which is different around different path lengths. Because the path length varies as the Earth moves around the Sun, this also causes an extra time delay, called the Shapiro delay. (5) The effective frequency of a radio receiver observed in an inertial frame varies because of the Doppler effect of the Earth's motion as the Earth moves around the Sun. Because the arrival time depends on the frequency that is due to dispersion in the ISM (see below), a correction may be needed for pulsars with high a dispersion measure. All these effects are now precisely calculable and are routinely taken into account in timing studies.

The variation in the arrival time of the pulse that is due to phenomena taking place in the ISM is of more importance in theoretical studies of the structure of the ISM. Two such effects are the plasma dispersion (arising from the dependence of the propagation velocity on the frequency of the electromagnetic radiation) and the Faraday rotation of the plane of polarisation that is due to the interstellar magnetic field. We shall now discuss these effects.

The first one, which was discussed in Vol. I, Chap. 9, Section 9.4, gives the dispersion relation for the electromagnetic wave propagating through the ISM plasma with electron density, n_e, as

$$\omega^2 = \omega_p^2 + c^2 k^2, \quad \omega_p^2 = \frac{4\pi e^2}{m_e} n_e. \tag{6.96}$$

Because of this dispersion, the time T for a wave to traverse a distance L turns out to be

$$T = \int_0^L \frac{dl}{v_g} \simeq \frac{L}{c} + \frac{e^2}{2\pi m c \nu^2} \int_0^L n_e\, dl = \left(\frac{L}{c} + 1.345 \times 10^{-3}\ \mathrm{s}\ \nu^{-2} \int_0^L n_e\, dl\right). \tag{6.97}$$

The second term on the right-hand side is the delay t that is due to plasma dispersion and is usually quoted as

$$t = 4.2 \times 10^3\ \mathrm{s} \left(\frac{\nu}{1\ \mathrm{MHz}}\right)^{-2} \mathrm{DM}, \quad \mathrm{DM} \equiv \int_0^L n_e\, dl, \tag{6.98}$$

where the dispersion measure DM is measured in units of cm^{-3} pc. Higher frequencies will arrive earlier and the pulse will traverse the radio spectral range at a rate of

$$\dot{\nu} = -1.205 \times 10^{-4} \frac{\nu_{\mathrm{MHz}}^3}{\mathrm{DM}}\ \mathrm{MHz\ s^{-1}}. \tag{6.99}$$

Therefore a receiver with a bandwidth B MHz will smear out a narrow pulse over an interval

$$\Delta t = 8.3 \times 10^3\ \mathrm{s}\ \nu_{\mathrm{MHz}}^{-3}\ B(\mathrm{DM}). \tag{6.100}$$

If a pulsar with a high dispersion measure is observed with a wide bandwidth, its pulse is stretched, causing a reduction in the peak intensity. To recover the lost sensitivity, a pulsar is usually observed over several narrow frequency channels whose outputs are then added with appropriate delays so that the pulse components are properly superposed. We shall now consider some details of this process.

A monochromatic wave of frequency ω and amplitude $A(\omega)$ can be represented as

$$S_0(\omega) = A(\omega)e^{-i(\omega t - kL)} = A(\omega)e^{i\phi(\omega)}e^{-i\omega t} \tag{6.101}$$

after it has traversed a distance L through the ISM. Here

$$\phi(\omega) \equiv k(\omega)L = \frac{\omega L}{c}\left(1 - \frac{\omega_p^2}{\omega^2}\right)^{1/2} \cong \frac{\omega L}{c}\left(1 - \frac{\omega_p^2}{2\omega^2}\right) \tag{6.102}$$

is the phase acquired by the wave and the last relation is valid for $\omega \gg \omega_p$. In the case of pulsar observations, we usually concentrate on a detector system operating in a range of frequencies $\omega_l < \omega < \omega_u$ and measure the frequencies with respect to the lower limit. Adding an unimportant phase, we can write the amplitude in Eq. (6.101) as

$$S_1(\omega) = A(\omega)e^{-i(\omega - \omega_l)t}e^{i\phi(\omega)} = A(\omega_b)e^{-i\omega_b t}e^{i\phi(\omega_b)}, \tag{6.103}$$

where $\omega_b \equiv (\omega - \omega_l)$. A given signal will have all the frequencies in the range $\omega_l < \omega < \omega_u$ and hence the signal in the time domain will be

$$V(t) = \int_0^B d\omega_b\, A(\omega_b)e^{-i\omega_b t}e^{i\phi(\omega_b)}, \tag{6.104}$$

where $B \equiv (\omega_u - \omega_l)$ is the bandwidth.

This equation clearly shows that the resulting pulse shape is distorted because of the frequency-dependent phase $\phi(\omega_b)$. To obtain the true signal $V_{\text{true}}(t)$ [corresponding to $\phi = 0$ in Eq. (6.104)], we need to correct for the phase $\phi(\omega)$. To do this systematically, let us write $\phi(\omega)$ in the form

$$\phi(\omega) = \phi(\omega_l) + \mathcal{R}(\omega_l)\omega_b + \mathcal{C}(\omega), \tag{6.105}$$

where $\mathcal{R}(\omega)$ denotes the rate of variation of phase with frequency and is given by

$$\mathcal{R}(\omega) = \frac{d\phi}{d\omega} = \frac{L}{c}\left(1 + \frac{\omega_p^2}{2\omega^2}\right). \tag{6.106}$$

The first term $\phi(\omega_l)$ in Eq. (6.105) is an unimportant constant phase; the second term is linear in ω and represents the first-order rate of change of phase; $\mathcal{C}(\omega)$ represents the rest of the corrections. When $\phi(\omega)$ is written in this form, it is very easy to see the different roles played by the second and the third terms. Note that

$\mathcal{R}(\omega)$ has the dimensions of time and, in fact, the quantity

$$\mathcal{D}(\omega) \equiv \mathcal{R}(\omega) - \mathcal{R}(\infty) = \frac{L}{c}\left(\frac{\omega_p^2}{2\omega^2}\right) = \left(\frac{2\pi e^2}{m_e c^2}\right)(Ln_e)\omega^{-2}$$

$$= 4.2 \times 10^3 \text{ s} \left(\frac{\text{DM}}{\text{cm}^{-3}\text{ pc}}\right)\left(\frac{\nu}{1 \text{ MHz}}\right)^{-2} \quad (6.107)$$

denotes the time delay in the arrival of a wave with frequency ω compared with a wave with infinite frequency represented in Eq. (6.98). If we retain just this term and ignore $\mathcal{C}(\omega)$ then the wave can be represented as

$$\exp\{-i[\omega t + \phi(\omega)]\} \simeq \exp{-i[\omega t + \mathcal{R}(\omega_l)(\omega - \omega_l)]}$$
$$\simeq e^{-i\omega_l t} \exp\{-i\omega_b[t - \mathcal{R}(\omega_l)]\}, \quad (6.108)$$

clearly showing that the effect of $\mathcal{R}(\omega)$ term is to produce a shift in the arrival time of waves of different frequencies. This effect is clearly seen by the shifting of the peak of the pulse profile in Fig. 6.9. In actual observations we can introduce a frequency-dependent delay to correct for this linear term; because the ISM disperses the wave, the observational technique used to reconstruct the original pulse profile is called dedispersion. When only the linear term is corrected, the process is called *incoherent dedispersion*.

The incoherent dedispersion, of course, cannot reproduce the original signal correctly as we have accounted for only the first two terms on the right-hand side of Eq. (6.105). To cancel the phase distortion $\phi(\omega)$ completely, we also need to correct for $\mathcal{C}(\omega)$. We can easily compute this term by writing Eq. (6.105) as

$$\mathcal{C}(\omega) = \phi(\omega) - [\phi(\omega_l) + (\omega - \omega_l)\mathcal{R}(\omega_l)]. \quad (6.109)$$

Straightforward algebra now gives

$$\mathcal{C}(\omega) = \frac{L}{c}\frac{\omega_p^2}{2}\left(\frac{\omega - \omega_l}{\omega_l^2}\right)\left(\frac{\omega_l}{\omega} - 1\right) = -\mathcal{D}(\omega_l)\omega_b\left(\frac{\omega_b/\omega_l}{1 + \omega_b/\omega_l}\right). \quad (6.110)$$

It is possible to devise detectors in which the phase of the wave can be modified by $\exp[-i\mathcal{C}(\omega)]$ so that the necessary correction is achieved; this process is called coherent dedispersion and is required for a more accurate restoration of the initial signal. The pulse profile in Fig. 6.9, in fact, has been obtained by a similar technique.

Exercise 6.3

Practice with pulsars I: A radio pulsar has a flux of 0.06 Jy at 400 MHz, a period $P = 0.6$ s, and a period derivative $\dot{P} = 3 \times 10^{-15}$. (The pulsar fluxes usually quoted are the averaged value over a period; the peak flux can be 3–20 times higher because it "lights up" for only a short time.) The pulses at 400 MHz arrive 2.5 s later than the pulses

6.6 Pulsar Timing

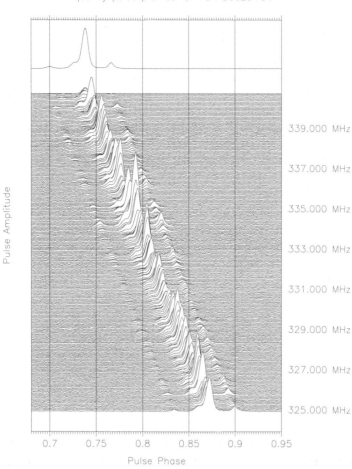

Fig. 6.9. Pulsar dispersion and dedispersion. (Figure courtesy Y. Gupta, based on the data from Giant Metre Radio Telescope, Pune, India.)

at 1 GHz. Taking the mean density of electrons in the ISM to be 0.02 cm^{-3}, estimate (a) the distance to the pulsar, (b) the luminosity of the pulsar, (c) the fraction of the total luminosity that is radiated in one octave of frequency around 400 MHz, and (d) the pulsar's surface magnetic field.

Exercise 6.4
Practice with pulsars II: A pulsar at $\alpha = 19^h\,57^m\,25^s$, $\delta = +20°\,40'$ (epoch 1950) has proper motion $v_\alpha = -0.00114$ s yr^{-1}, $v_\delta = -0.026''$ yr^{-1} and a parallax of 0.0007 arcsec. It has a companion star of $m_V = 19.0$, whose heliocentric radial velocity varies sinusoidally about a mean of $\gamma = 55$ km s^{-1}. Calculate (a) the galactic coordinates l and b of the pulsar, (b) its transverse velocity $\mathbf{v}_t$ relative to the Sun and the angle that the vector makes to the galactic plane, and (c) the absolute magnitude M_V of its companion star.

Exercise 6.5

Dispersion of pulsar signals: (a) The Crab pulsar has dispersion measure DM = 57 cm^{-3} pc and a pulse period of 0.033 s. Its pulses have been detected at radio, optical, x-ray and γ-ray energies. When they arrive at Earth, by how many seconds (and by how many pulses) are the 100-MHz radio pulses delayed with respect to the corresponding x-ray pulses? (b) For accurate pulsar timing, the corresponding pulses have to be aligned to within 1/100 of a pulse. How accurately must the dispersion measure of the Crab pulsar be determined to allow alignment of the 100-MHz radio pulses and the x-ray pulses to this accuracy? (c) It is desired to discover a millisecond pulsar with a pulse period of 0.001 s (the shortest known period is 0.0015 s) by performing a survey at 430 MHz with a receiver with a bandwidth of 250 kHz. What is the largest dispersion measure (in cm^{-3} pc) at which the survey is likely to succeed?

It has been found that the dispersion measure of pulsars satisfies the constraint (DM) $\sin b < 20$ cm^{-3} pc, where b is the galactic latitude of the pulsar. If we model the galaxy with z denoting the vertical height from the disk and r denoting the radial coordinate in the plane of the disk, then the square of the path length along the galactic latitude b is given by $ds^2 = dr^2 + dz^2 = \mathrm{cosec}^2\, b\, dz^2$. From the definition of the dispersion measure, it follows that

$$\mathrm{DM} = \int_0^s n_e\, ds = \mathrm{cosec}\, b \int_0^z n_e\, dz. \tag{6.111}$$

Therefore

$$\mathrm{DM} \sin b = \int_0^z n_e\, dz < \int_0^\infty n_e\, dz \approx 20 \text{ cm}^{-3} \text{ pc}. \tag{6.112}$$

If we take the electron distribution to be exponential, with $n_e = n_0 \exp-(z/h)$, then this constraint gives $n_0 h \approx 20$ cm^{-3} pc. For $n_0 \approx 0.02$ cm^{-3}, the scale height is ~ 1 kpc.

Let us next consider the Faraday rotation of pulsar radiation that allows us to estimate the interstellar magnetic field, because pulsar radio signals show a high degree of linear polarisation. It was shown in Vol. I, Chap. 9, Subsection 9.5.1 that the rotation of the plane of polarisation is essentially determined by the rotation measure RM that depends on the integral of $n_e B_\parallel$ along the line of sight to the pulsar, where n_e is the electron density and $B_\parallel$ is the component of the magnetic field parallel to the line of sight. If the rotation of the plane of polarisation at two adjacent frequencies ν and $\nu + \Delta\nu$ differ by an amount $\Delta\phi$, then

$$\Delta\phi = -2\,(\mathrm{RM}) \left(\frac{300 \text{ MHz}}{\nu}\right)^2 \left(\frac{\Delta\nu}{\nu}\right) \text{ rad} \tag{6.113}$$

[obtained by differentiation of Eq. (9.134) of Vol. I, Chap. 9, Section 9.5]. Because the dispersion measure is related to an integral of n_e, the ratio (RM/DM)

6.6 Pulsar Timing

for the same pulsar provides an average value of $B_\parallel$. Numerically,

$$\frac{\text{RM}}{\text{DM}} = 0.81 \frac{\int_0^L n_e B \cos\theta \, dl}{\int_0^L n_e \, dl}, \tag{6.114}$$

where RM is measured in radians per square meter, DM in cm^{-3} pc and the magnetic field in microgauss.

The extreme accuracy with which pulsar timings are possible allows us to detect any orbiting planet around a pulsar.[3] A planet of mass m orbiting a pulsar of mass M in an orbit with semimajor axis a and period P will produce a periodic timing residual of amplitude

$$\mathcal{A} = \frac{m}{M+m} \frac{a \sin i}{c} \approx 0.46 \text{ s} \frac{m}{M_{\text{Jup}}} \left(\frac{1.4 \, M_\odot}{M}\right)^{2/3} \left(\frac{P}{1 \text{ yr}}\right)^{2/3} \sin i. \tag{6.115}$$

That is, a planet like Jupiter orbiting around a neutron star of mass $1.4 \, M_\odot$ will produce a residual of 0.46 s if $P = 1$ yr and $i = 60°$; if the planet has a lower mass, say, that of the Earth, the corresponding value will be 1.5 ms and for the Moon, it will be $\sim 18 \, \mu$s. Because it is possible to measure timing residuals to an accuracy of 1 μs, any pulsar with an orbiting companion having a mass greater than the lunar mass is potentially detectable. In fact, a millisecond pulsar PSR 1257 + 12 indeed seems to have two planets orbiting it on nearly circular orbits of periods of 67 and 98 days.

The timing of pulsars also allows us to put bounds on the amount of stochastic gravitational wave background that can exist in our universe. We shall see in Vol. III that certain models for the evolution of an early universe predict a copious production of gravitons during the very early evolution of the universe. If these models are correct, these gravitons will be present as background radiation in the present-day universe, just as photons are present as cosmic microwave background radiation. There are, however, two significant differences between these. (1) The spectrum of the gravitons will not be thermal as the gravitons do not interact sufficiently strongly with matter in order to be thermalised. The spectrum needs to be computed specifically from the theory, and most models lead to a power spectrum in which every logarithmic band in frequency contributes the same amount of power. (2) The gravitons will have a very large wavelength in the present epoch and hence can be considered as a classical stochastic gravitational wave background characterised entirely by its power spectrum.

If the fluctuations in the background metric tensor g_{ij}, due to the gravitational wave, are taken to be h_{ik}, then the strength of the gravitational wave can be measured by the typical rms value of any nonvanishing component of h_{ik}, which we shall denote by the symbol h_{gw}. Theory usually predicts the ratio $\Omega_{\text{gw}}(\nu)$ between the energy density of gravitational waves per logarithmic frequency band and the critical energy density ρ_c needed to close the universe today, that is,

$$\Omega_{gw}(\nu) \equiv \frac{1}{\rho_c}\left(\frac{d\rho_{gw}}{d\ln\nu}\right). \qquad (6.116)$$

This quantity is related to the dimensionless amplitude h_{gw} by

$$h_{gw}(\nu) = 1.3 \times 10^{-20}\left(\frac{\nu}{100\text{ Hz}}\right)^{-1}\Omega_{gw}(\nu)^{1/2}\left(\frac{H}{100\text{ km s}^{-1}\text{ Mpc}^{-1}}\right), \qquad (6.117)$$

where H is the Hubble constant. Imagine now such a stochastic gravitational wave passing by us in a direction transverse to our line of sight to a pulsar. When one of the maxima of the gravitational wave is passing by the Earth, the pulses from the pulsar will be spaced slightly further apart because of the gravitational redshifting from the gravitational wave. Half a period of the gravitational wave later, the minimum of the gravitational wave will be passing by the Earth and we will see the pulses slightly more closely spaced together. During a measurement period of time T, the gravitational fluctuations will cause a time difference of the order of $h_{gw}T$. If a pulsar is being monitored for a stretch of time T and the timing residuals are less than ΔT, then it follows that $h_{gw} \leq (\Delta T/T)$, provided that the period of the wave is of the order of T. This last condition is required for obtaining the maximum sensitivity of the measurement. Observationally, $\Delta T \leq 10^{-6}$ s and $T \approx 8$ yr for some of the best investigated pulsars. This implies that

$$h_{gw} \leq \left(\frac{\Delta T}{T}\right) \leq 10^{-14}, \quad \Omega_{gw} < 10^{-8}, \qquad (6.118)$$

at the frequencies $\nu = 8 \text{ yr}^{-1} = 10^{-8}$ Hz. Such an extremely tight bound on Ω_{gw} is an illustration of the accuracy with which time residuals of pulsar arrival times are measured.

6.7 Pulsar Scintillation

The propagation of radiation from pulsars through the ISM is affected by the fluctuations in the effective refractive index of the ISM, leading to several interesting effects. Because this phenomenon – viz., the propagation of radiation through a turbulent medium – is very general and occurs in a variety of astrophysical contexts, we shall provide a general discussion, keeping the pulsar as a backdrop.

When a wave front arriving from a distant source encounters a turbulent medium with a refractive index varying randomly over space, two major effects occur: (1) The flux received by the observer will show spatial variation and also temporal scintillation if there is transverse motion of the source, observer, or scattering medium; (2) the image of the source detected by a telescope will be

6.7 Pulsar Scintillation

distorted. Both of these phenomena are well known at optical wavelengths. The twinkling of stars arises because of the first effect through the scattering in Earth's atmosphere. The second effect leads to the resolution of optical telescopes being limited by atmospheric seeing to ~ 1 arcsec (see Vol. I, Chap. 3, Section 3.16). Similar phenomena can, of course, occur at all other wavelengths when conditions are appropriate.

To discuss this scintillation in the simplest context, we begin by assuming that (1) the source is at infinity so that the incoming wave front can be assumed to be planar and (2) the entire effect of turbulent medium can be mimicked by a thin scattering screen at a distance D from the observer, whose effect is to produce a phase change $\phi(\mathbf{x})$ on a ray crossing the screen at a location with transverse coordinates $\mathbf{x} = (x, y)$. If the wave front from the source has unit amplitude, then the amplitude $\psi(\mathbf{X})$ at the plane of the receiver is given by

$$\psi(\mathbf{X}) = \frac{e^{-i\pi/2}}{2\pi r_F^2} \int \exp\left[i\phi(\mathbf{x}) + i\frac{(\mathbf{x} - \mathbf{X})^2}{2r_F^2}\right] d\mathbf{x}; \quad r_F = \sqrt{\frac{\lambda D}{2\pi}}, \quad (6.119)$$

where $\mathbf{X} = (X, Y)$ are the transverse coordinates in the observer plane, λ is the wavelength of the radiation, and r_F is called the *Fresnel length*. We obtain this result by multiplying the standard propagator in paraxial optics (discussed in Vol. I, Chap. 3, Section 3.13, and valid when $|\mathbf{x} - \mathbf{X}| \ll D$) by the random-phase factor $\exp i\phi(\mathbf{x})$. The intensity of radiation received by the observer is given by $F(\mathbf{X}) = |\psi(\mathbf{X})|^2$. In the absence of any scattering, $\phi(\mathbf{x}) = 0$ and, obviously, $\psi(\mathbf{X}) = 1$. In this case, the saddle-point limit to the integral is dominated by contributions from a circular zone of radius $\sqrt{2r_F}$ (called the *first Fresnel zone*) around the point $\mathbf{x} = \mathbf{X}$. In the presence of scattering by a turbulent medium, the phase fluctuation $\phi(\mathbf{x})$ will be a random variable and can be characterised by its moments. In the simplest case, $\phi(\mathbf{x})$ may be taken to be a Gaussian random field with zero mean and the *structure function*

$$D_\phi(\mathbf{x}) \equiv \langle [\phi(\mathbf{x}' + \mathbf{x}) - \phi(\mathbf{x}')]^2 \rangle_{\mathbf{x}'}, \quad (6.120)$$

where the averaging is over all points $\mathbf{x}'$. The structure function is related to the two-point correlation function $C(\mathbf{x}) \equiv \langle \phi(\mathbf{x} + \mathbf{x}')\phi(\mathbf{x}') \rangle$ by

$$D_\phi(\mathbf{x}) = 2[C(0) - C(\mathbf{x})]. \quad (6.121)$$

Expanding $\phi(\mathbf{x})$ in Fourier series, we find that

$$D_\phi(\mathbf{x}) = \langle [\phi(\mathbf{x}' + \mathbf{x}) - \phi(\mathbf{x}')]^2 \rangle$$
$$= \int \frac{d^2\mathbf{k}}{(2\pi)^2} \frac{d^2\mathbf{p}}{(2\pi)^2} \langle \phi_\mathbf{k} \phi_\mathbf{p}^* \rangle \, e^{i(\mathbf{k}-\mathbf{p})\cdot\mathbf{x}'}(e^{i\mathbf{k}\cdot\mathbf{x}} - 1)(e^{-i\mathbf{p}\cdot\mathbf{x}} - 1). \quad (6.122)$$

If the power spectrum of the random field is $P(\mathbf{k})$, then, by definition,

$$\langle \phi_\mathbf{k} \phi_\mathbf{p}^* \rangle = (2\pi)^2 \delta_D(\mathbf{k} - \mathbf{p}) P(\mathbf{k}). \quad (6.123)$$

Using this in Eq. (6.122), we get

$$D_\phi(\mathbf{x}) = 2 \int \frac{d^2\mathbf{k}}{(2\pi)^2} P(\mathbf{k})(1 - \cos \mathbf{k} \cdot \mathbf{x})$$
$$= 2 \int_0^\infty \frac{k\,dk}{(2\pi)^2} P(k) \int_0^{2\pi} d\phi\,[1 - \cos(kx \cos \phi)], \qquad (6.124)$$

where the second equality is valid if $P(\mathbf{k}) = P(k)$, which will be the case if the turbulence is isotropic. Using the expression for the Bessel function,

$$J_0(qx) = \int_0^{2\pi} \frac{d\phi}{2\pi} \cos(qx \cos \phi), \qquad (6.125)$$

we get

$$D_\phi(\mathbf{x}) = 2 \int_0^\infty \frac{k\,dk}{2\pi} P(k)[1 - J_0(kx)]. \qquad (6.126)$$

The corresponding expression for the two-point correlation function is given by

$$C(\mathbf{x}) = \int_0^\infty \frac{k\,dk}{2\pi} P(k) J_0(kx), \qquad (6.127)$$

which is essentially the second term in $D_\phi(x)$. For such a Gaussian random field, it is also easy to show that $\langle \exp[i\phi(\mathbf{x})] \rangle = 0$ whereas

$$\langle \exp[i(\phi(\mathbf{x}) - \phi(\mathbf{x}'))] \rangle = \exp\left\{-\frac{1}{2}\langle[\phi(\mathbf{x}) - \phi(\mathbf{x}')]^2\rangle\right\}$$
$$= \exp\left[-\frac{1}{2}D_\phi(\mathbf{x} - \mathbf{x}')\right]. \qquad (6.128)$$

Hence $\langle \psi(\mathbf{X}) \rangle = 0$ and

$$\langle \psi(\mathbf{X})\psi^*(\mathbf{Y}) \rangle = \exp\left[-\frac{1}{2}D_\phi(\mathbf{X} - \mathbf{Y})\right]. \qquad (6.129)$$

This result (which was proved in Vol. I, Chap. 3, Section 3.16 in a different context) shows that the correlation of the field amplitudes at different points on the observer plane is related to the structure function of the turbulence.

At the next-higher order, we can study the intensity correlations on the observer plane, $\langle F(\mathbf{X})F(\mathbf{X}') \rangle$, which will contain information about the scintillation. In the case of homogeneous, isotropic turbulence, this quantity can depend on only $|\mathbf{X} - \mathbf{X}'|$ and thus can be described by the power spectrum of the intensity correlation function, $W(q)d^2q$, where q is the magnitude of the spatial wave vector; $q^2 W(q)$ represents the amount of power per logarithmic wavelength scale.

6.7 Pulsar Scintillation

If the turbulence in the medium is isotropic and is described by the Kolmogorov spectrum at the scales of interest, then the structure function has the form

$$D_\phi(r) = \left(\frac{r}{r_{\text{diff}}}\right)^{5/3}, \quad (r^2 = x^2 + y^2). \tag{6.130}$$

The index 5/3 is derived in Vol. I, Chap. 8, Section 8.15, and the amplitude is expressed in terms of a diffractive length scale r_{diff}. In optical astronomy, it is conventional to write the same result in the form

$$D_\phi(r) = 6.88\left(\frac{r}{r_0}\right)^{5/3}, \quad (r_0 = 3.18\, r_{\text{diff}}), \tag{6.131}$$

where r_0 is called the *Fried length*.

Given the length scales r_F and r_{diff}, it is possible to discuss two limiting cases of scattering. (1) The first case, which may be called weak scattering, occurs if $D_\phi(r_F) \ll 1$ so that random-phase fluctuations within the Fresnel zone are small. In this case, the concept of the Fresnel zone remains useful with weak perturbations to the wave front. The condition for weak scattering requires that $r_{\text{diff}} \gg r_F$. (2) The second case is the opposite limit, with $r_{\text{diff}} \ll r_F$ implying $D_\phi(r_F) \gg 1$. Because the phase fluctuations now vary significantly over the Fresnel scale, r_F becomes irrelevant and, instead, r_{diff} is the characteristic size of a coherent patch. In this case there will be many points x, y on the scattering screen that will contribute to a given point X, Y on the observer plane.

Table 6.1 gives the characteristic values for these parameters in several astrophysical contexts. In the case of ionised media (corresponding to ionospheric, interplanetary, and interstellar scattering), the cold plasma dispersion relation, along with the Kolmogorov spectrum, gives $r_{\text{diff}} \propto \lambda^{6/5} D^{-3/5}$. (This relation is obtained in Vol. I, Chap. 3, Section 3.16.) Hence the strength of scattering increases with increasing distance and is fairly chromatic.

Table 6.1. *Strong and weak scattering by turbulent media*

Medium	λ cm	D cm	r_F cm	r_{diff} cm
Earth's atmosphere	5×10^{-5}	10^6	3	10
Troposphere	20	10^5	6×10^2	$\sim 10^5$
Ionosphere	3×10^2	3×10^7	4×10^4	$\sim 10^5$
Solar wind	10^2	10^{13}	10^7	$> 10^7$
ISM	10^2	10^{21}	10^{11}	$\sim 10^9$

6.7.1 Weak Scattering ($r_{\text{diff}} \gg r_F$)

In the case of weak scattering, the dominant influences arise from mild (de)focussing of the wave front that is due to phase fluctuations at the size of the Fresnel zone, characterised by $D_\phi(r_F)^{1/2}$. If the Fresnel zone coincides with the phase fluctuation that increases coherence (usually called the focussing fluctuation), the size of the coherent patch becomes larger than r_F and $F(X,Y) > 1$ at the observer; if the Fresnel zone aligns with a defocussing fluctuation, $F(X,Y) < 1$ at the observer. Because the phase variation $\langle (\Delta\phi)^2 \rangle^{1/2}$ is small over the Fresnel zone, the net-mean-square fluctuation in the intensity will be approximately $\langle \exp i(\phi - \phi') - 1 \rangle$ where ϕ and ϕ' are the two locations separated by one Fresnel length. Using Eq. (6.128) and (6.130) we get

$$\langle e^{i(\phi - \phi')} \rangle - 1 \cong \frac{1}{2} D_\phi(r_F) \cong \frac{1}{2} \left(\frac{r_F}{r_{\text{diff}}} \right)^{5/3}. \tag{6.132}$$

The rms fluctuation in the intensity scales as the square root of this expression, i.e., as $(r_F/r_{\text{diff}})^{5/6}$.

An alternative way of obtaining the above result is in terms of an effective focal length introduced by the scattering screen. We have seen in Vol. I, Chap. 3, Section 3.15 that a lens of focal length f will introduce a path difference of $(x_\perp^2/2f)$ at a lateral distance $x_\perp$ from the optical axis. The corresponding phase difference will be $\Delta\phi \approx (x_\perp^2/f\lambda)$. This relation allows us to define an effective focal length for the phase variations introduced by the screen as

$$f(x_\perp) = \frac{x_\perp^2}{\lambda \Delta\phi} \approx \frac{x_\perp^2}{\lambda D_\phi(x_\perp)^{1/2}}. \tag{6.133}$$

In arriving at the second approximation, we have estimated $\Delta\phi$ by the rms fluctuation at the relevant scale. In our case, the rms focal length $f(r_F)$ at the Fresnel scale is $f(r_F) \approx r_F^2/\lambda D_\phi(r_F)^{1/2}$. In any optical system involving a lens, the rms flux variation will be of the order of D/f after propagating through a distance D. Hence, in our case, the rms flux variation will be

$$\frac{D}{f(r_F)} \simeq \left(\frac{D}{r_F^2} \right) \lambda D_\phi^{1/2}(r_F) = \frac{1}{2\pi} \left(\frac{r_F}{r_{\text{diff}}} \right)^{5/6}, \tag{6.134}$$

which agrees with the previous result.

We can make the above analysis more rigorous by computing the power spectrum of intensity scintillations.[4] This is straightforward but tedious as it requires the evaluation of the four-point function for Gaussian random fields. We shall merely quote the final result. In the case of weak scintillation, governed by Kolmogorov turbulence, the power per unit logarithmic interval in two-dimensional Fourier space (with the wave vector denoted by q) is given by

$$q^2 W(q) = C \left(\frac{r_F}{r_{\text{diff}}}\right)^{5/3} (qr_F)^{-5/3} \sin^2\left(\frac{1}{2}q^2 r_F^2\right), \qquad (6.135)$$

where $C \approx 11.2$ is a constant. For length scales longer than r_F, the power decreases as $(q^{-1})^{-7/3}$ whereas for length scales smaller than r_F it increases as $(q^{-1})^{5/3}$; the largest fluctuations are on the Fresnel scale, as to be expected. The amplitude is indeed modified as predicted by relation (6.132).

If there is a transverse velocity v between the scattering medium and the line of sight, the scintillation will appear at a time scale $t_F \approx (r_F/v)$. These intensity variations are correlated over a wide bandwidth $\Delta \nu \approx \nu$.

The above result is strictly valid for only a point source but, of course, can be used whenever the angular size of the source θ_s is smaller than the Fresnel angle $\theta_F \equiv (r_F/D)$. In the case of an extended incoherent source, the net scintillation pattern will be the sum of flux variations that are due to all the elements in the source. Two elements separated by an angle θ will be shifted in the observer plane by an amount $(-D\theta)$. If, for example, the source is a Gaussian with a profile

$$I_s(\theta) = \left(\frac{1}{\pi \theta_s^2}\right) \exp\left[-\frac{|\theta|^2}{\theta_s^2}\right], \qquad (6.136)$$

then the power spectrum is modified to

$$W_e(q) = W(q) \exp\left[-\frac{1}{2} q^2 D^2 \theta_s^2\right], \qquad (6.137)$$

where the subscript e stands for 'extended' source. Because the source acts as a spatial filter of size $D\theta_s$ the flux variations are damped at a length scale of $D\theta_s$ rather than at r_F. The corresponding scintillation time scale is now $(D\theta_s/v)$. Further, because $q^2 W(q) \propto q^{7/3}$ for $q < r_F^{-1}$, the mean-square fluctuations are suppressed relative to the point source by a factor $(\theta_F/\theta_s)^{7/3}$.

In the scattering of optical radiation in Earth's atmosphere, the stars act as point sources whereas planets are extended sources. Hence planets exhibit much weaker scintillation and their flux variations, if any, occur at larger time scales than those for stars. The entries in Table 6.1 show that there are several examples of weak scattering that arise in astrophysical cases.

6.7.2 Strong Scattering ($r_{\text{diff}} \ll r_F$)

In the strong-scattering region, each point on the observer plane receives radiation from a large number of points on the scattering screen. To model this, we can divide the screen into coherent patches of size r_{diff} around each of the contributing rays. These patches are expected to be randomly distributed, with typical separation also given by r_{diff}. Each coherent patch will scatter the radiation into a diffraction cone of angle

$$\theta_{\text{scat}} \approx \frac{\lambda}{r_{\text{diff}}} \approx \frac{r_F^2}{r_{\text{diff}} D} \equiv \frac{r_{\text{ref}}}{D}, \qquad (6.138)$$

where we have defined a new length scale $r_{\text{ref}} \equiv (r_F^2/r_{\text{diff}}) \gg r_{\text{diff}}$. A given observer now receives radiation from all patches that lie within a region $D\theta_{\text{scat}} \approx r_{\text{ref}}$ on the screen. The image will now be scatter broadened by the angle θ_{scat}.

The number of rays contributing to a given point is approximately $(r_{\text{ref}}/r_{\text{diff}})^2 = (r_F/r_{\text{diff}})^4 \gg 1$, and each ray has a random phase that is uncorrelated with the phases of other rays. Because the angle spanned by the rays is $\sim \theta_{\text{scat}}$, the length scale of the flux variation is $(\lambda/\theta_{\text{scat}}) \approx r_{\text{diff}}$. The scintillation time scale will be $t_{\text{diff}} = (r_{\text{diff}}/v)$.

As in the case of weak scattering, we can define a critical angular size that distinguishes point sources from extended sources. Because the scintillation length scale is r_{diff}, this critical angle is essentially

$$\theta_{\text{diff}} = \frac{r_{\text{diff}}}{D} \simeq \left(\frac{r_{\text{diff}}}{r_F}\right)^2 \theta_{\text{scatt}} \ll \theta_{\text{scatt}}. \qquad (6.139)$$

The effect of a Gaussian image profile can be calculated as in the case of weak scattering and leads to an increase in the scintillation time scale and a reduction of rms fluctuation. One difference from the weak scattering is that the fractional bandwidth in this case is $(\Delta\nu/\nu) \approx (r_{\text{diff}}/r_F)^2 \ll 1$; diffractive scintillation is a narrow-band phenomenon. The diffractive scintillation of radio pulsars by a turbulent ISM is easily detectable. For a pulsar moving with a velocity $v \approx 10^7$ cm s^{-1} and $r_{\text{diff}} \approx 10^9$ cm, the time scale of variation is $t_{\text{diff}} \approx 10^2$ s. This provides a host of information about both the irregularities in the ISM and the velocity distribution of pulsars.

In this case of strong scattering, there occurs another effect, called *refractive scintillation*. This arises from the fact that coherent fluctuations of several patches of size r_{diff}, contained within a region of size r_{ref}, can (de)focus radiation at the observer. A phase fluctuation of size r_F, when it focusses radiation towards the observer, will tilt the phase profile of all the diffraction cones from individual coherent patches of size r_{diff}. This will allow the observer to see a larger flux arising from a larger number of coherent patches. When the phase fluctuation has a (de)focussing profile, the opposite effect occurs and the observer receives significantly less intensity. The rms focal length of the phase fluctuations with scales r_{ref} is

$$f(r_{\text{ref}}) \approx \frac{r_{\text{ref}}^2}{\lambda D_\phi^{1/2}(r_{\text{ref}})} \approx D \left(\frac{r_F}{r_{\text{diff}}}\right)^{1/2}. \qquad (6.140)$$

Because $f \gg D$, the resulting flux variations are weak and are scaled by the factor $D/f(r_{\text{ref}}) \approx (r_{\text{diff}}/r_F)^{1/2}$; the characteristic time scale is now $t_{\text{ref}} = r_{\text{ref}}/v$.

In contrast to diffractive scintillation, this is a broadband phenomenon, with $\Delta \nu \sim \nu$. The angular size of the source, in order to be treated as a point, is essentially $\theta_{\text{scat}} = (r_{\text{ref}}/D)$. This effect has also been observed in the case of pulsars; for a typical pulsar with $v \approx 10^7$ cm s^{-1} and $r_{\text{ref}} \approx 10^{13}$ cm, the time scale of variability is $t_{\text{ref}} \approx 10^6$ s, which is a few weeks. We shall see in Vol. III that a similar effect occurs in the case of extragalactic radio sources as well.

These results can again be verified by direct computation of the power spectrum of intensity correlations, as in the case of weak scintillation. Now we get the power per unit logarithmic interval in the two-dimensional Fourier space as

$$q^2 W(q) = C \left(\frac{r_F}{r_{\text{diff}}}\right)^{5/3} (q\, r_F)^{7/3} \exp\left[-(q\, r_{\text{ref}})^{5/3}\right], \quad q \lesssim r_{\text{ref}}^{-1},$$
$$= C'(q\, r_{\text{diff}})^2, \quad r_{\text{ref}}^{-1} \lesssim q < r_{\text{diff}}^{-1},$$
$$= C''(q\, r_{\text{diff}})^{-5/3}, \quad r_{\text{diff}}^{-1} < q, \qquad (6.141)$$

where C, C', and C'' are numerical constants. This spectrum has *two* peaks at widely separated length scales: the dominant peak at $qr_{\text{diff}} \approx 1$, which corresponds to the diffractive scintillation, and a second weaker peak at $qr_{\text{ref}} \approx 1$, which corresponds to refractive scintillation.

The above analysis is based on the assumption of Kolmogorov turbulence that, strictly speaking, can be valid only at scales larger than an inner cutoff scale r_{in}. If r_{diff} becomes smaller than r_{in}, then certain modifications in the analysis are necessary. The coherent patches on the scattering screen still have a size of r_{diff} but are now separated by a distance of $r \approx r_{\text{in}}$. This can lead to complex caustic phenomena whose observational relevance is not quite certain.

Finally, we comment on the effect of scattering on the actual images that are produced by the rays propagating through the turbulent medium. The weak-scattering limit is easy to analyse. In the case of a telescope observing through Earth's atmosphere, it is clear that random fluctuations will degrade the resolution of the image if the integration time t_{in} is larger than $t_{\text{diff}} = (r_{\text{diff}}/v)$. In that case, it was shown in Vol. I, Chap. 3, Section 3.16 that the resolution is limited by the angular size $\theta_{\text{scat}} = (\lambda/r_{\text{diff}})$, even if the aperture of the telescope is larger than r_{diff}. On the other hand, if $t_{\text{in}} < t_{\text{diff}}$ – corresponding to short exposures – the phase fluctuations remain frozen during the observation. The full aperture a of the telescope now is utilised and the image will be made of several speckles with individual sizes λ/a. In principle, by analysing the shapes of the speckles, we can study the source with the maximum angular resolution available for the given telescope aperture, but because the energy from the source is distributed over different speckles, the signal-to-noise ratio will be poorer than that in the case of an undistorted wave front.

The situation is more complicated in the case of strong scattering because of the existence of two different time scales, t_{diff} and t_{ref}. The different regimes are now conventionally called by the following names: *ensemble-averaged image* for $t_{in} > r_{ref}$, *average image* for $t_{ref} > t_{in} > t_{diff}$, and *snapshot image* for $t_{diff} > t_{in}$. Of these, the ensemble average and the average image corresponding to comparatively shorter integration time scales are easy to obtain. They have been used in several studies of radiation propagating through turbulent media. The snapshot regime offers the scope of probing a source with an angular resolution of $\theta_{diff} = (r_{diff}/D) \approx (\lambda/r_{ref})$. This is equivalent to working with a telescope of aperture r_{ref} that, in practice, can be much larger than the actual aperture of the telescope. This is possible because regions of size r_{ref} in the scattering screen are already acting as a giant lens, focussing the rays on the observer. A somewhat similar phenomenon occurs even in the case of atmospheric scintillation, causing the familiar twinkling of stars. The fact that stars twinkle but planets do not allows us to conclude that the angular sizes of the stars must be smaller than $\sim$1 arcsec, even though the aperature of the eye is too small to resolve images at a 1-arcsec scale. This is possible only because the atmospheric patches at scales of r_F provide an effective aperature (which is larger than that of the eye), improving the resolution. In the snapshot-image regime, something similar occurs, with r_{ref} acting as the basic scale size rather than r_F. Although this technique has yet to be demonstrated as a standard imaging tool, it suggests the possibility that strong scattering may actually help in *increasing* the resolution in some cases.

7
Binary Stars and Accretion

7.1 Introduction

This chapter deals with several observed phenomena in binary star systems and depends on the material developed in dynamics (Vol. I, Chap. 2) and Chaps. 3, 5, and 6 of this volume.[1] The material related to accretion disks developed here will be needed in the modelling of active galactic nuclei in Vol. III.

7.2 Overview

The discussion of stellar evolution in Chaps. 2–6 concentrated on the star as a single dynamical entity, uninfluenced by its surroundings. The evolutionary phenomena change significantly and a variety of new effects come into play if the star is a member of a binary system that consists of two stars gravitationally bound to each other. We saw in Chap. 3 that star formation takes place in giant molecular clouds in the ISM. The chances that a given star is gravitationally bound to another star is fairly high under such circumstances and – in fact – well over half of all the stars in the sky are members of binary or multiple star systems. It is therefore necessary to study the effect of a close companion on the evolution of a star.

Such an effect clearly depends on how close the two stars are. When the stars are reasonably far away (compared with the sum of their radii at any stage in their evolution) they are said to form a detached binary system, and the influence of one star on another is minimum. Such systems are still extremely important in astronomy, as the study of the orbits of the binary star allows us to estimate the masses of the stars.

The situation becomes more complicated if the separation between the stars is not far enough and – during the course of the evolution – the radius of one of the stars becomes significantly large. We can quantify this aspect by studying the effective equipotential curves, in the corotating frame of the two stars, which we obtain by assuming that the gravitational effect of stars is equivalent to that of two-point masses. These equipotential surfaces were worked out in Vol. I, Chap. 2, Section 2.3, and are characterised by five Lagrangian points. (See Fig. 2.1 of Vol. I, Chap. 2, Section 2.3). Of these five points, three (denoted $L1$, $L3$, and $L2$) are along the line connecting the centres of the stars. If the radius of the star becomes large enough to fill the Roche lobe around it, then the material that reaches the inner Lagrangian point $L3$ can flow from one star to the other. It is convenient to define a length scale $\mathcal{R}_{RL}$ as the radius of a sphere having the same volume as the region bounded by the Roche lobe. To a good approximation, this is given by

$$\mathcal{R}_{RL,1} \approx a \left[\frac{0.49\, q^{2/3}}{0.6\, q^{2/3} + \ln(1 + q^{1/3})} \right] \equiv a f(q), \qquad (7.1)$$

where a is the semimajor axis of the system and $q = (m_1/m_2)$ is the ratio of the masses; this expression is the Roche lobe radius of the first star, and, of course, a corresponding expression exists for the second star, with q being replaced with $1/q$. If the radius of the star becomes comparable with $\mathcal{R}_{RL}$ during some phase of the evolution, the mass transfer can occur between the stars.

As one of the stars in a binary system evolves, its radius changes during several phases. For example, in a star with $m_1 \approx 5\, M_\odot$ the radius increases marginally (by less than of a factor of 2) during the core hydrogen burning; it increases significantly (by more than a factor of 10) during the shell hydrogen burning before core helium burning and again during the phase preceding the core carbon ignition (by another factor of nearly 10). The first phase of expansion is somewhat slow whereas the next two are comparatively rapid.

As long as the radius of the first star R_1 is smaller than the Roche lobe radius $\mathcal{R}_{RL,1}$ the evolution of the star can be approximated as that of an isolated system. However, when $R_1 \gtrsim \mathcal{R}_{RL,1}$ the stellar material fills the Roche lobe and a significant mass transfer can take place from the first star to the second, decreasing m_1 and increasing m_2. Such a transfer changes the Roche lobe radius in relation (7.1) because of changes in q as well as in a. Mass transfer will also change the radius R_1 of the star because the radius of the star essentially scales as some power of its mass. If the net effect is to decrease R_1 below the new Roche lobe radius, then the mass transfer will cease and the system will be stable. On the other hand, if the net effect is to decrease the Roche lobe radius below the stellar radius, mass transfer will continue at a significant rate. It follows that the behaviour of the system depends essentially on how the Roche lobe radius and stellar radius vary in response to changes in m_1.

To investigate this effect in detail, let us consider the simplest possible situation in which the total mass $m = m_1 + m_2$, as well as the total angular momentum,

$$J = \frac{m_1 m_2}{m_1 + m_2} \sqrt{G(m_1 + m_2)a} = (Gm^3 a)^{1/2} \frac{q}{(1+q)^2}, \qquad q = \frac{m_1}{m_2}, \qquad (7.2)$$

is conserved during the evolution. In this case, we can express a and hence the

Roche lobe radius of relation (7.1) in terms of the conserved quantities J, m and the variable q as

$$\mathcal{R}_{RL,1} = R_c \frac{0.49(1+q)^4}{0.6q^2 + q^{4/3}\ln(1+q^{1/3})}, \quad R_c \equiv \left(\frac{J^2}{Gm^3}\right). \quad (7.3)$$

Note that R_c is defined such that $(Gm^2/R_c^2) = (J^2/mR_c^3)$; this is the radius of a circular orbit for two stars of mass m and orbital angular momentum J. From Eq. (7.3) it is clear that the Roche lobe radius increases as q^2 for $q \gg 1$ and as q^{-2} for $q \to 0$. It has a distinct minimum around $q_{min} \approx 1$. Because the mass of the primary star can be expressed as $m_1 = mq/(1+q)$, we can easily express $\mathcal{R}_{RL,1}$ as a function of the mass m_1 of the mass-losing star. The functional form is shown in Fig. 7.1(a) by a thick curve. The slope of the curve $\mathcal{R}_{RL,1}(m_1)$ varies continuously and increases as a function of m_1. It is straightforward to determine from Eq. (7.3) the logarithmic derivative of $\mathcal{R}_{RL,1}$ with respect to m_1, and we get

$$n_L = \frac{\partial \ln \mathcal{R}_{RL,1}}{\partial \ln m_1} = -2(1-q) + \frac{2}{3}(1+q)$$
$$\times \left[1 - \frac{0.6q^{2/3} + 0.49q^{1/3}(1+q^{1/3})^{-1}}{0.6q^{2/3} + \ln(1+q^{1/3})}\right]. \quad (7.4)$$

This slope n_L is a monotonically increasing function of q and has the limiting value of $-(5/3)$ for $q \to 0$ and increases as $2q$ for large q. This quantity represents how fast the Roche lobe radius shrinks or grows as m_1 decreases or increases.

To decide about the stability of mass transfer, we next need to determine how the radius of the star varies with m_1, i.e., we need to determine $n_* \equiv (\partial R/\partial m_1)$. This index depends on the quantity that is kept constant during the process of mass transfer. If the star has a fixed chemical composition and remains in thermal equilibrium, then for main-sequence stars over a mass range of $(0.1-10)$ $M_\odot$ we have $n_* \approx 0.7$. On the other hand, if the mass transfer maintains hydrostatic equilibrium but not necessarily thermal equilibrium, then the derivatives in n_* need to be evaluated at constant entropy. Lower-mass stars (with $M < 0.7 M_\odot$) are fully convective and thus have constant specific entropy throughout the interior. If the mass transfer occurs on a time scale less than the evolutionary time scale of the star, the condition $s = $ constant can be maintained throughout the mass transfer. Because this requires that $P \propto \rho^{5/3}$, with $P \propto (GM^2/R^4)$ and $\rho \propto (M/R^3)$, we have $R \propto M^{-1/3}$ or $n_{RM} = -(1/3)$. On the other hand, stars more massive than 10 $M_\odot$ have radiative envelopes and $n_{RM} \approx 2$ for these stars. This suggests that the critical values for n_* we are concerned with are $-1/3$, 0.7, and 2.

As an example, let us consider a star for which the mass–radius relationship is given by $R \propto m_1^{0.7}$. Depending on the constant of proportionality as well as on the value of R_c, three possible situations can emerge, as shown by the three lines

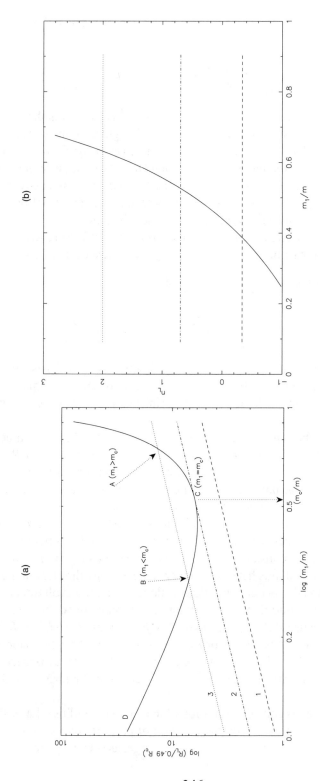

Fig. 7.1. (a) The behaviour of the Roche lobe radius is compared with the stellar radius as a function of the mass m_1 of the mass-losing star. (b) The slope of the Roche lobe radius as a function of the mass m_1 of the mass-losing star; $m = m_1 + m_2$ is the total mass of the binary system.

in Fig. 7.1(a). If the parameters are such that the curve marked 1 is relevant, then the radius of the star is smaller than the Roche lobe radius and no mass transfer takes place. If the parameters are such that the radius of the star varies, as shown by the curve marked 3, then the two radii intersect at points A and B, depending on the value of m_1. If m_1 is greater than a critical value m_c, then the radius of the star equals the Roche lobe radius at point A. When the mass transfer occurs, m_1 will decrease and R and $\mathcal{R}_{\text{RL},1}$ will decrease along curves AB and AC, respectively. This clearly has the effect of decreasing the Roche lobe radius below the stellar radius, making the mass transfer unstable. On the other hand, if mass m_1 of the star is less than a critical value m_c, the intersection occurs at B. When m_1 decreases, the stellar radius falls below the Roche lobe radius and the mass transfer ceases.

The critical value of the mass that distinguishes stable and unstable behaviour occurs when the curve giving the stellar radius just touches the curve for the Roche lobe. Given the relation $R \propto m_1^{n_*}$, we have to determine only the value of m_1/m at which n_L is equal to n_*. It can be easily verified that n_L crosses the critical values $n_* = -(1/3), 0.7$, and 2 (for low-mass convective stars, main-sequence stars, and high-mass stars dominated by the radiative envelope) when $q_c \approx 0.63, 1.12$, and 1.71, respectively. The corresponding critical values for m_1/m are 0.387, 0.528, and 0.631, respectively. For example, if $q > 0.63$ [giving $(m_1/m) > 0.387$], then $n_L > n_*$ for low-mass convective stars. The radius of the Roche lobe will shrink faster than the stellar radius and the mass transfer will be unstable on dynamical time scales. Similarly, $q > 1.12$ [giving $(m_1/m) > 0.528$] will represent instability for main-sequence stars in thermal time scales, and $q > 1.71$ [giving $(m_1/m) > 0.631$] will denote the instability of radiative stars on thermal time scales. Figure 7.1(b) denotes the slope n_L and the three critical values.

Once the mass loss starts, the Roche lobe radius also decreases until $m_1 \approx m_2$ and then increases again as the secondary becomes more massive than the primary. Because $R_1 > \mathcal{R}_{\text{RL},1}$ between A and B, it follows that the mass loss must reduce the primary's mass from the initial value corresponding to point A to a final value corresponding to point B. This process takes place at a thermal time scale

$$t_K \simeq \frac{Gm_1^2}{RL} \simeq 3 \times 10^7 \frac{(m_1/M_\odot)^2}{(R/R_\odot)(L/L_\odot)} \text{ yr} \tag{7.5}$$

if the stellar envelope is radiative; it could even occur on a dynamical time scale

$$t_{\text{ff}} \simeq \left(\frac{R^3}{Gm_1}\right)^{1/2} \simeq 5 \times 10^{-5} \frac{(R/R_\odot)^{3/2}}{(m_1/M_\odot)^{1/2}} \text{ yr} \tag{7.6}$$

if the envelope is convective.

Beyond point B, the radius of the primary is less than the Roche lobe, allowing for two subsequent possibilities. The first possibility is one in which the binary becomes detached with negligible further influence of one on the other. The second – and the more interesting possibility – is when a new phase of nuclear burning sets in and the star's radius increases at a nuclear time scale:

$$t_n \simeq \left(\frac{m_1}{M_\odot}\right)\left(\frac{L_\odot}{L}\right) \times 10^{10} \text{ yr.} \tag{7.7}$$

The star would gradually expand, with the stellar radius essentially following the Roche lobe radius along BD with a slow mass transfer. Both of these phases of mass transfer, with the first one occurring in as short as 10^4–10^5 yr and the second one taking approximately 10^7–10^{10} yr, can occur for the same binary system. The net amount of mass transfer can be obtained from the curve $R_1(m_1)$, which is available from stellar-evolution theory. Given m_1 and m_2 before the mass loss, the final values m'_1 and m'_2 corresponding to point B in Fig. 7.1(a) can be easily determined. Because the mass ratio (m_1/m_2) can become reversed during the mass transfer, the less-massive member of a system can often appear to be more evolved.

One elementary consequence of such a mass transfer will be the change in the orbital characteristics of the binary system, which are essentially determined by the masses of the stars. In fact, the system, becomes fairly distorted when any one of the stars fills its Roche lobe and the orbital characteristics change quite a bit. More importantly, the evolution of individual stars can be significantly modified by such a mass transfer. We saw in the previous chapters that the mass of the star is the single governing parameter that determines its entire evolution. However, this conclusion assumed that the mass remains constant (except possibly for mass loss by stellar winds which can be ignored during most of the evolution). If a significant amount of mass transfer occurs between two stars in a binary system, then the evolutionary history of a star can be very different from what happens in the case of isolated stars. As a simple example, consider a situation in which originally $m_1 > m_2$. Being more massive, the first star evolves faster, enters the red giant phase, and fills the Roche lobe. A significant amount of mass transfer from m_1 to m_2 can now reverse the mass ratios and make $m_1 < m_2$. Once m_2 becomes more massive, it will evolve faster and could, for example, end up in a supernova explosion, leaving behind a compact remnant – a white dwarf, a neutron star, a pulsar, or a black hole. Observations might now indicate a seemingly paradoxical situation in which the star with larger mass appears to be less evolved, contrary to what is expected from stellar-evolution theory.

The above discussion also illustrates the possibility of having binary systems of which one member (or even both) is a compact stellar remnant (white dwarf, neutron star, pulsar, or black hole). Such systems show a variety of interesting behaviour depending on the particular combination and, broadly speaking, we can distinguish the following possibilities:

(1) The compact remnant is a white dwarf, and the partner is a normal star. Such binaries are involved, for example, in systems called *cataclysmic variables, classical novas*, and *dwarf novas*. The gravitational force of the compact remnant pulls out the matter from the companion star that, under certain circumstances, could form an accretion disk around it. In many cases, the strong magnetic field of the remnant also plays a role in the dynamics of accretion and a very complex pattern of behaviour can arise.

(2) The compact remnant could be a neutron star, and the partner could be a low-mass star. Because the potential-well depth of a neutron star is significantly larger than that of a white dwarf, fairly high temperatures can be reached in the accretion disks, leading to radiation that peaks in the x-ray band. The *low-mass x-ray binaries* (LMXBs) belong to this class.

(3) In other situations, the neutron star could accrete matter from the stellar wind of the high-mass companion star. X-ray emission can again occur, and the system is called a *high-mass x-ray binary* (HMXB).

(4) Another interesting class of binary systems are those in which one star is a pulsar and the other one is a suitable compact remnant. The radiation from the pulsar allows precise measurement of orbital parameters of the system, and theoretical models can be tested accurately in such systems. In the vicinity of a neutron star or a black hole, general relativistic effects can be quite significant. When *both* the stars of a binary system are neutron stars or pulsars, then it is possible to use such a system to test the strong field predictions of the general theory of relativity.

7.3 Example of an Evolving Binary System

Let us now take a closer look at the evolution of a binary system made of a primary with 20 $M_\odot$ and a secondary with 6 $M_\odot$ in circular orbit of radius $a = 0.24 \times 10^{13}$ cm. The initial period is given by

$$P = 365.25 \frac{(a/a_\odot)^{3/2}}{(M/M_\odot)^{1/2}} \text{ days} = 0.116 \frac{(a/R_\odot)^{3/2}}{(M/M_\odot)^{1/2}} \text{ days}, \quad (7.8)$$

where $a_\odot = 1.5 \times 10^{13}$ cm is the astronomical unit and $M = m_1 + m_2$; for our case, the period is ~ 4.4 days. The following stages characterise the subsequent evolution.

(1) The more-massive star, with a mass of 20 $M_\odot$, evolves faster and exhausts the hydrogen core in approximately 6.16×10^6 yr. The scaling relations derived in Chap. 2, Section 2.4 show that the main-sequence lifetime of the 6 $M_\odot$ star is nearly 10 times longer. As the envelope of the primary expands, it fills the Roche lobe and matter is transferred on a dynamical time scale of approximately 2×10^4 yr to the secondary. The matter cannot fall directly onto the secondary because of its angular momentum; instead, it forms an accretion disk around

the secondary and is transferred slowly. The total time elapsed until now is approximately 6.18×10^6 yr.

The angular momentum of the system,

$$J = m_1 m_2 \left(\frac{aG}{M}\right)^{1/2} = 8.85 \times 10^{52} \frac{(m_1/M_\odot)(m_2/M_\odot)(a/a_\odot)^{1/2}}{(M/M_\odot)^{1/2}} \text{ ergs s}, \quad (7.9)$$

is conserved during the mass transfer, making the relative orbital radius scale with the masses as

$$a \propto \frac{1}{m_1^2 m_2^2} = \frac{1}{m_1^2 (M - m_1)^2}. \quad (7.10)$$

If $M = m_1 + m_2$ remains constant, the mutual separation of the stars changes as a result of mass transfer, with a reaching a minimum when the two masses are equal.

(2) In approximately 5×10^5 yr, the secondary has grown to a hydrogen-rich star of mass 20.6 $M_\odot$ whereas the primary is a helium burning star with $m_1 = 5.4 M_\odot$. The centre of gravity of the system shifts closer to the secondary, the Roche lobe curves change, and the binary period becomes ~5.2 days. The primary will now go through helium burning (in approximately 6×10^5 yr) and subsequent stages of carbon, oxygen, and silicon burning, each with a shorter lifetime, whereas the secondary will evolve as a young hydrogen burning star of a 20.6 $M_\odot$ star. We now have a less-massive star (primary) on the verge of becoming a supernova whereas the more-massive star is in the main sequence. The age of the system now is approximately 6.7×10^6 yr.

(3) The primary explodes as a supernova, expelling a shell of matter of ~3.4 $M_\odot$ and leaving behind a remnant of mass $m_R = 2 M_\odot$. The remnant collapses rapidly on a dynamical time scale, forming a neutron star (or a black hole, depending on the mass). The supernova explosion, in general, leads to an enlargement of the scale of the orbit and can also give a significant recoil velocity, producing some amount of orbital eccentricity. However, for the parameters given above, the explosion is not strong enough to disrupt the gravitational binding of the system. The condition for such a disruption can be easily estimated. The initial energy of two stars in circular orbit is given by

$$E_{\text{total}} = \frac{1}{2}\left(\frac{m_1 m_2}{m_1 + m_2}\right) v^2 - \frac{Gm_1 m_2}{a} = -\frac{Gm_1 m_2}{2a}. \quad (7.11)$$

Let the supernova explosion occur and eject a mass $\Delta m = m_1 - m_R$ in a spherically symmetric manner around the primary star. The expanding shell reaches the orbit of the secondary in a time scale of

$$t \simeq \frac{a}{v_{\text{ej}}} = \frac{R_\odot}{v_{\text{ej}}}\left(\frac{a}{R_\odot}\right) = 10^{-3} (P\sqrt{M/M_\odot})^{2/3} \text{ days}, \quad (7.12)$$

where P is in days and $v_{\text{ej}} \approx 10^4$ km s^{-1}. This time scale is usually quite small compared with the orbital period. Once the ejecta cross the orbit, the net

7.3 Example of an Evolving Binary System

gravitational force acting on the secondary will be decreased. However, the instantaneous velocity of m_1 is hardly affected. The total energy now is given by

$$E'_{\text{total}} = \frac{1}{2}\left(\frac{m_R m_2}{m_R + m_2}\right)v^2 - \frac{Gm_R m_2}{a}. \quad (7.13)$$

The speed of m_2 with respect to m_R is still given by $v^2 = G(m_1 + m_2)/a$. Using this, we find that

$$E'_{\text{total}} = \frac{Gm_R m_2}{2a}\left(\frac{m_1 + m_2}{m_R + m_2} - 2\right). \quad (7.14)$$

The binary will be disrupted if E'_{tot} is positive, which requires the condition that $m_R < (1/2)(m_1 - m_2)$, which can be stated in an equivalent form as

$$\Delta M > \frac{1}{2}(m_1 + m_2). \quad (7.15)$$

In other words, for stars in circular orbits, the system will be disrupted only if at least half the total mass is ejected. (It follows that the disruption cannot occur if the less-massive member explodes.) When this condition is not satisfied, the orbit readjusts to a new radius b, where

$$\frac{b-a}{a} = \frac{\Delta M}{m_1 + m_2 - 2\Delta M}, \quad (7.16)$$

and a new period is given by

$$P_b = P_a \left(\frac{b}{a}\right)^{3/2}\left(\frac{2b-a}{b}\right)^{1/2}. \quad (7.17)$$

Exercise 7.1
Extension of results to elliptical orbits: Show that the condition for disruption given in inequality (7.15) is multiplied by the factor $1 - \epsilon$, where ϵ is the eccentricity of the orbit, if the original orbit is elliptical. Also show that Eqs. (7.16) and (7.17) continue to be valid for initial elliptical orbits as well, provided that a and b are interpreted as semimajor axes.

---

(4) At this stage, we have a compact remnant with a mass of 2 $M_\odot$ and a main-sequence secondary with a mass of 20.6 $M_\odot$ forming a detached binary in an initially eccentric orbit. Tidal interactions between the two stars will dissipate energy and – because circular orbits have the minimum energy for a given angular momentum – the orbits will be circularised. (We will see more details of tidal interactions in Chap. 10, Section 10.7.) This typically has a time scale of approximately 7×10^6 yr.

(5) The secondary star completes core hydrogen burning and becomes a blue supergiant. During the next 3×10^4 yr or so, the secondary loses matter in the form of a strong stellar wind and the compact primary accretes the matter from this wind. Much of the gravitational potential energy of the accreting matter will

be converted into x rays, and the binary systems, in this phase of evolution, can be a strong x-ray source. We shall discuss such systems extensively in Section 7.4.

(6) Eventually (after approximately 3×10^4 yr or so), the expanding blue supergiant fills its own Roche lobe and another phase of mass exchange begins that is qualitatively similar to the wind accretion. An accretion disk can form around the compact star, once again converting the gravitational potential energy into x radiation. The form of radiation received from such an object will also depend strongly on the geometry, with most of the x rays emerging perpendicular to the plane of the accretion disks. The age of the system now is approximately 1.2×10^7 yr.

The above scenario, although extremely simplified, illustrates the basic features of binary evolution. Depending on the parameters of the individual stars and their evolutionary track, different variations in the basic theme are possible, leading to a rich variety of astrophysical phenomena. We shall now explore these features one by one.

Exercise 7.2

OB-runaway stars: Most of the stars in the disk of the Milky Way have speeds of less than 20 km s^{-1} compared with the local standard of rest. There are, however, a few stars with speeds that exceed 200 km s^{-1} but they belong to the older Pop I stars, having orbits that are not within the galactic plane. They can be accounted for as belonging to the halo of the galaxy and as having originated from the matter out of which galaxy has formed. There can be no young blue OB-type stars in this halo category, as they will have a lifetime of (10^6-10^7) yr, which is much shorter than the age of the galaxy (10^{10} yr). It has been observed, however, that there *do* exist a significant number of OB stars with speeds of a few hundred kilometres per second, called *OB-Runaway stars*. One possible origin for these stars is in the disruption of a binary orbit in a supernova explosion taking place in the binary. Show that, if the binary orbit is disrupted, the orbital speed of the OB star relative to the original centre of mass will be $\sim$400 km s^{-1} for reasonable values of the parameters. How could we test such a hypothesis?

We conclude this section with some general comments relevant to the evolution of binary orbits when mass transfer is significant.

To begin with, note that the Newtonian mechanics of the binary orbit using the Roche potential curve are based on the following approximations: (1) The gravitational field of the two stars can be described by that of two-point masses. Because stars tend to have mass distributions that are centrally concentrated, this assumption could provide a reasonable description of the interaction of the tenuous outer layers of the star in a given gravitational field. (2) The orbits are circular. (3) The stars are in synchronous rotation with orbital motion. Then the equation of motion for a gas element in the star, in the corotating frame, is given by

$$\frac{d^2\mathbf{r}}{dt^2} + 2\mathbf{\Omega} \times \frac{d\mathbf{r}}{dt} = -\frac{1}{\rho}\nabla P - \nabla\phi, \qquad (7.18)$$

where $\mathbf{\Omega}$ is the angular velocity of rotation and $\phi =$ constant gives the Roche equipotential surface (see Vol. I, Chap. 2, Section 2.3, for more details; it must be stressed that ϕ includes the term proportional to r^2 that will lead to the centrifugal force in this frame). For synchronously rotating stars that are not moving in the corotating frame, the left-hand side of the above equation vanishes, implying, that $\nabla P = -\rho \nabla \phi$. As usual (see Vol. I, Chap.10, Section 10.3), isobars coincide with equipotential surfaces and with surfaces of constant density. Under such circumstances, matter can flow through the Lagrange point $L2$ (located between the stars along the line joining the centres) when the star fills its Roche lobe. Although the analysis based on the Roche approximation gives a vivid picture of the phenomena, it is necessary to keep in mind certain limitations of this approach. (1) To begin with, if the stars are not corotating, their shape will not be identical to that given by the Roche equipotential surface and we must incorporate periodic tidal torquing. (2) The interpretation of the Roche potential *away* from the two stars along the line connecting them needs some care. Because the coordinate system is rotating, the centrifugal acceleration will increase as we move away and will dominate at large distances. This can give the (wrong) impression that the matter to the left of $L1$ or to the right of $L3$ will move away from the system. This can happen only if matter is kept in corotation even out to long distances from the stars so that the potential ϕ (which includes the centrifugal potential) continues to be relevant for the matter. This is, of course, unrealistic and – in physically relevant situations – the matter to the left of $L1$ or to the right of $L3$ will cease to corotate at some distance. It will feel only the gravitational attraction of the two stars and will remain bound to the system.

Exercise 7.3

Change of orbital parameters: To study the changes in the orbital parameters in a binary system, it is convenient to introduce the angular momentum and energy per unit reduced mass, defined as

$$h = [G(m_1 + m_2)a(1 - e^2)]^{1/2}, \quad E = \frac{G(m_1 + m_2)}{2a}. \tag{7.19}$$

Show that the logarithmic differentiation of the above equations with straightforward algebraic manipulations lead to

$$\frac{da}{a} = \frac{d(m_1 + m_2)}{(m_1 + m_2)} - \frac{dE}{E}, \tag{7.20}$$

$$\frac{dP}{P} = \frac{d(m_1 + m_2)}{(m_1 + m_2)} - \frac{3}{2}\frac{dE}{E}, \tag{7.21}$$

$$\frac{ede}{1 - e^2} = \frac{d(m_1 + m_2)}{(m_1 + m_2)} - \frac{1}{2}\frac{dE}{E} - \frac{dh}{h}. \tag{7.22}$$

These equations describe the changes in the orbital element in terms of loss of mass, energy, and angular momentum. [Unfortunately, they are only of formal values as they cannot be integrated unless (dE/E) and (dh/h) are given as functions of other variables.]

7.4 Low-Mass and High-Mass X-Ray Binaries

The transfer of mass from a star to a compact companion allows for the possible release of the gravitational potential energy in the form of radiation. The luminosity L arising from an accretion process at a rate of $\dot{M}$ onto a compact remnant of mass M and radius R will be $L \approx (GM\dot{M}/R)$. If this radiation is thermalised then the effective temperature will be determined by $L = 4\pi R^2(\sigma T_{\text{eff}}^4)$. For binary systems in which a star is losing mass onto a neutron star, this process can lead to the copious production of x rays. Such binary x-ray sources are of considerable interest and have been investigated extensively. We shall now discuss some general features of such systems elaborating on Section 1.6.

One general constraint that is applicable for gaseous accretion on to a compact object is that the resulting luminosity has to be less than the Eddington luminosity. For example, we obtain the maximum accretion rate possible for a compact object like a neutron star of mass M and radius R by equating the accretion luminosity

$$L_{\text{acc}} = \frac{GM\dot{M}}{R} = 10^{4.5} \left(\frac{L_\odot}{R_6}\right)\left(\frac{M}{M_\odot}\right)\left(\frac{\dot{M}}{1.5 \times 10^{-8}\ M_\odot\ \text{yr}^{-1}}\right), \quad (7.23)$$

(where R_6 is the radius of the compact star in units of 10^6 cm) to the Eddington luminosity

$$L_{\text{Edd}} = \frac{4\pi GMm_p c}{\sigma_T} = 10^{4.5}\left(\frac{M}{M_\odot}\right)L_\odot. \quad (7.24)$$

This gives the maximum accretion rate as

$$\dot{M}_{\text{Edd}} = (1.5 \times 10^{-8}\ M_\odot\ \text{yr}^{-1})\ R_6. \quad (7.25)$$

For a typical neutron star with $M = 1.4\ M_\odot$, $R_6 = 1$, the maximum accretion rate that can be sustained in steady state is approximately $1.5 \times 10^{-8}\ M_\odot\ \text{yr}^{-1}$. It is also clear from the above relations that an accretion rate of $(10^{-12}-10^{-8})\ M_\odot\ \text{yr}^{-1}$ will produce luminosities in the range $(3-10^4)\ L_\odot$. The question arises as to which kinds of stars can lead to such an accretion rate when they are part of a binary system in which the second star is a neutron star.

There are two different processes in which accretion onto a neutron star can take place in a binary system. In the first case, the companion star fills the Roche lobe and transfers mass through the Lagrangian point to the neutron star. In the second case, the companion star has a significant stellar wind from which the orbiting neutron star accretes the mass. We shall now show that the first process is effective only when the companion stars have masses less than $\sim 1\ M_\odot$, whereas the second process is effective only when the companion star is very massive, with $M \gtrsim 15\ M_\odot$.

In the first case of Roche lobe overflow, the transfer takes place at low speeds and all the transferred matter will be accreted by the companion. Let us further assume that the mass-losing star is the massive partner; that is, the stellar mass is larger than $\sim 1.4\ M_\odot$, which is the typical mass of the companion neutron star.

In this case, the orbit and the size of the Roche lobe will shrink, making the star larger than the Roche lobe and losing mass at an accelerated pace. This situation is unstable, and rapid mass transfer will ensue roughly at the thermal time scale of the mass-losing star:

$$\tau_{\text{th}} = \frac{GM^2}{RL} \simeq 3 \times 10^7 \text{ yr } \frac{(M/M_\odot)^2}{(R/R_\odot)(L/L_\odot)}. \tag{7.26}$$

Using the scaling relations $L \propto M^{3.5}$, $R \propto M^{0.5}$, we get

$$\tau_{\text{th}} = 3 \times 10^7 \text{ yr} \left(\frac{M}{M_\odot}\right)^{-2}. \tag{7.27}$$

Assuming that most of the stellar mass (say, $\sim 0.8\,M_\odot$) is transferred at this time scale, we get a mass-transfer rate of

$$\dot{M}_{\text{Roche}} \simeq 3 \times 10^{-8} \left(\frac{M}{M_\odot}\right)^3 M_\odot \text{ yr}^{-1}. \tag{7.28}$$

It follows that $\dot{M}_{\text{Roche}}$ is below $\dot{M}_{\text{Edd}}$ only for $M \lesssim M_\odot$. However, in this case, the mass-losing star is really not the more-massive partner of the binary and the mass transfer is no longer unstable, showing that the above case is *not* feasible.

For such a combination of parameters (that is, one star with $M \lesssim M_\odot$ accreting onto a neutron star of mass $1.4\,M_\odot$), the mass transfer will actually cause the orbit to widen and the mass loss will take place at nuclear time scales:

$$\tau_{\text{nucl}} \approx 10^{10} \left(\frac{M}{M_\odot}\right)\left(\frac{L}{L_\odot}\right)^{-1} \text{ yr}; \tag{7.29}$$

with $L \propto M^{3.5}$, we have $\dot{M} \approx 10^{-10}\,M_\odot \text{ yr}^{-1}(M/M_\odot)^{3.5}$. This is smaller than the Eddington rate only if $M \lesssim 3\,M_\odot$. However, our basic assumption requires that $M \lesssim 1.4\,M_\odot$, which is a more stringent condition. We thus conclude that Roche lobe overflow can produce steady accretion with $\dot{M} < \dot{M}_{\text{Edd}}$, provided that the mass of the star is less than $\sim 1.4\,M_\odot$.

The situation is somewhat different in the case of systems with massive $(M > 15\,M_\odot)$ stars accreting to a compact companion. Such stars have strong stellar winds, leading to a mass-loss rate of $\sim 10^{-9}\,M_\odot \text{ yr}^{-1}$ or more, with wind velocities reaching approximately $V_w = (1-2) \times 10^3 \text{ km s}^{-1}$. Most of this wind will flow past the compact neutron star companion but a fraction will be captured. We can obtain the critical radius r_{acc} within which accretion occurs by equating the gravitational potential depth of the compact star (GM_n/r_{acc}) at the accretion radius to the kinetic energy (per unit mass), $(V_w^2/2)$, of the material in the wind. Assuming that all the particle flux through an area πr_{acc}^2 is captured by the neutron star (which is orbiting at a distance a from the massive companion), we get the accretion rate as

$$\dot{M}_{\text{acc}} = \frac{\pi r_{\text{acc}}^2}{4\pi a^2} \dot{M}_w = \left(\frac{G^2 M_n^2}{V_w^4 a^2}\right) \dot{M}_w \equiv S \dot{M}_w. \tag{7.30}$$

Table 7.1. The two main types of strong galactic binary x-ray sources

HMXB	LMXB
Optical counterpart is a massive and luminous early-type star, O or early B; $(L_{\text{opt}}/L_x) > 1$	Faint blue optical counterparts $(L_{\text{opt}}/L_x) < 0.1$
Concentrated in the galactic plane; young stellar population, age $<10^7$ yr	Concentrated towards the galactic centre; fairly widespread around the galactic plane; old stellar population, age $(5$–$15) \times 10^9$ yr
Regular x-ray pulsations; no x-ray outbursts;	X-ray outbursts; regular x-ray pulsations only in a few cases.
Relatively hard-x-ray spectra with $k_B T \gtrsim 15$ keV	Softer-x-ray spectra; with $k_B T \lesssim 10$ keV

This expression depends strongly on V_w; for $a = 50\ R_\odot$ and $V_w = 10^3$ km s^{-1}, we have $S \approx 5 \times 10^{-5}$. To have significant accretion, say, $\dot{M}_{\text{acc}} \gtrsim 10^{-12}\ M_\odot$ yr^{-1}, we need $\dot{M}_w \gtrsim 2 \times 10^{-8}\ M_\odot$ yr^{-1}. Such large stellar winds are only found in main-sequence stars more massive than $(20$–$25)\ M_\odot$ and in blue supergiants of masses greater than approximately $(15$–$20)\ M_\odot$ (see Fig. 1.7 of Chap. 1.)

These results lead to an important conclusion. Assuming that compact relativistic remnants like neutron stars or black holes can be formed as companions to stars of any mass, persistent binary accretion systems can arise in only two windows: (1) The first one is made of a compact star and a low-mass ($\lesssim 1.4\ M_\odot$) companion in which the accretion is due to Roche lobe overflow. (2) The second set is made of a compact star and a high-mass ($\gtrsim 15\ M_\odot$) companion in which the accretion to the compact object occurs from the strong stellar wind of the companion. In the range $(1.4$–$15)\ M_\odot$ the winds are too weak to use the second mode and the Roche lobe overflow is quenched by super-Eddington luminosity. Hence stars in this mass gap are unable to produce persistent binary accreting systems.

These two classes of accreting systems are conventionally called *low mass x-ray binaries* (LMXB) and *high mass x-ray binaries* (HMXB). Some of the key properties of these two systems are listed in Table 7.1. These x-ray binaries are systems that contain either a neutron star or a black hole, accreting material from a stellar companion, and are the brightest class of x-ray sources in the sky. There are $\sim$175 known sources, of which $\sim$34 show x-ray pulsation and many others show transient x-ray variability.

In a HMXB the companion is an O or a B star ($M \gtrsim 10\ M_\odot$) whose light dominates the optical and the UV bands and the optical luminosity can be comparable with or even greater than x-ray luminosity. Such a companion loses mass in a

7.4 Low-Mass and High-Mass X-Ray Binaries

Table 7.2. Accretion parameters

Stellar Object ($M = M_\odot$)	R (km)	Kinetic Energy per gm (ergs)	dm/dt for $L_x = 10^{37}$ ergs s^{-1} ($M_\odot$ yr^{-1})	Column Density C to Stellar Surface (gm cm^{-2})
Sun	7×10^5	2×10^{15}	0.8×10^{-4}	190
White dwarf	10^4	1.4×10^{17}	10^{-6}	22
Neutron star	10	1.4×10^{20}	10^{-9}	0.7
Black hole	3	0.7×10^{21}	2×10^{-10}	0.4

stellar wind with $\dot{M} \simeq (10^{-10}$–$10^{-6})$ $M_\odot$ yr^{-1} and speed $v \simeq 2000$ km s^{-1}. The x-ray source is powered by mass accretion from the stellar wind. In general, HMXBs exhibit strong flare-ups, x-ray pulsations, transient outbursts, and variability in time scales of minutes. The hard-x-ray spectrum (1–10 keV) is a power law with index in the range of 0–1. The characteristic time scale for HMXBs is $\tau_{\text{hmxb}} \simeq 5 \times 10^4$ yr. Because there are ~50 HMXBs in our galaxy, the formation rate is ~10^{-3} yr^{-1}, which is 10% of the type II supernova rate seen in the galaxy. This also implies that 1–3 out of 10 neutron stars are in HMXBs in our galaxy.

In contrast, LMXBs contain companions that are later than spectral type A, with $M \lesssim 2$ $M_\odot$ and mass transfer occurring by means of Roche lobe overflow. The emission from the accretion disk can dominate the radiation from the system and LMXBs appear as faint blue objects in the optical. The mass-transfer time scale for LMXBs is $\tau_{\text{lmxb}} \gtrsim 10^8$ yr. Because there are ~100 LMXBs in the galaxy, the formation rate is ~10^{-6} yr^{-1}, which is a factor of $10^{4.5}$ lower than the type II supernova rate. This suggests that 1 out of $10^{4.5}$ neutron stars is in a LMXB in our galaxy.

If spherically symmetric accretion takes place onto a compact remnant of mass M and radius R at a rate $\dot{M}$, it would have accumulated an amount of matter $\Delta M = \dot{M}(2GM/R^3)^{-1/2}$ over a dynamical time scale. The column density of the accumulated gas will be $C = \Delta M/(4\pi R^2)$. Expressing $\dot{M}$ in terms of the accretion luminosity $L = (GM\dot{M}/R)$, we find the column density as $C = R^{1/2}L/[4\pi(GM)^{3/2}]$. Table 7.2 gives C as well as a few other parameters for accretion onto different compact remnants. Note that the column density is low enough for x rays to escape from the system in the case of neutron stars but not for white dwarfs, as x rays are typically stopped at column densities of a few grams per square centimetre.

Exercise 7.4

Mass transfer in semidetached binaries: A binary system is called *semidetached* when one of the stars has just filled its Roche lobe. Consider a semidetached binary made of two stars of equal mass M. Let the Roche lobe be a sphere of radius R. Assuming that the two Roche lobes have an overlap of D, estimate the rate of mass transfer as

$$\dot{M} \approx \pi R \, d\rho \sqrt{\frac{3kT}{m_H}}, \qquad (7.31)$$

where the density of stellar material near the intersecting point is ρ and its temperature is T. For a Sun-like star, we have $d \approx 1.42 \times 10^8$ cm, $T \approx 6600$ K, $\rho \approx 2.52 \times 10^{-10}$ gm cm^3, and $R \approx 7 \times 10^{10}$ cm. Estimate the mass loss in this case. [Answer: The mass loss can be expressed as $\dot{M} = \rho v A$, with $A = \pi x^2$ and $x \simeq \sqrt{Rd}$. Expressing v in terms of T gives the above expression. Plugging in numerical values gives $\dot{M} \simeq 1.6 \times 10^{-10} \, M_\odot \, \text{yr}^{-1}$.]

7.5 Accretion Disks

The specific angular momentum of matter orbiting a compact object is usually a few orders of magnitude higher than the specific angular momentum permitted for the stable, closed orbits around a black hole or a neutron star. Hence it is necessary that most of the angular momentum of the in-falling gas be somehow lost if accretion onto the compact object is to be effective. Given the fact that energy can be lost fairly easily by radiation, it is the effective loss of angular momentum that acts as a limiting factor in mass transfer in binaries.

In general, we expect the in-falling gas to fall into a common plane because gas particles travelling on an inclined orbit are likely to collide in the plane of intersection, which will mix the angular momentum of different streams. Consequently matter in any given radius will eventually acquire the same specific angular momentum and will orbit in a given plane. Because the orbit of minimum energy for a given angular momentum is a circle in any spherically symmetric potential, we can also expect the orbits of accreting gas to be a succession of circles, with the accretion occurring by means of the material's being transferred to circles of smaller and smaller radii. All these suggest that we should study in detail the dynamics of gaseous material orbiting in a plane around a compact object. Such a configuration, called an accretion disk, plays a vital role in high-energy astrophysics, especially in the study of binary systems and active galactic nuclei (see Vol. III). We shall now discuss several features of accretion disks.

7.5.1 Model for Thin Disks

The simplest of such configurations will be based on the assumption that the accretion disk is thin and two dimensional, with the vertical thickness H being much smaller than the radius R. This necessarily requires the accretion luminosity L to be much smaller than the Eddington luminosity L_{Edd}; if $L \gtrsim L_{\text{Edd}}$, then the radiation force is comparable with that of gravity so that the vertical gaseous

7.5 Accretion Disks

structure can be supported against gravity by radiative force, making $H \approx R$. However, if $(L/L_{\text{Edd}}) \ll 1$, radiation alone cannot support matter against gravity at $H \gtrsim R$ and we need to check only whether gas pressure can provide the necessary support. Because the vertical component of the acceleration at the height H is approximately $(GM/R^2)(H/R)$, vertical support requires that

$$\frac{GM}{R^2}\frac{H}{R} = \frac{1}{\rho}\left|\frac{\partial P}{\partial z}\right| \simeq \frac{P_c}{\rho_c H}, \quad (7.32)$$

where P_c and ρ_c are the central density and the pressure, respectively, and M is the mass of the central compact object. Using the azimuthal velocity of the fluid in Keplerian orbit, which is approximately $v_\phi = (GM/R)^{1/2}$, we can write this condition in the form

$$c_s^2 \simeq \frac{P_c}{\rho_c} \simeq v_\phi^2 \frac{H^2}{R^2}. \quad (7.33)$$

For the disk to be thin, we must have $H \ll R$, requiring that $c_s^2 \ll v_\phi^2$. In the same limit, the radial acceleration that is due to pressure is given by

$$\frac{1}{\rho}\frac{\partial P}{\partial R} \simeq \frac{P_c}{\rho_c R} \simeq \frac{c_s^2}{R} \simeq \frac{GM}{R^2}\frac{H^2}{R^2}, \quad (7.34)$$

which is much smaller than the gravitational acceleration when $H \ll R$. This shows that the fluid in a thin disk moves in a Keplerian orbit to a high degree of precision.

Let us next consider the radial structure of such a disk *in steady state*, taking into account the possibility of some mechanism for angular-momentum loss. By definition of a thin disk, the velocity of gas has only the radial component v_R and azimuthal component v_ϕ, with $v_R \ll v_\phi$. The amount of mass $\dot{M}$ crossing any radius R per unit time is given by

$$-2\pi R \Sigma v_R = \dot{M} \quad (7.35)$$

where $\Sigma(R)$ is the surface density of the disk obtained by integration of the mass density along the z direction. In a Keplerian disk, the angular velocity of rotation varies as $\Omega(R) = (GM/R^3)^{1/2}$, so that the specific angular momentum varies as $R^2\Omega \propto R^{1/2}$, which decreases with decreasing R. Therefore accretion through successive Keplerian orbits towards the central mass is possible only if the fluid can constantly lose angular momentum because of some viscous torque. Assuming that the viscous force *per unit length* around the circumference at any R is given by $\mathcal{F} = \nu \Sigma (R d\Omega/dR)$, where ν is some unspecified coefficient of viscosity, the viscous torque around the whole circumference will be

$$G(R) = (2\pi R \mathcal{F})R = \nu \Sigma 2\pi R^3 \left(\frac{d\Omega}{dR}\right). \quad (7.36)$$

The direction of the torque is such that the fluid at a radius less than R (which is rotating more rapidly) feels a backward torque and loses angular momentum whereas the fluid at radius larger than R gains the angular momentum. To determine the radial structure of the disk we have to equate this torque to the rate of loss of specific angular momentum. Consider an annular ring located between the radii R and $R + dR$. In unit time, a mass $\dot{M}$ enters the ring at $R + dR$ with specific angular momentum $(R + dR)^2[\Omega(R + dR)]$ and leaves at R with specific angular momentum $R^2[\Omega(R)]$. Thus the net angular momentum lost by the fluid per unit time in this ring is $\dot{M}[d(R^2\Omega)/dR]\, dR$. This angular momentum is lost because the torque is acting at both R and $R + dR$ whose net effect is $(dG/dR)\, dR$. This leads to

$$\dot{M}\frac{d(R^2\Omega)}{dR} = -\frac{d}{dR}\left[\nu\Sigma\, 2\pi R^3 \frac{d\Omega}{dR}\right]. \tag{7.37}$$

Using $\Omega(R) = (GM/R^3)^{1/2}$ and integrating, we get

$$\nu\Sigma\, R^{1/2} = \frac{\dot{M}}{3\pi} R^{1/2} + \text{constant}. \tag{7.38}$$

The constant of integration can be fixed by an appropriate boundary condition. One possible way is to assume the existence of an inner radius R_* at which the shear vanishes. Using the condition that the right-hand side of the above equation should vanish at R_* to determine the integration constant, we get

$$\nu\Sigma = \frac{\dot{M}}{3\pi}\left[1 - \left(\frac{R_*}{R}\right)^{1/2}\right]. \tag{7.39}$$

The actual value of R_* will depend on the specific case. If the compact object is a black hole, $R_* = 6(GM/c^2)$ is the radius of the marginally stable orbit. In the case of accretion onto an unmagnetised white dwarf or neutron star, R_* can be taken to be the radius of the compact object. In the case of a magnetised compact object, the situation is much more complicated and R_* is determined by the balance between the pressure of the magnetic field and the ram pressure. This will be discussed in detail in Section 7.6.

Given the radial structure of the disk, we can immediately compute the viscous dissipation rate per unit area, given by

$$D(R) = \nu\Sigma\left(R\frac{d\Omega}{dR}\right)^2 = \frac{3GM\dot{M}}{4\pi R^3}\left[1 - \left(\frac{R_*}{R}\right)^{1/2}\right]. \tag{7.40}$$

The energy released between R and $R + dR$ is approximately $(3GM\dot{M}/2R^2)\, dR$ for $R \gg R_*$, which is three times larger than the change in the orbital energy $(GM\dot{M}/2R^2)\, dR$. The excess actually comes from the energy released at smaller radii and transported to R by viscous forces. The corresponding total luminosity obtained by integrating $D(R)$ is

$$L = \int_{R_*}^{\infty} D(R) 2\pi R \, dR = \frac{GM\dot{M}}{2R_*}. \tag{7.41}$$

Note that L is half of the total gravitational potential energy lost in the drop from infinite distances to R_*. The other half of the potential energy is actually present in the form of kinetic energy of the fluid at R_* and is not available for dissipation as heat. In the case of accretion to a black hole, this energy is lost inside the hole whereas for accretion onto a compact object with a surface, this energy is released in a boundary layer or in an accretion column. In general, such dissipation is fairly complex and depends on the details.

A thin disk in steady state radiates energy through its top and bottom surfaces. If we assume that the spectrum is a blackbody with surface temperature $T_s(R)$, then we must have $2\sigma T_s^4(R) = D(R)$ (with the factor 2 arising from the existence of two surfaces), implying that

$$\begin{aligned}T_s &= \left\{\frac{3GM\dot{M}}{8\pi R^3 \sigma}\left[1 - \left(\frac{R_*}{R}\right)^{1/2}\right]\right\}^{1/4} \\ &= 6.8 \times 10^5 \eta^{-1/4} \left(\frac{L}{L_E}\right)^{1/2} L_{46}^{-1/4} \mathcal{R}^{1/4}(x) x^{-3/4} \text{ K},\end{aligned} \tag{7.42}$$

where $\eta \equiv (L_E/\dot{M}_E c^2)$ is the radiative efficiency in the rest-mass units (expected to be ~ 0.1), $x = (c^2 R/2GM)$, and $\mathcal{R} = [1 - (R_*/R)^{1/2}]$. Far from the inner edge, $\mathcal{R} \approx 1$, $T_s \propto x^{-3/4} \propto R^{-3/4}$, which is characteristic of thin accretion disks. The overall temperature scale varies as $(\dot{M}/M^2)^{1/4} \propto (L/L_{\text{Edd}})^{1/2} L^{-1/4}$.

Central regions of such disks can emit soft x rays and hard UV radiation. However, the theoretical prediction of the radiation from the accretion disk cannot always be directly observed because of several practical limitations. We should realise that the geometry of the binary system involving a compact remnant and, say, a large normal star plays a decisive role in what will be observed. Consider, for illustration, an accretion disk around a neutron star that is orbiting around a much larger normal star. In general, such a disk will be thicker at larger R. Figure 7.2 shows the relevant geometry, assuming that x rays are produced from the inner regions of the accretion disks. At large inclinations these x rays will be blocked by the thickness of the disks themselves, and hence well-defined x-ray eclipses that are due to the orbiting star may not be seen. At somewhat lower inclinations we will see dips in the x-ray intensity that are due to partial eclipsing by the companion star as well as the outer regions of the disk during certain phases of the orbit. (Such sources are called dippers.) Even during a (geometrical) eclipse it may be possible to see scattered x radiation in the form of an x-ray corona.

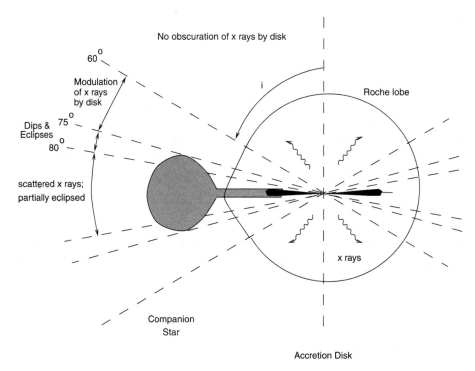

Fig. 7.2. Schematic diagram showing the influence of geometry in observing x rays from an accretion disk.

7.5.2 Nature of Disk Viscosity

The expressions derived in the last subsection, especially Eqs. (7.41) and (7.42), show that the final result is independent of the coefficient of viscosity and can be expressed entirely in terms of $\dot{M}$. Nevertheless, the entire mechanism depends on the existence of a suitable viscosity, and we shall now briefly discuss its possible source. Several models have been suggested in the literature as a possible source for the viscosity, and none of them are completely satisfactory; at present, we have no reliable model for the origin of viscosity and we shall limit our discussion to the various possibilities.

To begin with, it is easy to show that the standard fluid viscosity is inadequate in this context. It was shown in Vol. I, Chap. 8, Section 8.4 that the coefficient of kinematic viscosity that is due to molecular scattering is $\nu_{\rm mol} \approx \lambda_{\rm mfp} c_s$, where $\lambda_{\rm mfp}$ is the mean free path for the particles. For a plasma, $\lambda_{\rm mfp} \simeq (n\sigma)^{-1} \simeq (4\pi n b_c^2)^{-1}$, where b_c is the critical impact parameter determined by $(Ze^2/b_c) \simeq (3/2) k_B T$. Numerically,

$$\lambda_{\rm mfp} \approx 6.4 \times 10^4 \left(\frac{T^2}{n}\right) {\rm cm}, \quad c_s \approx 10^4 T^{1/2} {\rm cm\ s^{-1}}, \qquad (7.43)$$

where n is the number density of molecules and all quantities are in cgs units.

7.5 Accretion Disks

Therefore

$$\nu_{\text{mol}} \simeq 6.4 \times 10^8 \, T^{5/2} n^{-1} \text{ cm}^2 \text{ s}^{-1}. \tag{7.44}$$

We can estimate the importance of this viscosity by calculating the corresponding Reynold's number $Re \approx (Rv_\phi/\nu_{\text{mol}})$. Using

$$Rv_\phi \simeq 10^{18} \left(\frac{M}{M_\odot}\right)^{1/2} \left(\frac{R}{10^{10} \text{ cm}}\right)^{1/2} \text{ cm}^2 \text{ s}^{-1}, \tag{7.45}$$

we find that

$$Re_{\text{mol}} \simeq 2 \times 10^9 \left(\frac{M}{M_\odot}\right)^{1/2} \left(\frac{R}{10^{10} \text{ cm}}\right)^{1/2} (nT^{-5/2}). \tag{7.46}$$

For the typical values in accretion disks ($n \approx 10^{15}$ cm^{-3}, $T \approx 10^4$ K), the Reynolds number is very large, showing that molecular viscosity is irrelevant.

In laboratory fluids, the flow becomes turbulent for $Re \gtrsim 10^3$ and we may attempt to invoke turbulent viscosities in the case of accretion disks. In such a case, $\nu \approx \nu_{\text{turb}} l_{\text{turb}}$, where ν_{turb} is the velocity associated with turbulent eddies and l_{turb} is the size of the largest coherent turbulent cell. Assuming that $\nu_{\text{turb}} \lesssim c_s$ and $l_{\text{turb}} \lesssim H$ for a thin accretion disk, we can write

$$\nu_{\text{turb}} \simeq \alpha \, c_s \, H \quad (\alpha \lesssim 1). \tag{7.47}$$

This will lead to more reasonable results. Such a prescription is often used in modelling the viscosity in accretion disks with α parameterising our ignorance of the basic phenomena.

One possible way of obtaining the viscosity in the above form is due to tangled magnetic fields in the accretion disks along the following lines: The spatial part of Maxwell's stress tensor,

$$\sigma_{ij} = \frac{1}{4\pi} \left[E_i E_j + B_i B_j - \frac{1}{2}(E^2 + B^2) \delta_{ij} \right], \tag{7.48}$$

can lead to an electromagnetic torque on the plasma between two rings, given by

$$\mathcal{G}_m = \int d\phi \, r \int dz \, r \, \sigma_{r\phi}, \tag{7.49}$$

where the relevant component of the shear stress is $\sigma_{r\phi} = (B_r B_\phi/8\pi)$. If the magnetic-field lines are frozen to the plasma, they will become tangled because of fluid rotation, leading to $B_r \approx B_\phi \approx B$. Assuming that the magnetic-energy density $(B^2/8\pi)$ is comparable with ρc_s^2, we expect the shear stress to be proportional to ρc_s^2, say, $\alpha \rho c_s^2$. The kinematic viscosity, on the other hand, is defined such that the shear stress is $\nu \rho (Rd\Omega/dR) \approx \nu \rho \Omega \approx (\nu \rho c_s/H)$. This suggests that $\nu_{\text{mag}} \approx \alpha c_s H$, which matches with original prescription (7.47).

It is actually possible to realise such a magnetic viscosity through a generic instability that exists for axisymmetric fluid systems containing a magnetic field.

The origin of this instability lies in the following fact: Although the Rayleigh stability criterion, $d(R^2\Omega)/dR > 0$, holds for fluid systems (see Vol. I, Chap. 8, Section 8.13), it does not necessarily apply in the MHD context. In the latter, a radial perturbation of an initially poloidal magnetic field is unstable and grows rapidly. The energy for this instability arises from the shear but it is the magnetic field rather than the molecular diffusion that establishes the coupling between different radii.

Consider a disk in equilibrium with a dynamically *un*important (i.e., $B^2/8\pi \ll p$) magnetic field running parallel to the vertical z axis. To simplify the discussion further, let us suppose that the gas in the disk has the same specific entropy and that there is no radial pressure gradient. We will now perturb this system with the perturbed quantities, having the wave vector $\mathbf{k} = (0, 0, k_z)$ along the z axis and frequency ω, and restrict attention to perturbations with $\delta v_z = 0$ and $\delta B_z = 0$. The linearised MHD perturbation equations now become

$$-i\omega \frac{\delta \rho}{\rho} = 0, \tag{7.50}$$

$$-i\omega \delta v_r - 2\Omega \delta v_\phi - \frac{ikB}{4\pi\rho}\delta B_r = 0, \tag{7.51}$$

$$-i\omega \delta v_\phi + \frac{1}{2}\Omega \delta v_r - \frac{ikB}{4\pi\rho}\delta B_\phi = 0, \tag{7.52}$$

$$-i\omega \delta B_r - ikB\delta v_r = 0, \tag{7.53}$$

$$-i\omega \delta B_\phi - \frac{d\Omega}{d\ln r}\delta B_r - ikB\delta v_\phi = 0 \tag{7.54}$$

(see Vol. I, Chap. 9, Section 9.7, for a similar analysis). The adiabatic condition gives $\gamma \delta\rho/\rho = \delta p/p = 0$. Combining these equations leads, after straightforward algebra, to the linear dispersion relation

$$\frac{\omega^2}{\Omega^2} = q^2 + \frac{1}{2}(1 \pm \sqrt{1+16q^2}), \tag{7.55}$$

where $q = (kv_A/\Omega)$ and $v_A = (B/\sqrt{4\pi\rho})$. The upper sign leads to an oscillating stable wave whereas the lower sign leads to an instability provided that $q < \sqrt{3}$. For small q, the growth rate is imaginary, with $\mathrm{Im}\,\omega \approx \sqrt{3}q\Omega$. Because k cannot be much smaller than H^{-1}, we have $q > v_A/c_s$. The maximum growth rate of $(3/4)\Omega$ is achieved at $q = \sqrt{15/16}$. It is clear that the perturbations can grow in orbital time scales provided that $v_A < c_s$. It may appear paradoxical that a magnetic instability is important only when the magnetic field is weak (for $q < \sqrt{3}$). This arises because the magnetic field plays the role of only a coupling mechanism, and the energy for the instability arises from the orbital shear flow. In fact, the mode becomes damped when the energy densities of the magnetic field becomes comparable with the thermal energy because – at this stage – the energy needed to bend the field line is comparable with the energy of radial motion of fluid elements.

7.5 Accretion Disks

With such a large growth rate, nonlinear amplitude can be reached until the magnetic-energy density of perturbations becomes comparable with gas pressure. Because the entire instability is based on magnetic forces' pushing matter to rotate faster (or slower) than the local orbital speed, the torque across radial rings arises in an intrinsic manner. The Maxwell stresses are comparable with – if not greater than – normal fluid stresses when this perturbation exists.

Although any of these mechanisms may lead to a viscous coefficient that is proportional to local-pressure scale height, it must be stressed that the estimation of α from fundamental physics remains an unsolved problem. In particular, α could depend on the local conditions and vary across the disk.

7.5.3 Emergent Spectrum from a Thin Disk

For the sake of simplicity, we shall ignore the above complications and proceed with the assumption that there exists *some* suitable viscous dissipation with $\nu \approx \alpha c_s H$. In such a case, the basic set of equations governing the accretion disk can be summarised as follows:

$$\rho_c = \frac{\Sigma}{H}, \tag{7.56}$$

$$H = \frac{c_s R^{3/2}}{(GM)^{1/2}}, \tag{7.57}$$

$$c_s^2 = \frac{P_c}{\rho_c}, \tag{7.58}$$

$$\alpha c_s H \Sigma = \frac{\dot{M}}{3\pi}\left[1 - \left(\frac{R_*}{R}\right)^{1/2}\right], \tag{7.59}$$

$$P_c = \frac{\rho_c k_B T}{\mu m_p} + \frac{4\sigma}{3c}T^4, \tag{7.60}$$

$$\frac{4\sigma T^4}{3\Sigma \kappa_R} = \frac{3GM\dot{M}}{8\pi R^3}\left[1 - \left(\frac{R_*}{R}\right)^{1/2}\right]. \tag{7.61}$$

The first three equations arise from the assumption that the vertical structure of the gas can be approximated by a constant temperature T. In such a limit, we obviously have $\Sigma = H\rho_c$, $c_s^2 = (P_c/\rho_c)$, and $H = (Rc_s/v_\phi)$. The fourth equation is the same as Eq. (7.39) derived in Subsection 7.5.1 and expressed in terms of the α parameterisation for the viscosity. The fifth equation merely expresses the total pressure as the sum of gas and radiation pressures, with μm_p giving the mean molecular weight of the gas. The last equation arises from the assumption that energy transport is radiative, with the energy loss through each surface being equal to half of energy generation so that $(4\sigma T^4/3\tau) = (1/2)D(R)$, where τ, the optical depth, is $\tau = \Sigma \kappa_R$, with κ_R denoting Roseland mean opacity. Once an accretion scenario is specified by the quantities M, $\dot{M}$, and α, the above six equations can be solved for the six unknowns ρ_c, Σ, H, c_s, P_c, and T at each R. It is, however, not possible solve these equation in a general fashion because (1) the pressure has two contributions, one proportional to T and another proportional to T^4 and (2) different forms of opacity dominate at different locations. The simplest procedure is to use the most dominant process at each location, solve the equations, and check for consistency. This is easily done by dividing the disk into three regions, outer, middle, and inner, and solving the equations in each case. We shall now describe these solutions:

(1) *Outer disk:* Here the gas pressure dominates over radiation pressure, and we can set $P_{\rm rad} \approx 0$. The dominant opacity is due to free–free absorption rather than to electron scattering, and we can set

$$\kappa_R \approx \kappa_{\rm ff} = 6.6 \times 10^{22} \rho\, T^{-7/2} \text{ cm}^2 \text{ gm}^{-1}. \tag{7.62}$$

It is convenient to express the lengths in terms of Schwarzschild radius, the mass in terms of solar mass, and $\dot{M}$ in terms of $\dot{M}_{\rm Edd} = 2.2 \times 10^{-8}(M/M_\odot)M_\odot \text{ yr}^{-1}$; that is, we use the dimensionless variables

$$m \equiv \frac{M}{M_\odot}, \quad \dot{m} \equiv \frac{\dot{M}}{1.39 \times 10^{18} m \text{ gm s}^{-1}}, \tag{7.63}$$

$$r \equiv \frac{R}{2.95 \times 10^5 m \text{ cm}}, \quad h \equiv \frac{H}{2.95 \times 10^5 m \text{ cm}}, \tag{7.64}$$

and the notation

$$f = \left[1 - \left(\frac{R_*}{R}\right)^{1/2}\right]^{1/4}. \tag{7.65}$$

In terms of these, the solution for the outer disk is given by

$$\Sigma = 4.5 \times 10^5 \alpha^{-4/5} \dot{m}^{7/10} m^{1/5} r^{-3/4} f^{14/5} \text{ gm cm}^{-2}, \tag{7.66}$$

$$T = 1.8 \times 10^8 \alpha^{-1/5} \dot{m}^{3/10} m^{-1/5} r^{-3/4} f^{6/5} \text{ K}, \tag{7.67}$$

$$h = 7.2 \times 10^{-3} \alpha^{-1/10} \dot{m}^{3/20} m^{-1/10} r^{9/8} f^{3/5}. \tag{7.68}$$

The solution shows that $h \ll r$ for all reasonable values of the parameters, which is required for consistency; h is also only weakly dependent on α, which is not known with any level of accuracy. A minor problem arises because of the relation $h \propto r^{9/8}$, which implies that the thickness increases faster than the linear rate. It is now possible for the outer-disk portion to be irradiated by photons from the inner hot region, thereby making a local solution slightly suspect. It is, however, not possible to make clear statements about the ratio $(h/r) \propto \alpha^{-1/10} r^{1/8}$ without knowing the variation of α with r. Whether h/r will increase with r depends on whether the physical process behind the α parameterisation makes α increase with r more slowly than $r^{5/4}$ or not; because we have virtually no handle on α, it is difficult to answer such questions. The optical depth of the disk in the vertical direction, $\tau = \Sigma \kappa_R$, is given by

$$\tau_{f\!f} = 87\,\alpha^{-4/5}\dot{m}^{1/5}m^{1/5}f^{4/5}, \qquad (7.69)$$

which again shows that the disk is likely to be optically thick.

From the value of electron-scattering opacity, $\kappa_{R,es} = 0.4\text{ cm}^2\text{ gm}^{-1}$, we find the corresponding optical depth as

$$\tau_{es} = 1.8 \times 10^5\,\alpha^{-4/5}\dot{m}^{7/10}m^{1/5}r^{-3/4}f^{14/5}, \qquad (7.70)$$

which becomes important only as r decreases. At a sufficiently large r, our assumptions are consistent and the outer disk is dominated by free–free opacity. The transition from outer disk to middle disk may be taken to occur at a critical radius r_{om} at which the two optical depths are equal. This gives

$$r_{om} = 2.6 \times 10^4 \dot{m}^{2/3}f^{8/3}, \quad R_{om} = 1.1 \times 10^{10}\dot{m}^{2/3}f^{8/3}\text{ cm}, \qquad (7.71)$$

where the second expression is for a neutron star of mass $1.4\,M_\odot$. Note that this expression is independent of α. (We shall see later in Section 7.6 that the magnetic effects of a neutron star become important typically at a distance of $R_m \simeq 2 \times 10^8 B_{12}^{4/7} R_6^{10/7} m^{-1/7}\dot{m}^{-2/7}$ cm; for the usual parameters, $R_{om} \gg R_m$, showing that accreting neutron stars will definitely have outer and middle disks.)

(2) *Middle disk:* A similar analysis works for other regions as well. In the middle disk, $P_{\text{gas}} \gg P_{\text{rad}}$ but the opacity is dominated by electron scattering with $\kappa_{R,es} \gg \kappa_{R,f\!f}$. The solution will be

$$\Sigma = 9.7 \times 10^4\,\alpha^{-4/5}\dot{m}^{3/5}m^{1/5}r^{-3/5}f^{12/5}\text{ g cm}^{-2}, \qquad (7.72)$$

$$T = 8.1 \times 10^8\,\alpha^{-1/5}\dot{m}^{2/5}m^{-1/5}r^{-9/10}f^{8/5}\text{ K}, \qquad (7.73)$$

$$h = 1.6 \times 10^{-2}\,\alpha^{-1/10}\dot{m}^{1/5}m^{-1/10}r^{21/20}f^{4/5}. \qquad (7.74)$$

(3) *Inner disk:* This region is dominated by radiation pressure with $P_{\text{rad}} \gg P_{\text{gas}}$ and electron-scattering opacity. The solutions will be

$$\Sigma = 0.42\,\alpha^{-1}\dot{m}^{-1}r^{3/2}f^{-4}\text{ g cm}^{-2}, \qquad (7.75)$$

$$T = 3.7 \times 10^7\,\alpha^{-1/4}m^{-1/4}r^{-3/8}\text{ K}, \qquad (7.76)$$

$$P_{\text{rad}} = 4.8 \times 10^{15}\,\alpha^{-1}m^{-1}r^{-3/2}\text{ dyn cm}^{-2}, \qquad (7.77)$$

$$h = 7.46\,\dot{m}\,f^4. \qquad (7.78)$$

Obtaining $\rho = (\Sigma/H)$ we can calculate the gas pressure as

$$P_{\text{gas}} = 9.6 \times 10^8\,\alpha^{-5/4}\dot{m}^{-2}m^{-5/4}r^{9/8}f^{-8}\text{ dyn cm}^{-2}. \qquad (7.79)$$

It is clear that P_{gas} increases with r while P_{rad} decreases. Hence the transition radius from middle to inner disk occurs at the point where $P_{\text{gas}} = P_{\text{rad}}$. This gives

$$r_{mi} = 360\,\alpha^{2/21}\dot{m}^{16/21}m^{2/21}f^{64/21}, \quad R_{mi} = 1.5\times 10^8\,\alpha^{2/21}\dot{m}^{16/21}f^{64/21} \text{ cm}. \tag{7.80}$$

This radius is usually quite large compared with the Schwarzschild radius and hence an inner disk will exist in the case of a black hole accretion. For a neutron star, we need to compare R_{mi} with the magnetospheric radius R_m. Whether an inner disk exists for a magnetically active neutron star depends on the values of $\dot{m}$ and B. Also note that h in the inner region is independent of r, implying a disk of constant thickness. Further, when $\dot{m} \to 1$, $h \gtrsim 1$, implying the breakdown of the thin-disk approximation. This is merely a restatement of the fact that when the accretion rate approaches the Eddington rate, radiation pressure can thicken the disk. Figure 7.3 summarises the different regions.[2]

If the disk is optically thick in the z direction, then the radiation spectrum at each R will be a blackbody spectrum at the local temperature $T(R)$. The total flux can be obtained by integration over all radii, giving

$$F_\nu \propto \int_{R_*}^{R_{\text{out}}} \frac{2h\nu^3/c^2}{e^{h\nu/k_B T(R)} - 1}\, 2\pi R\, dR. \tag{7.81}$$

There are three regimes in the frequency space where a simple form of spectrum can be obtained. First, for $h\nu \ll k_B T(R_{\text{out}})$, the Planck spectrum is proportional to ν^2 at all R so that F_ν also integrates to a function proportional to ν^2. Second, in the range, $k_B T(R_{\text{out}}) \ll h\nu \ll k_B T_*$, where $T_* = (3GM\dot{M}/8\pi R_*^3 \sigma)^{1/4}$, we can simplify the radial dependence of the temperature to $T \propto R^{-3/4}$ and extend the region of integration from 0 to ∞. Simple rescaling then shows that

$$F_\nu \propto \nu^{1/3} \int_0^\infty \frac{x^{5/3}\, dx}{e^x - 1} \propto \nu^{1/3}. \tag{7.82}$$

Finally, for $h\nu \gg k_B T_*$, we are at the Wien end of the Planck function for all R and the spectrum varies exponentially. A schematic of the spectrum is shown in Fig. 7.4(a). It is essentially a Planck spectrum with a middle part stretched out to a $\nu^{1/3}$ plateau.

The above analysis is valid only when the opacities can be effective in thermalising the spectrum, which is indeed true in the outer disk. In the middle and the inner disks, the dominant opacity is electron scattering, which does not modify the spectrum of the radiation. The radiation actually will be emitted from a region deeper down at an optical depth of $\tau_{\text{es}} \approx (\kappa_{\text{es}}/\kappa_{\text{ff}})^{1/2}$. The net effect will be to flatten the $\nu^{1/3}$ part further to a ν^0 regime, with the exponential cutoff pushed to higher frequencies [see Fig. 7.4(b)].

Exercise 7.5

Diffusion of matter in a disk: (a) Show that, in the case of a time-dependent disk, the continuity equation for mass and the z component of angular momentum can be expressed

7.5 Accretion Disks

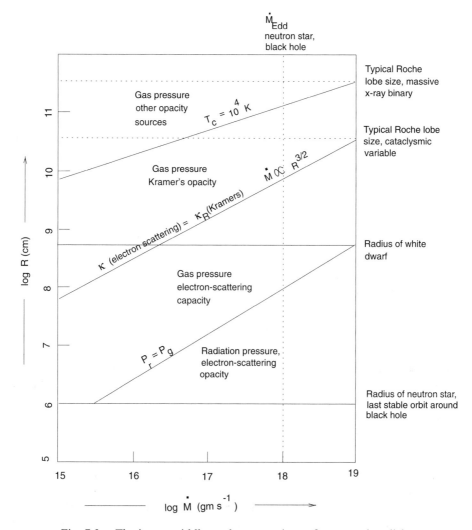

Fig. 7.3. The inner, middle, and outer regions of an accretion disk.

in the form

$$\frac{\partial \Sigma}{\partial t} + \frac{1}{R}\frac{\partial}{\partial R}(R\Sigma v_R) = 0, \tag{7.83}$$

$$\frac{\partial}{\partial t}(R^2 \Sigma \Omega) + \frac{1}{R}\frac{\partial}{\partial R}(R^3 \Sigma \Omega v_R) = \frac{1}{2\pi R}\frac{\partial G}{\partial R}. \tag{7.84}$$

(b) In the case of a Keplerian disk with a viscous torque parameterised by a constant-viscosity coefficient ν, reduce these equations to the form

$$\frac{\partial}{\partial t}(R^{1/2}\Sigma) = \frac{3\nu}{R}\left(R^{1/2}\frac{\partial}{\partial R}\right)^2 (R^{1/2}\Sigma). \tag{7.85}$$

Solve this equation with an initial condition corresponding to a ring of mass m at a

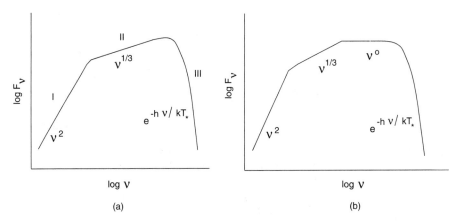

Fig. 7.4. Schematic diagram showing the emergent spectrum from a thin accretion disk.

radius $R = R_0$ and estimate how the mass evolves. {Answer. Equation (7.85) can be easily solved by the standard separation of variables techniques. The solution is given by

$$\Sigma(x, t) = \frac{m}{\pi R_0^2} \tau^{-1} x^{-1/4} \exp\left[-\frac{(1+x^2)}{\tau}\right] I_{1/4}(2x/\tau), \quad (7.86)$$

where $x = (R/R_0)$, $\tau = 12\nu t R_0^{-2}$, and $I_{1/4}(z)$ is the modified Bessel function. The solution represents the diffusion of matter both inwards and outwards. At very late times, almost all of the original mass would have accreted to the central star but almost all the angular momentum would have been carried away to a very large radii by a very small amount of matter.}

Exercise 7.6

Boundary layers in accretion: In the thin Keplerian accretion disk around a star of mass M and radius R_*, the angular velocity of matter at a radius R is given by $\Omega_K(R) = (GM/R)^{1/2}$. Near the star the flow must lead to a boundary layer of radial extent b such that the angular velocity decreases from $\Omega_K(R_* + b)$ to the angular velocity of the star Ω_*. [Of course, $\Omega_* < \Omega_K(R_*)$]. The simplified analysis given in the text ignored the boundary layer and assumed that $b \ll R_*$. To verify this assumption, start with the basic equation

$$v_R \frac{\partial v_R}{\partial R} - \frac{v_\phi^2}{R} + \frac{1}{\rho} \frac{\partial P}{\partial R} + \frac{GM}{R^2} = 0 \quad (7.87)$$

and estimate each of the terms. Argue that we must have

$$b \simeq \frac{R_*^2}{GM} c_s^2 \simeq \frac{H^2}{R_*}, \quad (7.88)$$

where H is the scale height of the disk in the z direction. The radiation emitted near the boundary layer emerges through a region of radial extent H on the two faces of the disk. Assuming it is optically thick and radiates roughly as a blackbody of area $2 \times 2\pi R_* H$, show that the effective temperature of the boundary layer is

$$T_{\rm BL} \simeq \left(\frac{R_*}{H}\right)^{1/4} T_*. \tag{7.89}$$

For accretion onto a white dwarf, estimate this expression numerically and show that

$$T_{\rm BL} \simeq 1 \times 10^5 \dot{M}_{16}^{7/32} M_1^{11/32} R_9^{-25/32} \,{\rm K}. \tag{7.90}$$

7.6 Magnetic Effects in Accretion

In the preceding discussion we have alluded several times to the effect of magnetic field. We shall now discuss some of these features in detail.

Fairly complicated phenomena can arise if the compact object (onto which the gas is accreting) is a source of large magnetic field, for example, in the case of a neutron star. A strong magnetic field can prevent the flow of plasma transverse to magnetic field lines and the matter can accrete onto such a system only along the field lines that terminate near the polar caps. To determine when such an effect will be important, we will first study the idealised case of a spherical accretion from infinity onto a magnetised neutron star.

For most of the compact remnants, the magnetic field can be taken to be that of a dipole so that the field at a distance R is given by $B(R) = (R_0/R)^3 B_0$, where B_0 is the surface magnetic field and R_0 is the radius of the compact object. The magnetic pressure that is due to this field at a distance R is

$$P_{\rm mag} = \frac{[B(R)]^2}{8\pi} = \frac{B_0^2}{8\pi} \frac{R_0^6}{R^6}. \tag{7.91}$$

If the in-falling matter is taken to have fallen from rest at infinity, its in-fall velocity at R will be $v_i \approx (GM/R)^{1/2}$ and the corresponding accretion rate will be $\dot{M} = (4\pi R^2 \rho v_i)$, where ρ is the density of the gas. This gas exerts a ram pressure $P_g = \rho v_i^2$, which is essentially the rate at which momentum is transported inwards per unit area at R. When the magnetic pressure is larger than the gas pressure, charged particles cannot penetrate transverse to the magnetic field lines and the dynamics of accretion is strongly affected. Expressing the ram pressure of the gas as

$$P_g \simeq \frac{\dot{M}}{4\pi R^2} \sqrt{\frac{GM}{R}} \tag{7.92}$$

and equating it to $P_{\rm mag}$, we can determine the radius of the magnetosphere at which a significant deviation from simple free-fall accretion will occur. This gives, except for factors of the order of unity, the radius

$$R_m \simeq G^{-1/7} B_0^{4/7} R_0^{12/7} M^{-1/7} \dot{M}^{-2/7}. \tag{7.93}$$

Expressing the magnetic field in units of 10^{12} G as $B_{12} = (B_0/10^{12}\text{ G})$ and using $R_6 = (R_0/10^6 \text{ cm})$, $m = (M/M_\odot)$, and $\dot{m} = (\dot{M}/\dot{M}_{\text{Edd}})$, where

$$\dot{M}_{\text{Edd}} = 9.4 \times 10^{17} R_6 \text{ gm s}^{-1} = 1.5 \times 10^{-8} R_6 \, M_\odot \text{ yr}^{-1} \qquad (7.94)$$

is the Eddington accretion rate defined in Eq. (7.25), we can express the radius of the magnetosphere as

$$R_m \simeq 1.9 \times 10^8 \, B_{12}^{4/7} R_6^{10/7} m^{-3/7} \dot{m}^{-2/7} \text{ cm}. \qquad (7.95)$$

For typical neutron star parameters, R_m is $\sim$100 times larger than the radius of the neutron star, showing that there is a significant region close to the compact remnant that is dominated by the magnetic field. In contrast, for a white dwarf with $R_6 \approx 500$ the condition $R_m > R_0$ can be satisfied only if $B > 10^6$ G. Only a fraction of white dwarfs have such strong magnetic fields.

It is clear that the only way matter can be accreted to a neutron star is if the matter falls down the poles of the magnetic field configuration. This suggests that there will be an accretion column associated with magnetically dominated remnants. Such a column can also exist for a white dwarf, provided it has a strong magnetic field.

The above analysis is somewhat simplified as we did not take into account the disk geometry involved in the accretion, which will, in general, depend on the angle between the magnetic field and the normal to the accretion disk. An approximate analysis of this case can be performed along the following lines. In the steady state, the flow of plasma, influenced by both the gravitational field and the magnetic field is governed by the equation

$$\rho(\mathbf{v}\cdot\nabla)\mathbf{v} = -\nabla P - \rho\nabla\phi + \frac{1}{4\pi}(\nabla\times\mathbf{B})\times\mathbf{B}. \qquad (7.96)$$

To obtain an equation for the conservation of angular momentum along the z axis, we need to consider the ϕ component of this equation. We first note the identities

$$\rho(\mathbf{v}\cdot\nabla)\mathbf{v} = \rho\left[\frac{1}{2}\nabla v^2 + (\nabla\times\mathbf{v})\times\mathbf{v}\right], \quad 0 = \nabla\cdot\mathbf{B} = \frac{1}{r}(rB_r)_{,r} + B_{z,z}, \qquad (7.97)$$

using which we can express

$$[(\nabla\times\mathbf{B})\times\mathbf{B}]_\phi = \frac{1}{r}B_r(rB_\phi)_{,r} + B_z B_{\phi,z} \qquad (7.98)$$

in the form

$$[(\nabla\times\mathbf{B})\times\mathbf{B}]_\phi = \frac{1}{r^2}(r^2 B_r B_\phi)_{,r} + (B_z B_\phi)_{,z}. \qquad (7.99)$$

Similarly, the term $(\nabla\times\mathbf{v})\times\mathbf{v}$ can be expressed in a simpler form, with $\mathbf{B}$ replaced with $\mathbf{v}$ and v_z set to zero. Finally, noting that all the gradient terms vanish

7.6 Magnetic Effects in Accretion

for the ϕ component because of axisymmetry, we get

$$\rho v_r r (r v_\phi)_{,r} = \frac{1}{4\pi}[(r^2 v_r v_\phi)_{,r} + r^2(v_z v_\phi)_{,z}]. \tag{7.100}$$

In the absence of the disk, the dipole field lines are poloidal and are dominated by the B_z component in the equatorial plane. The rotating plasma in the disk drags the magnetic field lines along the ϕ direction, producing a sizable B_ϕ component. Further, continuity across the disk requires B_ϕ to be equal in magnitude but opposite in direction above and below the disk. This implies that the magnetic field varies more rapidly with z than with r. Finally, because the disk is close to the equatorial plane and the gas moves in nearly circular orbits, B_r is negligible compared with the B_ϕ and B_z components; therefore the first term on the right-hand side of Eq. (7.100) can be ignored. Integrating the resulting equation along the vertical axis from $-H$ to H and using $\dot{M} = 2\pi r \Sigma v_r = 2\pi r (2H\rho) v_r$, we get

$$\dot{M}(r v_\phi)_{,r} = r^2 B_z B_\phi. \tag{7.101}$$

This equation can be used to determine the radius r_0 where the flow makes a transition from pure Keplerian motion to the one dominated by the magnetic field. Assuming that this transition takes place over a distance $\delta \ll r$, over which the torque exerted by the magnetic field on the disk alters the Keplerian flow, the above equation can be recast in the form

$$\dot{M}\left(\frac{r}{\delta}\right) v_\phi \approx r^2 B_z B_\phi. \tag{7.102}$$

Assuming now that $B_\phi \approx B_z \approx B(r)$ is the unperturbed dipole field in the equatorial plane and that $H \lesssim \delta \lesssim r$, Eq. (7.102) can be solved for r_0:

$$\left(\frac{H}{r}\right)_0^{2/7} R_m \lesssim r_0 \lesssim R_m. \tag{7.103}$$

Using our preceding result for (H/r) in the middle region of the accretion disk and taking $\alpha \approx 1$, we get, for reasonable values of the parameters,

$$0.3 R_m \lesssim r_0 \lesssim R_m. \tag{7.104}$$

This shows that even when the disk geometry is taken into account, the final result does not change significantly and we conclude that even in the case of an accretion disk, matter can rain down the compact object only near the polar region of the magnetic field (see Fig. 7.5).

The fraction of the area into which matter can flow can be easily estimated from the equation for field lines for dipole geometry, $r(\theta) = C \sin^2 \theta$ (see Chap. 6, Subsection 6.3.1). To have $r = R_m$ at $\theta = \alpha$ (see Fig. 7.5), we need $C = (R_m/\sin^2 \alpha)$. Hence

$$\sin^2 \beta = \frac{R_0}{C} = \left(\frac{R_0}{R_m}\right) \sin^2 \alpha. \tag{7.105}$$

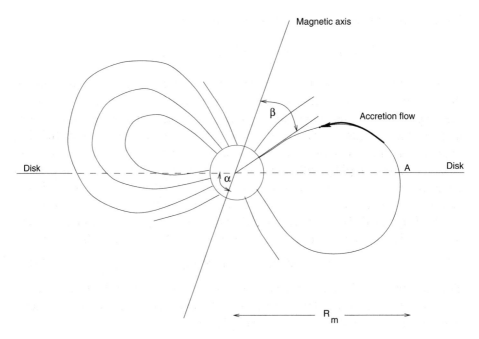

Fig. 7.5. Accretion of matter from a disk onto the polar cap of a magnetised compact remnant.

The fractional area of the accreting polar cap is

$$f_{\text{disc}} \simeq \frac{\pi R_0^2 \sin^2 \beta}{4\pi R_0^2} \cong \frac{R_0 \sin^2 \alpha}{4 R_m} \simeq 0.1 \frac{R_0}{R_m}. \qquad (7.106)$$

Exercise 7.7

Magnetic pressure and ram pressure: A steady, constant magnetic field exits along the z axis in the region $x < 0$; i.e., $\mathbf{B} = (0, 0, B_z)$ for $x < 0$ and $\mathbf{B} = 0$ for $x > 0$. Consider a beam of charged particles of density ρ and streaming velocity $v_x < 0$ moving from the force-free region ($x > 0$) towards $x \leq 0$. Describe the motion of the ions and the electrons near the $x = 0$ plane and determine how far the particles can penetrate into the magnetic field region. From this result, interpret the procedure of equating the magnetic pressure to the ram pressure in order to determine the influence of the magnetic field.

Exercise 7.8

One-dimensional accretion column: The simplest model for an accretion column is provided by gas falling along one direction (say, the z axis) onto the surface of a white dwarf. The one-dimensional flow will be supersonic at large distances and will settle on the surface of the white dwarf at subsonic speeds. These two regions of flow will be separated by a shock front at a height D. The energy released by accretion needs to be removed by suitable cooling mechanisms in the shock column. If we take the main cooling mechanism of the gas to be bremsstrahlung, then the equations describing steady accretion flow become

$$\rho v = \text{constant},$$

$$\rho v \frac{dv}{dz} + \frac{d}{dz}\left(\frac{\rho k_B T}{\mu m_H}\right) + g\rho = 0, \qquad (7.107)$$

$$\frac{d}{dz}\left[\rho v \left(\frac{3k_B T}{\mu m_H} + \frac{v^2}{2} + gz\right) + \frac{\rho v k_B T}{\mu m_H}\right] = -a\rho^2 T^{1/2}.$$

The first one is the conservation of mass; the second is the conservation of momentum that incorporates gravitational acceleration and the pressure gradient, with the gas pressure being given by $P = (\rho k_B T/\mu m_H)$; the last equation is the conservation law for energy, with the bremsstrahlung cooling expressed as $4\pi j_{\text{br}} = a\rho^2 T^{1/2}$, where a is a numerical factor. (a) Manipulate these equations to recast them in the form

$$\frac{3}{2}v\frac{dT}{dz} + T\frac{dv}{dz} = -\frac{a\mu m_H}{k_B}\rho T^{1/2}, \qquad P + \rho v^2 = P_{\text{ram}} + g\int_0^D \rho\, dz, \qquad (7.108)$$

where $z = 0$ is taken to be the base of the column at which $v \simeq 0$ and D is the height of the shock front. Explain why the ram pressure $P_{\text{ram}} = \rho_1 v_1^2$ will be much larger than the second term involving the integral. Hence the integral can be ignored. (b) Moreover, because P varies by only a factor of $4/3$ in the shock front, caused by the supersonic accretion flow hitting the stellar surface, we can make the approximation $P = \text{constant} = P_{\text{ram}}$ everywhere. Show that this leads to the solution

$$T^{5/2} = \frac{\mu m_H a}{k_B} \frac{\rho_2 T_s^2}{(-v_2)} z + \text{constant}, \qquad (7.109)$$

where T_s is the temperature at the shock front. Near the base of the column ($z = 0$) we expect T to be very small, allowing us to ignore the constant in the above equation and write

$$\frac{T}{T_s} = \left(\frac{z}{D}\right)^{2/5}, \qquad D = \frac{k_B T_s^{1/2}(-v_2)}{\mu m_H a \rho_2}. \qquad (7.110)$$

(c) Evaluate the term involving the integral that has been ignored and show that it is indeed negligible for the solution we have obtained. (d) Describe the accretion column physically, based on this solution.

7.6.1 Magnetic Torques and Pulsar Spin-Up

The accretion disk that is strongly coupled to the magnetosphere of the star exerts a torque on it, which eventually is transmitted to the star itself. To see the details of this process, we begin by noting that, in order for the angular momentum to be transferred inwards, the Keplerian angular velocity $\Omega_K(R) = (GM/R^3)^{1/2}$ at $R = R_m$ should be greater than the angular velocity Ω of the neutron star and the corotating magnetosphere. For the sake of simplicity, let us consider the case in which $\Omega_K(R_m) \gg \Omega$ so that the neutron star is a slow rotator. The rate of transfer of angular momentum at the radius $R = R_m$ to the star's magnetosphere is $\dot{M}R_m^2\Omega_K(R_m)$. Equating this to the angular acceleration, we get

$$I\frac{d\Omega}{dt} = \dot{M}R_m^2\Omega_K(R_m) = \left(\frac{L}{GM}\right)R_m^3\Omega_K(R_m). \tag{7.111}$$

The corresponding change in the period $P = (2\pi/\Omega)$ is governed by

$$\frac{\dot{P}}{P} = -\frac{1}{2\pi}\left(\frac{PL}{GMI}\right)R_m^3\Omega_K(R_m). \tag{7.112}$$

Substituting for $\Omega_K(R_m)$ and R_m, we find that simple algebra gives

$$\frac{\dot{P}}{P} \approx -\frac{G^{-3/7}}{2\pi I}\left(\frac{B_0 R_0^6}{M}\right)^{2/7}(L^{6/7}P). \tag{7.113}$$

For $M \approx M_\odot$, $R_0 \approx 10^6$ cm, and $B_0 \approx 10^{12}$ G, we get the numerical relation

$$\log_{10}\left(-\frac{\dot{P}}{P}\right) = -4.4 + \log_{10}\left(PL_{37}^{6/7}\right), \tag{7.114}$$

where L_{37} is in units of 10^{37} ergs s^{-1}. This relation is entirely among quantities that are directly measurable and has been verified observationally to a fairly high degree.

In the above analysis, we assumed that $\Omega \ll \Omega_K(R_m)$. As the accretion spins up the star, this will eventually cease to be valid, and, of course, this process can spin up the star only until Ω becomes equal to $\Omega_K(R_m)$. This leads to a limiting period for the star determined by the condition $\Omega = \Omega_K(R_m)$ and is given by

$$P_{\text{eq}} = \frac{2\pi}{\Omega_m} = 1.4 B_{12}^{6/7} R_6^{15/7} m^{-8/7} \dot{m}^{-3/7} \text{ s} = 3.8 B_9^{6/7} R_6^{15/7} m^{-8/7} \dot{m}^{-3/7} \text{ ms}. \tag{7.115}$$

Because $\dot{m} \lesssim 1$, the minimum period to which the stars can be spun up is approximately $4 \times B_9^{6/7}$ ms for reasonable values of parameters. Figures 6.1 and 6.6 in Chap. 6 show the positions of over 10^3 pulsars in the B–P plane with the constraint arising from the spin-up marked in it.

These features have important observational consequences when the star that is accreting mass is a pulsar. A radio pulsar in a binary system, which has undergone such a process, will end up on or to the right of a curve $P \propto B^{6/7}$ in the P–B plane. This result is also verified observationally. By taking $P_{\text{min}} \approx 4B_5^{6/7}$ ms and combining with the relation $B_0 = 3 \times 10^{19}$ G $(P\dot{P})^{1/2}$ [obtained from Eq. (6.4) of Chap. 6], we find that $\dot{P} \leq 2 \times 10^{-15} P^{4/3}$. This puts an equivalent constraint on the radio pulsars that are spun up by binary accretion in the P–$\dot{P}$ plane, shown in Figs. 6.1 and 6.6 as the "spin-up limit."

The spin-up of a pulsar by the accretion of matter occurs through a series of phases, each characterised by the relative values of R_m compared with the

7.6 Magnetic Effects in Accretion

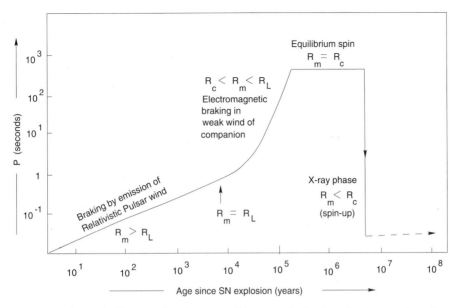

Fig. 7.6. Schematic diagram showing a possible scenario for the spinning up of a pulsar.

radius of the light cylinder ($R_L = c/\Omega$) and the radius of corotation [$R_c = (GM/\Omega^2)^{1/3}$]. When a neutron star first is formed in a binary, it will be spinning rapidly and will slow down like a normal pulsar. There is very little penetration of the magnetosphere and the light cylinder by the stellar wind of the companion star during this time, and $R_m > R_L$ in this phase. The second phase arises when $R_c < R_m < R_L$ and the matter is blocked by the magnetic pressure and forced to corotate with the star. However, when $R < R_m$ (during the still early phase in which stellar wind is weak and the pulsar is spinning fast) the matter cannot corotate with the star and will eventually be expelled from the magnetosphere. Nevertheless, the coupling of the plasma and the magnetic field around $R \approx R_m$ will exert an electromagnetic torque on the magnetosphere and will slow down the neutron star. As it slows down, the corotation radius will approach R_m and accretion of matter onto the neutron star becomes possible. Accretion, which brings in the angular momentum, will spin up the star, changing the corotation radius. This process will rapidly settle down when the period of the neutron star is such that $R_m = R_c$. In the next phase, as the stellar wind from the companion increases, the neutron star will become a strong x-ray source and will be rapidly spun up further, with R_m becoming smaller than R_c. Eventually the companion star could also explode, leaving behind a rapidly spinning neutron star.[3] The evolution of the period during these phases is shown schematically in Fig. 7.6.

To set this phenomenon in context, let us consider the evolution of a binary system made of two stars of masses of ∼25 $M_\odot$ (primary) and ∼10 $M_\odot$ (secondary). The more-massive primary star will evolve rapidly in a time scale of approximately 5×10^6 years. During the evolution it will fill its Roche lobe and transfer its hydrogen-rich envelope to the companion. This will lead to a binary system with a helium star of $\sim 8.5\ M_\odot$ and a main-sequence star of 26.5 $M_\odot$. Helium stars, being very luminous, evolve through helium and carbon burning within approximately 5×10^5 years and the degenerate core will collapse to form the neutron star. For the parameters we have chosen, the supernova explosion will eject less than half the total mass of the system and hence the binary will not be disrupted. The next interesting phase in the evolution occurs when the now-massive secondary becomes a supergiant. The mass transfer from the supergiant to the neutron star by means of strong stellar winds can make the system an x-ray source for a brief interval of $\sim 10^4$ yr. However, as the massive companion increases in radius, there will be a significant loss of mass from the binary system because the neutron star can capture only a fraction of the mass carried away by the stellar wind. The loss of mass will also be accompanied by a significant loss of angular momentum, causing the orbital size of the system to shrink rapidly. Such a loss of mass and the spiralling in of the neutron star will terminate when the outer envelope of the companion star is almost completely depleted, leaving only a helium core. This core (which, calculations suggest, will have a mass of $\sim 8.8\ M_\odot$) will now evolve rapidly and explode as a supernova. If the second supernova explosion does not disrupt the system, we will be left with two neutron stars orbiting around each other. It is entirely possible for at least one of them to be seen as a pulsar, provided its pulsar activity is going on and the geometry permits the emission cone to be in our line of sight.

In the above case, the ages of the two pulsars are vastly different. The second-born pulsar will have a shorter age and typically a high value of the magnetic field. But the first one, as it is an older pulsar, might have a much lower magnetic field because of the decay of the field in the course of time. We also saw in the last section that the accretion of matter by a magnetised neutron star transfers angular momentum to it, thereby spinning it up. In the above case, we would therefore expect the firstborn neutron star to appear as a low-period, low magnetic field, object.

The possible track of such a star is shown in Fig. 7.7. The evolution from A to B is the normal slowing down of the pulsar rotation due to radiation. Under normal circumstances, a pulsar would have faded away once it is to the right of the line marked "death line" (see Chap. 6, Section 6.3). The track along B to C indicates the decay of the magnetic field. (In reality, of course, the decay of the magnetic field as well as the slowing down of the pulsar is a continuous phenomenon, and the figure is schematic to the extent that it presents the two phenomena, one after the other.) In the case of an isolated neutron star, there will be no further significant evolutionary phase. However, if the neutron star is part of a binary system, then it can be spun up by the transfer of mass from the companion and can evolve along the line CD. Because there is a maximum limit to the spin-up of neutron stars [see Eq. (7.115)], these spun-up remnants are expected to lie to the right of the line marked "spin-up." By and large, they will

7.6 Magnetic Effects in Accretion

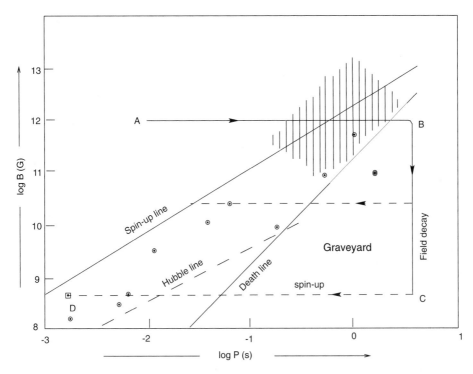

Fig. 7.7. Track of a spin-up pulsar.

appear as a low-period, low-magnetic-field, pulsar. Note that it is also possible to have low-period pulsars if they are very young; such pulsars are expected to be located around A in the figure and will have fairly high value of the magnetic field. It is the low magnetic field, in combination with the low period, that distinguishes the recycled pulsars in binaries from the young, isolated pulsars.

7.6.2 Millisecond Pulsars

The evolution of a pulsar in a binary system could also hold the key for a population of pulsars known as millisecond pulsars.[4] A pulsar with a period of ~ 1.5 ms will have a very low age of $\sim 10^2$ yr if its magnetic field has the standard value of $\sim 10^{12}$ G. However, in reality, the population of a millisecond pulsar also has a very low magnetic field of approximately $(10^8–10^9)$ G. This places them broadly in the same location as the spun-up binary pulsars discussed above. One possible model for such a population is based on the idea that the high-energy emission from the pulsar has completely evaporated the companion star. It is also possible to think of other cases in which the companion star could be missing but it is not yet very clear whether any of these models are as compelling and unique as an explanation for the millisecond pulsar population. Approximately 80% of observed millisecond pulsars are in binary systems with magnetic field $B \simeq (8 \times 10^7 - 3 \times 10^9)$ G; this is to be contrasted with the fact that only $\sim 1\%$ of the slower pulsars are in the binaries and have a much higher magnetic field. There seems to be a link between evolution in binary systems and a low magnetic field. Although the standard picture is attractive, there are several questions concerning the recycling models that remain unresolved. In particular, $\sim 20\%$ of all disk millisecond pulsars are isolated, and it is not very clear whether they were energetic enough to have, say, evaporated their companions.

It must be stressed that the real mystery of millisecond pulsars is not their periods (for example, the inferred initial period of the Crab pulsar is about ~ 16 ms) but their low magnetic fields. The standard model for the millisecond pulsar proceeds along the following lines. We begin with a pair of massive stars with small orbital separation in a binary configuration. The more-massive primary evolves, expands, transfers matter to the secondary, and explodes, leaving the neutron star behind. The secondary, now evolves on a time scale of less than $\sim 10^8$ yr and expands, overflowing the Roche lobe. This matter is accreted by the neutron star that is spun up with pulsed x-ray emission.

The mass transfer from the massive secondary to the neutron star of $M \simeq 1.4\,M_\odot$ results in orbit contraction. For initial orbital periods $P_i \gtrsim 1$ yr the secondary can swell up as a red giant before the over flow of the Roche lobe. The gravitational influence of the neutron star can eject the loosely bound outer envelope of the giant secondary, possibly leading to a common envelope evolution. Eventually the secondary core evolves into a degenerate remnant, leading to either a pair of neutron stars or a highly circular white dwarf–neutron star binary or an eccentric double-neutron system or possibly even a neutron star–black hole system. The process of accretion spins up the neutron star and is limited only when the magnetic pressure balances the accretion pressure, giving rise to the equilibrium period $P_{eq} \simeq 1.7$ ms $B_9^{6/7}(\dot{M}/\dot{M}_{Edd})^{-3/7}$. It is this process that spins up the pulsar, giving it a new lease of life.

The main arguments in favour of this model are the following: First, a large fraction of millisecond pulsars in a galactic disk have low-mass white dwarfs as their binary companions, indicating that mass transfer must have taken place. Because the initial orbit just after the formation of a neutron star in a supernova must have had large eccentricity, the reduction of eccentricity to the present value ($\ll 10^{-3}$) requires a long period of close contact. Second, the period–magnetic field combination of the recycled pulsar is expected to be confined to the right of the spin-up line; this is indeed true for practically all millisecond pulsars.

To form a millisecond pulsar along these lines, we need (1) a low value of the magnetic field, (2) large $\dot{M}$, and also (3) a long time scale over which accretion is sustained. Even at the Eddington rate, these conditions require a binary system with a low-mass secondary that evolves on nuclear time scale of 10^9 yr or longer. These three conditions are satisfied by LMXBs containing a neutron star and a low-mass ($M \lesssim 1 M_\odot$) companion. For an initial period of $P_i \gtrsim 2$ days, stable mass transfer is driven by nuclear evolution, with the companion evolving as a

7.6 Magnetic Effects in Accretion

giant and the orbit expanding by almost a factor of 10. The final orbital period can vary from 10 days to years but the mass that is transferred is approximately the same for $P_f \lesssim 1$ yr. Hence we expect the spin period to be essentially independent of P_f for $P_f \lesssim 1$ yr, and slow pulsars should be found in systems with $P_f \gtrsim 1$ yr. This is broadly what is seen with the widest low-mass binary pulsars also being the slowest ones. For 0.7 days $\lesssim P_i \lesssim 2$ days, the evolution is complicated because angular momentum is lost both by gravitational radiation and by mass loss by means of magnetised wind. Detailed computations indicate that P_f ranges from 1 to 30 days in such systems. When P_i is less than 0.7 days, the mass transfer begins just when the donor star is leaving the main sequence and will result in the decrease of its radius. The net result will lead to a shrinking of the orbit and further evolution is not well understood.

Given a particular formation rate for LMXBs, the ratio of the number of millisecond pulsars to the number of LMXBs in our galaxy will be of the order of the ratio of their lifetime. Granting that the lifetime of millisecond pulsars is approximately Hubble time and taking the lifetime of LMXBs to be (10^7-10^8) yr (based on the evolutionary calculations), we find that the ratio of millisecond pulsars to LMXBs will be approximately 10^2-10^3. The total number of binary millisecond pulsars in the galaxy is estimated to be $\sim 10^6$, suggesting that there could be other routes to the formation of pulsars than through the evolution of LMXBs. However, no other simple alternative route to the formation of millisecond pulsars exists. One possible way out of this difficulty will be to find a mechanism that will shorten the x-ray phase of low-mass binaries. Consider, for example, a case in which a low-mass companion is transferring mass to a neutron star at the Eddington rate. After $\sim 10^7$ yr, the neutron star would have accreted enough angular momentum to have spun up to a period of ~ 1 ms. If the pressure of low-frequency dipole radiation from such a rotating star prevents further accretion of matter, then the lifetime of the x-ray phase will be of the same order as the time required for spining up the neutron star to the period $P \approx 1$ ms, which is $\sim 10^7$ yr. Such a short x-ray lifetime can provide a case for reconciling the large number of millisecond pulsars in the galaxy with the number of LMXBs.

This entire picture depends on the original formation of a LMXB with a neutron star and low-mass companion. Even assuming that the supernova explosion did not disrupt the binary, it certainly would have given a velocity kick to the system. The existence of millisecond pulsars in globular clusters that have very low ($v \lesssim 60$ km s^{-1}) escape velocities requires that in at least a fraction of the cases the initial velocity kick should be fairly low in magnitude. The model also requires the low-mass binary pulsars to have evolved out of a LMXB phase, linking up the birthrates for the two systems. It is not clear whether this is consistent with observations.

It was mentioned in Chap. 6, Section 6.6, that at least one millisecond pulsar has two planets revolving around it. If these planets have been formed along with the pulsar's progenitor, they would also have been evaporated when the companion was destroyed, assuming that the above model for the formation of a millisecond pulsar is correct. It is possible that the planets have condensed out of the debris formed during the evaporation of the companion as the debris disk could be quite similar to a protosolar disk. In that case, it is somewhat intriguing that such planet formation does not occur more often.

7.7 General Relativistic Effects in Binary Systems

So far we have discussed the binary star systems by using nonrelativistic theory. However, in a binary star system with two compact remnants, $(GM/c^2 R)$ can be near unity and general relativistic effects can be important. We shall now discuss some of the features of binary systems that depend on the general relativistic effects.

7.7.1 Gravitational Radiation from Binary Pulsars

We begin by considering a situation in which one member of the binary is a radio pulsar from which periodic signals are received and the other member is a neutron star or black hole. Because the pulsar is orbiting near a strongly gravitating compact remnant, its orbit will not be an ellipse (as in the case of Newtonian gravity) but will precess (see Vol. I, Chap. 11, Section 11.9). Such a pulsar acts as an extraordinarily stable clock, located in a strong gravitational field in which the clock rate will be affected by general relativistic effects. By monitoring the orbit as well as the arrival time of pulses from such a pulsar, we can gain significant information about the general relativistic effects present in such a system.

Consider a pulsar of mass M_1 and a companion of mass M_2 moving about their common centre of mass. We will first relate the observed and the real periods of the pulsar to each other when general relativistic effects are important. We begin by writing

$$\frac{(\delta t)_{\text{obs}}}{(\delta t)_{\text{true}}} = \frac{(\delta t)_{\text{obs}}}{(\delta t)_{\text{stat}}} \frac{(\delta t)_{\text{stat}}}{(\delta t)_{\text{true}}} \tag{7.116}$$

where the subscript stat refers to an observer who is stationary with respect to the centre of mass. If the observer on Earth is also stationary with respect to the centre of mass, then

$$\frac{(\delta t)_{\text{obs}}}{(\delta t)_{\text{stat}}} = \left(1 - \frac{GM_2}{rc^2}\right)^{-1}, \tag{7.117}$$

where r is the distance between the masses M_1 and M_2 (see Vol. I, Chap. 11, Section 11.2). The second factor in Eq. (7.116) is given by the standard Doppler

7.7 General Relativistic Effects in Binary Systems

formula (see Vol. I, Chap. 3, Section 3.12)

$$\frac{(\delta t)_{\text{stat}}}{(\delta t)_{\text{true}}} = \left(1 - \frac{v_1^2}{c^2}\right)^{-1/2} \left(1 + \frac{\mathbf{v_1} \cdot \mathbf{n}}{c}\right), \quad (7.118)$$

where $\mathbf{n}$ is a unit vector pointing towards the pulsar from the Earth and $\mathbf{v}_1$ is the velocity of the pulsar. To the the lowest order in $\mathcal{O}(v^2/c^2)$, the net effect is given by

$$\frac{(\delta t)_{\text{obs}}}{(\delta t)_{\text{true}}} = 1 + \frac{\mathbf{v_1} \cdot \mathbf{n}}{c} + \frac{1}{2}\frac{v_1^2}{c^2} + \frac{GM_2}{rc^2}. \quad (7.119)$$

Let us now consider the various terms on the right-hand side. It is convenient to start with an analysis of the Newtonian situation in which the pulsar of mass M_1 and a companion of mass M_2 are moving in elliptical orbits about their common centre of mass, and then to introduce the necessary modifications from general relativity. We choose the x–y plane as the plane of the orbit with the origin at the centre of mass. The orbital plane is taken to be inclined at an angle i with respect to the line of sight with the x axis along the line of nodes (see Fig. 7.8). If ω is the angle between the periastron and the line of nodes, the position of the pulsar at any instant is given by

$$x = r_1 \cos\psi, \quad y = r_1 \sin\psi, \quad (7.120)$$

where

$$\psi = \omega + \phi, \quad r_1 = \frac{a_1(1-e^2)}{1 + e\cos\phi}. \quad (7.121)$$

If

$$\mathbf{n} = \mathbf{e}_{z'} = \cos i\, \mathbf{e}_z + \sin i\, \mathbf{e}_y \quad (7.122)$$

denotes the unit vector pointing from the Earth to the pulsar, then the line-of-sight component of the velocity vector of the pulsar is given by

$$\mathbf{v}_1 \cdot \mathbf{n} = (\dot{r}_1 \sin\psi + r_1\dot\psi \cos\psi) \sin i. \quad (7.123)$$

For the elliptical orbit, $\dot\phi$ can be expressed as

$$\dot\phi = \frac{2\pi}{P(1-e^2)^{3/2}} (1 + e\cos\phi)^2, \quad (7.124)$$

where P is the period of the orbit. Using this, we obtain, after some simple algebra,

$$\mathbf{v}_1 \cdot \mathbf{n} = K[\cos(\omega + \phi) + e\cos\omega], \quad K = \frac{2\pi a_1 \sin i}{P(1-e^2)^{1/2}}. \quad (7.125)$$

For any binary system in which the line-of-sight velocity of a star can be measured, we can use these equations to determine the following parameters: The eccentricity e and period P can be found from Eq. (7.124), which will provide

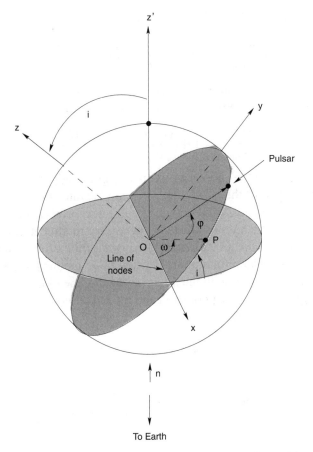

Fig. 7.8. Geometrical details of the binary pulsar orbit.

the function $\phi(t)$ when integrated. Further, by studying the coefficients of $\cos\phi$ and $\sin\phi$ in Eq. (7.125), we can obtain K and ω. Eliminating P and e from K, we determine $a_1 \sin i$, using which we can determine the mass function, defined as

$$f = \frac{(M_2 \sin i)^3}{(M_1 + M_2)^2} = \frac{(a_1 \sin i)^3}{G} \left(\frac{2\pi}{P}\right)^2. \qquad (7.126)$$

It might seem that Doppler shift can actually provide us with more information because of the additional terms of the order of $\mathcal{O}(v^2/c^2)$. This, however, is not true for elliptical orbits in Newtonian gravity. Using

$$v_1^2 = \dot{r}_1^2 + r_1^2 \dot{\psi}^2 = \left(\frac{2\pi}{P}\right)^2 \frac{a_1^2}{1-e^2}(1 + 2e\cos\phi + e^2), \qquad (7.127)$$

$$\frac{GM_2}{r} = \frac{GM_2^2}{(M_1 + M_2)r_1}, \quad \left(\frac{2\pi}{P}\right)^2 = \frac{GM_2^3}{(M_1 + M_2)^2 a_1^3}, \qquad (7.128)$$

7.7 General Relativistic Effects in Binary Systems

we can easily show that

$$\frac{1}{2}v_1^2 + \frac{GM_2}{r} = \beta \cos\phi + \text{constant}, \quad (7.129)$$

where

$$\beta \equiv \frac{GM_2^2(M_1 + 2M_2)e}{(M_1 + M_2)^2 a_1(1 - e^2)}. \quad (7.130)$$

Because the time dependence of this quantity is the same as the higher-order term $K \cos\omega \cos\phi$ in Eq. (7.125), β will not be independently measurable. Thus, in purely Newtonian gravity, we cannot do any better.

The situation is, however, different when general relativistic effects that cause the periastron to precess by the amount (see Vol. I, Chap. 11, Section 11.9)

$$\dot{\omega} = \frac{6\pi GM_2}{a_1(1-e^2)Pc^2} \quad (7.131)$$

are taken into account. To the lowest order, we can replace ω with $(\omega_0 + \dot{\omega}t)$ in Eq. (7.125), thereby obtaining four independent time-varying trigonometric combinations of ϕ and $\dot{\omega}t$. This allows us to determine separately from the observations the parameters K, ω_0, $\dot{\omega}$, and β. Because $\dot{\omega}$ and β involve different combinations of the four parameters M_1, M_2, a_1, and $\sin i$, compared with the mass function, the measurement of $\dot{\omega}$ and β allows complete determination of the parameters of the binary system.

In fact, the situation is better than that. It was shown in Vol. I, Chap. 11, Section 11.12 that two masses that are moving in an elliptic orbit with eccentricity e lose energy and angular momentum because of the radiation of gravitational waves. The rate of energy loss [given by Eq. (11.240) of Vol. I, Chap. 11] is

$$\dot{E} = -\frac{32G^4}{5c^5} \frac{m_1^2 m_2^2 (m_1 + m_2)}{a^5} \frac{1}{(1-e^2)^{7/2}} \left(1 + \frac{73}{24}e^2 + \frac{37}{96}e^4\right). \quad (7.132)$$

This causes the period to change by the amount $(\dot{P}/P) = (3\dot{a}/2a) = (3\dot{E}/2E)$. Using $|E| = (Gm_1m_2/2a)$, we get

$$\frac{1}{P}\frac{dP}{dt} = -\frac{96}{5}\frac{G}{c^5}\frac{M^2\mu}{a^4} f(e), \quad (7.133)$$

where μ is the reduced mass of the system, M is the total mass, and $f(e)$ is given by

$$f(e) \equiv \left(1 + \frac{73}{24}e^2 + \frac{37}{96}e^4\right)(1-e^2)^{-7/2}. \quad (7.134)$$

Because all the parameters in Eq. (7.133) are now determined, we can actually *predict* the value of $\dot{P}$ that is due to gravitational radiation from the system. Observing $\dot{P}$ will act as a consistency check on the model. Alternatively, we can determine the masses by any combination of the observed quantities and check whether the curves in the M_1–M_2 plane intersect at a single point. Such an exercise was carried out[5] for a particular pulsar PSR 1913+16 with remarkable success, verifying the predictions of general relativity. From the observed values for this pulsar, we can obtain a very precise estimate of the total mass,

$$M_1 + M_2 = (2.8278 \pm 0.0007)\, M_\odot, \tag{7.135}$$

in terms of solar mass. (In fact, the Newtonian gravitational constant is not known to a precision of four significant figures; hence the total mass can be quoted more accurately in terms of solar masses than in grams.) From other observed values we can also determine the combinations

$$\gamma = \frac{G^{2/3} M_2 (M_1 + 2M_2) e}{(M_1 + M_2)^{4/3} c^2} \left(\frac{P}{2\pi}\right)^{1/3} = (0.0007344\, \text{s}) M_2 (2.8278 + M_2), \tag{7.136}$$

$$\sin i = \left(\frac{2\pi}{P}\right)^{2/3} \frac{(M_1 + M_2)^{2/3} a_1 \sin i}{G^{1/3} M_2} = \frac{1.019}{M_2}, \tag{7.137}$$

$$\dot{P} = -\frac{192\pi}{5} \frac{G^{5/3} M_1 M_2 f(e)}{c^5 (M_1 + M_2)^{1/3}} \left(\frac{2\pi}{P}\right)^{5/3} = -1.202 \times 10^{-12} M_2 (2.8278 - M_2) \tag{7.138}$$

where the masses are measured in units of solar mass. The masses are approximately $M_1 = M_2 = (1.41 \pm 0.06)\, M_\odot$, leading to a prediction that $\dot{P} = -(2.40258 \pm 0.00004) \times 10^{-12}$. This should be compared with the observed value $\dot{P} = -(2.425 \pm 0.010) \times 10^{-12}$. These match within a ratio of 1.002 ± 0.005.

The remarkable timing accuracy available in the case of a few binary pulsars monitored for a couple of decades can be used to put rigorous bounds on any possible mechanism that could affect the arrival time of the pulses. One particular example of this is given by the possibility of putting bounds on the amount of stochastic gravitational wave background that can exist in our universe. This was discussed in Chap. 6, Section 6.6. The best bounds currently available are from the binary pulsars.

7.7.2 Black Holes in Binary Systems

In a number of binary x-ray sources, the amplitude of the radial velocity of the optical star is sufficiently large to imply that the mass of the companion is likely to be higher than $\sim 3.4\, M_\odot$. From the discussion in Chap. 3, Section 3.6, it follows that the compact remnant is very likely to be a black hole, as no stable configuration for a neutron star exists for such a high value of mass.

7.7 General Relativistic Effects in Binary Systems

Table 7.3. Possible black hole candidates

Candidate	L_x	d (kpc)	m_v	e	K (km s^{-1})	P (days)	$f(M)/M_\odot$
Cygnus X-1	2×10^{37}	2.5	9	0.001($\pm$0.01)	74.7 $\pm$ 1.0	5.6	0.25 $\pm$ 0.01
LMC X-3	3×10^{38}	55	17	0.13($\pm$0.05)	235 $\pm$ 11	1.7	2.3 $\pm$ 0.3
A0620-00	1×10^{38}	$\sim$1	18	0.01($\pm$0.01)	443 $\pm$ 4	0.32	2.91 $\pm$ 0.08
GS 2023+338	2×10^{38}	$\sim$2.7	18.4	0.00	211 $\pm$ 4	6.47	6.26 $\pm$ 0.31
GS 1124-68	1×10^{38}	$\sim$2.5	20	0.00	409 $\pm$ 18	0.433	3.07 $\pm$ 0.40

In fact, realistic equations of state imply that an upper limit to the mass of the neutron star is $\sim$2.2 $M_\odot$. There are several candidate systems in which it is likely that the mass of the compact remnant is higher than $\sim$2 $M_\odot$; all of them are actively investigated for possible detection of black-hole-related phenomena.

The x-ray sources with a black hole might show a number of characteristics that are different from normal LMXBs that contain neutron stars. Unfortunately, none of these characteristics can be thought of as a totally conclusive signature for a black hole. For example, the x-ray spectrum of a black hole candidate binary is expected to have either a very hard or supersoft tail up to energies beyond 100 keV; but it is not clear whether this feature is either necessary or sufficient for us to conclude that the binary system contains a black hole. Table 7.3 summarises the features of five black hole candidates, selected somewhat arbitrarily from a larger set; in the table, m_v is the visual magnitude of the companion, e is the orbital eccentricity, $K = \{2\pi a_1 \sin i /[P(1-e^2)^{1/2}]\}$, and P is the period.

Two of these systems – Cygnus X-1 and LMC X-3 – are HMXBs, and the other three are LMXBs, with the companions filling the Roche lobe. Even if we assume that the masses of optical stars are negligible in the case of these LMXBs, the radial-velocity curves require the masses of their compact companion to be larger than $\sim$3 $M_\odot$, making them strong candidates for black holes. All three systems exhibit strong x-ray transients that involve a flare-up as an x-ray source for several weeks (increasing by more than 6 magnitudes in optical) followed by a fading away.

Among the other candidates, Cygnus X-1 has been investigated extensively, and we will provide a brief discussion of this binary system. The key observations are as follows: This binary shows x-ray variability in all time scales from milliseconds to years with significant $t_{min} \approx$ 1-ms bursts. Assuming that such a phenomenon requires coherence of the underlying mechanism over a length scale $R \approx ct_{min}$, we can put an upper limit to the size of the compact object at $\sim$300 km. The optical star in Cygnus X-1 is a ninth-magnitude supergiant HDE 226868. Careful study based on the interstellar absorption gives the distance to HDE 226868 as $\sim$2.5 kpc with an absolute minimum distance of 2 kpc.

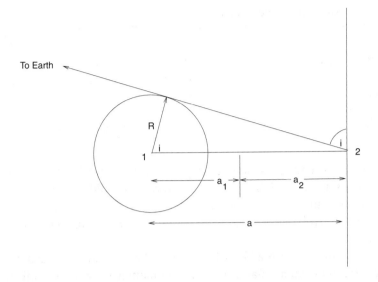

Fig. 7.9. Geometry of Cygnus X-1 illustrating the condition for the absence of an x-ray eclipse.

The Doppler shift of the spectrum of the star allows us to determine $P = 5.60$ days, $a_1 \sin i = (5.82 \pm 0.08) \times 10^{11}$ cm, and $f = (0.252 \pm 0.010) \, M_\odot$; the eccentricity is likely to be small with $e < 0.02$.

A first limit on the mass of companion star, based on these data, can be obtained as follows. A typical OB supergiant has a mass greater than 20 $M_\odot$ but stellar-evolution calculation shows that the mass should be at least 8.5 $M_\odot$ to provide the observed luminosity of HDE 226868 at a distance of $d > 2$ kpc. Because

$$f = \frac{(M_2 \sin i)^3}{(M_1 + M_2)^2} \tag{7.139}$$

(where 1 refers to the star and 2 to the compact remnant), the minimum value for M_2 is obtained by setting $\sin i = 1$. This gives $M_2 \geq 3.3 \, M_\odot$.

Alternatively, we can provide a bound on M_2 without assuming anything about M_1 by using the fact that no x-ray eclipse is seen in this system. From Fig. 7.9, it is clear that the absence of eclipse requires that

$$\cos i > \frac{R}{a}, \tag{7.140}$$

where R is the radius of the optical star and a is the separation between the two stars. Using $a = [1 + (M_1/M_2)]a_1$ we can convert this constraint into the condition

$$M_2 \sin i \cos^2 i \geq \frac{f R^2}{(a_1 \sin i)^3}. \tag{7.141}$$

7.7 General Relativistic Effects in Binary Systems

The combination $\sin i \cos^2 i$ has a maximum value of $2/3\sqrt{3}$; hence we must have

$$M_2 \geq \frac{3\sqrt{3} f R^2}{2(a_1 \sin i)^2}. \tag{7.142}$$

The radius R can be expressed in terms of the luminosity and the effective temperature of the star by $R^2 = (L/4\pi \sigma T_e^4)$. Determining this quantity from observations requires a careful analysis of the bolometric corrections and the interstellar absorption of light in the visual band. A conservative estimate turns out to be

$$R^2 = (6.62 \times 10^6 \text{ km})^2 \left(\frac{d}{1 \text{ kpc}}\right)^2. \tag{7.143}$$

Substituting Eq. (7.143) back into relation (7.142) gives the result

$$M_2 \geq 3.4 \, M_\odot \left(\frac{d}{2 \text{ kpc}}\right)^2. \tag{7.144}$$

This suggests that the companion star is sufficiently massive to be a black hole.

The Cygnus X-1 system also shows a two-state behaviour: in the low state, which occurs 90% of the time, the source emits a hard-x-ray component with a power-law spectrum in the range (1–250) keV. In the high state, which occurs in the time scale of days to months, an intense soft spectral component appears in the (3–6) keV band. It is possible to provide an explanation for such an x-ray spectrum based on gaseous accretion disks, although it is considerably more difficult to explain the peculiar transitions between the two states. It is generally believed that such complexities are irrelevant for the arguments given above for the estimate of the mass.

7.7.3 Gravitational Radiation from Coalescing Binaries

The discussion in Subsection 7.7.1 shows that gravitational radiation is emitted by the system PSR 1913+16. Over a period of time, this drain of energy will cause the two remnants to move towards each other and eventually coalesce together. Hence the final stages of evolution of a binary system made of two compact relativistic objects are of considerable interest, both theoretically and observationally. As the two members of the binary systems approach coalescence, large quantities of gravitational waves are expected to be emitted. Several planned experiments for the detection of gravitational waves rely on emission from such a coalescence of compact relativistic objects. For example, the coalescence of a neutron star–neutron star (NS–NS) system at a distance of 1000 Mpc or a black hole–black hole (BH–BH) system at 3000 Mpc will produce an effective gravitational wave amplitude of $h \simeq 10^{-22}$ in the frequency band of $(10-10^3)$ Hz; a NS–NS coalescence at 60 Mpc or a BH–BH coalescence at 200 Mpc will produce an amplitude that is ~30 times higher. The chances of such detection in a statistical sense, of course, depend on the rate of occurrence of such events, which are difficult to estimate reliably. We will now provide a simplified discussion of the energy spectrum from two coalescing compact objects, assuming that they move towards each other head on rather than in a spiralling orbit.

Taking the motion to be along the x axis with the centre of mass at the origin, we can describe the system in terms of the reduced mass $\mu = m_1 m_2/(m_1 + m_2) \equiv (m_1 m_2/M)$ and the separation $x = x_1 - x_2$, with $m_1 x_1 = -m_2 x_2 = \mu x$. We have seen in Vol. I, Chap. 11, Section 11.12 that the emission of gravitational radiation is determined by the quadrupole moment

$$D_{jk} \equiv \sum_A m_A \left[x_j^A x_k^A - \frac{1}{3} \delta_{jk} (x^A)^2 \right] \qquad (7.145)$$

{which differs from the definition used in Vol. I [see Eq. (11.225)] by a factor of 3}. In this particular case, the components of the quadrupole tensor are given by

$$D_{xx} = \frac{2}{3} \mu x^2, \quad D_{yy} = D_{zz} = -\frac{1}{3} \mu x^2. \qquad (7.146)$$

To estimate the total amount of energy that is emitted by the system, we need to compute the third time derivative of D_{ij}. From the equation of motion $\ddot{x} = -(GM/x^2)$ we get $\dot{x}^2 = (2GM/x)$ if the particles are falling freely from infinity, starting at rest. Using this, we obtain

$$\dddot{D}_{xx} = -\frac{4}{3} G \mu M \frac{\dot{x}}{x^2}, \qquad (7.147)$$

etc., giving the total amount of energy radiated per second as

$$\frac{dE}{dt} = \frac{1}{5} \frac{G}{c^5} \langle \dddot{D}_{xx}^2 + \dddot{D}_{yy}^2 + \dddot{D}_{zz}^2 \rangle = \frac{8}{15} \frac{G^3 \mu^2 M^2}{c^5} \left\langle \frac{\dot{x}^2}{x^4} \right\rangle. \qquad (7.148)$$

The total energy emitted when the stars approach each other from infinity, up to a separation $x_{\min}$, will be

$$\Delta E = \int \frac{dE}{dt} dt = \int \frac{dE}{dt} \frac{1}{\dot{x}} dx = \frac{8}{15} \frac{G^3 \mu^2 M^2}{c^5} (2GM)^{1/2} \int_{x_{\min}}^{\infty} \frac{dx}{x^{9/2}}, \qquad (7.149)$$

where we have used $\dot{x} = (2GM)^{1/2} x^{-1/2}$. This integral diverges as $x_{\min} \to 0$; however, the Newtonian treatment of the trajectory breaks down for $x_{\min} \lesssim (2GM/c^2)$. Cutting off the integral at $x_{\min} = (2GM/c^2)$, we get

$$\Delta E = \frac{2}{105} \frac{\mu^2 c^2}{M}. \qquad (7.150)$$

In spite of the approximations used, this result agrees with more exact treatments that incorporate general relativistic effects and are calculated numerically. The

7.7 General Relativistic Effects in Binary Systems

results shows that, if $m_1 \approx m_2 \approx \mu$, about 1% of the rest-mass energy of the stars can be emitted in the form of gravitational waves.

To determine the spectrum of gravitational radiation emitted in this case, we start by rewriting ΔE in Eq. (7.149) as

$$\Delta E = \frac{1}{5}\frac{G}{c^5} \int_{-\infty}^{\infty} d\omega |\hat{D}_{jk}^{(3)}(\omega)|^2 = \frac{2}{5}\frac{G}{c^5} \int_{0}^{\infty} d\omega |\hat{D}_{jk}^{(3)}(\omega)|^2, \qquad (7.151)$$

where we have used Parseval's theorem and limited the integration to positive values of frequency. The quantity $\hat{D}_{jk}^{(3)}(\omega)$ denotes the Fourier transform of the third derivative of the quadrupole tensor. This gives the spectral-energy distribution

$$\frac{dE}{d\omega} = \frac{2}{5}\frac{G}{c^5}|\hat{D}_{jk}^{(3)}(\omega)|^2. \qquad (7.152)$$

The necessary Fourier transform can be expressed, after one integration by parts, in the form

$$\hat{D}_{jk}^{(3)}(\omega) = \frac{1}{(2\pi)^{1/2}} \int_{-\infty}^{\infty} \dddot{D}_{jk}(t) e^{i\omega t} \, dt$$

$$= \frac{1}{(2\pi)^{1/2}} \left[\ddot{D}_{jk}(t) e^{i\omega t} \Big|_{-\infty}^{\infty} - i\omega \int_{-\infty}^{\infty} \ddot{D}_{jk}(t) e^{i\omega t} \, dt \right]. \qquad (7.153)$$

Because $\ddot{D}_{xx}(t) = (4/3)(GM\mu/x)$ we need the explicit form of the trajectory to evaluate the integrals. For a free fall from infinity, we can take it to be

$$x = \left(\frac{3}{2}\right)^{2/3} (2GM)^{1/3}(-t)^{2/3} \quad (x \to \infty \text{ as } t \to -\infty) \qquad (7.154)$$

for $x > x_{\min} \equiv (2GM/c^2)$. Assuming that $x = x_{\min}$ at $t = t_{\min}$ and taking the derivatives of the quadrupole tensor to be zero for $t > t_{\min}$, we see that the first term in Eq. (7.153) vanishes in both limits. Therefore

$$\hat{D}_{xx}^{(3)}(\omega) = -\frac{i\omega}{(2\pi)^{1/2}} \int_{-\infty}^{t_{\min}} \frac{2^{7/3}(GM)^{2/3}\mu}{3^{5/3}(-t)^{2/3}} e^{i\omega t} \, dt$$

$$= -\frac{i\omega^{2/3} 2^{7/3}(GM)^{2/3}\mu}{(2\pi)^{1/2} 3^{5/3}} \int_{|\omega t_{\min}|}^{\infty} e^{-iy} \frac{dy}{y^{2/3}}, \qquad (7.155)$$

where $y = -\omega t$. The integral can be expressed in terms of an incomplete gamma function; but for $|\omega t_{\min}| \ll 1$ we can replace the lower limit of the integration with zero, getting the value of the integral as $\Gamma(1/3)\exp(-i\pi/6)$. Evaluating the other components in a similar fashion and combining all the results, we get

$$\frac{dE}{d\omega} \approx \frac{2^{11/3}}{5\pi(3)^{7/3}} \Gamma^2[1/3] \frac{G}{c^5}(GM\omega)^{4/3}\mu^2 \quad (\omega \to 0). \qquad (7.156)$$

The $\omega^{4/3}$ dependence is quite generic in the limit of $\omega \to 0$ and arises from the behaviour of $D_{ij}(t)$ near $t \to -\infty$, when the Newtonian approximation is quite valid.

To calculate the chances of detecting such gravitational radiation from coalescing neutron stars, we need an estimate of the formation and the evolution rate of these binaries. By and large, the birthrate of NS–NS binaries is quite small and is probably $\sim 10^{-3}$ of the corresponding rate for a single pulsar. In spite of this, the fraction of steady population at any given time is comparable in the two cases because of the vastly differing time scales involved in the evolution. Single pulsars have a lifetime of less than $\sim 10^7$ yr, whereas in the NS–NS system the firstborn pulsar undergoes accretion, leading to the decrease in B and P; this leads to a longer radio emission lifetime of approximately 3×10^9 yr. This would lead to a conservative merger rate of ~ 3 yr^{-1} within a local volume of radius 200 Mpc.

It must, however, be stressed that these estimates are very tentative. For example, a similar analysis would suggest that the formation rate of NS–NS binaries is $\sim 1\%$ of the birthrate of massive x-ray binaries that are the progenitors of all HMXBs. Hence, for every NS–PSR binary, we would expect to see ~ 100 pairs of young pulsar–recycled pulsar combinations with similar B and P values. This is not seen, possibly suggesting that the original estimates for the birthrates are not quite accurate. There is also the intriguing possibility of BH–NS and BH–BH binaries. Because the minimum main-sequence mass for black hole formation is somewhere around 30 $M_\odot$ or so, it seems reasonable to assume that all stars with $M > 50 M_\odot$ will form a black hole. The high mass of the binary primary (say, $M \gtrsim 10 M_\odot$) will ensure that the binary remains bound even after the supernova explosion that forms the secondary neutron star. This would suggest that the birthrate of a BH–NS system should be comparable to a NS–NS system. The observational status regarding such pairs is still somewhat uncertain.

7.8 Varieties of Accreting Binary Systems

The mechanics of a star transferring a mass through an accretion disk onto a compact object occurs in a wide variety of situations, leading to different astrophysical phenomena. In what follows, we shall give a brief description of some of these systems (also see Table 7.4).

There are several classes of semidetached binary systems in which the primary component is a white dwarf. Depending on the characteristics, they may be called dwarf nova, classical nova, or supernova (type I) roughly in the order of increasing luminosity. The term *nova* that is common to these classes of objects is somewhat misleading as their outbursts are based on very different mechanisms.

Cataclysmic variables show periodic outbursts in which the luminosity increases by a factor between 10 (for most dwarf novas) and 10^6 (for classical novas), separated by long quiet intervals. The primary is a white dwarf with a mass of $\sim 0.85 M_\odot$ and the secondary is a main-sequence star of type G or later.

7.8 Varieties of Accreting Binary Systems

Table 7.4. Taxonomy of binary systems

Name	Description	Remarks
Algols	Two normal stars (main sequence or subgiants): semidetached binary	Provide checks on stellar evolution, information on mass loss
RS Canum Venaticorum	Chromospherically active binaries	Useful for studies of dynamo-based magnetic activity; exhibits starspot chromospheres, corona, and flares similar to the Sun
W Ursae Majoris	Short period (0.2–0.8 days) Contact binaries	High levels of magnetic activity, important for studying stellar dynamo model
Cataclysmic variables and novas	White dwarfs with cool M-type secondaries; short periods	Exhibits accretion phenomena and accretion disks
X-ray binaries	Neutron star or black hole as the compact component; powerful x-ray sources with $L_x > 10^{35}$ ergs s^{-1}	Study of structure and evolution of compact remnants; indirect evidence for black holes
ζ Aurigae/ VV Cephi	Long-period interacting binaries; Late-type supergiant plus a hot companion	Study of supergiant phase, especially atmospheres of supergiants

The periods of orbit vary between 76 min and 16 h. Dwarf novas and novas are essentially similar to cataclysmic variables, and their mechanism for release of energy recurs.

Approximately 300 dwarf novas have been discovered that brighten by between 2 and 6 magnitudes during the outbursts, which last for 5 to 20 days. These outbursts are separated by 30 to 300 days. To explain such an increased luminosity, the accretion has to change from $\dot{M} \approx (10^{-11}$–$10^{-10})\, M_\odot$ yr^{-1} during the quiet period to $(10^{-9}$–$10^{-8})\, M_\odot$ yr^{-1} during the outburst. The models describing the modulation of the mass transfer rate rely on the instability in the outer layers of the secondary star, which could be powered by the hydrogen partial-ionisation zone or by the lowering of the viscosity in the accretion disk.

In contrast, a classical nova requires higher accretion rates. About 30 such novas are detected in the Andromeda galaxy each year but only 2 or 3 per year are seen in our galaxy, possibly because of dust obscuration. These novas increase in brightness by 10–12 magnitudes in a span of only a few days with a subsequent decline taking place over several months. The typical absolute visual magnitude of a nova in the quiescent state is approximately $M_V = 4.5$, corresponding to a luminosity of $L = 1.3\, L_\odot$. The accretion rate needed to provide this luminosity is approximately $\dot{M} = (2RL/GM) \simeq 9 \times 10^{-10} \, M_\odot \, \text{yr}^{-1}$ for typical white dwarf parameters. The gas accreted at this rate accumulates on the surface of the white dwarf where it is compressed and heated. At the base of the layer, turbulent mixing enriches the gas with the carbon, nitrogen, and oxygen of the white dwarf. When approximately $(10^{-4}$–$10^{-5}) \, M_\odot$ of hydrogen has accumulated, the temperature at the base reaches a few million degrees and a shell of hydrogen burning involving the CNO cycle begins. Because the base is supported by electron degeneracy pressure, runaway thermonuclear reaction occurs, with the temperatures racing to $\sim 10^8$ K. When the luminosity exceeds the Eddington luminosity, radiation pressure can lift the accreted material and expel it into space, causing the eruption of the nova. Eventually the binary system reverts to its quiescent configuration and the accretion process begins again. With an accretion rate of $(10^{-8}$–$10^{-9}) \, M_\odot \, \text{yr}^{-1}$ it will take another $(10^4$–$10^5)$ yr to build up a surface layer with $10^{-4} \, M_\odot$ of accreted matter. We shall now discuss some of these objects in greater detail.

7.8.1 Novas

A nova represents a transient, possibly recurring, outburst of a low-luminosity star. The absolute visual magnitude of a star may change as much by 11 mag, reaching a maximum between -6 and -9, corresponding to luminosity in the range $(0.25$–$4) \times 10^5 \, L_\odot$. The light curves of a nova are broadly similar to each other and have the form shown schematically in Fig. 7.10.

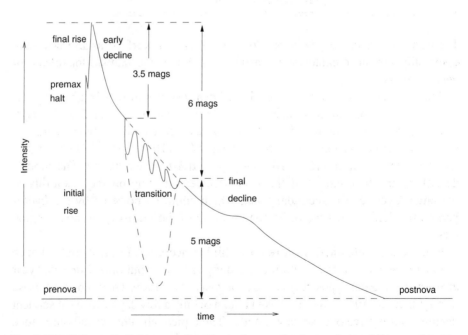

Fig. 7.10. Schematic light curve for a typical nova; the time axis is arbitrary and not to scale.

7.8 Varieties of Accreting Binary Systems

This could arise from the following scenario. Consider a binary system in which both the stars are approximately a few solar masses initially. The more-massive star of the two reaches the red giant stage first, transfers and loses mass and becomes a hot white dwarf. The secondary then evolves towards the red giant end, fills the Roche lobe, and begins to transfer mass to the hot white dwarf. An accretion disk will form, just as in the case of x-ray sources, except for the fact that the compact object is now a white dwarf, and the mass will rain on a hot white dwarf. This accreted material can be compressed and heated to a temperature higher than 10^7 K, and when the critical mass is reached, hydrogen burning through CNO cycle can take place at the base of the accreted layer. Further, if the matter is sufficiently degenerate, a thermonuclear runaway can occur for the CNO cycle, which has a strong dependence on temperature. The strong temperature dependence also leads to convective instabilities and can trigger a nova explosion involving $\sim 10^{-4}$ $M_\odot$ of mass.

Simple estimates confirm the feasibility of this scenario. For a white dwarf of mass M and radius R with a nondegenerate envelope of thickness $l < R$, we can estimate the heating rate as follows. If the gravitational potential energy of the in-falling gas is converted into the thermal-energy density of the envelope, $\epsilon_{th} \approx n_{env} k_B T$, where n_{env} is the number density of particles in the envelope, then the energy balance in the envelope requires that

$$\frac{dU}{dt} \simeq (4\pi R^2 l) n_{env} k_B \frac{dT}{dt} = (4\pi R^2 l) \left(\frac{M_{env}}{A_{env} m_p}\right) k_B \frac{dT}{dt}, \quad (7.157)$$

where A_{env} is the atomic weight of matter in the envelope and $M_{env} \approx (4\pi R^2 l) A_{env} n_{env} m_p$ is the mass of the envelope. If the accretion rate is $\dot{M}_a$, this gives the rate of change of temperature as

$$\frac{dT}{dt} \simeq \frac{A_{env} G m_p}{k_B R} \frac{M}{M_{env}} \dot{M}_a. \quad (7.158)$$

For a white dwarf with $M \approx M_\odot$, $R \approx 10^9$ cm, $A_{env} \approx 4$, $M_{env} \approx 10^{-4} M_\odot$, and $\dot{M}_a \approx (10^{-5} – 10^{-7}) M_\odot$ yr^{-1}, the heating rate is $\dot{T} \approx (10^7 – 10^9)$ K yr^{-1}. Such high temperatures can easily lead to the triggering of nuclear reactions and ejection of the envelope with speeds of $\sim 10^3$ km s^{-1}.

The change in the luminosity of such a nova can be estimated from the change in the energy density of the outer envelope by use of

$$\frac{dL}{dt} = -\frac{d}{dt} \int_{R_0}^{R(t)} u(r,t) 4\pi r^2 \, dr = -\frac{d}{dt} \int_0^{R(t)} 4\pi a T^4(r,t) r^2 \, dr, \quad (7.159)$$

where it is assumed that the temperature of the white dwarf at $r < R_0$ remains unchanged during the evolution. In the Eddington approximation (see Chap. 2, Section 2.4) the temperature is related to the optical depth by

$$T^4 = \frac{1}{2} T_{env}^4 \left(1 + \frac{3}{2}\tau\right) = \frac{L(t)}{8\pi \sigma R^2} \left(1 + \frac{3}{2}\tau\right). \quad (7.160)$$

In our case, we can estimate the optical depth by using $\kappa\rho \approx \kappa_0 \Delta M/(4\pi R^3/3)$, where ΔM is the mass ejected by the nova. This gives

$$\tau = \int_r^R \kappa\rho\,dr = \frac{3\kappa_0 \Delta M}{4\pi R^3}(R-r). \qquad (7.161)$$

By using this result in Eq. (7.160), we can convert Eq. (7.159) into a differential equation for $L(t)$ with one unknown function $R(t)$. Assuming uniform expansion with velocity $v = (R/t)$, this equation can be integrated to determine the rise time of the nova. After some algebra (see Exercise 7.9) we get

$$t_{max} = \left(\frac{3\kappa_0 \Delta M/16\pi vc}{1 + 2v/3c}\right)^{1/2}. \qquad (7.162)$$

For $\Delta M \approx 10^{-5}\,M_\odot$, $v \approx 10^3$ km s^{-1}, and $\kappa_0 \approx 0.4$, the rise time is $t_{max} \approx 3.5$ h and the size of the envelope is $R(t_{max}) \approx v t_{max} \approx 10^{12}$ cm. These values are typical for the nova. Unlike in a supernova, this does not lead to complete disruption of the star and hence can reoccur.

Approximately 50 novas occur each year within a massive spiral galaxy; this frequency is greater than the formation rate of white dwarfs. The expansion velocities of nova shells are approximately (500–2000) km s^{-1}. Because the mass ejected is approximately $(10^{-4}$–$10^{-6})\,M_\odot$, the kinetic energies involved in the process are $(10^{43}$–$10^{44})$ ergs. The relatively small masses of the ejecta show that approximately 10^4–10^5 outbursts can arise because of mass accretion in a single binary system. In other words, approximately 10^7–10^8 close binary stars can give rise to 10^{11}–10^{12} nova outbursts in the $(12$–$15) \times 10^9$ yr lifetime of our galaxy. The energy radiated as photons is ~ 100 times larger than the kinetic energy of the ejecta; this is in contrast to the supernova shells in which the kinetic energy of bulk motion is 100 times larger than the radiated photon energy.

Exercise 7.9
Filling in the gaps: Derive Eq. (7.162).

7.8.2 Supernova Type I

On the other hand, it is possible to imagine a case in which the accretion of matter onto a carbon–oxygen white dwarf causes its core mass to exceed the Chandrasekhar limit of 1.4 $M_\odot$. This will lead to a catastrophic collapse of the central region, making the carbon burn and causing a deflaggeration wave to pass through the star. It is likely that no remnant is left behind from such an explosion and the whole star is disrupted. The end products of the carbon burning will eventually be iron group elements, and the energy released will go into the kinetic energy of the expansion of the supernova. Such an explosion triggered by accretion in a binary system is called *type I supernova*. The decay of ^{56}Ni provides the characteristic decline of the luminosity of the supernova as in the case of type II. The type I supernovas are characterised by the absence of hydrogen lines in their spectra but reveal the presence of other elements, with atomic masses ranging from helium through iron. The absence of hydrogen is understandable from the fact that the pregenitors are highly evolved stars that have lost almost all of their hydrogen before the explosion. The two crucial parameters that govern the explosion are the rate of mass accretion and the mass of the white dwarf, with different combinations leading to different events. In some cases it is possible to disrupt the white dwarf completely, leaving behind no stellar remnant.

Observationally, supernovas that do not show prominent hydrogen lines are called type I whereas those that do show hydrogen spectra are called type II. All type I supernovas show similar rates of decline of their brightness after the phase of maximum light at the rate of approximately 0.065 ± 0.007 mag day^{-1} at 20 days. After $\sim$50 days the dimming rate slows and becomes $\sim$0.01 mag day^{-1}. This is similar to that observed for a type II supernova and can be explained in terms of the β decay of $^{56}_{27}$Co to $^{56}_{26}$Fe. This decay has a half-life of $\tau = 77.7$ days so that the number of cobalt atoms varies with time as $N(t) = N_0 e^{-\lambda t}$, with $\lambda = (\ln 2/\tau)$. Because the rate at which energy is being deposited onto the supernova remnant is proportional to dN/dt, the slope of the light curve is $d(\ln L)/dt = -0.434\lambda$ or, equivalently, $dM_{\text{bol}}/dt = 1.086\lambda$. For $\tau = 77.7$ days, this gives a slope of 0.01 mag day^{-1}. A typical type I supernova curve is shown in Fig. 7.11. Some of the observational details distinguishing type I and II supernovas are shown in Fig. 7.12.

There is general agreement that the light curves of type I supernovas are sufficiently similar for them to be used as standard candles to determine intergalactic distances. We shall see in Vol. III that this method can be of use in constraining cosmological models.

7.8.3 Cataclysmic Variables

These are fairly generic classes of binary systems, usually containing a white dwarf and a late-type main-sequence star or a red giant star or, in some rare cases, another white dwarf. The periods of these binaries vary from $\sim$80 min to $\sim$15 h, and the light output shows abrupt increases in a time scale of a day followed by a decline that can last from weeks to a year. The differences between the individual systems arise mostly because of the geometry in which accretion takes place. The nova discussed above can in fact be thought of as one specific example of a general class of cataclysmic variable.

There also exist dwarf novas in which a star brightens by 2 to 5 mag repeatedly, with the interval between the outbursts varying from tens to hundreds of

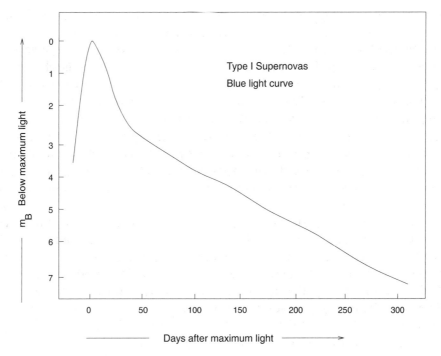

Fig. 7.11. Light curve of a type I supernova in a blue band with magnitudes relative to the value at maximum.

days. (This is to be contrasted with a change of ∼20 mag that can occur in classical novas.) Nearly all the 300 or so dwarf novas now known have evidence for a stream of matter flowing from a cool star to the white dwarf. There is indirect evidence for the stream of matter to be heating an accretion disk around the white dwarf and causing a hot spot. In some of the cases we can directly see Doppler-shifted atomic lines from the accretion disk that indicate the rotation (see Fig. 7.13). It is even possible to actually map the temperature distribution of the accretion disk and compare it with the theoretically predicted variation of $T \propto R^{-3/4}$. The exact mechanism for the periodic outbursts is not known but several models exists, all of which rely on the mass-transfer rate varying because of complex physical phenomena.

When the white dwarf has a strong enough magnetic field, the flow of matter from the accretion disk has to be funneled near the magnetic poles of the compact remnants. This leads to an accretion column near the poles of the white dwarf and several associated phenomena. As the material in the accretion column falls on the white dwarf, temperatures of $k_B T \approx (GMm_p/3R) \approx 50$ keV can be reached if all the kinetic energy is dissipated as radiation. A fair amount of this energy goes into heating the plasma, which will have a shock front propagating through it because the velocity in the accretion column will be supersonic. In many cases we do see evidence for x rays but with temperatures of $(1–5)$ keV.

7.8 Varieties of Accreting Binary Systems

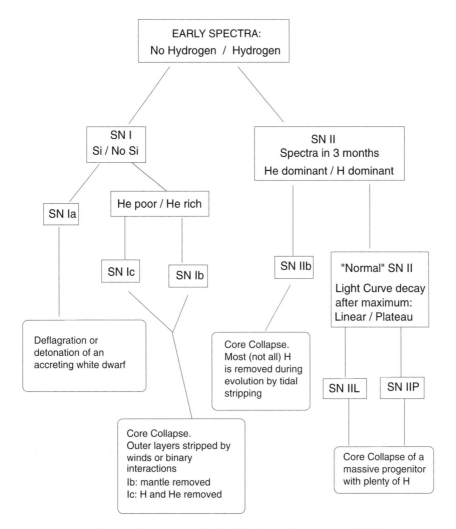

Fig. 7.12. Observational classification of supernovas (SNs).

There is also an intermediate type of star called a *recurrent nova* that is made of binary systems with giant stars and white dwarfs rather than a main-sequence star.

7.8.4 X-Ray Transients and Be Stars

Another class of sources believed to be related to binary phenomena is that of the x-ray transients, which show abnormally increased x-ray activity in an aperiodic manner. There are ∼30 such sources known that can be divided into two classes. The first class has softer-x-ray spectra and are found in low-mass systems in which the optical companion is a small late-type K-M star. They

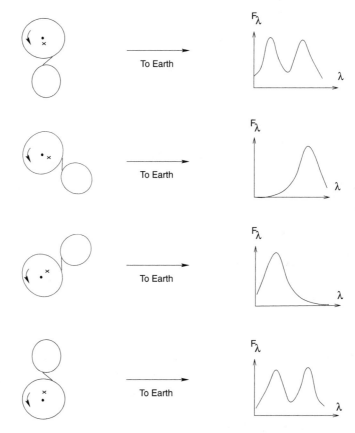

Fig. 7.13. Doppler shift from an accretion disk and the effect of an eclipse.

brighten by approximately 6–8 mag during x-ray outbursts and are also called soft-x-ray transients. The second class of objects have very hard-x-ray spectra and are associated with HMXBs. The optical companion seems to be a mid-B giant/main-sequence star; these are conventionally called Be stars. These stars were previously known and classified, with the "e" standing for emission lines that vary in intensity and profile. The exact nature of Be systems is unknown, and some models suggest that these stars eject mass, probably around the Equator, forming a ringlike material. Any compact object orbiting such a star will show increased x-ray output when it encounters the ring. Further, if the orbit is eccentric then the x-ray output would be modulated by a factor that is significantly larger than what can be accounted for by a stellar wind from an ordinary B star. One well-studied example of such a system is A0538-66, whose behaviour could be modelled along the lines shown in Fig. 7.14. The rapid rotation of the Be star has created around it an atmosphere and possibly an equatorial range of matter through which the eccentric orbit of the neutron star passes. If the accretion takes place, an increased outburst of x rays and optical activity can

7.8 Varieties of Accreting Binary Systems

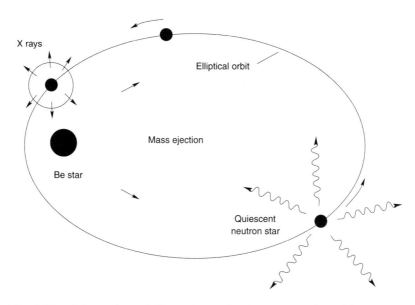

Fig. 7.14. Schematic modelling of accretion in an eccentric Be binary.

start. What is more, the high x-ray luminosity can cause the destruction of the tenuous material around the neutron star orbit, thereby ending the activity at least temporarily. When the outburst and the mass-transfer rate fall, so does the external pressure on the magnetosphere of the neutron star, making R_m increase. Once R_m is larger than the corotation radius, the centrifugal barrier will prevent matter from accreting to the neutron star. Clearly, the faster the spin of the neutron star, the higher will be the x-ray luminosity at which the system will turn on. The system A0538-66 has the neutron star with the fastest period of all the Be systems and hence has the largest centrifugal barrier to the incoming material.

7.8.5 X-Ray Bursters

Certain x-ray sources that show dramatic increases in their x-ray luminosity over a short duration are called x-ray bursters. In these sources $\sim 10^{39}$ ergs of x-ray energy is radiated in ~ 10 s. The start of the burst is very rapid (within a time scale of ~ 1 s), and the energy output increases in all bands of the spectrum. The decay of the burst, however, is different in different spectral bands with the lowest-energy band (corresponding to coolest material) showing a long tail and the higher energies showing sharper decay. This will happen if the temperature of the burst that starts off at a high value drops as the burst progresses, changing from 3×10^7 K to 1.5×10^7 K in ~ 10 s. One possible model for such bursts is based on unstable nuclear burning that takes place on the surface of the neutron star along the following lines.

The accretion of hydrogen from the disk onto the neutron star will produce a surface layer of hydrogen that will undergo nuclear burning, thereby producing an underlying layer of helium. Eventually, if the densities and the temperatures in the helium layer reach a critical point for fusion, the surface layer will undergo rapid, unstable, helium burning, producing a thermonuclear flash peaked at gamma-ray energies. The steady accretion of fresh hydrogen can produce another burst at a later time; actual energetics of the burst will vary with the gaps between the bursts because a longer interval will lead to more helium's being produced on the surface. The rapid rise of the temperature during the thermonuclear event followed by its subsequent cooling can be correctly reproduced in this model. For typical accretion rates, the interval between accretion rates is $\sim$3 h. It may be noted that a hydrogen atom falling in the potential well of a typical neutron star will gain $\sim$100 MeV of kinetic energy that could be eventually made available for steady x-ray emission; in contrast, a nuclear reaction involving the same hydrogen atom will release $\sim$1 MeV of energy. Hence we expect the steady emission averaged over time to be $\sim$2 orders of magnitude higher in energy production compared with the burst emission that is indeed seen in many sources.

There is another class of bursting x-ray sources, usually called rapid bursters, that operates through a different mechanism. In this case, the bursts recur very rapidly, typically once every 10 s. The strengths of the bursts vary significantly, with the ratio between the strongest and the weakest burst being as high as 10^3. Contrary to the x-ray burster, the rapid burster shows no evidence for cooling in the tail of the spectra, suggesting that the effective temperature varies very little. It is generally believed that the rapid bursters arise because of a complex plasma process that may be roughly described as follows. Consider a neutron star that has a sufficiently strong magnetic field so that its magnetosphere exerts strong enough pressure to hold back the surrounding gas. However, as the gas builds up around the magnetospheric boundary, it will start influencing the magnetic field lines and could lead to a rupture in the boundary. This causes the matter to fall directly onto the neutron star surface, producing a series of rapid x-ray bursts. Once the matter has fallen in, the magnetosphere regains the original shape and the process can repeat again.

The above two examples indicate how the variations on the basic theme of accretion can lead to widely different physical phenomena. In particular, note that even though matter must be flowing into the poles of the bright x-ray pulsars, not all of them show x-ray bursts from time to time. This is related to the fact that pulsars and bursters have widely different ages. All the x-ray pulsars have massive early-type stars as companions; such stars have very small lifetimes and hence the system must have formed fairly recently. On the other hand, the bursters are associated with LMXBs in which the companion is less massive than the Sun and hence is likely to be very old. The magnetic field of the neutron star would have decayed much more in the older systems compared with that of the younger systems. Because a strong field is needed to channel the accreting matter, x-ray pulsations are expected in only young neutron stars associated with massive stellar companions. Further, the helium burning instability does not occur at high-mass-transfer rates, which is the case in all the x-ray pulsars associated with HMXBs. Hence x-ray bursts are inhibited in these systems.

8
The Sun and the Solar System

8.1 Introduction

This chapter deals with the physics of the Sun and the constituents of the solar system. It draws heavily on the material developed in Chaps. 2 and 3 and on Vol. I, Chaps. 2, 8, and 9.

8.2 The Standard Solar Model

Given the mass of the Sun, its initial composition, and its current age, we should be able to develop a model for the Sun by using the equations described in Chaps. 2 and 3. Such an evolutionary calculation will predict all other structural properties of the Sun at the present time, which may then be compared with observations. Among the input variables, the mass of the Sun, $M_\odot = (1.9891 \pm 0.0004) \times 10^{33}$ gm, is known quite accurately. The age of the Sun has to be estimated indirectly and is expected to be approximately $(4.5 \pm 0.1) \times 10^9$ yr. The initial composition of the Sun is not well known but the ratio $Z/X = 0.02739 - 0.02765$ is thought to be well determined.[1] Because $X + Y + Z = 1$ and Z/X are given, the initial composition can be parameterised by a single variable, say, the value of helium fraction Y. By varying the value of Y, we can construct a class of solar models and choose the one that fits best with the observations. In reality, there arises (at least) one more parameter in modelling the solar structure because of theoretical uncertainty in the description of convection. The mixing-length parameter α, which is the ratio between the convective mixing length and the local-pressure scale height, is a variable in the model building (see Vol. I, Chap. 8, Section 8.14). It is usual to vary both α and Y in order to obtain the model for the Sun. We shall now describe such a *standard solar model* (SSM), providing the connections with the discussion of stellar structure in Chap. 2, wherever possible.

The physical parameters for the present-day Sun based on the standard solar model are shown in Fig. 8.1. Numerical integration of the standard solar model

8.2 The Standard Solar Model

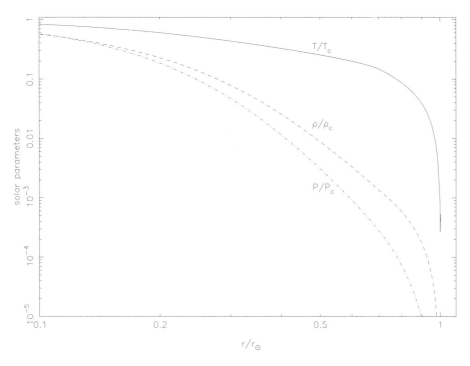

Fig. 8.1. Temperature, pressure, and density inside the Sun scaled with respect to the central values. (Figure courtesy of H.M. Antia.)

provides the following numbers for the Sun: (1) central pressure $p_c = 2.4 \times 10^{17}$ dyn cm^{-2}; (2) central density $\rho_c = 154$ gm cm^{-3}; (3) the pressure falls to half its central value at $0.12\, R_\odot$, the temperature falls to half its central value at $0.26\, R_\odot$, the luminosity falls to half its value at $0.11\, R_\odot$, and half the hydrogen is consumed virtually at the origin. Here $R_\odot = 6.96 \times 10^{10}$ cm is the radius of the visible disk of the Sun. We shall first try to understand these values.

Because the density is approximately constant near the centre, $(dp/dr) \approx -G(4\pi/3)\rho_c^2 r$. Integration gives $p(r) \approx p_c - (2\pi/3)G\rho_c^2 r^2$; hence $p = p_c/2$ at $r_{1/2} = (3p_c/4\pi G\rho_c^2)^{1/2} \approx 0.1\, R_\odot$. This is fairly close to the value of $0.12\, R_\odot$ obtained from numerical integration.

Because the Sun is radiative near the core, $\nabla_{\rm ad} = 0.4$ and the temperature must vary more shallowly than $T \propto p^{0.4}$. Thus, at the radius where T has dropped by a factor of 2, the pressure would have dropped at least by $2^{2.5} = 5.7$. In reality, the ratio is more like 13, which suggests a scaling law $T \propto p^{0.27}$. This has to be expected because the gradient ∇ varies from 0.34 at the centre to 0.23 at $r = 0.26\, R_\odot$ and to 0.19 at $0.5\, R_\odot$.

The energy generation in the Sun is primarily due to the p–p chain and the thermonuclear reactions take place in the core region with $r \lesssim 0.2\, R_\odot$.

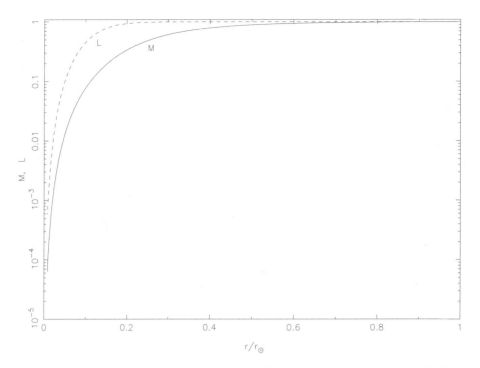

Fig. 8.2. The fractional luminosity and mass inside the Sun as functions of radius. (Figure courtesy of H.M. Antia.)

(See Fig. 8.2.) In the region $0.2\,R_\odot \lesssim r \lesssim 0.7\,R_\odot$, the energy transport is radiative whereas in most of the outer regions $0.7\,R_\odot \lesssim R_\odot$, the energy transport is due to convection. This is clearly seen in Fig. 8.3, which shows the adiabatic and radiative temperature gradients inside the Sun; the crossover occurs around $0.71\,R_\odot$. These features agree well with the theoretical description given in Chap. 2, Section 2.3. (It must be noted that there are several other interesting physical phenomena that take place just outside the optical disk of the Sun; these aspects are described later in Section 8.6.) The p–p reaction rate that generates solar energy varies as

$$\epsilon_{pp} \propto X^2 T^4 \rho^2 \propto X^2 p^2 T^2 \propto X^2 p^{2.5}, \qquad (8.1)$$

where the last relation follows from $T \propto p^{0.27}$ obtained above. If $q \equiv r/(\sqrt{2}r_{1/2})$, then the pressure varies near the centre as $p \propto (1 - q^2)$, leading to an energy generation $\epsilon_{pp} \propto (1 - q^2)^{5/2}$. Using this, we can easily find the value of q at which half the total luminosity is generated. We get $q_{1/2} = 0.52$, predicting that half the luminosity should come from within $0.074\,R_\odot$. This is smaller than the true value of $0.1\,R_\odot$ obtained from numerical integration because we have neglected the variation of X. Figure 8.4 gives the variation of the hydrogen fraction inside the Sun.

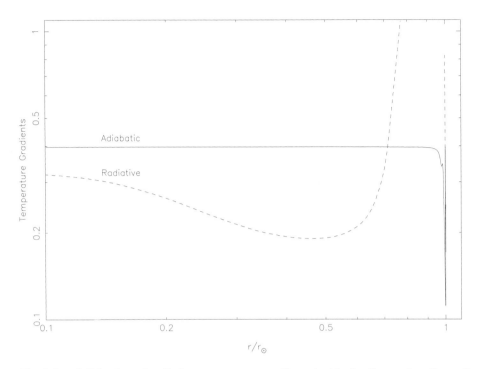

Fig. 8.3. Adiabatic and radiative temperature gradients inside the Sun as functions of radius. (Figure courtesy of H.M. Antia.)

Because hydrogen is half depleted at the centre, more energy generation is occurring at larger radii. Since Sun is $t_\odot = 4.5$ Gyr old, the amount of hydrogen that must have fused is approximately $(L_\odot t_\odot/0.0067c^2) = 0.046\,M_\odot$. The mass within the half-luminosity radius is $0.097\,M_\odot$ and thus originally contained $0.097X\,M_\odot = 0.068\,M_\odot$ of hydrogen. In this region, $(0.046\,M_\odot/2) = 0.023\,M_\odot$ of hydrogen must have been burnt. This is a fraction $0.023/0.068 \approx 1/3$ of the total hydrogen mass in this region.

At the centre of the Sun, $\mu = 0.829$, $\rho = 153$ gm cm^{-3}, and $T = 1.54 \times 10^7$ K. The gas-energy density is $\epsilon_{\rm gas} = (3/2)(\rho/\mu m_p)k_B T \approx 3.6 \times 10^{17}$ ergs cm^{-3}, and the radiation-energy density is $\epsilon_{\rm rad} = aT^4 \approx 4.25 \times 10^{14}$ ergs cm^{-3}. The ratio between them is approximately 1.2×10^{-3}. At half the solar radius, $\mu = 0.605$, $\rho = 1.43$, and $T = 3.9 \times 10^6$ K, giving $\epsilon_{\rm gas} = 1.16 \times 10^{15}$ ergs cm^{-3} and $\epsilon_{\rm rad} \approx 1.75 \times 10^{12}$ ergs cm^{-3}, giving a ratio of 1.5×10^{-3}. In fact, this ratio is nearly constant in the Sun in the range $0 < r < 0.8\,R_\odot$. This constancy implies that

$$\mathcal{R} \equiv \left(\frac{\rho}{1\,{\rm gm\,cm^{-3}}}\right)\left(\frac{T}{10^6\,{\rm K}}\right)^{-3} = {\rm constant}, \qquad (8.2)$$

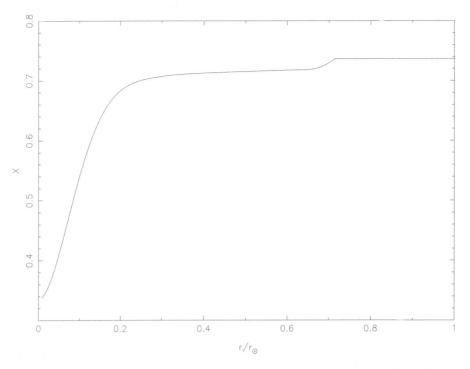

Fig. 8.4. Mass fraction of hydrogen inside the Sun as a function of radius. (Figure courtesy of H.M. Antia.)

giving $p \propto \rho T \propto \rho^{4/3}$; thus most of the Sun can be described by an $n = 3$ polytrope.

At the centre, $\mathcal{R} \approx 0.042$ and the opacity is $\kappa_R = 1.3$ cm^2 gm^{-1}. At half the solar radius, $\mathcal{R} = 0.024$ and $\kappa_R = 10$ cm^2 gm^{-1}. The mean free path of the photon in the Sun is $\lambda = (\kappa\rho)^{-1}$. Plugging in the previous values, we get $\lambda_c \approx 5 \times 10^{-3}$ cm and $\lambda_{0.5} \approx 7 \times 10^{-2}$ cm. Diffusing a distance D by random walking with mean free path λ will take $n = (D/\lambda)^2$ steps, which in turn takes a time $t = (n\lambda/c) = (D^2/\lambda c)$. To diffuse out from the half-radius of the Sun, $D = (R_\odot/2)$ so that $t_{0.5} = R_\odot^2/(4\lambda_{0.5}c) \approx 6 \times 10^{11}$ s. For diffusing from the centre, we can repeat the calculation with $D = R_\odot$ and $\lambda = \lambda_c$. However, this is inaccurate because λ varies fairly rapidly near the centre with $\kappa = (5, 8, 14, 30, 50) \times 10^{-3}$ cm^2 gm^{-1} at $r = (0, 0.1, 0.2, 0.3, 0.4)R_\odot$. Estimating the diffusion time per cell by $t_n = (r_{n+1}^2 - r_n^2)/(\lambda_{n+1}c)$ gives $t_n = (2, 3, 3, 2, 2) \times 10^{11}$ s. Adding all these up and $t_{0.5}$, we estimate the diffusion time as $t_{\text{diff}} \approx 1$–$2 \times 10^{12}$ s.

The gravitational binding energy of a polytrope of index n is

$$U_{\text{grav}} = -\frac{3}{(5-n)}\left(\frac{GM^2}{R}\right) \approx -\frac{3}{2}\left(\frac{GM_\odot}{R_\odot}\right) \tag{8.3}$$

8.2 The Standard Solar Model

because we saw above that, for the Sun, $n \approx 3$. Virial theorem gives $U_{\text{kinetic}} = -U_{\text{grav}}/2$. To relate thermal energy U_{th} to U_{kinetic}, note that

$$\frac{U_{\text{kinetic}}}{U_{\text{th}}} = \frac{(3/2)Nk_BT}{C_VT} = \frac{(3/2)(C_P - C_V)}{C_V} = \frac{3}{2}(\gamma - 1). \tag{8.4}$$

Because the Sun is mostly ionised particles, $\gamma = 5/3$ and $U_{\text{th}} \approx U_{\text{kinetic}}$. This gives the thermal time scale as $t_{\text{KH}} = (U_{\text{th}}/L_\odot) = 7 \times 10^{14}$ s. This is a factor of 350–700 larger than t_{diff}. The reason is quite simple. All the Sun's radiation leaks out at $t \approx t_{\text{diff}}$, but it is replenished by the thermal energy of matter. Therefore, for all the Sun's thermal energy to be carried off in radiation, it has to leak out by a larger factor of $(\epsilon_{\text{gas}}/\epsilon_{\text{rad}}) \approx 700$. This result is a direct consequence of the equation for radiative diffusion $(dp_{\text{rad}}/dr) = -\kappa\rho L/(4\pi cr^2)$. Writing

$$\frac{dp_{\text{rad}}}{dr} \approx \frac{p_{\text{rad}}}{p_{\text{gas}}} \frac{dp_{\text{gas}}}{dr} \approx \frac{p_{\text{rad}}}{p_{\text{gas}}} \frac{2\epsilon_{\text{gas}}}{3r}, \tag{8.5}$$

we find that $L \approx (\epsilon_{\text{rad}}/\epsilon_{\text{gas}})(4\pi r^3/3)\epsilon_{\text{gas}}(c/\kappa\rho r^2)$. Noting that $t_{\text{diff}} \approx (\kappa\rho r^2/c) = r^2/c\lambda$, we can write this equation as $t_{\text{KH}} \approx (\epsilon_{\text{gas}}/\epsilon_{\text{rad}})t_{\text{diff}}$.

Figure 8.3 gives the temperature gradient relevant for determining radiative-energy versus convective-energy transport in the interior of the present Sun. For an ideal monotonic gas, the onset of convection occurs when $(d\ln T/d\ln P) > 0.4$. The figure shows that this occurs around $r \approx 0.7\,R_\odot$. Also note that, very close to the surface, radiative conditions will prevail and hence the solar atmosphere has to be treated separately.

It was shown in Vol. I, Chap. 8 that the adiabatic temperature gradient is given by

$$\left.\frac{dT}{dr}\right|_{\text{ad}} = -\left(1 - \frac{1}{\gamma}\right)\frac{\mu m_H}{k}\frac{GM_r}{r^2} = -\frac{g}{C_P} \tag{8.6}$$

and the convective-gradient difference is given by

$$\delta\left(\frac{dT}{dr}\right) = \left[\frac{L_r}{4\pi r^2}\frac{1}{\rho C_P\alpha^2}\left(\frac{\mu m_H}{k}\right)^2\left(\frac{g}{T}\right)^{3/2}\beta^{-1/2}\right]^{2/3}. \tag{8.7}$$

For the Sun, we can take $\alpha = \beta = 1$, $M_r \simeq 1\,M_\odot$, $L_r \simeq 1\,L_\odot$, $r \simeq 0.75\,R_\odot$, $C_P \simeq (5/2)n\mathcal{R}$, $P \simeq 3 \times 10^{13}$ dyn cm^{-2}, $\rho \simeq 0.1$ gm cm^{-3}, $\mu \simeq 0.6$, and $T \simeq 1.8 \times 10^6$ K. Then $(dT/dr)_{\text{ad}} \simeq 1.4 \times 10^{-4}$ K cm^{-1} and $\delta(dT/dr)_{\text{ad}} \simeq 8 \times 10^{-11}$ K cm^{-1}. The ratio between the two is 6×10^{-7}. Clearly we can approximate the temperature gradient in the convection zone as given by the adiabatic gradient. We also derived in Vol. I, Chap. 8 the convective velocity that is needed to carry the flux:

$$\bar{v}_c = \left(\frac{\beta g}{T}\right)^{1/2}\left[\delta\left(\frac{dT}{dr}\right)\right]^{1/2}l = \beta^{1/2}\left(\frac{T}{g}\right)^{1/2}\left(\frac{k}{\mu m_H}\right)\left[\delta\left(\frac{dT}{dr}\right)\right]^{1/2}\alpha. \tag{8.8}$$

With the above numbers, this becomes $\bar{v}_c \approx 7 \times 10^3$ cm s$^{-1} \approx 10^{-4}\,c_s$.

The central composition of the Sun at present corresponds to $X \approx 0.34$ and $Y \approx 0.64$ if the initial composition is taken to be $X \approx 0.71$ and $Y \approx 0.27$. The decrease in the hydrogen fraction and the increase in the helium fraction are natural consequences of stellar evolution. The evolution also increases the luminosity by $\sim 40\%$ and the radius by $\sim 10\%$.

We can understand evolution by using the scaling relations developed in Chap. 2. We have seen in Chap. 2 that the luminosity is a strong function of μ with $L \propto \mu^\alpha$, where $\alpha \simeq 7.5\text{--}8$. From virial theorem, we have the scaling $T \propto \mu M^{2/3} \rho^{1/3}$, where μ is the mean molecular weight. Further, when radiative diffusion controls the energy flow, we must have $L \propto (RT^4/\kappa \rho)$. Using the relations $\kappa = \kappa_0 \rho T^{-3.5}$ and $R^3 \propto (M/\rho)$, it is straightforward to obtain the scaling

$$L \propto \frac{M^{5.33} \rho^{0.117} \mu^{7.5}}{\kappa_0}. \tag{8.9}$$

Because L varies slowly with ρ and because κ_0 is only a weak function of X or Y, the major evolutionary dependence of L on the composition arises through the $\mu^{7.5}$ factor. If we also model the variation of ρ with μ, by using the energy-generation law, we get a still steeper dependence of $L \propto \mu^8$. Hence

$$\frac{L(t)}{L(0)} \cong \left[\frac{\mu(t)}{\mu(0)}\right]^\alpha, \quad \alpha = 7.5\text{--}8, \tag{8.10}$$

where

$$\mu(t) \cong \frac{4}{3 + 5X(t)} \tag{8.11}$$

[see Chap. 2, Eq. (2.6)]. We have to determine only the time evolution of $X(t)$. Because hydrogen burning releases approximately $Q = 6 \times 10^{18}$ ergs per gram of hydrogen, it follows that $\dot{X}(t) = -[L(t)/MQ]$. Hence

$$\frac{d\mu(t)}{dt} = -\frac{5}{4}\mu^2(t)\frac{dX}{dt} = \frac{5}{4}\mu^2(t)\frac{L(t)}{MQ}. \tag{8.12}$$

Integrating this equation along with relation (8.10) and fixing the initial conditions, we find that

$$L(t) = L(0)\left[1 - \frac{5(\alpha+1)}{4}\frac{\mu(0)L(0)t}{MQ}\right]^{-\alpha/(\alpha+1)}. \tag{8.13}$$

Substituting the numbers, expressing the results in terms of $L_\odot = (3.847 \pm 0.003) \times 10^{33}$ ergs s^{-1}, and $t_\odot = 4.5 \times 10^9$ yr, taking $\mu(0) \approx 0.6$, $\alpha \simeq 7.4$ gives

$$\frac{L(t)}{L_\odot} \approx \frac{L(0)}{L_\odot}\left[1 - 0.3\frac{L(0)}{L_\odot}\frac{t}{t_\odot}\right]^{-15/17}. \tag{8.14}$$

To get $L(t) = L_\odot$ at $t = t_\odot$, we need $L(0) \approx 0.79\, L_\odot$; this is fairly close to the value of $L(0) \approx 0.73\, L_\odot$ obtained by numerical integration of the equations.

8.3 Solar Neutrinos

The p–p and CNO cycles taking place in the core of the Sun produce large quantities of neutrinos in some parts of the chain. Listed below are some of the specific steps in which this occurs in the standard nuclear cycles, with the energy range of the neutrinos given in parentheses[2]:

$$\begin{aligned}
p + p &\to {}^2\text{H} + e^+ + \nu_e & (0\text{--}0.42\,\text{MeV}), \\
p + e^- + p &\to {}^2\text{H} + \nu_e & (1.4\,\text{MeV}), \\
{}^3\text{He} + p &\to {}^4\text{He} + e^+ + \nu_e & (0\text{--}18.8\,\text{MeV}), \\
e^- + {}^7\text{Be} &\to {}^7\text{Li} + \nu_e & (0.383, 0.861\,\text{MeV}), \\
{}^8\text{B} &\to {}^8\text{Be} + e^+ + \nu_e & (0\text{--}15\,\text{MeV}), \\
{}^{13}\text{N} &\to {}^{13}\text{C} + e^+ + \nu_e & (0\text{--}1.2\,\text{MeV}), \\
{}^{15}\text{O} &\to {}^{15}\text{N} + e^+ + \nu_e & (0\text{--}1.7\,\text{MeV}), \\
{}^{17}\text{F} &\to {}^{17}\text{O} + e^+ + \nu_e & (0\text{--}1.7\,\text{MeV}).
\end{aligned} \quad (8.15)$$

The neutrinos from the first reaction have a continuous spectrum of energies up to a cutoff value of ~ 0.42 MeV. The decay of ^{8}B can similarly lead to neutrinos with energies up to 15 MeV. The ^{7}Be reaction leads to neutrinos at two different but specific values of energies, 0.861 or 0.383 MeV, depending on the final state of ^{7}Li. The decay leading to the 0.861 MeV neutrino is more probable and occurs 90% of the time. The three reactions in the CNO cycle lead to neutrinos with energies up to a maximum of 1.2, 1.73, and 1.74 MeV, respectively. There are also some other additional reactions that can lead to neutrinos, but these are rarer than the above reactions.

These neutrinos react with matter through only weak interactions and have scattering cross sections of $\sigma \approx 10^{-44}$ cm^2 when their energies are of the order of mega electron volts. Hence the mean free path for the absorption of neutrinos in the solar interior is $\lambda \approx (\sigma n)^{-1} \approx 10^9\, R_\odot$, where n is the number density of particles in the Sun. It is obvious that these neutrinos will stream out of the Sun with negligible scattering and hence should be detectable by suitable techniques on Earth.

We can easily estimate the mean solar neutrino flux on Earth by taking the dominant reaction channel. The cumulative effect of a p–p chain is to convert four protons into one helium nucleus, releasing two positrons, two neutrinos, and ~ 28 MeV of energy, which is eventually radiated as photons. Because two neutrinos come out with the release of 28 MeV of energy in the form of radiation, the total number of neutrinos released per second will be approximately $2(L_\odot/28\,\text{MeV})$. Hence the neutrino flux on Earth will be approximately

$$F = \frac{2L_\odot}{4\pi R^2 (28\,\text{MeV})} = \frac{2 \times 4 \times 10^{33}\,\text{ergs s}^{-1}}{4\pi (1.5 \times 10^{13}\,\text{cm})^2 \times 28\,\text{MeV}} = 6 \times 10^{10}\,\text{cm}^{-2}\,\text{s}^{-1}. \quad (8.16)$$

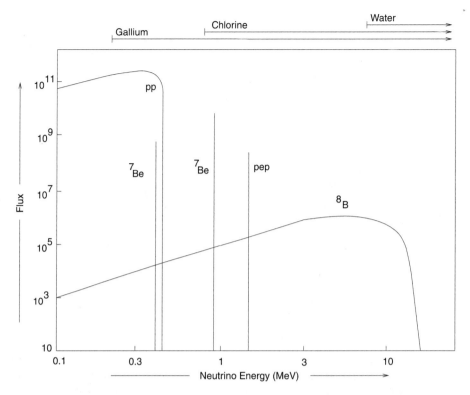

Fig. 8.5. Energy spectrum of solar neutrinos expected from the standard solar model.

A more detailed calculation, based on the standard solar model, leads to the spectrum of neutrinos expected on Earth, as shown in Fig. 8.5.

Several experiments have attempted to detect and measure the flux of solar neutrinos arriving on Earth. One of these procedures involves allowing the solar neutrinos to react with ^{37}Cl, thereby producing an electron and ^{37}Ar. The product ^{37}Ar is radioactive with a half-life of 35 days. By measuring the amount of argon nuclei that are produced, we can estimate the number of neutrino captures per second that took place in the target. This reaction requires a threshold energy of 0.81 MeV for the neutrino and hence is sensitive only at higher-energy neutrinos. With an absorption cross section of $\sim 10^{-44}$ cm^2 and an effective flux of $\sim 10^9$ neutrinos cm^{-2} s^{-1}, we would expect 10^{-35} reactions per second per target atom. It is convenient to introduce a unit called the solar neutrino unit, abbreviated as SNU, which stands for 10^{-36} captures per second per target atom. From the above discussion it is clear that we expect a detection at the level of ~ 10 SNU. A more precise calculation, based on the standard solar model, gives $7.7^{+1.2}_{-1.0}$ SNU, where the error bar covers uncertainties in nuclear-reaction rates, opacities, and model-building techniques and is fairly conservative. Of these, ~ 5.9 SNU [corresponding to a flux of $5.15 \times 10^6 (1.00^{+0.19}_{-0.14})$ cm^2 s^{-1}] is from the ^{8}B decay and 1.15 SNU is from the ^{7}Be reaction. Observations, however, give a much lower value for the solar neutrino flux, approximately 2.55 ± 0.25 SNU. This discrepancy has been called the *solar neutrino problem* and possibly indicates some unknown aspects of neutrino physics.[3]

The stability of the theoretical results essentially arises from the following facts. For stars like the Sun, powered by the p–p chain, $L \propto T_c^4$ (see Chap. 2, Section 2.3) and the neutrino flux ϕ_ν varies as a fairly high power of the central temperature: $\phi_\nu \propto T_c^n$, with $n \approx 18$ for the ^{8}B neutrino flux and $n \approx 8$ for the ^{7}Be neutrino flux. Combining the dependence of ϕ_ν and L, we can write $(\delta\phi/\phi) = (n/4)(\delta T_c/T_c)$. Because the luminosity $L \propto T_c^4$ of the Sun is known to an accuracy of 0.4%, the fractional uncertainty in T_c is like $(0.4/4) = 0.1\%$. The most uncertain neutrino flux corresponding to the largest n will be for the ^{8}B case, and even here it can only be $18 \times 0.1 \approx 1.8\%$. The uncertainty is significantly lower for the neutrino flux from ^{7}Be and p–p chains (for which $n \approx -1.2$).

Because the procedure for detecting the neutrinos with ^{37}Cl has a threshold of ~ 0.81 MeV, it misses most of the neutrinos from the p–p chains. It is possible to use other procedures for the detection of neutrinos that have lower thresholds. For example, scattering of electrons by neutrinos, leading to Čerenkov radiation from electrons in H_2O has a threshold of ~ 7.3 MeV for neutrino energy and is used in another experimental setup. Yet another procedure relies on the reaction ^{71}Ga$(\nu_e, e^-)^{71}$Ge, which has a very low threshold of 0.23 MeV. Even these experiments show that the observed neutrino flux is significantly lower than that predicted by theory. For example, the prediction from SSM for the gallium-based experiments is $(129)^{+8}_{-6}$SNU whereas the experimentally observed values are approximately 73.4 ± 5.7 SNU. What is more, these experiments suggest that the key reason for the discrepancy could possibly lie in the fundamental physics of neutrinos rather than in stellar modelling. We shall now briefly review these results.

The experiment based on Čerenkov radiation is also mostly sensitive to the electron neutrinos arising from the ^{8}B reaction. The currently observed flux is $\phi_K(^8\text{B}) = [2.80 \pm 0.19(\text{stat}) \pm 0.33(\text{syst})] \times 10^6 \text{ cm}^{-2} \text{ s}^{-1}$, which is $\sim 44\%$ of the ^{8}B flux predicted by the standard solar model. This shows that completely different experimental setups based on different procedures confirm the deficit of solar neutrinos compared with that of the theoretical predictions. In fact, it is possible to combine the results of the chlorine experiment and the ones based on Čerenkov radiation to draw some important conclusions. The neutrinos arising from the ^{8}B reaction in the Sun and detected in the Čerenkov detector experiment alone should have led to the following capture rate in the chlorine experiment: $\phi_K(^8\text{B})\sigma_{\text{Cl}} \approx [3.11^{+0.24}_{-0.23} \pm 0.39(\text{syst})]$ SNU. This is *higher* than the observed rate in the chlorine experiment, namely, 2.55 ± 0.25 SNU. In other words, the chlorine experiment is missing some of the neutrinos that it should have seen based on the observed neutrino flux in the H_2O experiment. This conclusion is based essentially on neutrino physics, which predicts the spectrum of ^{8}B neutrinos, shown in Fig. 8.5. Both the ^{37}Cl and the H_2O experiments are sensitive to these neutrinos. Given the actual detection rate in the latter experiment, we can predict the former. We find that the results are not compatible.

It is actually possible to do a more sophisticated analysis of the two experiments and show that the chlorine experiment is actually seeing less than 0.46 SNU of neutrinos from ^{7}Be and CNO reactions at a 95% confidence limit. This result is in strong disagreement with the theoretical prediction for ^{7}Be neutrinos alone, which is 1.15 ± 0.11 SNU from the standard solar model calculations. The spread is a conservative limit arising from theoretical uncertainties in the model. In other words, combining the results of Čerenkov radiation experiments and chlorine experiments shows that something must be happening to the electron neutrinos *after* they are created in the solar interior.

These conclusions are reinforced by the gallium-reaction experiments. The prediction from the standard solar model in this case is 129^{+8}_{-6} SNU, where the quoted errors are 1σ uncertainties. More than half of these, $\sim$70 SNU, arise from p–p reactions whose flux can be predicted to an accuracy of better than 1%. Another 30% ($\sim$34.4 SNU) arises from the ^{7}Be flux that can again be calculated to a fairly high degree of accuracy. A wide class of solar models leads to 70 ± 1 SNU for the p–p chain and 34 ± 4 SNU from ^{7}Be reactions. The ^{8}B neutrinos that dominate the chlorine and the Čerenkov experiments contribute less than 15% in the gallium experiment. The experimentally observed rate currently available in two separate gallium experiments (combined together) is 73.4 ± 5.7 SNU. By performing an analysis similar to the one mentioned above, we can combine these results and get the bound on the observed contribution from ^{7}Be in the gallium experiment to be less than $\sim$19 SNU at 95% confidence limit. This is significantly lower than the theoretical value of 34 ± 4 SNU. The seeming exclusion of everything but the p–p neutrinos in the gallium experiment is a problem independent of the preceding ones because the p–p neutrinos are not observed in the other experiments. The missing ^{7}Be neutrinos in the gallium experiment cannot be explained away by any change in solar physics. Note that the ^{8}B neutrinos observed in H_2O experiments are produced in competition with the missing ^{7}Be neutrinos; any solar model explanation that reduces the predicted ^{7}Be flux also reduces the observed ^{8}B flux to unacceptably low values.

A possible solution to the solar neutrino problem might lie in a phenomenon called *neutrino oscillation*. In the simplest models describing electroweak interactions, it is assumed that there are three different species of neutrinos, electron neutrinos (ν_e), muon neutrinos (ν_μ), and tau neutrinos (ν_τ), all of which are massless. In that case, each neutrino will retain its identity, and the solar neutrino experiment essentially measures the flux of electron neutrino ν_e; the experiments are insensitive to the other two types of neutrinos. Laboratory experiments, however, allow for the neutrinos to be massive; in fact, there is fair amount of evidence that the neutrinos are massive. If that is the case, then we can construct particle-physics models in which neutrino oscillations can exist. In this phenomenon, the nature of the neutrino changes periodically in time so that the identity of the neutrino can oscillate between, say, the ν_e state and the ν_μ state. It is possible to arrange the masses of the neutrinos in such a way that the deficit of electron neutrinos seen in terrestrial experiments is explained as being due to the conversion of ν_e to ν_μ, say.

8.4 Solar Oscillations

In the study of stellar oscillations in Chap. 3, it was pointed out that sustained oscillations can be classified as p modes (in which the restoring force is due to pressure) or g modes (in which the restoring force is due to gravity). Given the internal structure of the star, we can compute all the eigenvalues and eigenfunctions for these oscillations. In the case of the Sun, it is possible to directly measure the frequency of the p modes and compare them with theoretical predictions. This approach has become an extremely sensitive probe of the internal structure of the Sun. We shall now discuss some essential features of this analysis.[4]

The normal modes of oscillation of any star like the Sun can be classified by three quantum numbers, n, l, and m, where n denotes the radial order of the mode, l denotes the total angular momentum, and m relates to the azimuthal angle. The fundamental modes will have $n = 0$ and the overtones will have $n > 0$. The acoustic modes have an approximate dispersion relation of $\omega^2 \approx (n+1)(gl/R_\odot)$, which is a fair approximation for $l \gg 1$. These modes have a cutoff frequency determined essentially by the minimum-pressure scale height $H_{\min}$ as $\omega_{ac}^2 \approx (g/2H_{\min})$. At $\omega \lesssim \omega_{ac}$, the period of oscillation $P_{ac} \approx 3$ min. The maximum number of l values is given by $l_{\max} \approx (R_\odot/2H_{\min}) \simeq 3500$. The surface velocities range between 0.2 cm s^{-1} and 20 cm s^{-1} for individual modes, and the total rms velocity that is due to $\sim 10^7$ is ~ 0.4 km s^{-1}. The total energy in all the modes together is $\sim 10^{34}$ ergs, which is equivalent to solar radiation emitted in a couple of seconds.

The general behaviour of the modes can be characterised in terms of two critical frequencies S_l^2 and N^2 introduced in Chap. 3, Section 3.7. As $r \to 0$, $S_l^2 \approx l(l+1)(v_s^2/r^2)$ goes to infinity; near the surface of the star, $S_l^2 \to 0$ because the temperature of the ideal gas, and hence the sound speed v_s, vanishes. The general behaviour of S_l^2 (for $l = 2$) in the solar interior is shown by the dashed curve in Fig. 8.6. On the other hand, $N^2 \to 0$ near the centre because it is proportional to the local acceleration that is due to gravity, which vanishes near the centre. Hence N^2 starts at zero, increases near the origin, reaches a maximum, levels off, and drops to negative values around $r \approx 0.7 R_\odot$, indicating the onset of convective instability. The behaviour of N^2 is shown by the solid curve in Fig. 8.6. (To be precise, N^2 again rises to a positive value very close to the surface, where radiative transport dominates over convection again; this is not shown in the figure.)

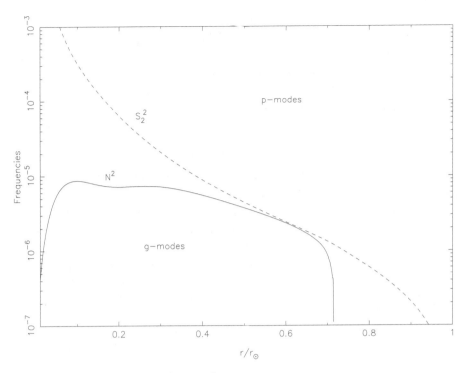

Fig. 8.6. The variation of N^2 and S_l^2 for $l = 2$ inside the Sun. (Figure courtesy of H.M. Antia.)

The discussion in Chap. 3 showed that undamped propagating modes exist only when the square of the frequency, σ^2, is larger than both S_l^2 and N^2 or when σ^2 is less than both S_l^2 and N^2. The first set corresponds to p modes, and the second set corresponds to g modes. These regions are also indicated in the figure. It is clear that as we go to higher and higher frequencies, the complexity of the p mode increases with the occurrence of an increasing number of nodes. The corresponding g modes occur in the region that is below both the dashed and the solid curves. Quite clearly, the g modes penetrate deep inside the Sun and can probe the internal structure whereas the p modes are confined to the outer region.

For the p modes that propagate mostly near the outer regimes, it is possible to obtain an approximate formula for the turnaround radius at which the modes stop propagating inwards and turn around. If the turnaround radius is r_t, then the condition $\sigma^2 = S_l^2(r_t)$ becomes

$$\frac{v_s(r_t)}{r_t} = \frac{2\pi f_{nl}}{[l(l+1)]^{1/2}}, \qquad (8.17)$$

where we have used the conventional frequency f_{nl}, defined as $2\pi f_{nl} = \sigma$. The velocity of sound may be taken to be $v_s = (\Gamma_1 N_A k_B T/\mu)^{1/2}$, with the temperature $T(r)$ given by

$$T(r) = \frac{1}{1+n_{\text{eff}}} \frac{GM\mu}{N_A k_B} \left(\frac{1}{r} - \frac{1}{R}\right)$$

$$= \frac{2.293 \times 10^7}{1+n_{\text{eff}}} \mu \left(\frac{M}{M_\odot}\right) \left(\frac{R}{R_\odot}\right)^{-1} \left(\frac{R}{r} - 1\right) \text{K} \quad (8.18)$$

[see Eq. (2.95) of Chap. 2]. Taking $\mu = 0.6$, $\Gamma_1 = 5/3$, and $n_{\text{eff}} \approx 3.25$ (corresponding to Kramer's opacity), we find that Eq. (8.17) becomes

$$\frac{2.56 \times 10^8 f_{nl}^2}{l(l+1)} x_t^3 + x_t - 1 = 0, \quad x_t = \frac{r_t}{R}. \quad (8.19)$$

This cubic equation determines the turnaround radius. For example, consider the $l = 2$ p mode (with a single node) that has a frequency of $f_{1,2} \approx 4 \times 10^{-4}$ Hz (with a period of ~ 2500 s). Equation (8.19) gives the turning point as $r_t \approx 0.44 R$, which agrees well with numerical estimates. This shows that even a low-order p mode can probe the interior up to $\sim 40\%$ of the radius.

Because a large number of eigenfrequencies for solar oscillations are known observationally, it is possible to put severe constraints on the theoretical models by use of the information provided by them. All these analyses require varying parameters of the standard solar model in such a way that the difference $\Delta\sigma^2 \equiv \sigma_{\text{obs}}^2 - \sigma_{\text{theory}}^2$ is minimised in some statistical sense. One possible way of addressing this problem, of course, would be to construct a wide variety of solar models slightly differing from a fiducial one and determine which modifications reduce $\Delta\sigma^2$. A more interesting and challenging approach would be to use the observed values of $\Delta\sigma^2$ to determine the modification of parameters in the solar model. This latter approach proceeds along the following lines: In general, the nonradial oscillations of the Sun are governed by an eigenvalue problem of the type $\mathcal{L}(\mathbf{q}) = \sigma^2 \mathbf{q}$, where $\mathbf{q}$ is the linearised displacement field and $\mathcal{L}$ is a linear differential operator. Such a problem can be reformulated in terms of a variational principle and σ^2 can be expressed in the form

$$\sigma^2 = \frac{\int \mathbf{q}^* \mathcal{L}(\mathbf{q})\, dM_r\, d\Omega}{\int \mathbf{q}^* \mathbf{q}\, dM_r\, d\Omega}, \quad (8.20)$$

where $d\Omega$ is the element of solid angle and $M_r = M(r)$ is the mass contained within a radius r. Let us suppose that a fiducial solar model with a density run $\rho_0(r)$ leads to the eigenfunctions $\mathbf{q}_0$ and eigenvalues σ_0^2. If we now vary the density slightly, then to the lowest order, the variation in the eigenfrequencies will be given by an expression of the form

$$\frac{\Delta\sigma^2}{\sigma_0^2} = \int_0^M K(\mathbf{X}_0, \mathbf{q}_0, r) \frac{\Delta\rho}{\rho} dM_r, \quad (8.21)$$

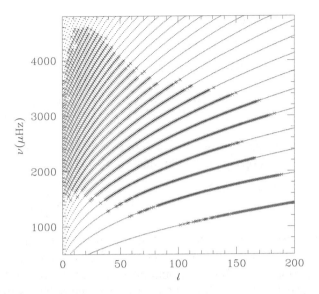

Fig. 8.7. The observed and the theoretical oscillation frequencies based on the standard solar model. (Figure courtesy of H.M. Antia.)

where $\mathbf{X}_0$ are the original model parameters. This integral equation can, in principle, be numerically inverted to provide $(\Delta\rho/\rho)$ if the left-hand side is known from observations. Figure 8.7 shows the theoretical and the observed frequencies for a range of modes based on the standard solar model; it is obvious that there is remarkable agreement between the two.

As an illustration of the above procedure in a very simple context, consider the eigenfrequencies in the WKB limit given by

$$\int_{r_t}^{R_\odot} \frac{1}{c}\left(\sigma^2 - \frac{l(l+1)c^2}{r^2}\right)^{1/2} dr = (n+\alpha)\pi, \qquad (8.22)$$

where α allows us to impose the boundary conditions between the the propagation regime and the evanescent regime. This relation has the structure

$$\pi\frac{n+\alpha}{\sigma} = \int_{r_t}^{R_\odot} \sqrt{\frac{r^2}{c^2} - \frac{1}{x^2}} \frac{dr}{r} \equiv F(x), \quad x = \frac{\sigma}{\sqrt{l(l+1)}}. \qquad (8.23)$$

Plotting the left-hand side as a function of x, we can determine the value of α by fitting. For the standard solar model, we find that $\alpha \approx 1.58$. Once the form of the function $F(x)$ is known from the observed eigenfrequency distribution, this equation can be inverted to give

$$r = R_\odot \exp\left[-\frac{2}{\pi}\int_{c(R_\odot)/R_\odot}^{c(r)/r}\left(\frac{1}{x^2} - \frac{r^2}{c^2}\right)^{-1/2}\frac{dF}{dx}dx\right]. \qquad (8.24)$$

8.4 Solar Oscillations

Given $F(x)$, this provides a relation of the form $r = f(c/r)$, which in turn gives $c(r)$. Thus the observed eigenfrequencies can directly provide information about the variation of sound speed in solar interior.

Exercise 8.1

Dispersion relations and eigenfrequencies: Equation (8.22) is based on the dispersion relation for sound waves of the form $\sigma^2 = c^2 k^2$. Suppose this dispersion relation changes into the form $\sigma^2 = c^2 k^2 + \delta f(r)$. Show that the change in the eigenfrequencies are given by the relation

$$S \frac{\delta \sigma}{\sigma} \simeq \frac{1}{2\sigma^2} \int_{r_1}^{R} \left(1 - \frac{l(l+1)c^2}{\sigma^2 r^2}\right)^{-1/2} \delta f \frac{dr}{c} + \pi \frac{\delta \alpha}{\sigma}, \qquad (8.25)$$

where

$$S = \int_{r_1}^{R} \left[1 - \frac{l(l+1)c^2}{\sigma^2 r^2}\right]^{-1/2} \frac{dr}{c} - \pi \frac{d\alpha}{d\sigma}. \qquad (8.26)$$

When the sound speed changes from c to $c + \delta c$, the change in the dispersion relation can be characterised by a function $\delta f = 2\sigma^2 (\delta c / c)$. On the other hand, the effect of the local gravitational field (which was ignored in the approximation used in Chap. 3, Section 3.7) changes the dispersion with $\delta f = -4\pi G \rho$. Evaluate the change $\delta \omega$ in both of these cases.

The eigenfrequency spectrum also contains information about the rotation of matter inside the Sun. Surface observations indicate that the equatorial rotational velocity of the Sun is ~ 2 km s^{-1}, corresponding to a period of ~ 25 days and an angular velocity of $\Omega \approx 2.9 \times 10^{-6}$ s^{-1}. The rotation, however, is differential, with the period increasing with increasing latitude. Because of this rotation, the eigenfrequencies depend not only on n and l but also on m and must be indicated as σ_{nlm}. However, because the value of σ is ~ 0.02 s^{-1}, which is considerably higher than the surface value of Ω, we expect the $(2l+1)$ frequencies of σ_{nlm} for a given value of n and l to be quite close to the original value. Thus the rotational splitting of the eigenfrequencies removes the degeneracy only weakly, and we can write $\sigma_{nlm} = \sigma_{nl,0} + \Delta \sigma_{nlm}$, where the second term is a small perturbation to the first. In such a case, we can use standard techniques of quantum-mechanical perturbation theories to evaluate the correction. For example, in the case of a uniformly rotating Sun, such an analysis gives

$$\Delta \sigma_{nlm} = -m\Omega \left\{ 1 - \frac{\int_0^M (2\xi_r \xi_t + \xi_t^2) \, dM_r}{\int_0^M \left[\xi_r^2 + l(l+1)\xi_t^2\right] dM_r} \right\} \equiv -m\Omega + m\Omega C_{nl}, \qquad (8.27)$$

where the eigenfunctions $\xi_r(r)$ and $\xi_t(r)$ correspond to the nonrotating model and depend on only n and l and the quantity C_{nl} denotes the value of the second term inside the braces. The first term on the right-hand side, $-m\Omega$, arises purely because of the rotational Doppler effect of the wave modes as viewed from the inertial frame. The dynamical effects, arising for example because of the Coriolis force, are contained in the second term $m\Omega C_{nl}$. In the case of a uniform rotation, we see that the splitting of the eigenmodes is uniformly spaced and varies linearly with m. Because, in reality the Sun rotates differentially, the actual splitting will be different from the one predicted above. Once again this deviation can be used to model the rotational structure of the Sun, treating it as an inversion problem. Such an analysis suggests that the rotation from the surface down to $r \approx 0.7\, R_\odot$ is similar to that on the surface, whereas the rotation at $r \lesssim 0.6\, R_\odot$ closely resembles a rigid rotation.

8.5 The Atmosphere and the Corona of the Sun

The study of stellar structure in Chaps. 2 and 3 concentrated mostly on the internal structure and the evolution of the stars. During this study, it was noted that the physical processes near the outer boundary of the star can be fairly complicated and hence these regions have to be investigated separately, as was done in Chap. 2, Section 2.5, for the stellar atmosphere. In the case of the Sun, it is possible to observe a host of physical phenomena that take place near its surface, and even outside, at $r \gtrsim R_\odot$. We shall now describe some of these physical phenomena.[5]

Figure 8.8 shows the variation of density and temperature in the outer regions of the Sun. The origin is chosen to be the base of the photosphere at which the optical depth is approximately unity. From the analysis in earlier chapters we know that this point corresponds to the surface from which (near) Planckian radiation is emitted by the star. For the Sun, the effective temperature of this surface is ~5770 K. The figure also shows that the temperature remains constant as we go farther away from the surface of the photosphere, although the density of the gas falls rapidly. This region is called the *chromosphere*. (During the eclipse of the Sun by the Moon, this region appears as a thin ring of coloured light, which explains the origin of the terminology.) Above the chromosphere there is a narrow *transition zone* in which the temperature rises to a very high value of approximately 3×10^5 K. Outside the transition zone, the temperature continues to rise, but more gradually, and has a maximum value of approximately 1.5×10^6 K at about 3×10^5 km above the surface. At temperatures $T \gtrsim 10^6$ K, the plasma emits most of its radiation in soft x rays. The study of this x-ray emission from this region, called the *solar corona*, by different x-ray satellite missions, has helped to diagnose different physical processes in this region.

One major feature that is important in the study of outer regions of the Sun is the existence of a reasonably strong magnetic field. To the lowest order of approximation, the Sun's magnetic field is dipolar in character and is axisymmetric. The strength of the field on a typical point on the solar surface is approximately a few Gauss. There is, however, a significant variation in this value and there are localised regions (called *sunspots*) in which the field can be much higher.

8.5 The Atmosphere and the Corona of the Sun

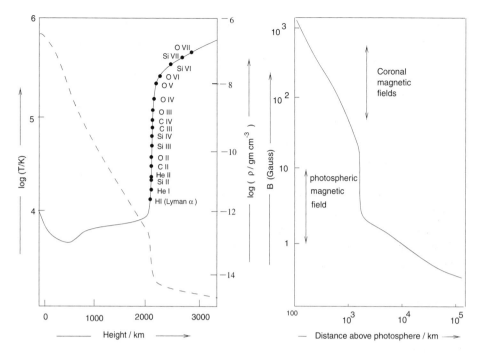

Fig. 8.8. (a) Variation of temperature and density outside the solar photosphere; (b) critical magnetic field B_c at which the magnetic pressure equals the gas pressure.

In general, the magnetic field will exert significant dynamical effect when the pressure that is due to magnetic field $(B^2/8\pi)$ is of the order of the gas pressure $nk_B T$. Equating the two pressures, we can obtain a critical value of the field $B_c = (8\pi n k_B T)^{1/2}$. Given n and T, we can work out the critical magnetic field as a function of the distance from the solar surface, which is plotted in Fig. 8.8(b). The observed values of the magnetic field at some locations are also indicated on the same graph; it is clear that magnetic effects cannot be ignored in the study of physical phenomena in the outer region.

At the base of the photosphere, we can see a patchwork of bright and dark regions with a spatial extent of ∼700 km and a characteristic lifetime of 5 to 10 min. These features, called *solar granulations*, are expected to arise because of the convective currents below the photosphere. The cell-like structure has brighter regions that are blue shifted (indicating that the fluid motion is radially outwards) and darker regions that are red shifted (indicating that the fluid motion is downwards). This agrees with the expectation of hotter fluid elements rising in convective flow and forming cell-like structures, as discussed in Vol. I, Chap. 8, Section 8.14.

The convection is driven by the increase in the opacity near the surface of the Sun. As discussed in Chap. 2, Subsection 2.4.3, the predominant source of opacity in these regions is the H^- ions. In addition to the continuum opacity, discrete absorption lines (called Fraunhofer lines) are also produced in the photosphere with the darkest part of the lines originating from regions higher in the photosphere where the gas is cooler. These absorption lines allow very precise measurement, by means of the Doppler effect, of the gaseous motion in this region. Because the peak of the blackbody spectrum is near 5900 Å, the strength of the UV continuum decreases rapidly at shorter and much longer wavelengths. Hence the Fraunhofer lines, which are swamped by the Planckian continuum in the visible and the near-UV portions of the spectrum, appear as emission lines outside this band.

One of the complex features seen in the solar atmosphere that is transient in nature is a sunspot, a dark spotlike region seen on the Sun's disk. Observations over the past two centuries show that the number of sunspots is a periodic function of time with a period of nearly 11 yr. Individual sunspots have a lifetime of approximately a month or so and remain at a constant latitude; however, succeeding sunspots tend to form at progressively lower latitudes and as the last of the sunspots of one cycle vanishes near the equator, a new cycle begins near the latitudes of approximately $\pm 40°$. The darkest portion of the sunspots (called the *umbra*) is $\sim$30,000 km in diameter and is surrounded by a region called the *penumbra*, which has a filamentlike structure. The temperature of the umbra may be as low as 3900 K compared with the Sun's effective temperature of 5770 K; this leads to a decrease of a factor of $\sim$5 in the bolometric flux from the local sunspot region. A study of spectral lines arising from the sunspots shows that they contain localised strong magnetic fields of several thousand Gauss. The field direction is vertical in the umbra, becoming horizontal across the penumbra. During the 11-yr cycle, the leading sunspots will have the same polarity in one hemisphere while the corresponding sunspots will have opposite polarity in the other hemisphere. The trailing sunspots have opposite polarity compared with the leading sunspots, and the entire set of polarity will be reversed during the next 11-yr cycle. In fact, the overall dipole magnetic field of the Sun reverses over this time scale, with the magnetic north pole switching from the geographic north pole to the geographic south pole.

Outside the photosphere lies the chromosphere, which extends for $\sim$2000 km in height and has an intensity of $\sim 10^{-4}$ of the photosphere. The density drops by a factor of 10^4, and the temperature increases from 4400 K to $\sim$25,000 K. From the analysis of stellar atmosphere, based on the Boltzmann and the Saha equations done in Chap. 2, Section 2.5, it is clear that lines that are not produced at lower temperatures and higher densities can arise in chromospheric conditions. In fact, in addition to the hydrogen Balmer line, the lines of HeII, FeII, SiII, CrII, and CaII (in particular the CaII doublet of 3968 Å and 3933 Å) can appear in this spectrum.

Just above the chromosphere, the temperature rises rapidly to 10^6 K over a region (called the *transition zone*) of only few hundred kilometres. The emission from this region is essentially in the UV and the extreme UV. For example, the

Lyman-α line is produced at 20,000 K, the CIII 977Å line originates at a level where $T=$ 90,000 K, the OVI 1032Å line originates at $T=$ 30,000 K, and the MgX 625Å line occurs at 1.4×10^6 K. It must be noted that the narrow size of the transition zone is vital for its stability. We have seen in Vol. I, Chap. 8, Section 8.13 that the gas within the temperature range of approximately (10^4-10^6) K can be thermally unstable. Such a gas does exist in the transition zone, and the thermal runaway is prevented by the conductive flux of heat from the hot outer region to the cool solar photosphere. The size of the transition zone is smaller than the minimum length scale over which thermal instability can overcome heat conduction.

The coronal region is located outside the transition zone and extends into space for several solar radii. The total energy output of the corona is $\sim 10^{-6}$ times that of photosphere and its density is very low, with only $\sim 10^5$ particles cm^{-3}. The emission from the corona arises because of different physical processes. Of these, the central sources of the emission lines are the highly ionised atoms located throughout the corona. (It is conventional to call this structure the E corona.) Because the temperatures are high, the exponential factor of the Saha equation encourages ionisation when ionisation potentials are comparable with the thermal energy; the low density effectively decreases the recombination rate, keeping the atoms ionised. In addition to the E corona, we can also observe the F corona, which arises from the scattering of photospheric light by dust grains that are located at $r \gtrsim 2.3\ R_\odot$. The Doppler broadening of Fraunhofer line radiation scattered by such dust is detectable, and the F corona actually merges with the zodiacal light that is the faint glow along the ecliptic arising from the reflection of the Sun's light by interplanetary dust. Finally, there is also a K corona, which refers to a continuum white-light emission arising from photospheric radiation scattered by free electrons located between $1\ R_\odot \lesssim r \lesssim 2.3\ R_\odot$. The spectral lines of the photosphere are broadened by the large thermal velocities of the electrons and blend together in this region. Although the corona is optically thin for visible photons, the situation is different for radio waves in low frequencies. In this band, the corona has an effective photosphere where the optical depth is approximately unity. For this reason, the radio band has been useful in the study of the solar corona.

The major issue in the study of the corona is related to the question as to which physical processes supply the energy necessary to keep the corona hot and radiating. Below the photosphere of the Sun, there exist significant convective motions carrying a large amount of kinetic energy. It is, however, difficult to transform the kinetic energy of convective motion into a heat source for the corona directly. In fact, many other stars that do not have strong convective motion do exhibit significant coronal x-ray emission, thereby suggesting that these two processes may not be directly related. It is, however, possible to use magnetic fields and MHD wave propagation to transfer energy from the inner regions of the Sun to the corona. It is generally believed that such is the case, although a completely satisfactory model is not available at present.

Exercise 8.2

Zeeman effect: One of the standard astronomical procedures to determine magnetic fields – especially in sunspots, for example – is based on the splitting of atomic-energy levels in the presence of a magnetic field. The purpose of this exercise is to derive this splitting in a simplified context.

(a) The Hamiltonian for an atom in a uniform magnetic field **B** can be taken to be

$$\hat{H} = \frac{1}{2m}\sum_a \left[\hat{\mathbf{p}}_a + \frac{|e|}{c}\mathbf{A}(\mathbf{r}_a)\right]^2 + U + \frac{|e|\hbar}{mc}\mathbf{B}\cdot\hat{\mathbf{S}}, \tag{8.28}$$

where $\mathbf{A} = (1/2)(\mathbf{B}\times\mathbf{r})$ is the vector potential, U is the electrostatic interaction energy of the system, $\hat{\mathbf{S}} = \sum\hat{\mathbf{s}}_a$ is the total electron spin of the atom, and the charge of the electron is expressed as $-|e|$. Simplify this expression and show that it can be written in the form

$$\hat{H} = \hat{H}_0 + \mu_B(\hat{\mathbf{L}} + 2\hat{\mathbf{S}})\cdot\mathbf{B} + \left(\frac{e^2}{8mc^2}\right)\sum_a(\mathbf{B}\times\mathbf{r}_a)^2, \tag{8.29}$$

where $\hat{H}_0$ is the atomic Hamiltonian in the absence of the magnetic field, $\hbar\hat{\mathbf{L}} = \sum \mathbf{r}_a\times\mathbf{p}_a$ is the total angular momentum of the system, and $\mu_B \equiv (|e|\hbar/2mc)$ is the Bohr magneton.

(b) To the lowest order this will induce an energy-level splitting given by

$$\Delta E = \mu_B B(\langle L_z\rangle + 2\langle S_z\rangle) = \mu_B B(\langle J_z\rangle + \langle S_z\rangle). \tag{8.30}$$

Show that this can be expressed in the form

$$\Delta E = g\mu_B B M_J, \tag{8.31}$$

where

$$g = 1 + \frac{J(J+1) - L(L+1) + S(S+1)}{2J(J+1)}, \tag{8.32}$$

and J, L, S, and M_J denote the eigenvalues of total angular momentum, orbital angular momentum, spin, and the z component of the total angular momentum, respectively. {Hint: In expression (8.30) we can replace $\langle J_z\rangle$ with the eigenvalue M_J. To determine the average value of S_z we note that $\langle \mathbf{S}\rangle$ must be in the direction of **J** in the formal sense, giving $\langle \mathbf{S}\rangle = k\mathbf{J}$. Hence $\langle S_z\rangle = kJ_z = kM_J$; further, $\langle \mathbf{S}\cdot\mathbf{J}\rangle = \langle k\mathbf{J}^2\rangle = kJ(J+1)$. Combining these two, we can write $\langle S_z\rangle = M_J(\mathbf{J}\cdot\mathbf{S}/\mathbf{J}^2)$. Argue that we can replace $\mathbf{S}\cdot\mathbf{J}$ with $(1/2)[J(J+1) - L(L+1) + S(S+1)]$ to obtain the final answer.}

(c) The splitting in the above case is linear in the field. If, however, $S = L = 0$, this splitting vanishes and the lowest-order correction is due to the B^2-dependent term in the Hamiltonian. Compute the splitting in this case.

8.6 Solar Flares

Near the surface of the Sun, there exists several other energetic phenomena, all of which require release of some form of internal energy as radiation. Of particular interest are the *solar flares*, which are probably the most energetic events observed in the Sun. A typical large solar flare has a diameter of $L \approx 3 \times 10^9$ cm and a height of 2×10^9 cm, thereby occupying a volume of approximately 2×10^{28} cm^3. Such a solar flare releases an energy of $\sim 10^{32}$ ergs in a time scale of $\sim 10^3$ s. The thermal energy contained in the volume of the flare is $E_{\text{thermal}} \approx (3\rho_{\text{col}} L^2 / m_H) k_B T$, where ρ_{col} is the column density of matter and T is the corresponding temperature. This expression corresponds to any single component of gas, and if electrons and protons make equal contributions in a plasma, then the total thermal energy will be twice the above value. The column density varies in the range $(0.1-3 \times 10^{-6})$ gm cm^{-2} as we go from the chromosphere to the corona; the temperature T increases from 10^4 to 3×10^6 K in the same region. The maximum value of $\rho_{\text{col}} T$ occurs in the chromosphere, but even here the thermal energy is only $E_{\text{therm}} \approx 2 \times 10^{29}$ ergs, which is insufficient to power the flares. On the other hand, the energy in the magnetic field contained in the same region is $E_{\text{mag}} = (VB^2/8\pi) \approx 10^{27}(B/1\text{ G})^2$ ergs. Hence, if the magnetic field strength in the flare is approximately (300–1000) G, then the field energy will be adequate to supply the energy in the flare.

Although the energetics are satisfactory, it is also necessary to provide a mechanism by which the energy trapped in the magnetic field can be released. In the limit of infinite conductivity for a plasma, the magnetic field is frozen in the fluid and is transported along with the fluid particles. The dissipation of the magnetic field in a plasma, because of finite conductivity σ, operates at a time scale $\tau \approx 4\pi L^2 \sigma$ (see Vol. I, Chap. 9, Section 9.6). For a fully ionised plasma, the conductivity is given by

$$\sigma = \frac{2(2k_B T)^{3/2}}{Ze^2 m_e^{1/2} \pi^{3/2} \ln \Lambda} = 2.63 \times 10^{-2} \text{ s}^{-1} \frac{T^{3/2}}{Z \ln \Lambda}, \qquad (8.33)$$

where T is in degrees Kelvin and $\log \Lambda \approx 10$ (see Vol. I, Chap. 9, Section 9.3). Adopting this value, we can estimate the time scale of dissipation in a solar flare as $\tau \approx 5 \times 10^{11}$ s, which is significantly larger than the natural time scale of the flare. Hence we need a better mechanism for converting magnetic field energy into the kinetic energy of the plasma.

Two physical processes, called *magnetic buoyancy* and *magnetic reconnection*, can act as components of such a mechanism. The basic idea behind these processes is illustrated in Fig. 8.9. Figure 8.9(a) shows a magnetic flux tube inside the Sun. This is a piece of plasma approximately cylindrical in shape threaded by a strong magnetic field. It is shown below that such a plasma element will show a tendency to rise up from the interior of the Sun, stretching the field lines. This could eventually lead to a configuration like the one shown in Fig. 8.9(b),

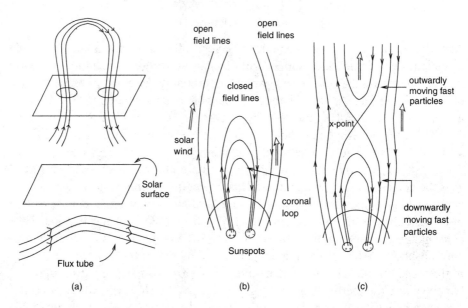

Fig. 8.9. Schematic diagram showing magnetic buoyancy and reconnection.

with an arclike structure containing the magnetic field and the plasma anchored in the base. When two such loops come close to each other, there arises a region in which the magnetic field direction reverses over a short distance. When this happens there will be a tendency for the magnetic field lines to annihilate each other, releasing the energy in the form of a flare. This process, which occurs within many different contexts, is called *magnetic reconnection*. In fact, magnetic reconnection can arise in more complex field configurations and lead to solar flares quite generically. We shall now consider some simple estimates for these two processes, starting with the magnetic buoyancy.

8.6.1 Magnetic Buoyancy

Consider a flux tube containing plasma with internal density n_i and located in a region with density n_o. When the plasma in the flux tube is in pressure balance with the outside matter, we must have $p_o = p_i$. Further, if the matter is in local thermodynamic equilibrium, then $T_0 = T_i = T$ (say). In the absence of a magnetic field, the equality of pressure and temperature will lead to the equality of densities $n_o = n_i$ and no bulk motion will occur. However, the situation is different when the flux tube is threaded by a strong magnetic field, contributing a pressure $(B^2/8\pi)$. In that case, the pressure-balance equation gives

$$2n_o k_B T = \frac{B^2}{8\pi} + 2n_i k_B T, \tag{8.34}$$

where it is assumed that the outside magnetic field is negligible small. The factor

2 for the gas pressure arises because both electrons and protons contribute to the pressure in the plasma, and n denotes the number density of the protons. This gives $n_o - n_i = (B^2/16\pi k_B T)$. The gravitational buoyancy force acting on this flux tube is therefore

$$F = (n_o - n_i) m_H g V = \frac{B^2 m_H g V}{16 \pi k_B T}, \qquad (8.35)$$

where V is the volume of the flux tube and g is the local acceleration that is due to gravity. The combination $H \equiv (2 k_B T / m_H)$ denotes the pressure scale height of the atmosphere in terms of which the buoyancy force can be written as

$$F = \frac{B^2 V}{8 \pi H}. \qquad (8.36)$$

When the flux tube rises through a distance of the order of H, the work done by this force, FH, will appear as kinetic energy:

$$\frac{1}{2} M u^2 = \frac{1}{2} \rho V u^2 = FH = \frac{B^2 V}{8\pi}. \qquad (8.37)$$

The resulting velocity of the flux tube is therefore

$$u = \left(\frac{B^2}{4\pi} \frac{V}{M} \right)^{1/2} = \left(\frac{B^2}{4\pi \rho} \right)^{1/2} = v_A, \qquad (8.38)$$

which is the same as the local Alfven speed. We thus find that a flux tube containing a strong magnetic field can rise through the atmosphere roughly at the Alfven speed. However, because it will be tied to the material of the solar atmosphere at its extremities, it is natural for looplike structures to be formed in such a system. For example, when a magnetic tube rises to the surface, its edges anchored on the solar atmosphere can give rise to the sunspots. The polarity of the sunspots arises from the direction of the magnetic field in the magnetic tube. It must be emphasised that magnetic buoyancy is a very general feature and arises in a host of astrophysical systems and is not confined to solar flares.

8.6.2 Magnetic Reconnection

We shall next see how the magnetic field energy in the coronal loops can be converted into the energy released in a solar flare. This is possibly done through a process called *magnetic reconnection*, which is also a general phenomenon that can occur in many different contexts in plasmas threaded by magnetic fields. The generic configuration in which magnetic reconnection occurs is shown in Fig. 8.10. Near the origin of the xy plane, the direction of the magnetic field reverses over a length scale L_y (say). For $y > 0$, the field lines are pointing more or less towards the positive x axis whereas for $y < 0$ they are pointing essentially towards the negative x axis. The fluid motion along the y direction is such as to

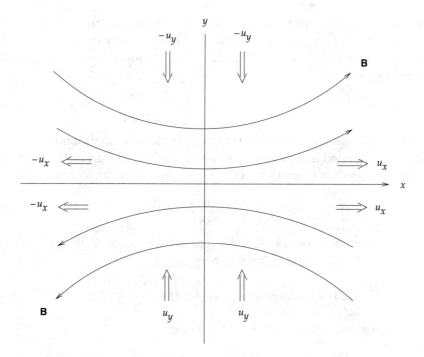

Fig. 8.10. The geometry for magnetic reconnection.

bring the oppositely directed magnetic field lines towards each other. It is clear from the geometry that there will exist a very large current,

$$\mathbf{j}_e = (c/4\pi)\nabla \times \mathbf{B}, \tag{8.39}$$

in the neighbourhood of the $y = 0$ plane. This is true even if B is small, provided that L_y is significantly smaller. Such a large current will lead to ohmic dissipation of the magnetic field energy at a rate, per unit volume, given by

$$\frac{|\mathbf{j}|^2}{\sigma} = \frac{\eta}{4\pi}|\nabla \times \mathbf{B}|^2 \cong \frac{\eta}{4\pi}\frac{B^2}{L_y^2}, \quad \eta \equiv \frac{c^2}{4\pi\sigma}. \tag{8.40}$$

This energy will accelerate the charged particles in the plasma, thereby energising the flare. We shall try an ansatz in which the velocity field along the x axis is away from the origin, as shown in Fig. 8.10, and make an estimate of different quantities.

We shall assume that the motion is sufficiently slow for the fluid to be treated as incompressible. (This assumption can be verified a posteriori.) Let the flow along the x axis gather momentum over a distance of the order of $x \approx \pm L_x$. From Bernouli's theorem applied for motion along the x axis, we have

$$\frac{1}{2}u_x^2 + \frac{P}{\rho} = \text{constant}. \tag{8.41}$$

Taking $u_x = 0$ at $x = 0$, the speed at $x = \pm L_x$ is given by $u_x = (2\Delta P/\rho)^{1/2}$, where ΔP is the pressure drop across the length scale L_x. Because the approximate force balance along the y axis suggests that

$$P + \frac{B^2}{8\pi} = \text{constant}, \quad (8.42)$$

the pressure drop has to be

$$\Delta P \cong \frac{B^2}{8\pi} \quad (8.43)$$

over a distance of the order of L_x along the x axis for the matter to be squirted along this axis. This shows that

$$u_x = (2\Delta P/\rho)^{1/2} = v_A; \quad (8.44)$$

that is, the plasma is squirted out with a speed equal to the local Alfven speed.

In steady state, mass conservation requires that the vertical inward flow along the y axis be compensated for by the horizontal outward flow along the x axis for every strip of width Δz along the z axis. Hence

$$2\rho u_x(2L_y \Delta z) = 2\rho u_y(2L_x \Delta z), \quad (8.45)$$

giving

$$u_y = (L_y/L_x)u_x = (L_y/L_x)v_A. \quad (8.46)$$

This relation also shows that the rate at which the reconnection of field lines occurs, decided by u_y, is smaller than the Alfven speed by a factor (L_y/L_x).

Energy conservation gives yet another relation among the various parameters. The rate at which ohmic dissipation produces heat in a volume $L_x L_y \Delta z$ will be $(\eta B^2/4\pi L_y)L_x \Delta z$, and the rate of annihilation of magnetic field energy that is due to fluid motion along the y axis is $(B^2/8\pi)u_y L_x \Delta z$. Equating these two expressions, we obtain

$$u_y = \frac{2\eta}{L_y}. \quad (8.47)$$

Eliminating L_y between Eqs. (8.46) and (8.47) we find that the rate of reconnection is given by

$$u_y = 2Re_{\text{mag}}^{-1/2} v_A, \quad (8.48)$$

where Re_{mag} is the magnetic Reynolds number of the plasma (see Vol. I, Chap. 9, Section 9.6):

$$Re_{\text{mag}} \equiv \frac{2L_x u_x}{\eta} = \frac{2L_x v_A}{\eta}. \quad (8.49)$$

The corresponding thickness of the neutral sheet is $L_y = (2L_x/Re_{\text{mag}}^{1/2})$. This rate in Eq. (8.48) is significantly faster than the diffusive dissipation of energy over a length scale $L \approx L_x$, which is governed by $\tau \approx 4\pi\sigma L^2$. The diffusive velocity for this process is $v \approx (L/\tau) \approx (4\pi\sigma L)^{-1} \approx (v_A/Re_{\text{mag}})$. Clearly, magnetic reconnection is faster than the standard plasma dissipation by a factor $Re_{\text{mag}}^{-1/2}$.

Unfortunately, this gain is not sufficient to explain the energetics involved in a solar flare. Taking $L_x = 10^9$ cm, $B = 300$ G, $n = 10^{10}$ cm^{-3}, and $T = 2 \times 10^6$ K, we find that the Alfven speed works out to $v_A \approx 6 \times 10^8$ cm s^{-1}. In the solar context $Re_{\text{mag}} \gtrsim 10^{14}$, implying that the reconnection takes place at a speed of the order of $10^{-6} v_A$. The characteristic reconnection time scale $\tau_R \approx (L_x/v_A) \approx 1$ s. The energy in the neutral sheet is approximately 3×10^{23} ergs, leading to a luminosity of $L_{\text{flare}} \approx 3 \times 10^{23}$ ergs s^{-1}. The energy contained within the flare is $E_{\text{flare}} \approx 3 \times 10^{30}$ ergs, which can be liberated in a time scale $\tau_{\text{flare}} \approx (E_f/L_f) \approx 10^7$ s. This rate, which corresponds to months, is slower than the rate required for explaining the observations, which suggests that flares take place at a time scale of a few hours.

The difficulty can be traced to the fact that the reconnection rate was $(v_A/Re_{\text{mag}}^{1/2})$. The factor $Re_{\text{mag}}^{1/2}$, however, is crucially dependent on the geometry of the reconnection region and the boundary condition. By modifying these conditions, we can obtain considerably slower functions of Re_{mag} in the denominator, for example, $\log Re_{\text{mag}}$. In such a case, the reconnection rate can be significantly higher and the solar flare can be explained by the existence of such reconnecting regions. These results, while plausible, are highly model dependent.

Exercise 8.3
Electric-field acceleration in the reconnection region: The reconnection of the magnetic field will produce large values for $\nabla \times \mathbf{B}$ and $(\partial \mathbf{E}/\partial t)$ in localised regions. The resulting electric field can be a source of acceleration for the electrons. Model the motion of the electron by an equation of the form $m_e \dot{\mathbf{v}} = -e\mathbf{E} - \gamma m_e \mathbf{v}$, where γ is the rate of collisions which acts as a damping term. Show that a runaway situation results if the electric field is larger than a critical value (known as the *Dreicer field*) that is given by $E_d \approx (2/3)(e/\lambda_D^2) \ln \Lambda$, where λ_D is the Debye length of the plasma. Estimate E_d for a typical solar flare.

Exercise 8.4
X-ray emission from solar flares: The plasma in the solar flare will emit bremsstrahlung radiation in the x-ray band that will be nonthermal in character because the electron distribution could possibly be a power law rather than a Maxwellian. A simple model for this process can be constructed based on the following assumptions. (a) Nonrelativistic electrons are injected into the flare with a spectrum that varies with energy E as E^{-x}. (b) They lose energy (predominantly) through ionisation so that the rate of loss of energy of any single electron is inversely proportional to its velocity. Show that these assumptions, along with the standard expression for bremsstrahlung radiation, lead to a number spectrum of emitted photons $n(\omega) \propto \omega^{\alpha_x} \propto \omega^{-(x-1)}$. [Also show that, if the

ionisation losses are ignored, then the x-ray spectrum of the photons has the power-law index $\alpha_x = -(x + 1/2)$.] On the other hand, the same electrons gyrating in the magnetic fields will lead to synchrotron radiation with a power-law index α_R. Ignoring ionisation losses, show that these two indices are related by $\alpha_R = 0.75 - 0.5\,\alpha_x$.

8.7 Generation of the Solar Magnetic Field

We shall now turn to the question of how the solar magnetic field is generated. It may appear at first sight that an axisymmetric magnetic field for the Sun can be maintained by a corresponding axisymmetric fluid flow in steady state. This is, however, not possible, as can be seen by the following argument.

Consider any plane passing through the axis of symmetry, like the one shown in Fig. 8.11. The magnetic-field lines on this meridional plane must be closed curves; hence there must exist a neutral point (marked N in the figure) that is encircled by all the field lines. Joining all the neutral points together will give a simple closed curve around the axis of symmetry. From Maxwell's equation $\nabla \times \mathbf{B} = (4\pi/c)\mathbf{j}$, it is clear that the component j_ϕ should be nonzero at any

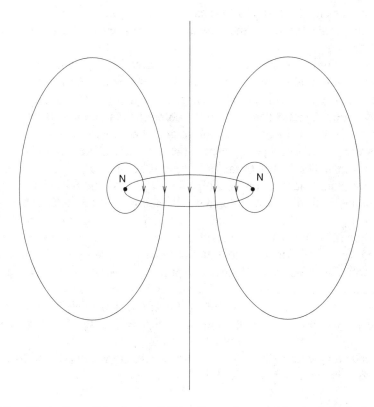

Fig. 8.11. Illustration of the impossibility of an axisymmetric steady-state fluid flow producing an axisymmetric magnetic field.

neutral point, where ϕ is the azimuthal angle around the axis of symmetry. Consider now the line integral of $\mathbf{j} = j_\phi \mathbf{e}_\phi$ along the closed curve passing through the neutral points. Using the relation

$$\mathbf{j} = \sigma\left(\mathbf{E} + \frac{\mathbf{v}}{c} \times \mathbf{B}\right), \qquad (8.50)$$

we get

$$\oint j_\phi \, dl = \sigma\left(\oint \mathbf{E} \cdot d\mathbf{l} + \frac{1}{c} \oint \mathbf{v} \times \mathbf{B} \cdot d\mathbf{l}\right). \qquad (8.51)$$

Because $\mathbf{B}$ can have a ϕ component at only the neutral point, the magnetic field is parallel to $d\mathbf{l}$, making the second term on the right-hand side vanish identically. The first term on the right-hand side can be converted to the surface integral of $\partial \mathbf{B}/\partial t$, which vanishes in steady state. Thus the right-hand side of Eq. (8.51) vanishes identically, but the left-hand side is equal to the product of j_ϕ and the length of the closed loop along the neutral points and hence is nonzero. This clearly shows that we cannot have a purely axisymmetric steady-state configuration that obeys the generalised Ohm's law to generate the magnetic field of the Sun.

A field configuration of the type $\mathbf{B} = B_\phi \mathbf{e}_\phi$ is called *toroidal*, whereas a magnetic field configuration with only r and θ components (and vanishing ϕ component) is called *poloidal*. The above argument shows that, in order to maintain a steady-state magnetic field, a toroidal component alone will not do. A more complicated scheme for generating the solar magnetic field requires a configuration in which the poloidal component of the field generates a toroidal component and vice versa. The first part can be achieved because of the differential rotation of the Sun. The rotation of the plasma will drag the poloidal component differentially and will stretch it because the Sun is rotating faster at the equator than at the poles. If there are turbulent convective motions inside the Sun, then the radially moving plasma blobs will also stretch the toroidal field in the radial direction. In such a case, turbulent motions can generate a poloidal field to produce a poloidal component. If these two processes coexist, then it is possible to generate a self-sustaining scenario for the solar magnetic field.

Because we need to study turbulent fluid flow, it is convenient to treat the fluid velocity as being made of two parts with $\mathbf{v} = \bar{\mathbf{v}} + \mathbf{v}'$, where $\bar{\mathbf{v}}$ is the mean velocity field and $\mathbf{v}'$ is the turbulent fluctuating part. Corresponding to such a motion, the magnetic field can also be separated into mean ($\bar{\mathbf{B}}$) and fluctuating ($\mathbf{B}'$) parts. Averaging Ohm's law [Eq. (8.50)], we get

$$\bar{\mathbf{j}} = \sigma\left(\bar{\mathbf{E}} + \frac{1}{c}\bar{\mathbf{v}} \times \bar{\mathbf{B}} + \frac{1}{c}\mathcal{E}\right), \qquad (8.52)$$

8.7 Generation of the Solar Magnetic Field

where

$$\mathcal{E} = \overline{\mathbf{v}' \times \mathbf{B}'}. \tag{8.53}$$

The extra term shows that when the fluctuating velocity field is correlated with the magnetic field in a nontrivial fashion, then an extra term arises in Ohm's law, thereby invalidating the original argument regarding the impossibility of an axisymmetric steady-state solution. The central equation governing the evolution of magnetic field now is modified to

$$\frac{\partial \bar{\mathbf{B}}}{\partial t} + \frac{\partial \mathbf{B}'}{\partial t} = \nabla \times (\bar{\mathbf{v}} \times \bar{\mathbf{B}} + \mathbf{v}' \times \bar{\mathbf{B}} + \bar{\mathbf{v}} \times \mathbf{B}' + \mathbf{v}' \times \mathbf{B}') + \lambda \nabla^2 (\bar{\mathbf{B}} + \mathbf{B}'), \tag{8.54}$$

where $\lambda = (c^2/4\pi\sigma)$. Averaging this equation, we get

$$\frac{\partial \bar{\mathbf{B}}}{\partial t} = \nabla \times (\bar{\mathbf{v}} \times \bar{\mathbf{B}}) + \nabla \times \mathcal{E} + \lambda \nabla^2 \bar{\mathbf{B}}, \tag{8.55}$$

which governs the evolution of the mean magnetic field in the presence of an extra term characterised by $\mathcal{E}$.

To proceed further, we need to estimate $\mathcal{E}$, which, unfortunately, cannot be done from fundamental principles. One possible, approximate, method is to evaluate $\mathcal{E}$ as follows. Subtracting Eq. (8.55) from Eq. (8.54), we see that the fluctuating part of the magnetic field evolves according to

$$\frac{\partial \mathbf{B}'}{\partial t} = \nabla \times (\mathbf{v}' \times \bar{\mathbf{B}} + \bar{\mathbf{v}} \times \mathbf{B}' + \mathbf{v}' \times \mathbf{B}' - \mathcal{E}) + \eta \nabla^2 \mathbf{B}' \approx \nabla \times (\mathbf{v}' \times \bar{\mathbf{B}}). \tag{8.56}$$

In arriving at the last equality, we have ignored all the terms that are linear in $\mathbf{B}'$ and have retained the leading-order term. If the coherence time of the fluctuating velocity field is τ, then the typical magnetic-field fluctuation, generated in this time scale, is given by

$$\mathbf{B}' \approx \tau \nabla \times (\mathbf{v}' \times \bar{\mathbf{B}}). \tag{8.57}$$

We can now use this expression in the definition of $\mathcal{E}$ to make an estimate of this quantity. An elementary calculation gives

$$\mathcal{E}_i = \alpha_{ij} \bar{\mathbf{B}}_j + \beta_{ijk} \frac{\partial \bar{\mathbf{B}}_k}{\partial x_j}, \tag{8.58}$$

with

$$\alpha_{ij} = \epsilon_{ilk} \overline{v'_l \frac{\partial v'_k}{\partial x_j}} \tau, \quad \beta_{ijk} = -\epsilon_{ilk} \overline{v'_l v'_j} \tau. \tag{8.59}$$

These expressions become simpler if the turbulence is isotropic so that it does not contain any preferred vectorial quantities. In that case, we must have, from symmetry,

$$\alpha_{ij} = \alpha \delta_{ij}, \quad \beta_{ijk} = -\lambda_T \epsilon_{ijk}, \qquad (8.60)$$

where

$$\alpha = -\frac{1}{3}\overline{\mathbf{v}' \cdot (\nabla \times \mathbf{v}')}\,\tau, \quad \lambda_T = \frac{1}{3}\overline{\mathbf{v}' \cdot \mathbf{v}'}\,\tau, \qquad (8.61)$$

leading to

$$\mathcal{E} = \alpha \bar{\mathbf{B}} - \lambda_T \nabla \times \bar{\mathbf{B}}. \qquad (8.62)$$

Substituting this into Eq. (8.55), we get the evolution equation for a turbulent, magnetoactive plasma as

$$\frac{\partial \bar{\mathbf{B}}}{\partial t} = \nabla \times (\bar{\mathbf{v}} \times \bar{\mathbf{B}}) + \nabla \times (\alpha \bar{\mathbf{B}}) + (\lambda + \lambda_T)\nabla^2 \bar{\mathbf{B}}. \qquad (8.63)$$

The effect of turbulence is to add two additional terms: a term $\lambda_T \nabla^2 \bar{\mathbf{B}}$ that leads to an extra diffusion of the magnetic field that is due to turbulence and, further, a new source term for the magnetic field in the form $\nabla \times (\alpha \bar{\mathbf{B}})$. It turns out that this equation can have steady-state axisymmetric magnetic-field configurations as solutions. To see this more concretely, we will solve this equation in a simplified context.

To begin with, we can ignore λ compared with λ_T, as in most astrophysical contexts the turbulent diffusion acts as the main dissipative term. In that case, our equation reduces (in a slightly simplified notation) to

$$\frac{\partial \mathbf{B}}{\partial t} = \nabla \times (\mathbf{v} \times \mathbf{B}) + \nabla \times (\alpha \mathbf{B}) + \lambda_T \nabla^2 \mathbf{B}. \qquad (8.64)$$

Further, we shall solve this equation in a locally Cartesian coordinate system rather than in spherical coordinates. Taking the x axis to be radially outwards, the y axis to be in the toroidal (ϕ direction), and the z axis in the direction of increasing latitude, we can provide a local coordinate grid in the northern hemisphere of the Sun. In such a coordinate system, axisymmetry implies that all quantities should be independent of y. For the magnetic field we will choose the ansatz

$$\mathbf{B} = B_y(x,z)\hat{\mathbf{e}}_y + \nabla \times [A(x,z)\hat{\mathbf{e}}_y], \qquad (8.65)$$

where the first term is the toroidal component of the magnetic field and the second one is the poloidal component. The velocity field of the plasma is essentially generated by the differential rotation of the Sun and hence can be modelled with just the y component: $\mathbf{v} = (0, v_y, 0)$. In general, the y component of the velocity can vary both in the radial direction (along the x axis) and along the latitude (along the z axis). To simplify the mathematics we shall, however, take it to vary along only the radial direction and write $\mathbf{v} = v_y(x)\mathbf{e}_y$. The shear for this velocity

8.7 Generation of the Solar Magnetic Field

field is given by $S = (\partial v_y/\partial x)$. Finally we shall make the approximation that α, λ_T, and S are approximately constant. In that case, ansatz Eq. (8.65) will satisfy Eq. (8.64), if the following conditions are satisfied:

$$\frac{\partial B_y}{\partial t} = -S\frac{\partial A}{\partial z} - \alpha \nabla^2 A + \lambda_T \nabla^2 B_y, \qquad (8.66)$$

$$\frac{\partial A}{\partial t} = \alpha B_y + \lambda_T \nabla^2 A. \qquad (8.67)$$

These equations clearly illustrate the physical origin of the magnetic field in our model. Equation (8.67) will become a plain diffusion equation if $\alpha = 0$ and any poloidal field will diffuse away. It is the additional term αB_y, which is essentially due to turbulence, that acts as a source to this poloidal field. Equation (8.66) shows how a toroidal field is generated because of the differential rotation [given by $(-S(\partial A/\partial z))$] as well as by turbulence (through the $-\alpha \nabla^2 A$ term). For an astrophysical situation with strong differential rotation, we can ignore $-\alpha \nabla^2 A$ term as compared with $-S(\partial A/\partial z)$. In that case, we need to solve Eq. (8.67) along with

$$\frac{\partial B_y}{\partial t} \cong -S\frac{\partial A}{\partial z} + \lambda_T \nabla^2 B_y. \qquad (8.68)$$

The magnetic-field configuration, which is a solution to these equations, is said to be generated by an $\alpha\Omega$ dynamo. (The nomenclature arises from the fact that the rotation of the star is usually denoted by the symbol Ω.) These equations are straightforward to solve with the standard ansatze

$$A = \hat{A}\exp(\omega t + ikz), \quad B_y = \hat{B}\exp(\omega t + ikz), \qquad (8.69)$$

which lead to the dispersion relation of the form

$$\omega = -\lambda_T k^2 \pm \left(\frac{i-1}{\sqrt{2}}\right)\sqrt{k\alpha S}. \qquad (8.70)$$

For the magnetic field to grow with time, the real part of ω must be positive. Irrespective of whether αS is positive or negative, this can arise only in the solution in which the negative sign is taken in Eq. (8.70). In that case, it is easy to see that the condition for the growth of dynamo is given by

$$\frac{|\alpha S|}{2k^3 \lambda_T^2} \geq 1. \qquad (8.71)$$

This result shows that, given a helical turbulence characterised by an α term and a differential rotation giving rise to the shear S, it is possible to have sustained growth of the magnetic field.

The eigenmodes are, however, different in the case of $\alpha S > 0$ and $\alpha S < 0$. This is most easily seen for the marginally stable mode corresponding to equality in relation (8.71). When $\alpha\mathcal{S} > 0$, this mode is given by

$$(A, B_y) \propto \exp\left[-i\left(\frac{k\alpha\mathcal{S}}{2}\right)^{1/2} t + ikz\right], \qquad (8.72)$$

which corresponds to a wave propagating towards the pole. In the case $\alpha\mathcal{S} < 0$, the corresponding modes are given by

$$(A, B_y) \propto \exp\left[i\left(\frac{k|\alpha\mathcal{S}|}{2}\right)^{1/2} t + ikz\right], \qquad (8.73)$$

which propagates towards the equator. There is some evidence for the equatorward propagation of the solar magnetic field, which suggests that $\alpha\mathcal{S}$ must be negative in the northern hemisphere of the Sun.

8.8 Solar Wind

We saw in Chap. 3 there can be mass loss from the surface of a star in the form of a stellar wind. In the case of the Sun, such a wind will be detectable in the solar system and is studied extensively.

The spherically symmetric flow of a nondissipative fluid in steady state, under the gravitational force of the Sun, can be described by the equations

$$4\pi r^2 \rho v = \text{constant}, \qquad \rho v \frac{dv}{dr} = -\frac{dp}{dr} - \frac{GM\rho}{r^2}. \qquad (8.74)$$

Eliminating p and ρ, we can rewrite this in the form

$$\left(v - \frac{c_s^2}{v}\right)\frac{dv}{dr} = \frac{2c_s^2}{r} - \frac{GM}{r^2}, \qquad (8.75)$$

where $c_s = (\gamma p/\rho)^{1/2}$ is the local sound speed, which is a function of r. If the equation of state is taken to be $p \propto \rho^\gamma$ with $\gamma \neq 1$, these equations can be integrated to give

$$\frac{1}{2}v^2 + \frac{c_s^2}{\gamma - 1} - \frac{GM}{r} = C, \qquad (8.76)$$

where C is a constant of integration. This result was derived and discussed in detail in Vol. I, Chap. 8, Subsection 8.9.2. In the case of solar wind, however, the isothermal assumption with $p = \rho\mathcal{R}T$ is more relevant, which makes Eq. (8.76) invalid. Instead, Eq .(8.75) now integrates to

$$\frac{v^2}{c_s^2} - \log\frac{v^2}{c_s^2} = 4\log\frac{r}{r_c} + \frac{2GM}{rc_s^2} + \text{constant}, \qquad (8.77)$$

where $r_c \equiv (GM/2c_s^2)$ is the critical radius at which the velocity of flow is equal to the isothermal sound speed $c_s \equiv (k_B T)^{1/2}$, which is now a constant independent of r. We can evaluate the constant of integration by using $v = c_s$ at $r = r_c$, and we get its numerical value as 3. Substituting back and rearranging the terms, we can write the solution as

$$\frac{\mathcal{M}^2}{2} - \ln \mathcal{M} = \frac{2r_c}{r} + 2\ln\frac{r}{r_c} - \frac{3}{2}, \tag{8.78}$$

where $\mathcal{M} = (v/c_s)$ is the Mach number. The relevant solution is the one in which v is small near $r = R_\odot$ and increases to supersonic values for $r \gg r_c$. At such large radii, the first and the third terms of this equation balance each other, giving

$$v \approx 2c_s\sqrt{[\log(r/r_c)]}. \tag{8.79}$$

The corresponding pressure and the density fall as

$$p \propto \rho \propto \frac{1}{r^2\sqrt{(\log r)}} \tag{8.80}$$

at large distances.

At the base of the corona, $v \ll c_s$, and we can ignore the first term on the left-hand side of Eq. (8.78) and obtain

$$\dot{M} = 4\pi R_\odot^2 \rho_b v(R_\odot) = 4\pi R_\odot^2 \rho_b c_s \mathcal{M}(R_\odot) \approx 4\pi r_c^2 \rho_b c_s \exp\left(\frac{3}{2} - \frac{2r_c}{R_\odot}\right). \tag{8.81}$$

At the base of the corona, $\rho \approx 10^{-15}$ gm cm^{-3} and $c_s \approx 1.3 \times 10^7$ cm s^{-1}, giving $\dot{M} \approx 1.4 \times 10^{12}$ gm s$^{-1} \approx 2 \times 10^{-14} M_\odot$ yr^{-1}. From the solution we can now estimate v and ρ at the location of the Earth at $r = 1$ AU $= 1.5 \times 10^{13}$ cm. This gives $\mathcal{M} \approx 3.8$ and $v \approx 500$ km s^{-1}. From the conservation of $\dot{M}$, it follows that the density of particles in a solar wind at the location of the Earth is $\rho \approx 10^{-23}$ gm cm^{-3}. This matches reasonably well with the observed features of the solar wind.

The above description does not take into account the magnetic-field structure of the Sun. Because the corona and solar wind are made of ionised plasma, magnetic-field lines will be frozen to the fluid. Because the Sun rotates about its axis, the magnetic-field lines will be twisted around in the coronal region. We shall now describe breifly the modifications that arise because of the existence of a magnetic field.

In the lowest order of approximation, we can ignore the effect of magnetic field on the flow of the plasma and assume that the field lines are dragged by the fluid. In that case, the field lines will end up in a configuration as shown schematically in Fig. 8.12. At any given location, the component of $\mathbf{v}$ perpendicular to magnetic field $\mathbf{B}$ must be equal to the speed of the field line in that direction. At a given radius R, the field line rotates at a speed $\Omega(R - R_\odot)$. Because the velocity component perpendicular to the field is $v \sin \psi$ and the

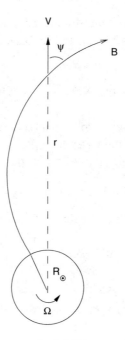

Fig. 8.12. A spiralling magnetic-field line attached to the rotating Sun.

speed of the magnetic-field component is $\Omega(R - R_\odot)\cos\psi$, we get the condition $v\sin\psi = \Omega(R - R_\odot)\cos\psi$ or

$$\tan\psi = \frac{\Omega(R - R_\odot)}{v}. \tag{8.82}$$

At the solar surface, ψ vanishes and the field line is radially outwards. At $R \approx 1$ AU, the inclination is $\psi \approx 45°$. If we take v to be a constant, then the equation to the field line is given by $\phi = \Omega t = \Omega(R_\odot - R)/v$ so that $R = R_\odot - (v\phi/\Omega)$, which represents an *Archimedes spiral*.

To the next order of approximation, we have to take into account the effect of the magnetic field on the fluid flow. To do this, we have to solve the steady-state MHD equation for both B and v consistently. In this particular context, we shall take the magnetic field and velocity to be of the form

$$\mathbf{B} = [B_r(r), 0, B_\phi(r)], \quad \mathbf{v} = [v_r(r), 0, v_\phi(r)]. \tag{8.83}$$

The equation $\nabla\cdot\mathbf{B} = 0$, corresponding to the flux conservation, determines B_r as

$$B_r = \frac{B_\odot R_\odot^2}{r^2}. \tag{8.84}$$

The conservation of mass, $\nabla\cdot(\rho\mathbf{v}) = 0$, implies the condition

$$\rho v_r r^2 = \text{constant}. \tag{8.85}$$

8.8 Solar Wind

The evolution of the magnetic field in an infinitely conductive plasma in steady state is determined by the simple equation $\nabla \times (\mathbf{v} \times \mathbf{B}) = 0$. In the present case, this can be integrated to give

$$r(v_r B_\phi - v_\phi B_r) = \text{constant} \equiv C. \tag{8.86}$$

Using the boundary conditions $B_\phi = 0$ and $v_\phi = \Omega R_\odot$ at $r = R_\odot$ we can determine the constant as $C = -\Omega R_\odot^2 B_\odot = -\Omega r^2 B_r$, where the second equality follows from flux conservation. Using this in Eq. (8.86) we find that

$$\frac{B_\phi}{B_r} = \frac{v_\phi - r\Omega}{v_r}, \tag{8.87}$$

which is a condition for the magnetic field to flow along with the fluid, conserving the flux. To determine the actual variations of the field and fluid velocities along the radial direction, we have to use the equation of motion for the fluid, which can be written in the form

$$\rho(\mathbf{v} \cdot \nabla)\mathbf{v} = -\nabla\left(p + \frac{B^2}{8\pi}\right) + \mathbf{B} \cdot \nabla \frac{\mathbf{B}}{4\pi} - \frac{GM_\odot \rho}{r^2}\hat{\mathbf{r}}. \tag{8.88}$$

On the right-hand side, the first term gives the gradient of total pressure made of ordinary fluid pressure p and the magnetic pressure $(B^2/8\pi)$; the second term is the magnetic tension of the field lines, and the third term is the gravitational force that is due to the Sun. The ϕ component of this equation is

$$\rho(\mathbf{v} \cdot \nabla)v_\phi = (\mathbf{B} \cdot \nabla)\frac{B_\phi}{4\pi} \tag{8.89}$$

or

$$4\pi \rho v_r \frac{d}{dr}(rv_\phi) = B_r \frac{d}{dr}(r B_\phi). \tag{8.90}$$

However, from the conservation law for magnetic flux and mass it follows that $(\rho v_r / B_r)$ is a constant, allowing this equation to be integrated to give

$$rv_\phi - \frac{B_r}{4\pi \rho v_r} r B_\phi = \text{constant} \equiv L. \tag{8.91}$$

This quantity represents the total angular momentum per unit mass carried away by the solar wind that is due to the torque's arising from the $\mathbf{J} \times \mathbf{B}$ force on the solar surface. Substituting for B_ϕ from Eq. (8.87), we can express the tangential velocity in the form

$$v_\phi = \Omega r \frac{M_A^2 L/(r^2 \Omega) - 1}{M_A^2 - 1}, \tag{8.92}$$

where $M_A = v_r (4\pi \rho / B_r^2)^{1/2}$ is called the *Alfven–Mach number* and represents the ratio between the radial velocity and the Alfven velocity. For v_ϕ to remain finite at the Alfven critical point $r = r_A$ corresponding to $M_A = 1$, it is

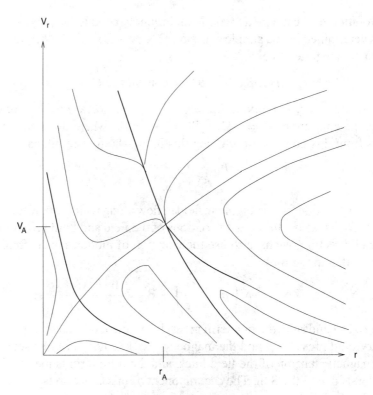

Fig. 8.13. The radial velocity for solar wind in the presence of a magnetic field.

necessary that $L = \Omega r_A^2$. Equation (8.84) determines B_r, and Eqs. (8.87) and (8.91) determine B_ϕ and v_ϕ in terms of v_r. To solve the problem completely, we need to determine only $v_r(r)$. This is governed by the radial part of Eq. (8.88), which reads

$$\rho v_r \frac{dv_r}{dr} - \frac{\rho v_\phi^2}{r} = -\frac{dp}{dr} - \frac{B_\phi}{4\pi r}\frac{d}{dr}(rB_\phi) - \frac{GM_\odot \rho}{r^2}. \quad (8.93)$$

Given an equation of state connecting p and ρ and the mass conservation law $\rho r^2 v_r = $ constant, this equation can be integrated to determine $v_r(r)$. The nature of the solutions to this equation is sketched in Fig. 8.13. In this case, there are three critical points, and the outer two are very close to each other. The solar-wind solution passes through all the critical points.

The above analysis also clarifies the terminology of supersonic motion as applied to a solar wind. In conventional gas dynamics, the flow is considered supersonic if the speed is larger than sound speed $c_s \approx (k_B T/m)^{1/2}$. In the case of solar wind, with $T \approx 10^6$ K, $c_s \approx 170$ km s^{-1}. Thus the solar wind is definitely supersonic compared with normal sound speed. However, in the case in which a magnetic field contributes significantly to the energy density, then the relevant speed of propagation for disturbances is the Alfven velocity $v_A = B/(4\pi\rho)^{1/2}$.

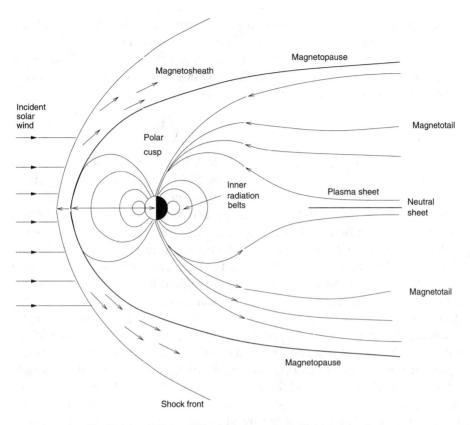

Fig. 8.14. Effect of Earth's magnetic field on solar wind.

In the case of a solar wind, the Alfven speed works out to ~ 35 km s^{-1}, and hence the solar wind is also supersonic with respect to the propagation speed of disturbances in the magnetised plasma.

The above discussion concentrated on the effect of a solar magnetic field on the wind. As the wind approaches Earth, it is also affected by the Earth's magnetic field. In this case, the energy density of the Earth's magnetic field acts as an obstacle to the flow of the plasma in a solar wind and leads to a formation of a bow shock, very much like that in the case of a spherical obstacle moving supersonically through a fluid. The overall structure of the solar wind when affected by Earth's magnetic field is as shown in Fig. 8.14. For purposes of visualisation, the surface marked *magnetopause* can be taken to be the surface of a solid obstacle. The plasma flows around this, and the bow shock is located at a standoff distance slightly ahead of this surface. The magnetic field in the region facing the solar wind is squashed inwards while the field in the direction away from the solar wind is stretched out and dragged along the plasma.

Although the above picture is intuitively clear, there are several subtleties that need to be pointed out in this context. In the case of normal fluid shocks (discussed in Vol. I, Chap. 8, Section 8.11), particle collisions play a vital role in converting steady-streaming motion into heat. In such a situation, the size of a shock front is essentially determined by the mean free path of the fluid. In the case of plasma in a solar wind, the collisional time scale for momentum transfer by proton–proton collisions is given by

$$t_c(p-p) = 11.7\left(\frac{T^{3/2}}{n_p \ln \Lambda}\right) \text{ s}, \qquad (8.94)$$

where all quantities are in cgs units and $\ln \Lambda \approx 40$ for the solar wind. With $T \approx 10^6$ K and $n_p \approx 5$ cm^{-3}, this time scale is $t_c \approx 6 \times 10^7$ s. For a plasma velocity of $v \approx 5 \times 10^7$ cm s^{-1}, the mean free path is $vt_c \approx 3 \times 10^{15}$ cm. This is far larger than the size of the magnetosphere of the Earth. It is clear that direct particle collisions play no role in the formation of the shock in this particular case.

This is a general phenomenon that happens whenever magnetic fields are dominant. In the presence of a magnetic field, particles in a plasma are forced to gyrate around the field lines and follow the field lines. In such a case, the collisional mean free path is not the relevant parameter that determines the region over which energy and momentum are transferred. The relevant parameter is usually the gyroradius of the particle in the magnetic field. To see in detail the role played by the magnetic field, consider the motion of any one proton incident upon the $x = 0$ plane when there is zero magnetic field for $x < 0$ and a field of approximately $B = 10^{-3}$ G for $x > 0$. (This is similar to the values involved in the case of Earth's magnetosphere.) For this field strength, the gyrofrequency is $\omega_L = (eB/m_p c) \approx 10$ Hz; if the speed of the proton is $v = 5 \times 10^7$ cm s^{-1}, the gyroradius is $r_L = (B/\omega_L) \approx 5 \times 10^6$ cm. This length scale is much smaller than the size of the magnetosphere. In the idealised case, such a proton will make half a circular orbit in the magnetic-field regime and will exit back to the field-free region; that is, the magnetic field will reflect back the proton completely. The electrons will also undergo a similar effect but with a much smaller radius of gyration. Such a magnetic reflection will produce a substantial charge separation and potential difference that, of course, cannot be sustained in the presence of a copious number of free charged particles. Hence the interface between the two regions will be unstable to a generation of several plasma instabilities, and the local electromagnetic fields will violently scatter the ordered stream of particles in the solar wind. This will cause a significant transfer of energy and momentum between the particles over a region of the size of the gyroradius. It is this process that leads to a collisionless shock in the presence of magnetic fields.

Although the details regarding the origin of such shock fronts are quite different from those of a normal fluid shock, the mathematical description of flow across the shock front remains essentially the same. We again can study shocks in the presence of a magnetic field by using the continuity equations for energy, momentum, etc., taking into account the contribution from the magnetic field. In general, the tangential components of the fluid flow remain continuous whereas the normal components are governed by the continuity equation. The density enhancements are also typically by a factor of 2–4 in the case of a solar wind that is flowing past the magnetopause.

The overall size of the solar wind is determined by the balance of the ram pressure $P \approx \rho v^2 \approx 10^{-8}(r/1\,\text{AU})^{-2}$ dyn cm^{-2} of the wind and the pressure $P = nk_B T + (B^2/8\pi) \approx 3 \times 10^{-13}$ dyn cm^{-2} of the ambient ISM. Equating the two, we find that $r \approx 158$ AU. Closer to the planet, the size of the magnetopause is determined by the balance between the ram pressure of the solar wind and the pressure of the planet's magnetic field $(B^2/8\pi)$, with $B = B_0(R/r)^3$, where R is the radius of the planet. This pressure balance gives $(r/R) \approx 13(B/1\,\text{G})^{1/3}(a/1\,\text{AU})^{1/3}$, where a is the distance of the planet from the Sun. For Earth, $B_0 = 0.3$ G and $a = 1$ AU, giving $r \simeq 9R$, whereas for Jupiter, $B_0 = 10$ G and $a = 5$ AU, giving $r \simeq 50\,R$.

The distorted magnetic-field structure around the Earth, shown in Fig. 8.14, leads to several interesting physical phenomena. To begin with, the magnetic field prevents the solar-wind plasma and other charged particles from hitting the Earth directly. Instead, the particles spiral around the magnetic-field lines, which are stronger near the poles and weaker near the equator. Such a constant gyration of particles trapped in a magnetically reflecting structure (as discussed in Vol. I, Chap. 3, Section 3.8) leads to the existence of regions with a high concentration of charged particles known as *Van Allen radiation belts*. The innermost belt is composed of protons and is at a height of ~4000 km above the Earth's surface. This also contains a region made of atomic nuclei that could have been part of the ISM at some stage. The outermost belt is composed of electrons at an altitude of ~16,000 km. Highly energetic particles confined in these belts can enter the Earth's atmosphere near the pole and strike atoms and molecules, causing collisional excitation, ionisation, and dissociation. When the atoms or molecules undergo recombination or deexcitation, they emit radiation that leads to the physical phenomena called the *aurora borealis* and the *aurora australis*. In particular, the excitation of oxygen atoms can lead to the green 5580 Å and the red 6300 Å lights, which are characteristic of the auroras. The Earth is also a source of very long kilometre wave radiation, the intensity of which increases with the occurrence of polar auroras.

The magnetic fields of other planets and their interaction with the surrounding plasma, notably solar wind, lead to similar phenomena. For example, Jupiter is the strongest radio emitter of all the planets. In the wavelength range of 100 MHz to 10 GHz, this radiation is essentially due to synchrotron emission from high-energy electrons trapped in Jupiter's radiation belts. In this particular case, one of the closest satellites to Jupiter, called Io, acts as a strong source of ionised particles. Io is tidally affected by Jupiter very strongly and the dissipation of mechanical energy in Io's interior arising from tidal friction acts as a continuous source of heat for the satellite, leading to a great deal of volcanic activity. The volcanic eruption throws out heavy elements like sulphur and oxygen into Io's atmosphere, where they are ionised by the Sun's UV radiation. This forms an ion torus around Jupiter and injects charged particles into Jupiter's magnetosphere. There are magnetic-field lines connecting Io and Jupiter, and the spiralling of charges along these lines leads to a steep spectrum of radio emission at longer wavelengths (approximately 10–40 MHz) that is highly variable but is strongly correlated with the position of Io in the orbit.

Another aspect of the distorted magnetic-field structure is the existence of a neutral sheet in the direction opposite to the solar wind (see Fig. 8.14). Because the magnetic-field lines of the Earth point in different directions in the northern and the southern hemispheres, there arises a sheetlike two-dimensional region in between where the magnetic field vanishes with the field lines oriented in an opposite direction just above and below the field. From the relation $\nabla \times \mathbf{B} = (4\pi \mathbf{J}/c)$, it follows that there is an electric current that flows in the plasma sheet. If the plasma moves in such a way as to bring together regions with oppositely directed magnetic fields, it is possible for the magnetic fields to be annihilated, transferring the magnetic-field energy into particle energy through the acceleration of particles by the electric field that is created when the magnetic-field strength changes. This process of magnetic reconnection was discussed previously in Section 8.6 in the context of solar flares. Here we see that it can serve as an important particle-acceleration mechanism in any magnetoactive plasma.

Exercise 8.5

Earth's magnetic field as a particle diagnostic: The trajectory of a charged particle of given energy in the magnetic field of the Earth can contain valuable information about the propagation of interplanetary charged particles and their energy spectra. To understand this, begin from the classical problem of the trajectory of a charged particle in the magnetic field of Earth, which, for simplicity, may be approximated as a dipole field:

$$B_r = -\frac{2\mu}{r^3} \sin\theta, \quad B_\theta = -\frac{\mu}{r^3} \cos\theta. \tag{8.95}$$

Show that the solution to the equations of motion for a particle with charge q and momentum p in such a magnetic field can be expressed in the form

$$r \sin\lambda \cos\theta + \left(\frac{R_S^2}{r}\right) \cos^2\theta = -2b, \quad R_S^2 = \frac{q\mu}{p}, \tag{8.96}$$

where b is the impact parameter of the charged particle and λ is the angle between the instantaneous velocity vector of the particle and the meridian plane that traces the particle in its orbit. Because $\sin\theta$ must lie in the interval $[-1, 1]$, this sets limits to the accessible range of (r, θ) for particles with a given value of b. Plot the accessible and excluded regions in polar coordinates for different values of b and show that there is a critical value corresponding to $b = -R_S$ at which the behaviour changes. Show how this feature can be used to put bounds on the momentum of the charged particle when its trajectory near Earth is known.

8.9 Brief Description of the Solar System

We have seen in Chap. 3, Section 3.3, that stars have originated within collapsing interstellar gas and dust clouds when certain local condensations acquire high enough central temperatures to ignite nuclear reactions. The environment of such a star will contain the debris from original gas clouds, and it is conceivable that physical processes can lead to further condensations of objects of different sizes around the star. It is generally believed that such physical processes have led to planetary systems (possibly) around several stars. The Sun, for example, is one such star around which a planetary system exists.[6]

The origin and dynamics of such planetary systems are extremely complex and are not well understood. Observationally, however, we have very detailed information about the constituents of the solar system, although it is not very clear what fraction of this information is fundamental and pertinent and which aspects are incidental. We give below a very brief summary of the nature of the solar system and a rough description of the processes that could have led to its formation.

The planets around the Sun, with the possible exception of the Pluto–Charon system, can be classified into *terrestrial* and *jovian* types. The terrestrial planets are rocky, closer to the Sun, have a higher mean surface temperature, lower mass and radius, higher density, fewer or no satellites, and no rings around them. The jovians, on the other hand, are either gaseous or made of liquid or ice, located a larger distance from the Sun, have lower mean surface temperature, higher mass and radius, lower densities, and a larger number of satellites. Some of these properties can be understood from the fact that farther planets will receive less radiation from the Sun and hence will have, on the average, lower temperatures. Table 8.1 summarises these characteristics. The planets have different number of satellites orbiting them, and some of the larger satellites of the jovian planets are similar in structure to the terrestrial planets themselves.

Table 8.1. Properties of planets

Property	Terrestrial	Jovian
Basic form	Rocky	Gas/liquid/ice
Mean orbital distance (AU)	0.39–1.52	5.2–30.1
Mean surface temperature (K)	200–750	75–170
Mass ($M_\oplus$)	0.055–1.0	14.5–318
Equatorial radius ($R_\oplus$)	0.38–1.0	3.88–11.2
Mean density (gm cm^{-3})	3.95–5.52	0.69–1.64
Sidereal rotation period (equator)	23.9 h–243 days	(9.8–19.2) h
Number of known moons	0–2	8–20
Ring systems	No	Yes

The orbits of the major and (most) minor planets are approximately coplanar, and this plane is close to the Sun's equatorial plane. The planets mostly orbit the Sun in the same direction as the rotation of the Sun. The mean radial separation between the orbits increases with the distance from the Sun, and the orbital paths of the eight major planets do not cross or even approach closely. The orbit of Pluto crosses that of Neptune peripherally.

In addition to the planets and satellites, the solar system contains several rocky objects orbiting around the Sun, most of which are contained at distances between 2 and 3.5 AU in a structure called the *asteroid belt*. The size distribution of asteroids is a power law with an index between -2.5 and -3. The eccentricities and inclination are exponentially distributed with mean values of $0.14°$ and $15°$, respectively. These orbital parameters correspond to random velocities of ~ 5 km s^{-1}, which is quite high compared with the escape velocity, 1.7 km s^{-1}, from the largest asteroid Ceres.

Another important class of objects orbiting the Sun is that of the *comets*, which are essentially made of ice and dust. The orbital periods of the comets vary tremendously from $\sim 10^2$ yr to more than 10^6 yr. Approximately 10^{12} comets with radii bigger than 1 km orbit the Sun at distances beyond 10^4 AU. This swarm of particles, called an *Oort cloud*, could be the source of the longer-period comets. At ~ 30 AU from the Sun, there exists a flattened disklike region containing a collection of cometary nuclei (called the *Kuiper belt*) from which most of the short-period comets are expected to originate.

The Sun contains nearly 99.9% of the mass of the solar system but only ~ 1% of the angular momentum. In fact, most of the angular momentum of the solar system arises because of the orbital motion of Jupiter; this orbital angular momentum is nearly 20 times the rotational angular momentum of the Sun. In contrast, the orbital angular momenta of satellites of jovian planets are far less than the spin angular momenta of the planets themselves. Of the planets, Venus, Uranus, and Pluto exhibit retrograde rotation; that is, they rotate around their axes in a direction opposite to that in which they revolve around the Sun. In general, the axial tilt of the rotation of the planets is less than $30°$.

The terrestrial planets are primarily made of dense, refractory (that is, with high condensation temperature), rocky material. The jovian planets have hydrogen and helium as the dominant components. Jupiter is 99% hydrogen and helium by mass, whereas the corresponding figure for Saturn is ~ 77%. Most of the satellites of the outer planets are made of rock and ice whereas the asteroids are rocky.

The age of the solar system seems to be few billion years. Radio isotope dating of meteorites leads to an age of $(4.56 \pm 0.02) \times 10^9$ yr, whereas the determination of lunar rocks and terrestrial rocks lead to ages in the range $(3.1–4.4) \times 10^9$ and 4.1×10^9 yr, respectively. Isotopic ratios of elements are remarkably uniform across the solar system.

It is generally believed that the origin of such a planetary system can be understood along the following lines:

8.9 Brief Description of the Solar System

(1) As a primordial interstellar gas cloud collapses, the most-massive condensates evolve rapidly into stars belonging to the upper end of the main sequence while the less-massive pieces are still in the process of collapsing. The massive stars evolve over a time scale of few million years and explode as supernova, enriching the surrounding region. As the material from the expanding nebula of the supernova moves out, at a high speed, the gas cools and some of the refractory elements like calcium, aluminium, and titanium condense out. Such an expanding nebula also collides with smaller clouds around it, compresses them, and triggers their collapse.

(2) Because the original solar nebula would have possessed some initial angular momentum, the cloud will spin up as it collapses and will produce a protosun, surrounded by a disk of gas and dust. Such a case, containing a centrally condensed gravitating object surrounded by a disk of material, arises in several astrophysical situations. Three separate processes, among others, can contribute towards the transfer of angular momentum from smaller to larger radii in such a disk. The first is the viscous torque acting between particles moving at higher velocities at smaller Keplerian orbits and particles with smaller velocities at larger radii. The second is the drag on the plasma that will attempt to corotate with the magnetic field. Finally, the third possibility will arise because of gravitational torques that can originate because of deviations from the axial symmetry in the disk.

(3) Such an accretion disk will have a radially decreasing temperature profile (which was derived to be $T \propto r^{-3/4}$ for a simple case in Chap. 7, Section 7.5) whose detailed form depends on the various physical processes. It is, however, clear that the condensation temperature from water to ice will be reached at some distance from the protosun, thereby dividing the inner region (where terrestrials form) from the outer region (where jovians form).

(4) The slow relative velocity of orbiting material, made of silicates and others, will lead to low-energy collisions in which material sticks together, forming larger aggregates called planetesimals. In the outer jovian regions, planetesimals made of water and ice would form and – at still larger radii – methane–ice structures can originate. Eventually the inner planets grow out of the planetesimals because of progressive accretion. Numerical simulations suggest that this process could take approximately $(10^7–10^8)$ yr.

(5) Around this time, the Sun would have triggered the nuclear reactions and entered the T-Tauri phase. A violent stellar wind would have driven away matter from the inner solar system, increasing the density of matter at outer regions – especially near the water–ice condensation zone around which the water vapour swept up by the stellar wind would have condensed as ice. Such an increased density dramatically accelerates the local condensation process and could have led to the formation of Jupiter.

(6) Once a protoplanet like Jupiter has accreted a mass of approximately $(10–15) M_{earth}$, its gravitational field will dominate the local region and it will generate a local accretion disk. This will lead to the formation of the Galilean

moons of Jupiter along lines similar to the formation of terrestrial planets around the protosun. The formation of such a giant planet at ∼5 AU will also lead to a strong gravitational perturbation on matter in the protodisk, in general. For example, such a gravitational perturbation will evolve material near it (in the asteroid belt today) to progressively eccentric orbits that will eventually crash on Jupiter or the Sun or escape from the solar system. The loss of mass in the vicinity of the asteroid belt and Mars could account for the rather low mass in these objects. The same perturbation leads to comparatively high velocities to the remaining members of the asteroid belt, preventing them from effectively condensing as a single object.

(7) At larger radii from the Sun, the accretion process takes longer because of lower density and longer orbital periods. Therefore, by the time planets like Uranus and Neptune have accreted rock–ice cores (taking ∼10^7 years), which are capable of exerting significant gravitational influence, the T-Tauri phase of the Sun would have been over and most of the gas would have been swept out of the solar system. This will naturally lead to large core sizes for Uranus and Neptune compared with those of Jupiter and Saturn. The gravitational influence of Neptune and Uranus would have eventually driven away the icy debris in the local region to the location of the present-day Oort cloud where their orbits would have been randomised by passing stars and interstellar clouds. Other planetesimals that formed beyond Neptune's orbit would have survived this gravitational influence and would still be in orbit in the Kuiper belt.

(8) Such a process would still leave some amounts of large planetesimals that will collide with the planets and satellites. It is believed that the removal of the low-density mantel of Mercury, the flipping of the rotation axis of Venus, the forming of the Earth–Moon system, the formation and enrichment of ring systems of jovian planets, etc., could have arisen from such collisions.

In the past decade, planets have been discovered around several nearby, normal, stars. More than a dozen stars are now known to have planets with masses ranging from a few tenths of to a few times the mass of Jupiter and with orbits at distances ranging from a few hundredths of an astronomical unit to a few astronomical units from their parent stars. These are detected essentially through very precise measurements of the radial velocity or positions of the parent stars. There has also been a detection of a brown dwarf GL 229B at ∼7 arcsec (∼50 AU) from its parent star. The brown dwarf has a surface temperature of ∼10^3 K, and its spectrum shows methane absorption lines as in the giant planets of the solar system. These discoveries suggest that planetary systems are a rule rather than exception in our galaxy.

It must be stressed that the detection of planets around stars is a technologically challenging task. The small radii as well as small T_{eff} for the planets (or, for that matter, brown dwarfs) make them very subluminous and certainly $L < 10^{-4} L_\odot$. Detecting the photons from such an object from near a much brighter star requires special techniques and careful analysis. An alternative procedure will be to measure the motion of the star of mass M_* around the common centre of mass of the star–planet system. The Keplerian motion of a planet of mass M_p induces an angular displacement θ for the star in the sky, where

$$\theta(\text{arcsec}) \simeq \frac{10^{-3}}{d(\text{pc})} \left[\frac{P(\text{yr})}{(M_*/M_\odot)} \right]^{2/3} \frac{M_p}{M_{\text{Jupiter}}}. \tag{8.97}$$

The corresponding wobbling in the radial velocity along the line of sight is

$$v_{\text{rad}} \simeq \frac{30\,\text{m s}^{-1}}{[P(\text{yr})]^{1/3}} \frac{M_p \sin i}{M_{\text{Jupiter}}} \left(\frac{M_*}{M_\odot} \right)^{-2/3}, \tag{8.98}$$

where d is the distance of the object from the Earth, P is the orbital period of the planet, and i is the inclination of the plane of the orbit to the plane of the sky. The direct astrometric method of measuring θ avoids the problem of knowing i and also gives the direction around which the planet should be searched for by direct imaging techniques. However, relation (8.97) clearly shows that even giant planets induce only a fraction of milliarcsecond displacement over periods of several years even for stars a few parsecs from us. These displacements are also usually of the order of the physical dimensions of the stellar disk and hence bright or dark spots on the star can mimic the displacement that is due to a planet, producing similar shifts in the centre of gravity of the light received from the star. The radial-velocity method, on the other hand, can provide only $(M_p/\sin i)$ and hence a lower limit on the companion mass. The current sensitivity of ~ 5 m s^{-1} could detect Jupiters or Saturns around the other stars but not Uranuses, Neptunes, or minor planets.

8.10 Aspects of Solar System Dynamics

To the lowest order of approximation, a planet like Mercury will move around the Sun in an elliptical orbit with the Sun at one focus. Such orbits are planar, and in the plane of the orbit the trajectory is given by

$$r = \frac{a(1-e^2)}{1+e\cos f}, \tag{8.99}$$

where a is the semimajor axis, e is the eccentricity, and f is the angular coordinate measured from the perihelion (called *true anomaly*). Because the axis of rotation of the Earth is inclined at an angle of $\sim 23.5°$ with respect to the normal of the orbital plane (called the *ecliptic plane*), the equatorial plane of the Earth cuts the ecliptic at the same angle of $23.5°$. The axis of the Earth's rotation will be perpendicular to the line connecting the centres of the Earth and the Sun at two points in its orbit, which are called the *equinoxes*. The line connecting the equinoxes is called the line of nodes (see Fig. 8.15).

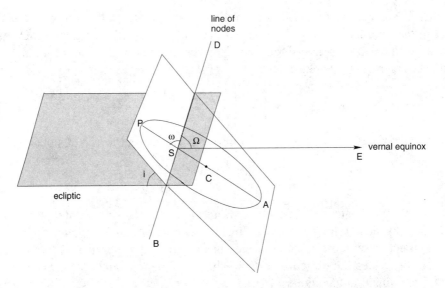

Fig. 8.15. Geometry corresponding to orbital elements.

For any other celestial object orbiting the Sun, it is necessary to specify six different quantities to uniquely characterise the orbit and the position of the celestial object. The first two parameters [(1) and (2)] specify the exact orientation and location of the orbital plane. This is done by specifying two angles: (1) the angle of inclination i of the orbital plane of the object with respect to the ecliptic; (2) the angle in the plane of the ecliptic between the direction of the vernal equinox and the line of nodes of the celestial object in question, denoted by Ω. In the orbital plane, we have to specify the orientation of the ellipse on which the object moves. We can achieve this by giving the third parameter (3) as the direction of, say, the major axis. (3) This angle in the orbital plane, between the line of nodes and the direction of the perihelion, is usually denoted by ω. The next two parameters [(4) and (5)] specify the orbit in terms of two parameters, viz., (4) the semimajor axis a of the orbit and (5) the eccentricity e of the orbit. (6) Finally, to specify the location of the object in its orbital path, we need to know its position at some time. This is given by the epoch T of the last instant of perihelion passage.

Incidentally, the same procedure can be used to specify the location of any other celestial body in a elliptic orbit, e.g., a star in a binary star system (see Chap. 7, Section 7.7). The parameters ω, a, and e have intrinsic meanings for any elliptic orbit; i is taken to be the inclination of the orbital plane with respect to the plane of the sky for a general case; Ω becomes the angle in the plane of the sky between the line of sight and the line of nodes; T is taken with respect to any conveniently specified point in the orbit.

In reality, the motion of a planet like Mercury will be affected by several forces in addition to the approximate inverse-square-law force of the Sun in the

8.10 Aspects of Solar System Dynamics

point particle approximation. The study of perturbation to the planetary orbit arising from corrections to the inverse-square-law force is of great practical and theoretical importance.

In general, the solar system is described by a Hamiltonian H that is not integrable; but H can be separated into a dominant term that is integrable and a series of perturbations. The description of such Hamiltonians was given in Vol. I, Chap. 2, where it was pointed out that the phase-space structure of orbits of an integrable Hamiltonian can be significantly altered by nonintegrable perturbations. In general, such systems can lead to chaotic motions and long-term predictability will be impossible for a wide class of initial conditions. It is conceivable that the solar system belongs to this class of dynamical systems. Some features of the long-term predictability and chaos in the solar system will be discussed in Subsection 8.10.5 below.

The situation, however, is different for short-term predictability. Over shorter time scales, it is indeed possible to treat the effects of perturbations in a systematic manner and to study their effects on planetary orbits. We will first describe a general procedure that is powerful enough to handle such perturbation calculations and then illustrate it with a simple example.

8.10.1 Perturbation Theory

Let us assume that the Hamiltonian for a system with f degrees of freedom can be written in the form $H = H_0 + V(p, q, t)$, where H_0 is the part of the Hamiltonian for which the exact solution is known and V is a small perturbation. (The set of coordinates q_i and momenta p_i, with $i = 1, 2, \ldots, f$ will be denoted as q, p without explicit subscripts when no confusion is likely to arise.) A precise characterisation of the smallness is a difficult (and often intractable) problem in complete generality; we shall assume that this issue can be settled in some well-defined manner over the time scale of interest. Let the complete solution to the unperturbed problem be denoted by the functions $[p_n(\alpha, \beta, t), q_n(\alpha, \beta, t)]$, where (α_i, β_i) are the $2f$ integration constants, with $i = 1, 2, \ldots, f$ and $\alpha_1 = E_0$. Expressing the constants α_n and β_n in terms of p and q, we obtain relations of the form

$$C_i = C_i(p, q, t) \quad (i = 1, \ldots, 2f) \tag{8.100}$$

for a system with f degrees of freedom. To provide a formal solution to the problem, we can try to obtain a canonical transformation that will produce α's as the new momenta and β's as the new coordinates. Of the $2f$ constants (α_n, β_n), one constant corresponds to the choice of origin of time for the closed system. Let this be β_1, which we shall take to be the same as C_1. In that case, it follows that

$$\dot{C}_i = [C_i, H_0] = \delta_{i1}, \tag{8.101}$$

where $[A, B]$ is the Poisson bracket between A and B (see Vol. I, Chap. 2). These equations for $i \neq 1$ merely state that C_i's are constants. For $i = 1$, Eq. (8.101) leads to a simple translation in time, as expected.

When the perturbation is switched on, the functions C_i's are no longer constants and their time variation is given by

$$\dot{C}_i = [C_i, H_0 + V] = [C_i, V] + \delta_{i1}. \tag{8.102}$$

We can evaluate the Poisson bracket $[C_i, V]$ by noting that $V(p, q, t)$ can be equivalently written as $V(C_j, t)$. Then

$$[C_i, V] = \frac{\partial C_i}{\partial q_n} \frac{\partial V}{\partial C_j} \frac{\partial C_j}{\partial p_n} - \frac{\partial C_i}{\partial p_n} \frac{\partial V}{\partial C_j} \frac{\partial C_j}{\partial q_n}, \tag{8.103}$$

so that the time variation is given by

$$\dot{C}_i = [C_i, C_j] \frac{\partial V}{\partial C_j} + \delta_{i1}. \tag{8.104}$$

This equation is exact and forms a convenient starting point for perturbation theory. We assume, as usual, that C's can be expressed as a sum of the form

$$C_i = C_i^{(0)} + C_i^{(1)} + \cdots, \tag{8.105}$$

where each successive term is of one higher order in the strength of perturbation. In that case, the lowest-order corrections are given by

$$\dot{C}_i^{(1)} = [C_i^{(0)}, C_j^{(0)}] \frac{\partial V^{(0)}}{\partial C_j^{(0)}} + \cdots, \tag{8.106}$$

where $V^{(0)}$ denotes $V[C^{(0)}]$, etc. This equation shows that if the Poisson brackets of the original integrals of motion are known and the perturbing potential can be expressed in terms of the original constants characterising the system, then the perturbation theory can be formulated in a general manner in terms of the time variation of the original constants. We shall now illustrate this procedure in terms of a specific example.

8.10.2 Example: Precession of Mercury

As an example, let us consider the long-term secular changes in the orbit of Mercury that are due to the perturbing influence of Jupiter, under the assumptions that (1) both Mercury and Jupiter orbit the Sun in the same plane and (2) the orbit of Jupiter is circular. In the absence of Jupiter's perturbation, Mercury's motion at any time t can be calculated in terms of the four constants a, A_x, A_y, and t_0, where a is the semimajor axis, A_x and A_y are the components of the Runge–Lenz vector, $\mathbf{A} \equiv \mathbf{L} \times \dot{\mathbf{r}} + GM_\odot m\hat{\mathbf{r}}$, which gives the direction of the major axis in the plane of the orbit, and t_0 is the time chosen such that the planet is at aphelion (see Vol. I, Chap. 2, Section 2.3). If $\gamma \equiv GM_\odot m$, then the magnitude of the Runge–Lenz vector is given by $A = \gamma e$, where e is the eccentricity of the orbit. Thus, given a and $\mathbf{A}$, we know the exact shape and orientation of the orbit while t_0 fixes the location of the planet *in* the orbit. We can now choose the constants α's and β's for the problem as

$$\alpha_1 = E = -\frac{\gamma}{2a}, \quad \alpha_2 = L = \sqrt{\gamma a(1-e^2)}, \quad \beta_1 = -t_0, \quad \beta_2 = \vartheta_{a0}, \quad (8.107)$$

where ϑ_{a0} denotes the angle between $\mathbf{A}$ and the y axis at $t = 0$. These relations can be inverted to express the parameters of the orbit in terms of α's and β's as

$$A_x = -\gamma e \sin \beta_2, \quad A_y = \gamma e \cos \beta_2, \quad (8.108)$$

$$a = -\frac{\gamma}{2\alpha_1}, \quad e^2 = 1 + \frac{2}{\gamma^2}\alpha_1\alpha_2^2 = \frac{1}{\gamma^2}(A_x^2 + A_y^2). \quad (8.109)$$

We can now calculate all the relevant Poisson brackets $[C_i^{(0)}, C_j^{(0)}]$ needed in Eq. (8.106). We get, for example,

$$[A_x, A_y] = \frac{\partial A_x}{\partial \beta_2}\frac{\partial A_y}{\partial \alpha_2} - \frac{\partial A_y}{\partial \beta_2}\frac{\partial A_x}{\partial \alpha_2} = -2\alpha_1\alpha_2 = \frac{\gamma}{\alpha}\sqrt{\gamma a(1-e^2)} \quad (8.110)$$

and, similarly,

$$[a, t_0] = \frac{2a^2}{\gamma}, \quad [A_x, t_0] = \frac{a(1-e^2)}{\gamma e^2} A_x,$$

$$[A_y, t_0] = \frac{a(1-e^2)}{\gamma e^2} A_y, \quad [a, A_x] = [a, A_y] = 0. \quad (8.111)$$

The analysis up to this point is independent of the nature of the perturbation. The effect of Jupiter's perturbation will be to make the perihelion of Mercury precess so that $\mathbf{A}$ will no longer be constant (see Fig. 8.16). To estimate this effect, we will first consider the gravitational potential corresponding to the perturbation of Jupiter (with mass m_J) on Mercury, which can be written in the form

$$V = -\frac{\delta}{R}, \quad \delta = Gm_J, \quad (8.112)$$

where

$$\frac{1}{R} = \frac{1}{\sqrt{r_1^2 - 2r_1 a_2 \cos\vartheta + a_2^2}} = \frac{1}{a_2}\sum_{n=0}^{\infty}\left(\frac{r_1}{a_2}\right)^n P_n(\cos\vartheta). \quad (8.113)$$

Here $\mathbf{r}_1$ is the radius vector from the Sun to Mercury, a_2 is the radius of Jupiter's orbit, and ϑ is the angle between the lines connecting the Sun to Jupiter and Mercury. Treating r_1/a_2 as a small parameter, we can retain the first three terms

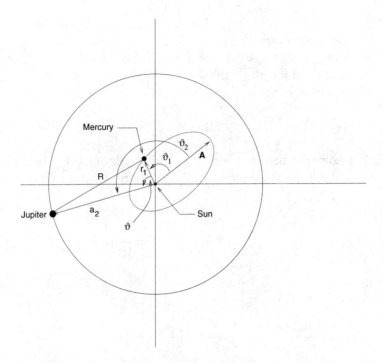

Fig. 8.16. Coordinate system for calculating the perturbation of Jupiter on Mercury's orbit.

in the above expansion and use the values

$$P_0(\cos \vartheta) = 1, \quad P_1(\cos \vartheta) = \cos \vartheta, \quad P_2(\cos \vartheta) = \frac{1}{2}(3 \cos^2 \vartheta - 1). \tag{8.114}$$

We now note that $\cos \vartheta$ and r_1 are periodic functions of time. As we are interested in only the secular effect, we can average these quantities over the orbits. For example, the $n = 1$ term in the sum of Eq. (8.113) contains the combination

$$r_1(t) \cos \vartheta(t) = r_1(t)(\cos \vartheta_1 \cos \vartheta_2 + \sin \vartheta_1 \sin \vartheta_2), \tag{8.115}$$

in which $r_1 \cos \vartheta_1$ and $r_1 \sin \vartheta_1$ vary with the orbital period T_1 of Mercury whereas $\cos \vartheta_2$ and $\sin \vartheta_2$ vary with the orbital period T_2 of Jupiter. Because T_1 and T_2 are different *and* incommensurable, this term averages to zero. Similarly, $P_2(\cos \vartheta)$ averages to $\bar{P}_2 = 1/4$. To compute the second term, we need the time average of r_1^2 over an elliptic orbit. This can be done in a fairly straightforward manner (see Exercise 8.7), and we get

$$\overline{r_1^2} = \left(1 + \frac{3}{2}e_1^2\right)a^2. \tag{8.116}$$

Collecting all these results together, we find that the average perturbing potential energy is given by

$$V = \bar{V} = -\frac{\delta}{a_2}\left[1 + \frac{1}{4}\left(1 + \frac{3}{2}e_1^2\right)\left(\frac{a}{a_2}\right)^2\right]. \qquad (8.117)$$

We can now compute the right-hand side of Eq. (8.106); we find that

$$\overline{\dot{A}_x} = [A_x, \bar{V}] = -\frac{3\delta a^2}{8a_2^3}[A_x, e_1^2], \qquad (8.118)$$

which becomes, on use of Eqs. (8.109),

$$\overline{\dot{A}_x} = -\frac{3}{8}\frac{\delta a^2}{\gamma^2 a_2^3}[A_x, A_x^2 + A_y^2]$$

$$= -\frac{3}{4}\frac{\delta a^2}{\gamma^2 a_2^3}[A_x, A_y]A_y = -\frac{3}{4}\frac{\delta a}{\gamma a_2^3}\sqrt{\gamma a(1-e_1^2)}\,A_y. \qquad (8.119)$$

During one revolution of Mercury with a period $T_1 = (2\pi a^{3/2}/\gamma^{1/2})$, A_x changes by the amount

$$\Delta A_x = \overline{\dot{A}_x}T_1 = -\frac{3}{4}2\pi\frac{\delta}{\gamma}\left(\frac{a}{a_2}\right)^3\sqrt{1-e_1^2}\,A_y. \qquad (8.120)$$

The forward angular precession in one orbit is given by

$$-\frac{\Delta A_x}{A_y} = \left(\frac{3\pi}{2}\right)\left(\frac{\delta}{\gamma}\right)\left(\frac{a}{a_2}\right)^3(1-e_1^2)^{1/2}. \qquad (8.121)$$

Using the values $(\delta/\gamma) = 9.548 \times 10^{-4}$, $(a/a_2) = 7.441 \times 10^{-2}$, and $e_1 = 0.2056$, we find that the precession per century that is due to Jupiter's perturbations is $\sim 155''$.

This procedure illustrates the manner in which perturbation theory can be applied to planetary dynamics. In a similar manner we can compute the perturbation that is due to all other planets, and the total comes to $\sim 531''$ per century. The most dominant effect in the precession of the Mercury – when viewed from Earth – arises from the precession of the Earth's rotational axis (see Subsection 8.10.3 below), which adds $\sim 5026''$ per century. The observed value is $5600''$, giving a difference of $43''$ per century, which agrees well with the correction that is due to general relativistic effects (see Vol. I, Chap. 11, Section 11.9).

Exercise 8.6

Precession of Moon's orbital angular momentum: As seen from Earth, the elliptical path of the Moon lies in a plane that is slightly inclined to Earth's orbital plane around the Sun. This causes a wobbling of the angular-momentum vector of the Moon that is due to the perturbing action of the Sun. Show that if the Earth's orbit around the Sun is taken to be a circle, then the perturbing potential is given by

$$V = \frac{GM_\odot \mu}{2R} \left[3\left(\frac{\mathbf{R} \cdot \mathbf{r}}{R^2}\right)^2 - \frac{r^2}{R^2} \right], \quad (8.122)$$

where μ is the reduced mass of the Earth and the Moon, $\mathbf{R}$ is the position of the Earth, and $\mathbf{r}$ is a vector from the Earth to the Moon. Using this expression, calculate the precession of the angular-momentum vector of the Moon's orbit and show that the period of precession T_p is given by

$$\frac{T_p}{T_{\text{moon}}} = \frac{4}{3} \frac{m_{\text{earth}} + m_{\text{moon}}}{M_\odot} \left(\frac{R}{a_{\text{moon}}}\right)^3 \sec \alpha, \quad (8.123)$$

where $T_{\text{moon}} = 27.32$ days is the orbital period of the Moon and $\alpha = 5°8'43''$ is the inclination of the angular-momentum vector from the normal to Earth's orbital plane. Estimate T_p numerically and show that the precession is at the rate of one revolution for 18.6 yr.

Exercise 8.7
Keplerian averages: Show that the average of r^n over a period of Keplerian orbit with semimajor axis a and eccentricity e can be expressed in the form

$$\overline{r^n} = \frac{a^n}{2\pi} \oint (1 - e \cos \psi)^{n+1} d\psi, \quad (8.124)$$

where the integral is from 0 to 2π. Evaluate the integral for $n = 1, 2, 3$ explicitly.

8.10.3 Precession of the Equinoxes

To the lowest order of approximation, Earth can be treated as a symmetric rotator, with two moments of inertia about the principle axis being nearly equal and the third one being different from the other two: $I_x = I_y \neq I_z$. The deviation of Earth's shape from a sphere can be attributed, in large part, to the flattening arising from the rotation of the Earth. Using the description of rotating fluids, developed in Vol. I, Chap. 10, Section 10.3, we can estimate this deviation as being characterised by an eccentricity of $e \approx (5/4)(\Omega^2 R_\oplus^3 / GM_\oplus) \approx 1/300$. This deviation in the shape of the Earth from a perfect sphere leads to some interesting effects.

To begin with, because the Earth rotates with an angular velocity $\omega_z = \Omega = 7.5 \times 10^{-4}$ s^{-1} corresponding to a period of 24 h, its axis of rotation will undergo a free precession at the rate

$$\omega_p = \frac{I_z - I_x}{I_x} \Omega \simeq \frac{\Omega}{304} \simeq 2.5 \times 10^{-6}. \quad (8.125)$$

This effect has been derived in Vol. I, Chap. 2, Subsection 2.3.2, and we have used the observed values for the moments of inertia to arrive at the numerical estimate. This precession corresponds to a period of ~ 300 days, which turns out

8.10 Aspects of Solar System Dynamics

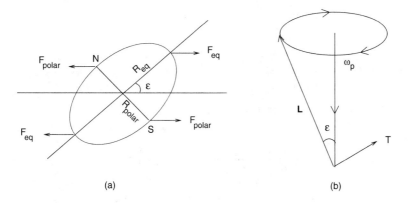

Fig. 8.17. Torque and precession of the Earth.

to be ~50% smaller than the observed period. It is believed that the discrepancy arises because of complicating factors such as the tidal distortion of the Earth.

A more interesting precession of the axis of rotation of the Earth is the forced precession, which arises because of the torque exerted by the Sun and the Moon on Earth. The gravitational force of the Sun on the Earth is, of course, is significantly larger than that of the Moon on the Earth. However, the tidal force that is due to the Sun and the Moon has the ratio

$$\frac{(M_\odot/d_\odot^3)}{(M_{\text{moon}}/d_{\text{moon}}^3)} \simeq \frac{(M_\odot/R_\odot^3)(R_\odot/d_\odot)^3}{(M_{\text{moon}}/R_{\text{moon}}^3)(R_{\text{moon}}/d_{\text{moon}})^3}$$

$$= \frac{\rho_\odot}{\rho_{\text{moon}}} \left(\frac{\theta_\odot}{\theta_{\text{moon}}}\right)^3 \simeq \frac{\rho_\odot}{\rho_{\text{moon}}} \simeq \frac{1.4}{3.3} \simeq 0.42, \quad (8.126)$$

where $\theta \equiv R/d$ is the angle subtended by the Sun or the Moon at Earth and we have used the fact that $\theta_\odot \simeq \theta_{\text{moon}}$. Consider an axisymmetric ellipsoidal figure of the Earth that is acted on by the tidal force of the Sun and the Moon (see Fig. 8.17). The equatorial and polar tidal forces are

$$F_{\text{eq}} \simeq GQR_{\text{eq}}\cos\epsilon, \quad F_{\text{polar}} \simeq GQR_{\text{polar}}\sin\epsilon,$$

$$Q \approx \left[\left(\frac{M_\odot}{d_\odot^3}\right) + \left(\frac{M_{\text{moon}}}{d_{\text{moon}}^3}\right)\right] M_{\text{earth}}. \quad (8.127)$$

The corresponding tidal torques are $T_{\text{eq}} = \lambda F_{\text{eq}} R_{\text{eq}} \sin\epsilon$ and $T_{\text{polar}} = \lambda F_{\text{polar}} R_{\text{polar}} \cos\epsilon$, where λ is a numerical factor of ~0.3. Because these act in opposite directions, the net torque is given by

$$T = T_{\text{polar}} - T_{\text{eq}} = \lambda G M_{\text{earth}} \left(\frac{M_\odot}{d_\odot^3} + \frac{M_{\text{moon}}}{d_{\text{moon}}^3}\right)(R_{\text{eq}}^2 - R_{\text{polar}}^2)\sin\epsilon\cos\epsilon.$$

$$(8.128)$$

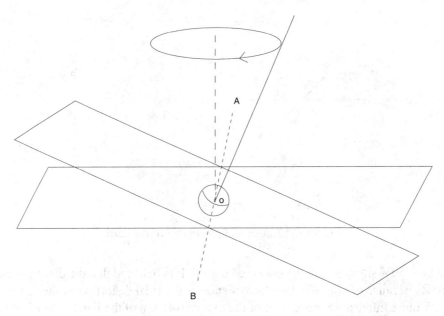

Fig. 8.18. Geometry corresponding to the precession of equinoxes.

Taking the angular momentum of the Earth to be $L \simeq (2/5)M_{\text{earth}}R_{\text{eq}}^2\Omega$, we can estimate the precession frequency by the rule $\mathbf{\Omega}_p \times \mathbf{L} = \mathbf{T}$. This gives

$$\Omega_p \simeq \frac{5\lambda}{2} G \left(\frac{M_\odot}{d_\odot^3} + \frac{M_{\text{moon}}}{d_{\text{moon}}^3} \right) \left(1 - \frac{R_{\text{polar}}^2}{R_{\text{eq}}^2} \right) \frac{\cos \epsilon}{\Omega}. \qquad (8.129)$$

The corresponding precessional period $P_p = (2\pi/\Omega_p)$ turns out to be $\sim$26,000 yr. We shall now provide a more rigorous derivation of this result.

The geometry relevant for the forced precession that is due to the torque from the Sun is shown in Fig. 8.18. The horizontal plane in the figure is the orbital plane of the Earth around the Sun (called the ecliptic). The axis of rotation of the Earth is inclined with respect to the normal to this plane by an angle $\vartheta = 23°27'$. This makes the equatorial plane inclined by the same angle with respect to the ecliptic. These two planes intersect along the line of nodes denoted by AB. When viewed from Earth, the Sun moves on the ecliptic plane in a clockwise direction; at A, the Sun passes above the equatorial plane and the direction OA is called the ascending node. At B the Sun passes below the equatorial plane, and the direction OB is called the descending node. By averaging the Sun's orbit around the Earth, we can approximate it as a ring of material kept on the ecliptic plane. Because the Earth is not a perfect sphere, such a ring of material will exert a torque on the Earth, trying to align the axis of rotation towards the normal to the ecliptic plane. This will cause the axis of rotation to precess around the normal, thereby making the equinoxes precess in the sky. We shall now estimate the rate of this precession.

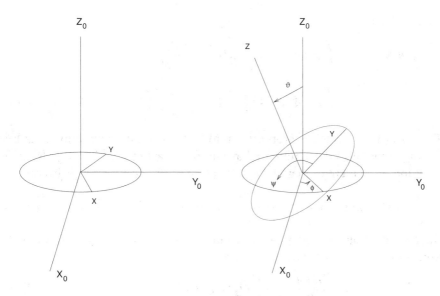

Fig. 8.19. Definition of Euler angles used for the analysis in this section.

To illustrate the general principles involved in the study of such a forced precession, we shall work out this result by first developing the equations of motion for a symmetric rotator from a suitable Lagrangian. To determine the kinetic-energy term of the Lagrangian, we need to express the angular velocity of the rotator in terms of suitably defined angles that characterise its orientation. This is most easily done with the Euler angles defined as in Fig. 8.19. The angular velocity, in the case of a symmetric rotator, can be immediately written down by inspection:

$$\omega_x = \dot{\vartheta}; \quad \omega_y = \dot{\phi} \sin \vartheta; \quad \omega_z = \dot{\phi} \cos \vartheta + \dot{\psi}. \tag{8.130}$$

The kinetic-energy term now becomes

$$T = \frac{1}{2}\left[I_x\left(\omega_x^2 + \omega_y^2\right) + I_z\omega_z^2\right] = \frac{1}{2}[I_x(\dot{\vartheta}^2 + \dot{\phi}^2 \sin^2 \vartheta) + I_z(\dot{\phi} \cos \vartheta + \dot{\psi})^2]. \tag{8.131}$$

To estimate the gravitational potential energy of interaction between the Sun and the Earth, we have to evaluate the integral

$$V = \int \rho(\mathbf{x})\Phi(\mathbf{x})\,d^3\mathbf{x} = \rho_0 \int \Phi(\mathbf{x})\,d^3\mathbf{x}, \tag{8.132}$$

where $\Phi(\mathbf{x})$ is the gravitational potential at $\mathbf{x}$ that is due to the Sun located at $\mathbf{R}(t)$ and the integral is over the body of the Earth, which is taken to have a uniform density ρ_0. Because the gravitational potential varies as $|\mathbf{x}|^{-1} \equiv |\mathbf{R}+\mathbf{l}|^{-1}$ with $l \ll R$, it is easy to evaluate this quantity by Taylor expanding $\Phi(\mathbf{x})$ around the origin. By expanding $\Phi(\mathbf{x}) = \Phi(\mathbf{R}+\mathbf{l})$ in Taylor series in $\mathbf{l}$ up to the quadratic term, we have

$$\int d^3\mathbf{x}\Phi(\mathbf{x}) \cong \int d^3\mathbf{l}\,\Phi(R) + \int d^3\mathbf{l}\,\frac{1}{2}\frac{\partial^2\Phi}{\partial R^a\partial R^b}l^a l^b$$

$$\cong \int d^3\mathbf{l}\,\Phi(R) + \int d^3\mathbf{l}\,\frac{1}{2}\frac{\partial^2\Phi}{\partial R^a\partial R^b}\left(l^a l^b - \frac{1}{3}l^2\delta^{ab}\right). \quad (8.133)$$

(The linear term, which is proportional to l^a, vanishes on integration over $\mathbf{l}$; in arriving at the second equality, we have used the fact that $\nabla^2\Phi = 0$.) Using the definition of moment of inertia $I_{ab} = \text{dia}(I_x, I_x, I_z)$, the potential V becomes

$$V = m\Phi(R) + \frac{1}{2}I_{ab}\frac{\partial^2\Phi}{\partial R^a\partial R^b} = m\Phi(R) + \frac{1}{2}(I_z - I_x)\frac{\partial^2\Phi}{\partial z^2}, \quad (8.134)$$

where we have again used $\nabla^2\Phi = 0$. Thus the potential energy is given by the expression

$$V(r) = -\frac{GMm}{R} - \frac{GM}{2R^5}(I_z - I_x)(X^2 + Y^2 - 2Z^2) + \cdots. \quad (8.135)$$

We will assume that the Sun moves around the Earth in a circular orbit of radius R with a constant angular velocity $-n$ (the negative sign indicates the clockwise motion of the Sun). In that case, the position of the Sun at any given instant can be expressed as

$$X = R\cos(\phi + nt), \quad Y = -R\sin(\phi + nt)\cos\vartheta, \quad Z = R\sin(\phi + nt)\sin\vartheta. \quad (8.136)$$

Substituting Eq. (8.136) into expression (8.135), we can express the potential in terms of the Euler angles:

$$V(r) = -\frac{GMm}{R} - \frac{GM}{2R^3}(I_z - I_x)\left[1 - 3\sin^2(\phi + nt)\sin^2\vartheta\right]. \quad (8.137)$$

Because the Sun's motion is very rapid, we can average this expression over time to get the mean potential that is acting on the rotating Earth. Replacing $\sin^2(\phi + nt)$ with $1/2$, we find that the mean potential becomes

$$\bar{V}(r) = -\frac{GMm}{R} - \frac{GM}{2R^3}(I_z - I_x)\left(1 - \frac{3}{2}\sin^2\vartheta\right). \quad (8.138)$$

The first term in the potential does not affect the rotational motion; hence the rotation of the Earth is governed by the Lagrangian of the form

$$L = \frac{1}{2}[I_x(\dot{\vartheta}^2 + \sin^2\vartheta\,\dot{\phi}^2) + I_z(\dot{\phi}\cos\vartheta + \dot{\psi})^2] + \frac{GM}{2R^3}(I_z - I_x)\left(1 - \frac{3}{2}\sin^2\vartheta\right). \quad (8.139)$$

We can obtain the equations of motion for the Euler angles by varying this Lagrangian with respect to the three angles. Because ϕ and ψ are cyclic, we get two conservation laws,

$$\frac{\partial L}{\partial \dot\phi} = (I_x \sin^2 \vartheta + I_z \cos^2 \vartheta)\dot\phi + I_z \Omega \cos \vartheta = p_\phi = \text{constant}, \quad (8.140)$$

$$\frac{\partial L}{\partial \dot\psi} = I_z(\dot\phi \cos \vartheta + \Omega) = p_\psi = \text{constant}, \quad (8.141)$$

where $\Omega \equiv \dot\psi$ is the daily rotational rate of the Earth. Because $\Omega \gg \dot\phi$, these constants can be approximated as

$$p_\phi \simeq I_z \Omega \cos \vartheta, \quad p_\psi \simeq I_z \Omega, \quad (8.142)$$

which in turn show that ϑ is nearly constant. Varying the Lagrangian with respect to ϑ and ignoring $\dot\phi$ compared with Ω, we get

$$I_x \ddot\vartheta + I_z \Omega \dot\phi \sin \vartheta + \frac{3GM}{2R^3}(I_z - I_x) \sin \vartheta \cos \vartheta \simeq 0. \quad (8.143)$$

Using the fact that ϑ is approximately constant, we can ignore the first term and obtain the precessional velocity as

$$\dot\phi = -\frac{3GM}{2R^3}\left(\frac{I_z - I_x}{I_z}\right)\frac{\cos \vartheta}{\Omega} = -6\pi^2 \frac{\Omega}{(\Omega T)^2}\left(\frac{I_z - I_x}{I_z}\right) \cos \vartheta, \quad (8.144)$$

where the second equality follows from use of the result $(GM/R^3) = (2\pi/T)^2$, with T denoting the orbital period of the Earth.

This, however, is the contribution that is due to the Sun alone. A similar effect arises because of the Moon's torque on the Earth as well, which can be computed exactly along the same lines. Adding both the results, we find that the precession has a period of ~26,000 yr. This corresponds to ~50″ per year for the motion of the equinoxes.

The analysis also shows that, in general, the precession rate is given by the formula

$$\omega_p \approx \left(\frac{4n^2}{n'}\right)\left(\frac{\Delta I}{I}\right) \cos \theta, \quad (8.145)$$

where n is the angular velocity of revolution of the object around the central body, n' is the angular velocity of rotation of the object on its own axis, $(\Delta I/I)$ is the fractional difference in the moment of inertia of the body that is due to asphericity, and θ is the inclination of the axis of rotation to the normal of the orbital plane.

Exercise 8.8

Zodiacal signs: As the Sun is seen to move clockwise around the Earth in the plane of the ecliptic, it passes above the equatorial plane at the ascending node and passes below the equatorial plane at the descending node. The ascending node at present points towards Pisces and the Age of Aquarius is to begin when it enters Aquarius. (a) Estimate when the Age of Aquarius will begin. (b) In some of the horoscopes published in magazines, the

Sun is taken to be in Aries from 21 March to 19 April. Actually it is in Aries around 17 April to 17 May. Assuming that the astrologers have not taken into account the precession, when was the original zodiacal table devised?

Because the Moon's orbital plane is inclined at $\sim 5°$ to the ecliptic, there is another torque on the Earth, which varies with the precessional rate of the Moon's orbital spin. This was estimated in Exercise 8.6 to be ~ 18 yr. This will lead to an extra wobbling of the precession of the equinoxes with an 18-yr periodicity.

Finally, it must a noted that a similar effect will occur in the case of artificial satellites orbiting a nonspherical body like the Earth. There will be a torque acting between the planet and the satellite, tending to align the orbital and equatorial planes, if the planet is treated as a symmetric rotator. This will cause a drift in the orbit of artificial satellites orbiting Earth and actually provides a very precise way of measuring the asphericity of earth. There is a similar effect on the Moon's orbital plane that is due to the Earth but it is subdominant to the effect that is due to the Sun.

Exercise 8.9
Cathedrals and pinholes: Consider a tall building like a cathedral with a small pinhole somewhere on its roof. The motion of the Sun in the sky will lead to a path of the sunlight on the floor of the building, which will have an annual variation that is due to the motion of the Earth around the Sun. Can such an arrangement be used realistically to measure the precession of the equinoxes and similar phenomena? Make an estimate of the accuracy that can be achieved.

8.10.4 Tidal Friction

The planets and satellites of the solar system are not rigid bodies and hence undergo a deformation in shape under the action of forces. This is particularly true in the case of the Earth, in which a large mass of water exists on the surface. The action of the Moon's gravitational force can deform the surface of the Earth, thereby leading to tidal bulges on the surface of the Earth. These tidal bulges will be located directly under the Moon if (1) there is no other force (such as friction) acting on the system and (2) the rotational period of the Earth is identical to the orbital period of the Moon around the Earth. In reality, the surface deformation of the Earth is affected by fairly strong frictional forces and the rotation of the Earth is quite fast (compared with the orbital rate of the moon around the Earth). Hence the tidal bulge will not be aligned along the direction of the Moon. The fast rotation of the planet will drag the axis of the bulge ahead of the Earth–Moon line, leading to several consequences: (i) This causes a torque between the Moon and the Earth that tends to align the axis of the bulge along the Earth–Moon line; an equal and opposite torque acts on the moon, increasing its orbital angular momentum. (ii) Because frictional forces are dissipative, the loss of energy causes the rotational velocity of the Earth to decrease (thereby increasing the length of the day). (iii) This also causes the angular momentum of the Earth's rotation to decrease. It follows that the orbital angular momentum of the Moon around the Earth should increase in order to conserve the total angular momentum, which in turn has the effect of making the Earth–Moon distance increase. (iv) Because circular orbits have the lowest energy for a given angular momentum, this process also leads to the circularisation of orbits. Eventually the configuration should stabilise, with the orbital period of the Moon around the Earth matching the rotational period of the Earth on its axis. (v) A corresponding effect, of course, occurs because of the tides produced on the Moon by the Earth. If the initial position of the Moon was closer to the Earth and its orbital period was different from the rotational period about its own axis, then the tides raised by the Earth on the Moon will have the effect of synchronising the orbital and rotational periods of the Moon. Because the Moon is a smaller body, this effect on the Moon occurs at a shorter time scale and has already caused the Moon to be in such a synchronous, tidally locked orbit with respect to the Earth. In fact, most of the satellites in the solar system have this synchronous feature.

We will now consider some details of the tidal interaction by using a fairly simple model. Let us choose a coordinate system centred on Earth with a point near the surface of the Earth having the coordinates (a, θ), where a measures the radial distance and θ measures the angle from the instantaneous Earth–Moon line. If M_1 and M_2 are the masses of the Earth and the the Moon and R is the distance between their centres, then the gravitational potential at any given point with coordinates (a, θ) near the surface of the Earth is given by

$$\Phi(a, \theta) = -\frac{GM_1}{a} - \frac{1}{2}\frac{Ga^2}{R^3}(3M_2 \cos^2\theta + M_1) + \text{constant}. \quad (8.146)$$

The first term is the gravitational potential that is due to the Earth and the second term gives the correction that is due to the tidal forces. The equipotential surfaces, given by $\Phi(a, \theta) = \text{constant}$, will provide the lowest-order deformation of the spherical shape of the Earth that is due to tidal forces. To get a better feeling for this shape, we can set $a = R_\oplus + h$ (where $R_\oplus$ is the radius of the Earth) and expand $\Phi(a, \theta)$ in a Taylor series in h. A simple calculation now gives, to the lowest nontrivial order in h, the expression

$$\Phi(h, \theta) \approx g\left[h - \frac{3}{2}\left(\frac{M_2}{M_1}\right)\left(\frac{R_\oplus}{R}\right)^3 R_\oplus \cos^2\theta\right] + \text{constant}, \quad (8.147)$$

where g is the acceleration that is due to gravity on the surface of the Earth and higher-order terms in $(R_\oplus/R)$ and (h/R) are ignored. Setting Φ to constant shows that the deformation of the surface h is given by

$$h = \frac{3}{2}\left(\frac{M_2}{M_1}\right)\left(\frac{R_\oplus}{R}\right)^3 R_\oplus \cos^2\theta + \text{constant}. \quad (8.148)$$

The difference between $h(\theta = 0)$ and $h(\theta = \pi/2)$ is ~ 54 cm for $M_2/M_1 \approx 1/80$; $R_\oplus \approx 6400$ km and $R = 3.8 \times 10^5$ km. The corresponding calculation for the Sun's gravitational field on the Earth produces a shift of ~ 25 cm. It should be noted that the tidal force has a $\cos^2\theta = (1/2)(\cos 2\theta + 1)$ dependence, where θ is the angle measured from the location of the Moon. This is characteristic of a tidal force.

To model the tidal friction, it is necessary to study the effect of the Moon's tidal force on a deformable structure like the oceans on Earth. To a lowest order of approximation, this can be done as follows. Consider a satellite that orbits a planet in the equatorial plane with an orbital angular velocity Ω_s. Let the angular velocity of rotation of the planet about its own axis be Ω_p. Any given point P on the equator of the planet will move past the planet–satellite line with an angular velocity $\Omega_m = \Omega_p - \Omega_s$. If the instantaneous location of a point on the equator is denoted by the angle ϕ with respect to some axis, then the angle between the planet–satellite line and this point P will vary as $(\Omega_m t - \phi)$. We have seen above that the tidal force acting on point P will vary as $F = A\cos 2(\Omega_m t - \phi)$ with a suitable choice for the origin of t. The displacement ξ of a deformable structure to such a tidal force can then be modelled by a damped-wave equation of the following kind:

$$\frac{1}{v^2}\frac{\partial^2 \xi}{\partial t^2} - \frac{1}{R^2}\frac{\partial^2 \xi}{\partial \phi^2} + \frac{\gamma}{v^2}\frac{\partial \xi}{\partial t} = A\cos 2(\Omega_m t - \phi). \tag{8.149}$$

The first two terms on the left-hand side govern the propagation of waves with speed v on the surface of the planet, with the spatial distance dz replaced with $R\,d\phi$, where R is the radius of the planet. The third term is a simple model for the frictional damping of these waves with a characteristic time constant γ^{-1}. The right-hand side is the tidal-force term. [In the absence of the right-hand side, the equation can represent the propagation of (damped) waves in the ocean. Such waves, except for damping, were considered in Vol. I, Chap. 8, Section 8.13]. Equation (8.149) can be solved by elementary techniques, and the solution is given by

$$\xi = \frac{AR^2}{4\left[\left(1 - \frac{R^2\Omega^2}{v^2}\right)^2 + \frac{1}{4}\frac{\gamma^2\Omega^2 R^4}{v^4}\right]^{1/2}} \cos 2(\Omega_m t - \phi - \epsilon),$$

$$\tan 2\epsilon = \frac{\gamma \Omega_m R^2}{2v^2}\frac{1}{\left(1 - \frac{\Omega_m^2 R^2}{v^2}\right)}. \tag{8.150}$$

Note that the key effect of the friction is to introduce a phase shift ϵ between the maximum of the force and the maximum of the displacement. The sign of ϵ determines whether the tidal bulge (the maximum of displacement) leads or lags behind the planet–satellite line. In the case of the Earth and the Moon, the tidal bulge is dragged ahead of the Moon with an angle $\epsilon \approx 3°$. It is clear therefore that the tidal torque will try to spin up the orbital velocity of the Moon and decrease the rotational rate of the Earth on its own axis.

The final length of the day (which will be the same as the final orbital period of the Moon) can be easily estimated from the conservation of angular momentum. If ω_1 is the rotational speed of Earth, ω_2 is the orbital speed of Moon, and ω_f is the final values for these two in synchronous rotation, then the conservation of angular momentum gives

$$I_1 \omega_1 + I_2 \omega_2 + M_2 \omega_2 R^2 = I_1 \omega_f + I_2 \omega_f + M_2 \omega_f R_f^2. \qquad (8.151)$$

Further, orbital dynamics requires that

$$\frac{R_f}{R} = \left(\frac{\omega_2}{\omega_1}\right)^{2/3}. \qquad (8.152)$$

(In arriving at the above equation, we have ignored the spin angular momentum of the Moon at the *present* epoch and the spin angular momenta of the Earth and the Moon in the *final* state; it can be easily verified that these are ignorable.) Taking $I_1 = (2/5)M_1 R_1^2$, we get

$$\frac{\omega_2}{\omega_f} = \left[1 + \frac{2}{5}\left(\frac{M_1}{M_2}\right)\left(\frac{R_1}{R}\right)^2 \left(\frac{\omega_1}{\omega_2}\right)\right]^3 = 1.99. \qquad (8.153)$$

Hence the final day (*and* month) will be ~ 54 days and the final Moon's orbital radius will be approximately 5.9×10^5 km.

At present, the fractional increase in the distance to the Moon is 1.3×10^{-10} yr^{-1} and the lengthening of the day is by 7×10^{-4} s/century. Precise calculation of the tidal friction, described above, suggests that – although it could be the dominant effect – there could be other contributions to the lengthening of day, etc. A completely satisfactory model, accounting for all the observations, is not yet available.

Exercise 8.10

Effect of tides: Give simple physical arguments to show that (a) the tide raised on the satellite by the planet has the following effects: (1) It despins the satellite towards synchronous rotation, (2) it acts to damp the satellite's orbital eccentricity, and (3) it can also tidally heat up the satellite. (b) Also show that the tide raised on the planet by the satellite has the following effects. (1) It increases the satellite's semimajor axis if $\omega_{\rm rot} > \omega_{\rm rev}$ and $\omega_{\rm rev} > 0$, where $\omega_{\rm rot}$ is the angular velocity of rotation of the planet and $\omega_{\rm rev}$ is the angular velocity of the satellite around the planet. If $\omega_{\rm rot} < \omega_{\rm rev}$ (as in the case of the satellite Phobos of Mars), the semimajor axis will decrease and the satellite can eventually crash onto the planet. (2) It decreases the planet's spin rate if $\omega_{\rm rot} < \omega_{\rm rev}$ and $\omega_{\rm rev} > 0$.

8.10.5 Long-Term Evolution of the Solar System

The solar system is an example of a nearly integrable system in the sense that – to the lowest order – the orbits of the planets around the Sun, treated as a Kepler problem, are integrable. The phase-space structure of such a system is made of invariant tori, described in Vol. I, Chap. 2, Section 2.7. When nonintegrable pieces are added to the Hamiltonian, the nature of the system changes and perturbation theory will, in general, be divergent and cannot be relied on to predict long-term behaviour. As was described in Vol. I, Chap. 2, Section 2.7, such a situation leads to chaos, which is associated with the trajectories of the original, unperturbed problem in which the ratios of the characteristic frequencies are rational numbers. Such a situation is called a *resonance*, and there are three main kinds of resonance phenomena that take place in the solar system. The first is the spin-orbit resonance which arises because of the commensurability of the period of rotation of a satellite and the period of its orbital revolution. The external driving force is the gravitational tidal torque on the satellite from the planet that is due to the nonsphericity of the satellite. The second may be called secular resonance, which arises from the commensurability of the frequencies of precession of the orbital orientation as described by the direction of the perihelion and the direction of the normal. Finally, there can be the mean motion resonance, which occurs when the orbital periods of two bodies are in the ratios of small numbers. The most familiar example of the spin-orbit resonance is in the Earth–Moon system, in which the Moon's rotation period is equal to its orbital period around the Earth. Indeed, most satellites in the solar system are locked into such a state.

Let us begin with the spin-orbit resonance, the mathematics of which can be reduced to that of a common pendulum. In an idealised model of an ellipsoidal satellite with the principle moments of inertial $A < B < C$ and spinning about the axis of largest moment of inertia, the equation of motion for the spin is given by

$$\ddot{\theta} = -\epsilon \frac{GM}{r^3} \cos 2(\theta - \psi), \quad \epsilon \equiv \frac{3(B - A)}{2C}, \quad (8.154)$$

where θ is the orientation of the satellite's long axis relative to the perihelion of the orbit, $\psi(t)$ is the angular distance of the orbit measured from the direction of the perihelion, and $r(t)$ is the radial position of the planet. [The derivation of Eq. (8.154) is along the same lines as that of relation (8.143).] It is assumed that the spin axis is normal to the orbital plane. If we further take the orbit to be circular, then the equation reduces to that of a pendulum, with

$$\ddot{\phi} = -2\epsilon\omega^2 \sin\phi, \quad \phi \equiv 2(\theta - \omega t), \quad (8.155)$$

where ω is the orbital frequency. These equations admit solutions in which the satellite's long axis is liberated around the planet–satellite direction, and its mean spin rate is the same as the orbital frequency. This is clearly the 1:1 spin-orbit resonance whose width is determined by the parameter $\omega(2\epsilon)^{1/2}$.

8.10 Aspects of Solar System Dynamics

It is, however, possible to have other kinds of resonances. When the orbit is noncircular but has a small eccentricity e, we can expand Eqs. (8.154) in a Taylor series in e to get

$$\ddot{\theta} = -\epsilon\omega^2\left\{\sin 2(\theta - \omega t) - \frac{1}{2}e[\sin(2\theta - \omega t) - 7\sin(2\theta - 3\omega t)] + \mathcal{O}(e^2)\right\}.$$
(8.156)

In addition to the original resonance, there are two new terms corresponding to 1:2 and 3:2 spin-orbit resonances. The planet Mercury is a well-known example of a 3:2 spin-orbit resonance with an orbital period of 88 days and a rotational period of 59 days. The width of the 3:2 resonance is a factor $(7e/2)^{1/2}$ smaller than the 1:1 resonance. For Mercury, with $e = 0.2$, this factor is ~ 0.84 so that the 3:2 resonance is approximately as strong as the 1:1 case. In many other satellites this is not the case, and hence higher-order resonances are not easy to observe.

There are also several orbital resonances that arise in the solar system, and we will discuss some of them. There exists a set of asteroids called trojan asteroids that are at 1:1 resonance with the orbit of Jupiter. The trojan asteroids occupy the same orbit as Jupiter but are located at a position that forms an equilateral triangle with respect to Jupiter and the Sun. These are clearly the stable Lagrangian points of the reduced planar three-body problem (see Vol. I, Chap. 2, Section 2.3), and they correspond to a location with an overabundance of asteroidal density.

There are also regions in the asteroid belts called *Kirkwood gaps*, the most prominent being at 3.3 AU (at 2:1 resonance) and at 2.5 AU (at 3:1 resonance). Resonances like the 3:1 orbital case have to be investigated numerically in order to obtain reliable conclusions. In this particular case, it is found that an orbit near the resonance maintains a low eccentricity ($e < 0.1$) for nearly 10^6 yr and then dramatically jumps over to a high eccentricity orbit with $e > 0.3$. There are also chaotic regions, in which the trajectories separate from each other at an exponential rate characterised by a factor $\exp(\Gamma t)$, where Γ (called the *Lyapunov exponent*) is fairly independent of the initial conditions and is $\sim 10^6$ yr. The outer boundaries of the chaotic zone in a 3:1 resonance coincide well with the boundaries of the Kirkwood gap, suggesting a possible explanation for the gap structure in the asteroid belt.

The resonances are also expected to play a key role in emptying the solar system of the debris. Numerical investigations show that there exists a band in the semimajor axis, centred on each planet's semimajor axis, within which all test-particle trajectories are chaotic because of orbital resonances. These test particles will quickly undergo a close encounter with the relevant planet. This process can perturb particles between the giant planets in a time scale of $\sim 10^6$ yr. The same effect also seems to lead to some dynamical erosion of the inner edge of the Kuiper belt.

The fact that regions near major planets can lead to chaotic orbits raises questions about the long-term stability of the planetary orbits themselves. This issue is plagued by the computational complexity and the necessity to perform very stable numerical integration over a long period of time. Such investigations suggest that the solar system does have large Lyapunov exponents and is chaotic at time scales of possibly tens of millions of years. However, the apparent regularity of the motion of the Earth shows that the solar system has indeed survived for 4.5×10^9 yr and hence the chaotic regions must be narrow. In practise, it could mean that, although we may not be able to calculate the exact location of the Earth in its orbit after, say, 20×10^6 yr, it is very likely that the Earth will continue to be in a low-inclination low-eccentricity orbit, with a semimajor axis of ~ 1 AU. These orbital elements could possibly evolve over only a much longer time scale, unlike the case of many minor bodies whose orbital parameters have evolved significantly within the lifetime of the solar system.

9
The Interstellar Medium

9.1 Introduction

This chapter discusses the physical features of the material that exists *between* the stars in our galaxy.[1] It draws heavily from Vol. I, Chaps. 6–9.

9.2 Overview

We have seen in Chap. 3 that stars form out of clouds of gas in the galaxy. This process of star formation from the protostellar cloud is never totally efficient and will certainly lead to the existence of a residual, ambient medium around the stars. We also saw that the there is transfer of material from the stars to the surrounding region; stellar winds of high-mass stars, ejection of the outer mantle in the formation of planetary nebulas, and supernova explosions are three processes that lead to such a mass transfer. These phenomena couple the stars directly with the medium around them. This medium is generically called the interstellar medium (ISM).

The physics of the ISM is extremely complex because the medium is very inhomogeneous and is made of regions with fairly diverse physical conditions. We shall first provide a general overview and a description of the ISM and then take up specific topics for discussion.

The composition of our galaxy is made of stars that provide a mass of approximately $(10^{10}$–$10^{11})$ $M_\odot$ and the ISM that provides a mass of $\sim 10^9$ $M_\odot$. Both stars and the ISM are distributed predominantly on the disk of the galaxy, with a typical radius of 10 kpc and a thickness of 250 pc. (In addition, there is a halo of invisible matter around the galaxy making up $\sim 10^{12}$ $M_\odot$, which is distributed more spherically; see Vol. III.) Hydrogen (90%) and helium (10%) in gaseous state make up most of the ISM mass, with a small fraction being contributed by silicate and graphite dust. Hydrogen exists as neutral (HI), singly ionized (HII), and in molecular (H_2) form, with HI dominating at larger distances from the centre of the galaxy. The conversion of the ISM *to* stars occurs typically at the

Table 9.1. The local ISM

Component	n (cm^{-3})	T (K)	Volume Fraction	Mass Fraction	Size (pc)
Hot ICM	0.005	5×10^5	0.5	0.001	—
Warm ICM	0.3	8000	0.5	0.1	900
HI clouds	5–20	10–100	0.05	0.4	300
H$_2$ diffuse	300	10–30	0.005	0.5	300
HII	10–10^4	10^4	0.001	0.02	30–80

star-formation rate of $\dot{M}_* \approx 3\ M_\odot\ \mathrm{yr}^{-1}$. The contribution of mass *from* the stars *to* the ISM occurs through supernovas (0.03 $M_\odot\ \mathrm{yr}^{-1}$), stellar winds of OB stars (0.08–0.5 $M_\odot\ \mathrm{yr}^{-1}$), and stellar winds from red giants (0.3–1 $M_\odot\ \mathrm{yr}^{-1}$).

Most of the mass in the ISM is distributed in the form of clouds of cold neutral material (with $T \approx 80$ K, $n \approx 10\text{--}100$ cm^{-3}) and in giant molecular clouds [(GMCs) with $T \approx 10$ K, $n \approx 10^2\text{--}10^5$ cm^{-3}]. However, most of the volume is occupied by hotter, gaseous material, which itself can be divided broadly into three components: (1) a hot diffuse component (with $T \approx 10^6$ K, $n \approx 10^{-3}$ cm^{-3}), (2) a warm ionised component (with $T \approx 8000$ K, $n \approx 0.1$ cm^{-3}), and (3) a warm neutral component (with $T \approx 5000$ K, $n \approx 0.1$ cm^{-3}). These three components as well as the cold neutral medium and the GMCs approximately satisfy the condition for pressure equilibrium:

$$P = nk_B T \approx 10^{-13}\ \mathrm{dyn\ cm}^{-2} = \mathrm{constant}. \tag{9.1}$$

These are summarized in Table 9.1 and Fig. 9.1.

Further evidence for the inhomogeneity of the ISM is provided by the existence of structures usually called *loops*, *spurs*, and *bubbles*. These are characteristic inhomogeneities with shapes described by their names. For example, the so-called loop I is a feature that leaves the galactic plane around $l = 30°$ and is seen up to $b = 30°$ and can be thought of as part of a circular loop with diameter of $\sim$116° centred at $l = 30°$, $b = 17°$. The radio emission from loop I is linearly polarised with a steep spectrum and can be described as synchrotron radiation. The magnetic field is mainly along the line of the loop (as it would have been if the field were swept up by an expanding shell) and has a value of approximately 5.5×10^{-6} G. This structure is very likely to be an outer shell of a supernova remnant. Assuming that the supernova is located above the galactic plane, we could explain the absence of the lower part of the loop as being due to the fact that the concentration of mass in the galactic plane has halted the expansion of remnant in that direction. The most striking example of a bubble is the *local bubble*, which is a region with dimensions of $\sim$100 pc located around the Sun with $T \approx 10^6$ K and $n \approx 5 \times 10^{-3}$ cm^{-3}. There is x-ray emission coming from the hot ionised gas and Lyman-α absorption along the line of sight

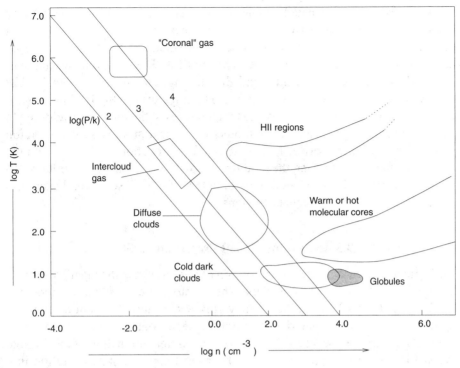

Fig. 9.1. Different physical regions in the ISM.

of OB stars. Structures like bubbles and loops are fairly transient compared with the time scale of the galaxy but are important in the detailed modelling of the ISM.

The ISM also displays a variety of thermal and bulk motion. The thermal motion is essentially characterised by the speed $c_s \approx (k_B T/m)^{1/2}$, which becomes $c_s = 10 T_4^{1/2}$ km s^{-1} for hydrogen atoms and $c_s = 400 T_4^{1/2}$ km s^{-1} for electrons (T_4 is the temperature in units of 10^4 K, etc.). There is a wide variation in the thermal speeds because of the corresponding variation in the temperature. The bulk flow of gas arises from the motion of GMCs, propagation of pressure disturbances at sound speeds, peculiar velocities of matter, stellar winds and supernova blast waves, and varies by ∼4 orders of magnitude.

The dynamical processes in the ISM can be broadly divided into those pertaining to local disturbances and those pertaining to the global, large-scale nature of the ISM. The first category consists of regions around OB stars or supernovas in which the ionising radiation (or the blast wave of the supernova) violently disturbs the background ISM. Of these, some aspects of the physics of the supernova remnants have already been discussed in Chap. 4, Section 4.9. The regions around hot stars will be made of ionised ISM and are conventionally called HII regions. The physics of such regions as well as the propagation of the ionisation front through the ISM will be discussed in Sections 9.3 and 9.4.

In addition to these isolated regions, the global characteristics of the ISM are decided by the balance between different physical processes. The first among them is the condition for mechanical balance, which can be stated as pressure equality between different components of the ISM. The second is thermal balance, which requires the heating and the cooling rates to match locally. The condition for this can lead to several interesting features in the ISM because of rather large variations in densities and temperatures of the regions. In addition to this, we also expect balance between other dynamical processes like ionisation and recombination or between excitation and deexcitation of different atomic and molecular levels. In fact, the requirement of all these balances is often sufficient to determine the local condition of the ISM. We shall see several examples of these processes in subsequent sections.

9.3 Ionisation of the ISM Around a Star

We begin our discussion with a description of regions of the ISM that are ionised by hot UV radiation from OB stars. These stars have effective temperatures of $T_{\text{eff}} \approx (2\text{--}6) \times 10^4$ K and hence they emit a substantial amount of photons, capable of ionising the material around it. In steady state, the rate of ionisation in the region around the star will match the rate of recombination of the plasma. We shall first work out a simple model for such a steady-state condition in a region having only hydrogen. When the star begins to ionise the ISM, it drives an ionisation front through it, and the dynamics of the propagation of this front is much more complicated than the steady-state situation. The dynamics will be discussed in Section 9.4.

Because the steady-state condition is determined by the balance between photoionisation and recombination, we briefly recall the relevant physics for these processes discussed in detail in Vol. I, Chap. 6, Section 6.12. The photoionisation cross section for neutral hydrogen can be expressed in the form

$$\sigma(\nu) \cong \sigma_{\text{PI}} \left(\frac{\nu}{\nu_I}\right)^{-s} \quad (\nu > \nu_I,\ s \simeq 3), \tag{9.2}$$

where $h\nu_I \approx 13.6$ eV is the ionisation threshold and $\sigma_{\text{PI}} \approx 2\pi(e^2/m_e c^2)(c/\nu_I) \approx 10^{-17}$ cm^2. We can determine the cross section for recombination σ_{rec} by relating it to σ_{PI} by the principle of detailed balance. The relevant parameter governing recombination is the recombination rate $\alpha(T) \equiv \langle v\sigma_{\text{rec}} \rangle$, where the averaging is over a thermal distribution of electrons. This is given by

$$\alpha(T) \equiv \langle v\sigma_{\text{rec}} \rangle \simeq 4 \times 10^{-13} \left(\frac{T}{10^4\ \text{K}}\right)^{-1/2} \text{cm}^3\ \text{s}^{-1} \tag{9.3}$$

if recombination to all states of hydrogen atom including the ground state is considered. Because direct recombination to the ground state can produce a photon that is very likely to ionise another atom immediately, it may be necessary to exclude the $n=1$ state from the computation of recombination rate in some situations. The expression is modified in such a case (Vol. I, Chap. 6, Section 6.12) but the numerical value around $T = 10^4$ K does not change significantly.

Consider now an ionising source (such as an OB star, although the discussion is quite general) with a luminosity $L(\nu)$. The flux of photons from the source with frequency ν, available at a distance r from the source, is given by

$$n_\gamma(\nu) = \frac{L(\nu)}{h\nu} \left[\frac{e^{-\tau_\nu(r)}}{4\pi r^2} \right], \tag{9.4}$$

where $\tau_\nu(r)$ is the optical depth in the medium up to the radius r:

$$\tau_\nu = \int_0^r n_H(r')\sigma(\nu)\,dr', \tag{9.5}$$

where n_H is the number density of neutral hydrogen atoms. The ionisation rate is given by the integral of $\sigma(\nu)n_\gamma(\nu)n_H$ over frequencies $\nu > \nu_I$, and the recombination rate is given by $n_e n_p \alpha(T)$. Equating the two, we get the condition for the ionisation equilibrium as

$$n_e n_p \alpha(T) = n_H \int_{\nu_I}^\infty \frac{L(\nu)\sigma(\nu)e^{-\tau_\nu(r)}}{4\pi r^2 h\nu}\,d\nu. \tag{9.6}$$

This can be rewritten more conveniently as

$$\frac{x^2}{1-x} = \frac{c\sigma_{PI}}{\alpha(T)}\Gamma, \quad x \equiv \frac{n_e}{n_0} = \frac{n_e}{n_e + n_H}, \tag{9.7}$$

where

$$\Gamma = \frac{1}{4\pi r^2 c n_0} \int_{\nu_I}^\infty \left[\frac{L(\nu)}{h\nu}\right] \left(\frac{\nu_I}{\nu}\right)^3 e^{-\tau_\nu}\,d\nu \tag{9.8}$$

is a dimensionless number, giving essentially the ratio between the number densities of ionising photons and neutral hydrogen atoms at a radius r (sometimes called the *ionisation parameter*). Given any ionising source, Eq. (9.7) determines $x(r)$. Strictly speaking, this is an implicit equation because τ_ν depends on $x(r)$ through $n_H(r) = n_0(1-x)$. However, it is possible to understand several features of the ionisation region by suitable approximations without solving this equation exactly.

We shall see later (towards the end of this section) that the region around such ionising sources are fully ionised up to some radius ($r = r_s$, say) with a transition to neutral gas occurring over a small region of thickness Δ with $\Delta \ll r_s$. Anticipating this result, we may take the gas to be almost fully ionised up to $r < r_s$ and fully neutral for $r > r_s$. It follows that τ_ν is small throughout the ionised nebula and rises rapidly near the edge; hence the factor $\exp(-\tau_\nu)$ in the definition of Γ can often be replaced with unity for $r < r_s$. As a specific numerical example, consider a region at $r \approx 5$ pc from an O6 star with an effective temperature $T_* = 4.2 \times 10^4$ K and $R_* = 6 \times 10^{11}$ cm. Because $(h\nu_I/k_B T_*) \approx 3.8$, we can take the relevant part of the blackbody spectrum of the star to be given by the Wien limit:

$$B_{\nu \geq \nu_I}(T_*) \simeq \frac{2h\nu^3}{c^2} \exp\left(-\frac{h\nu}{k_B T_*}\right). \tag{9.9}$$

In this case, the flux of the effective number of ionising photons (with $e^{-\tau} \simeq 1$) is

$$n_\gamma^{\text{ion}} \cong \int_{\nu_I}^\infty \frac{L(\nu)}{h\nu} \left(\frac{\nu_I}{\nu}\right)^3 d\nu = \frac{8\pi R_*^2}{c^2} \left(\frac{k_B T_*}{h}\right)^3 x_I^3 \int_{x_I}^\infty \frac{dx}{x} e^{-x}$$
$$\simeq 9.6 \times 10^{48} \text{ s}^{-1}, \tag{9.10}$$

and $n_0 \approx 10 \text{ cm}^{-3}$, giving $\Gamma \approx 10^{-2}$. Using $T \approx 10^4$ K, we find that $(c\sigma_{\text{PI}}/\alpha(T)\Gamma \approx 4.5 \times 10^3$ and $(1-x) \approx 2 \times 10^{-4}$; the region is almost completely ionized.

To study the variation of $x(r)$ in more detail, we approximate Eq. (9.7) by ignoring the ν dependence of $\tau(\nu)$ in the $\exp[-\tau(\nu)]$ factor and setting $\sigma(\nu) \approx \sigma_{\text{PI}}$ in this factor. This gives

$$\frac{x^2}{1-x} = \left(\frac{\sigma_{\text{PI}}}{4\pi \alpha n_0}\right) \left(\frac{1}{r^2}\right) \int_{\nu_I}^\infty d\nu \frac{L(\nu)}{h\nu} \left(\frac{\nu_I}{\nu}\right)^3 \exp\left\{-n_0 \sigma_{\text{PI}} \left(\frac{\nu_I}{\nu}\right)^3\right.$$
$$\left. \times \int_0^r [1-x(r')] dr'\right\}$$
$$\cong \left(\frac{\sigma_{\text{PI}} n_\gamma^{\text{ion}}}{4\pi \alpha n_0}\right) \left(\frac{1}{r^2}\right) \exp\left\{-n_0 \sigma_{\text{PI}} \int_0^r [1-x(r')] dr'\right\}$$
$$\equiv \left(\frac{n_0 \sigma_{\text{PI}} r_s^3}{3r^2}\right) \exp\left\{-n_0 \sigma_{\text{PI}} \int_0^r [1-x(r')] dr'\right\}, \tag{9.11}$$

where we have defined a radius r_s (called the *Stromgen* radius) by

$$r_s = \left(\frac{3 n_\gamma^{\text{ion}}}{4\pi} \frac{1}{n_e n_p \alpha}\right)^{1/3} \simeq \left(\frac{3 n_\gamma^{\text{ion}}}{4\pi n_0^2 \alpha}\right)^{1/3} \tag{9.12}$$

if $n_e = n_p = n_0$. This quantity can be interpreted as the effective, overall size of such an ionised region. To see this, note that we can determine the size of the ionised region by equating the total number of recombinations taking place per second to the total number of ionising photons emitted by the source per second. [This assumes that each photon is eventually capable of ionising a hydrogen atom and $\exp(-\tau_\nu) \approx 1$ in the region under consideration.] The condition

$$n_e n_p \alpha(T) \left(\frac{4\pi}{3} r^3\right) \cong \int_{\nu_I}^\infty \frac{L(\nu)}{h\nu} \left(\frac{\nu_I}{\nu}\right)^3 d\nu \equiv n_\gamma^{\text{ion}} \tag{9.13}$$

gives $r = r_s$. The number n_γ^{ion} is completely fixed by the properties of the ionising source; for an O star with $n_0 \simeq 10 \text{ cm}^{-3}$ and $n_\gamma^{\text{ion}} \approx 3 \times 10^{48}$, we have $r_s \approx 10$ pc.

9.3 Ionisation of the ISM Around a Star

The total optical depth for ionisation through the Stromgen sphere, *if* it is made of neutral hydrogen, would be

$$\tau_s \equiv n_0 \sigma_{\text{PI}} r_s \simeq 1.6 \times 10^3 \left(\frac{n_0}{10 \text{ cm}^{-3}}\right) \left(\frac{r_s}{10 \text{ pc}}\right), \tag{9.14}$$

which is quite large for the relevant range of parameters.

We can now rewrite Eq. (9.11) in terms of the dimensionless variables $l = r/r_s$ and τ_s as

$$\frac{x^2}{1-x} = \frac{\tau_s}{3l^2} \exp\left\{-\tau_s \int_0^l [1 - x(\bar{l})] \, d\bar{l}\right\}, \tag{9.15}$$

which in turn can be converted to an equation for the variable $q(l) \equiv \tau_s[1 - x(l)]$:

$$\frac{1}{q}\left(1 - \frac{q}{\tau_s}\right)^2 = \frac{1}{3l^2} \exp\left(-\int q \, dl\right). \tag{9.16}$$

The asymptotic forms of the solution to this equation can be easily determined. For large τ_s, the leading-order behaviour of q is $q(l) \approx 3l^2$; substituting into the exponential, we get the next-order solutions, again valid for $\tau_s \gg 1$, as

$$q(l) \cong \frac{3l^2}{1 - l^3}, \quad x(l) = 1 - \frac{q}{\tau_s} = 1 - \frac{3}{\tau_s}\frac{l^2}{1 - l^3}. \tag{9.17}$$

These solutions show that $x \lesssim 1$ for small l and large τ_s, as originally claimed; the material is fully ionised in this region. It is also clear that the above form of the solution breaks down when $l \approx 1 - \mathcal{O}(\tau_s^{-1})$. In that region x approaches zero and Eq. (9.15) can be approximated as

$$x^2 \cong \frac{\tau_s}{3l^2} \exp\left\{-\tau_s \int_0^{l_0} [1 - x(l')] \, dl'\right\} \exp[-\tau_s(l - l_0)], \tag{9.18}$$

where $l_0 = 1 - \mathcal{O}(\tau_s^{-1})$ and we have set $x = 0$ in the range $l_0 < l < 1$. To the same degree of accuracy, we can replace the upper limit on the integral inside exponential with 1 and we get the solution

$$x^2 = \frac{A^2}{l^2} \exp[-\tau_s(l - 1)], \quad A^2 \equiv \frac{\tau_s}{3} \exp\left[-\tau_s \int_0^1 (1 - x) \, dl\right]. \tag{9.19}$$

This solution shows that x approaches zero exponentially in the region of fractional thickness τ_s, that is, the actual thickness Δ of the region in which the transition from ionised to neutral gas is made is given by $\Delta \approx (r_s/\tau_s)$. This vindicates our previous approximations and establishes the Stromgen radius as the characteristic size of the ionised region. Some of the properties of the ionised regions are given in Table 9.2.

We have assumed so far that the ISM is made of pure hydrogen whereas in reality it has other components, especially helium. The analysis can be easily extended to cover helium, which has two ionisation potentials. The relevant

Table 9.2. Properties of HII regions

Spectral Type	$T_{\rm eff}$ (K)	$R_*/R_\odot$	$n_\gamma \times 10^{-48}$ (s^{-1})	$r_s(n_e n_H)^{1/3}$ (pc cm^{-2})	$\tau_s n_H^{-1/3}$ (cm)
O5	47,000	13.8	51	110	0.69
O6	42,000	11.5	17.4	77	0.48
O7	38,500	9.6	7.2	57	0.36
O8	36,500	8.5	3.9	47	0.29
O9	34,500	7.9	2.1	38	0.24
B0	30,900	7.6	0.43	22	0.14
B1	22,600	6.2	0.0033	4.4	0.028

parameters for helium are listed in Table 9.3. The Stromgen radius for HeIII will be given by the corresponding expression

$$r_s({\rm HeIII}) = \left[\frac{3}{4\pi} \frac{n_\gamma^{\rm ion}}{n_e n({\rm HeIII})\alpha}\right]^{1/3} \simeq \left[2.5 \frac{n_\gamma^{\rm ion}(E_0^{\rm He})}{n_\gamma^{\rm ion}(E_0^{\rm H})}\right]^{1/3} r_s({\rm H}), \quad (9.20)$$

where we have used the fact that $\alpha({\rm HeIII}) \approx 4\alpha(H)$ and $n({\rm HeIII}) \approx 0.1 n_p$. In the case of stars with steep spectra, the ratio between the available number of ionising photons, $n_\gamma^{\rm ion}$, at $E_0^{\rm He}$ and $E_0^{\rm H}$ can be small and the resulting structure of ionisation around a star will be made of several layers. The innermost layer will be made of HeIII, followed by a region of HeII and He, all of which will also contain ionised hydrogen; finally there will be an outer region where both helium and hydrogen are neutral. (We will see in Vol. III that the situation is different for ionised regions around quasars.)

The kinetic energy of a photoelectron emitted in an ionisation event is approximately $E_0 = 13.6$ eV, and this energy will be quickly thermalised among the constituent particles. The electron–electron-scattering cross section in a plasma is of the order of $\sigma_{ee} \approx 4\pi b_{\rm crit}^2$ with $b_{\rm crit} \approx (2e^2/m_e v^2)$ (see Vol. I, Chap. 9, Section 9.3). Therefore the collision rate among electrons is given by

$$R_{ee} \simeq \langle v\sigma_{ee}\rangle \simeq 4\pi\langle vb_c^2\rangle \simeq 1.4 \times 10^{-5} \text{ cm}^3 \text{ s}^{-1} \left(\frac{\epsilon_{\rm kin}}{1 \text{ eV}}\right)^{-3/2}$$
$$\simeq 3 \times 10^7 \alpha (T = 10^4 \text{ K}), \quad (9.21)$$

Table 9.3. Ionisation parameters for helium

Parameter	HeI $\to$ HeII	HeII $\to$ HeIII
E_0 (eV)	24.6	54.4
$\sigma_{\rm PI}(\nu_I)$	8×10^{-18}	$\frac{1}{4}\sigma_H \simeq 2 \times 10^{-18}$
α	$3 \times 10^{-13} T_4^{-1/2}$	$Z^2 \alpha_H \simeq 10^{-12} T_4^{-1/2}$

9.3 Ionisation of the ISM Around a Star

where $\alpha(T)$ is the recombination rate defined in Eq. (9.3). This shows that the electron scattering occurs at a significantly high rate compared with that of recombination, allowing thermalisation of the energy. Even electron–ion collisions and ion–ion collisions occur at sufficiently high rates to ensure a well-defined temperature to the system. Therefore the rate of photoelectric heating of the region is $\mathcal{R}_{\text{ion}} E_0$, where $\mathcal{R}_{\text{ion}}$ is the ionisation rate. The recombination of an electron and a proton releases an energy of $\sim k_B T$ in the form of photons that escape from the system so that the rate of recombination cooling is $\sim \mathcal{R}_{\text{rec}} k_B T$, where $\mathcal{R}_{\text{rec}}$ is the recombination rate. If these are the only heating and cooling processes, then the thermal-energy balance requires that $\mathcal{R}_{\text{ion}} E_0 = \mathcal{R}_{\text{rec}} k_B T$. Because $\mathcal{R}_{\text{ion}} = \mathcal{R}_{\text{rec}}$, it follows that $k_B T = E_0$; that is, the temperature of the ionised region should be $k_B T = 13.6\,\text{eV} \approx 10^5$ K, which is an order of magnitude higher than the observed values. We shall see, however, in Section 9.5 that there are other cooling processes in the ionised region that can operate effectively to lower the temperature to $\sim 10^4$ K.

Exercise 9.1

Ionisation in planar approximation: The question of the sharpness of the boundary of the ionisation layer, discussed above, was complicated by the fact that we needed to use spherical geometry. It is, however, possible to illustrate the key features by approximating the discussion to a planar geometry, which we shall do. (a) Show that, in this case, the ionisation fraction $x(r)$ is determined by the equation

$$\frac{dx}{dr} = -\alpha_0 n \frac{x(1-x)^2}{2-x}. \tag{9.22}$$

(b) Integrate this equation with an arbitrary boundary condition that $x = (1/2)$ at $r = 0$ and show that the ionisation fraction $x(r)$ and the ionising flux $J(r)$ are determined by the conditions

$$r \sigma_{\text{PI}} n = 2 - 2\ln\left(\frac{x}{1-x}\right) - \frac{1}{1-x}, \tag{9.23}$$

$$J_x = \left[\frac{2x^2}{(1-x)}\right] J_{1/2}, \tag{9.24}$$

where $J_{1/2} \equiv (\alpha n / 2\sigma_{\text{PI}})$ represents the photon flux at $x = 0.5$. (c) Show that the key conclusions derived in the text continue to be valid. {Answers: Let the flux of ionising photons at a distance r from the star be $J(r)$. The flux changes by an amount $dJ = -\sigma_{\text{PI}} n_H J\, dr = -\sigma_{\text{PI}} n(1-x) J\, dr$ while traversing a distance dr. This gives

$$\frac{dJ}{dr} = -\sigma_{\text{PI}} n(1-x) J. \tag{9.25}$$

The ionisation balance equation (9.7) can be reexpressed in terms of J as

$$\frac{x^2}{1-x} = \frac{J}{n}\left[\frac{\sigma_{\text{PI}}}{\alpha(T)}\right], \tag{9.26}$$

allowing (dJ/dr) to be expressed in terms of (dx/dr). This leads to differential equation (9.22) for $x(r)$. Integrating the equation, we get the variation of $x(r)$ to be given implicitly by

$$2\ln\left(\frac{x}{1-x}\right) + \frac{1}{1-x} = A - \lambda, \qquad (9.27)$$

where A is an integration constant and $\lambda = \sigma_{\text{PI}} nr$ is the mean free path of an ionising photon in neutral hydrogen of density n. Because the equation becomes singular near $x = (0, 1)$ there is some ambiguity in fixing the integration constant. For simplicity, let us choose $\lambda = 0$ at $x = 1/2$, which is equivalent to taking $A = 2$. This gives

$$\lambda = \sigma_{\text{PI}} nr = 2 - 2\ln\left(\frac{x}{1-x}\right) - \frac{1}{1-x}. \qquad (9.28)$$

We can also define a quantity $J_{1/2} \equiv (\alpha n/2\sigma_{\text{PI}})$ that represents the photon flux at $x = 0.5$. The photon flux at a point where the degree of ionisation x is given by

$$J_x = \left[\frac{2x^2}{(1-x)}\right] J_{1/2}. \qquad (9.29)$$

Equations (9.28) and (9.29) provide an implicit relation for determining $J(r)$. From these, it is easy to show that the flux drops exponentially for $r \gtrsim (\sigma_{\text{PI}} n)^{-1}$. Comparing with the radius of the Stromgen sphere, we find that $\Delta \ll r_s$, showing that ionisation nebulas have fairly sharp edges.}

9.4 Propagation of Ionisation Fronts

The discussion in the last section focussed on the steady-state situation around an ionising source that arises at late times. Initially, when the star is formed and starts emitting UV photons, it will ionise the region around it in a progressive manner with an ionisation front propagating outwards and converting neutral hydrogen to ionised hydrogen fairly rapidly. The radius $R(t)$ of the ionised region increases with time but there is very little mass motion. However, ionisation also increases the kinetic energy of the particles in the region to a higher value compared with that of the ambient medium. This, in turn, causes a pressure imbalance between the hot inner HII region and the cold outer HI region and will cause the gas outside to be compressed. Eventually there will be mass motion, caused by the higher pressure of the ionised region, that drives a shock front though the ISM. We will now analyse the physics of this process by using a simplified model.

Let us first consider the initial propagation of the ionisation front through the ambient medium that has an unperturbed density $\rho = m_H n_0$. The production of electrons by ionisation will be governed by the equation

$$\frac{\partial n_e}{\partial t} + \nabla \cdot (n_e \mathbf{u}) = -\alpha n_e^2 - \nabla \cdot \left[\frac{n_\gamma^{\text{ion}}}{4\pi r^2}\hat{\mathbf{r}}\right], \qquad (9.30)$$

where n_e is the number density of electrons, α is the recombination coefficient [see Eq. (9.3)], and n_γ^{ion} is the effective number of ionising photons released by the central source per second, defined by relation (9.10). (We have ignored the optical-depth correction, which makes very little difference in the analysis.) In the initial stage, the ionisation front will propagate rapidly [that is, (dR/dt) will be much larger than the thermal velocity v_T of the ionised region] through the neutral gas, converting hydrogen into protons and electrons. Because this involves very little mass motion, we can obtain the growth rate of the ionisation region by integrating Eq. (9.30) over the volume of the ionised region. The last term on the right-hand side gives just n_γ^{ion}; the first term on the right-hand side gives

$$-\int \alpha n_e^2 d^3x \simeq -\alpha n_0^2 \frac{4\pi}{3} R^3 \qquad (9.31)$$

if we take $n_e \simeq n_0$. Integrating the left-hand side of Eq. (9.30), we obtain $(d/dt)(4\pi R^3 n_0/3)$. This leads to the equation

$$\frac{d}{dt}\left(\frac{4\pi}{3}n_0 R^3\right) = n_\gamma^{ion} - \frac{4\pi}{3} R^3 \alpha n_0^2. \qquad (9.32)$$

Changing the variable of integration to R^3, we can easily integrate this to provide the solution

$$R(t) = r_s\left(1 - e^{-t/t_R}\right)^{1/3}, \qquad (9.33)$$

where $r_s = (3n_\gamma^{ion}/4\pi n_0^2 \alpha)^{1/3}$ is the Stromgen radius defined above and $t_R \equiv (\alpha n_0)^{-1}$, giving the characteristic time scale for recombination. For $n_0 \approx 10\,\text{cm}^{-3}$, this is $t_R \approx 10^4$ years.

Equation (9.33) shows that the ionisation front is moving into the medium with a speed $U(t) = (dR/dt)$, which is decreasing with time. For sufficiently large t, $R \to r_s$ and $U \to 0$. However, our derivation was based on the assumption that U is far greater than the thermal velocity of the ionised medium and hence must break down long before this. In other words, we cannot ignore the bulk flow of matter at later stages when the ionisation front slows down.

During the initial stages, the propagation of the ionisation front will not cause any bulk flow in the medium and will merely convert the neutral hydrogen into a plasma at a temperature $T \simeq 10^4$ K. (This equilibrium temperature is due to the efficient cooling mentioned earlier at the end of last section; we will discuss the processes in detail in Section 9.5.) The ambient ISM, however, is usually at a temperature $T \simeq 10^2$ K. Very soon, this temperature difference will make the gas flow relative to the ionisation front, thereby causing a shock front to move through the neutral medium. We may assume that the radiative processes exist in both regions, maintaining the constancy of the temperature. We now have a situation involving both an ionisation front and a shock front and we need to determine the flow of gas based on the jump conditions. This can be done as follows.

Let us consider the phenomena in the rest frame of the shock with the neutral HI region on the right and the ionised HII region on the left. In this frame, the gas is flowing at the speed v_1 on the right-hand side and v_2 on the left-hand side, with both flows being directed from right to left. On the right-hand side, we have no radiating flux, whereas the flux of radiation on the left is given by

$$J_2 = \frac{n_\gamma^{\text{ion}}}{4\pi r^2} e^{-\tau(r)}, \tag{9.34}$$

where we have included the optical depth $\tau(r)$ for the photons from the source to the surface of the sphere of radius r. This quantity J_2 can be treated as a given prespecified number. Assuming that each ionising photon ionises a hydrogen atom, we get two equations

$$\frac{\rho_1 v_1}{m_{\text{H}}} = J_2, \quad \rho_1 v_1 = \rho_2 v_2. \tag{9.35}$$

In addition, we have the jump condition on the energy in the form of the constancy of $p + \rho v^2$; writing $p = c^2 \rho$, where c is the speed of sound, we get one more condition:

$$\rho_1 \left(v_1^2 + c_1^2\right) = \rho_2 \left(v_2^2 + c_2^2\right). \tag{9.36}$$

These equations can be combined to give a quadratic equation for ρ_2/ρ_1:

$$c_2^2 \left(\frac{\rho_2}{\rho_1}\right)^2 - \left(v_1^2 + c_1^2\right)\left(\frac{\rho_2}{\rho_1}\right) + v_1^2 = 0, \tag{9.37}$$

which has the solution

$$\frac{\rho_2}{\rho_1} = \frac{v_1}{v_2} = \frac{1}{2c_2^2}\left\{\left(c_1^2 + v_1^2\right) \pm \left[\left(c_1^2 + v_1^2\right)^2 - 4c_2^2 v_1^2\right]^{1/2}\right\}. \tag{9.38}$$

Because we want the solutions to be real, this puts an interesting constraint on the velocity v_1. The quantity within brackets in expression (9.38) has to be positive for the solutions to be real; treating this as a quadratic in v_1^2, we find that the expression vanishes for the values

$$v_\pm = \left[c_2 \pm \left(c_2^2 - c_1^2\right)^{1/2}\right]. \tag{9.39}$$

Because $c_2 \gg c_1$, the two roots can be approximated as

$$v_R \approx 2c_2, \quad v_D \approx \frac{c_1^2}{2c_2}, \tag{9.40}$$

where the larger root has the subscript R denoting "rarefied" and the smaller root has the subscript D denoting "dense." This nomenclature arises from the fact that higher speeds lead to lower densities in the medium and vice versa. We can now approximate the term within the brackets in Eq. (9.38) as $(v_1^2 - v_R^2)(v_1^2 - v_D^2)$.

9.4 Propagation of Ionisation Fronts

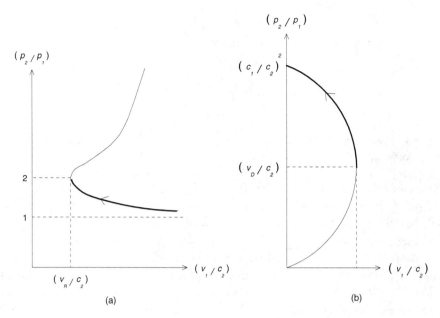

Fig. 9.2. The density ratios as functions of velocity for the R- and the D-type solutions.

Then the original solution for ρ_2/ρ_1 in Eq. (9.38) can be written as

$$\frac{\rho_2}{\rho_1} \cong \frac{1}{2c_2^2}\left\{(v_R v_D + v_1^2) \pm \left[(v_1^2 - v_R^2)(v_1^2 - v_D^2)\right]^{1/2}\right\} = \frac{v_1}{v_2}. \quad (9.41)$$

It is now clear that the real solution will exist only when one of the following two conditions are satisfied:

$$v_1 \geq v_R \text{ or } v_1 \leq v_D. \quad (9.42)$$

If $v_1 > v_R$, we call the ionisation front an R type. If $v_1 < v_D$, it is called a D type. For each of these types, we have two possible roots corresponding to the plus or the minus sign in relation (9.41). Figure 9.2(a) shows ρ_2/ρ_1 as a function of v_1/c_2 for an R-type front; Figure 9.2(b) shows the corresponding quantities for the D-type front. From the figure and relation (9.41), we can easily conclude the following:

For the R-type front, when $v_1 \gg c_2$, the density ratio approaches unity as $(\rho_2/\rho_1) \to [1 + (c_2^2/v_1^2)] \approx 1$. This corresponds to the initial stages of the ionisation front's moving rapidly through the medium without changing the density too much. As the speed v_1 decreases, we are moving along the lower branch of the curve in Fig. 9.2(a). Because v_1 has to be greater than v_R, the ionisation front can at most get to $v_1 = v_R$ along this branch, shown by the thick lower part of the curve in Fig. 9.2(a).

At this juncture, the ionised gas is moving into the neutral gas with a speed $c_2 \gg c_1$. This will initiate a shock wave in the neutral gas, provided the pressure

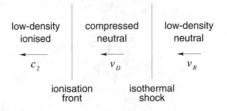

Fig. 9.3. The flow of gas when an isothermal shock exists in conjunction with the ionisation front.

wave can catch up with the ionisation front. This becomes possible when the ionisation front slows down to the critical value $v_R \approx 2c_2$. (The pressure wave will be moving at a speed c_2, relative to the bulk flow speed of c_2, thereby attaining a *net* speed of $2c_2$.) This allows the critical R-type front to make a transition to a D-type front through an isothermal shock. We saw in Vol. I, Chap. 8, Section 8.11 that across an isothermal shock, the product of the upstream velocity of the undisturbed medium (say, v_0) and the downstream velocity of the compressed gas v_1 is equal to c_1^2. We can now visualise the situation shown in Fig. 9.3. Neutral low-density gas flows into an isothermal shock with an R-critical speed of $v_0 = v_R$ from the extreme right. From the matching conditions for the isothermal shock, we know that the velocity in the middle region will be $v_1 = (c_1^2/v_R) = (c_1^2/2c_2)$ when $v_R = 2c_2$ – which is identical to v_D. Thus in the middle region we have a D-type flow of compressed neutral gas flowing with critical speed. This gas flows through a D-critical ionisation front, changing its speed from v_D to c_2, and emerges as a low-density ionised gas.

Figure 9.2(b) shows that, for a D-type ionisation front corresponding to the upper branch in the figure, the density ratio has the asymptotic value $(\rho_2/\rho_1) = (c_1/c_2)^2$; that is, $(\rho_2 c_2^2) = (\rho_1 c_1^2)$, signifying pressure equilibrium between the two regions. This is what we expect in the late stages of evolution. The transition from an R-critical ionisation front to a D-critical ionisation front occurs when the isothermal shock and the ionisation front are located in the same place. In this type of evolution, we have a shock front running ahead of the ionisation front through the neutral gas.

9.5 Heating and Cooling of the ISM

We saw in Section 9.3 that if the only heating and cooling processes that operate in an ionised nebula are due to photoionisation and recombination, then the temperature of the ionised region will be very high, $\sim 10^5$ K. Observations suggest that the temperature is actually lower by more than an order of magnitude. This arises because there are many other coolants in the ISM that contribute to the cooling of the nebula. These coolants are usually the heavier atoms, such as oxygen, carbon, etc., which are present in only small amounts and hence have

9.5 Heating and Cooling of the ISM

a negligible *direct* influence on the overall dynamical behaviour of the ISM. Nevertheless, they play an extremely vital role in the thermodynamical aspects of the ISM and hence could *indirectly* influence the dynamical characteristic of the ISM in an interesting manner. To understand the thermal balance that exists in ISM, we shall discuss each of the heating and the cooling processes that contribute significantly.

The first heating process that acts in the ISM is the heating that is due to the photoionisation of elements – especially the most abundant one, viz., hydrogen. This process operates not only near an ionising source such as an O star, but even throughout the ISM, if sufficiently high energy photons or cosmic rays, capable of ionising the hydrogen, permeate the medium. The net heating of the ISM, because of a flux J_ν of photons, is given by

$$\mathcal{H}_{\text{ion}} = n_{\text{H}} \int_{\nu_I}^{\infty} \frac{4\pi J_\nu}{h\nu}(h\nu - h\nu_I)\sigma_{\text{ion}}(\nu)\,d\nu. \tag{9.43}$$

The flux of photons is given by the first factor inside the integral, $(4\pi J_\nu/h\nu)$, and the amount of energy acquired by the electron is given by the second factor $h(\nu - \nu_I)$; $\sigma_{\text{ion}}(\nu)$ is the photoionisation cross section given by relation (9.2). Because equilibrium between ionisation and recombination requires the condition

$$n_e n_p \alpha(T) = n_{\text{H}} \int_{\nu_I}^{\infty} \frac{4\pi J_\nu}{h\nu}\sigma_{\text{ion}}(\nu)\,d\nu \tag{9.44}$$

[which is essentially Eq. (9.6) written in terms of the photon flux], Eq. (9.43) can be expressed in the form

$$\mathcal{H}_{\text{ion}} = n_e n_p \alpha(T)\langle h(\nu - \nu_I)\rangle_I \tag{9.45}$$

where the averaging symbol $\langle\cdots\rangle_I$ is defined as

$$\langle F(\nu)\rangle_I \equiv \frac{\int_{\nu_I}^{\infty} \frac{4\pi J_\nu}{h\nu} F(\nu)\sigma_{\text{ion}}(\nu)\,d\nu}{\int_{\nu_I}^{\infty} \frac{4\pi J_\nu}{h\nu}\sigma_{\text{ion}}(\nu)\,d\nu} \tag{9.46}$$

for any function $F(\nu)$. For most of the realistic ionising sources $[\langle h(\nu - \nu_I)\rangle/h\nu_I] \approx 0.2$–$2$. This implies that typical energies of the photoelectrons produced during the ionisation are ~ 10 eV. As we saw above, this energy will be rapidly thermalised by electron–electron and electron–ion collisions.

The inverse process of photoionisation, which involves recombination of an electron and an ion, will lead to the cooling of the medium whenever the photons can escape from that region. Direct recombination to ground state, leading to an emission of a photon with energy $h\nu = 13.6$ eV, does not contribute significantly to this process because such a photon will be absorbed by another neutral hydrogen atom with a very high probability. Recombination to excited states of hydrogen followed by progressive radiative deexcitation with the emission of low-energy photons is more effective for cooling the system. Because the kinetic energy $(1/2)m_e v^2$ is lost in each recombination, the cooling rate that is due to recombination can be expressed in the form

$$\mathcal{C}_{\rm rec} = n_e n_p k_B T \alpha(T) \left[\frac{\langle (1/2)m_e v^2 \rangle_R}{k_B T} \right], \qquad (9.47)$$

where the averaging $\langle \cdots \rangle_R$ is defined as

$$\left\langle \frac{1}{2}m_e v^2 \right\rangle_R \equiv \frac{\int_0^\infty dv\, v f(v) \sigma_{\rm rec} (1/2) m_e v^2}{\int_0^\infty dv\, v f(v) \sigma_{\rm rec}}, \qquad (9.48)$$

where $f(v)$ is the standard Maxwell–Boltzmann distribution, and $\sigma_{\rm rec}$ is the cross section for recombination. The advantage of writing $\mathcal{C}$ in this particular form of Eq. (9.47) is that the quantity in the brackets is a very weak function of temperature and can be usually taken to be a constant of the order of unity. The recombination coefficient $\alpha(T)$ should be computed excluding the $n = 1$ state in most of the situations. It is obvious from Eqs. (9.45) and (9.47) that, if these are the only two processes operating, then $\mathcal{H} = \mathcal{C}$ will lead to $k_B T \approx h\nu_I \approx 10$ eV. We shall next consider other cooling processes that operate in the system.

Exercise 9.2

Carbon lines and ionisation parameter: Consider an optically thin gas cloud embedded in an ionising spectrum with F_ν (erg cm^{-2} s^{-1} Hz^{-1}) $\propto \nu^{-1}$. The cloud contains carbon atoms in different states of ionisation. The ionisation potential χ, the photoionisation cross section at the threshold $\sigma_{\rm PI}$, and the recombination coefficient α_R for different states of ionisation are as follows: $(\alpha_R, \sigma_{\rm PI}, \chi) = (6.5 \times 10^{-13}$ cm^3 s^{-1}, 1.2×10^{-17} cm^2, 11.3 eV) for CI $\Leftrightarrow$ CII, $(8.5 \times 10^{-12}$ cm^3 s^{-1}, 4.6×10^{-18} cm^2, 24.4 eV) for CII $\Leftrightarrow$ CIII, $(1.8 \times 10^{-11}$ cm^3 s^{-1}, 1.6×10^{-18} cm^2, 47.9 eV) for CIII $\Leftrightarrow$ CIV, and $(3.4 \times 10^{-11}$ cm^3 s^{-1}, 6.8×10^{-19} cm^2, 64.5 eV) for CIV $\Leftrightarrow$ CV. Compute (a) the fraction of carbon in each of the ionisation states I–V and (b) the ratio of densities $n({\rm CII})/n({\rm CIV})$ when the ionisation parameter is $F_{\rm H}/n_e = 4 \times 10^8$ cm s^{-1}, 4×10^6 cm s^{-1}; $F_{\rm H}$ is the flux of ionising photons in units of per square centimetres per second. [Hint: The ionisation balance between two states i and $i+1$ is determined by

$$n_{i+1} n_e \alpha_R = n_i \int_{\nu_I}^\infty \frac{F_\nu}{h\nu} \sigma_{\rm PI}(\nu)\, d\nu, \quad F_{\rm H} = \int_{\nu_I}^\infty \frac{F_\nu}{h\nu}\, d\nu. \qquad (9.49)$$

Taking the form of F_ν and using $\sigma_{\rm PI}(\nu) = \sigma_{\rm PI}(\nu_I)(\nu/\nu_I)^{-3}$, we get $(n_{i+1}/n_i) = (\sigma_{\rm PI}/4\alpha_R)(F_{\rm H}/n_e)$. The rest of the calculation can now be done in a straightforward manner. The ratio of the number of atoms in the levels $i+1$ and i for $i = 1, 2, 3, 4$ is given by $(7.4 \times 10^3, 2.2 \times 10^2, 35.6, 2)$ when the ionisation parameter is 4×10^8 cm s^{-1} and is 100 times smaller for the second case. The fraction $n({\rm CII})/n({\rm CIV})$ is $2.08 \times 10^{13}(F_{\rm H}/n_e)^{-2}$, which is 1.3×10^{-4} for the first case and 1.3 for the second case.]

The most important cooling process, other than recombination, is through radiative line cooling, which occurs as follows. Consider an atom that collides

9.5 Heating and Cooling of the ISM

with an electron, say, and gets excited to a higher-energy state. The collisional cross section for such processes is $\sigma_{\text{coll}} \approx 10^{-15}$ cm^2 [see relation (9.52) below], giving a collision rate of

$$\tau_{\text{coll}}^{-1} = n\sigma_{\text{coll}} v \cong \left(\frac{k_B T}{m_H}\right)^{1/2} n\sigma_{\text{coll}} \approx 9 \times 10^{-12} n T^{1/2} \text{s}^{-1}. \quad (9.50)$$

Such an excited atom can get back to the ground state either through collisional deexcitation or through spontaneous radiative transition, in which it decays radiatively, emitting the extra energy in the form of a photon. The rate for such radiative transitions, determined by the Einstein coefficient A_{21} (see Vol. I, Chap. 4, Section 4.5), is $A_{21} \approx 10^8$ s^{-1} for permitted electric dipole radiation whereas the rate is much smaller for transitions that are forbidden to occur through the dominant channels. Whenever a particular radiative transition rate is high compared with the collisional rate computed in Eq. (9.50), that transition will act as a dominant cooling process. The original collision would have transferred some kinetic energy of the electron to the atom, which eventually is lost from the system in the form of photons that escape from the region. This process can be expressed symbolically as

$$\text{atom} + e^- \rightarrow \text{atom}^* + e^- \rightarrow \text{atom} + e^- + \gamma. \quad (9.51)$$

Because of the rather low rate of collisions in the ISM, the excited atom is likely to persist in the excited state until it decays by means of a radiative transition in several cases. (For comparison, it may be noted that, in the lower regions of the Earth's atmosphere, the deexcitation is usually collisional and not radiative because of the relatively high density.)

The cross section for the collision between an electron and an atom that causes a transition between levels i and j is given by

$$\sigma_{ij} \cong \pi \left(\frac{\hbar}{m_e v}\right)^2 \frac{\Omega_{ij}}{g_i} \simeq 10^{-15} \text{ cm}^2 \left(\frac{\Omega_{ij}}{g_i}\right) \left(\frac{T}{10^4 \text{ K}}\right)^{-1}. \quad (9.52)$$

This expression was discussed in Vol. I, Chap. 6, Section 6.13 and arises essentially from the fact that the effective cross section of the electron is decided by its de Broglie wavelength ($\hbar/m_e v$). Hence we expect Ω_{21} to be of the order of unity, which is usually true. (For a precise estimate, tabulated values of Ω_{ij} need to be used.) It should also be noted that $\Omega_{ij} = \Omega_{ji}$ in this parameterisation, giving

$$g_2 v_2^2 \sigma_{21}(v_2) = g_1 v_1^2 \sigma_{12}(v_1), \quad \frac{1}{2} m_e \left(v_1^2 - v_2^2\right) = E_{12}, \quad (9.53)$$

so that both the collisional excitation and the deexcitation rates per unit volume can be expressed in terms of the cross section σ_{ij}. Let us consider two levels, 1 and 2, where level 1 is the ground state and level 2 is an excited state with an energy gap E_{12}. The collisional deexcitation rate per unit volume between these two levels is

$$n_e n_2 \mathcal{R}(2 \to 1) \equiv n_e n_2 \mathcal{R}_{21} \equiv n_e n_2 \int_0^\infty v \sigma_{21} f(v) \, dv, \quad (9.54)$$

where $f(v)$ is the distribution function for electrons and this equation defines the rate coefficient $\mathcal{R}_{21}$. The corresponding excitation rate per unit volume is given by

$$n_e n_1 \mathcal{R}(1 \to 2) \equiv n_e n_1 \mathcal{R}_{12} \equiv n_e n_1 \mathcal{R}_{21} \left(\frac{g_2}{g_1}\right) e^{-\beta E_{12}}, \quad (9.55)$$

which follows from the principle of detailed balance.

We are now in a position to determine the population of atoms in the two levels when steady state is reached through collisional excitation, collisional deexcitation, and radiative deexcitations, with the last process given by the rate $n_2 A_{21}$, where A_{21} is the spontaneous decay rate. The steady-state condition requires the net rate of upward transitions to be balanced by the corresponding rate of downward transitions, leading to

$$n_e n_1 \mathcal{R}(1 \to 2) = n_e n_2 \mathcal{R}(2 \to 1) + n_2 A(2 \to 1) \quad (9.56)$$

or

$$\frac{n_2}{n_1} = \frac{n_e \mathcal{R}_{12}}{A_{21}} \left(1 + \frac{n_e \mathcal{R}_{21}}{A_{21}}\right)^{-1} = \frac{\mathcal{R}_{12}}{\mathcal{R}_{21}} \left(1 + \frac{A_{21}}{n_e \mathcal{R}_{21}}\right)^{-1}$$

$$= \frac{g_2}{g_1} e^{-\beta E_{12}} \left(1 + \frac{A_{21}}{n_e \mathcal{R}_{21}}\right)^{-1}, \quad (9.57)$$

with $A_{21} \equiv A(2 \to 1)$. When $n_e \gg (A_{21}/\mathcal{R}_{21})$, this leads to the standard population ratio expected in thermodynamic equilibrium. When $n_e \lesssim (A_{21}/\mathcal{R}_{21})$ the excited state is depleted compared with the value in thermodynamic equilibrium because of spontaneous decay. Note that the photons emitted in the spontaneous decay are allowed to escape from the system and hence do not participate in the inverse process.

The corresponding cooling rate that is due to spontaneous decay will be $\mathcal{C} = n_2 A_{21} E_{21}$; writing $n = n_1 + n_2$ and $n_2 = n[1 + (n_1/n_2)]^{-1}$, we find that this cooling rate becomes

$$\mathcal{C} \cong n_2 A_{21} E_{21} = \frac{n A_{21} E_{21}}{1 + \left(\frac{n_1}{n_2}\right)} = \frac{n A_{21} E_{21}}{1 + (g_1/g_2) e^{\beta E_{21}} \left(1 + \frac{A_{21}}{n_e \mathcal{R}_{21}}\right)}. \quad (9.58)$$

This simple analysis already allows us to draw some interesting conclusions. The ratio $n_c \equiv (A_{21}/\mathcal{R}_{21})$ defines a critical density of electrons for any two particular levels under consideration. The behaviour of n_2/n_1 depends on how the actual density of electrons n_e compares with the critical density n_c. When $n_e \gg n_c$ the

9.5 Heating and Cooling of the ISM

level populations are given by

$$\frac{n_2}{n_1} \simeq \frac{\mathcal{R}_{12}}{\mathcal{R}_{21}} = \left(\frac{g_2}{g_1}\right) e^{-\beta E_{12}}, \tag{9.59}$$

and the cooling rate

$$\mathcal{C}_{\text{line}}(n_e \gg n_c) \simeq n_2 A_{21}(h\nu_{21}) \simeq n_1 \left(\frac{g_2}{g_1}\right)(A_{21}h\nu_{21})e^{-\beta E_{12}} \tag{9.60}$$

is proportional to the spontaneous decay probability A_{21}. These level populations represent standard thermodynamic equilibrium. On the other hand, for low densities ($n_e \ll n_c$) Eq. (9.57) gives $(n_2/n_1) \approx (n_e \mathcal{R}_{12}/A_{21})$. In this case, the rate of cooling that is due to photon emission is given by

$$\mathcal{C}_{\text{line}}(n_e \ll n_c) \simeq n_2 A_{21}(h\nu_{21}) \simeq n_1 n_e \mathcal{R}_{12}(h\nu_{12}) \tag{9.61}$$

and is independent of A_{21}.

To see the importance of the low-density result (when the level populations are not in thermodynamic equilibrium) let us compare the cooling rate obtained above, with the recombination cooling rate that is due to hydrogen. As an illustrative example, we shall consider the [OIII] line with a wavelength $\lambda = 5007$ Å, corresponding to the transition $^1D_1 \rightarrow {}^3P_2$ with the energy difference of $E_{12} \approx 2.5$ eV. The two rates [computed from relation (9.61) by use of relation (9.59) with $\mathcal{R}_{21} = \langle \sigma_{21} v \rangle$] are given by

$$\mathcal{C}_{\text{line}} \simeq n_{\text{ion}} n_e \langle \sigma_{21} v \rangle E_{12} \exp(-\beta E_{12}), \tag{9.62}$$

$$\mathcal{C}_{\text{Hy}} \simeq n_H n_e \langle \sigma_{21}^H v \rangle E_0 \exp(-\beta E_0). \tag{9.63}$$

Because $\langle \sigma v \rangle$ is nearly same for both, this ratio becomes

$$\frac{\mathcal{C}_{\text{line}}}{\mathcal{C}_{\text{Hy}}} \simeq \left(\frac{n_{\text{ion}}}{n_H}\right)\left(\frac{E_{12}}{E_0}\right) e^{\beta(E_0 - E_{12})}. \tag{9.64}$$

At $T = 10^4$ K, corresponding to $k_B T \approx 1$ eV, the argument of the exponential is quite large and is $(E_0 - E_{12})/k_B T \approx 11.1$. The prefactor $(E_{12}/E_0) \approx$ (2.5 eV/13.6 eV) ≈ 0.183 is insufficient to offset the effect of the large exponential. This implies that $\mathcal{C}_{\text{line}} \gg \mathcal{C}_{\text{hy}}$ even when $(n_{\text{ion}}/n_H) \gtrsim 5 \times 10^{-5}$. In the ISM the abundance of heavy ions is of the order of (C/H) $\approx 3 \times 10^{-4}$, (N/H) $\approx 10^{-4}$, (Ne/H) $\approx 7 \times 10^{-5}$, and (O/H) $\approx 6 \times 10^{-4}$. Most of these ions have energy levels that differ by typically a few electron volts, for which the critical density n_c is fairly high. Hence even a small fractional contamination of heavy ions with $n_{\text{ion}}/n_H \approx 10^{-4}$ can have dominant effect on the cooling process. At $T \approx 10^4$, the collisional rate $\mathcal{R} \approx \sigma v \approx 10^{-9}$ cm^3 s^{-1} so that the critical density is given by

$$n_c \simeq \frac{A_{21}}{\mathcal{R}_{21}} \simeq 10^9 \text{ cm}^{-3} \left(\frac{T}{10^4 \text{ K}}\right)^{1/2} \left(\frac{A_{21}}{1 \text{ s}^{-1}}\right). \tag{9.65}$$

For the OIII 5007Å line, $A_{12} \approx 2 \times 10^{-2}$ s^{-1} giving $n_c \approx 2 \times 10^7$ cm^{-3}. Whenever $n_e \ll n_c$, the line radiation can be an effective coolant.

These cooling processes are also responsible for lowering the temperature of the ionised HII regions. To see this effect, let us consider some ion with abundance $A_{\text{ion}} \approx 6 \times 10^{-4}$ and two energy levels with a gap of $E_{12} = 4$ eV. We will assume that these levels are connected by a forbidden transition such that the line cooling rate is given by low-density expression (9.62). If this line cooling is matched by photoionisation heating, then the thermal balance between heating and cooling requires $\mathcal{C}_{\text{line}}$ to be equal to the product of the rate of ionisation and the energy $E_0 \approx 13.6$ eV injected per ionisation. Because the rate of ionisation is equal to the rate of recombination in steady state, the net heating can be expressed as

$$\mathcal{H}_{\text{ion}} = n_p n_e \alpha(T) E_0 \cong n_p^2 \alpha(T) E_0 \qquad (9.66)$$

for a fully ionised nebula with $n_e = n_p$. Equating this to $\mathcal{C}_{\text{line}}$ we get

$$a_{\text{ion}} E_{12} \langle \sigma v \rangle e^{-E_{12}/k_B T} \cong \alpha(T) E_0. \qquad (9.67)$$

Both $\langle \sigma v \rangle$ and $\alpha(T)$ vary approximately as $T^{-1/2}$ if recombinations to all states including $n = 1$ are taken into account and the ratio $\langle \sigma v \rangle / \alpha(T) \approx 2.3 \times 10^5$ is independent of T. [If the $n = 1$ state is excluded while the recombination rate is computed, then $\alpha(T) \propto T^{-0.75}$ while $\langle \sigma v \rangle \propto T^{-1/2}$. This will give a weak T dependence to the ratio $\langle \sigma v \rangle / \alpha(T)$, which we shall ignore for simplicity.] The above equation can be solved for T to give

$$T \cong \frac{E_{12}}{k_B} \left\{ \ln \left[\frac{\langle \sigma v \rangle}{\alpha(T)} \frac{E_{12}}{E_0} a_{\text{ion}} \right] \right\}^{-1} \simeq 10^4 \text{ K}. \qquad (9.68)$$

If only recombinations up to the $n = 2$ state are included in computing the recombination rate, the expressions are modified slightly but the final numerical value is not changed significantly.

The above results are of rather a general nature. From relations (9.62) and (9.52) it is clear that the line cooling rate is given by

$$\mathcal{C}_{\text{line}} \simeq n_i n_e \left(\frac{\pi \Omega_{21}}{g_2} \right) \frac{\hbar^2}{m_e^2} \left(\frac{k_B T}{m_e} \right)^{-1/2} E_{12} \exp\left(-\frac{E_{12}}{k_B T}\right)$$

$$\propto T^{-1/2} \exp(-E_{12}/k_B T), \qquad (9.69)$$

where we have indicated only the temperature dependence in the last part. At low temperatures ($k_B T \ll E_{12}$), the colliding particles do not have sufficient energy to populate the excited state significantly and the exponential factor suppresses the cooling. At high temperatures ($k_B T \gg E_{12}$), the line cooling rate decreases as $T^{-1/2}$ because the collision rate σv falls as $T^{-1/2}$. It follows that this effect is most significant when $k_B T \approx E_{12}$. For different regions of the ISM with different temperatures, efficient radiative cooling will arise from different kinds of lines.

9.5 Heating and Cooling of the ISM

It is also necessary that the density of particles inducing the collision be small compared with the critical density for the given transition. The existence of forbidden lines in OIII, NII, etc., with E_{12} of a few electron volts, dominates the cooling and keeps the temperature of the HII regions at $T \simeq 10^4$ K.

As an example of the above phenomena, in a different context, we will discuss the transitions that can occur in a molecular gas cloud at temperature $T = 10$ K, containing CO and NH_3. The lowest rotational transition $J = 1$–0 for the CO molecule has an Einstein A coefficient $A_{CO} = 6 \times 10^{-8}$ s^{-1} and a frequency ν_{CO} of 115 GHz. The corresponding $J = 1$–0 transition in NH_3 has $A_{NH_3} = 1.5 \times 10^{-3}$ s^{-1} and a frequency ν_{NH_3} of 570 GHz. The molecular abundances by number are CO/H $= 3 \times 10^{-4}$ and NH_3/H $\approx 3 \times 10^{-7}$. The transitions between $J = 0$ and $J = 1$ can be due to collisional excitation with H_2 molecules followed by radiative deexcitation, with the collisional cross section being roughly $(1/10)$ of the geometrical cross section. Ignoring all other levels except $J = 0$ and 1, we find that the collisional excitation and deexcitation rates are related by

$$\mathcal{R}_{01} = \frac{g_1}{g_0} \mathcal{R}_{10} e^{-E_{10}/k_B T}, \quad \mathcal{R}_{10} = \int_0^\infty (\sigma v) f(v) dv \simeq \sigma \left(\frac{8 k_B T}{\pi m_{H_2}} \right)^{1/2}. \tag{9.70}$$

In this particular case, the degeneracy factor for the Jth level is given by $g_J = (2J + 1)$ so that $(g_1/g_0) = 3$. Hence

$$\mathcal{R}_{01} = 3\sigma \left(\frac{8 k_B T}{\pi m_{H_2}} \right)^{1/2} \exp\left(-\frac{E_{10}}{k_B T}\right). \tag{9.71}$$

The cooling rates that are due to CO and NH_3 can be directly read off from relation (9.58), noting that – in this particular case – it is the collisions with H_2 molecules that are causing the excitation. Using the numerical values given above and $\mathcal{R}_{10} \approx 2 \times 10^{-12}$ cm^3 s^{-1}, we find that the cooling rates per hydrogen atom that are due to CO and NH_3 are given by

$$\left(\frac{\mathcal{C}}{n_H}\right)_{CO} \simeq 10^{-26} \left(1 + \frac{3 \times 10^4 \text{ cm}^{-3}}{n_{H_2}}\right)^{-1} \text{ ergs cm}^3 \text{ s}^{-1}, \tag{9.72}$$

$$\left(\frac{\mathcal{C}}{n_H}\right)_{NH_3} = 2.7 \times 10^{-25} \left(1 + \frac{4 \times 10^7 \text{ cm}^{-3}}{n_{H_2}}\right)^{-1} \text{ ergs cm}^3 \text{ s}^{-1}. \tag{9.73}$$

These cooling rates are plotted in Fig. 9.4. It is clear from the expressions and the figure that CO dominates the cooling at low densities whereas NH_3 dominates at high densities, with the transition occurring around $n_H = 1.3 \times 10^6$ cm^{-3}. This result is also understandable in terms of the critical density introduced above. For this system, $n_{cr} \approx 3 \times 10^4$ cm^{-3} for CO whereas it is $n_{cr} \approx 7.5 \times 10^8$ cm^{-3} for NH_3. The limiting forms of the cooling rate in the two limits $n_H \ll n_{cr}$ and

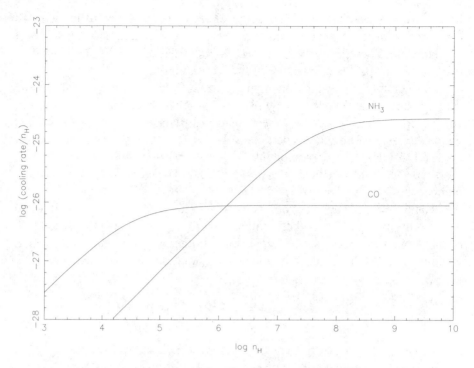

Fig. 9.4. Cooling rate by CO and NH$_3$ in a molecular cloud with $T = 10$ K.

$n_\text{H} \gg n_\text{cr}$ are given by

$$\mathcal{C} = n_1 A_{10} E_{10} = \begin{cases} n_H n E_{10} \mathcal{R}_{01} & (n_\text{H} \ll n_\text{cr}) \\ \dfrac{n A_{10} E_{10}}{1 + (g_0/g_1)\exp(\beta E_{10})} & (n_\text{H} \gg n_\text{cr}) \end{cases}. \quad (9.74)$$

In the first case, when $n_\text{H} \ll n_\text{cr}$, the cooling rate is low because there are insufficient numbers of collisions exciting the system and the rate is essentially set by the rate of collisions between the two levels. In the second case, when $n_\text{H} \gg n_\text{cr}$, there are sufficient numbers of collisions, which guarantee a thermal distribution of population between the two levels. Because each collisional excitation is balanced by a collisional deexcitation, the rate of collisions $\mathcal{R}$ has no effect on the cooling rate and $\mathcal{C}$ is independent of n_H; the only way to change the cooling rate is to change the temperature in this case.

9.6 Global Structure of the ISM

The balance between various sources of heating and cooling is of importance not only in localised regions like ionised nebulas, but also in determining the global equilibrium of the ISM. In particular, the existence of coolants of different energy scales and spontaneous decay rates leads to the possibility that matter can exist in different phases in the ISM. We shall now discuss how such a possibility can arise by using a fairly simple model.

It is necessary to satisfy several conditions for the global equilibrium of the ISM, especially if it is made of matter in different phases existing together. To begin with, mechanical equilibrium requires the pressure to be approximately constant throughout the ISM except in those regions where active bulk flows are occurring (say, around a supernova that has just exploded). Roughly speaking, this requires constancy of the product of the local number density of particles and the local temperature so that denser regions of the ISM have lower temperatures. Second, the maintenance of steady state in populating all the energy levels requires upwards and downwards transitions to be balanced for every process. This condition would include the balance between ionisation and the recombination of atoms, in particular hydrogen. Finally, energy conservation requires the net heating rate of the ISM to be matched by the net cooling rate. Let us now see what kind of structures are permitted, subject to these constraints.

We will model the ISM as an optically thin medium for all the radiative processes, which we shall discuss below. As regards composition, the ISM will contain hydrogen, helium, and a small fraction of metals that are important in their role as coolants. It turns out that the overall structure is not seriously affected by helium, and hence we shall ignore it in our model. All atoms, in particular hydrogen, can exist in ionised as well as in neutral form because there are different sources of ionisation in the ISM. For our model, we shall parameterise the ionisation source as a flux of photons with a power-law spectra. There could, of course, be other sources of ionisation for example, high-energy particles in cosmic rays, that can be dealt with in a similar manner.

Because the pressure is given by $P = nk_BT$, it will be contributed mostly by the species that have the largest value of n in any local region. Hence, in the ISM, the pressure is provided mostly by neutral and ionised hydrogen and electrons. The total pressure will be

$$P = (n_H + n_p + n_e)k_BT = n(1 + x_H)k_BT, \qquad (9.75)$$

where n_H, n_p, and n_e are the number densities of the hydrogen atom, protons, and electrons, respectively, $n = n_H + n_p$ is the the total number of protons contributing to the mass density, and $x_H \equiv (n_e/n) = (n_p/n)$ is the ionisation fraction. The condition of pressure balance demands the constancy of P in Eq. (9.75).

Let us next consider the equation for ionisation balance that determines x_H. We take the flux of ionising protons to be given by

$$F(\epsilon)d\epsilon = F_H \left(\frac{\epsilon}{E_H}\right)^{-s} d\epsilon \quad (E_H < \epsilon \lesssim 10^4 E_H) \qquad (9.76)$$

in the range of energies indicated and zero outside; $E_H = 13.6\,\text{eV}$ is the ionisation potential for hydrogen and s is the spectral index that indicates how steeply the photon energy falls. For numerical estimates we will use $s \simeq 1.5$. The unit for $F(\epsilon) d\epsilon$ is ergs cm^{-2} s^{-1}, and F_H has the units cm^{-2} s^{-1}. The corresponding total energy density of photons is given by

$$U = \int_{E_H}^{\infty} \frac{F(\epsilon)}{c} d\epsilon = \frac{F_H E_H}{(s-1)c}. \tag{9.77}$$

The dimensionless number

$$\Sigma = \frac{U}{P} = \frac{F_H E_H}{P(s-1)c} \tag{9.78}$$

is a convenient parameter giving three times the ratio between radiation pressure and gas pressure in the ISM, which we shall use extensively in our discussion. Given the above flux, the photoionisation rate per unit volume is given by

$$\mathcal{R}_{\text{photo}} = n_H \int_{E_H}^{\infty} \sigma_{\text{PI}}(\epsilon) \frac{F(\epsilon)}{\epsilon} d\epsilon = \frac{n(1-x_H)\sigma_0 F_H}{(s+\mu)}, \tag{9.79}$$

where we have taken the photoionisation cross section to be

$$\sigma_{\text{PI}} = \sigma_0 \left(\frac{\epsilon}{E_H}\right)^{-\mu} \quad (\epsilon > E_H, \mu \cong 3). \tag{9.80}$$

The index μ varies between 2.7 to 3 when a realistic cross section for photoionisation is used and $\sigma_0 \approx 6 \times 10^{-18}$ cm^2. The recombination rate is

$$\mathcal{R}_{\text{rec}} = \alpha_H(T) n_e n_p = n^2 x_H^2 \alpha_0 \left(\frac{T}{10^4 \text{ K}}\right)^{-1/2}, \quad \alpha_0 \simeq 10^{-13} \text{ cm}^3 \text{ s}^{-1}. \tag{9.81}$$

Equating the two, $\mathcal{R}_{\text{photo}} = \mathcal{R}_{\text{rec}}$, and expressing n in terms of pressure P, we get the ionisation fraction as given by

$$\frac{1}{x_H^2} = 1 + \frac{(s+\mu)\alpha_H(T)P}{\sigma_0 F_H k_B T}, \tag{9.82}$$

or, in terms of Σ,

$$x_H = \left[1 + \left(\frac{s+\mu}{s-1}\right)\left(\frac{E_H}{k_B T}\right)\left(\frac{\alpha_H(T)}{c\sigma_0}\right)\frac{1}{\Sigma}\right]^{-1/2}. \tag{9.83}$$

This equation determines the ionisation fraction of hydrogen in terms of temperature and Σ. Note that the pressure and the radiation flux enter into this expression only through Σ.

We now consider the equilibrium of other species of atoms in steady state. Just as we have modelled the hydrogen ionisation as a two level process, we will also assume that each species of the ions (labelled $i = 1, 2, \ldots$) can be characterised by two-energy levels separated by an energy gap E_i. The two levels are connected by collisional excitation, collisional deexcitation, and spontaneous decay, as in

Section 9.5. In that case, the equilibrium requires that

$$n_1 C_i(1 \to 2) = n_2 [C_i(2 \to 1) + A_i(2 \to 1)], \tag{9.84}$$

where the collisional rates are given by

$$C_i(2 \to 1) = n_e \gamma(T) = n_e \gamma_0 \left(\frac{T}{10^4 \text{ K}}\right)^{-1/2},$$

$$C_i(1 \to 2) = C_i(2 \to 1) e^{-E_i/k_B T}, \tag{9.85}$$

with $\gamma_0 \approx 10^{-7}$ cm^3 s^{-1}. Hence the fractional abundance of ions in the excited states is

$$x_i \equiv \frac{n_2}{n_1 + n_2} = \left\{ 1 + e^{E_i/k_B T} \left[1 + \frac{A_i k_B T}{\gamma(T)} \frac{(1 + x_H)}{x_H P} \right] \right\}^{-1}, \tag{9.86}$$

where we have expressed n in terms of the equilibrium pressure. This expression can be used for different ions $i = 1, 2, 3$, etc., with corresponding values for E_i, A_i, etc.

Finally, let us consider the equilibrium between the heating and the cooling rates for the system. The photoionisation heating rate is given by the product of ionisation rate by a photon of energy E and the kinetic energy given to the electron in the process of ionisation. Integrating over all photon energies, we get

$$\mathcal{H}_{\text{phot}} = n_H \int_{E_H}^{\infty} (\epsilon - E_H) \sigma_{\text{PI}}(\epsilon) \frac{F(\epsilon)}{\epsilon} d\epsilon = \frac{E_H \sigma_0 F_H n_H}{(s + \mu)(s + \mu - 1)}. \tag{9.87}$$

Using the equality between $\mathcal{R}_{\text{photo}}$ and $\mathcal{R}_{\text{rec}}$ [which gives $\sigma_0 F_H n_H (s + \mu)^{-1} = \alpha_H n^2 x_H^2$], the heating rate per baryon can be written as

$$n^{-1} \mathcal{H}_{\text{phot}} = \frac{n E_H \alpha_H(T) x_H^2}{(s + \mu - 1)} = \frac{E_H \alpha_H(T)}{(s + \mu - 1)} \left(\frac{P}{k_B T}\right) \frac{x_H^2}{(1 + x_H)}. \tag{9.88}$$

If there are other ionising processes, then they have to be added to this heating rate. For simplicity, however, we shall assume that photoionisation is the only source of heat.

The cooling is contributed by different ions as well as through the process of recombination. The line cooling rate that is due to an ion i is

$$\mathcal{C}_{\text{line}} = n_2 A_i E_i = n \left(\frac{n_{\text{tot}}}{n}\right) x_i A_i E_i = n a_i A_i E_i x_i, \tag{9.89}$$

where a_i is the abundance of the ith ion. Using Eq. (9.86), we find that the line cooling rate per baryon becomes

$$n^{-1} \mathcal{C}_{\text{line}} = \frac{a_i A_i E_i}{1 + e^{E_i/k_B T} \left[1 + \left(\frac{1 + x_H}{x_H}\right) \frac{A_i}{\gamma(T)} \frac{k_B T}{P} \right]}. \tag{9.90}$$

We obtain the net cooling by summing over the different ions with suitable values for the parameters. The recombination cooling rate per baryon is given by

$$n^{-1}\mathcal{C}_{\text{rec}} = nx_H^2 k_B T \alpha_H(T) = \left(\frac{x_H^2}{1+x_H}\right) P\alpha_H(T). \quad (9.91)$$

Equating the net heating rate per baryon [Eq. (9.88)] to the net cooling rate per baryon [sum of Eq. (9.90) for each ion and Eq. (9.91)], we get an equation of the form $f(P, T) = 0$, which determines the relationship between equilibrium pressure and temperature for the ISM. The phases that are allowed for matter in the ISM are essentially determined by the nature of this curve in the P–T plane. If, for a given value of pressure, there exists only one value of temperature, then the solution is unique and the matter will exist with that particular value of temperature and pressure. It is, however, possible – given the complexity of the cooling and the heating processes – to have multiple values of temperature for the same value of pressure. In that case, there can exist more than one phase in the ISM with high temperatures compensated for by low densities and vice versa.

To obtain the nature of such a curve, it is necessary to solve the equation $f(P, T) = 0$ numerically, taking into consideration all the different line cooling processes that are available. It is, however, possible to obtain a simple qualitative picture in two limits. The first limit is the one in which cooling is dominated by a set of lines in which the collisional deexcitation is more dominant than radiative deexcitation. In that case, the expression for the line cooling rate per baryon given in Eq. (9.90) can be approximated as

$$n^{-1}\mathcal{C}_{\text{line}} \simeq \left(\frac{x_H}{1+x_H}\right)\left(\frac{P}{k_B T}\right)\gamma(T)\sum_{i=1}^N a_i E_i e^{-E_i/k_B T}, \quad (9.92)$$

where the sum is over all the dominant lines relevant to the problem. Equating this cooling rate to the photoionisation heating rate given in Eq. (9.88), we can solve for x_H, obtaining

$$x_H(T) \simeq \frac{(s+\mu-1)}{E_H}\left[\frac{\gamma(T)}{\alpha_H(T)}\right]\sum_i a_i E_i e^{-E_i/k_B T}. \quad (9.93)$$

[Of course, the approximation used should be in such a range as to ensure that $x_H < 1$ in the relation (9.93).] On the other hand, the equilibrium between photoionisation and recombination rates lead to Eq. (9.83), which can be rewritten as

$$\frac{1-x_H^2}{x_H^2} = \frac{(s+\mu)}{(s-1)}\left(\frac{E_H}{k_B T}\right)\left[\frac{\alpha_H(T)}{c\sigma_0}\right]\left(\frac{P}{U}\right), \quad (9.94)$$

where U is the energy density of the radiation field. Solving for pressure, we obtain

$$\frac{P}{U} = \left(\frac{c\sigma_0}{\alpha_H(T)}\right)\left(\frac{k_B T}{E_H}\right)\frac{(s-1)}{(s+\mu)}\left(\frac{1-x_H^2}{x_H^2}\right) = f(T), \quad (9.95)$$

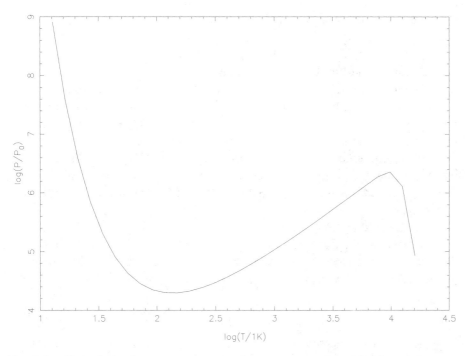

Fig. 9.5. The relation between pressure and temperature of the ISM in steady state, based on a toy model.

in which $x_H(T)$ is treated as a known function fixed by relation (9.93). This equation expresses pressure directly as a function of temperature, given the line cooling processes in the ISM subject to the condition that collisional deexcitation dominates the cooling.

This function can be multivalued for a suitable range of parameters. As an example, in Fig. 9.5 a model for the ISM is shown that involves just two lines with energy spacings $E_1 = 10^2$ K and $E_2 = 9 \times 10^4$ K; the fractional abundances of the ions are 10^{-4}. The ionising radiation flux is chosen to be such that it induces photoionisations at a rate $\mathcal{R}$ and imparts an energy ϵ_h to the electron during each ionisation. The figure shows the pressure in units of

$$P_0 \equiv 1 \text{ K cm}^{-3} \left(\frac{\mathcal{R}}{10^{-15} \text{ s}^{-1}}\right)\left(\frac{\epsilon_h}{13.6 \text{ eV}}\right). \tag{9.96}$$

It is obvious from the figure that a pressure of $10^5 P_0$, for example, can be maintained by matter in three different phases. The temperatures for the three phases will be $T = 40$ K, 10^3 K, and 1.6×10^4 K. The corresponding densities will be $n = 2.5 \times 10^3 \, n_0$, $10^2 \, n_0$, and $3.16 \, n_0$, where

$$n_0 \equiv 1 \text{ cm}^{-3} \left(\frac{\mathcal{R}}{10^{-15} \text{ s}^{-1}}\right)\left(\frac{\epsilon_h}{13.6 \text{ eV}}\right). \tag{9.97}$$

Of the three equilibrium points, the middle one is clearly unstable, as pressure increases with increasing temperature in this region. This model therefore predicts a simple, stable two-phase ISM containing a low-density high-temperature gas in pressure equilibrium with high-density low-temperature clouds. The photoionisation heating is balanced by the line cooling of the first ion in the low-temperature phase and the line cooling by the second ion at the high-temperature phase. The equilibrium temperatures are essentially characterised by the energy levels of the two ions. In fact, the physics near the equilibrium points is identical to the cooling process in an ionised nebula, discussed earlier (see page 487), where it was found that line cooling can act as an effective thermostat in the ISM. The absolute value of the pressure and density in this model can be rescaled by adjusting $\mathcal{R}$ and ϵ_h.

For a more exact modelling, we must take into account all the different line cooling processes and heating sources. The mathematical reason for the existence of three phases can be traced to the fact that, in the intermediate range of temperatures, the density

$$n = \frac{P}{(1+x_H)k_B T} = \left(\frac{U}{k_B}\right) \frac{f(T)}{T[1+x_H(T)]} \tag{9.98}$$

drops rapidly with increasing temperature, thereby making the pressure a decreasing function of temperature.

The effect of radiative decay can be incorporated near any equilibrium point by considering just one line as the main coolant. We will illustrate this process, in the context of standard photoionisation, by a spectrum of photons characterised by an index s. Balancing the photoionisation heating rate by recombination cooling and line cooling, we get the equation

$$\alpha(T)n x_H^2 \chi = \alpha(T) n x_H^2 k_B T + \mathcal{C}_{\text{line}}, \quad \chi = \frac{E_H}{s+\mu-1}, \tag{9.99}$$

which can be written in full as

$$a_i A_i E_i \left[\left(1 + e^{E_i/k_B T}\right) + e^{E_i/k_B T} \frac{A_i}{\gamma(T) n x_H} \right]^{-1} = \alpha(T) n x_H^2 \left(\frac{\chi}{k_B T} - 1\right) k_B T. \tag{9.100}$$

By rearranging this expression, we can solve for n as a function T and x_H, obtaining $n(T, x_H)$ as

$$n = \frac{\mathcal{A}(T)}{x_H^2} - \frac{\mathcal{B}(T)}{x_H}, \tag{9.101}$$

where $\mathcal{A}(T)$ and $\mathcal{B}(T)$ are given by

$$\mathcal{A} = \frac{a_i A_i (E_i/k_B T)}{\alpha(T)[(\chi/k_B T) - 1](1+e^{E_i/k_B T})}, \quad \mathcal{B} = \frac{A_i}{\gamma(T)} \frac{e^{E_i/k_B T}}{(1+e^{E_i/k_B T})}. \tag{9.102}$$

9.6 Global Structure of the ISM

The pressure now becomes

$$P(x_H, T) = k_B T \left(\frac{1+x_H}{x_H^2}\right) [\mathcal{A}(T) - x_H \mathcal{B}(T)]. \tag{9.103}$$

Because we expect the pressure to be a constant, this quadratic equation in x_H can be solved to express x_H as a function of temperature for a given value of pressure. This gives

$$x_H = \frac{(q_1 - 1) + [(q_1 + 1)^2 + (4P/k_B T \mathcal{B})q_1]^{1/2}}{2[(P/k_B T \mathcal{B}) + 1]} = x_H(T, P), \tag{9.104}$$

with q_1 given by

$$q_1 = \frac{\mathcal{A}}{\mathcal{B}} = a_i \left(\frac{\gamma_0}{\alpha_0}\right) \frac{(E_i/\chi)}{(1 - k_B T/\chi)} e^{-E_i/k_B T}. \tag{9.105}$$

The parameters governing the line cooling, of course, must satisfy the constraint that $x_H < 1$ in order for the range of values to be physically meaningful.

So far we have not used the condition for ionisation equilibrium. Condition (9.83) gives, as usual, an expression for $x_H(T, P)$ given by

$$x_H(T, P) = \left\{ 1 + \left(\frac{s+\mu}{s-1}\right) \frac{(s+\mu-1)\chi}{k_B T} \left[\frac{\alpha_H(T)}{c\sigma_0}\right] \frac{P}{U} \right\}^{-1/2}. \tag{9.106}$$

We obtain the equilibrium values of T and P by equating the two expressions for x_H in Eqs. (9.104) and (9.106). Figure 9.6 shows the curves for $x_H(T, P)$ in Eq. (9.104) as a function of T for different values of P (dashed curves with P expressed in units of nT). Superimposed on it are the curves given by Eq. (9.106) for different values of the parameter P/U (solid curves). The intersection of the curves determines the temperature and the pressure and implicitly the radiation-energy density U under which such an equilibrium is possible. The fact that the curves intersect for a wide class of n, T, and P illustrates the nonuniqueness of the solution. The curves are drawn for a single ion with $s = 1.5$, $\mu = 2.7$, $E_H = 13.6$ eV, $a_i = 10^{-4}$, $A_i = 10^{-2}$ s^{-1}, and $E_i \simeq 10^{-2}$ eV, just for illustrative purposes. We generate the continuous curves by changing P/U from 3×10^3 to 3×10^4 [in equally spaced intervals in $\log(P/U)$] and the broken curves are generated by changing the value of nT from 10^2 to 10^4 K cm^{-3} [in equally spaced intervals in $\log(nT)$].

Exercise 9.3
Evaporation of ISM clouds: A spherical cloud of cold interstellar gas of radius R_c, $T_c = 100$ K, and $n_c = 20$ atoms cm^{-3} is immersed in the hot coronal gas with $T_h = 10^6$ K and $n_h = 0.003$ electrons cm^{-3}. Heat is conducted with a conductivity given by $\kappa = 0.06 \times 10^{-6} T_h^{5/2}$ in cgs units. Estimate the time scale for the evaporation of the cloud. {Answer: Using the approximation $T_h \gg T_c$ and assuming that other cooling

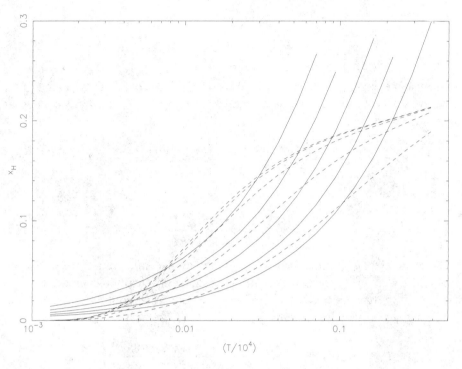

Fig. 9.6. The curves based on Eqs. (9.104) and (9.106) illustrating the variety of equilibria possible for the ISM.

processes are subdominant, we can estimate the heat input into the cloud to be $\dot{E}_{\text{cond}} \simeq 4\pi R_c^2 \kappa (T_h/R_c) \simeq 2.3 \times 10^{34} (R_c/1 \text{ pc})$ ergs s^{-1}. The time scale for evaporation will be $t_{\text{evap}} \simeq (4\pi R_c^3/3)[(3/2) n k_B T_h / \dot{E}_{\text{cond}}] \simeq 7 \times 10^5$ yr $(R_c/1 \text{ pc})^2$.}

Finally, it must be stressed that the two-phase model of the ISM is somewhat oversimplified, although the thermal instability mechanism has a much wider domain of validity. In fact, such thermal instabilities could be important in several other localised regions of the ISM made of shocked gas shells, spurs, loops, etc.

9.7 Interstellar Electron Density and Magnetic Field

Because the ISM has widely varying physical parameters and conditions, it needs to be probed by very different techniques in order to obtain the overall picture. We shall begin with the probes of electron density and magnetic field in the ISM.

9.7.1 Electron Density from the Dispersion Measure

We have seen in Vol. I, Chap. 9, Section 9.5 that the dispersion relation for transverse electromagnetic waves in the high-frequency limit can be expressed

9.7 Interstellar Electron Density and Magnetic Field

in the form

$$\omega^2 = c^2 k^2 + \Omega_e^2, \qquad (9.107)$$

where $\Omega_e^2 = [4\pi n_e (Ze)^2/m_e]$. The group velocity corresponding to this dispersion relation is $c_g \equiv (\partial \omega/\partial k) = c(1 - \Omega_e^2/\omega^2)^{1/2}$, so that the time taken for a group of waves travelling from a cosmic source towards Earth, through a plasma, will be

$$t_\omega = \int_0^r \frac{ds}{c_g} = \int_0^r \frac{ds}{c} \left(1 - \frac{\Omega_e^2}{\omega^2}\right)^{-1/2}. \qquad (9.108)$$

For $\omega \gg \Omega_e$, we can use the Taylor expansion of the square root and the definition of Ω_e^2 to get

$$t_\omega = \frac{r}{c} + \frac{2\pi e^2}{m_e c \omega^2} \text{DM}, \qquad (9.109)$$

where the quantity DM, called the *dispersion measure* of the plasma, is defined as

$$\text{DM} \equiv \int_0^r n_e \, ds. \qquad (9.110)$$

This quantity gives the column density of electrons along the line of sight. By using Eq. (9.109) we can relate the difference in the time of arrival $\Delta \tau_D$ of a signal at two different frequencies to the dispersion measure. Numerically,

$$\Delta \tau_D = 1.34 \times 10^{-9} \mu s \left(\frac{\text{DM}}{\text{cm}^{-2}}\right) \left[\left(\frac{\nu_1}{1 \text{ MHz}}\right)^{-2} - \left(\frac{\nu_2}{1 \text{ MHz}}\right)^{-2}\right]. \qquad (9.111)$$

More conventionally, n is measured in inverse cubic centimeters and l in parsecs, so that

$$\frac{\text{DM}}{\text{cm}^{-3} \text{ pc}} = 2.410 \times 10^{-10} \left(\frac{\Delta \tau_D}{\mu s}\right) \left[\left(\frac{\nu_1}{1 \text{ MHz}}\right)^{-2} - \left(\frac{\nu_2}{1 \text{ MHz}}\right)^{-2}\right]^{-1}. \qquad (9.112)$$

If the time delay at the two frequencies can be measured, then this equation allows for the dispersion measure to be determined along any given direction.

With the pulsars used as sources of radiation, the above technique allows us to determine the electron density along different directions in our galaxy. The currently available data provide estimates for the dispersion measure in ~ 500 different directions through the interstellar gas. A typical value of the dispersion measure is ~ 30 cm^{-3} pc, giving the column density of electrons as 10^{20} cm^{-2}. At $\nu = 100$ MHz the delay will be ~ 12 s compared with a wave with infinite frequency. The warm ionised component is probably responsible for the majority of electrons that cause the dispersion in the radiation from pulsars (see Exercise 9.7). Because the distances to many pulsars are known either from direct parallax measurements or by other means, the dispersion measure allows us to model the electron density in part of the galaxy. A fit for the electron density in the galaxy that is in reasonable agreement with observation is given by

$$n_e = \frac{2 \text{ cm}^{-3}}{1 + (R/10 \text{ kpc})} \left[0.025 + 0.015 \exp\left(-\frac{|z|}{70 \text{ pc}}\right) \right], \quad (9.113)$$

which is useful for numerical estimates.[2]

9.7.2 Faraday Rotation and the Interstellar Magnetic Field

We saw in Vol. I, Chap. 9, Section 9.5 that a linearly polarised electromagnetic wave travelling through a plasma containing the magnetic field will have its plane of polarisation rotated by an angle

$$\psi = \int_0^L \Delta k \, dz = \frac{2\pi e^3}{m_e^2 c^2 \omega^2} \int_0^L n_e B_\parallel \, dz \equiv \frac{2\pi e^3}{m_e^2 c^2 \omega^2} \langle \text{RM} \rangle \quad (9.114)$$

while propagating a distance L along the magnetic field. The last equality defines a quantity RM, called the *rotation measure*, that is an integral of the product of electron density and the component of the magnetic field along the line of propagation. Numerically, in conventional astronomical units,

$$\frac{\Delta \psi}{\text{rad}} = 0.81 \left(\frac{\lambda}{10^2 \text{ cm}}\right)^2 \int_0^{L/\text{pc}} \left(\frac{B_\parallel}{\mu\text{G}}\right) \left(\frac{n_e}{\text{cm}^{-3}}\right) d\left(\frac{z}{\text{pc}}\right). \quad (9.115)$$

As in the case of the dispersion measure, observations of Faraday rotation at two different frequencies can be used to determine the rotation measure by use of the above result in the form

$$\frac{\text{RM}}{\text{rad m}^{-2}} = 0.81 \int_0^{L/\text{pc}} \left(\frac{B_\parallel}{\mu\text{G}}\right) \left(\frac{n_e}{\text{cm}^{-3}}\right) d\left(\frac{z}{\text{pc}}\right)$$

$$= \left[\left(\frac{\Delta\psi_1}{\text{rad}}\right) - \left(\frac{\Delta\psi_2}{\text{rad}}\right) \right] \left[\left(\frac{\lambda_1}{\text{m}}\right)^2 - \left(\frac{\lambda_2}{\text{m}}\right)^2 \right]^{-1}. \quad (9.116)$$

Measurement of the rotation measure provides the integral of the magnetic-field component along the line of sight to the source. There are many galactic and extragalactic radio sources that emit linearly polarised radiation, thereby allowing this method to be used as an effective probe of the interstellar magnetic field.

If both rotation and dispersion measures are known along the same (or nearby) directions, then their ratio will provide the mean value of the magnetic field along the line of sight:

$$\langle B_\parallel \rangle \propto \frac{\text{rotation measure}}{\text{dispersion measure}} \propto \frac{\int n_e B_\parallel \, dl}{\int n_e \, dl}. \quad (9.117)$$

The typical magnetic fields obtained by this method vary in the range (0.6 to 3) $\times 10^{-6}$ G in our galaxy.

The detailed structure of the magnetic field in our galaxy has been investigated extensively, and the following features emerge from these studies: (1) When averaged over kiloparsec scales, **B** usually points within the plane, although, on smaller scales, plumes of the magnetic field rise above the plane to tens of parsecs in a manner reminiscent of magnetic loops on the surface of the Sun. (2) The mean field runs azimuthally around the galaxy, and at solar radius and beyond it is in the direction of galactic rotation. At ~500 pc inside the solar radius, the direction of the field reverses, and it seems to reverse still again at a radius of ~5.5 kpc from the centre of the galaxy. (3) At small scales, **B** wiggles significantly and it is usually interpreted as being due to the superposition of an ordered large-scale component and a random small-scale component. The dispersion and rotation measures are sensitive to only the ordered field, which is approximately 2×10^{-6} G near the Sun, rising to 6×10^{-6} G at a galactocentric radius of 4.5 kpc. (4) The magnitude of the random component is quite hard to estimate. One possible method is to relate it to the galactic synchrotron emission. This gives a value that decreases from 10^{-5} G at $R = 5$ kpc, through 6×10^{-6} G at $R = 10$ kpc, to 3×10^{-6} G at $R = 20$ kpc. It appears that the random field is at least a few times larger than the ordered component. (5) The strongest interstellar magnetic fields are in the filamentary structures of the ISM from which synchrotron emission is detected; they give rise to fields of $\sim 10^{-3}$ G.

Exercise 9.4

Numbers for the ISM: The hot phase of the interstellar gas may be taken to be fully ionised plasma with $T \simeq 3 \times 10^5$ K, $n_e = n_p = 0.003$ cm^{-3}, and magnetic field $B = 3 \times 10^{-6}$ G. Calculate (a) the Debye length, (b) the Coulomb mean free path for electrons and protons, (c) the average electron-cyclotron radius, (d) the average proton-cyclotron radius, and (e) the average electron-cyclotron frequency.

Exercise 9.5

Paradox?: The measurements of the Zeeman effect at 21 cm and the measurement of Faraday rotation along some line of sight in the ISM give different values for the interstellar magnetic field. How can you explain this? (Hint: Which region of the ISM does the 21-cm Zeeman effect probe? Which region does the Faraday rotation probe?)

Exercise 9.6

Simple magnetic configuration for the ISM: The aim of this exercise is to investigate a simple configuration for the disk of the galaxy containing both the magnetic field and ionised ISM gas. We will first provide a general formulation of magnetostatics applicable to such a context and then work out one specific case. In steady state, we can set the fluid velocity and all time derivatives to zero in the equations of MHD. We further assume that the heating and the cooling balance each other, leading to an isothermal configuration in which the equation of state for gas is $P = a^2 \rho$, with $a^2 = (k_B T/m)$.

In the case of a galaxy, we take the magnetic field to be in the y–z plane with a vector potential $\mathbf{A} = A(y, z)\hat{\mathbf{x}}$ and $\mathbf{B} = -\hat{\mathbf{x}} \times \nabla A$. (a) Show that the force balance equation

$$-\rho\nabla\phi - \nabla P + \frac{1}{4\pi}(\nabla \times \mathbf{B}) \times \mathbf{B} = 0, \qquad (9.118)$$

where ϕ is the gravitational potential that is due to matter in the plane of the galaxy, can be rewritten in the form

$$-\frac{1}{4\pi}(\nabla^2 A)\nabla A = e^{-\phi/a^2}\nabla q, \quad q \equiv Pe^{\phi/a^2}. \qquad (9.119)$$

This equation implies that the gradients of q and A are parallel, allowing us to express q as a function of A; $q = q(A)$. Hence Eq. (9.119) becomes

$$\nabla^2 A = -4\pi e^{-\phi/a^2}\frac{dq}{dA}. \qquad (9.120)$$

(b) The above equation suggests that for any given function $q(A)$, we can determine a particular configuration of magnetic field and matter. This is, however, not quite correct for the following reason. In steady-state magnetostatics, both mass and magnetic flux are conserved and hence we must have a particular value for the mass to flux distribution in the system. This distribution should be an invariant as the system evolves and settles on the final state, which puts a constraint on $q(A)$. To incorporate this constraint, consider any two field lines labelled by A and $A + dA$. Show that the amount of matter dM located between these field lines in the range $-Y$ to $+Y$ in the y direction and per unit length in the x direction is given by

$$dm = dA \int_{-Y}^{+Y} \rho \frac{\partial z}{\partial A} dy. \qquad (9.121)$$

Here we have used the locus of the field line $z = z(y, A)$ corresponding to a given value of A to convert z integration to A integration. Show that this gives

$$q(A) = a^2 \frac{dm}{dA}\left(\int_{-Y}^{+Y} e^{-\phi/a^2}\frac{\partial z}{\partial A} dy\right)^{-1}. \qquad (9.122)$$

(c) As an application of the above formalism, consider a magnetic-field configuration with $\mathbf{B} = B(z)\hat{\mathbf{y}}$ with the additional condition that $(B^2/8\pi P) \equiv \alpha_0$ is a constant. All physical variables can now be taken to depend on only the vertical direction z. Show that, in this case, $A(z)$ is determined by

$$\frac{dA}{dz} = B_0(0)\exp\left[-\frac{\phi(z)}{2(1+\alpha_0)a^2}\right]. \qquad (9.123)$$

Given the potential $\phi(z)$, this can be integrated to determine $A(z)$ and thus will solve the problem completely. (d) Describe the solution quantitatively for a reasonable $\phi(z)$.

9.8 Radiation from Ionised Interstellar Gas

The HII regions exhibit a rich spectra of emission in both lines and in continuum and have been studied all the way from radio to UV bands. The continuum part of the radio spectrum is mostly from thermal bremsstrahlung. The line spectrum is made of several forbidden lines (e.g., those of oxygen), recombination lines of hydrogen in Balmer series, and radio recombination lines. We first discuss the continuum radiation and then the lines.

We have seen in Vol. I, Chap. 6, Section 6.9 that the thermal bremsstrahlung emissivity can be expressed in the form

$$4\pi j_\nu = 6.8 \times 10^{-38} n^2 T^{-1/2} e^{-h\nu/k_B T} \bar{g}_{ff} \text{ ergs cm}^{-3} \text{ s}^{-1} \text{ Hz}^{-1} \quad (9.124)$$

if all variables are expressed in cgs units. The total emissivity, integrated over all frequencies, is

$$4\pi J \simeq 1.4 \times 10^{-27} n^2 T^{1/2} \bar{g}_B \text{ ergs cm}^{-3} \text{ s}^{-1}. \quad (9.125)$$

In these expressions, $\bar{g}_{ff}$ and $\bar{g}_B$ represent the velocity-averaged Gaunt factor and the frequency-averaged Gaunt factor, respectively. The relevant frequency-dependent Gaunt factor is given by the fitting formula

$$\bar{g}_{ff}(T,\nu) \cong 11.69 + 0.83 \ln\left(\frac{T}{10^4 \text{ K}}\right) - 0.55 \ln\left(\frac{\nu}{100 \text{ MHz}}\right), \quad (9.126)$$

whereas $\bar{g}_B(T)$, which is a frequency average of $\bar{g}_{ff}(\omega)$, varies between 1.1 and 1.5. Choosing a value of $\bar{g}_B \approx 1.2$ usually gives an accuracy of ∼20%.

It is clear that bremsstrahlung emissivity from an optically thin region is a probe of the integral of n_e^2 along the line of sight. This suggests the definition of the quantity called the *emission measure*,

$$\frac{\text{EM}}{\text{pc cm}^{-6}} = \int_0^{s/\text{pc}} \left(\frac{n_e}{\text{cm}^{-3}}\right)^2 d\left(\frac{s}{\text{pc}}\right), \quad (9.127)$$

which provides a convenient parameterisation. This bremsstrahlung process is most important in radio and x-ray bands. Diffuse regions of ionised hydrogen at $T = 10^4$ K are strong sources of bremsstrahlung in radio wavelengths. The shells of supernova remnants at much higher temperatures of approximately $T = 10^7$ K will be a source of bremsstrahlung in x-ray bands. As an example, consider the thermal bremsstrahlung emission from the ionised gas cloud of Orion A, located at a distance of ∼0.5 kpc and having an angular size of 4 arc min. The plasma temperature of the cloud is $T_e \approx 7000$ K and radio observations at 5 GHz give a brightness temperature of $T_b \approx 273$ K. From the relation $T_b \approx \tau T_e$, valid for an optically thin emission, we find the optical depth to be $\tau = 0.039$. From the relation $\tau = 0.082 T_e^{-1.35} \nu^{-2.1}$ (EM), we get the emission measure as EM $= 2.2 \times 10^6$ pc cm^{-6}. Taking the transverse size of the cloud $l = 0.5$ kpc tan(4') ≈ 0.6 pc to be the same as the size along the line of sight and using EM $= n_e^2 l$, we get $n_e = 1900$ cm^{-3}.

When the region that is emitting bremsstrahlung is optically thick, the bremsstrahlung spectrum is self-absorbed at low frequencies. It was seen in Vol. I,

Chap. 6, Section 6.9 that the free–free absorption coefficient, α_ν^{ff}, in the radio-frequency regime is given by

$$\alpha_\nu^{ff} = 5.5 \times 10^{-2} \left(\frac{n_e}{1~\text{cm}^{-3}}\right)^2 \left(\frac{T}{10^4~\text{K}}\right)^{-3/2} \left(\frac{\nu}{100~\text{MHz}}\right)^{-2} \text{kpc}^{-1}. \quad (9.128)$$

When the region of size R is optically thick with $\alpha_\nu R \gg 1$, the radiation will take the low-frequency form of the Planckian spectrum and we will have $I_\nu \propto \nu^2$. This occurs for frequencies $\nu < \nu_c$, where

$$\nu_c \cong 23~\text{MHz} \left(\frac{n_e}{1~\text{cm}^{-3}}\right) \left(\frac{T}{10^4~\text{K}}\right)^{-3/4} \left(\frac{R}{1~\text{kpc}}\right)^{1/2}. \quad (9.129)$$

The form of the spectrum from such a region will allow us to measure ν_c and j_ν and thus determine both T and n_e, provided the source is homogeneous.

Exercise 9.7

Inhomogeneous ISM: A realistic model for the ISM in the disk of our galaxy involves gas in three different phases: (1) A 1-kpc layer of ionised coronal gas with $T \simeq 10^6$ K and $n_H \simeq 3 \times 10^{-3}$ cm^{-3}. (2) Warm ionised matter (WIM) in the form of clouds with a radius of ~ 5 pc and $n_H = 0.3$ cm^{-3}, $T = 8000$ K; the number density of these clouds is $n_{\text{wim}} \simeq 3 \times 10^{-4}$ pc^{-3}. (3) Clouds of cold neutral matter (CNM) with $n_H = 30$ cm^{-3}, $T = 100$ K, and radius 2.5 pc; the number density of these clouds is comparable with n_{wim} because these clouds are often at the dense cores of warm ionised clouds. (a) Along the line of sight of a pulsar 1 kpc away, estimate the number of WIM clouds and the number of CNM clouds that are intercepted. (b) Compute the contribution to the dispersion measure from each of these constituents along the line of sight. (c) Consider a line of sight leaving the disk of the galaxy in the vertical direction. Compute the contributions to the emission measure from each of these phases. (d) Compare the mean contribution to the gas density and the fraction of volume occupied by each of these phases. [Answers: (a) The number of clouds intercepted will be given roughly by cloud cross section × number density of clouds × path length. For WIM, this is $(\pi \times 5^2)(3 \times 10^{-4}) \times 10^3 \simeq 24$ and for CNM this is $(\pi \times 2.5^2)(3 \times 10^{-4}) \times 10^3 \simeq 6$. (b) Because DM $\simeq \int n_e$, we need to compute the fraction of the 1-kpc path length that is occupied by the three phases of the ISM. The average path length intercepted through a randomly placed spherical cloud of radius R can be easily shown to be $(\pi/2)R$. Hence the total path length from WIM is $(24)(\pi/2)(5) \simeq 188.5$ pc, and the corresponding CNM path length will be $(6)(\pi/2)(2.5) \simeq 23.6$ pc. The remaining distance of $(1000 - 188.5 - 23.6) \simeq 788$ pc will be through the hot coronal gas. For coronal gas and fully ionised WIM, $n_e = n_H$, DM$_{\text{WIM}} \simeq 56.5$ pc cm^{-3}, and DM$_{\text{coronal}} \simeq 2.36$ pc cm^{-3}. For the CNM $n_e \simeq 0$, as the medium is essentially neutral. In reality, cosmic rays induce a small nonzero ionisation fraction of approximately 7.7×10^{-4}, which gives DM$_{\text{CNM}} \simeq 0.54$ pc cm^{-3}. (c) The disk of the galaxy is only ~ 500 pc thick on either side of the midplane. Therefore all the path lengths computed in part (b) become smaller by a factor of 2. The emission measure is EM $\simeq \int n_e^2$. Plugging in the numbers, we get $EM_{WIM} \simeq 8.48 \, \text{pc cm}^{-6}$, $EM_{corona} \simeq 3.55 \times 10^{-3} \, \text{pc cm}^{-6}$, and $EM_{CNM} \simeq 6.23 \times 10^{-3} \, \text{pc cm}^{-6}$. (d) The fractional volume occupied by WIM and CNM clouds will be the ratio (volume of one cloud/volume of space in which one cloud is formed) that is equal to $(4\pi/3)R^3 n_{cloud}$. This gives 0.157 for WIM and 0.02 for CNM. The remaining fraction, 0.823, is occupied by coronal gas. The mean gas number density will be $\bar{n} = (0.157 \times 0.3) + (0.02 \times 30) + (0.823 \times 3 \times 10^{-3}) \simeq 0.639 \, \text{cm}^{-3}$. The fraction of the gas density contributed by WIM will therefore be $(0.157 \times 0.3)/0.639 \simeq 0.074$. Similarly, the fraction from CNM will be 0.922 and the fraction from coronal gas will be 3.87×10^{-3}.]

Let us next consider the line radiation from ionised gas. We saw in Section 9.5 that forbidden and permitted lines of different atoms act as a coolants in ionised regions. The line-cooling process that results in the emission of photons of specific frequency helps us determine the physical conditions of the nebula and the ISM. This is best done with elements for which there are three energy levels that satisfy the following conditions: (1) Let the energy levels of the elements be E_1, E_2, and E_3. We will assume that only collisional excitation, deexcitation, and spontaneous decay occur between the ground and either excited states. (2) Further, we will neglect direct transitions between levels 2 and 3. Then the steady-state condition for transitions among these three levels can be expressed by the equations

$$n_3(\mathcal{R}_{31} + A_{31}) = n_1 \mathcal{R}_{13} = n_1 \frac{g_3}{g_1} \mathcal{R}_{31} e^{-(E_{13}/k_B T)}, \qquad (9.130)$$

$$n_2(\mathcal{R}_{21} + A_{21}) = n_1 \mathcal{R}_{12} = n_1 \frac{g_2}{g_1} \mathcal{R}_{21} e^{-(E_{12}/k_B T)}. \qquad (9.131)$$

Hence the ratio of the two line intensities will be given by

$$\frac{j(3 \to 1)}{j(2 \to 1)} = \frac{n_3 h \nu_{31} A_{31}}{n_2 h \nu_{21} A_{21}} = \frac{g_3 A_{31} \nu_{31}}{g_2 A_{21} \nu_{21}} \frac{[1 + (A_{21}/\mathcal{R}_{21})]}{[1 + (A_{31}/\mathcal{R}_{31})]} e^{-(E_{13} - E_{12})/k_B T}. \qquad (9.132)$$

There are two limiting forms of this expression that are of interest. If the element is such that E_{12} and E_{13} are quite different, then the argument of the exponent $(E_{13} - E_{12})/k_B T = E_{23}/k_B T$ will vary rapidly with the temperature T. Hence the main dependence of the line strength on the temperature will arise from the exponential factor, and the ratio of the line strengths can be used as a probe of the temperature. On the other hand, if $E_{13} \approx E_{12}$ so that $E_{23} \ll k_B T$, the exponential factor is irrelevant and the temperature dependence of the line ratios essentially arises from $\mathcal{R}_{ij} \propto n T^{-1/2}$. This corresponds to a situation in which the upper energy level is split by a small amount having $E_3 \approx E_2$, with the excited-state energies satisfying $(E_3, E_2) \gg E_1$.

An example of the first situation is provided by the transitions in OIII that are due to $^1S_0 \to {}^1D_2$ ($\lambda = 4363$ Å), $^1D_2 \to {}^3P_2$ ($\lambda = 5007$ Å), and $^1D_2 \to {}^3P_1$

($\lambda = 4959$ Å). The first transition corresponds to an energy of 2.84 eV, whereas the other two are close together with energies of ~ 2.51 eV. In this case, our equations can provide the ratio

$$\frac{j(4363)}{j(5007)+j(4959)} = \frac{n_3 h \nu_{32} A_{32}}{n_2 h \nu_{21} A_{21}} = \frac{g_3 \mathcal{R}_{31} \nu_{32}}{g_2 \mathcal{R}_{21} \nu_{21}} \frac{A_{32}}{A_{32}+A_{31}} e^{-E_{32}/k_B T}. \tag{9.133}$$

The coefficient in front of the exponential for OIII is ~ 0.12, thereby giving

$$\frac{j(4363)}{j(5007)+j(4959)} = 0.12 \, \exp\left(-\frac{3.29 \times 10^4 \text{ K}}{T}\right). \tag{9.134}$$

The assumption that collisional deexcitation is subdominant is valid for $n_e \lesssim 10^5$ cm^{-3}. Because of the exponential factor, the ratio varies by ~ 2 orders of magnitude when the temperature varies from 6000 to 20,000 K and thus can be a sensitive probe of the temperature. A similar analysis can be performed with other lines, for example, NII lines. If all the lines are produced from the same region of the nebula, then they should provide the same value for the temperature. Any variation observed in the temperature can be used as a further probe of the structure of the nebula.

As an example of the second case, let us consider the transitions in OII that are due to $^2D_{3/2} \to {}^4S_{3/2}$ ($\lambda = 3726$ Å) and $^2D_{5/2} \to {}^4S_{5/2}$ ($\lambda = 3729$ Å). In this case, $E_{23} \ll k_B T$ and the line-intensity ratio is given by

$$\frac{j(3729)}{j(3726)} = \frac{g_3 A_{31}}{g_2 A_{21}} \left[\frac{1+(A_{21}/\mathcal{R}_{21})}{1+(A_{31}/\mathcal{R}_{31})}\right]. \tag{9.135}$$

Level 3 is $^2D_{5/2}$, corresponding to $g_3 = 2J+1 = 2(5/2)+1 = 6$, whereas level 2 is $^2D_{3/2}$, with $g_2 = 2J+1 = 2(3/2)+1 = 4$. The collisional rate for this transition can be expressed in the form

$$\mathcal{R}_{ij} = 8.6 \times 10^{-6} \frac{n_e}{T^{1/2}} \frac{\Omega_{ij}}{g_j}, \tag{9.136}$$

with $\Omega_{31} = 0.88$ and $\Omega_{21} = 0.59$. The spontaneous transition rates are $A_{31} = 4.2 \times 10^{-5}$ s^{-1} and $A_{21} = 1.8 \times 10^{-4}$ s^{-1}. In the dilute regime (as $n_e \to 0$), the line intensity takes the limiting value

$$\frac{j(3729)}{j(3726)} \approx \frac{g_3 \mathcal{R}_{31}}{g_2 \mathcal{R}_{21}} = \frac{\Omega_{31}}{\Omega_{21}} \approx 1.5. \tag{9.137}$$

In the high-density limit (as $n_e \to \infty$), the line ratio drops to

$$\frac{j(3729)}{j(3726)} \approx \frac{g_3 A_{31}}{g_2 A_{21}} \approx 0.35. \tag{9.138}$$

The transition occurs over a band of 2 orders of magnitude when $n_e T^{-1/2}$ varies from 1 to 10^2 in cgs units. In this range the line ratio can be a sensitive probe of the parameter $n_e T^{-1/2}$. If T can be independently estimated from the methods described above, we one can have an independent estimate of n_e.

Exercise 9.8

Clumping in HII region: Assume that a HII region contains a large number of small clumps, each of which has uniform electron density n_e. These clumps occupy a small fraction ϵ of the total volume, with the rest of the region empty. The average electron density is defined as $n_{\rm rms} \equiv \langle n_e^2 \rangle^{1/2}$, where the averaging is over the entire volume of the region. (a) How does the luminosity of an emission line like Hα (for which $n_c \gg n_e$) depend on $n_{\rm rms}$ and ϵ? (b) How about another emission line for which $n_c \ll n_e$? (c) How does the optical depth of a resonance absorption line depend on $n_{\rm rms}$ and ϵ? [Answers: If V_c is the total volume of all clumps and V is the overall volume of the region, then $n_{\rm rms}^2 = n_e^2 (V_c/V)$; therefore $n_e = n_{\rm rms} \epsilon^{-1/2}$. (a) In this case, the line emissivity goes as $j \propto n^2$. Therefore total line luminosity is $L \propto n_e^2 V_c = n_{\rm rms}^2 V \propto n_{\rm rms}^2$ and is independent of ϵ. (b) Here j is proportional to n, implying that L is proportional to $n_e V_c \propto (n_{\rm rms} \epsilon^{-1/2})(\epsilon V)$ so that $L \propto n_{\rm rms} \epsilon^{1/2}$. (c) The optical depth that is due to one clump of size l is $\tau_c \propto n_e l \propto n_{\rm rms} \epsilon^{-1/2} l$. Because the cross-sectional area of the clump is $\sim l^2$ and the number density of the clump is N_c/V, where N_c is the total number of clumps, the number of clumps intercepted along any line of sight will be $\mathcal{N} = (N_c/V) l^2 R$, if R is the size of the region. Taking $V \simeq R^3$, we get $\mathcal{N} \simeq N_c (l/R)^2$. The total optical depth will be $\tau = \tau_c \mathcal{N} \simeq (n_{\rm rms} \epsilon^{-1/2} l) N_c (l/R)^2 \simeq n_{\rm rms} R \epsilon^{-1/2} (V_c/V) \propto n_{\rm rms} \epsilon^{1/2}$.]

We also detect, in addition to forbidden lines, different hydrogen lines from the ionised regions. For example, the Hβ line, occurring in the transition $n = 4$ to $n = 2$ corresponding to the wavelength $\lambda = 4861$ Å, is one of the strongest lines from the ionised region. Its intensity is given by

$$L(H\beta) = 2.28 \times 10^{-19} n_e^2 T_e^{-3/2} V \exp(9800/T_e) \text{ ergs s}^{-1}, \quad (9.139)$$

where V is the volume of the source. Clearly, this intensity provides information about the integral of $n_e^2 T^{-3/2}$ along the line of sight. It is not easy to disentangle the n_e and T dependences without other independent observations.

Another important class of line radiation observed from this region is the radio recombination lines, discussed in Vol. I, Chap. 7, Section 7.4, which arise because for transitions between high n states of hydrogenlike atoms. Lines corresponding to $(n+1) \to n$ are strongest and are called α *lines*; those for transitions $(n+2) \to n$ are called β *lines*, etc. In the standard notation used by astronomers, H 109 α is a line corresponding to the transition $110 \to 109$ of hydrogen; He 127 β corresponds to the transition $129 \to 127$ of helium, etc. The α transitions with $n > 160$ in most of the abundant elements produce lines in the radio band with $\lambda > 1$ cm. In this case, the linewidth is essentially determined by Doppler broadening and the shape of the line is a Gaussian with the full Doppler width at half-intensity being given by

$$\frac{\Delta \nu}{\nu} = \frac{2}{c} \left[2 \left(\frac{k_B T_e}{M} + v_t^2 \right) \ln 2 \right]^{1/2}, \qquad (9.140)$$

where v_t^2 is the velocity dispersion that is due to turbulence, if it is present. For such a line shape, the value of the line profile at the centre of the line will be [Vol. I, Chap. 7, Section 7.4, Eq. (7.44)]

$$\phi(0) = \left(\frac{\ln 2}{\pi} \right)^{1/2} \left(\frac{2}{\Delta \nu} \right). \qquad (9.141)$$

Given the number densities n_e and n_i of electrons and ions, respectively, in a given temperature T_e, we can compute the optical depth τ_L (at the centre of the line) from a plasma in which such radio recombination lines are produced. The result [Vol. I, Chap. 7, Eq. (7.45)] can be expressed in terms of the emission measure of the plasma as

$$\tau_L = 1.92 \times 10^3 \left(\frac{T_e}{K} \right)^{-5/2} \left(\frac{EM}{cm^{-6}\,pc} \right) \left(\frac{\Delta \nu}{kHz} \right)^{-1}, \qquad (9.142)$$

where EM is the emission measure of the plasma defined in Eq. (9.127). Because $\tau_L \ll 1$ in most contexts, the brightness temperature of the line [Vol. I, Chap. 7, Eq. (7.47)] is given by $T_L \approx T_e \alpha_\nu R \simeq T_e \tau_L$. Numerically

$$T_L = 1.92 \times 10^3 \left(\frac{T_e}{K} \right)^{-3/2} \left(\frac{EM}{cm^{-6}\,pc} \right) \left(\frac{\Delta \nu}{kHz} \right)^{-1}. \qquad (9.143)$$

Such a plasma will also be emitting continuum radiation that can be modelled as a thermal bremsstrahlung, as described above. For a pure hydrogen plasma, we can obtain the brightness temperature of the line, compared with that of the continuum, by dividing the above expression by the corresponding results for thermal bremsstrahlung, which is given by

$$\left(\frac{T_L}{T_c} \right) \simeq 2.3 \times 10^4 \left(\frac{\Delta \nu}{kHz} \right)^{-1} \left(\frac{\nu}{GHz} \right)^{2.1} \left(\frac{T_e}{K} \right)^{-1.15}, \qquad (9.144)$$

where we have used the fitting function in Eq. (6.240) of Vol. I, Chap. 6, Section 6.9. Note that the ratio between the brightness temperatures of a given radio recombination line and the free–free emission depends on the radio frequency and the temperature of the region but is independent of the electron density. Hence it can be used as a probe of the temperature of the HII region. This leads to observed temperatures of (8000–10,000) K. If both the line and the continuum are emitted by the same cloud with some temperature $T_e = T_c$, the brightness temperature at the line centre will be

$$T_{bL} = T_e \left[1 - e^{-(\tau_L + \tau_c)} \right], \qquad (9.145)$$

where τ_L and τ_c are the corresponding optical depths. At frequencies adjacent to the line, only continuum contributes, and we have

$$T_{bc} = T_e(1 - e^{-\tau_c}). \qquad (9.146)$$

Hence the brightness temperature of the line alone is given by

$$T_L = T_{bL} - T_{bc} = T_e e^{-\tau_c}(1 - e^{-\tau_L}). \qquad (9.147)$$

This shows that even the *lines* will disappear in the range at which the *continuum* optical depth is large, that is, at low frequencies $\nu < \nu_c$ [see relation (9.129)]. T_L, however, drops at high frequencies, and hence these observations have to be carried out at some intermediate frequency for optimal results.

Radio recombination lines are also capable of providing information about the elemental abundance, especially the He/H ratio. To illustrate this procedure, let us consider the detection of 109 α line of hydrogen and the recombination lines of singly ionised helium. If the nuclear mass is Am_p, the reduced mass μ for a hydrogen like configuration will be

$$\mu = \frac{Am_p m_e}{(m_e + Am_p)} = m_e \left[1 + \left(\frac{m_e}{Am_p}\right)\right]^{-1}. \qquad (9.148)$$

Because the energy levels scale as $E_n \propto \mu Z^2$, the energy levels will change slightly with A. For the 109 α line of hydrogen and helium, we get the frequency difference as $\Delta \nu = \nu_{He} - \nu_H \approx 2.07$ MHz. The corresponding Doppler velocity $v = c(\Delta \nu / \nu) \approx 124$ km s^{-1}. At a temperature of $\sim 10^4$ K, the thermal velocities are $b \approx 12$ km s^{-1}, which will determine the widths of the lines. Because $v \gg b$, the lines are easily resolved and can be observed, thereby providing a measure of He/H.

9.9 21-cm Observations of Neutral Hydrogen

The most important probe of the distribution of neutral hydrogen in the ISM is the 21-cm line emitted by HI observed in emission or absorption, which we shall now discuss.[3]

We have seen in Vol. I, Chap. 6, Section 6.2 that the absorption coefficient for any process can be expressed, in general, in the form

$$\alpha_\nu = \left(\frac{\pi q^2}{m_e c}\right) f_{ji} \phi_\nu n \left(1 - e^{-h\nu/k_B T}\right) = \left(\frac{n\lambda^2}{8\pi}\right)\left(\frac{g_j}{g_i}\right)(A_{ji}\phi_\nu)\left(1 - e^{-h\nu/k_B T}\right)$$
$$\equiv \alpha_{01}\left(1 - e^{-h\nu/k_B T}\right), \qquad (9.149)$$

where $\phi(\nu)$ is the line profile and we have used the general result between the absorption coefficient and the line profile. The temperatures in the ISM vary in the range of $(10-10^4)$ K, and the factor $(h\nu/k_B)$ is approximately 2×10^4 K in the optical band. Hence the exponential can be ignored, and almost all the atoms can be treated as being in the ground state. The measurement of α_{01} will then lead directly to the densities, irrespective of the value of T. In contrast, for the 21-cm radio line, $(h\nu/k_B) \equiv T_* \simeq 0.068$ K; hence the exponential is important and the absorption is determined by the small difference in the population between the singlet and the triplet states, which in turn is governed by the unknown spin temperature $T = T_s$. Expanding the exponential in a Taylor series for $T_s \gg T_*$ and using $A_{12} = 2.9 \times 10^{-15}$ s^{-1}, we get

$$\alpha \simeq \alpha_{01}\left(\frac{0.068 \text{ K}}{T_s}\right) \simeq 1.03 \times 10^{-14}\left[\frac{n_H \phi(\nu)}{T_s}\right] \quad (9.150)$$

in cgs units. The brightness temperature of the source will vary as

$$T_b = T_{b0}\, e^{-\tau_\nu} + T_s(1 - e^{-\tau_\nu}), \quad (9.151)$$

so that the change in T_b that is due to absorption is

$$\Delta T_b \equiv T_b - T_{b0} = (T_s - T_{b0})(1 - e^{-\tau_\nu}) = (T_s - T_{b0})(1 - e^{-\alpha_\nu s}), \quad (9.152)$$

where s is the path length. For an optically thin medium ($\alpha_\nu s \ll 1$) in which emission dominates ($T_{b0} \ll T_s$), this relation becomes

$$T_b \simeq T_s \alpha s \simeq 1.03 \times 10^{-14} \text{ K} \, (n_H \phi_\nu s) \simeq 3 \times 10^4 \text{ K}\left(\frac{n_H}{1 \text{ cm}^{-3}}\right)\left(\frac{s}{1 \text{ pc}}\right)\phi_\nu, \quad (9.153)$$

which is independent of the (unknown) value of T_s, that is, the spin temperature is irrelevant as long as the optically thin clouds are studied in emission.

More usefully, we can relate the column density $\mathcal{C} \equiv n_H s$ of neutral hydrogen to the integral of the brightness temperature across the line by integrating relation (9.153) over ν and recalling that ϕ_ν is normalised to unity. Then

$$\mathcal{C}_H \simeq 10^{14} \text{ cm}^{-2} \int T_B\, d\nu. \quad (9.154)$$

The value of the integral depends on the broadening mechanism. If the line is Doppler broadened and optically thin, then $T_B(\nu) = T_B(0) \exp\{-[(\nu - \nu_0)/\Delta\nu_D]^2\}$, giving

$$\int T_B\, d\nu = T_B(0)\sqrt{\pi}\,(\Delta \nu_D). \quad (9.155)$$

We can convert the integral over ν into one over the velocity v by using $(\Delta\nu/\nu) = (v/c)$:

$$\int T_B(\nu)\, d\nu = \int T_B(\nu)\left(\frac{d\nu}{\nu}\right)\nu = \int \frac{T_B(\nu)}{\lambda}\, dv = \frac{T_s}{\lambda}\int \tau(v)\, dv. \quad (9.156)$$

This gives

$$\frac{C_H}{\text{cm}^{-2}} \simeq 10^{18} \left(\frac{T_s}{K}\right) \int_{-\infty}^{\infty} \tau(v)\, d\left(\frac{v}{\text{km}^{-1}\,\text{s}}\right). \quad (9.157)$$

Let us next consider the situation in which the neutral hydrogen has a radial bulk velocity $U(s)$ along the line of sight. To handle this case, we note that by definition $d\tau = \alpha\, ds$, with α given by relation (9.150). Therefore, in velocity space,

$$d\tau(v) = -w \frac{n_H(s)}{T(s)} \phi[v - U(s)]\, ds, \quad (9.158)$$

where w is a numerical coefficient that depends on the units used for s and v; $w = 5.5 \times 10^{-19}$ if s is in centimetres and v is in kilometres per second, and $w = 1.69$ if s is in parsecs and v is in kilometres per second. The total optical depth at a velocity v is then given by

$$\tau(v,s) = w \int_0^s \frac{n_H(x)}{T(x)} \phi[v - U(x)]\, dx. \quad (9.159)$$

Converting the integral to one over the velocity field, we get

$$\tau(v, U) = w \int_{U(0)}^{U(s)} \frac{n_H[s(U)]}{T[s(U)]} \phi[v - U] \frac{dU}{|dU/ds|}. \quad (9.160)$$

This relation shows that the observed optical depth contains direct information about the bulk flow of neutral hydrogen gas. Because most of the ISM is made of neutral hydrogen, this can act as an effective probe of the bulk state of motion in our galaxy. In particular, it provides information about the overall rotation of our galaxy.

As an example of use of this formula, consider a velocity field in the form

$$U(s) = U_c + b(s - s_c)^2 \quad (9.161)$$

in the galaxy, with n_H and T remaining constant. In the integral in Eq. (9.159) we can extend the limits of integration from 0 to ∞ in considering the total neutral hydrogen in the galaxy. For the integral in the ranges 0–s_c and s_c–∞, the relationship between s and U is given by

$$s = s_c - \frac{1}{\sqrt{|b|}} |U - U_c|^{1/2}, \quad s = s_c + \frac{1}{\sqrt{|b|}} |U - U_c|^{1/2}. \quad (9.162)$$

The integral can therefore be written as

$$\tau(v) = \frac{1}{2} \frac{w n_H}{T} \frac{1}{\sqrt{|b|}} \int_{U_c}^{\infty} \phi(v - U) \frac{dU}{\sqrt{|U - U_c|}} - \frac{1}{2} \frac{w n_H}{T} \frac{1}{\sqrt{|b|}}$$

$$\times \int_{U_c + bs_c^2}^{\infty} \phi(v - U) \frac{dU}{\sqrt{|U - U_c|}}. \quad (9.163)$$

Taking the linewidth to be due to the Doppler effect with a velocity dispersion

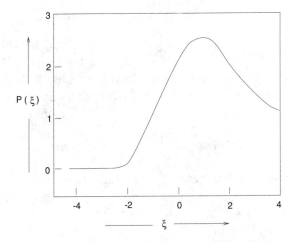

Fig. 9.7. Integral determining the relation between velocity and optical depth.

σ and assuming that $bs_c^2 \gg \sigma$, we can extend the limit $(U_c + bs_C^2)$ to ∞ so that the two terms on the right-hand side of Eq. (9.163) can be combined. Simple manipulation now allows us to express the final result as

$$\tau(v) = \frac{w}{\sqrt{2\pi\sigma}} \frac{n_H}{T} \frac{1}{\sqrt{|b|}} P\left(\frac{v - U_c}{\sigma}\right), \qquad (9.164)$$

where

$$P(\xi) = \int_0^\infty \frac{1}{\sqrt{x}} \exp\left[-\frac{1}{2}(\xi - x)^2\right] dx. \qquad (9.165)$$

This relation allows us to extract information about U_c and b from the observed form of $\tau(v)$, thereby reconstructing the velocity field $U(s)$ in the regions where Eq. (9.161) is satisfied. Figure 9.7 shows the behaviour of the function $P(\xi)$.

In our galaxy, the motion of hydrogen gas around the equatorial plane can be described, to the first approximation, by an angular velocity of rotation $\Omega(R)$, which depends on only the distance from the galactic centre. Figure 9.8 shows the geometry relating the Sun to an element of neutral hydrogen gas at the point marked A. The semicircle drawn with the line connecting the centre of the galaxy to the Sun as the diameter gives the locus of points at which the gas will be moving radially away from the Sun. At all other points, it is only the component v of the gas velocity in the radial direction that will contribute to the measurement of the Doppler shift. From the geometry it is clear that

$v = $ component of ΩR along line of sight – component of $\Omega_0 R_0$
along line of sight,
$= \Omega R \cos(90° - l - \theta) - \Omega_0 R_0 \cos(90° - l),$
$= \Omega R(\sin\theta \cos l + \cos\theta \sin l) - \Omega_0 R_0 \sin l. \qquad (9.166)$

9.9 21-cm Observations of Neutral Hydrogen

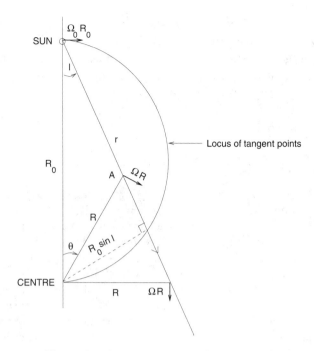

Fig. 9.8. Geometry illustrating the measurement of galactic rotation from the Doppler shift.

Because $r \sin l = R \sin \theta$ and $R \cos \theta = R_0 - r \cos l$, this relation becomes

$$v = R_0 [\Omega(R) - \Omega_0] \sin l. \qquad (9.167)$$

Along any direction l in the range $0 < l < \pi/2$, the radial distance R first decreases with increasing r, reaches a minimum when the line of sight cuts the semicircle, and then increases with increasing R. Assuming that $\Omega(R)$ decreases with R in the relevant range, this will cause v to first increase with r, then to reach a maximum, and finally decrease. The measurement of v at a specified value will then lead to two possible solutions for r, and this ambiguity needs to be resolved by some other input. Once this is done, information about v can be converted to information about galactic rotation $\Omega(R)$. It is also obvious that, near the maximum, the velocity field has a quadratic dependence on the line-of-sight distance, allowing us to use the results obtained above.

Such a mapping of the galaxy by use of the 21-cm line has been instrumental in deciphering the spiral-arm structure of the Milky Way. The spiral arms located at different radii from the centre of the galaxy will contain neutral hydrogen rotating at different speeds. When the line of sight cuts through these spiral arms, each of these will produce its own 21-cm feature, Doppler shifted by the corresponding speeds of rotation. By deciphering these features, we can obtain the actual spiral structure of the galaxy. These studies have also shown that the total mass of neutral hydrogen in the galaxy is approximately $(3\text{–}5) \times 10^9 \ M_\odot$, with the major part confined to the plane of the galaxy. In the range of radii $R = (4\text{–}20)$ kpc, the neutral hydrogen density is effectively independent of radius but falls exponentially in the vertical direction, with a scale length of 250 pc for the warm component and 130 pc for the cold component; this gives a projected density on the plane of 3×10^{20} atoms cm^{-2}. The cold component, however, is very nonuniform, with clouds of hydrogen filling less than one-tenth of the volume of the disk.

Exercise 9.9

Practice with 21-cm observations: (a) A neutral hydrogen cloud located at $d = 30$ pc emits 21-cm radiation. If the observed flux is $f = 4.5 \times 10^{-15}$ ergs cm^{-2} s^{-1}, what is the mass of the neutral hydrogen in the cloud? (b) Along some direction in the galaxy, 21-cm emission is seen in a narrow line (N) centred at radial velocity 100 km s^{-1} and a broad line (B) centred at velocity 50 km s^{-1}. The brightness temperature measurements across the lines give a central brightness of 25 K and a FWHM of 5 km s^{-1} for N and a central brightness of 6 K and a FWHM of 34 km s^{-1} for B. There is a bright radio source with brightness temperature $T_0 = 10^9$ K in the same direction. The absorption of this source shows that the difference in the brightness temperature of the background source and the brightness of the source with absorption is 3×10^8 K for N and 10^6 K for B. Determine (1) the Doppler velocity widths, (2) the optical depth at the line centre, (3) the kinetic temperature of the absorbing systems, and (4) the column densities. {Answers: (a) The luminosity of the source is $4\pi d^2 f = 4.85 \times 10^{26}$ ergs s^{-1}. Equating this to $L_{21 \ \text{cm}} = (3/4) N_H A_{21} h\nu$ with $A_{21} = 2.869 \times 10^{-15}$ s^{-1}, we get $N_H = 2.39 \times 10^{58}$; this corresponds to $M_H = m_p N_H \approx 20 \ M_\odot$. (b) (1) Because b is related to the FWHM by $0.5 = \exp[-(\text{FWHM}/2b)^2]$, we have $b = \text{FWHM}/(4\ln 2)^{1/2}$. Given FWHM $= (5, 34)$ km s^{-1} for (N, B), we get $b = (3, 20)$ km s^{-1}, respectively. (2) Because T_0 is much higher than the gas temperature, we can write the brightness temperature as $T_B = T_0 e^{-\tau}$. This gives $\log(1 - e^{-\tau}) = \log(T_0 - T_B) - \log T_0 = (-0.52, -3)$ for N, B. Hence the optical depths are $\tau = (0.36, 10^{-3})$ for N, B. (3) If T_k is the kinetic temperature, then at the line centre, $T_{B,\text{centre}} = T_k(1 - e^{-\tau_c})$. Using the known values of $T_{B,\text{centre}}$ and τ_c, we get $T_k = (83, 6000)$ K for N, B. (4) The central optical depth is given by $\tau_c = 5.49 \times 10^{-14} \ (N_H / \sqrt{\pi} b T_k)$ when all quantities are in cgs units. This gives $N_H = 3.23 \times 10^{18} \tau_{\text{centre}} (b/1 \ \text{km s}^{-1})(T_k/1 \ \text{K})$. Substituting the known values, we get $N_H = (3, 4) \times 10^{20}$ cm^{-2} for N, B.}

9.10 Molecular Lines from the ISM

The ISM also contains a very large class of organic and inorganic molecules, from simple H_2 to extremely complicated hydrocarbons. In general, the molecular line emission provides information about a denser region of the ISM than, for example, the 21-cm line emission. This is because the fragile molecules will be dissociated by the optical and the UV photons and can exist within only denser molecular clouds [$n \approx (10^3\text{–}10^4)$ cm^{-3}], where they are shielded from

9.10 Molecular Lines from the ISM

high-energy photons by dust and molecular hydrogen at the periphery of the cloud. The molecules therefore provide a direct probe of the conditions of these clouds.[4]

Some of these molecules, such as CH, CN, etc., possess electronic transitions in the optical band and can be detected by absorption features against bright stars. More importantly, many molecules possess transitions in the radio band that can be observed more easily and with virtually no obscuration from interstellar dust. The molecules OH, H_2O, and H_2CO were among the early set of molecules to be detected in the radio band with very high brightness temperature. (We shall see in Section 9.11 that this could possibly be explained by maser action.)

Because hydrogen is the most abundant element, we would have expected molecular hydrogen H_2 to be relatively common. However, as described in Vol. I, Chap. 7, Section 7.7, such a homonuclear diatomic molecule does not possess a permanent electric dipole moment and hence has no simple rotational transition. It is possible to detect H_2 molecules in the UV region through electronic transitions and these (satellite-based) observations confirm that H_2 is very abundant in the ISM.

The next most abundant molecule is CO; it has a permanent electric dipole moment and a strong rotational transition corresponding to $J = 1 \to 0$ at 115 GHz ($\lambda = 2.6$ mm) and the next few transitions of the rotational ladder at 230 GHz, 345 GHz, ... (Vol. I, Chap. 7, Section 7.7). The CO observations at radio wavelength nicely complement the surveys of the 21-cm neutral hydrogen. It also indirectly helps us to surmise the existence of H_2 molecules, as the latter must exist wherever CO molecules can exist. In fact, the excitation mechanism for CO molecules is usually collisions with H_2, and hence this correlation is fairly strong.

The absorption coefficient for the rotational transition in the CO molecule can be expressed by the general formula [see Eq. (9.149)]:

$$\alpha_\nu = \frac{g_i}{g_k} n_k \frac{c^2}{8\pi \nu^2} A_{ik} \left(1 - e^{-\Delta E/k_B T_{\text{ex}}}\right) \phi(\nu), \qquad (9.168)$$

where $g_i = (2J+3)$, $g_k = (2J+1)$, $\Delta E = 2hB(J+1)$, and T_{ex} is the excitation temperature. For the case of CO, $2B = 115$ GHz and the Einstein coefficient for molecular rotational transitions from $(J+1)$ to J can be expressed in the form [see Eq. (7.91) of Vol. I, Chap. 7]

$$A_{ik} = \frac{64\pi^4 \nu^3}{3hc^3} \langle d^2 \rangle, \quad \langle d^2 \rangle = \frac{J+1}{2J+3} d_p^2, \qquad (9.169)$$

where d_p is the permanent dipole moment. (For CO, this has the value $d_p = 0.11$ debye, where 1 debye $= 10^{-18}$ cgs units.) The number density is $n_k = f_k x_{\text{CO}} n_{\text{tot}}$, where x_{CO} is the abundance of CO molecules relative to n_{tot} and

$$f_k = \frac{n_k}{n_{\text{tot}}} = \frac{g_k}{Z_{\text{rot}}} e^{-E_J/k_B T}, \qquad (9.170)$$

where the partition function can be approximated by the integral

$$Z \simeq \int_0^\infty (2J+1) \exp\left[-\frac{hBJ(J+1)}{k_B T_{ex}}\right] dJ \simeq \frac{k_B T_{ex}}{hB}. \qquad (9.171)$$

Several useful results can be now obtained from this basic result. Consider the $J = 1$ to 0 transition first. In this case, $A_{10} \simeq 6 \times 10^{-8}$ s^{-1}; taking the linewidth at the centre of the line to be $\phi(\nu_0) \approx (1/\Delta\nu)$ and substituting the numerical values in Eq. (9.168) we get

$$\alpha_\nu \equiv 1.5 \times 10^{-15} n_0 \left(\frac{\Delta\nu}{1 \text{ km s}^{-1}}\right)^{-1} \left(1 - e^{-h\nu/k_B T_{ex}}\right). \qquad (9.172)$$

For the CO transition with $\nu = 115$ GHz, $(h\nu/k_B) \approx 5.6$ K and so $h\nu \ll k_B T_{ex}$. This simplifies the absorption coefficient to

$$\alpha_\nu \approx 8.3 \times 10^{-18} n_0 \left(\frac{\Delta\nu}{1 \text{ km s}^{-1}}\right)^{-1} \left(\frac{T_{ex}}{10^3 \text{ K}}\right)^{-1}. \qquad (9.173)$$

Using $f_{J=0} \simeq 2.8 \times 10^{-3} T_3^{-1}$, we get the corresponding optical depth at the line centre as

$$\tau_{\nu_0} \cong 0.23 \left(\frac{N_c}{10^{23} \text{ cm}^{-2}}\right) \frac{\langle x_{CO}\rangle}{10^{-4}} \left(\frac{T_{ex}}{10^3 \text{ K}}\right)^{-2} \left(\frac{\Delta\nu}{1 \text{ km s}^{-1}}\right)^{-1}, \quad N_c = n_{tot} s. \qquad (9.174)$$

This suggest that denser molecular clouds can be optically thick for CO rotational transitions. The measurement of τ will provide direct information about the column density of CO.

Similar considerations apply for all other transitions between any two adjacent levels $J+1$ and J. Because for $h\nu \ll k_B T$, we have the brightness temperature varying as $T_B(\nu) \simeq T\alpha_\nu s \simeq T\tau_\nu \propto N_c$, we can repeat the analysis exactly as in the case of 21-cm radiation. In particular, the integral over frequency for the brightness temperature gives

$$\int T_B \, d\nu = \left(\frac{hc^2}{8\pi k_B}\right) \left(\frac{A_{ik} g_i}{g_k}\right) \left(\frac{g_k}{\nu}\right) \frac{1}{Z} e^{-E/k_B T_{ex}} (N_c)$$

$$= \frac{32\pi^3 h}{3ck_B^2} \frac{B^3(J+1)^3}{T_{ex}} d_p^2 \exp\left[-\frac{hBJ(J+1)}{k_B T_{ex}}\right] N_c x_{CO}. \qquad (9.175)$$

Because the optical depth is $\tau = (T_B/T_{ex})$ in the optically thin case, this result can also be written as an integral over τ. Further, it is customary to write the integral over the Doppler velocity V so that $d\nu = \nu(dV/c)$. Then we get [with $\nu = 2B(J+1)$]

$$\int \tau_\nu \, dV = \frac{16\pi^3 h}{3k_B^2} \frac{B^2(J+1)^2}{T_{ex}^2} d_p^2 N_c x_{CO} \exp\left[-\frac{hBJ(J+1)}{k_B T_{ex}}\right]. \qquad (9.176)$$

When the long-wavelength approximation is not valid, the results have to be suitably modified by taking the full Planck spectrum. These results are of course applicable for the rotational transitions of any heteroatomic dipolar molecule.

Incidentally, the existence of the isotope of ^{13}C also helps in the diagnostics of the ISM conditions. In solar systems ^{12}C is ~90 times more abundant than ^{13}C. The ratio of intensities from the ^{12}C^{16}O line and the ^{13}C^{16}O line is often much less than 90, suggesting that the ^{12}C^{16}O lines are saturated whereas the ^{13}C^{16}O line is optically thin. If the ^{12}C^{16}O line is indeed saturated, then the brightness temperature at the line centre will give the excitation temperature, which is expected to be the same for both isotopes. We can now use the relation between an optically thin line intensity and the column density to determine the column density of ^{13}C^{16}O molecules and then multiply by the relative abundance to get the abundance of ^{12}C^{16}O molecules.

Another standard procedure used for determining the abundance of different elements in the ISM is based on the curve of growth. We have seen in Vol. I, Chap. 7, Section 7.3 that the curve of growth provides a relation between (W_λ/λ) and $(N_j \lambda f_{jk})$, where W_λ is the equivalent width and f_{jk} is the relevant oscillator strength. If the linewidth $\phi(\Delta \nu)$ is due to Doppler broadening with a velocity dispersion b, the relationship between these two quantities is given by

$$\left(\frac{W_\lambda}{\lambda}\right) = \frac{2bF(\tau_0)}{c}; \quad F(\tau_0) = \int_0^\infty \left[1 - \exp\left(-\tau_0 e^{-x^2}\right)\right] dx, \quad (9.177)$$

where the optical depth at the line centre is τ_0. For small τ_0 the relationship is

$$\left(\frac{W_\lambda}{\lambda}\right) = \frac{\pi e^2}{m_e c^2} N_j \lambda f_{ik} = 8.8 \times 10^{-13} N_j \lambda f_{ik}, \quad (9.178)$$

wheras for intermediate τ_0 we have the asymptotic form $F(\tau_0) = [\ln(\tau_0)]^{1/2}$.

Considering the coupling between stars and the ISM through different processes (star formation, stellar winds, supernova, etc.), it is interesting to ask how the abundances of elements in the ISM compare with, say, those in the Sun. By and large, the ISM abundances follow the broad pattern among the various elements, similar to the stellar abundances, but there is some amount of systematic depletion of the ISM abundance of metals compared with the metal abundance in the Sun. This is usually interpreted as being due to metals being locked up in molecular clouds and dust grains, although a detailed modelling of the depletion has not yet been done successfully.

It is important, however, to remember that, given the fairly inhomogeneous composition of the ISM, the abundances will depend crucially on the region of the ISM that is probed. The abundance of the tenuous ISM, away from molecular clouds, is usually determined by spectroscopy. The technique essentially involves looking for the absorption features that are due to elements in the ISM on the spectrum of a sufficiently distant star. In practice, however, such determination of the abundances in the ISM is complicated by several factors. To begin with, we need to be certain that the feature is due to the ISM and not from elements in the atmosphere of the star itself, which can be done by examination of the linewidth; there will be more thermal broadening in the stellar atmosphere than in the ISM. Second, it is necessary to choose lines for which observations fall on the linear regime of the curve of growth so that the line is not beginning to be saturated. This requires using fainter lines with a consequent loss in the signal-to-noise ratio. Third, lines could arise because of elements in the ionised state and can provide direct information about the number of atoms of the elements in only that particular state of ionisation. This is particularly true for elements such as carbon, calcium, and magnesium, with ionisation potentials of less than those of hydrogen, which could be ionised even when hydrogen remains neutral. It is necessary to correct for this feature while estimating the total abundance of the element. Fourth, this technique requires the use of sufficiently bright stars that exist along only ~100 or so specific directions in the sky. Even along these lines of sight, the stars we use are within few hundred parsecs of the Sun, and hence the abundances are only representative of the local regions of the ISM. Finally, this method clearly cannot be used to determine the abundances of dense interstellar clouds because the dust grains in these clouds will block too much light from the background stars.

9.11 Maser Action in the ISM

There are compact regions in the ISM from which radiation at brightness temperatures as high as 10^{13} K is observed. One example is a region with angular size of 10^{-6} to 10^{-4} times that of the HII region from which line emissions with widths of approximately (0.1–3.0) km s^{-1} [compared with the linewidths of approximately (2.0–100) km s^{-1} in the normal cases] are detected. The second example is emission in the IR band from dust shells around cool supergiants. These must be due to some nonequilibrium phenomenon and are usually explained as being due to maser action in some molecules. Even though emission from a region that is *not* in local thermal equilibrium is common in astronomy, maser action is unique in the sense that population inversion is required.

The basic idea behind this phenomenon is illustrated in Fig. 9.9. Consider the transitions along the levels $0 \to k \to 1$ and $1 \to 0$. If the rate for the second process ($1 \to 0$) is small compared with the first ($0 \to k \to 1$), then level 1 will be populated in significant excess of what is allowed in thermodynamic equilibrium. This will lead to maser action between 1 and 0, and the intensity of radiation will be much more than what is expected in equilibrium. In fact, the actual mechanism for a maser can be more complicated than what is illustrated in Fig. 9.9 in the sense that k may actually represent a set of energy levels rather than a single one – all of which contribute to populating level 1, provided that direct

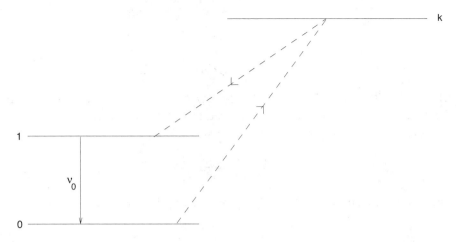

Fig. 9.9. The energy levels that lead to maser action.

radiative decay from these levels to ground state is forbidden by some selection rules.

Because the situation is far from local thermodynamic equilibrium, it is best to analyse it from basic principles. The formula for the absorption coefficient,

$$\alpha_\nu = \frac{h\nu}{c} B_{10} \phi_\nu (n_0 - n_1), \tag{9.179}$$

shows that α_ν becomes negative if $n_1 > n_0$, which arises in population inversion. The corresponding transfer equation,

$$\frac{dI}{dl} = \frac{h\nu}{4\pi} \phi_\nu [(n_1 - n_0) B_{10} I_\nu + n_1 A_{10}], \tag{9.180}$$

now needs to be solved with the population ratio n_1/n_0 determined by

$$n_1 \left(C_{10} + A_{10} + B_{10} I_\nu \frac{\Omega_b}{4\pi} + P_{10} \right) = n_0 \left(C_{01} + B_{01} I_\nu \frac{\Omega_b}{4\pi} + P_{01} \right), \tag{9.181}$$

where A characterises the spontaneous processes, C denotes the collisional processes, and P_{01} and P_{10} are the net rates of excitation and deexcitation of level 1, respectively, through all states, generically labelled as k. The induced emission coefficients B are same as in the case of thermodynamic equilibrium, but the following points need to be noted: Because the radiation is not in thermal equilibrium, I_ν is unknown. Further, because the photons are emitted in the direction of the incident one, this term is reduced by the factor $\Omega_b/4\pi$, where Ω_b is the beam's solid angle. In all practical cases, we may take

$$C_{10} = C_{01} = C; \quad A_{10} \ll \left(\frac{B I_\nu \Omega_b}{4\pi} \text{ or } C \right). \tag{9.182}$$

Then the population ratio is given by

$$\frac{n_1}{n_0} = \frac{C + \hat{B}I_\nu + P_{01}}{C + \hat{B}I_\nu + P_{10}}, \qquad (9.183)$$

where $\hat{B} = B\Omega_b/4\pi$. This result can be rewritten in a more transparent form. We first note that if collisions and induced transitions are completely ignored, then the corresponding ratio is $(n_1/n_0) = (P_{01}/P_{10})$, giving

$$\frac{\Delta n_0}{n} = \frac{(n_1 - n_0)_0}{n} = \frac{P_{01} - P_{10}}{P_{01} + P_{10}}. \qquad (9.184)$$

Therefore our result in Eq. (9.183) can be expressed in the equivalent form:

$$\left(\frac{\Delta n}{n}\right) = \frac{(\Delta n_0/n)}{1 + (C + \hat{B}I_\nu)P_{10}^{-1}}. \qquad (9.185)$$

[We have used $(n_1/n_0)_0 P_{10} = P_{01}$ to eliminate P_{01}.] Two features are obvious from this equation: First, the collisions and stimulated emission reduce the value of Δn compared with the value in their absence. Second, the population inversion decreases with increasing I_ν, which in turn will be high if the inversion is large. This suggests the existence of some feedback control over an otherwise runaway process. Substituting Eq. (9.185) into Eq. (9.180),

$$\frac{dI_\nu}{dl} = \frac{h\nu_0}{4\pi}\phi_\nu(\Delta n B I_\nu + n_1 A_{10}) = \frac{aI_\nu}{1 + (I_\nu/I_s)} + b, \qquad (9.186)$$

where

$$a = \frac{h\nu_0}{4\pi}\phi_\nu\frac{B\Delta n_0}{1 + C/P_{10}}, \quad b = \frac{h\nu_0}{4\pi}\phi_\nu n_1 A_{10}, \quad I_s \equiv \frac{P_{10} + C}{B\Omega_b}4\pi. \qquad (9.187)$$

For consistency with relations (9.182), we will again ignore b (which scales as A) and find the solution

$$\ln\frac{I_\nu(l)}{I_\nu(0)} = \frac{I_\nu(0) - I_\nu(l)}{I_s} + al. \qquad (9.188)$$

In the range where $I_\nu \ll I_s$, the intensity grows exponentially as

$$I_\nu(l) = I_\nu(0)e^{al} \qquad (9.189)$$

when the beam traverses the cloud. The initial $I_\nu(0)$ could easily arise from the emission from, say, a nearby region of the cloud. When I_ν becomes comparable with I_s the exponential growth saturates, and we get the behaviour

$$I_\nu(l) \approx I_s al. \qquad (9.190)$$

The maximum value of the brightness temperature can be $\sim 10^{12}$ K for OH$^-$ molecules acting as masers and $\sim 10^{15}$ K for H$_2$O molecules. If the intrinsic kinetic temperature of the clouds is $\sim 10^2$ K, then such an enhancement can arise when the beam traverses a distance of $al = 20$ or 30.

Finally, we comment on the effect of maser action on the linewidth. In the context of the ISM, linewidths are usually due to the Doppler effect. The line profile near the centre of the line varies as $\phi_\nu \propto \exp -(\Delta \nu / \Delta \nu_D)^2 \approx [1 - (\Delta \nu / \Delta \nu_D)^2]$, so that the optical depth is

$$\tau \equiv al \approx \tau_c [1 - (\Delta \nu / \Delta \nu_D)^2], \tag{9.191}$$

where τ_c is the line-centre optical depth. Equation (9.189) now becomes

$$I_\nu \propto \exp(\tau_c) \exp[-\tau_c (\Delta \nu / \Delta \nu_D)^2], \tag{9.192}$$

showing that the FWHM is reduced by the factor $[\ln(2/\tau_c)]^{1/2}$ during maser action, explaining why the lines are considerably narrow.

Although the above model contains the basic idea behind the maser action, the modelling of actual systems can be much more complicated. For example, H_2O masers have been observed from regions of size $l = 10^{13-14}$ cm, where the theoretical value for a is around 2×10^{-14} cm^{-1}. This will produce an optical depth of only $al \approx 0.2\text{–}2.0$, which is inadequate to explain the observed brightness temperatures. To circumvent this difficulty, it is necessary to enhance the optical path length by a factor of 10 without changing the geometrical path length (which can be done in models that contain filamentary geometry for the gas clouds); but the results in these cases are highly model dependent.

Exercise 9.10

Condition for maser action: Consider a three-level atom in which the only dominant processes are (1) collisional excitation and deexcitation between levels 1 and 2, with the deexcitation level rate being $n_e \gamma_{21}$; (2) collisional excitation and deexcitation between levels 1 and 3 with the corresponding rate $n_e \gamma_{31}$; (3) the spontaneous radiative decay from level 2 to level 1 with the rate A_{21}; (4) spontaneous radiative decay from level 3 to level 2 with rate A_{32}; take $A_{21} \ll A_{32}$ and $g_3 \gamma_{31} \leq g_2 \gamma_{21}$. (a) Define the critical electron densities n_{c2}, n_{c3} for levels 2 and 3. (b) Solve for the level populations in terms of the various parameters. (c) Under what conditions could one produce a maser action in the 2-to-1 emission line? {Answers: (a) The critical densities are defined as $n_{c2} = (A_{21}/\gamma_{21}), n_{c3} = (A_{32}/\gamma_{31})$. (b) The equilibrium condition for level 3 is given by

$$n_3(\gamma_{31} n_e + A_{32}) = n_1 \gamma_{13} n_e = n_1 n_e \gamma_{31} \frac{g_3}{g_1} e^{-\Delta E_{13}/k_B T}. \tag{9.193}$$

The departure from local thermodynamic equilibrium is determined by the ratio

$$\left(\frac{b_3}{b_1}\right) \equiv \frac{n_3}{n_1}\left(\frac{g_1}{g_3}\right) e^{\Delta E_{13}/k_B T} = \frac{\gamma_{31} n_e}{\gamma_{31} n_e + A_{32}} = \left(1 + \frac{n_{c3}}{n_e}\right)^{-1}. \tag{9.194}$$

A corresponding calculation for level 2 gives, after straightforward algebra,

$$\frac{b_2}{b_1} = \left(1 + \frac{n_{c2}}{n_e}\right)^{-1} \left[1 + \left(\frac{g_3}{g_2}\right)\left(\frac{\gamma_{31}}{\gamma_{21}}\right)\left(1 + \frac{n_e}{n_{c3}}\right)^{-1} e^{-\Delta E_{23}/k_B T}\right]. \quad (9.195)$$

(c) For maser action we would need $(n_2/n_1) > (g_2/g_1)$, which requires $(b_2/b_1) > e^{\Delta E_{12}/k_B T}$. From the expression obtained above, it is clear that if $k_B T \ll \Delta E_{23}$, then $(b_2/b_1) \leq 1$, and hence maser action is not possible. Therefore it is necessary for the gas to be hot with $k_B T \gtrsim (\Delta E_{23}, \Delta E_{12})$. To obtain the condition on the density, we note that $A_{21} \ll A_{32}$ and $g_3 \gamma_{31} \lesssim g_2 \gamma_{21}$ imply $n_{c2} \ll n_{c3}$. Now if $n_e \ll n_{c2}$ then $(b_2/b_1) \to (n_e/n_{c2})[1 + (g_3\gamma_{31}/g_2\gamma_{32})] \ll 1$, implying that there is no maser action. If $n_e \gg n_{c3}$ then a similar analysis shows that (b_2/b_1) is only approximately of the order of unity at best; so we cannot get a strong maser. The ideal condition therefore will be $n_{c2} \ll n_e \ll n_{c3}$, giving $(b_2/b_1) \to [1 + (g_3\gamma_{31}/g_2\gamma 21)] > 1.\}$

9.12 Interstellar Dust Grains

Nearly 1% of the mass contained in the ISM is in the form of dust grains whose dimensions are between 100 Å and 1000 Å. These dust grains, made of graphite, silicate, and other components, exist throughout the interstellar gas in our galaxy as well as in external galaxies. Although dynamically not significant (contributing less than 1% to mass) they do play an important role in the scattering and extinction of starlight, thereby affecting the visual appearance of objects in a marked manner.

There are several processes that can lead to the formation of such interstellar grains. For example, matter in the expanding outer atmosphere of a cool red giant, a planetary nebula, or even in the mass shells ejected by a nova can lead to the condensation of dust grains. The atoms of a gas in thermal equilibrium will undergo random fluctuations in any region of space and can form larger aggregates temporarily. At high temperatures such aggregates are unstable and will disperse away in a short time. For a stable aggregate of atoms to make a dust grain, it is necessary that the energy does change by some amount, E_{cond}, usually called condensation energy. Because the probability for a random energy fluctuation in thermal equilibrium varies as $\exp(-E_{\text{cond}}/k_B T)$, it is clear that lower temperatures favour the condensation. The temperature T_s at which a heavy element can condense as a grain is called the sublimation temperature. This quantity varies from $\sim 10^3$ K (for aluminium and iron) to rather low values of 10^2 K (for carbon). Some other compounds, such as H_2O and NH_4, also condense around 100 K, often onto preexisting dust grains. As the expanding envelope cools, heavier elements will condense first, followed by lighter ones, if other conditions remain invariant. It should, however, be noted that as the system expands the density also decreases so that the fraction of the element that condenses will be smaller for elements with lower T_s.

9.12 Interstellar Dust Grains

Several properties of such dust grains have been discussed in Vol. I, Chap. 6, Section 6.4. If the radius of the grain is r_d, so that the geometrical cross section is $\sigma_d = \pi r_d^2$, the actual cross section for photon scattering can be parameterised as $Q\sigma_d$, where Q takes into account both the absorption and the scattering of light by the dust. As discussed in Vol. I, Chap. 6, Section 6.4, Q essentially depends on the ratio between the grain size and the wavelength of the radiation parameterised by $x = (2\pi r_d/\lambda)$. The factor Q increases with x as a power law x^α, with $\alpha \approx 1$ for $x \ll 1$; for $x \gg 1$, Q reaches a constant value asymptotically. If $x \gg 1$ (when the grain size is much bigger than the wavelength of radiation), there will be very significant absorption of photons followed by reradiation as a continuous spectrum. On the other hand, if $x \ll 1$, the scattering will follow the Rayleigh law and will vary as x^4. The actual behaviour ($Q \propto x$) is a strong indication that the grain diameters are comparable with the wavelength of visible light.

The observed flux of radiation $F(\lambda)$ from a star at a wavelength λ will be related to its intrinsic flux at the same wavelength $F_0(\lambda)$ by $F(\lambda) = F_0(\lambda)\exp[-\tau(\lambda)]$, where $\tau(\lambda)$ is the optical depth. Taking logarithms and converting to magnitudes we get

$$m(\lambda) = m_0(\lambda) - 2.5\log\{\exp[-\tau(\lambda)]\} = m_0(\lambda) + 1.086\tau(\lambda). \quad (9.196)$$

The quantity $A(\lambda) \equiv 1.086\tau(\lambda)$ describes the effect of dust on the change of the magnitude of a star. Observations suggest that $A(\lambda) \approx 3n_H$ mag kpc^{-1}, where n_H is the average number density of hydrogen along the line of sight. If the magnitudes are measured in the blue (4500 Å) and the visible (5500 Å) we can compare the difference $m_B - m_V$ for distant stars and nearby stars. This quantity E_{B-V} is called *colour excess*, and it is found (empirically) that $A_V \approx 3.2\,E_{B-V}$.

The absorption and scattering of light lead to a loss of intensity along the line of sight, thereby causing extinction. Figure 9.10 indicates schematically the observed extinction that is due to interstellar dust as a function of the wavelength. The curve is normalised (rather arbitrarily) such that the extinction in the visible band is by 1 mag. Several features can be noted in this curve. The first one is at 2200 Å (0.22 μ), which is the only strong dust absorption in the UV band. The most likely candidate producing this absorption is graphite arising from the abundant element carbon. The second strong absorption feature is in the IR peaked at 9.7 μ but having a width that extends from 8 to 12 μ. This is attributed to silicate particles that are chemical compounds based around silicon and oxygen atoms in conjunction with common elements such as iron, aluminium, calcium, and magnesium. This feature is likely to have been caused by a combination of many different types of silicate particles, and in fact the absorption feature does not match any known single mineral. Spectra of IR sources located inside dense clouds show absorption features that are due to frozen molecules. The most prominent ones at 3.1 and 6 μ are probably due to H_2O. Evidence for frozen CO, NH_3, and CH_3OH has also been found in several

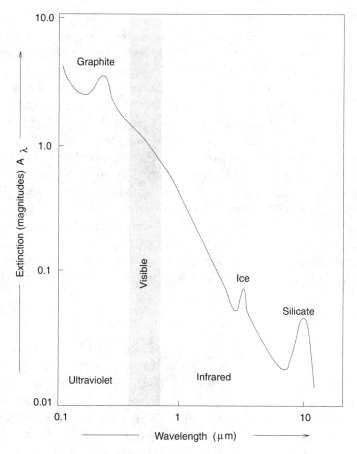

Fig. 9.10. Extinction curve that is due to interstellar dust.

cases. All these chemicals sublimate from solid to gaseous state at fairly low transition temperatures and hence can exist only in the cool dark interiors of molecular clouds.

In addition to these, the absorption curves also show a series of over 50 diffuse interstellar bands (not shown in Fig. 9.10), the strongest ones being at 4430 Å, 6284 Å, and 6177 Å. These bands are nearly 10 to 100 times wider than atomic absorption line features and cannot be explained by standard Doppler broadening. The net absorption, however, is still less than 10^{-3} of that produced by dust. It is possible that these are also associated with some kind of dust grains but actual modelling has so far been inconclusive.

The ISM spectroscopy also reveals strong IR emission features at, for example, 3.3 μ, 6.2 μ, 7.7 μ, 8.6 μ, and 11.3 μ, which are never seen in absorption. In fact they are seen only in regions where dust is heated by the UV radiation of stars, as in the HII regions or in planetary nebulas. They are identified with groups of molecules called polycyclic aromatic hydrocarbons (PAHs). They consist of 10 to 50 carbon atoms bonded together in rings that are extremely stable. When one of the molecules absorbs a UV photon, it can be excited to a higher energy level; decay to ground state occurs through a series of jumps involving IR photons. PAHs having ~50 atoms form an intermediate structure between atoms and solid dust grains.

The scattering of light will also lead to polarisation of the scattered light. We have seen in Vol. I, Chap. 6, Exercise 6.9 that the degree of polarisation for a spherical scatterer with $r_d \ll \lambda$ is given by

$$P = \frac{1 - \cos^2 \theta}{1 + \cos^2 \theta}, \quad (9.197)$$

where θ is the angle of scattering. The polarisation is total when $\theta = (\pi/2)$. Such polarised, scattered light is observed in dusty regions around stars.

In steady state, the absorption of ambient radiation energy by the grain must match the loss of energy by the grain, provided all other sources of heating for the grain are negligible. The gain in energy from the ambient field is given by

$$\Gamma_R = c \int_0^\infty u_\lambda Q(\lambda) \sigma_d \, d\lambda, \quad (9.198)$$

where u_λ is the energy density of radiation. The emission from the grain will be thermal and will be characterised by the temperature T_d of the dust grain. Using the standard relation between absorption and emission, we easily see that the rate of loss of energy by the grain will be

$$\Lambda_d = 4\pi \int_0^\infty Q(\lambda) \sigma_d \, B_\lambda(T_d) \, d\lambda. \quad (9.199)$$

Equating the two, we get the condition

$$4\pi \int_0^\infty Q(\lambda) B_\lambda(T_d) \, d\lambda = c \int_0^\infty u_\lambda Q(\lambda) \, d\lambda. \quad (9.200)$$

If it is assumed that the ambient energy density that is due to a star with a temperature T_{eff} and radius $R_* \simeq 6 \times 10^{11}$ cm and that the dust grain is at $r \simeq 5$ pc so that the a dilution factor is $W = (R_*/r)^2 \simeq 10^{-15}$, then

$$u_\lambda = \frac{4\pi}{c} B_\lambda(T_{\text{eff}}) W. \quad (9.201)$$

Equation (9.200) can be easily solved to give the relation between the effective temperature and the grain temperature as

$$T_d = T_{\text{eff}} \, W^{1/5}. \quad (9.202)$$

If $T_{\text{eff}} \approx 10^4$ K and $W \approx 10^{-15}$, then the grain temperature is $T_d \approx 10$ K, which is a reasonable estimate for most grains irradiated by stars.

There are other sources of heating of the grain that may be important in certain cases. The collision of the grains with molecules can result in the transfer of the kinetic energy of the molecule to the grain; when atoms combine to form molecules at the site of a grain, it may deposit a significant fraction of the binding energy of the grain. These two processes are usually subdominant to radiative heating in HI regions. On the other hand, in HII regions, the absorption of Lyman-α radiation from the central star can be an effective heating process. Because the emission of Lyman-α photons that is due to recombination depends on the local hydrogen density n_H, the grain temperature in these regions will also depend on n_H. We get typically $T_d \approx 20$ K for $n_H \approx 10$ cm^{-3}, rising to ~ 50 K for $n_H \approx 10^3$ cm^{-3}.

Dust grains in HII regions are continuously hit by free electrons and protons and hence can end up having a substantial amount of electric charge accumulated on that. As the charge accumulates, it tends to repel the like charges, thereby leading to a saturation effect. This has already been worked out in Vol. I, Chap. 9, Exercise 9.1, where it was shown that the charge on the grain is determined by the solution to the equation

$$1 - x_0 = \left(\frac{m_p}{m_e}\right)^{1/2} \frac{n_e}{n_i} \exp(x_0), \qquad (9.203)$$

where $x_0 = (|Ze^2|/r_d k_B T)$. (The notation has been changed slightly from Vol. I, taking into account the sign of x_0.) For $n_e \approx n_i$, this leads to a solution $(Ze^2/r_d k_B T) \approx 2.51$. For $r_d \approx 2 \times 10^{-5}$ cm and $T \approx 10^4$ K, we get $Z \approx 300$. This shows that electrical properties of grains can be significantly affected by the fact that they are charged.

The grains also interact strongly with the interstellar magnetic field. This arises partly because the grains are not strictly spherical but are elongated in some direction. When such a grain is in thermal equilibrium, its mean-square angular velocity $\langle \omega_j^2 \rangle$ about any one principle axis of symmetry (labelled by $j = 1, 2, 3, \ldots,$) will be related to the temperature by $(1/2) I_j \langle \omega_j^2 \rangle = (1/2) k_B T$, giving

$$\langle \omega_j^2 \rangle^{1/2} = \left(\frac{k_B T}{I_j}\right)^{1/2} \approx \left(\frac{k_B T}{\rho r_d^5}\right)^{1/2} \approx 4 \times 10^5 \text{ s}^{-1} \qquad (9.204)$$

if $T = 10^2$ K and $r_d = 10^{-6}$ cm. The lowest value for ω_j is obtained when the rotation is around one of the shorter axes. Further, if the grains are mildly paramagnetic, it is preferable for them to have the rotational axis aligned along the local magnetic-field direction for the following reason. When the rotational axis is along the direction of the magnetic field, the induced magnetic moment in the material does not change during the rotation; but if, for example, the rotation axis is perpendicular to the magnetic field, then the direction of the induced magnetic field continuously changes during the rotation, causing internal torques to damp out the motion. Such an alignment is indeed seen in the grains as it will produce transmitted light polarised in the direction of the local magnetic field. The details of the grain alignment are, however, quite complicated, and a final picture has not yet emerged.

9.13 Giant Molecular Clouds and Star Formation

A typical *giant molecular cloud* (GMC) will have a diameter of ~ 400 pc, a number density of molecular hydrogen $n(H_2) \simeq 300$ cm^{-3} with $[n(H_2)/n(HI)] \gtrsim 10$. The temperature is $T \approx 10$ K, corresponding to the sound speed of $c_s \simeq 0.2$ km s^{-1}. The free-fall collapse time of a denser core of such a gas cloud, in the absence of any form of pressure support, will be about

$$t_{\text{ff}} \simeq 5 \times 10^5 \text{ yr} \left(\frac{n}{10^4 \text{ cm}^{-3}} \right)^{-1/2}. \qquad (9.205)$$

The GMCs, however, live for a much longer time, $\sim 10^{7.5}$ yr. Further, the widths of the lines emitted from the molecular clouds are also much larger than the thermal widths corresponding to $T \simeq 10$ K. (See Table 9.4 for some properties of interstellar clouds.) One possible explanation for these phenomena is that the clouds are magnetically supported against the collapse, and the eventual collapse occurs only through the leaking of magnetic-field lines through the gas. To obtain such a magnetic support, we would require the condition (see Vol. I, Chap. 9, Section 9.6) $(3/5)(GM^2/R) \leq (B^2/8\pi)R^3$ or $M \lesssim M_{\text{crit}} = (B^2 R^4/8\pi G)^{1/2}$. Numerically, this translates into the bound

$$M_{\text{crit}} \simeq 200 \, M_\odot \left(\frac{B}{3 \times 10^{-6} \text{ G}} \right) \left(\frac{R}{3 \text{ pc}} \right)^2. \qquad (9.206)$$

In such a magnetically supported system, the Alfven speed will be comparable will the virial speed: $v_{\text{Alv}}^2 = (B^2/4\pi\rho) \simeq (5 \, GM/R)$. The diffusion of the magnetic-field line occurs because of the finite conductivity of the cloud arising from the electron density $(n_e/n_{\text{neutral}}) \simeq 10^{-9}$, which in turn arises because of cosmic-ray ionisation. For a cloud with $M \simeq M_{\text{crit}} \propto (BR^2)$ the surface density $\sigma = (M/R^2)$ is a constant whereas $\rho \propto (M/R^3) \propto (1/R)$. The velocity dispersion will scale as $\Delta v \simeq v_{\text{Alv}} \propto B\rho^{-1/2} \propto R^{1/2}$, which is close to the observed result.

Table 9.4. Representative interstellar clouds

Cloud Type	Globule	IR/HII Cloud	Dark Cloud	Diffuse Cloud	Cloud Complex
A_V (mag)	4	30	4	0.2	4
N_H (cm^{-2})	8×10^{21}	6×10^{22}	8×10^{21}	4×10^{20}	8×10^{21}
n_H (cm^{-3})	7×10^3	4×10^4	2×10^3	20	2×10^2
R (pc)	0.3	0.4	1	5	10
$T(^\circ \text{K})$	10	50	10	80	10
$M/M_\odot$	30	400	300	400	3×10^4

The magnetic force also affects the neutral atoms because of the ambipolar diffusion, discussed in Vol. I, Chap. 9, Section 9.6, where it was shown that the force on the neutral particles is given by

$$f \simeq n_i n_n m_n v_{\text{drift}} \langle \sigma v \rangle, \quad \langle \sigma v \rangle \simeq 10^{-9} \text{ cm}^3 \text{ s}^{-1}. \tag{9.207}$$

If this force is supported by the gravitational potential gradient

$$\rho \nabla \phi \simeq n_n m_n \frac{GM}{r^2}, \tag{9.208}$$

then we get the relation

$$\frac{GM}{r^2} \simeq n_i \langle \sigma v \rangle v_{\text{drift}}. \tag{9.209}$$

(In arriving at these results, we have taken $\rho \simeq n_n m_n$, as $n_n m_n \gg n_i m_i$.) Rearranging relation (9.209),

$$\frac{GM}{r^3} \simeq \frac{4\pi G}{3} n_n m_n \simeq n_i \langle \sigma v \rangle \left(\frac{v_{\text{drift}}}{r} \right), \tag{9.210}$$

so that the time scale for the drift to be significant becomes

$$t_{\text{drift}} \equiv \frac{r}{v_{\text{drift}}} \simeq \frac{n_i}{n_H} \frac{\langle \sigma v \rangle}{4 G m_n} \simeq 5 \times 10^{13} \left(\frac{n_i}{n_H} \right) \text{ yr}. \tag{9.211}$$

The ionisation of H_2 molecules by cosmic rays leads to a value of $(n_i/n_H) \simeq 10^{-7}[n(H_2)/10^4]^{-1/2}$, thereby giving $t_{\text{drift}} \simeq 5 \times 10^6$ yr $[n(H_2)/10^4]^{-1/2}$. For comparison, note that HI gas has $(n_i/n_H) \simeq 10^{-3}$, giving $t_{\text{drift}} \simeq 5 \times 10^{10}$ yr, which is much larger than the lifetime of the cloud, thereby making ambipolar diffusion irrelevant. The actual collapse of such a cloud – as the magnetic field leaks out – is quite complicated with ionisation, magnetic support, and gravitational dynamics all playing a significant role.

It should also be noted that a density of one hydrogen atom/cm^3 over a scale of 1 kpc will lead to an attenuation by 1 mag. This density of one hydrogen atom cm^{-3} over 3×10^{21} cm is equivalent to a column density of 3×10^{21} hydrogen atom/cm^2. A cloud with $M \simeq M_{\text{crit}}$, $B \simeq 20 \times 10^{-6}$ G will have a surface density $\Sigma \simeq 0.015$ gm cm$^{-2} \simeq 50\, M_\odot$ pc$^{-2} \simeq 10^{22}$ atoms cm^{-2}. This will lead to an attenuation of ~ 3 mag ($\sim 90\%$), thereby shielding the inside regions quite well.

Exercise 9.11

Cosmic-ray ionisation: One of the sequences of reactions that leads to the ionisation of H_2 molecules is the following: (1) $H_2 + p \rightarrow H_2^+ + e^- + p$, in which a high-energy cosmic-ray proton ejects an electron from the H_2 molecule. The rate for this can be expressed in the form $Jn(H_2)$, where $J \simeq 10^{-17}$ s^{-1} if the density is in cgs units. (2) $H_2^+ + H_2 \rightarrow H_3^+ + H$; the rate for this can be expressed in the form $\alpha_2 n(H_2) n(H_2^+)$, where α_2 is a suitable recombination coefficient. (3) $H_3^+ + e^- \rightarrow 3H$. The rate for this can be expressed in the form $\alpha_3 n(H_3^+) n_e$. (a) Show that, in steady state, this will give

$[n_e/n(H_2)] \propto [J/\alpha_3 n(H_2)]^{1/2}$. (b) Also show that the ratio between the drift time scale and the free-fall time scale of a molecular cloud is $\sim 10 \, (J/10^{-17} \, s^{-1})^{1/2}$, independent of the density of the cloud.

The molecular clouds also contain several regions of star formation that are studied extensively in order to understand the early phases of stellar evolution. We saw in Chapter 3 that massive O- and B-type stars will be first to form inside a collapsing cloud undergoing fragmentation. Initially, the protostar will appear as an IR source embedded within the molecular cloud. With rising temperature the dust in the cloud will vaporise, molecules will dissociate and eventually – when the main-sequence star is formed – the UV photons will ionize the cloud, creating a HII region. This will lead to the driving of an ionisation and shock front through the cloud along the lines described in Section 9.4. If several OB stars form at the same time, the shock fronts can expel much of the loosely bound mass, thereby halting further star formation. This suggests that star formation can have a complex feedback reaction on the cloud. If the loss of mass from the cloud is significant, resulting in the loss of gravitational binding energy, the newly formed cluster of stars will become unbound and will tend to drift apart. Such groups of newly formed OB main-sequence stars are called OB associations, one example being the trapezium cluster in Orion, which is estimated to be less than 10^7 yr old. It is densely populated with more than 2000 stars parsec^{-3}, most of which have masses in the range of $(0.5-2) \, M_\odot$. Doppler-shift measurements of CO molecules show that the gas is very turbulent and the stars themselves are moving in a manner that will disperse the cluster.

Another class of low-mass pre-main-sequence stars in the transition stage are called T-Tauri stars named after the first star of their class located in the constellation Taurus. These stars exhibit strong Balmer lines of emission as well CaII and iron lines. The shape of some of the lines in the T-Tauri stars, especially the Hα lines, can be interpreted as being due to the existence of an expanding shell of matter around the star arising from the significant mass loss. The shape, usually called the *P-Cygni profile* is shown in Fig. 9.11 (see Exercise 3.5 in Chap. 3). The absorption part of the spectrum arises from the continuum radiation from the star passing through a cooler diffuse gas, whereas the emission peak arises from the part of the shell on the back and in front of the star. The shift that is due to the Doppler effect corresponds to speeds of ~ 80 km s^{-1} for the expanding gas.

The collapse of a gas cloud leading to star formation will not, in general, be spherically symmetric – especially if the collapsing region had an initial angular momentum. The arguments given in Chap. 7 suggest that the most natural consequence will then be the formation of a disk of material. In some cases (called *Herbig-Haro* objects) it has actually been noted that there are narrow beams of jet ejected in opposite directions perpendicular to the accretion disk. These jets expand supersonically into the ISM and excite the gas collisionally,

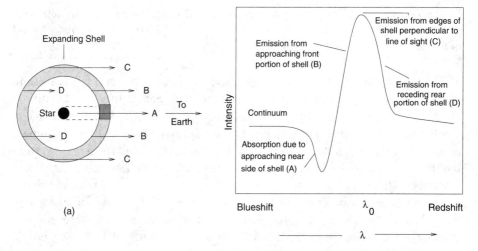

Fig. 9.11. The geometry of a star with an expanding shell leading to the P-Cygni profile.

thereby resulting in relatively bright regions with emission-line spectra. Further, the light from the parent star can also be reflected in these objects, leading to continuous emission. A typical structure of a T-Tauri star with an accretion disk and the jet is shown in Fig. 9.12. The existence of accretion disks around the protostar is important in the theories of planetary formation.

Exercise 9.12

Bulk motions in a nebula: The Hα line and the [NII] 6584-Å line emitted by a nebula are seen to have FWHMs of 0.5 and 0.4 Å, respectively. Provide an interpretation of this result in terms of bulk motions and estimate the electron temperature and velocity of random bulk motion. {Answers: In a nebula, Doppler broadening is the only relevant mechanism and the FWHM for Doppler broadening scales as $m^{-1/2}$. Therefore we would expect the Hα line to be $\sqrt{14} \approx 3.7$ times broader than the [NII] line, but observationally it is only 1.25 times broader. This shows that some other process is in operation, which could be the

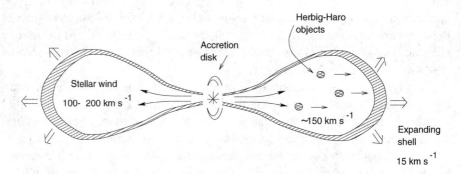

Fig. 9.12. Schematic diagram showing a T-Tauri star with an accretion disk.

bulk motion of the region from which the emission takes place. Assuming that FWHMs that are due to bulk thermal motions add in quadrature, we have $F_{\text{bulk}}^2 + F_{\text{th,H}}^2 = F_{\text{tot,H}}^2 = (0.5 \text{ Å})^2$; $F_{\text{bulk}}^2 + F_{\text{th,N}}^2 = F_{\text{tot,N}}^2 = (0.4 \text{ Å})^2$; and $F_{\text{th,H}} = 14 \times F_{\text{th,N}}^2$. Solving these three equations, we get $F_{\text{th,H}} = 0.31$ Å, $F_{\text{th,N}} = 0.08$ Å and $F_{\text{bulk}} = 0.39$ Å . The corresponding temperature and bulk velocity are given by $k_B T = (m_H c^2)(\Delta\lambda/\lambda)^2(1/8 \ln 2)$; $v_{\text{bulk}} \simeq c(\Delta\lambda/\lambda)$. Numerically $T = 4400$ K and $v_{\text{bulk}} = 18$ km s^{-1}.}

Exercise 9.13

Angular-momentum loss through Alfven waves: Consider a spherically symmetric cloud of radius R rotating with an angular velocity ω in a region containing an axially symmetric magnetic field **B** that is parallel to the axis of rotation. Surrounding this spherical cloud is the ISM with $\rho = 4 \times 10^{23}$ gm cm^{-3} and $T = 50$ K. We will solve the equations of MHD under the following assumptions: (1) Conductivity is very high so that flux freezing occurs in the medium. (2) The ISM plasma surrounding the rotating cloud is initially at rest and later on picks up only a v_θ component of velocity. (3) The magnetic field has only θ and z components at all times. (a) Show that the relevant equations are

$$\frac{\partial B_\theta}{\partial t} = B_z \frac{\partial v_\theta}{\partial z}, \quad \frac{\partial v_\theta}{\partial t} = \frac{B_z}{4\pi\rho} \frac{\partial B_\theta}{\partial z}, \tag{9.212}$$

which have the general solution

$$v_\theta = f(t - z/v_A, r) + g(t + z/v_A, r), \tag{9.213}$$

$$B_\theta = -(4\pi\rho)^{1/2}[f(t - z/v_A, r) - g(t + z/v_A, r)] \tag{9.214}$$

where f and g are arbitrary functions. (b) Argue that boundary conditions require that $g = 0$ for propagation along the positive z axis and $f = 0$ for propagation along the negative z axis. As the Alfven waves propagate through the plasma, they set the plasma in rotation, thereby transferring angular momentum from the cloud to the plasma. Show that the rate of loss of the cloud's angular momentum is

$$\frac{dJ_{\text{cloud}}}{dt} = -4\pi v_A \rho \int_0^{d=v_A t} v(t - z/v_A, r) r^2 \, dr. \tag{9.215}$$

To make an estimate, assume that other physical effects stop this corotation for $d \approx R$ and that the velocity structure is of the form $v(t - z/v_A, r) = \omega(t)r$. Taking $J = I\omega(t) = (2MR^2/5)\omega(t)$, integrate the equation for angular momentum loss to obtain

$$\omega(t) = \omega(0) e^{-t/\tau}, \tag{9.216}$$

where

$$\tau = \frac{4M}{5\sqrt{\rho\pi} \, BR^2}. \tag{9.217}$$

(c) A condensation with $M = 10 \ M_\odot$ has an average density of $\rho \simeq 4 \times 10^{-17}$ gm cm^{-3} when the fragmentation stops. At this time its angular velocity is approximately $\omega(0) = 2 \times 10^{-15}$ s^{-1}. How long will it take for the angular velocity to increase by 1 order of magnitude, which is typical for stars with $R \simeq R_\odot$? How does it compare with the relevant free-fall time?

10
Globular Clusters

10.1 Introduction

This chapter deals with the kinematics and the dynamics of systems containing large-number (above 10^6) stars called *globular clusters*.[1] It draws heavily on Chap. 10 of Vol. I and the basic ideas of stellar evolution described in the earlier chapters of this volume.

Globular clusters play an important role as systems in which many aspects of stellar-evolution theory can be directly tested and, in the process, can be used to provide significant information about the age and the mass function of stars in our galaxy. They also are examples of systems dominated by gravitational many-body interactions that hence undergo evolution in a manner quite different from other – simpler – systems. Finally, their dynamical evolution provides important points of similarity with central regions of galaxies, elliptical galaxies, and galaxy clusters. The kinematical aspects are discussed in the next section, and the rest of the chapter is devoted to dynamical issues.

10.2 Stellar Distribution and Ages of Globular Clusters

We have seen in the earlier chapters that the formation of stars from gaseous clouds is a fairly complex process that is not completely understood. Irrespective of the details, it seems reasonable to expect that, when stars form in a cloud, there will be a tendency for a large number of them to be close together, thereby forming a cluster of stars. Two broad categories of such star clusters have been seen in our galaxy, usually called *open clusters* and *globular clusters*.

Open clusters are made of population I stars and are found in the galactic disk. A typical open cluster contains 100–1000 stars in a region of size (1–10) pc. These are relatively young objects with a median age of $\sim 10^8$ yr. There is a significant paucity of open clusters with ages greater than $\sim 10^9$ yr and there is some observational evidence suggesting that the formation of open clusters is an ongoing process. There are $\sim 10^5$ open clusters in our galaxy. The luminosity

10.2 Stellar Distribution and Ages of Globular Clusters

functions of open clusters seem to increase towards the fainter end to the extent that observations can determine; in that sense, the above number should be thought of as a lower bound.

Globular clusters, on the other hand, are made of population II stars. A typical globular cluster contains 10^4–10^6 stars confined within a median radius of ~ 10 pc. They are significantly older systems with an age of $\sim 10^{10}$ yr and are distributed in a spherically symmetric manner about the galactic centre. Because the formation of an initial bunch of stars in the galaxy is a spatially inhomogeneous process (see Vol. III) occurring over a period of time, it is possible that some regions would have produced copious amount of stars fairly early on. In such a case, we would find a fair number of isolated objects, each of which might contain a large number of very old stars, gravitationally bound together by their self-gravity. The globular clusters seem to be such systems. By and large, they appear to be approximately circular in projection and contain virtually no diffuse gas or dust. Because almost all the stars in a globular cluster have formed at the same epoch, they serve as a testing ground for studying different aspects of stellar evolution and stellar dynamics.

Although the above description is broadly correct, globular clusters do vary significantly in terms of the details. The brightest cluster in our galaxy, for example, has an absolute visible magnitude of $M_V \simeq -10.4$ contributed by $\sim 10^6$ stars. The faintest clusters contain only $\sim 10^3$ stars, with $M_V \simeq -3$. Most of the globular clusters are distributed around an absolute magnitude of $M_V \simeq -7$. Globular clusters have also been detected in other galaxies but their properties are not specifically correlated in any way with the properties of the host galaxy. These aspects need to be explained in any complete theory for the formation of globular clusters.

During the bulk of this chapter, we shall concentrate on globular clusters. Brief comments will be made about open clusters in Section 10.8.

The most important diagnostic of stellar-evolution theory that we obtain from globular clusters is the H-R diagram or – equivalently – the colour-magnitude (CM) diagram for the cluster. Figure 10.1(a) shows a schematic structure of the CM diagram for a typical globular cluster. This differs quite a bit from the corresponding CM diagram of field stars in, say, the solar neighbourhood. In the latter, we find a well-defined main-sequence band from very high-mass stars to very low-mass stars with a red giant branch connecting to it (see left panel of Fig. 1.3 of Chap. 1). The CM diagram of a globular cluster has very few stars in the upper-mass end of the main sequence and shows a distinct *turn-off point* along the main sequence. The reason for this is easy to understand if we assume that stars (with a distribution of initial masses) would have formed almost at the same time in a globular cluster. We have seen in Chap. 2 that high-mass stars have significantly higher luminosities and that they evolve faster. Hence the high-mass end of the main sequence would have been depleted in the globular clusters – which is precisely the reason for the existence of the turn-off point.

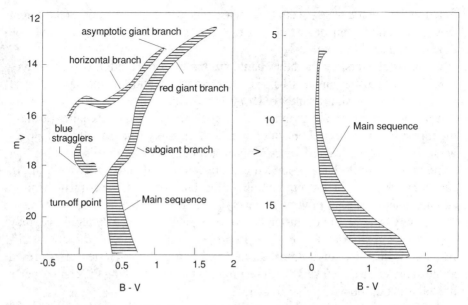

Fig. 10.1. (a) Schematic CM diagram for globular clusters; (b) schematic CM diagram for open clusters for comparison.

The other branches seen in Fig. 10.1 have all been explained in the standard stellar-evolution theory. For comparison, Fig. 10.1(b) gives the H-R diagram for an open cluster. It is clear that it has a extensive main sequence, spreading towards fainter magnitude, suggesting that open clusters are relatively young.

The precise determination of the location of the turn-off point, along with accurate modelling of stellar evolution, will allow us to determine the ages of the globular clusters. In principle this essentially involves calculating the age of the stars that are just moving away from the main sequence.[2] In practise, however, this requires a more careful treatment of stellar evolution along the following lines. To begin with, we need to choose the initial abundance of the chemical elements that is appropriate for the globular cluster. It has been noted that the abundance of metals in the globular clusters shows a fair amount of variation. Because the accurate determination of the evolutionary track of stars requires the correct estimate of metal abundance, it is necessary to use the values appropriate for the globular cluster that is being modelled. (For example, a relative overabundance of oxygen is seen in a class of clusters, which can affect the hydrogen burning efficiency in the CNO cycle.) Given the abundances, we can locate the initial positions of stellar population on the zero-age main-sequence band. The next step involves numerically evolving the stellar population forward in time by the solving the stellar structure evolution as accurately as possible. We can thus calculate the luminosities and colours for all the stars in the population at any given later time; that is, we know the location of any given star in the CM

10.2 Stellar Distribution and Ages of Globular Clusters

diagram at any future time. Finally, we draw a line connecting all the stars in the CM diagram at any given time. Such a line, called an *isochrone*, provides a theoretical image of the steller population when the cluster has a specified age. It is possible to draw isochrones corresponding to different ages as well as those for the same age but for different input parameters and stellar-evolution models. A comparison of these isochrones with the observed sequence of stars in the globular clusters (by a statistical analysis for the best fit) should provide us not only the age of the globular cluster but also information on input parameters and assumptions regarding stellar evolution.

Several theoretical studies suggest that the absolute luminosity at the turn-off point L_{TO} is related to the age and other parameters by

$$\log\left(\frac{L_{TO}}{L_\odot}\right) \approx [0.019(\log Z)^2 + 0.065 \log Z + 0.41\, Y - 1.179] \log\left(\frac{t}{\text{Gyr}}\right)$$
$$+ 1.246 - 0.028(\log Z)^2 - 0.272 \log Z - 1.073\, Y, \qquad (10.1)$$

where t_9 is the age in 10^9 yr and luminosity is measured in $L_\odot$. The absolute magnitude at the turn-off point for most clusters is ~ 4 and the metallicity is $\sim 10^{-1.5}$, giving an age of ~ 13 Gyr.

This procedure, of course, requires the knowledge of the distance to the cluster, as we need to convert the apparent magnitude to the absolute magnitude or luminosity. This is usually done by a process called *main-sequence fitting*. In this process, we first establish a unique relationship between the colour of a main-sequence star and its absolute magnitude from some other known sample and then use it to obtain the absolute magnitude from the colour of the star in the cluster. It must, however, be noted that a 10% uncertainty in the distance to the cluster will result in an error of 0.2 in its distance modulus and in the absolute scale for its CM diagram, which in turn could lead to a 20% error in the cluster's age. Similarly, uncertainties in the correction for extinction, which are important in connecting absolute and apparent magnitudes, can also lead to sizable errors in the age.

We can circumvent some of these difficulties by using the difference in the absolute magnitude between the turn-off point and the plateau region of the horizontal branch. This difference, of course, is independent of the distance modulus. Further, the brightness of the horizontal branch depends only weakly on the mass of the stars that populate it (and hence on the age of the cluster) whereas the magnitude of the turn-off point depends strongly on the age. Using this technique, we get an age of 15 ± 1 Gyr. We can devise similar techniques by taking the difference between the colours or magnitudes of any other pair of unique points in the CM diagram.[3]

One interesting aspect of the evolutionary sequence of stars in the globular clusters is the propensity of stellar remnants in these objects. We have seen in Chap. 3 that stars with initial masses in the range $(1-8)\, M_\odot$ would have become white dwarfs. This suggests that in a typical globular cluster, white dwarfs make up nearly 10% of the stellar population for the standard Salpeter initial mass function. Although these are fairly faint (stretching in U magnitude between 22 and 26) this population has indeed been observed from the Hubble Space Telescope. Similarly, we would have expected stars with initial masses $M_i \gtrsim 8\ M_\odot$ to have become compact remnants, of which the vast majority will be neutron stars. Although an individual neutron star would be unobservably dim, it will certainly be seen if it is a part of an x-ray-emitting binary system. In fact globular clusters contain 12 of the 100 brightest x-ray sources, although they contribute only $\sim 10^{-4}$ of the total luminosity of the galaxy. Eventually we would expect a fraction of these binaries to end up as millisecond pulsars, as discussed in Chap. 7. (We had already noted in Chap. 7 that there is some discrepancy in relating number densities of millisecond pulsars and x-ray binaries.) In a sample of $\sim 10^3$ pulsars, ~ 44 are in globular clusters, of which 35 are millisecond pulsars with $P < 25$ ms. Among these, 19 are in binary systems.

10.3 Time Scales in the Evolution of Globular Clusters

We shall next turn our attention to the dynamical evolution of globular clusters. To the lowest order of approximation, treated as an isolated system, globular clusters are examples of systems dominated by long-range gravitational forces. As we have seen in Vol. I, Chap. 10, the statistical description of such systems is quite different compared with that of systems dominated by short-range forces. To the next order, we need to incorporate the fact that globular clusters do interact with other constituents of the galaxy and hence cannot always be treated as completely isolated systems. Finally, the stars in the globular clusters themselves evolve in time and interact with each other during close encounters; it is therefore not strictly correct to treat the constituents of the globular cluster as static point particles.

Fortunately these processes are governed by vastly different time scales and hence can be separated in the discussion of globular-cluster evolution. It is also possible to determine different spatial scales over which different gravitational phenomena dominate in the dynamics of the system. To appreciate these features, we start with an order-of-magnitude estimate of various spatial and temporal scales relevant for globular clusters.

The spatial distribution of stars in a globular cluster is fairly nonuniform and is centrally concentrated. This allows us to define three different radii that are of use. The first is the *core radius*, which is observationally determined as the radius at which the projected surface brightness of the globular cluster falls to half its central value. The second is the *median radius*, which is the radius of the sphere that contains half of all the light emitted by the cluster. Finally, the *tidal radius* determines the outer edge of the cluster beyond which a member of the cluster will be influenced more by the external gravitational field of other matter

10.3 Time Scales in the Evolution of Globular Clusters

in the galaxy rather than by the self-gravity of the cluster. For a typical globular cluster, these three radii are 1.5, 10, and 50 pc, respectively.

The first two of these radii can also be characterised in terms of the mass distribution in the globular cluster and hence are properties of the cluster itself. The tidal radius, on the other hand, needs to be defined, taking into account the background gravitational potential of the galaxy in which the cluster is moving. Let $\Phi(\mathbf{X})$ denote the gravitational potential of the smooth distribution of matter in the galaxy, where the origin is taken to be the centre of the galaxy; let $\phi(\mathbf{x})$ denote the gravitational potential of the stars in the globular cluster itself, with $\mathbf{x} = 0$ denoting the cluster centre. The net gravitational potential felt by any star in the globular cluster will be the sum of these potentials evaluated at the point $\mathbf{R} + \mathbf{x}$ (with respect to the centre of the galaxy), where $\mathbf{R}$ denotes the position of the centre of the cluster at any given time. A star at this position in the cluster will feel a differential acceleration with respect to the centre of the globular cluster because of the background galactic potential Φ. This differential acceleration is given by

$$g_{\text{bg}}^i = -\left[\frac{\partial \Phi}{\partial x^i}\bigg|_{\mathbf{R}+\mathbf{x}} - \frac{\partial \Phi}{\partial x^i}\bigg|_{\mathbf{R}}\right] \approx -x^j \frac{\partial^2 \Phi}{\partial R^i \partial R^j}, \quad (10.2)$$

as the size of globular cluster is small compared with the scale of variation of the galactic potential. This differential acceleration can tidally distort the cluster and even pull stars out of it unless the self-gravity is sufficiently strong. The gravitational acceleration of the star that is due to the matter's belonging to the cluster is given by $g_{\text{self}}^i = -(\partial \phi / \partial x^i)$. The condition for tidal stability of the cluster is $g_{\text{self}}^i \gg g_{\text{bg}}^i$. For an order-of-magnitude estimate, we can take

$$g_{\text{bg}} \simeq x \nabla^2 \Phi \simeq x(4\pi G \rho_{\text{bg}}), \quad g_{\text{self}} \simeq x \nabla^2 \phi = x(4\pi G \rho_{\text{cluster}}). \quad (10.3)$$

The condition for tidal stability becomes $\rho_{\text{cluster}}(x) \gg \rho_{\text{bg}}(R)$, where the densities have to be interpreted as the mean densities within spheres of radius R (for the galaxy) and radius x (for the cluster). The tidal radius can be estimated from $\bar{\rho}_{\text{cluster}}(r_t) = \bar{\rho}_{\text{bg}}$. Given the background density and the density profile of the cluster, this equation determines the tidal radius.

Let us next consider the time scales involved in the description of a globular cluster, which are more complicated. To the lowest order of approximation, globular clusters represent a system of bodies evolving under the collective gravitational field and, as such, they form a gravitationally interacting many-body system. We saw in Vol. I, Chap. 10 that the statistical mechanics of a system dominated by gravity will be very different from more conventional systems in which the interaction is through short-range collisions. Many of the complications and new features discussed in Vol. I, Chap. 10 become relevant for the study of globular-cluster evolution, as can be seen from a discussion of the relevant time scales over which different processes operate in a typical globular cluster.

The first time scale that characterises a gravitational many-body system is the orbital time scale $t_{\text{cross}} \approx (R/v)$, where R is the size of the system and v is the typical velocity. If the system is in steady state, so that virial theorem is applicable and $v^2 \approx (GM/R) = (GNm/R)$, where m is the mass of the individual stars in the globular cluster containing a total of N stars, then t_{cross} is also of the same order as $t_{\text{dyn}} \equiv (GM/R^3)^{-1/2} \approx (G\rho)^{-1/2}$. For a globular cluster $N \approx 10^5$, $t_{\text{cross}} \approx 10^5$ yr, and the age of the system is $t_{\text{age}} \approx 10^{10}$ yr. In other words, stars would have made several ($\sim 10^5$) orbits during the lifetime of the system. This time scale t_{cross} also decides the duration over which the initial phase of violent relaxation operates in a self-gravitating system.

The second important time scale for a gravitational many-body problem is the relaxation time that is due to gravitational collisions t_{relax}. This is the time scale over which the granularity of the system can make the orbits deviate significantly from the paths that stars would have followed under the action of the gravitational field generated by the smooth density of matter. It has been shown in Vol. I, Chap. 10, Section 10.7 that

$$t_{\text{relax}} \approx \alpha \left(\frac{N}{\ln N} \right) t_{\text{cross}}, \quad (10.4)$$

where α is a numerical coefficient that depends on the precise definition used for the time scales, usually $\alpha \simeq 0.1$. For a globular cluster, $(N/\ln N) \approx 8.7 \times 10^3$ and $t_{\text{relax}} \simeq 10^9$ yr $< t_{\text{age}}$; hence relaxation effects would have played a major role in determining the present structure. This is in marked contrast to systems like galaxies with $N \approx 10^{11}$ and $t_{\text{cross}} \approx 10^8$ yr, in which $t_{\text{relax}} \gg t_{\text{age}}$ and the gravitational collisions are irrelevant over the age of the system. As we shall see, it is the collisional relaxation that makes the dynamical evolution of globular clusters different from other systems. In passing, we may note that a similar situation exists in the case of galactic nuclei and clusters of galaxies, which we will encounter in Vol. III.

There are three key effects that arise because $t_{\text{relax}} \lesssim t_{\text{age}}$ for the globular cluster. The first one is the obvious feature that the distribution of stars will tend towards a state of maximum entropy over this time scale. We have seen in Vol. I, Chap. 10, Section 10.3 that such a state is highly inhomogeneous for a self-gravitating system. Further, it can also lead to certain special kinds of instabilities in the system, which we shall discuss in Section 10.4.

Second, the system will tend towards a state with equipartition of kinetic energy in a time scale $t \approx t_{\text{relax}}$. Stars with larger kinetic energy will, on the average, lose energy to stars with less kinetic energy. If the initial distribution function is determined by violent relaxation, then the initial velocity distribution of stars will be independent of stellar mass and hence more-massive stars will have larger kinetic energy. Over a time of the order of t_{relax}, more-massive stars will lose energy to less-massive stars and will sink towards the centre of the system. This will lead to mass segregation in the system.

10.3 Time Scales in the Evolution of Globular Clusters

Finally, the escape of high-velocity stars in the tail of the Maxwellian distribution, from the gravitational potential well of the system, will become significant over a time scale t_{evap}, which is somewhat larger than t_{relax}. An estimate for t_{evap} can be made along the following lines: The escape speed v_{esc} for a star at a position $\mathbf{x}$ is given by $v_{\text{esc}}^2 = -2\Phi(\mathbf{x})$. The mean-square escape speed of the system is therefore

$$\langle v_{\text{esc}}^2 \rangle = \frac{\int \rho(\mathbf{x}) v_{\text{esc}}^2 \, d^3\mathbf{x}}{\int \rho(\mathbf{x}) \, d^3\mathbf{x}} = -2 \frac{\int \rho(\mathbf{x})\Phi(\mathbf{x}) \, d^3\mathbf{x}}{M} = -\frac{4W}{M}, \qquad (10.5)$$

where W is the potential energy of the system and M is the total mass. In virial equilibrium, $(1/2)M\langle v^2 \rangle = -(W/2)$, giving $\langle v_{\text{esc}}^2 \rangle = 4\langle v^2 \rangle$. For a Maxwellian distribution, the fraction f of particles with speeds higher than v_{esc} can be easily computed to be

$$f = \frac{4\pi}{N} \int_{2v}^{\infty} F(u) u^2 \, du = \frac{4}{\sqrt{\pi}} \int_{2.45}^{\infty} e^{-x^2} x^2 \, dx = 7.38 \times 10^{-3}, \qquad (10.6)$$

where $F(u)$ is Maxwellian distribution function for the velocity. Modelling the evaporation of stars from the cluster by the assumption that this fraction f escapes over a time scale t_{relax}, we find that the rate of loss is $(dN/dt) = -f(N/t_{\text{relax}})$. This gives the time scale for evaporation as

$$t_{\text{evap}} \equiv \frac{t_{\text{relax}}}{f} \approx 136\, t_{\text{relax}}, \qquad (10.7)$$

which defines the third time scale for the cluster.

Exercise 10.1

Example of equipartition: Consider a cluster containing two classes of stars with individual masses m_1 and m_2. Assume further that (1) $m_2 \gg m_1$ but (2) the total mass contributed by heavy stars m_2 is far less than the core mass contributed by the lighter-mass stars. (a) Show that the virial theorem for the heavier stars can be written in the form

$$2K_2 + W_2 - G \int_0^{\infty} \frac{\rho_2(r) M_1(r)}{r} 4\pi r^2 \, dr = 0, \qquad (10.8)$$

where the symbols have the standard meanings. Assuming that $\rho_1(r) \equiv \rho_{c1} \simeq$ constant and $W_2 = -f(GM_2^2/r_{h2})$, where f is a dimensionless constant and r_{h2} is the median radius of the heavy stars, deduce that

$$\langle v_2^2 \rangle = f \frac{GM_2}{r_{h2}} + \frac{4\pi G \rho_{c1}}{3} \langle r_2^2 \rangle, \qquad (10.9)$$

where $\langle r_2^2 \rangle$ can be expressed as $\langle r_2^2 \rangle = g^2 r_{h2}^2$, where g is another dimensionless constant of the order of unity. (b) It is also conventional to introduce a quantity called the *King radius* r_{c1} by the definition $r_{c1}^2 = (9\sigma^2/4\pi G \rho_{c1})$. Rewrite Eq. (10.9) in terms of these quantities as

$$\frac{4\pi}{3}\frac{m_1}{m_2}G\rho_{c1}r_{c1}^2 = f\frac{GM_2}{r_{h2}} + \frac{4\pi}{3}g^2 G\rho_{c1}r_{h2}^2 \qquad (10.10)$$

and show that equipartition state $m_2\langle v_2^2\rangle = m_1\langle v_1^2\rangle = 3m_1\sigma^2$ can be reached only if the following condition is satisfied:

$$\frac{M_2}{\rho_{c1}r_{c1}^3} \le \frac{1.61}{fg}\left(\frac{m_1}{m_2}\right)^{3/2}. \qquad (10.11)$$

What happens if this condition is violated? Explain the physical origin of this result, which has sometimes been called the *equipartition instability*.

The finite radius r_* of the star introduces another characteristic time scale, viz., the one in which a star suffers a physical collision with another. This is given by $t_{\text{coll}} \approx (n\sigma v)^{-1}$, where $n \approx (N/R^3)$ is the number density of stars, $\sigma \approx (4\pi r_*^2)$ is the collisional cross section, r_* is the radius of the star, and v is the rms velocity of stars. Therefore

$$\frac{t_{\text{coll}}}{t_{\text{cross}}} \approx \left(\frac{1}{4\pi N}\right)\left(\frac{R}{r_*}\right)^2. \qquad (10.12)$$

Using the virial theorem $v^2 = (GNm/R)$, we can eliminate R in terms of v; similarly, we can eliminate r_* in terms of the escape speed from the stellar surface $v_* = (2Gm/r_*)^{-1/2}$. Then we get

$$\frac{t_{\text{coll}}}{t_{\text{cross}}} \approx 0.02 N \left(\frac{v_*}{v}\right)^4, \qquad (10.13)$$

or, in terms of the relaxation time in relation (10.4) with $\alpha \simeq 0.1$,

$$\frac{t_{\text{coll}}}{t_{\text{relax}}} \approx 0.2 \left(\frac{v_*}{v}\right)^4 \ln N. \qquad (10.14)$$

This shows that, for globular clusters, t_{coll} is larger than t_{relax} and hence physical collisions do not significantly change the evolution over a few relation times. (As an aside, note that the situation could be different for other systems if $v_* \ll v$. One such example is the rich cluster of galaxies in which v_* is approximately 300 km s^{-1} and $v \approx 1500$ km s^{-1}.)

One effect of close physical encounters between stars will be the formation of binaries. A binary star cannot form in a two-body encounter of point masses if the particles are not initially bound to each other. However, if the close interaction between the two stars can induce tidal distortions on each other, so that some of the kinetic energy can be dissipated as internal energy of the system, then binary formation is indeed possible in such an encounter. The above analysis shows that such a formation rate is negligible for globular clusters.

10.3 Time Scales in the Evolution of Globular Clusters

Binaries can also be formed in three-body encounters, the rate for which can be estimated along similar lines. It can be easily shown (see Exercise 10.2) that the total number of binaries formed over a time $t \simeq t_{\rm relax}$ is $\sim 0.1 (N \ln N)^{-1}$ so that this process can be ignored if $N \gtrsim 100$.

Exercise 10.2

Binary formation in three-body encounters: Estimate the number of binaries formed over one relaxation time in a gravitating system through three-body interaction. [Answer: The change in the velocity Δv in an encounter between two stars of mass m, speed v, and impact parameter b is $\Delta v \approx (Gm/bv)$. To form a binary by a three-body encounter requires $\Delta v \approx v$ or $b \approx (GM/v^2)$. For any given star, such an encounter occurs at a mean rate of $\mathcal{R} \approx (nvb^2)^{-1}$. During such an encounter, there is a probability $b^3 n$ that a third star will also be within a distance b. This shows that the time scale for a three-body encounter $t_{3-\rm body} \approx (n^2 b^5 v)^{-1}$. Using $b \approx (GM/v^2)$ and $v^2 \approx (GNm/R)$ we find $(t_{3-\rm body}/t_{\rm relax}) \approx 10 N^2 \ln N$. Therefore the total number of binaries formed per relaxation time is $N(t_{\rm relax}/t_{3-\rm body}) \approx 0.1 (N \ln N)^{-1}$.]

Exercise 10.3

Potential energy from strip brightness: One of the earliest methods used to determine the gravitational potential energy of a globular cluster is the following. Let $S(x)\, dx$ be the total luminosity in a strip of width dx that passes at a distance x from the centre of the globular cluster when projected onto the sky plane and let W be the gravitational potential energy of the system. If the globular cluster has a constant mass-to-light ratio Q, then show that

$$W = -2GQ^2 \int_0^\infty S^2(x)\, dx. \qquad (10.15)$$

[Hint: Let $I(R)$ be the surface brightness and $j(r)$ be the luminosity density of the cluster, where r and R are the radial coordinates in three and two dimensions, respectively. Then we have the relation

$$I(R) = 2 \int_r^\infty \frac{j(r)\, r\, dr}{\sqrt{r^2 - R^2}}. \qquad (10.16)$$

This equation can be solved for $j(r)$ by Abell's integral formula:

$$j(r) = -\frac{1}{\pi} \int_r^\infty \frac{dI(R)}{dR} \frac{dR}{\sqrt{R^2 - r^2}}. \qquad (10.17)$$

The strip brightness, on the other hand, is related to $I(R)$ by

$$S(x) = 2 \int_x^\infty \frac{I(R) R\, dR}{\sqrt{R^2 - x^2}}, \qquad (10.18)$$

which allows us to relate $j(r)$ to $S(x)$ by

$$j(x) = -\frac{1}{2\pi x} \frac{dS(x)}{dx}. \qquad (10.19)$$

For a constant mass-to-light ratio Q, the total mass inside a radius r will be

$$M(r) = -2Q \int_0^r \frac{dS}{dx} x \, dx. \tag{10.20}$$

We can express the gravitational potential energy in terms of $M(r)$ as

$$W = -4\pi G \int_0^\infty \rho M r \, dr, \tag{10.21}$$

using which we can obtain the required result.]

10.4 Fokker–Planck Description of Globular-Cluster Dynamics

The existence of vastly different time scales for a globular cluster suggests the utility of different physical descriptions during different epochs. Initially, over a time scale $t \lesssim t_{\rm cross}$, the system of stars undergoes violent relaxation along the lines described in Vol. I, Chap. 10, Section 10.6. This phase, however, is not of much observational relevance in the real universe. During the interval $t_{\rm cross} \ll t \ll t_{\rm relax}$, the system would have reached a quasi steady state and is best described by the collisionless Boltzmann equation. The steady-state solution to the collisionless Boltzmann equation can be taken to be a function of the isolating integrals of motion for the system. Several such models were described in Vol. I, Chap. 10, Section 10.4, of which the King model is found to be a reasonable description of the globular cluster. The distribution function for the King model is given by

$$f(E) = \begin{cases} 0 & (E > E_0) \\ K[e^{-\beta(E-E_0)} - 1] & (E < E_0) \end{cases}, \tag{10.22}$$

where K, β, and E_0 are constants and $E = (1/2)v^2 + \phi(r)$ is the energy per unit mass. For such a model, the core radius is defined as

$$r_c = \sqrt{\frac{9}{4\pi G \rho_0 \beta}}, \tag{10.23}$$

where ρ_0 is the central density and the tidal radius r_t is taken to be the point at which $\phi = 0$. In addition to these, we often define a quantity called concentration by $c \equiv \log(r_t/r_c)$. This is convenient because the King models form a monotonic sequence specified by c, with different models for a given c being generated by different values of the core radius. Observations suggest that the globular clusters obey a scaling relation among core radius, central velocity dispersion σ, and the central surface brightness I_0 of the form[4]

$$r_c \propto \sigma^{2.2 \pm 0.15} I_0^{-1.1 \pm 0.1}. \tag{10.24}$$

10.4 Fokker–Planck Description of Globular-Cluster Dynamics

We will see in Vol. III that similar relations (defining what is called the *fundamental plane*) exist for elliptical galaxies, suggesting a certain dynamical similarity. This relation is close to what is expected from virial theorem, $r_c \propto (\sigma^2/I_0)(M/L)^{-1}$, provided that M/L ratios do not vary significantly among the globular clusters.

The spatial structure of the globular cluster during this phase can be described in terms of three different regions. The innermost region in which violent relaxation is reasonably complete will be described by an isothermal sphere and will have a particular core radius and velocity dispersion. The escape of stars in the Maxwellian tail from the inner core will define the outer halo of the globular cluster. Regions at still larger radii will be strongly dominated by the escape of stars from the globular cluster and will also be influenced by the tidal gravitational effects of the surrounding matter.

The really interesting features of globular-cluster evolution arise for $t \gtrsim t_{\text{relax}}$, when the gravitational collisions exert a significant influence over the structure of the system. We shall now study these features.

Systems involving a large number of particles, such as those of globular clusters, can be described by a distribution function $f(\mathbf{x}, \mathbf{v}, t)$ such that $dN = f(d^3x d^3v)$ denotes the number of stars at time t in the range $(\mathbf{x}, \mathbf{x} + d^3\mathbf{x})$ and $(\mathbf{v}, \mathbf{v} + d^3\mathbf{v})$. (It is assumed that we can interpret the element $d^3x d^3v$ in some sensible manner so that it is small enough to be treated as infinitesimal but large enough to provide a statistical description; we can circumvent such problems by reinterpreting f as the relative probability of finding a star at some location in phase space. We shall not bother to do so because it does not affect any of the results.) In general, the distribution function will evolve by an equation of the form $(df/dt) = C$, where C denotes the collision terms. For systems like stars in a globular cluster, physical collisions are negligible but the gravitational collisions are significant for $t \gtrsim t_{\text{relax}} \approx (N/\ln N)t_{\text{cross}}$. Expanding out the time derivative and using $\dot{\mathbf{v}} = -\nabla \phi$, we can express the equations describing such a gravitating system as

$$\frac{df}{dt} = \frac{\partial f}{\partial \mathbf{x}} + \dot{\mathbf{v}} \cdot \frac{\partial f}{\partial \mathbf{v}} + \dot{\mathbf{x}} \cdot \frac{\partial f}{\partial \mathbf{x}} = \frac{\partial f}{\partial t} + \mathbf{v} \cdot \frac{\partial f}{\partial \mathbf{x}} - \nabla \phi \cdot \frac{\partial f}{\partial \mathbf{v}} = C, \quad (10.25)$$

with the gravitational potential ϕ, which is due to smoothed density, determined by

$$\nabla^2 \phi = 4\pi G \rho, \quad \rho(\mathbf{x}, t) = m \int f(\mathbf{x}, \mathbf{v}, t) d^3v. \quad (10.26)$$

In the absence of gravitational two-body relaxation, $C = 0$, and Eq. (10.25) reduces to the collisionless Boltzmann equation studied in Vol. I, Chap. 10, Section 10.4. However, for $t \gtrsim t_{\text{relax}}$, we cannot ignore the right-hand side of Eq. (10.25).

In general, the collision term C represents the scattering of particles from a given velocity $\mathbf{v}$ to another velocity $\mathbf{v}'$. In the case of gaseous systems, these collisions are of short range and the velocity change occurs when two particles are close to each other. For systems dominated by a long-range force like gravity, even distant encounters can play an important role in determining the form of C. We saw in Vol. I, Chap. 10, Section 10.7 that it is indeed the distant encounters that produce the most dominant contribution to C. Because each of these encounters introduces only a small change in the velocity (so that $|\mathbf{v} - \mathbf{v}'| \ll v$), the effect of distant encounters can be treated as a diffusion in the velocity space. The form of the collision term will then be $C = -(\partial J^\alpha/\partial p^\alpha)$, where J^α is a current in the momentum space, with $p^\alpha = mv^\alpha$; its explicit form was determined in Vol. I, Chap. 10, Section 10.7 as

$$J_\alpha(p) = B_0 f(\mathbf{p}) \frac{\partial \eta}{\partial p_\alpha} - \frac{B_0}{2} \frac{\partial}{\partial p_\beta}\left(f \frac{\partial^2 \psi}{\partial p_\alpha \partial p_\beta}\right) \equiv a_\alpha(\mathbf{p}) f(\mathbf{p}) - \frac{1}{2} \frac{\partial}{\partial p_\beta}(\sigma^2_{\alpha\beta} f), \quad (10.27)$$

where $B_0 = (4\pi G^2 m^5 \ln \Lambda)$, $a_\alpha = (\partial \eta/\partial p_\alpha)$, and $\sigma^2_{\alpha\beta} = (\partial^2 \psi/\partial p_\alpha \partial p_\beta)$, with

$$\psi(\mathbf{p}) = \int d\mathbf{p}' f(\mathbf{p}')|\mathbf{p} - \mathbf{p}'|, \quad \eta(\mathbf{p}) \equiv \nabla^2 \psi(\mathbf{p}) = 2 \int d\mathbf{p}' \frac{f(\mathbf{p}')}{|\mathbf{p} - \mathbf{p}'|}. \quad (10.28)$$

The quantity $\ln \Lambda$ is the standard Coulomb logarithm for a gravitating system with

$$\Lambda \approx \frac{Rv^2}{2Gm} \approx N. \quad (10.29)$$

The collision term therefore has the form

$$C = -\frac{\partial}{\partial p^\alpha}\left[a_\alpha(\mathbf{p}) f(\mathbf{p}) - \frac{1}{2}\frac{\partial}{\partial p_\beta}\{\sigma^2_{\alpha\beta} f\}\right] = -\frac{\partial}{\partial p^\alpha}\{a_\alpha f\} + \frac{1}{2}\frac{\partial}{\partial p_\alpha \partial p_\beta}\{\sigma^2_{\alpha\beta} f\}. \quad (10.30)$$

The functions $a_\alpha(\mathbf{p})$ and $\sigma_{\alpha\beta}(\mathbf{p})$ can be thought of as diffusion coefficients in the momentum (or, equivalently, velocity) space.

The above description is a special case of a more general procedure based on the Fokker–Planck approach to diffusive phenomena. Because distant gravitational encounters are uncorrelated in the diffusion approximation, we can think of such encounters as changing the value of the velocity of a particle in a stochastic manner. Let $P(y_1, t_1 \mid y_2, t_2)$ denote the probability for any one component of the velocity to have a value y_2 at time t_2 if it has a value y_1 at time t_1. Because the distant encounters governing the evolution are *stationary* and depend on only $t_2 - t_1$ we must have $P(y_1, t_1 \mid y_2, t_2) = P(y_1, 0 \mid y_2, t_2 - t_1)$. Transitivity

10.4 Fokker–Planck Description of Globular-Cluster Dynamics

in time now requires the condition

$$P(y_1 \mid y_3, t_3) = \int P(y_1 \mid y_2, t_2) \, P(y_2 \mid y_3, t_3 - t_2) \, dy_2, \qquad (10.31)$$

where the 0 in $P(y_1, 0 \mid y_2, t_2 - t_1)$ is omitted for simplicity of notation. Taking two instants of time separated infinitesimally and expanding the above equation in Taylor series, we can compare the probabilities at times t and $t + \Delta t$ and show that, in the limit of $\Delta t \to 0$, the probability satisfies the equation

$$\frac{\partial}{\partial t} P(y_0 \mid y, t) = \sum_{n=1}^{\infty} \frac{(-1)^n}{n!} \frac{\partial^n}{\partial y^n} [M_n(y) P(y_0 \mid y, t)], \qquad (10.32)$$

with

$$M_n(y) \equiv \lim_{\Delta t \to 0} \frac{1}{\Delta t} \int (y' - y)^n P(y \mid y', \Delta t) \, dy'. \qquad (10.33)$$

For a diffusive random walk taking place in momentum space, the mean-square displacement $(\Delta v)^2$ will grow in proportion to Δt whereas higher moments of Δv will grow faster than Δt. So when the limit $\Delta t \to 0$ is taken in Eq. (10.33), only the first two moments, $M_1 \equiv A$ and $M_2 \equiv B$, will have nonzero limits, with the rest of the moments vanishing as $\Delta t \to 0$. In that case, Eq. (10.32) reduces to the *Fokker–Planck* equation:

$$\frac{\partial}{\partial t} P = -\frac{\partial}{\partial y}[A(y)P] + \frac{1}{2}\frac{\partial^2}{\partial y^2}[B(y)P], \qquad (10.34)$$

where $P(y_0 \mid y, t)$ is treated as a function of y and t with some fixed value for y_0.

The structure of this equation is identical to the one obtained above in Eq. (10.30) except for obvious modifications in dimensionality, etc. This shows that the collision term can be interpreted in terms of diffusion coefficients for Δv_i and $\Delta v_i \Delta v_j$. To bring out this connection explicitly we will change the description from momentum space to velocity space and write the same collision term C in a changed notation as

$$C[f] = -\frac{\partial}{\partial v_i}[f(\mathbf{w})D(\Delta v_i)] + \frac{1}{2}\frac{\partial^2}{\partial v_i \partial v_j}[f(\mathbf{w})D(\Delta v_i \Delta v_j)]. \qquad (10.35)$$

Using the explicit form of a_α and $\sigma_{\alpha\beta}$ obtained above, we can write the diffusion coefficients as

$$D(\Delta v_i) = 4\pi G^2 m_a(m + m_a) \ln \Lambda \frac{\partial}{\partial v_i} h(\mathbf{v}),$$

$$D(\Delta v_i \Delta v_j) = 4\pi G^2 m_a^2 \ln \Lambda \frac{\partial^2}{\partial v_i \partial v_j} g(\mathbf{v}), \qquad (10.36)$$

with

$$h(\mathbf{v}) = \int \frac{f_a(\mathbf{v}_a)\, d^3\mathbf{v}_a}{|\mathbf{v} - \mathbf{v}_a|}, \quad g(\mathbf{v}) = \int f_a(\mathbf{v}_a)\, |\mathbf{v} - \mathbf{v}_a|\, d^3\mathbf{v}_a. \quad (10.37)$$

To be precise, the diffusion coefficients depend on f, which in turn is the quantity we are trying to determine. In many situations, however, we can approximate the form of the diffusion coefficient by evaluating it for some suitable equilibrium distribution of f and study the evolution by using these coefficients.

Of particular importance are the diffusion coefficients in the case of a distribution function that depends on only the magnitude of the velocity $v = |\mathbf{v}|$ so that both g and h are only functions of v. It is then convenient to choose a coordinate system such that one of the axes, say, the z axis, is along the direction of $\mathbf{v}$ with the other two axes, say, x and y, being perpendicular to $\mathbf{v}$. From spherical symmetry it follows that

$$D[(\Delta v_x)^2] = D[(\Delta v_y)^2], \quad (10.38)$$

$$D(\Delta v_x) = D(\Delta v_y) = D(\Delta v_x \Delta v_y) = D(\Delta v_x \Delta v_z) = D(\Delta v_y \Delta v_z) = 0, \quad (10.39)$$

leaving only three independent diffusion coefficients that are conventionally denoted by the symbols

$$D(\Delta v_\parallel) \equiv D(\Delta v_z),$$
$$D(\Delta v_\parallel^2) \equiv D[(\Delta v_z)^2], \quad (10.40)$$
$$D(\Delta v_\perp^2) \equiv 2D[(\Delta v_x)^2] = 2D[(\Delta v_y)^2].$$

The factor 2 in the last definition takes into account the fact that there are two directions orthogonal to the preferred axis. Using the identity $(\partial/\partial v_i) = (v_i/v)(\partial/\partial v)$, we get

$$D(\Delta v_i) = \frac{v_i}{v} D(\Delta v_\parallel),$$

$$D(\Delta v_i \Delta v_j) = \frac{v_i v_j}{v^2}\left[D(\Delta v_\parallel^2) - \frac{1}{2}D(\Delta v_\perp^2)\right] + \frac{1}{2}\delta_{ij} D(\Delta v_\perp^2), \quad (10.41)$$

where

$$D(\Delta v_\parallel) = 8\pi G^2 m^2 \ln \Lambda \frac{dh(v)}{dv}, \quad (10.42)$$

$$D(\Delta v_\parallel^2) = 4\pi G^2 m^2 \ln \Lambda \frac{d^2 g(v)}{dv^2}, \quad D(\Delta 0 v_\perp^2) = \frac{8\pi G^2 m^2 \ln \Lambda}{v} \frac{dg(v)}{dv}. \quad (10.43)$$

10.4 Fokker–Planck Description of Globular-Cluster Dynamics

To obtain more explicit forms for g and h in this case, we begin by noting that Eqs. (10.37) are equivalent to the differential equations

$$\nabla^2 h = \frac{1}{v^2}\frac{\partial}{\partial v}\left(v^2 \frac{\partial h}{\partial v}\right) = -4\pi f, \tag{10.44}$$

$$\nabla^2 g = \frac{1}{v^2}\frac{\partial}{\partial v}\left(v^2 \frac{\partial g}{\partial v}\right) = h. \tag{10.45}$$

Because Eq. (10.44) has the same form as the equation connecting the gravitational potential to a spherically symmetric mass density, its solution can be immediately written as

$$h(v) = 4\pi \left[\frac{1}{v}\int_0^v v_a^2 f_a(v_a)\,dv_a + \int_v^\infty v_a f_a(v_a)\,dv_a\right]. \tag{10.46}$$

This function $h(v)$ in turn acts as a source for another Poisson equation in Eq. (10.45). The integration can now be performed in a straightforward manner and, after some algebra, we obtain

$$g(v) = \frac{4\pi v}{3}\left[\int_0^v \left(3v_a^2 + \frac{v_a^4}{v^2}\right) f_a(v_a)\,dv_a + \int_v^\infty \left(3\frac{v_a^3}{v} + vv_a\right) f_a(v_a)\,dv_a\right]. \tag{10.47}$$

Having determined $g(v)$ and $h(v)$, we can calculate the diffusion coefficient from Eqs. (10.36). This completely solves the problem.

As an example, consider the case in which the distribution function is Maxwellian with a velocity dispersion σ^2:

$$f(v) = \frac{\rho}{m(2\pi\sigma^2)^{3/2}}e^{-v^2/2\sigma^2}. \tag{10.48}$$

The integrals can now be expressed in terms of the error function $\mathrm{erf}(X)$ and we get

$$D(\Delta v_\parallel) = -\frac{8\pi G^2 \rho m \ln\Lambda}{\sigma^2}G(X),$$

$$D(\Delta v_\parallel^2) = \frac{4\sqrt{2\pi}\,G^2 \rho m \ln\Lambda}{\sigma}\left[\frac{G(X)}{X}\right], \tag{10.49}$$

$$D(\Delta v_\perp^2) = \frac{4\sqrt{2\pi}\,G^2 \rho m \ln\Lambda}{\sigma}\left[\frac{\mathrm{erf}(X) - G(X)}{X}\right],$$

with $X \equiv (v/\sqrt{2}\sigma)$ and

$$G(X) \equiv \frac{1}{2X^2}\left[\mathrm{erf}(X) - X\frac{d\,\mathrm{erf}(X)}{dX}\right] = \frac{1}{2X^2}\left[\mathrm{erf}(X) - \frac{2X}{\sqrt{\pi}}e^{-X^2}\right]. \tag{10.50}$$

Given the diffusion coefficients for the velocity, it is possible to calculate the corresponding diffusion coefficient for any other function of velocity, for example, the kinetic energy. We will also get, in addition to the contributions from the transverse direction, a cross term between the mean velocity and $\Delta v_\parallel$ in computing the diffusion in kinetic energy. This gives

$$D(\Delta E) = m \left[v D(\Delta v_\parallel) + \frac{1}{2} D(\Delta v_\parallel^2) + \frac{1}{2} D(\Delta v_\perp^2) \right]$$
$$= 32\pi^2 G^2 m^3 \ln \Lambda \left[\int_v^\infty v_a f_a(v_a) dv_a - \int_0^v \frac{v_a^2}{v} f_a(v_a) dv_a \right]. \quad (10.51)$$

The first term represents the increase in the velocity dispersion of the stars that is due to collisions, and the second term, with a minus sign, describes the cooling that is due to dynamical friction. As has been emphasised in Vol. I, Chap. 10, Section 10.7, both these processes should operate simultaneously in order to provide Maxwellian distribution as an equilibrium configuration.

The above analysis can be easily generalised to the case in which a star of a given mass m ("test star") moves through a collection of other stars each of mass m_a ("field stars"), with the latter described by a distribution function f_a. The velocity dispersion terms arising from $\sigma_{\alpha\beta}$ do not change in any manner, and we can compute it by using the mass m_a and distribution function f_a of the field stars. In the case of coefficients arising from a_α, the factor $2m^2$ has to be replaced with $m_a(m + m_a)$. This changes the coefficient $D(\Delta v_\parallel)$. Because $D(\Delta E)$ picks up a contribution from $D(\Delta v_\parallel)$, it also changes to the form

$$D(\Delta E) = 16\pi^2 G^2 m m_a \ln \Lambda \left[m_a \int_v^\infty v_a f_a(v_a) dv_a - m \int_0^v \frac{v_a^2}{v} f_a(v_a) dv_a \right]. \quad (10.52)$$

Note that the heating rate is proportional to the mass m_a of the field star, whereas the cooling that is due to dynamical friction is proportional to the mass m of the test star. The mean-square speed of the population at which these integrals cancel is proportional to m^{-1}, which is a manifestation of equipartition of energy.

Finally, we note that the approach used in this section, based on the Fokker–Planck equation, is quite general and is independent of the variables used to describe the system. In several cases $\mathbf{x}$ and $\mathbf{v}$ may not be the most convenient variables to describe the evolution of the system. We have seen in Vol. I, Chap. 2, Section 2.4 that the most natural variables to describe the evolution of a system will be the action-angle variables (J_a, θ_b), $a, b = 1, 2, 3$. In terms of these variables, the equation of motion for the system will have the form

$$\frac{\partial f}{\partial t} + \dot{J}_i \frac{\partial f}{\partial J_i} + \dot{\theta}_i \frac{\partial f}{\partial \theta_i} = C[f]. \quad (10.53)$$

10.4 Fokker–Planck Description of Globular-Cluster Dynamics

For a collisionless system with $C = 0$, the actions are conserved ($\dot{J}_i = 0$) and we can take $f(\mathbf{J}, \boldsymbol{\theta}, t) = f(\mathbf{J})$, thereby satisfying Eq. (10.53) with $C = 0$. In this collisionless limit, all the stars are moving in some given smooth gravitational potential and the phase-space structure of the orbits will be determined by the orbital tori labelled by the action $\mathbf{J}$. When $C \neq 0$, the actions $\mathbf{J}$ will not be conserved and the stars will move from one torus to another. However, because the time taken for this movement of stars across the tori (t_{relax}) is large compared with the time in which the star explores the entire surface of the torus (t_{cross}), we can assume that the distribution function still remains a function of $\mathbf{J}$ and t with a slow variation in time; that is, $f = f(\mathbf{J}, t)$. To the same order of accuracy, $\dot{J}_i \approx 0$, and Eq. (10.53) reduces to

$$\frac{\partial f}{\partial t} = C[f], \qquad (10.54)$$

where the right-hand side involves derivatives with respect to θ_i and J_i in a manner similar to Eq. (10.27) The advantage of this approach arises from the fact that the particle moves around the torus in the phase space at a much more rapid rate compared with the rate at which it diffuses from one torus to another; i.e., θ's vary much more rapidly compared with the J's. Using this fact, we can average Eq. (10.54) over θ's to obtain a simpler evolution equation for $f(\mathbf{J}, t)$. On integrating both sides of Eq. (10.54) over $(2\pi)^{-3} \int \cdots d^3\theta$, we find that the left-hand side remains unchanged. On the right-hand side, all the terms involving derivatives with respect to θ vanish because the motion is periodic in θ_i. Therefore the Fokker–Planck equation simplifies to

$$\frac{\partial f(\mathbf{J}, t)}{\partial t} = -\frac{\partial}{\partial J_i}[f \bar{D}(\Delta J_i)] + \frac{1}{2}\frac{\partial^2}{\partial J_i \partial J_j}[f \bar{D}(\Delta J_i \Delta J_j)], \qquad (10.55)$$

where

$$\begin{aligned}\bar{D}(\Delta J_i) &= \frac{1}{(2\pi)^3} \int D(\Delta J_i) \, d^3\theta, \\ \bar{D}(\Delta J_i \Delta J_j) &= \frac{1}{(2\pi)^3} \int D(\Delta J_i \Delta J_j) \, d^3\theta\end{aligned} \qquad (10.56)$$

are called orbit-averaged diffusion coefficients. (This entire procedure is called *orbit averaging*.) In general, orbit averaging of the Fokker–Planck equation reduces it from an equation having $6 + 1$ independent variables to one with $3 + 1$ independent variables.

If the system possesses additional symmetries, we can reduce the degree still further by a judicious choice of variables. One of the simplest contexts in which such an orbit-averaged equation can be used is when the distribution function depends on only energy $E = (1/2)v^2 + \phi$ per unit mass. In that case Eq. (10.55) takes the simpler form

$$\left(\frac{\partial N}{\partial t}\right)_{\text{enc}} = -\frac{\partial}{\partial E}\{N\langle\Delta E\rangle_V\} + \frac{1}{2}\frac{\partial^2}{\partial E^2}[N\langle(\Delta E)^2\rangle_V], \quad (10.57)$$

where $N(E)\,dE$ is the number of stars in the range $(E, E+dE)$ defined through the relation

$$N(E)\,dE = \int dV f(E)\, 4\pi v^2\, dv, \quad dV \equiv d^3\mathbf{x}. \quad (10.58)$$

Because $dE = v\,dv$, it follows that

$$N(E) = 4\pi f(E) \int v\, dV. \quad (10.59)$$

The diffusion coefficients are now for the energy and can be determined by averaging over all the stars of energy E within some volume V. For example,

$$\langle\Delta E\rangle_V = \frac{\int \langle\Delta E\rangle v r^2\, dr}{\int v r^2\, dr}. \quad (10.60)$$

A more useful form of the orbit-averaged Fokker–Planck equation is available if the cluster is spherically symmetric and the gravitational potential ϕ depends on only r. Then any particle of energy E has a spatial volume $(4\pi/3)r_{\text{max}}^3$ accessible to it, where r_{max} is determined by the condition $\phi(r_{\text{max}}) = E$. In this case, Eq. (10.59) becomes $N(E) = 16\pi^2 \mathcal{P} f(E)$, where

$$\mathcal{P}(E,t) \equiv \int_0^{r_{\text{max}}} v r^2\, dr = \int_0^{r_{\text{max}}} [2(E-\phi)]^{1/2} r^2\, dr \quad (10.61)$$

measures the phase-space volume per unit energy interval available to the particle. Then the collision term governing the evolution of $N(E) \propto \mathcal{P} f$ can be expressed in the form

$$C = \mathcal{P}\left(\frac{\partial f}{\partial t}\right) = \frac{\partial}{\partial E}\left\{-\mathcal{P} f \langle(\Delta E)\rangle_V + \frac{1}{2}\frac{\partial}{\partial E}[\mathcal{P} f \langle(\Delta E)^2\rangle_V]\right\}, \quad (10.62)$$

with

$$\langle\Delta E\rangle_V = \frac{1}{\mathcal{P}} \int_0^{r_{\text{max}}} \langle\Delta E\rangle\, v r^2\, dr, \quad (10.63)$$

etc. On the left-hand side, in evaluating the time rate of change of $f(E,t)$, we need an expression for $\langle dE/dt\rangle$. Because orbit averaging is based on the assumption that ϕ varies at a time scale much longer than the orbital time scale, we can estimate $\langle dE/dt\rangle$ by $\langle(\partial\phi/\partial t)\rangle$, where the averaging is over the accessible phase volume. Then

$$\frac{dE}{dt} = \left\langle\frac{\partial\phi}{\partial t}\right\rangle_V = \frac{1}{\mathcal{P}}\int_0^{r_{\text{max}}} v\frac{\partial\phi}{\partial t} r^2\, dr = -\frac{1}{\mathcal{P}}\frac{\partial \mathcal{Q}}{\partial t}, \quad (10.64)$$

10.4 Fokker–Planck Description of Globular-Cluster Dynamics

where $\mathcal{Q}(E, t)$ is defined by

$$\mathcal{Q}(E, t) \equiv \frac{1}{3}\int_0^{r_{\max}} v^3 r^2 \, dr = \frac{1}{3}\int_0^{r_{\max}} [2(E-\phi)]^{3/2} r^2 \, dr, \quad (10.65)$$

and, clearly, $\mathcal{P} = (\partial \mathcal{Q}/\partial E)$. The theory of adiabatic invariance developed in Vol. I, Chap. 2, Section 2.6 shows that $\mathcal{Q}$ can be treated as an adiabatic invariant during the evolution of the cluster. Incorporating all these results, we can write the Fokker–Planck equation in the form

$$\frac{\partial \mathcal{Q}}{\partial E}\frac{\partial f}{\partial t} - \frac{\partial \mathcal{Q}}{\partial t}\frac{\partial f}{\partial E} = 4\pi\Gamma\frac{\partial}{\partial E}\left[\frac{m}{m_f}f\int_{-\infty}^{E} f_f \frac{\partial \mathcal{Q}_f}{\partial E_f}\, dE_f \right.$$
$$\left. + \frac{\partial f}{\partial E}\left(\int_{-\infty}^{E} f_f \mathcal{Q}_f \, dE_f + \mathcal{Q}\int_{E}^{\infty} f_f \, dE_f\right)\right], \quad (10.66)$$

where f_f and $\mathcal{Q}_f$ indicate quantities that are functions of E_f. In general, this equation has to be solved numerically. We shall discuss in the next section several simple results that follow from the analysis of this equation.

It should, however, be noted that this approach has certain limitations of which the following two are most important. Processes that depend strongly on the eccentricity or angular momentum of the orbit cannot be properly modelled in the above approach. The Fokker–Planck equation also assumes that encounters are uncorrelated; hence binaries and other close correlations among stars can be treated only crudely at best.

Exercise 10.4

Dynamical friction from the Boltzmann equation: A simple and elegant derivation of dynamical friction force can be obtained along the following lines. Consider a spatially uniform distribution of field stars of mass m and spatial density n_0 with a Maxwellian distribution of velocities characterised by a velocity dispersion σ^2. A star of mass M moves along the trajectory $\mathbf{x}(t) = \mathbf{x}_0 + \mathbf{v}_0 t$ through the field stars. The gravitational potential Φ_M of M perturbs the distribution function $f_0(\mathbf{v})$ of the field stars, changing it to $f_0(\mathbf{v}) + f_1(\mathbf{x}, \mathbf{v}, t)$. We can determine the perturbation f_1 by linearising the collisionless Boltzmann equation. (a) To the linear order in Φ_M, show that

$$f_1(\mathbf{x}, \mathbf{v}, t) = \frac{GMn_0}{(2\pi\sigma^2)^{5/2}\pi}\int \frac{d^3\mathbf{k}}{k^2}\frac{\mathbf{k}\cdot\mathbf{v}}{\mathbf{k}\cdot\mathbf{v} - \omega}e^{-(1/2)v^2/\sigma^2}e^{i[\mathbf{k}\cdot(\mathbf{x}-\mathbf{x}_0)-\omega(\mathbf{k})t]}, \quad (10.67)$$

where $\omega(\mathbf{k}) = \mathbf{k}\cdot\mathbf{v}_0$. (b) The force on M that is due to the perturbed part of the distribution function is given by

$$\mathbf{F} = \frac{4G^2M^2n_0m}{(2\pi\sigma^2)^{3/2}}i\int\frac{d^3\mathbf{k}}{k^3}\mathbf{k}\int_{-\infty}^{\infty}\frac{e^{-(1/2)u^2/\sigma^2}u\,du}{ku - \omega(\mathbf{k})}, \quad (10.68)$$

where u is the component of $\mathbf{v}$ in the direction of $\mathbf{k}$. (c) Interpret the origin of the singularity in the integration over u. Using the $i\epsilon$ prescription described in Vol. I, Chap. 9,

Section 9.4, show that the force can be written in the form

$$\mathbf{F} = -\frac{4\pi G^2 M^2 n_0 m}{(2\pi\sigma^2)^{3/2}} \int \frac{d^3\mathbf{k}}{k^5} \mathbf{k}\omega \exp\left(-\frac{\omega^2}{2k^2\sigma^2}\right). \quad (10.69)$$

This integral is still divergent logarithmically. Explain the origin of this divergence and regularise it by limiting the integration to a range $(k_{\min}, k_{\max})$. Compare the resulting t_{relax} with the one obtained by the Fokker–Planck equation.

10.5 Aspects of Globular-Cluster Evolution

Because the two-body relaxation time is the most important parameter determining the cluster evolution, we shall begin by estimating this quantity from the approach developed in the last section. One possible definition of relaxation time could be

$$t_{\text{relax}} \equiv \frac{v^2}{D(\Delta v_\parallel^2)}. \quad (10.70)$$

Because $v_{\text{rms}} = \sqrt{3}\sigma$, the dimensionless variable in Eq. (10.50) is $X \equiv (v/\sqrt{2}\sigma) = 1.225$ for $v = v_{\text{rms}}$. Replacing v^2 with σ^2 in the diffusion equation and evaluating $G(X)/X$ in Eq. (10.50) at $X = 1.225$, we find

$$t_{\text{relax}} = 0.34 \frac{\sigma^3}{G^2 m \rho \ln \Lambda}$$

$$= \frac{1.8 \times 10^{10} \text{ yr}}{\ln \Lambda} \left(\frac{\sigma}{10 \text{ km s}^{-1}}\right)^3 \left(\frac{1 M_\odot}{m}\right) \left(\frac{10^3 M_\odot \text{ pc}^{-3}}{\rho}\right). \quad (10.71)$$

Because the density and σ can vary across a system, t_{relax} has no deep significance for an inhomogeneous system. The best agreement with numerical estimates is obtained if we replace (1) ρ in the above equation with the density inside the system's median radius r_h, which contains half the total mass; (2) $3\sigma^2$ with $\langle v^2 \rangle \approx 0.4(GM/r_h)$; and (3) Λ with $(r_h \langle v^2 \rangle / Gm) \approx 0.4 N$. This yields

$$t_{\text{rh}} = \frac{0.14 N}{\ln(0.4 N)} \sqrt{\frac{r_h^3}{GM}} = \frac{6.5 \times 10^8 \text{ yr}}{\ln(0.4 N)} \left(\frac{M}{10^5 M_\odot}\right)^{1/2} \left(\frac{1 M_\odot}{m}\right) \left(\frac{r_h}{1 \text{ pc}}\right)^{3/2}. \quad (10.72)$$

The subscript is changed from 'relax' to 'rh' as a reminder of these substitutions.

This revised – and more precise – estimate of relaxation time can be used to provide a simple model for the evaporation of stars from a cluster. The equilibrium solution to the Fokker–Planck equation, at which $(\partial f/\partial t)$ vanishes, is given by the Maxwell–Boltzmann distribution with $f \propto \exp(-\beta E) \propto \exp[-\beta(v^2/2 + \phi)]$. The evolution therefore proceeds towards this equilibrium function. It is, however, impossible for any realistic self-gravitating system to achieve this steady

10.5 Aspects of Globular-Cluster Evolution

state because the Maxwell–Boltzmann distribution contains particles of arbitrarily high energy. In any realistic case, particles with energies higher than the depth of the potential well will escape from the system, thereby preventing the system from reaching the above equilibrium configuration. If we choose the boundary condition for the gravitational potential in such a way that it vanishes at large distances, then particles with positive values of energy will not be bound and will escape from the cluster. The simplest model for a cluster is based on the assumption that the cluster evolves self-similarly and shrinks as the stars escape from it. Because the stars that escape are predominantly due to weak encounter, they have negligible energy, and we might assume that the total energy of the cluster remains constant as it shrinks. Taking the energy of the cluster to be $E \propto -(GM^2/R)$, we get the scaling that $R(t) = R_0[M(t)/M_0]^2$, where M_0 and R_0 are the initial values. Assuming that a fixed fraction λM of all mass is lost because of evaporation in a time scale $t \approx t_{\rm rh}$, the mass-loss rate is governed by the equation $(dM/dt) = -(\lambda M/t_{\rm rh})$. Our analysis above shows that $t_{\rm rh}^2 \propto MR^3$, so that $t_{\rm rh} = t_{\rm rh}^0 (MR^3/M_0 R_0^3)^{1/2}$. Combining these results, we get

$$\frac{dM}{dt} = -\frac{\lambda M_0^{7/2}}{t_{\rm rh}^0 M^{5/2}}, \tag{10.73}$$

which has the solution

$$M(t) = M_0 \left(1 - \frac{7\lambda t}{2t_{\rm rh}^0}\right)^{2/7}. \tag{10.74}$$

This suggests that the cluster evaporates completely in a time scale $t_{\rm ev} = (2t_{\rm rh}^0/7\lambda) \simeq 10^2 t_{\rm rh}^0$ if $\lambda \simeq 0.007$. The scaling for all other quantities can be easily determined from this result. In particular, the density varies as $\rho \propto M^{-5}$ and the velocity dispersion scales as $\sigma^2 \propto M^{-1}$. Observations suggest that there are no globular clusters with $t_{\rm rh} \lesssim 3 \times 10^7$ yr, which is $\sim 1\%$ of the currently estimated age of the galaxy. This is consistent with the hypothesis that if there were many more globular clusters initially, those with shorter relaxation time would have evaporated by now.

For a more detailed modelling of the evolution of a globular cluster, it is necessary to integrate the orbit-averaged Fokker–Planck equation numerically. Several such studies have been attempted, the results of which can be understood along the following lines.

We begin by noting that the Fokker–Planck equation conserves the total energy and the mass of the system but increases the entropy, defined as the integral of $f \ln f$ over the phase space. The configuration that maximises the entropy for a given energy and mass is an isothermal sphere for which $f \propto \exp(-\beta E)$. If such a solution is permissible, then it seems reasonable to assume that the evolution will proceed towards such a solution. It is, however, clear that such a solution does *not* exist for our system. To begin with, the high-energy tail of the Maxwellian distribution will be constantly depleted by the process of evaporation, and in any realistic case we cannot assume that the total mass of the system is conserved. When evaporation plays a role, it is not possible for a system to reach a steady state with the Maxwell–Boltzmann distribution. In fact, a self-gravitating system may not be able to achieve the maximum entropy configuration, even if there is no evaporation. To see this, we recall from Vol. I, Chap. 10, Section 10.8 that even if the system is confined within a sphere of radius R, an isothermal solution cannot exist if $(RE/GM^2) < -0.335$. Even when $(RE/GM^2) > -0.335$, the isothermal solution is unstable if $\rho(0) > 709\,\rho(R)$. In other words, collisional evolution of a self-gravitating system of particles is inherently unstable and does not possess the Maxwell–Boltzmann distribution as an equilibrium solution. (This effect is called *Anatonov instability*.)

The above analysis also indicates the nature of the configuration towards which the system is evolving. If we treat the system strictly by the Fokker–Planck equation (which conserves mass), then the particles that are "evaporating" should also be treated as a part of the system at any given time. In other words, the system now consists of a diffuse halo of particles spreading to large distances and a dense core that is shrinking to higher and higher densities. Such a highly inhomogeneous configuration involving a shrinking dense core of bound particles and an evaporating halo of high-energy particles does increase the entropy of the system without bounds for a given energy and mass. To see this, let us divide the cluster of total mass M and binding energy $-|E|$ into a core of mass M_1 and radius R_1 and a shell-like envelope of mass $M_2 \ll M_1$ and radius R_2. The binding energy of the core is $E_1 \approx -(GM_1^2/R_1)$; the binding energy of the envelope is essentially due to its interaction with the core and is given by $E_2 \approx -(GM_1 M_2/R_2)$; the corresponding velocity dispersion in the envelope is $\sigma_2^2 \approx (GM_1/R_2)$. Suppose we now shrink the radius of the core from R_1 to $(1-\epsilon)R_1$, release the gravitational energy $\Delta E \approx (\epsilon GM_1^2/R_1)$, and feed it to the halo so as to change its radius to R_2' such that $|E_2'| = |E_2 + \Delta E| \approx (GM_1 M_2/R_2')$. The phase volume available to the envelope now is $\mathcal{V} \cong (\sigma_2' R_2')^3 \approx (GM_1 R_2')^{3/2}$. The contribution to the entropy that is due to the envelope will be

$$S = -\int f \ln f\, d^3x\, d^3v \approx -\left[\frac{C}{\mathcal{V}} \ln\left(\frac{C}{\mathcal{V}}\right)\right]\mathcal{V} + \text{constant}, \quad (10.75)$$

where C is a normalisation constant and $f \approx (C/\mathcal{V})$. Using our expression for $\mathcal{V}$ and R_2', we get

$$S \simeq \frac{3}{2} \ln R_2' + \text{constant} \simeq -\frac{3}{2} C \ln |E_2 + \Delta E| + \text{constant}. \quad (10.76)$$

This contribution can increase without bounds as $\Delta E \to |E_2|$, even though the entropy of the core changes by only a finite amount by this transfer. It follows that a configuration with a dense core and a tenuous expanding envelope can increase the entropy of the self-gravitating system without bounds.

10.5 Aspects of Globular-Cluster Evolution

Given the above tendency of the gravitating system, one would expect the core of a globular cluster to collapse and shrink without bound eventually reaching a singularity with infinite density. It is indeed possible to obtain self-similar solutions to the orbit-averaged Fokker–Planck equation that describes such a core collapse. In such a solution, the density $\rho(r, t)$, for example, varies as

$$\rho(r, t) = \rho_c(t)\rho_*(r_*), \quad r_* = \frac{r}{r_c(t)}, \quad r_c = \sqrt{\frac{9\sigma^2}{4\pi G \rho_c}}. \quad (10.77)$$

Well outside the core, $r \gg r_c$, the particles are governed by the core mass acting as an external gravitating source and we expect the radius to be independent of time. This requires that

$$0 = \frac{\partial \rho(r, t)}{\partial t} = \frac{d\rho_c}{dt}\rho_* - \rho_c \frac{d\rho_*}{dr_*} \frac{r}{r_c^2} \frac{dr_c}{dt} \quad (\text{for } r \gg r_c) \quad (10.78)$$

or

$$\frac{r_*}{\rho_*} \frac{d\rho_*}{dr_*} = \frac{r_c}{\rho_c} \frac{dt}{dr_c} \frac{d\rho_c}{dt} \quad (\text{for } r_* \gg 1). \quad (10.79)$$

Because the left-hand side is dependent on only r_c whereas the right-hand side depends explicitly on t, Eq. (10.79) can be satisfied only if both sides are equal to a constant, say, $-\beta$. Hence, for $r_* \gg 1$,

$$\rho_*(r_*) \propto r_*^{-\beta}, \quad \rho_c(t) \propto r_c^{-\beta}(t). \quad (10.80)$$

To determine the explicit form of $r_c(t)$ we can proceed as follows: Because the core density and radius are to be related in a self-similar manner with $\rho_c \propto r_c^{-\beta}$, it follows that in the core

$$M_c \propto \rho_c r_c^3 \propto r_c^{3-\beta}, \quad (10.81)$$

$$E_c \propto -\frac{GM^2}{r_c} \propto -r_c^{5-2\beta} \propto -M_c^{(5-2\beta)/(3-\beta)}, \quad (10.82)$$

$$v_c^2 \propto \frac{GM_c}{r_c} \propto r_c^{2-\beta}. \quad (10.83)$$

If the core evolves through a series of quasi-static configurations so that stars of higher energy leave the core, then we would expect the core to grow hotter in the course of evolution. This requires the velocity dispersion to increase as the core shrinks, and hence we need $\beta > 2$. Further, if the absolute value of the energy of the core has to decrease, then relation (10.82) requires that $\beta < 2.5$. A mean value of $\beta \simeq 2.25$ agrees well with simulations. Further, because core collapse occurs because of two-body relaxation, we must have

$$\frac{1}{r_c} \frac{dr_c}{dt} \propto \frac{1}{t_{\text{relax}}} \propto r_c^{\beta/(2-3)}, \quad (10.84)$$

giving immediately

$$r_c(t) \propto (t_0 - t)^{2/(6-\beta)} \propto \tau^{0.53}, \quad \tau \equiv t_0 - t, \quad (10.85)$$

where the second proportionality follows from using $\beta = 2.2$. Correspondingly

$$\rho_c(t) \propto \tau^{-2\beta/(6-\beta)} \propto \tau^{-1.17}, \quad \sigma^2(t) \propto \tau^{(4-2\beta)/(6-\beta)} \propto \tau^{-0.11},$$
$$M_c(t) \propto \tau^{(6-2\beta)/(6-\beta)} \propto \tau^{0.42}, \quad E_c \propto \tau^{-2(5-2\beta)/(6-\beta)} \propto \tau^{-0.32}. \quad (10.86)$$

The time for core collapse is a fixed multiple of the central relaxation time, and simulations suggest that $\tau = 330 \, t_{\text{relax}}(0)$. This predicts that the number of globular clusters with the central relaxation time in the range $(0, t_{\text{relax}})$ should be proportional to t_{relax} as long as the relaxation time is small compared with the age of the galaxy. There seems to be some confirmation for this prediction observationally.

It was mentioned above that both evaporation and Anatonov instability have the effect of driving a cluster to a core–halo configuration. In the initial stages, the dominant effect is evaporation, which, acting alone, would have caused the core collapse in a time scale of $t \approx 10^2 \, t_{\text{rh}}$. However, as the central density increases because of this process, the Anatonov instability comes into play at $t \approx 3 \, t_{\text{rh}}$, and the core collapse proceeds at a more rapid pace. The singular behaviour at late stages, of course, is an artificiality of the continuum approximation that is used. For a realistic cluster with $\rho_c \approx 10^4 \, M_\odot \, \text{pc}^{-3}$, $\sigma \approx 10 \, \text{km s}^{-1}$, and $r_c \approx 1.5$ pc, our scalings can be used to express the radius and the velocity dispersion in terms of the number of stars N remaining in the cluster as

$$r_c \simeq 3 \times 10^{-6} \, N^{1.26} \, \text{pc}, \quad \sigma \simeq 40 \, N^{-0.13} \, \text{km s}^{-1}. \quad (10.87)$$

Clearly, our approximations break down when $N \approx \mathcal{O}(1)$, which occurs when $\sigma \simeq 40 \, \text{km s}^{-1}$ and $r_c \simeq 10^{13}$ cm. These values are fairly large compared with the dimensions of the stars, and hence the final configuration of the core is essentially determined by the formation of binary stars. We shall now discuss some aspects of binary formation in globular-cluster dynamics.

10.6 Binary Stars in Globular Clusters

The binary stars inside a globular cluster can be classified into different types based on either (1) their process of formation or (2) their binding energies. In the first classification, we can think of binaries as three-body binaries if they have been formed through a three-body interaction of stars, as tidal-capture binaries if they formed through a tidal interaction of stars, and as primordial binaries if they had always existed in the globular cluster. This classification, however, is not of much dynamical significance.

In a system made of a large number of stars and a certain number of binaries, there is another natural classification of the latter as hard binaries and soft

10.6 Binary Stars in Globular Clusters

binaries. If the magnitude of the binding energy of the binary star $|E| = (Gm_1m_2/2a)$, (where m_1, m_2 are the masses of the stars and a is the semi-major axis) is small compared with the random kinetic energy $(1/2)m_a\sigma^2$ of the field stars of mass m_a, then the binary is called *soft*. On the other hand, if $|E|/(m_a\sigma^2) > 1$, then the binary is called *hard*. In a random encounter between a field star and a binary, a hard binary is likely to get harder and transfer energy to the field star whereas a soft binary is likely to be disrupted. Hence the interaction of hard and soft binaries with other stars is quite different and needs to be discussed separately.

The fact that soft binaries become softer can be understood in terms of energy equipartition. Any encounter between a field star and a soft binary can be thought of as a perturbation to the binary orbit of a particle with reduced mass. Because the binary is very soft, the kinetic energy of the effective (reduced-mass) particle is much less than the kinetic energy of the field star. As the system moves towards energy equipartition, encounters will increase the energy of the particle with reduced mass, that is, the internal energy of the binary. Hence soft binaries do not play any vital role in the evolution of the cluster.

The evolution, on the other hand, is significantly affected by the encounter between field stars and hard binaries. Unfortunately this process is quite complex and needs to be studied numerically. The relative velocity of the binary and the field star is $\sim\sigma$, which is much smaller than the orbital speeds in a hard binary. In an encounter, if the magnitude of the internal energy $|E|$ of the hard binary is increased (so that the binding becomes harder), then the field star gains energy and escapes back to infinity. However, if $|E|$ decreases, the field star is likely to become bound to the binary, thereby forming a temporary triplet. Eventually one of the stars will be ejected from the triple system with an escape speed typically of the same order as that of the orbital speed. It follows that the energy of the ejected star is higher than that of the incoming single star and hence, by energy conservation, the binding energy E must decrease, making $|E|$ increase. The actual rate at which energy transfer occurs, because the hard binaries are becoming harder, has been estimated from numerical studies and is given by

$$\langle \dot{E} \rangle \simeq -2.5 \frac{n\, G^2 m^3}{\sigma}, \qquad (10.88)$$

where n is the number density of stars. This is equivalent to the relation $\langle \Delta E \rangle t_{\text{relax}} \approx -0.1\, m\sigma^2$, which is understandable except for the numerical coefficient, which is determined from the simulation.

In the encounters between the binary star and the field star, the first level of energy transfer occurs between the hard binary and one of the three stars involved in the encounter, although it need not necessarily be the incoming star. If this star escapes, then the acquired energy is lost to the cluster. However, if the star interacts with other cluster stars through further encounters, then the energy gained by the star from the hard binary will be shared by other field stars. [Note that when a star of mass m is ejected from a radius r it carries a negative potential energy $m\phi(r)$, thereby increasing the energy of the cluster by $-m\phi(r)$. For an ejection occurring from or near the core, $\phi(r) \simeq \phi_c \simeq -3(GM/r_h) \simeq 14(E_{\text{cluster}}/M)$. This clearly suggests that hard binaries can act as an energy source in a cluster.] When the formation of hard binaries in the core reaches a critical limit, they will stabilise the core and can possibly stop the core collapse.

To work out the details of this process, it is necessary to calculate the formation rate of the binaries that is due to all processes (three-body encounters, N-body encounters, tidal interactions, etc.), which is a fairly hard task. Simple estimates suggest that the tidal capture is the most dominant single process leading to the formation of binaries. Assuming that tidal capture is important when the impact parameter for collision between two stars is approximately twice the stellar radius r_*, the rate of tidal capture can be estimated as

$$\frac{1}{t_{\text{coll}}} = 16\sqrt{\pi} n\sigma r_*^2 \left(1 + \frac{v_*^2}{4\sigma^2}\right) = 16\sqrt{\pi} n\sigma r_*^2 (1+q), \quad (10.89)$$

where $v_* \equiv (2Gm/r_*)^{1/2}$, and

$$q = \frac{v_*^2}{4\sigma^2} = \frac{Gm}{2\sigma^2 r_*} \quad (10.90)$$

is called the *Safronov number*. Numerically,

$$\frac{t_{\text{coll}}}{t_{\text{relax}}} = 0.8 \ln \Lambda \left(\frac{q^2}{1+q}\right) \quad (10.91)$$

(see Exercise 10.5). When the core has shrunk to approximately $N = 100$ stars, relations (10.87) give $\sigma \simeq 20$ km s^{-1} and $q \simeq 240$. If we take $\ln \Lambda \simeq 10$, then $t_{\text{coll}} \simeq 2 \times 10^3 \, t_{\text{relax}}$; in a characteristic evolution time of $t \simeq 300 \, t_{\text{relax}}$, roughly 10% of the stars will be tidally captured into hard binaries.

It is easy to see that this process is more dominant than three-body encounters. The latter produces binaries at the rate of $0.1(N \ln N)^{-1}$ per relaxation time. Taking the evolution time to be $\sim 300 \, t_{\text{relax}}$, the three-body encounter will be important when $N \ln N \simeq 30$, giving $N \simeq 10$.

Exercise 10.5

Rate of inelastic encounters: Consider an encounter between two stars with initial relative velocity V_0, impact parameter b, angular momentum per unit reduced mass $L = bV_0$, and distance of closest approach r_{coll}. (a) Using the conservation of energy and angular momentum, show that

$$b^2 = r_{\text{coll}}^2 + \frac{4Gmr_{\text{coll}}}{V_0^2}. \quad (10.92)$$

(b) Let $f(\mathbf{v}_a) d^3\mathbf{v}_a$ be the number of stars per unit volume in the velocity range $(\mathbf{v}_a, \mathbf{v}_a + d^3\mathbf{v}_a)$. Show that the mean collision rate $(1/t_{\text{coll}})$, which denotes the average number of encounters a given star has per unit time with a distance of closest approach r_{coll}, is given by

$$\frac{1}{t_{\text{coll}}} = \frac{\pi}{n} \int f(\mathbf{v}) f(\mathbf{v}_a) b^2 |\mathbf{v} - \mathbf{v}_a| \, d^3\mathbf{v} \, d^3\mathbf{v}_a, \qquad (10.93)$$

where n is the number density of stars. (c) Evaluate the integrals to obtain the result

$$\frac{1}{t_{\text{coll}}} = 4\sqrt{\pi}\, n\sigma r_{\text{coll}}^2 + \frac{4\sqrt{\pi}\, Gmnr_{\text{coll}}}{\sigma}. \qquad (10.94)$$

Show that this result can be rewritten in the manner given in the text. Interpret the two terms in Eq. (10.94).

10.7 Interaction Between Clusters and Disk

The discussion so far has treated clusters as isolated physical entities. This is, however, not quite true, and there exist several interesting effects that arise because of the interaction between a cluster of stars and other constituents of our galaxy. We shall now describe one of these effects, called *disk shocking*.

The stellar distribution in solar neighbourhood can be parameterised by a mean density $\rho \simeq 0.18\, M_\odot\, \text{pc}^{-3}$ and a vertical scale height of $z_0 \simeq 350\,\text{pc}$. This scale height is much larger than the typical tidal radius $r_t \simeq (30\text{–}100)\,\text{pc}$ of a cluster, and hence a cluster will be totally immersed in the disk while it passes through it. The mean density of stars within the cluster decreases rapidly from the centre to the edge of the cluster, whereas the mean density of the disk material is reasonably uniform. It follows that stars sufficiently away from the centre of the cluster will be influenced strongly by the gravitational force of the background material in the disk rather than by the self-gravity of the cluster. If the normal to the orbital plane of a particular globular cluster is inclined at an angle θ with respect to the symmetry axis of our galaxy, then the z component of the velocity of the cluster while crossing the disk of the galaxy will be $v_\perp = v_c \sin\theta$, where v_c is the velocity of the circular orbit on the disk at the location of the crossing. Because the globular clusters are distributed in a spherically symmetric manner in our galaxy, the average value of $v_\perp$ will be

$$V_\perp = \langle v_\perp \rangle = v_c \int_0^{\pi/2} \sin^2\theta \, d\theta = \frac{\pi}{4} v_c \simeq 170\,\text{km s}^{-1} \qquad (10.95)$$

if $v_c = 220\,\text{km s}^{-1}$. On the other hand, the internal velocity dispersion of the cluster is $\sigma_* \approx 5\,\text{km s}^{-1}$, implying that – as the cluster moves through the disk – a typical star moves in its orbit a distance of $(2z_0/V_\perp)\sigma_* \simeq 20\,\text{pc}$.

In studying the effect of the gravitational perturbation of the disk on such a cluster, we can consider two extreme limits. Stars in the cluster that are very close to the cluster centre, say, at $r \lesssim 2$ pc, will have their orbits evolving adiabatically and will be hardly affected by the perturbation. However, the stars in the radii 20 pc $\lesssim r \lesssim r_t$ will be strongly affected by the perturbation. As a crude first approximation, we can treat the perturbation that is due to the disk by a procedure known as impulse approximation, which involves the assumption that the stars do not significantly change their position with respect to the galactic centre during the time of the encounter. In that case, the net change $\Delta \mathbf{V}$ in the velocity of a cluster star located at $\mathbf{x}$, with respect to cluster centre ($\mathbf{x} = 0$), is caused by the differential gravitational acceleration of the disk $[\mathbf{g}(\mathbf{x}) - \mathbf{g}(0)]$ acting while the cluster crosses the disk. Expanding the acceleration in a Taylor series, we get for the z component of the induced velocity,

$$\Delta v_z(\mathbf{x}) = \int \mathbf{x} \cdot (\nabla g_z)_{\mathbf{x}=0} \, dt. \tag{10.96}$$

From the expression

$$\mathbf{x} \cdot \nabla g_z = x \frac{\partial g_z}{\partial R} + z \frac{\partial g_z}{\partial Z}, \tag{10.97}$$

it is clear that the second term will dominate over the first for a thin disk. When only the dominant contribution is retained, the net change in v_z when the cluster centre moves from $Z = -Z_1$ to $Z = Z_1$ is

$$|\Delta v_z(\mathbf{x})| \simeq \left| z \int_{-Z_1}^{Z_1} \left(\frac{\partial g_z}{\partial Z} \right)_{Z=v_\perp t} dt \right| = \left| \frac{z}{v_\perp} \int_{-Z_1}^{Z_1} \frac{\partial g_z}{\partial Z} dZ \right| \simeq \frac{2z}{v_\perp} |g_z(R, Z_1)|. \tag{10.98}$$

This suggests that the mean-square velocity $\langle v^2 \rangle$ of a star will increase in a time scale

$$t_{\text{shock}} \equiv \left(\frac{1}{\sigma_*^2} \frac{d\langle v^2 \rangle}{dt} \right)^{-1} \simeq \frac{(1/2) P_\psi \sigma_*^2}{\langle |\Delta \mathbf{v}|^2 \rangle} \simeq \frac{P_\psi \sigma_*^2 V_\perp^2}{8 z^2 g_z^2}, \tag{10.99}$$

where P_ψ is the azimuthal period of the cluster's orbit and we have set $v_\perp = V_\perp$ [see Eq. (10.95)]. The average of z^2 can be estimated as $(1/3)r_h^2$; all other quantities are averages over the distribution of matter in the disk. In the case of the Milky Way, $P_\psi = 10^8$ yr, $\sigma_* = 5$ km s^{-1}, $V_\perp = 170$ km s^{-1}, and $r_h = 10$ pc, giving $t_{\text{shock}} \approx 6 \times 10^9$ yr. It is obvious from this result that disk shocking can enhance the rate at which globular clusters lose stars and thereby undergo core collapse.

Exercise 10.6
Decay of globular-cluster orbits: Dynamical friction was used in this chapter to study the internal effects in a globular cluster, but the same process also has implications for the orbital motion of the globular cluster in the galaxy. (a) Estimate the dynamical friction

force acting on a globular cluster of mass M moving on an initially circular orbit of radius r through a galaxy having a density profile $\rho(r) = (v_c^2/4\pi G r^2)$. (b) The tangential dynamical friction force causes the cluster to lose angular momentum. Hence show that the orbital radius decreases at a rate determined by $r(dr/dt) \propto -(GM/v_c)\ln \Lambda$. Integrate this equation and show that the globular cluster will spiral to the centre in a time scale $t_{\text{spiral}} \propto (r_i^2 v_c/GM)(\ln \Lambda)^{-1}$. Estimate this time scale for globular clusters in our galaxy.

10.8 Open Clusters

The globular clusters are reasonably spherical in shape and show marked central concentration. There are also collections of stars that have a more diffuse morphology and are called open clusters. We shall briefly mention some aspects of open clusters for the sake of completeness.

Open clusters occur prominently near the galactic plane (and hence are more likely to be obscured by dust) and contain a few tens of stars to many thousands with densities ranging from 10^{-1} to 10^3 stars pc^{-3}. Their integrated magnitudes vary between $M_V = -3$ and $M_V = -9$. The lowest-density systems are also sometimes called associations, especially in the case of star-forming regions like OB associations and T-Tauri associations.

These clusters exhibit diffuse emission of starlight reflected by dust grains around the stars. Thus, unlike globular clusters, open clusters contain a significant amount of ISM. Further, many of the open clusters contain bright blue stars, and in some of them we can directly observe the dense gaseous cores from which stars form. The spectral studies of open clusters also show that their metallicities are in the range $-0.75 \lesssim$ [Fe/H] $\lesssim 0.25$, thereby suggesting the hypothesis that open clusters are relatively young star-forming regions. The detailed study of the H-R diagram of the open clusters described earlier (see Fig. 10.1) confirms this view. By and large, the ages of open clusters vary from 10^6 yr all the way to 10^9 yr. This estimate, of course, is subject to significant uncertainties, as in the case of globular clusters. Nevertheless, the spread in the ages suggests that they are continuously forming in the disk of the galaxy. From a nearby sample of open clusters, it is estimated that the formation rate of young open clusters is probably ~ 80 kpc^{-2} Gyr^{-1} to 1 kpc^{-2} Gyr^{-1}. The older globular clusters have been depleted, most probably, because of the encounters with the galactic disk as described earlier. Evidence in favour of this idea comes from the observations that old clusters are found preferentially at larger distances from the galactic centre and far away from the galactic plane.[5]

Studies of open clusters also provide a direct handle on the stellar IMF for these clusters. The high-mass end of the stellar luminosity function can be studied before the OB stars have fully evolved and are still found in the OB associations. Such studies indicate[6] that the IMF for stars with $M > 7\,M_\odot$ is a power law $M^{-\alpha}$ with index $\alpha = 2.1 \pm 0.1$. These high-mass stars evolve quickly and hence die in the cluster region, with very few escaping as field stars. In fact, the IMF index for the corresponding range among field stars is $\alpha \approx 4.5$. For lower masses ($M \lesssim 3\ M_\odot$), the field and the open-cluster IMFs seem to indistinguishable, within observational uncertainties.

We have seen in the last chapter that the galactic plane contains several giant molecular cloud (GMC) complexes with masses $m_{\text{cloud}} \gtrsim 10^5\ M_\odot$ and radius $r_{\text{cloud}} \approx 10$ pc. Any cluster of stars that moves through the galactic disk will also encounter the GMCs and will be affected by them gravitationally. Because both the clusters as well as the GMCs move with velocities that are Maxwellian with a dispersion $\sigma \approx 7$ km s^{-1}, the relative velocity of encounter between the GMC and the cluster will be $\sqrt{2}\sigma \approx 10$ km s^{-1}. For clusters with internal velocity dispersion $\sigma_* \ll 10$ km s^{-1}, these effects will be quite significant. Globular clusters usually have velocity dispersions that are comparable with ~ 10 km s^{-1}, and hence are not strongly affected by these encounters. However, the galaxy also contains a large number of open clusters that contain typically 10^2–10^3 stars in a radius of approximately (1–10) pc. They have a much lower velocity dispersion of $\sigma_* \lesssim 1$ km s^{-1} and will be more significantly affected by their encounters with GMCs. Because the stars in an open cluster move very little during the time over which the gravitational perturbation of a GMC acts, we may estimate the effect of the encounter in impulse approximation.

Consider a perturber of mass M_2 moving with a velocity $\mathbf{V} = (0, 0, V)$ in the rest frame of the open cluster. If the impact parameter is b, any given star will feel an impulsive acceleration $\Delta \mathbf{g}$ relative to the centre of the cluster in the x–y plane of magnitude $(GM_2/b^3)\mathbf{x}_\perp$, where $\mathbf{x}_\perp$ is the coordinate of the star relative to the centre of the cluster in the x–y plane. This relative acceleration, acting over a time $2(b/V)$, will produce a net relative velocity for a cluster star of the amount

$$\Delta \mathbf{V}(\mathbf{x}) = \left(\frac{2GM_2}{b^2 V}\right)\mathbf{x}_\perp + \mathcal{O}(x_\perp^2). \qquad (10.100)$$

The net energy gained by this star will be $(1/2)m(\Delta V)^2$, and we obtain the total energy fed into the cluster by integrating over all the stars in the cluster. This gives

$$\Delta E = \frac{2G^2 M_2^2}{b^4 V^2}\int \rho(\mathbf{x})(x^2 + y^2)\, d^3\mathbf{x} = \frac{4G^2 M_2^2 M_1}{3 b^4 V^2}\overline{r^2}. \qquad (10.101)$$

We now have to take into account different impact parameters and relative velocities by averaging over V and integrating over b. For a Maxwellian velocity distribution, the probability for relative speed V is distributed as

$$dP = \frac{4\pi V^2\, dV}{[2\pi(\sqrt{2}\sigma)^2]^{3/2}} \exp\left[-\frac{V^2}{2(\sqrt{2}\sigma)^2}\right]. \qquad (10.102)$$

10.8 Open Clusters

Hence the rate of encounters $\mathcal{R}$ at a relative speed V between any given cluster and a distribution of clouds with impact parameters in the range $(b, b + db)$ will be

$$\mathcal{R} = n_{\text{cloud}}\, V\, 2\pi b\, db\, dP = \frac{n_{\text{cloud}} 8\pi^2 b\, db}{[2\pi(\sqrt{2}\sigma)^2]^{3/2}} \exp\left(-\frac{V^2}{4\sigma^2}\right) V^3\, dV, \quad (10.103)$$

where n_{cloud} is the number density of clouds. Using expression (10.101) for the energy transfer per encounter, we can estimate the rate of increase of the cluster's energy for all encounters with impact parameters in the range $(b_{\min}, b_{\max})$ as

$$\dot{E} = \frac{32\pi^2 G^2 (M^2 n)_{\text{cloud}} (M\overline{r^2})_{\text{cluster}}}{3[2\pi(\sqrt{2}\sigma)^2]^{3/2}} \int_0^\infty \exp\left(-\frac{V^2}{4\sigma^2}\right) V^3 dV \int_{b_{\min}}^{b_{\max}} \frac{db}{b^3}$$

$$= \frac{4\sqrt{\pi}\, G^2 (M^2 n)_{\text{cloud}} (M\overline{r^2})_{\text{cluster}}}{3\sigma b_{\min}^2} \quad (10.104)$$

if $(b_{\min}/b_{\max})^2 \ll 1$. A reasonable estimate for $b_{\min}$ will be the radius of the cloud r_{cloud}. Then $\dot{E}$ can be expressed in the form

$$\dot{E} = \frac{4\sqrt{\pi}}{3\sigma} G^2 \left(\frac{M^2 n}{r^2}\right)_{\text{cloud}} (M\overline{r^2})_{\text{cluster}}. \quad (10.105)$$

We can obtain the time scale t_d over which such encounters can disrupt the cluster by dividing the binding energy of the cluster by $\dot{E}$. This is, however, an overestimate as it does not take into account encounters in which $b < r_{\text{cloud}}$, e.g., head-on collisions. It is easy to see that if we assume that $\dot{E} \propto b^{-3}$ for $b > r_{\text{cloud}}$ and $\dot{E} = \text{constant}$ for $b < r_{\text{cloud}}$, the net result for $\dot{E}$ is multiplied by a factor $\mu = 3$. This will reduce the time scale by the same factor μ. Then

$$t_d \approx \frac{E_{\text{bind}}}{\dot{E}} = \frac{0.085\sigma}{\mu G} \left(\frac{M}{r^2 r_h}\right)_{\text{cluster}} \left(\frac{r^2}{M^2 n}\right)_{\text{cloud}}. \quad (10.106)$$

It may be noted that the parameters of the cloud enter in terms of only its projected surface density $\Sigma_{\text{cloud}} \equiv (M/\pi r^2)_{\text{cloud}}$ and the mean density $\rho_{\text{mol}} \equiv (Mn)_{\text{cloud}}$. Taking $\sigma_{\text{cloud}} = 250\, M_\odot\, \text{pc}^{-2}$ and $\rho_{\text{mol}} = 0.025\, M_\odot\, \text{pc}^{-3}$, $\sigma = 7$ km s^{-1}, and $\mu = 3$ we get

$$t_d \approx 5.7 \times 10^8 \left(\frac{M_{\text{cluster}}}{250\, M_\odot}\right) \left(\frac{1\, \text{pc}}{r_h}\right)^3 \left(\frac{r_h^2}{r^2}\right)\, \text{yr}. \quad (10.107)$$

Although this estimate is quite uncertain, it does predict that the population of open clusters with ages greater than 10^9 yr should decline rapidly. This is indeed seen observationally, although determining the exact half-life of open clusters is a difficult task.

Notes and References

Chapter 1

Some of the general references relevant for Chaps. 1 and 2 are the following.

1. C.J. Hansen and S.D. Kawaler, *Stellar Interiors* (Springer-Verlag, New York, 1994).
2. R. Kippenhahn and A. Weigert, *Stellar Structure and Evolution* (Springer-Verlag, New York, 1991).
3. D. Arnett, *Supernova and Nucleosynthesis* (Princeton Univ. Press, Princeton, NJ, 1996).
4. D.D. Clayton, *Principles of Stellar Evolution and Nucleosynthesis* (Univ. of Chicago Press, Chicago, IL, 1983).
5. R.L. Bowers and T. Deeming, *Astrophysics* (Jones and Bartlett, Boston, 1984), Vols. I and II.
6. Many of the introductory textbooks in astrophysics provide simple derivations of physical processes similar to the ones given here. In particular, see F.H. Shu, *The Physical Universe* (University Science, Mill Valley, CA, 1982); B.W. Caroll and D.A. Oslie, *Modern Astrophysics* (Addison-Wesley, New York, 1996). The derivation of masses and sizes of planets, stars, galaxies, etc., has been usually attempted in the context of anthropic principle. See, for example, B.J. Carr and M.J. Rees, *Nature* (*London*) **278**, 605 (1979); J.D. Barrow and F.J. Tipler, *The Anthropic Principle* (Oxford Univ. Press, Oxford, England, 1982); also see L.M. Celnikier, *Basics of Cosmic Structures* (Editions Frontiers, Paris, 1989). Most of the astronomical data mentioned in this chapter can be found in M.V. Zombeck, *Handbook of Space Astronomy and Astrophysics*, 2nd ed. (Cambridge Univ. Press, New York, 1990); also available on line from http://adsbit.harvard.edu/books/hsaa/index.html.
7. The data for the graph are taken from C.W. Allen, *Astrophysical Quantities*, (Athlone, London, 1973); H. Jahreiss and R. Wielen, in *The Nearby Stars and the Stellar Luminosity Function*, IAU Coll. **76**, A.G. Davis Phillip et al., eds. (Davis, Philadelphia, 1983); and P. Kroupa et al., *Mon. Not. R. Astron. Soc.* **244**, 76 (1990); *Mon. Not. R. Astron. Soc.* **262**, 545 (1993). Also see, J. Binney and M. Merrifield, Galactic Astronomy (Princeton Univ. Press, Princeton, NJ; 1998), Chap. 3.

8. Compared with theoretical topics, observational astronomy suffers from a dearth of good books. One excellent text is P. Lena et al., *Observational Astrophysics*, 2nd revised ed. (Springer, New York, 1998).

Chapter 2

1. Some of the general references relevant for this chapter are Refs. 1–5 cited above in Chap. 1 references.
2. Books devoted to stellar structure discuss these assumptions in much greater detail. See, for example, J.P. Cox, *Principles of Stellar Structure* (Gordon & Breach, New York, 1968); M. Schwarzschild, *Structure and Evolution of Stars* (Princeton Univ. Press, Princeton, NJ, 1958). Tabulated opacities are available on the web. One key reference is F.J. Rogers and C.A. Iglesias, *Astrophys. J. Suppl.* **79**, 507 (1992); some attempts to produce analytic fits to opacity can be found, for example, in R.F. Stellingwerf, *Astrophys. J.* **195**, 441 (1995); **199**, 705 (1995).
3. Several codes are available in literature and on the worldwide web for integrating the stellar-evolution equations. A simple code is given in B.W. Carroll and D.A. Ostlie, *Modern Astrophysics* (Addison-Wesley, Reading, MA, 1996) and a more sophisticated version is available in C.J. Hansen and S.D. Kawaler, *Stellar Interiors* (Springer-Verlag, New York, 1994). The results presented here are based on the code adapted and developed from these sources by the author.
4. See, for example, G. Fontaine et al., *Astrophys. J. Suppl.* **35**, 293 (1977).
5. Most of the books cited earlier discuss stellar atmospheres; a monograph devoted to this subject is D. Mihalas, *Stellar Atmospheres* (Freeman, San Francisco, 1978).
6. Many aspects of astronomical spectra including line and continuum radiation from stars are discussed in D. Emerson, *Interpreting Astronomical Spectra* (Wiley, New York, 1996).
7. Nuclear abundances in different astrophysical systems are discussed in detail in D. Arnett, *Supernova and Nucleosynthesis* (Princeton Univ. Press, Princeton, NJ, 1996). The data for the table are from A.N. Cox et al., ed., *Solar Atmosphere and Interior* (University of Arizona Press, Tucson, 1991).

Chapter 3

1. Most of the textbooks cited for Chap. 2 are also relevant for the study of stellar evolution. Results of numerical integration of equations of stellar evolution are available at, for example, the Centre de Donnees de Strasbourg (CDS) website, which can be accessed through http://cdsads.u-strasbg.fr/. The description in this chapter makes extensive use of these results; also see D. Schaerer et al., *Astron. Astrophys. Suppl. Ser.* **102**, 339 (1993).
2. The subject of star formation is still an active research area and hence is not adequately discussed in many textbooks. Some of the references relevant to this are the following: C. Hayashi, *Ann. Rev. Astron. Astrophys.* **3**, 171 (1965); S. Stahler, *Astrophys. J.* **332**, 804 (1988); F.H. Shu et al., *Ann. Rev. Astron. Astrophys.* **25**, 23 (1987); R. Bachiller, *Ann. Rev. Astron. Astrophys* **34**, 1111 (1996).
3. See, e.g., R.B. Larson, *Mon. Not. R. Astron. Soc.* **145**, 271 (1969); I. Appenzeller and W. Tscharnuter, *Astron. Astrophys.* **40**, 397 (1975).

4. The data are from I. Iben, *Astrophys. J.* **141**, 993 (1965).
5. This initial mass function is originally from E.E. Salpeter, *Astrophys. J.* **121**, 161 (1955). See S.L. Shapiro and S.A. Teukolsky, *Black Holes, White Dwarfs and Neutron Stars* (Wiley, New York, 1983); J. Binney and M. Merrifield, *Galactic Astronomy* (Princeton Univ. Press, Princeton, NJ, 1998); G. Gilmore and D. Howell, eds., *The Stellar Initial Mass Function*, Astronomical Society of the Pacific Series No. 142, for a discussion of initial mass function.
6. Semianalytic models for brown dwarfs are discussed in A. Burrows and J. Liebert, *Rev. Mod. Phys.* **65**, 301 (1993). This article also provides many of the fitting formulas used in this section and references to earlier work.
7. Several stellar evolution codes and data are available in literature, one of the more comprehensive sets being the one cited in Ref. 1 above; also see R. Kippenhahn et al., *Z. Astrophys.* **61**, 241 (1965) and the book by R. Kippenhahn and A. Weigert, *Stellar Structure and Evolution* (Springer-Verlag, New York, 1991).
8. This issue is still fairly controversial. Some of the recent papers that have references to earlier literature are A. Renzini et al., *Astrophys. J.* **400**, 280 (1992); R. Bhaskar and A. Nigam, *Astrophys. J.* **372**, 592 (1991); I. Iben and A. Renzini, *Phys. Rep.* **105**, 329 (1984).
9. B. Paczynski, *Acta Astron.* **20**, 47 (1970).
10. These are discussed, e.g., in P. Giannone et al., *Z. Astrophys.* **68**, 107 (1968) and D. Lauterborn et al., *Astron. Astrophys.* **10**, 97 (1971); **13**, 119 (1971).
11. A good reference for late-stage evolution of stars is I. Iben, *Astrophys. J. Suppl.* **76**, 55 (1991). This paper also contains extensive references to earlier work.
12. The fitting formula is from C.D. Garmani and S. Conti, *Astrophys. J.* **284**, 705 (1984).
13. See Ref. 3 and Ref. 7 of Chap. 1.
14. The fitting formulas are from I. Iben, and G. Laughlin, *Astrophys. J.* **341**, 312 (1989); I. Iben, *Ann. Rev. Astron. Astrophys.*, **12**, 215 (1974). Also see I. Iben and A. Renzini, *Phys. Rep.* **105**, 330 (1984).
15. This fit is from I. Iben, *Publ. Astron. Soc. Pac.* **83**, 697 (1971). The fit in Eq. (1.30) is from I. Iben and R.S. Tuggle, *Astrophys. J.* **173**, 135 (1972).

Chapter 4

1. Most books on stellar evolution also discuss supernova. Among the specialized texts, the following contain fairly detailed discussions: D. Arnett, *Supernova and Nucleosynthesis* (Princeton Univ. Press, Princeton, NJ, 1996); A.G. Petschek, ed., *Supernovae* (Springer, New York, 1990); S.A. Bludman et al., eds., *Les Houches (1990) on Supernova* (North-Holland, Amsterdam, 1994). There is also a host of papers dealing exclusively with SN 1987A, two of which are H.A. Bethe, *Rev. Mod. Phys.* **62**, 801 (1990); *Astrophys. J.* **412**, 192 (1993).
2. The figure is based on J.C. Wheeler, *Rep. Prog. Phys.* **44**, 85 (1991) and D. Sugimoto et al., eds., *Fundamental Problems in the Theory of Stellar Evolution* (Reidel, Dordrecht, The Netherlands, 1992).
3. Most of the books cited above as well as those dealing with stellar structure discuss neutrino processes, especially the articles in *Les Houches (1990)*. Another useful reference is H. Mitter and F. Widder, eds., *Particle Physics and Astrophysics*

(Springer, New York, 1989). The name Urca has no astrophysical significance and is purely historical; it refers to a casino in Rio de Janeiro in which the net effect is the loss of money by the customer – yet another example of naming conventions used in astronomy that have historic rather than scientific significance.
4. The figures are based on the articles of Z. Barkat, *Ann. Rev. Astron. Astrophys.* **13**, 45 (1975) and G. Beaudet et al., *Astrophys. J.* **174** (1972).
5. A more detailed discussion can be found, e.g., in D.Q. Lamb et al., *Phys. Rev. Lett.* **41**, 1623 (1978); *Nucl. Phys. A.*, **360**, 459 (1981).
6. A brief but adequate discussion of asymmetric supernova explosions is given in J.N. Bahcall and J.P. Ostriker, eds. *Unsolved Problems in Astrophysics* (Princeton Univ. Press, Princeton, NJ, 1997). A good review of the theory of gamma-ray bursts is by T. Piran, *Phys. Rep.* **134**, 575 (1999).
7. A detailed discussion of stellar nucleosynthesis can be found in D. Arnett, *Supernova and Nucleosynthesis* (Princeton Univ. Press, Princeton, NJ, 1996) as well as in D.D. Clayton, *Principles of Stellar Evolution and Nucleosynthesis* (Univ. of Chicago Press, Chicago, IL, 1983).

Chapter 5

1. In addition to the books on stellar structure already cited in the references of previous chapters, the following three books are of particular relevance: S.L. Shapiro and S.A. Teukolsky, *Black Holes, White Dwarfs and Neutron Stars* (Wiley, New York, 1983); E. Kawaler et al., eds., *Stellar Remnants* (Springer, New York, 1997); J.M. Irvine, *Neutron Stars* (Oxford Univ. Press, Oxford, England, 1978).
2. Based on the data in D.E. Winget et al., *Astrophys. J. Lett.* **315**, L77 (1987).
3. See, for example, Section 39 of L.D. Landau and E.M. Lifshitz, *Statistical Physics, Part 2* (Pergamon, New York, 1980).
4. See, for example, L.D. Landau and E.M. Lifshitz, *The Classical Theory of Fields* (Pergamon, New York, 1975) or C.W. Misner, K.S. Thorne, and J.A. Wheeler, *Gravitation* (Freeman, San Francisco, 1973).

Chapter 6

1. Some useful monographs dealing with pulsars are A.G. Lyne and F. Graham-Smith, *Pulsar Astronomy*, 2nd ed. (Cambridge Univ. Press, New York, 1998); R.D. Blandford et al., *Pulsars as Physics Laboratories* (Oxford Univ. Press, Oxford, England, 1993); F.C. Michel, *Theory of Neutron Star Magnetospheres* (Chicago Univ. Press, Chicago, IL, 1991); M. Bailes et al., eds., *Pulsars: Problems and Progress*, IAU Coll. **160**, ASP Series No. 105 (Astronomical Society of the Pacific, California, 1996).
2. See, for example, P. Meszaros, *High Energy Radiation From Magnetized Neutron Stars* (Cambridge Univ. Press, New York, 1992).
3. See, for example, J.A. Philips et al., *Planets around Pulsars*, ASP Conf. Series No. 36.
4. See V. I. Tatarskii and U.V. Zavorotnyi, *Prog. Optics*, **18**, 207 (1980); J. Goodman and R. Narayan, *Mon. Not. R. Astr. Soc.* **214**, 519 (1985) and references cited therein, for a discussion of Eqs. (6.135) and (6.141).

Chapter 7

1. Some general texts relevant for this chapter are P.A. Charles and F.D. Seward, *Exploring the X-Ray Universe* (Cambridge Univ. Press, New York, 1995); C.W.H. de Loore and C. Doom, *Structure and Evolution of Single and Binary Stars* (Kluwer, Dordrecht, The Netherlands, 1992); J. Frank, A. King, and D. Raine, *Accretion Power in Astrophysics* (Cambridge Univ. Press, New York, 1992); W.H.G. Lewin et al., *X-Ray Binaries* (Cambridge Univ. Press, New York, 1995). In addition, most books on stellar evolution do discuss binaries to some extent. Also see D. Bhattacharya and E.P.J. van den Heuvel, *Phys. Rep.* **203**, 1 (1991) and references cited therein.
2. This figure follows directly from the analytic expressions derived in the text. Also see J. Frank, A. King and D. Raine, *Accretion Power in Astrophysics* (Cambridge Univ. Press, New York, 1992) for a more detailed discussion and figure.
3. For a detailed discussion, see for example, the contribution by G. Srinivasan in *Stellar Remnants*, Saas-Fee Course 25 (Springer, New York, 1995).
4. For a detailed discussion of several aspects of millisecond pulsars, see A.A. Fruchter et al., *Millisecond Pulsars: A Decade of Surprise*, ASP Conf. Series 72.
5. A nice discussion of PSR 1913+16 is available in the article by J.H. Taylor in *The Universe Unfolding*, H. Bondi and M. Weston-Smith, eds. (Oxford Univ. Press, Oxford, England, 1998).

Chapter 8

1. For a discussion of solar abundances, see L.H. Aller, in *Spectroscopy of Astrophysical Plasmas*, A. Dalgarno and D. Layzer, eds. (Cambridge Univ. Press, New York, 1986); J.N. Bahcall and R.K. Ulrich, *Rev. Mod. Phys.* **60**, 297 (1988); for a discussion of solar age, see, for example, D.B. Guenther, *Astrophys. J.* **339**, 1156 (1989).
2. There is extensive literature on this subject. See, for example, J.N. Bahcall, *Neutrino Astrophysics* (Cambridge Univ. Press, New York, 1989); J.N. Bahcall et al., eds., *Solar Neutrinos, the First 30 Years* (Addison-Wesley, Reading, MA, 1995); see also the article by J.N. Bahcall, in *Unsolved Problems in Astrophysics*, J.N. Bahcall and J.P. Ostriker, eds. (Princeton Univ. Press, Princeton, NJ, 1997).
3. The theoretical predictions have a tendency to change over a time scale of approximately a year in spite of the fact that researchers are trying to do a careful job at any given time. The results quoted here are mostly from J.N. Bachall et al., *Phys. Letts* **B 433**, 1 (1998).
4. There are several reviews on helioseismology; see, for example, K.G. Libbrecht, *Space Sci. Rev.* **47**, 275 (1998); K.G. Libbrecht and M.F. Woodard, *Science* **253**, 152 (1991).
5. Several plasma physics aspects related to the Sun are discussed in the article by E.R. Priest in J.G. Kirk et al., eds., *Plasma Astrophysics* (Springer, New York, 1994). This article also contains extensive references to earlier work. Also see R.D. Bentley and J.T. Mariska, eds., *Magnetic Reconnection in Solar Atmosphere*, ASP Conf. Series No. 111.
6. Two monographs providing a contemporary discussion of planetary formation are Y.J. Pendleton and A.G.G.M. Tielens, eds., *From Stardusts to Planetesimals*, ASP

Series No. 122; R. Rebolo et al., *Brown Dwarfs and Extra Solar Planets*, ASP Series No. 134.

Chapter 9

1. The following books deal extensively with the physics of the interstellar medium: D.E. Osterbrock, *Astrophysics of Gaseous Nebulae and AGN* (University Science, Mill Valley, CA, 1989); H. Scheffler and H. Elsasser, *Physics of the Galaxy and Interstellar Matter* (Springer-Verlag, New York, 1987); J.E. Dyson and D.A. Williams, *The Physics of the Interstellar Medium* (Institute of Physics, London, 1997); C.G.C. Wynn-Williams, *The Fullness of Space* (Cambridge Univ. Press, New York, 1992); L. Spitzer, *Physical Processes in the Interstellar Medium* (Wiley, New York, 1978); W.B. Burton et al., *The Galactic Interstellar Medium*, Saas-Fee 21 (Springer-Verlag, New York, 1992); A.R. Taylor et al., *New Perspectives on the Interstellar Medium*, ASP Series No. 168; also see the relevant chapters in F.H. Shu, *The Physics of Astrophysics* (University Science, Mill Valley, CA, 1992), Vol. II. Some of the Exercises are adapted from the Web resources provided by C. L. Sarazin.
2. See, e.g., J.H. Taylor and J.M. Cordes, *Astrophys. J.* **411**, 674 (1993).
3. All books in radio astronomy discuss 21-cm observations of the ISM. In particular, see K. Rohlfs and T.L. Wilson, *Tools of Radio Astronomy* (Springer-Verlag, New York, 1996); B.F. Burke and F. Graham-Smith, *An Introduction to Radio Astrophysics* (Cambridge Univ. Press, New York, 1996); and the articles in G.L. Verschuur and K.I. Kellermann, eds., *Galactic and Extragalactic Radio Astronomy* (Springer-Verlag, New York, 1988).
4. Several aspects of spectroscopy of the ISM are discussed in D. Emerson, *Interpreting Astronomical Spectra* (Wiley, New York, 1996).

Chapter 10

1. The gravitational dynamics of globular clusters is well discussed in L. Spitzer, *Dynamical Evolution of Globular Clusters* (Princeton Univ. Press, Princeton, NJ, 1987) and in J. Binney and S. Tremaine, *Galactic Dynamics* (Princeton Univ. Press, Princeton, NJ, 1988). A recent monograph dealing primarily with the observational aspects of globular clusters is K.M. Ashman and S.C. Zepf, *Globular Cluster Systems* (Cambridge Univ. Press, New York, 1998). Also see the articles in S.G. Djorgovski and G. Meylan, *Structure and Dynamics of Globular Clusters*, ASP Series No. 50.
2. See, for example, J. Binney and M. Merrifield, *Galactic Astronomy* (Princeton Univ. Press, Princeton, NJ, 1998). The fitting functions given in this section are also discussed in detail in this book.
3. See, for example, B. Chaboyer, P. Demarque, and A. Sarajedini, *Astrophys. J.* **459**, 558 (1996).
4. See, for example, S.G. Djorgovski, *Astrophys. J. Lett.* **438**, L29 (1995).
5. J.C. Mermilliod, in J.E. Hesser, ed., *Star Clusters*, IAU Symposium 85 (Reidel, Dordrecht, The Netherlands, 1980). Also see the database at http://obswww.unige.ch/bda/ugp1/ugp1.html.
6. P. Massey et al., *Astrophys. J.* **454**, 151 (1995).

Index

21-cm radiation
   column density, 520
   mapping of galaxy, 521

accretion
   introduction to, 28
accretion disk
   Alfven radius, 372
   boundary layers, 370
   emergent spectrum, 365, 368
   magnetic field, 363, 371
   opacity, 366
   temperature profile, 361
   thin, 358
   viscosity, 362
atmospheric absorption, 36
   order of magnitude, 39
atmospheric emission, 39

binary accretion
   stability, 347
   varieties of, 392
binary star, 5
   cataclysmic variable, 397
   coalescence, 390
   example of evolution, 349
   general relativistic effects, 382
   globular clusters, 566
   gravitational radiation, 390
   novas, 394
   Roche lobe, 343
   supernova, 394
   time scales, 347
   x-ray bursters, 401
   x-ray transients, 400
black hole, 287
   binary star, 387
   kerr metric, 294
   Kruskal coordinates, 291
   rotation, 294
brown dwarf, 129
   maximum mass, 134
Brunt–Vassala frequency
   white dwarf, 267

Chandrasekhar mass, 139
   core collapse, 192
   white dwarf, 241
Chandrasekhar–Schoenberg mass, 144
colour index
   definition of, 13
compact remnants
   spatial density, 237
coordinate systems
   definition of, 4
core collapse
   self-similar solution, 205

dispersion measure
   pulsar, 328
dust grains
   charge, 536
   extinction, 533
dynamical friction, 561

Eddington luminosity
  order of magnitude, 30
energy content
  Synchrotron source, 223
equation of state, 268

Faraday rotation
  pulsar, 331
Fokker–Planck description
  orbit averaging, 558

galactic structure
  order of magnitude, 4
globular clusters
  age of, 542
  binary stars, 551
  diffusion coefficient, 556
  disk shocking, 569
  disruption of, 572
  evolution of, 563
  Fokker–Planck description, 553
  H-R diagram, 543
  main-sequence turn-off, 545
  orbits in galaxy, 570
  self-similar solution, 564
  time scales, 546

H-R diagram
  introduction to, 17
  order of magnitude, 7
  theoretical slope, 88
Hamiltonian dynamics
  resonance, 466
Hertzsprung gap, 149
high-mass stars
  AGB, 150
  core contraction, 144
  core hydrogen burning, 140
  evolution, 139
  horizontal branch, 150
  red giant phase, 150
  shell burning of hydrogen, 142
HII region, 483
  ionisation front, 488
  radiation from, 513

HMXB
  binary evolution, 354
H-R diagram
  main-sequence turn-off, 545

initial mass function, 127
interstellar medium, 34
  21-cm radiation, 519
  cooling, 493
  description of, 480
  dispersion measure, 509
  dust grains, 532
  extinction, 533
  Faraday rotation, 510
  heating, 493
  introduction to, 32
  ionisation front, 488
  magnetic field, 511
  maser, 528
  molecular lines, 525
  probes of, 515
  structure of, 501
  two-phase model, 504

limb darkening
  Sun, 102
line radiation
  chemical abundance, 109
  stellar atmosphere, 105
  stellar classification, 108
LMXB
  binary evolution, 354
low-mass stars
  core hydrogen burning, 151
  helium Aash, 157
  red giant phase, 154
  shell hydrogen burning, 152
luminosity function
  stars, 25

magnetic field
  effect on atomic structure, 319
  generation of, 431
  particle diagnostic, 444
magnetosphere
  of Earth, 442

magnitudes
   definition of, 11
Malmquist bias
   stellar statistics, 53
mass–radius relation
   brown dwarf, 132
   neutron star, 271
   white dwarf, 245
millisecond pulsar, 380
molecular clouds
   description of, 537
   star formation, 537
molecular lines
   equivalent width, 527
   maser, 528

neutrino processes
   opacity, 199
neutron star
   mass bound, 274
   superfluidity, 277

opacity
   photosphere, 92
open clusters, 571

pair production
   magnetosphere, 314
period luminosity relation
   variable stars, 177
planets
   tidal friction, 462
plasma
   dispersion measure, 509
plasma frequency, 38
precession
   of equinoxes, 456
   perturbation theory, 452
pulsar
   aligned rotator, 304
   death line, 317
   glitches, 323
   gravitational radiation, 300
   introduction to, 26
   light cylinder, 309

magnetic field, 300
magnetosphere, 299, 304
pair production, 314
scintillation, 334
spin-up, 376
surface structure, 306, 318
timing, 326
pulsar spin-up
   magnetic torques, 376
pulsar timing
   gravitational wave background, 333

recombination line, 518
recombination rate
   order of magnitude, 8
resolution
   spatial, 41
   spectral, 43
   temporal, 44
Roche lobe
   general comments, 353

scintillation
   strong, 340
   structure function, 335
   weak, 338
solar corona
   description of, 422
solar flare
   energetics, 425
   magnetic buoyancy, 426
   magnetic reconnection, 428
solar granulation, 421
solar neutrinos
   neutrino oscillations, 414
   spectrum, 413
solar oscillations
   frequencies, 415
   rotation, 419
solar system
   description of, 445
   Hamiltonian dynamics, 466
   Hamiltonian system, 450
   Kirkwood gaps, 467
   perturbation theory, 451

solar wind, 437
  magnetic field, 437
  van Allen belt, 443
star formation
  angular momentum, 128
  gravitational instability, 115
  H-R diagram, 122
  Hayashi track, 125
  Henyey track, 127
  magnetic field, 128
  mass scales, 117
  spherical collapse, 119
  T-Tauri stars, 540
stars
  age of, 542
  introduction to, 2
stellar atmosphere
  absorption lines, 103
  basic equations, 91
  chemical abundance, 110
  continuum radiation, 102
stellar evolution
  characteristic masses, 168
  introduction to, 15
  late stages, 161
  mass loss, 163
  nucleosynthesis, 230
  p-Cygni profile, 165
  planetary nebula, 164
  summary of, 15
stellar luminosity
  introduction to, 8
stellar mass
  order of magnitude, 3
stellar oscillations
  adiabatic, nonradial, 181
  adiabatic, spherical, 172
  Brunt–Vassala frequency, 185
  Lamb frequency, 185
  polytropes, 175
  spherical, nonadiabatic, 179
  Sun, 415
  variable stars, 176
stellar remnants
  introduction to, 17

stellar spectra
  introduction to, 22
stellar statistics
  initial mass function, 172
stellar structure
  basic assumptions, 65
  boundary conditions, 64
  central density, 79
  central temperature, 79
  convection, 62
  dependence on molecular weight, 87
  energy generation, 63
  equations of, 58
  fitting functions, 68
  opacity, 60
  scaling solutions, 83
  solutions of, 66
stellar wind
  Sun, 437
Sun
  internal structure, 405
  luminosity evolution, 409
  magnetic field, 431
  rotation of, 419
  solar neutrinos, 411
sunspots, 420
  Zeeman effect, 424
supernova
  binary evolution, 350
  bounce and shock, 207
  core collapse, 202
  light curve, 398
  luminosity, 210
  neutrino process, 192
  neutronisation, 195
  nucleosynthesis, 231
  photodisintegration, 195
supernova remnants, 21
  different phases, 216
  radiation from, 221
  Sedov phase, 218
  shock acceleration, 224
surveys
  bias, 49

confusion limit, 55
dynamical range, 49

telescopes
   resolution of, 40
   signal-to-noise, 45

variable stars
   instability strip, 180
   introduction to, 23

wave bands
   definition of, 9
white dwarf
   accretion column, 374
   cooling, 260
   Coulomb corrections, 244
   degeneracy pressure, 238
   general relativistic effects, 250
   high-density corrections, 246
   magnetic field, 257
   order of magnitude, 17
   oscillations, 265
   rotation, 257
   surface structure, 260

x-ray binaries
   introduction to, 29